Probability Theory and Stochastic Modelling

Volume 63

Probability Theory and Stochastic Modelling publishes cutting-edge research monographs in probability and its applications, as well as postgraduate-level textbooks that either introduce the reader to new developments in the field, or present a fresh perspective on fundamental topics.

Books in this series are expected to follow rigorous mathematical standards, and all titles will be thoroughly peer-reviewed before being considered for publication.

Probability Theory and Stochastic Modelling covers all aspects of modern probability theory including:

· Gaussian processes
· Markov processes
· Random fields, point processes, and random sets
· Random matrices
· Statistical mechanics, and random media
· Stochastic analysis
· High-dimensional probability

as well as applications that include (but are not restricted to) :

· Branching processes, and other models of population growth
· Communications, and processing networks
· Computational methods in probability theory and stochastic processes, including simulation
· Genetics and other stochastic models in biology and the life sciences
· Information theory, signal processing, and image synthesis
· Mathematical economics and finance
· Statistical methods (e.g. empirical processes, MCMC)
· Statistics for stochastic processes
· Stochastic control, and stochastic differential games
· Stochastic models in operations research and stochastic optimization
· Stochastic models in the physical sciences

Probability Theory and Stochastic Modelling is a merger and continuation of Springer's Stochastic Modelling and Applied Probabilityand Probability and Its Applications series.

Hai-Dang Nguyen • George Yin • Chao Zhu

Hybrid Switching Diffusions

Properties and Applications

Second Edition

Springer

Hai-Dang Nguyen
Department of Mathematics
University of Alabama
Tuscaloosa, AL, USA

George Yin
Department of Mathematics
University of Connecticut
Storrs, CT, USA

Chao Zhu
Department of Mathematical Sciences
University of Wisconsin-Milwaukee
Milwaukee, WI, USA

ISSN 2199-3130 ISSN 2199-3149 (electronic)
Probability Theory and Stochastic Modelling
ISBN 978-3-031-93302-8 ISBN 978-3-031-93303-5 (eBook)
https://doi.org/10.1007/978-3-031-93303-5

This Springer imprint is published by the registered company Springer Nature Switzerland AG
The registered company address is: Gewerbestrasse 11, 6330 Cham, Switzerland

Preface of the Second Edition

Almost 15 years have passed since the publication of the first edition of the book, during which the study of hybrid switching diffusions has made significant strides. Numerous new results have been obtained. In particular, we mention the work on switching diffusions in which the switching takes values in a countable state space, and the associated generator of the switching process depends on the past history. These advancements have enriched the theoretical framework of switching diffusions. In parallel with such theoretical developments, emerging applications have surfaced in fields such as ecology, population biology, social network modeling, and other related disciplines. These applications underscore the need to review and incorporate the latest progress. With this in mind, we have substantially revised the book.

Several chapters have been refined, reworked, and rewritten. Some chapters have been much expanded, whereas some other chapters have been shortened or removed. A new chapter that is devoted to switching processes with generators being past dependent and the state space of the switching processes taking values in countably infinite sets is added. In addition, a chapter focusing on applications in mathematical biology has been included.

Our colleague Hai-Dang Nguyen joined us for this new project. We hope that the reader finds the new edition worthwhile and useful and that our concerted effort makes valuable contributions to the scientific community.

The research presented in the book has been supported in part by the National Science Foundation and the Simons Foundation, respectively, for which we are grateful. Finally, we acknowledge the support of our respective departments throughout the years. We also thank the Springer professionals for helping us to finalize the book.

Tuscaloosa, AL, USA — Hai-Dang Nguyen
Storrs, CT, USA — George Yin
Milwaukee, WI, USA — Chao Zhu
March 2025

Preface of the First Edition

This book encompasses the study of hybrid switching diffusion processes and their applications. The word "hybrid" signifies the coexistence of continuous dynamics and discrete events, which is one of the distinct features of the processes under consideration. Much of the book is concerned with the interactions of the continuous dynamics and the discrete events. Our motivations for studying such processes originate from emerging and existing applications in wireless communications, signal processing, queueing networks, production planning, biological systems, ecosystems, financial engineering, and modeling, analysis, and control and optimization of large-scale systems, under the influence of random environments.

Displaying mixture distributions, switching diffusions may be described by the associated operators or by systems of stochastic differential equations together with the probability transition laws of the switching actions. We either have Markov-modulated switching diffusions or processes with continuous state-dependent switching. The latter turns out to be much more challenging to deal with. Viewing the hybrid diffusions as a number of diffusions joined together by the switching process, they may be seemingly not much different from their diffusion counterpart. Nevertheless, the underlying problems become more difficult to handle, especially when the switching processes depend on continuous states. The difficulty is due to the interaction of the discrete and continuous processes and the tangled and hybrid information pattern.

A salient feature of the book is that the discrete event process is allowed to depend on the continuous dynamics. Apart from the existence and uniqueness of solutions, we treat a number of basic properties such as regularity, the Feller property, the strong Feller property, and continuous and smooth dependence on initial data of the solutions of the associated stochastic differential equations with switching.

A large part of this work is concerned with the stability of switching diffusion processes. Here stability is meant in the broad sense including both weak and strong stability. That is, we focus on both a "neighborhoods of infinity" and neighborhoods of equilibrium points; the stability corresponding to the former is referred to as weak stability, whereas that of the latter is stability in the usual sense.

In studying deterministic dynamic systems, researchers used Lagrange stability to depict systems that are ultimately uniformly bounded. Treating stochastic systems, one would still hope to adopt such a notion. Unfortunately, the boundedness excludes many important cases. Thus, we replace this boundedness by a weaker notion known as recurrence (returning to a prescribed compact region in finite time). When the expected returning time is finite, we have so-called positive recurrence. A crucial question is: Under what conditions will the systems be recurrent (resp. positive recurrent). We demonstrate that positive recurrence implies ergodicity and provide criteria for the existence of invariant distributions together with their representation. In addition to studying asymptotic properties of the systems in the neighborhood of infinity, we examine the behavior of the systems at the equilibria. Also considered are invariance principles and stability of differential equations with random switching but without diffusions.

Because the systems are rarely solvable in closed form, numerical methods become a viable alternative. We construct algorithms to approximate solutions of such systems with state-dependent switching, and provide sufficient conditions for convergence for numerical approximations to the invariant measures.

In real-world applications, hybrid systems encountered are often of large scale and complex leading to intensive-computation requirement. Reduction of computational complexity is thus an important issue. To take this into consideration, we consider time-scale separation in hybrid switching-diffusion and hybrid-jump-diffusion models. Several issues including recurrence of processes with switching having multiple-weak-connected ergodic classes, two-time-scale modeling of stochastic volatility, and weak convergence analysis of systems with fast and slow motions involving additional Poisson jumps are studied in details.

This book is written for applied mathematicians, probabilists, systems engineers, control scientists, operations researchers, and financial analysts among others. The results presented in the book are useful to researchers working in stochastic modeling, systems theory, and applications in which continuous dynamics and discrete events are intertwined. The book can be served as a reference for researchers and practitioners in the aforementioned areas. Selected materials from the book may also be used in a graduate-level course on stochastic processes and applications.

This book project could not have been completed without the help and encouragement of many people. We are deeply indebted to Wendell Fleming and Harold Kushner, who introduced the wonderful world of stochastic systems to us. We have been privileged to have the opportunity to work with Rafail Khasminskii on a number of research projects, from whom we have learned a great deal about Markov processes, diffusion processes, and stochastic stability. We express our special appreciation to Vikram Krishnamurthy, Ruihua Liu, Yuanjin Liu, Xuerong Mao, John Moore, Qingshuo Song, Chenggui Yuan, Hanqin Zhang, Qing Zhang, and Xunyu Zhou, who have worked with us on various projects related to switching diffusions. We also thank Eric Key, Jose Luis Menaldi, Richard Stockbridge, and Fubao Xi for many useful discussions. This book, its contents, its presentation, and exposition have benefited a great deal from the comments by Ruihua Liu, Chenggui Yuan, Hanqin Zhang, and by several anonymous reviewers, who read early versions of the drafts and offered many insightful comments. We thank the series editor, Boris Rozovsky, for his encouragement and consideration. Our thanks also go to Springer senior editor, Achi Dosanjh, for her assistance and help, and to the production manager, and the Springer professionals for their work in finalizing the book. During the years of our study, the research was supported in part by the National Science Foundation, and the National Security Agency. Their continuing support is greatly appreciated.

Detroit, Michigan
Milwaukee, Wisconsin
July 2009

George Yin
Chao Zhu

Declarations

Competing Interests The authors have no competing interests to declare that are relevant to the content of this manuscript.

Conventions

We clarify the numbering system and cross-reference conventions used throughout the book. Equations are numbered consecutively within a chapter. For example, (2.10) indicates the tenth equation in Chapter 2. Corollaries, definitions, examples, lemmas, propositions, remarks, and theorems are numbered sequentially throughout each chapter. For example, Definition 3.1, Theorem 3.2, Remark 3.3, and so on. Assumptions are marked consecutively within a chapter, for example, (H3.1) and (H3.2), etc. For cross reference, an equation is identified by the chapter number and the equation number; similar conventions are used for theorems, remarks, assumptions, and so on.

Throughout the book, we assume that all deterministic processes are Borel measurable and all stochastic processes are measurable and progressively measurable with respect to a given filtration. A subscript generally denotes either a finite or an infinite sequence. However, the ε-dependence of a sequence is designated in the superscript. To facilitate reading, we provide a glossary of symbols used in the subsequent chapters.

Contents

About the Authors

Hai-Dang Nguyen received the B.S. degree in mathematics from the VNU, Hanoi University of Science, Hanoi, Vietnam, in 2008, and the Ph.D. in applied mathematics from Wayne State University, Detroit, MI, USA, in 2018. He joined the University of Alabama, Tuscaloosa, AL, USA, in 2018 as an Assistant Professor and was promoted to Associate Professor, in 2022. His research interests include stochastic systems and applications, mathematical biology, and stochastic control. Dr. Nguyen was the recipient of SIAM Activity Group on Control and Systems Theory Prize in 2019. He is an Associate Editor of *Nonlinear Analysis: Hybrid Systems* and *SIAM Journal on Control and Optimization*

George Yin received the B.S. degree in mathematics from the University of Delaware in 1983, the M.S. degree in electrical engineering, and the Ph.D. degree in applied mathematics from Brown University in 1987. He joined the Department of Mathematics, Wayne State University in 1987, and became Professor in 1996 and University Distinguished Professor in 2017. He moved to the University of Connecticut in 2020. He was the Editor-in-Chief of *SIAM Journal on Control and Optimization* from 2018-2023, and is/was on editorial board of many other journals. He is a Fellow of IEEE, a Fellow of IFAC, and a Fellow of SIAM.

Chao Zhu received his B.S. and M.S. degrees in mathematics from East China Normal University, Shanghai, China, in 1999 and 2002, respectively, and his Ph.D. in mathematics from Wayne State University in 2007, under the supervision of Prof. George Yin. He joined the Department of Mathematical Sciences, the University of Wisconsin-Milwaukee in 2007 and became a Professor in 2018. He is currently serving on the editorial boards of three journals. His research interests include stochastic analysis, and stochastic control and their applications in areas such as mathematical finance, mathematical biology, and risk management.

Glossary of Symbols

A'	transpose of A (either a matrix or a vector)
$\mathrm{Cov}(\xi)$	covariance of a random variable ξ
$\mathbb{C}$	space of complex numbers
$C(D)$	space of real-valued continuous functions defined on D
$C_b^k(D)$	space of real-valued functions with bounded and continuous derivatives up to the order k
C_b^k	$C_b^k(D)$ with $D = \mathbb{R}^r$
C_0^k	C^k-functions with compact support
$\mathscr{C}$	$= \mathscr{C}([-r, 0], \mathbb{R}^n)$, $\mathbb{R}^n$-valued continuous functions defined on $[-r, 0]$
D^c	complement of a set D
$\overline{D}$	closure of a set D, $\overline{D} = D \cup \partial D$
∂D	boundary of a set D
$D([0, T]; \mathsf{S})$	space of S-valued functions being right continuous and having left-hand limits
$D \subset\subset E$	$D \subset \overline{D} \subset E$ and $\overline{D}$ is compact
$\mathbb{E}_{x,i}$	expectation with $X(0) = x$ and $\alpha(0) = i$
$\mathbb{E}\xi$	expectation of a random variable ξ
$\mathcal{F}$	σ-algebra
$\{\mathcal{F}_t\}$	filtration $\{\mathcal{F}_t, t \geq 0\}$
I	identity matrix of suitable dimension
I_A	indicator function of a set A
K	generic positive constant with convention $K + K = K$ and $KK = K$
$\mathcal{M}$	state space of the switching process $\alpha(t)$
$N(x)$	neighborhood of x centered at the origin
$O(y)$	function of y such that $\sup_y \|O(y)\|/\|y\| < \infty$
$\mathbb{P}_{x,i}$	probability with $X(0) = x$ and $\alpha(0) = i$
$\mathbb{P}(\xi \in \cdot)$	probability distribution of a random variable ξ
$Qf(\cdot)(i)$	$= \sum_{j \neq i} q_{ij}(f(j) - f(i))$ where $Q = (q_{ij})$

$Q(x)$	x-dependent generator of the switching process
$\mathbb{R}$	space of real numbers
$\mathbb{R}^r$	r-dimensional real Euclidean space
$S(r)$ or S_r	ball centered at the origin with radius r
$X^{x,i}(t)$	$X(t)$ with initial data $X(0) = x$ and $\alpha(0) = i$
a^+	$= \max\{a, 0\}$ for a real number a
a^-	$= \max\{-a, 0\}$ for a real number a
a.s.	almost surely
$\langle a, b\rangle$	inner product of vectors a and b
$a_1 \wedge \cdots \wedge a_l$	$= \min\{a_1, \ldots, a_l\}$ for $a_i \in \mathbb{R}$, $i = 1, \ldots, l$
$a_1 \vee \cdots \vee a_l$	$= \max\{a_1, \ldots, a_l\}$ for $a_i \in \mathbb{R}$, $i = 1, \ldots, l$
$\mathrm{diag}(A^1, \ldots, A^l)$	diagonal matrix of blocks $A^1, \ldots, A^l$
$\exp(Q)$	e^Q for a matrix Q
f_x or $\nabla_x f$	gradient of f with respect to x
f_{xx} or $\nabla_x^2 f$	Hessian of f with respect to x
i.i.d.	independent and identically distributed
$\ln x$ or $\log x$	natural logarithm of x
$\mathfrak{m}(\cdot)$	Lebesgue measure on $\mathbb{R}$
$o(y)$	a function of y such that $\lim_{y\to 0} o(y)/\lvert y\rvert = 0$
$\mathfrak{p}(dt, dz)$	Poisson random measure with intensity $dt \times \mathfrak{m}(dz)$
$\mathrm{tr}(A)$	trace of matrix A
w.p.1	with probability one
$\lfloor x\rfloor$	integer part of x
$(\Omega, \mathcal{F}, \mathbb{P})$	probability space
$(\Omega, \mathcal{F}, \{\mathcal{F}_t\}, \mathbb{P})$	filtered probability space
$\alpha(t)$	continuous-time process with right-continuous sample paths
δ^{ij}	$= 1$ if $i = j$; $= 0$ otherwise
ε	positive small parameter
$\lambda_{\max}(A)$	maximal eigenvalue of a symmetric matrix A
$\lambda_{\min}(A)$	minimal eigenvalue of a symmetric matrix A
$\mathbb{1}$	a column vector with all entries being 1
$:=$ or $\stackrel{\text{def}}{=}$	defined to be equal to
$\Box$	end of a proof
$\lvert\cdot\rvert$	norm of an Euclidean space or a function space
$\lVert\cdot\rVert$	essential sup-norm

Chapter 1
Introduction and Motivation

1.1 Introduction

This book focuses on switching diffusion processes involving both continuous dynamics and discrete events. Before proceeding to the detailed study, we address the following questions. Why should we study such hybrid systems? What are typical examples arising from applications? What are the main properties we wish to study? This introductory chapter provides motivations of our study, delineates switching diffusions in a simple way, presents a number of application examples, and gives an outline of the entire book.

1.2 Motivation

Owing to their wide range of applications, hybrid switching diffusions, also known as switching diffusion systems, have become more popular recently and have drawn growing attention, especially in the fields of control and optimization. Because of the presence of both continuous dynamics and discrete events, such systems are capable of describing complex systems and their inherent uncertainty and randomness in the environment. The formulation provides ample opportunities for realistic models; it however adds more difficulties in analyzing the underlying systems. Resurgent efforts have been devoted to learning more about the processes and their properties. Much of the study originated from applications arising in control engineering, manufacturing systems, estimation, filtering, two-time-scale systems, and financial engineering; see [138, 143, 208, 254, 255, 267, 292, 295, 314], among others. In these applications, random-switching processes are used to model demand rate or machine capacity in production planning, to describe the volatility changes over time to capture discrete shifts such as market trends and interest rates and the like in finance and insurance, and to model time-varying parameters for

H.-D. Nguyen et al., *Hybrid Switching Diffusions*, Probability Theory and Stochastic Modelling 63, https://doi.org/10.1007/978-3-031-93303-5_1

network problems. In reference to recent developments, this book emphasizes the development of basic properties of switching diffusion processes.

The coexistence of continuous dynamics and discrete events and their interactions reflect the salient features of the systems under consideration. Such systems have become increasingly important for formulation, analysis, and optimization in many applications. Perhaps, one of the reasons is that many real-world applications in the new era require sophisticated models, in which the traditional dynamic system setup using continuous dynamics given by differential equations alone is inadequate. In addition, there have been increasing demands for modeling large-scale and complex systems, designing optimal controls, and conducting optimization tasks. In the traditional setting, the design of a feedback controller is based on a plant having fixed parameters, which is inadequate when the actual system differs from the assumed nominal model. As a consequence, the controller is not able to serve its purposes and cannot attenuate disturbances or perturbations. Much effort has been directed to the design of more "robust" controls in recent years. Studies of hybrid systems with Markov regime switching contribute significantly to this end. Various regime-switching models have been proposed and examined. The so-called jump linear systems, widely used in engineering, have been studied in Mariton [208]; controllability and stabilizability of such systems are treated in Ji and Chizeck [143]. Estimation problems are considered in Sworder and Boyd [267]. Manufacturing and production planning under the framework of hierarchical structure are covered in Sethi and Zhang [255] and Sethi et al. [254]. Financial engineering applications and the use of hybrid geometric Brownian motion models can be found in [295, 314], among others. To reduce the complexity, effort has been made to deal with hybrid systems with regime switching by means of time-scale separation in Yin and Zhang [303]; see also [292, 316]. Hybrid systems have received increasing attention in recent years. Other than the switching diffusion systems mentioned above, one may find the formulation of a somewhat different setup in Bensoussan and Menaldi [23]. An introductory text on stochastic control for jump diffusions can be found in Hanson [113], whereas a treatment of stochastic hybrid systems with applications to communication networks is in Hespanha [121]. For references on stochastic control and controlled Markov processes, we refer the reader to Fleming and Rishel [86], Fleming and Soner [87], Krylov [167], and Yong and Zhou [309].

A special feature of switching diffusion processes is: In the systems, continuous dynamics and discrete events are intertwined. For example, one of the early efforts of using such hybrid models for financial applications can be traced back to [10], in which both the appreciation rate and the volatility rate of a stock depend on a continuous-time Markov chain. In the simplest case, a stock market may be considered to have two "modes" or "regimes," up and down, resulting from the state of the underlying economy, the general mood of investors in the market, and so on. The rationale is that in the different modes or regimes, the volatility and return rates are very different. The introduction of hybrid models makes it possible to describe stochastic volatility in a relatively simple manner (simpler than the so-called stochastic volatility models). Another example is concerned with a wireless communication network. Consider the performance analysis of an adaptive linear multiuser detector in a cellular direct-sequence code-division multiple-access wireless network with changing user activity due to an admission or access controller at the base station. Under certain conditions, an associated optimization problem leads to a switching diffusion limit; see [294]. To summarize, a centerpiece in the applications mentioned above is a two-component Markov process: a continuous component and a discrete-event component.

Long-time behavior of such hybrid systems is one of the major concerns in stochastic processes, systems theory, control, and optimization. Nowadays, there is a fairly well-known theory for stability of diffusion processes. The basic setup of the stability study and the foundational work were originated in the work of Khasminskii [156] and Kushner [173]. The origin of the study may be traced back to Kac and Krasovskii [146] for differential equations perturbed by Markov chains. Comprehensive treatment of Markov processes is in Dynkin [76], whereas that of diffusion processes may be found in Gihman and Skorohod [102], Liptser and Shiryayev [190], and Stroock and Varadhan [266] and references therein. Concerning jump processes and Markov chains, we mention the work of Chung [52], Cox and Miller [56], and Doob [65]. Davis's piecewise deterministic viewpoint adds a new twist to Markov models. Ethier and Kurtz [81] examine weak convergence of Markov processes and provide an in-depth study on characterization of the limit process. Chen's book [42] presents an approach from the angle of coupling methods. Recently, stability of diffusion processes with Markovian switching has received much attention (see, e.g., [11, 12, 164, 199, 204, 311, 321, 324] and the references therein). Some of the recent developments in Markov modulated switching diffusions (where the Markov chain and the Brownian motion are independent) are found in Mao and Yuan [205]. Accommodating the many applications, although these systems are more realistically addressing the demands, the nontraditional setup makes the analysis of switching diffusions more difficult. To be able to effectively treat these systems, it is of foremost importance to have a thorough understanding of the underpinning of the systems. This is the main theme of the current work.

1.3 What Is a Switching Diffusion

In this book, we consider switching diffusion models, in which the discrete events are modeled by a finite-state or a countable-state process. The following diagram provides a visualization of a typical example. Consider a switching diffusion as given in Fig. 1.1 that consists of three diffusions sitting on three parallel plans. The discrete event is a three-state jump process. We denote the pair of processes by

$$\text{(continuous process, discrete event)} = (X(t), \alpha(t)).$$

Suppose that initially, the process is at $(X(0), \alpha(0)) = (x, 1)$. The discrete-event process sojourns in discrete state 1 for a random duration; during this period, the continuous component evolves according to the diffusion process specified by the

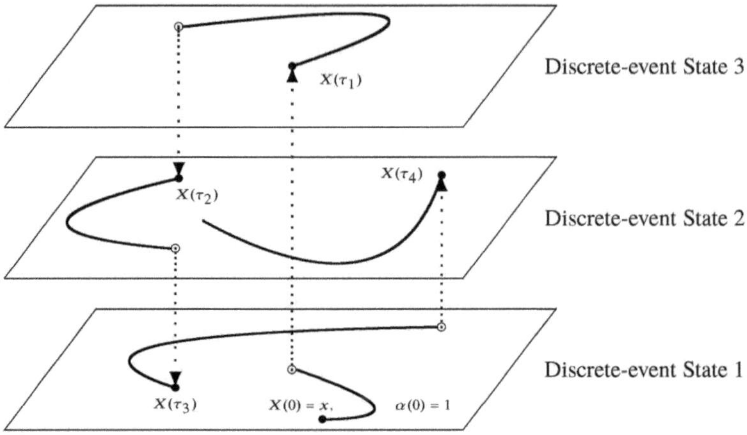

Fig. 1.1 A "sample path" of regime-switching diffusion $(X(t), \alpha(t))$

drift and diffusion coefficients associated with discrete state 1 until a jump of the discrete component takes place. At random moment τ_1, a jump to discrete state 3 occurs. Then the continuous component evolves according to the diffusion process whose drift and diffusion coefficients are determined by discrete event 3. The process wanders around in the third plan until another random jump time τ_2. At τ_2, the system switches to the second parallel plan and follows another diffusion with different drift and diffusion coefficients and so on.

At a first glance, one may feel that the process is not much different from a diffusion because the switching is a finite-state process. Nevertheless, even if the switching is a finite-state Markov chain independent of the Brownian motion (subsequently referred to as Markov-modulated switching diffusions), the switching diffusion is still more difficult to handle. The main reason is the coupling and interactions due to the switching process make the analysis much more complex. For example, when we study recurrence or stability, we have to deal with systems of coupled partial differential equations.

One of the main features of this book is that a large part of it deals with the switching component depending on the continuous component. This is often referred to as a state-dependent switching process in what follows; more precise statements about which will be made later. In this case, the analysis is much more difficult than that of the case in which the switching is independent of the continuous states. As shown in Chapter 2, some of the time-honored characteristics such as continuous dependence of initial data become highly nontrivial; even Feller properties are not straightforward to obtain.

1.4 Examples of Switching Diffusions

To demonstrate the utility of the switching diffusion models, this section provides a number of examples. Some of these examples are revisited in later chapters.

Example 1.1 (Lotka–Volterra Model) Concerning ecological population modeling, proposed by Lotka and Volterra, the well-known Lotka–Volterra models have been investigated extensively in the literature. When two or more species live in proximity and share the same basic requirements, they usually compete for resources, food, habitat, or territory. Both deterministic and stochastic models of the Lotka–Volterra systems have been studied extensively.

It has been noted that the growth rates and the carrying capacities are often subject to environmental noise. Moreover, the qualitative changes of the growth rates and the carrying capacities form an essential aspect of the dynamics of the ecosystem. These changes usually cannot be described by the traditional (deterministic or stochastic) Lotka–Volterra models. For instance, the growth rates of some species in the rainy season are much different from those in the dry season. Moreover, the carrying capacities often vary according to the changes in nutrition and food resources. Similarly, the interspecific or intraspecific interactions differ in different environments.

It is natural to consider a stochastic Lotka–Volterra ecosystem in a random environment that can be formulated by use of an additional factor process. Consider the following stochastic differential equation with regime switching:

$$\begin{aligned} dx(t) = {} & \operatorname{diag}(x_1(t), \dots, x_n(t)) \\ & \times [(b(\alpha(t)) - A(\alpha(t))x(t))dt + \Sigma(\alpha(t)) \circ dw(t)], \end{aligned} \tag{1.1}$$

or equivalently, in componentwise form

$$\begin{aligned} dx_i(t) = x_i(t)\Big\{&\Big[b_i(\alpha(t)) - \sum_{j=1}^{n} a_{ij}(\alpha(t))x_j(t)\Big]dt \\ &+\sigma_i(\alpha(t)) \circ dw_i(t)\Big\}, \ i = 1, \dots, n, \end{aligned} \tag{1.2}$$

where $w(\cdot) = (w_1(\cdot), \dots, w_n(\cdot))'$ is an n-dimensional standard Brownian motion, and for $\alpha \in \mathcal{M}$, $b(\alpha) = (b_1(\alpha), \dots, b_n(\alpha))'$, $A(\alpha) = (a_{ij}(\alpha))$, $\Sigma(\alpha) = \operatorname{diag}(\sigma_1(\alpha), \dots, \sigma_n(\alpha))$ represent different growth rates, community matrices, and noise intensities in different external environments, respectively, and z' denotes the transpose of z. Assume that $b_i(\alpha) > 0$ for each $\alpha \in \mathcal{M}$ and each $i = 1, \dots, n$, and the Markov chain $\alpha(\cdot)$ and the Brownian motion $w(\cdot)$ are independent. Without loss of generality, we also assume that the initial conditions $x(0)$ and $\alpha(0)$ are nonrandom. The above formulation is seen to be in the form of Stratonovich. As explained in [157], this form is more suitable for environment modeling. It is well-known that (1.2) is equivalent to the following stochastic differential equation in the

Itô sense:

$$
\begin{aligned}
dx_i(t) = x_i(t)\Big\{\Big[r_i(\alpha(t)) - \sum_{j=1}^{n} a_{ij}(\alpha(t))x_j(t)\Big]dt \\
+\sigma_i(\alpha(t))dw_i(t)\Big\},\quad i = 1, 2, \ldots, n,
\end{aligned}
\tag{1.3}
$$

where $r_i(\alpha(t)) := b_i(\alpha(t)) + \frac{1}{2}\sigma_i^2(\alpha(t))$ for each $i = 1, 2, \ldots, n$.

Regime-switching stochastic Lotka–Volterra models have received much attention lately. For instance, the study of trajectory behavior of Lotka–Volterra competition bistable systems and systems with telegraph noises was considered in [68], stochastic population dynamics under regime switching was treated in [197], the dynamics of a population in a Markovian environment were studied in [263], and the evolution of a system composed of two predator–prey deterministic systems described by Lotka–Volterra equations in a random environment was investigated in [269]. In the absence of regime switching, the system is completely modeled by stochastic time evolution in a fixed environment. The results in a fixed environment correspond to (1.2) or (1.3) in the case when the Markov chain has only one state or the Markov chain always stays in the fixed state (environment). When random environments are considered, the system's qualitative behavior can be drastically different. In a recent paper of Zhu and Yin [322], we considered several asymptotic properties of the models given above.

In the later part of the book, we, in fact, provide a rather detailed discussion on ecological and biological models. In depth study will be devoted to the permanence and extinction of switching diffusion processes, which is a crucial concern in mathematical biology. We leave further discussions to Chapter 12.

Example 1.2 Here, we consider the problem of balanced realizations, a concept that started gaining popularity in the fields of systems and control in the early 1980s, and has still attracted much attention lately. Balanced realizations have been studied for finite-dimensional linear systems for nearly three decades. The original problem is concerned with linear deterministic systems. In the 1980s, to bridge the gap between minimal realization theory and to treat the problem of finding lower-order approximations, a particular realization was first introduced in the seminal paper by Moore [216], where the problem was studied with principal component analysis of linear systems. The term "balanced" was used because the realizations have a certain symmetry between the input and the output maps characterized by the controllability and observability Grammians. Owing to their importance and their wide range of applications, balanced realizations have attracted much attention. One of the application areas is model reduction. Asymptotic stability of the reduced-order systems was studied in [216], and error bounds between the reduced-order model and the original system were obtained in [106] in terms of the associated singular values. There have been substantial extensions of the theory to time-varying linear systems. Key existence results concerning balanced realizations were contained in [259] and [271]. Subsequent work can be found in [116, 139, 233, 251] and references therein.

We generalize the ideas to include random switching to represent random perturbations under a stochastic environment. We suppose that for each instant t, the state consists of a pair $(X(t), \alpha(t))$ representing the continuous-state component and discrete-event component, respectively. Let $\alpha(t)$ be a continuous-time Markov chain with finite-state space $\mathcal{M} = \{1, 2, \ldots, m_0\}$ and generator:

$$Q = (q_{ij}) \in \mathbb{R}^{m_0 \times m_0} \text{ such that } q_{ij} \geq 0 \text{ for } i \neq j, \text{ and } \sum_{j=1}^{m_0} q_{ij} = 0. \tag{1.4}$$

Our proposed set of equations for balanced realizations includes coupling terms due to the Markov switching of the model. In the trivial case of no switching, the equations are decoupled and are standard balanced realization equations. We loosely carry the term "balanced realization" over to our more general setting. As in the classical theory for the case of no Markovian switching, the condition for balanced realization is represented by a system of algebraic equations. Our system of equations is a set of Riccati equations but with additional couplings between them owing to the discrete events not evident in the classical case. The underlying systems are modulated by a continuous-time Markov chain with state space $\mathcal{M}$. The matrix coefficients of the linear systems depend on the states of the Markov chain. At any given instant t, the Markov chain takes one of the values (e.g., i) in $\mathcal{M}$. Then the system dynamics are determined by the matrix coefficients associated with state i. After a random time, the Markov chain switches to a new state $j \neq i$ and stays there until the next jump. The system dynamics are then determined by the matrix coefficients associated with j. Such random jump linear systems arise frequently in applications where regime changes are utilized to model the random environment. Assume that for each $i \in \mathcal{M}$, $A(i, \cdot)$, $B(i, \cdot)$, and $C(i, \cdot)$ are bounded and continuously differentiable matrix-valued functions with suitable dimensions. Consider the following system:

$$\begin{aligned} \frac{d}{dt} X(t) &= A(\alpha(t), t) X(t) + B(\alpha(t), t) u(t), \quad X(t_0) = x_0, \\ y(t) &= C(\alpha(t), t) X(t), \end{aligned} \tag{1.5}$$

where the state $X(t) \in \mathbb{R}^{n \times 1}$, input $u(t) \in \mathbb{R}^{r \times 1}$, and output $y(t) \in \mathbb{R}^{m_0}$. Note that because a finite-state Markov chain is used, effectively, (1.5) can be written as

$$\begin{aligned} \frac{d}{dt} X(t) &= \sum_{i=1}^{m_0} A(i, t) X(t) I_{\{\alpha(t)=i\}} + \sum_{i=1}^{m_0} B(i, t) u(t) I_{\{\alpha(t)=i\}}, \quad X(t_0) = x_0, \\ y(t) &= \sum_{i=1}^{m_0} C(i, t) X(t) I_{\{\alpha(t)=i\}}, \end{aligned}$$

where I_S is the indicator function of the set S. In what follows, if a square matrix D is positive definite (resp., nonnegative definite), we often write it as $D > 0$ (resp., $D \geq 0$). For $D_1 \in \mathbb{R}^{\iota \times \ell}$ for some $\iota, \ell \geq 1$, D_1' denotes its transpose. For a suitable function $f(t, i)$, we denote

$$Qf(t,\cdot)(i) = \sum_{j=1}^{m_0} q_{ij} f(t, j) = \sum_{j \neq i} q_{ij}(f(t, j) - f(t, i)) \text{ for each } i \in \mathcal{M}.$$

For each $t \geq 0$ and $i \in \mathcal{M}$, a realization $(A(i, t), B(i, t), C(i, t))$ is said to be uniformly completely controllable if and only if there is a $\delta > 0$ such that for some positive $L_c(\delta)$ and $U_c(\delta)$,

$$\infty > U_c(\delta)I \geq G_c(t - \delta, t, i) \geq L_c(\delta)I > 0, \tag{1.6}$$

where $G_c(t - \delta, t, i)$ is the controllability Grammian

$$G_c(t - \delta, t, i) = \int_{t-\delta}^{t} \Phi(t, \lambda, i)B(i, \lambda)B'(i, \lambda)\Phi'(t, \lambda, i)d\lambda, \tag{1.7}$$

and $\Phi(t, \lambda, i)$ is the state transition matrix (see [2, p. 349]) of the equation

$$\frac{dz(t)}{dt} = A(i, t)z(t).$$

For each $t \geq 0$ and $i \in \mathcal{M}$, a realization $(A(i, t), B(i, t), C(i, t))$ is said to be uniformly completely observable if and only if there is a $\delta > 0$ such that for some positive $L_o(\delta)$ and $U_o(\delta)$,

$$\infty > U_o(\delta)I \geq G_o(t, t + \delta, i) \geq L_o(\delta)I > 0, \tag{1.8}$$

where $G_o(t, t + \delta, i)$ is the observability Grammian:

$$G_o(t, t + \delta, i) = \int_{t}^{t+\delta} \Phi'(\lambda, t, i)C'(i, \lambda)C(i, \lambda)\Phi(\lambda, t, i)d\lambda. \tag{1.9}$$

System (1.5) is said to have a balanced realization if there are nonsingular coordinate transformations $T(t, i)$ with

$$P(t, i) \overset{\text{def}}{=} T'(t, i)T(t, i) > 0$$

such that

$$P(t, i)G_c(t, i)P(t, i) = G_o(t, i) + QP(t, \cdot)(i), \quad i = 1, 2, \ldots, m_0. \tag{1.10}$$

Note that regime switching is one of the main features (fully degenerate regime-switching diffusion with the second-order term missing in the system) of such systems compared with the traditional setup. To study the behavior of such systems, it is crucial to have a thorough understanding of the switching diffusion processes. For modeling and analysis of the balanced realizations, we refer the reader to [193] for further reading.

Example 1.3 Continuing with the setup of the hybrid linear system,

$$\dot{x}(t) = A(\alpha(t))x(t) + B(\alpha(t))u(t),$$

where $\alpha(t)$ is a continuous-time Markov chain taking values in a finite set $\mathcal{M} = \{1, \ldots, m_0\}$, $A(i)$ and $B(i)$ for $i \in \mathcal{M}$ are matrices with compatible dimensions, and $u(\cdot)$ is the control. In lieu of one linear system, we have a number of systems coupled through the Markov chain. Such systems have enjoyed numerous applications in emerging application areas such as financial engineering and wireless communications, as well as in existing applications. A class of important problems concerns the asymptotic behavior of such systems when they are in operation for a long time. Very often, in many engineering problems, one is more interested in whether the system is stable. Much interest lies in finding admissible controls so that the resulting system will be stabilized. There has been continuing interest in dealing with hybrid systems under a Markov switching. In [208], stabilization for robust controls of jump linear quadratic (LQ) control problems was treated. In [143], both controllability and stabilizability of jump linear LQ systems were considered. In [34], adaptive LQG problems with finite-state process parameters were treated. Additional difficulties come when the switching process $\alpha(t)$ cannot be observed, but can only be observed in white noise. That is, we can observe

$$dX(t) = [A(\alpha(t))X(t) + B(\alpha(t))u(t)]dt + dw(t).$$

For such partially observed systems, it is natural to use nonlinear filtering techniques. The associated filter is the well-known Wonham filter [277], which is one of a handful of finite-dimensional filters in existence. Stabilization of linear systems with hidden Markov chains was considered in [35, 71]. In both of these references, averaging criteria were used for the purpose of stabilization under

$$\limsup_{t\to\infty} \mathbb{E}[|X(t)|^2 + |u(t)|^2] < \infty,$$

in [71], whereas stabilization under

$$\limsup_{t\to\infty} \frac{1}{t}\mathbb{E}\int_0^t [|X(s)|^2 + |u(s)|^2]ds < \infty$$

was considered in [35]. A question of considerable practical interest is: Can we design controls so that the resulting system will be stable in the almost sure sense? Using Wonham filters and converting the partially observed system to an equivalent fully observed system, almost sure stabilizing controls were found under the criterion

$$\limsup_{t\to\infty} \frac{1}{t} \log |X(t)| \le 0 \text{ almost surely} \tag{1.11}$$

in [26]. The main idea lies in analyzing the sample path properties using a suitable Lyapunov function. We refer the reader to the reference given above for further details.

Example 1.4 Let $\{\theta_n\}$ be a discrete-time Markov chain with finite-state space:

$$\mathcal{M} = \{\overline{\theta}_1, \ldots, \overline{\theta}_{m_0}\} \tag{1.12}$$

and transition probability matrix:

$$P^\varepsilon = I + \varepsilon Q, \tag{1.13}$$

where $\varepsilon > 0$ is a small parameter, I is an $m_0 \times m_0$ identity matrix, and $Q = (q_{ij}) \in \mathbb{R}^{m_0 \times m_0}$ is a generator of a continuous-time Markov chain (i.e., Q satisfies $q_{ij} \ge 0$ for $i \ne j$ and $\sum_{j=1}^{m_0} q_{ij} = 0$ for each $i = 1, \ldots, m_0$). For simplicity, suppose that the initial distribution $\mathbb{P}(\theta_0 = \overline{\theta}_i) = p_{0,i}$ is independent of ε for each $i = 1, \ldots, m_0$, where $p_{0,i} \ge 0$ and $\sum_{i=1}^{m_0} p_{0,i} = 1$.

Let $\{X_n\}$ be an S-state conditional Markov chain (conditioned on the parameter process). The state space of $\{X_n\}$ is $\mathcal{S} = \{e_1, \ldots, e_S\}$, where e_i, for $i = 1, \ldots, S$, denotes the ith standard unit vector with the ith component being 1 and the rest of the components being 0. For each $\theta \in \mathcal{M}$, $A(\theta) = (a_{ij}(\theta)) \in \mathbb{R}^{S\times S}$, the transition probability matrix of X_n, is defined by

$$a_{ij}(\theta) = \mathbb{P}(X_{n+1} = e_j | X_n = e_i, \theta_n = \theta) = \mathbb{P}(X_1 = e_j | X_0 = e_i, \theta_0 = \theta),$$

where $i, j \in \{1, \ldots, S\}$. Assume that for $\theta \in \mathcal{M}$, $A(\theta)$ is irreducible and aperiodic.

Note that the underlying Markov chain $\{\theta_n\}$ is in fact ε-dependent. We suppress the ε-dependence for notational simplicity. The small parameter ε in (1.13) ensures that the entries of the transition probability matrix are nonnegative, because $p^\varepsilon_{ij} = \delta_{ij} + \varepsilon q_{ij} \ge 0$ for $\varepsilon > 0$ small enough, where δ_{ij} denotes the Kronecker δ satisfying $\delta_{ij} = 1$ if $i = j$ and is 0 otherwise. The use of the generator Q makes the row sum of the matrix P be one. Although the true parameter is time varying, it is piecewise constant. Moreover, due to the dominating identity matrix in (1.13), $\{\theta_n\}$ varies slowly in time. The time-varying parameter takes a constant value $\overline{\theta}_i$ for a random duration and jumps to another state $\overline{\theta}_j$ with $j \ne i$ at a random time.

The assumptions on irreducibility and aperiodicity of $A(\theta)$ imply that for each $\theta \in \mathcal{M}$, there exists a unique stationary distribution $\pi(\theta) \in \mathbb{R}^{S\times 1}$ satisfying

$$\pi'(\theta) = \pi'(\theta)A(\theta), \text{ and } \pi'(\theta)\mathbb{1}_S = 1,$$

where $\mathbb{1}_\ell \in \mathbb{R}^{\ell \times 1}$ with all entries being equal to 1. We use a stochastic approximation algorithm to track the time-varying distribution $\pi(\theta_n)$ that depends on the underlying Markov chain θ_n.

We use the following adaptive algorithm of least mean squares (LMS) type with constant stepsize in order to construct a sequence of estimates $\{\widehat{\pi}_n\}$ of the time-varying distribution $\pi(\theta_n)$

$$\widehat{\pi}_{n+1} = \widehat{\pi}_n + \mu(X_{n+1} - \widehat{\pi}_n), \tag{1.14}$$

where μ denotes the stepsize. Define $\widetilde{\pi}_n = \widehat{\pi}_n - \mathbb{E}\pi(\theta_n)$. Then (1.14) can be rewritten as

$$\widetilde{\pi}_{n+1} = \widetilde{\pi}_n - \mu\widetilde{\pi}_n + \mu(X_{n+1} - \mathbb{E}\pi(\theta_n)) + \mathbb{E}(\pi(\theta_n) - \pi(\theta_{n+1})). \tag{1.15}$$

Note that $\widehat{\pi}_n$, $\pi(\theta_n)$, and hence $\widetilde{\pi}_n$ are column vectors (i.e., they take values in $\mathbb{R}^{S\times 1}$).

For $0 < T < \infty$, we construct a piecewise constant interpolation of the stochastic approximation iterates $\widehat{\pi}_n$ as

$$\widehat{\pi}^\mu(t) = \widehat{\pi}_n, \quad t \in [\mu n, \mu n + \mu). \tag{1.16}$$

The process $\widehat{\pi}^\mu(\cdot)$ so defined is in $D([0, T]; \mathbb{R}^S)$, which is the space of functions defined on $[0, T]$ taking values in $\mathbb{R}^S$ that are right continuous, have left limits, and are endowed with the Skorohod topology.

In one of our recent works [294], using weak convergence methods to carry out the analysis, we have shown that $(\widehat{\pi}^\mu(\cdot), \theta^\mu(\cdot))$ converges weakly to $(\widehat{\pi}(\cdot), \theta(\cdot))$, which is a solution of the following switching ordinary differential equation:

$$\frac{d}{dt}\widehat{\pi}(t) = \pi(\theta(t)) - \widehat{\pi}(t), \quad \widehat{\pi}(0) = \widehat{\pi}_0. \tag{1.17}$$

The above switching ODE displays a very different behavior from the trajectories of systems derived from the classical ODE approach for stochastic approximation. It involves a random element because $\theta(t)$ is a continuous-time Markov chain with generator Q. Furthermore, we can show that $\{(\widehat{\pi}_n - \mathbb{E}\pi(\theta_n))/\sqrt{\mu}\}$ is tight for $n \ge n_0$ for some positive integer n_0. In an effort to determine the rate of variation of the tracking error sequence, we define a scaled sequence of the tracking errors $\{v_n\}$ and its continuous-time interpolation $v^\mu(\cdot)$ by

$$v_n = \frac{\widehat{\pi}_n - \mathbb{E}\{\pi(\theta_n)\}}{\sqrt{\mu}}, \quad n \ge n_0, \quad v^\mu(t) = v_n \text{ for } t \in [n\mu, n\mu + \mu). \tag{1.18}$$

We have shown in [294] that $(v^\mu(\cdot), \theta^\mu(\cdot))$ converges weakly to $(v(\cdot), \theta(\cdot))$ satisfying the switching diffusion equation:

$$dv(t) = -v(t)dt + \Sigma^{1/2}(\theta(t))dw(t), \tag{1.19}$$

where $w(\cdot)$ is a standard Brownian motion and $\Sigma(\theta)$ is the almost sure limit given by

$$\begin{aligned} &\lim_{\mu\to 0} \frac{1}{n} \sum_{k_1=m}^{n+m-1} \sum_{k=m}^{n+m-1} (X_{k+1}(\theta) - \mathbb{E}X_{k+1}(\theta))(X_{k_1+1}(\theta) - \mathbb{E}X_{k_1+1}(\theta))' \\ &\quad = \Sigma(\theta) \text{ a.s.} \end{aligned} \tag{1.20}$$

Note that for each θ, $\Sigma(\theta)$ is an $S \times S$ deterministic matrix and in fact,

$$\begin{aligned} &\frac{1}{n} \sum_{k_1=m}^{n+m-1} \sum_{k=m}^{n+m-1} \mathbb{E}\left\{(X_{k+1}(\theta) - \mathbb{E}X_{k+1}(\theta))(X_{k_1+1}(\theta) - \mathbb{E}X_{k_1+1}(\theta))'\right\} \\ &\quad = \Sigma(\theta) \text{ as } \mu \to 0. \end{aligned} \tag{1.21}$$

Because of the regime switching, the system is qualitatively different from the existing literature on stochastic approximation methods; see Kushner and Yin [177].

Example 1.5 Suppose that $w(\cdot)$ is a d-dimensional Brownian motion and $\alpha(\cdot)$ is a continuous-time Markov chain generated by Q taking values in $\mathcal{M} = \{1, \ldots, m_0\}$. Define $\mathcal{F}_t = \sigma\{w(s), \alpha(s) : 0 \le s \le t\}$, and denote by $L^2_{\mathcal{F}}(0, T; \mathbb{R}^{m_0})$ the set of all $\mathbb{R}^{m_0}$-valued, measurable stochastic processes $f(t)$ adapted to $\{\mathcal{F}_t\}_{t\ge 0}$ satisfying $\mathbb{E}\int_0^T |f(t)|^2 dt < \infty$.

Consider a financial market in which $d+1$ assets are traded continuously. One of the assets is a bank account whose price $P_0(t)$ is subject to the following stochastic ordinary differential equation with Markovian switching:

$$\begin{cases} dP_0(t) = r(t, \alpha(t))P_0(t)dt, & t \in [0, T], \\ P_0(0) = p_0 > 0, \end{cases} \tag{1.22}$$

where $r(t, i) \ge 0$, $i = 1, 2, \ldots, m_0$, are given as the interest rate processes corresponding to different market modes. The other d assets are stocks whose price processes $P_m(t)$, $m = 1, 2, \ldots, d$, satisfy the following system of stochastic differential equations with Markov switching:

$$\begin{cases} dP_m(t) = P_m(t)\left\{b_m(t, \alpha(t))dt + \displaystyle\sum_{n=1}^{d} \sigma_{mn}(t, \alpha(t))dw_n(t)\right\}, & t \in [0, T], \\ P_m(0) = p_m > 0, \end{cases} \tag{1.23}$$

where for each $i = 1, 2, \ldots, m_0$, $b_m(t, i)$ is the appreciation rate process and $\sigma_m(t, i) := (\sigma_{m1}(t, i), \ldots, \sigma_{md}(t, i))$ is the volatility or the dispersion rate process of the mth stock, corresponding to $\alpha(t) = i$.

Define the volatility matrix as

$$\sigma(t, i) := \begin{pmatrix} \sigma_1(t, i) \\ \vdots \\ \sigma_d(t, i) \end{pmatrix}, \quad \text{for each } i = 1, \ldots, m_0. \tag{1.24}$$

We assume that the following non-degeneracy condition

$$\sigma(t, i)\sigma'(t, i) \geq \delta I, \quad \forall t \in [0, T], \quad \text{and} \quad i = 1, 2, \ldots, m_0, \tag{1.25}$$

is satisfied for some $\delta > 0$ and that all the functions $r(t, i)$, $b_m(t, i)$, $\sigma_{mn}(t, i)$ are measurable and uniformly bounded in t. Inequality (1.25) is in the sense of positive definiteness for symmetric matrices; that is, $\sigma(t, i)\sigma'(t, i) - \delta I$ is a positive definite matrix.

Suppose that the initial market mode $\alpha(0) = i_0$. Consider an agent with an initial wealth $x_0 > 0$. Denote by $x(t)$ the total wealth of the agent at time $t \geq 0$. Assuming that the trading of shares takes place continuously and that there are no transaction cost and consumptions, then one has (see, e.g., [309, p.57])

$$\begin{cases} dx(t) = \Big\{r(t, \alpha(t))x(t) + \displaystyle\sum_{m=1}^{d} \big[b_m(t, \alpha(t)) - r(t, \alpha(t))\big]u_m(t)\Big\}dt \\ \qquad\qquad + \displaystyle\sum_{n=1}^{d}\sum_{m=1}^{d} \sigma_{mn}(t, \alpha(t))u_m(t)dw_n(t), \\ x(0) \;\; = x_0 > 0, \;\; \alpha(0) = i_0, \end{cases} \tag{1.26}$$

where $u_m(t)$ is the total market value of the agent's wealth in the mth asset, $m = 0, 1, \ldots, d$, at time t. We call $u(\cdot) = (u_1(\cdot), \ldots, u_d(\cdot))'$ a portfolio of the agent. Note that once $u(\cdot)$ is determined, $u_0(\cdot)$, the asset in the bank account is completely specified inasmuch as

$$u_0(t) = x(t) - \sum_{i=1}^{d} u_i(t).$$

Thus, we need only consider $u(\cdot)$ and ignore $u_0(\cdot)$.

Setting

$$B(t, i) := (b_1(t, i) - r(t, i), \ldots, b_d(t, i) - r(t, i)), \;\; i = 1, 2, \ldots, m_0, \tag{1.27}$$

we can rewrite the wealth equation (1.26) as

$$\begin{cases} dx(t) = [r(t,\alpha(t))x(t) + B(t,\alpha(t))u(t)]dt + u'(t)\sigma(t,\alpha(t))dw(t), \\ x(0) \ = x_0, \ \ \alpha(0) = i_0. \end{cases} \tag{1.28}$$

A portfolio $u(\cdot)$ is said to be admissible if $u(\cdot) \in L^2_{\mathcal{F}}(0, T; \mathbb{R}^d)$ and the stochastic differential equation (1.28) has a unique solution $x(\cdot)$ corresponding to $u(\cdot)$. In this case, we refer to $(x(\cdot), u(\cdot))$ as an admissible (wealth, portfolio) pair. The agent's objective is to find an admissible portfolio $u(\cdot)$, among all the admissible portfolios whose expected terminal wealth is $\mathbb{E}x(T) = z$ for some given $z \in \mathbb{R}^1$, so that the risk measured by the variance of the terminal wealth

$$\text{Var}\, x(T) \equiv \mathbb{E}[x(T) - \mathbb{E}x(T)]^2 = \mathbb{E}[x(T) - z]^2 \tag{1.29}$$

is minimized. Finding such a portfolio $u(\cdot)$ is referred to as the mean-variance portfolio selection problem. An interested reader can find further details of the study on this problem in [320].

Example 1.6 This problem arises in the context of insurance and risk theory. Suppose that there is a finite set $\mathcal{M} = \{1, \dots, m_0\}$, representing the possible regimes (configurations) of the environment. At each $i \in \mathcal{M}$, assume that the premium is payable at rate $c(i)$ continuously. Let $U(t,i)$ be the surplus process given the initial surplus $u > 0$ and initial state i:

$$U(t,i) = u + \int_0^t c(\alpha(s))ds - S(t),$$

where $S(t)$, as in the classical risk model, is a compound Poisson process and $\alpha(t)$ is a continuous-time Markov chain with state space $\mathcal{M}$ representing the random environment. Under suitable conditions, we obtained Lunderberg-type upper bounds and nonexponential upper bounds for the ruin probability and treated the renewal-type system of equations for ruin probability when the claim sizes were exponentially distributed. To proceed further, we consider a class of jump diffusions with regime switching to prepare us for treating applications involving more general risk models.

One of the main features of [287] is that there is an additional Markov chain, which enables the underlying surplus to vary in accordance with different regimes. Consider jump diffusions modulated by a continuous-time Markov chain. Because the dynamic systems are complex, it is of foremost importance to reduce the complexity. Taking into consideration the inherent hierarchy in a complex system [260], and different rates of variations of subsystems and components, we use the two-time-scale method leading to systems in which the fast and slow rates of change are in sharp contrast. Then we proceed to reduce the system complexity by aggregation/decomposition and averaging methods. We demonstrate that under broad conditions, associated with the original systems, there are limit or reduced systems, which are averages with respect to certain invariant measures. Using weak convergence methods [175, 176], we obtain the limit system via martingale problem formulation.

Motivated by risk theory applications [72], we consider a switching jump diffusion model. We formulate the problem in a general way, which allows the consideration of other problems where switching jump diffusions are involved.

Let $\Gamma \subset \mathbb{R}^r - \{0\}$ so that Γ is the range space of the impulsive jumps, $w(\cdot)$ be a real-valued standard Brownian motion, and $N(\cdot, \cdot)$ be a Poisson measure such that $N(t, H)$ counts the number of impulses on $[0, t]$ with values in the set H. Let $f(\cdot, \cdot, \cdot) : [0, T] \times \mathbb{R} \times \mathcal{M} \mapsto \mathbb{R}$, $\sigma(\cdot, \cdot, \cdot) : [0, T] \times \mathbb{R} \times \mathcal{M} \mapsto \mathbb{R}$, $g(\cdot, \cdot, \cdot) : \Gamma \times \mathbb{R} \times \mathcal{M} \mapsto \mathbb{R}$, and $\alpha(\cdot)$ be a continuous-time Markov chain having state space $\mathcal{M}$ and generator $Q(t)$. A brief description of the jump-diffusion process with a modulating Markov chain can be found in the appendix. Consider the following jump-diffusion processes with regime switching:

$$\begin{aligned} X(t) = x_0 &+ \int_0^t f(s, X(s), \alpha(s))ds + \int_0^t \sigma(s, X(s), \alpha(s))dw(s) \\ &+ \int_0^t \int_\Gamma g(\gamma, X(s^-), \alpha(s^-))N(ds, d\gamma). \end{aligned} \tag{1.30}$$

We assume that $w(\cdot)$, $N(\cdot)$, and $\alpha(\cdot)$ are mutually independent. Note that compared with the traditional jump-diffusion processes, the coefficients involved in (1.30) all depend on an additional switching process—the Markov chain $\alpha(t)$.

In the context of risk theory, $X(t)$ can be considered as the surplus of the insurance company at time t, x_0 is the initial surplus, $f(t, X(t), \alpha(t))$ represents the premium rate (assumed to be ≥ 0), $g(\gamma, X(t), \alpha(t))$ is the amount of the claim if there is one (assumed to be ≤ 0), and the diffusion is used to model additional uncertainty of the claims and/or premium incomes. Similar to the volatility in stock market models, $\sigma(\cdot, \cdot, i)$ represents the amount of oscillations or volatility in an appropriate sense. The model is sufficiently general to cover the traditional as well as the diffusion-perturbed ruin models. It may also be used to represent security price in finance (see [209, Chapter 3]). The process $\alpha(t)$ may be viewed as an environment variable dictating the regime. The use of the Markov chain results from consideration of a general trend of the market environment as well as other economic factors. The economic and/or political environment changes lead to the changes of surplus regimes resulting in markedly different behavior of the system across regimes.

Defining a centered Poisson measure and applying generalized Ito's rule, we can obtain the generator of the jump-diffusion process with regime switching and formulate a related martingale problem. Instead of a single process, we have to deal with a collection of jump-diffusion processes that are modulated by a continuous-time Markov chain. Suppose that λ is positive such that $\lambda\Delta + o(\Delta)$ represents the probability of a switch of regime in the interval $[t, t+\Delta)$, and $\pi(\cdot)$ is the distribution of the jump. Then the generator of the underlying process can be written as

$$\begin{aligned}\mathcal{G}F(t,x,\iota) = &\left(\frac{\partial}{\partial t} + \mathcal{L}\right) F(t,x,\iota)\\ &+ \int_{\Gamma} \lambda[F(t,x+g(\gamma,x,\iota),\iota) - F(t,x,\iota)]\pi(d\gamma)\\ &+ Q(t)F(t,x,\cdot)(\iota), \quad \text{for each } \iota \in \mathcal{M},\end{aligned} \tag{1.31}$$

where

$$\begin{aligned}\mathcal{L}F(t,x,\iota) &= \frac{1}{2}\sigma^2(t,x,\iota)\frac{\partial^2}{\partial x^2}F(t,x,\iota) + f(t,x,\iota)\frac{\partial}{\partial x}F(t,x,\iota),\\ Q(t)F(t,x,\cdot)(\iota) &= \sum_{\ell=1}^{m_0} q_{\iota\ell}(t)F(t,x,\ell) = \sum_{\ell\neq\iota} q_{\iota\ell}(t)[F(t,x,\ell) - F(t,x,\iota)].\end{aligned} \tag{1.32}$$

Of particular interest is the case that the switching process and the diffusion vary at different rates. To reduce the amount of computational effort, we propose a two-time-scale approach. By concentrating on time-scale separations, we treat two cases in Chapter 9. In the first one, the regime switching is significantly faster than the dynamics of the jump diffusions, whereas in the second case, the diffusion varies an order of magnitude faster than the other processes. As shown, averaging plays an essential role in these problems.

Example 1.7 Consider the process $Y(\cdot) = (X(\cdot), \alpha(\cdot))$ that has two components, the diffusion component $X(\cdot)$ and the pure jump component $\alpha(\cdot)$. The state space of the process is $\mathbb{S} \times \mathcal{M}$, where $\mathbb{S}$ is the unit circle and $\mathcal{M} = \{1, \ldots, m_0\}$ is a state space with finitely many elements for the jump process. By identifying the endpoints 0 and 1, let $x \in [0, 1]$ be the coordinates in $\mathbb{S}$. We assume that the generator of the process $(X(t), \alpha(t))$ is of the form (1.31) with the jump part missing (i.e., $\lambda = 0$) and with $(x, i) \in [0, 1] \times \mathcal{M}$.

The probability density $p(x, t) = (p(x, t, 1), \ldots, p(x, t, m_0))$ of the process $Y(\cdot)$, with

$$\int_{\Gamma} p(x,t,i)dx = \mathbb{P}(X(t) \in \Gamma, \alpha(t) = i),$$

satisfies the adjoint equation, namely, the system of forward equations:

$$\frac{\partial p(x,t,i)}{\partial t} = \mathcal{D}^* p(x,t,i) + \sum_{j=1}^{m_0} p(x,t,j)q_{ji}(x),$$

where for a suitable function $f(x, t, i)$,

$$
\begin{aligned}
\mathcal{D}^* f(x,t,i) &= \mathcal{D}^*(x,t,i) f(x,t,i) \\
&= \frac{1}{2}\frac{\partial^2}{\partial x^2}(a_i(x,t) f(x,t,i)) - \frac{\partial}{\partial x}(b_i(x,t) f(x,t,i)), \\
p(x,0,i) &= g_i(x),
\end{aligned}
$$

for $i = 1, \ldots, m_0$, and $g(x) = (g_1(x), \ldots, g_{m_0}(x))$ is the initial distribution for $Y(t)$. The existence and properties of such switching diffusion processes can be found in [102, Section 2.2]. Suppose that all conditions of [93, Theorem 16, p. 82] are satisfied. Then the system of forward equations has a unique solution.

Suppose that $q_{ij}(x) > 0$ for each $i \neq j$. Then the transition density $p(x,t)$ converges exponentially fast to

$$
\nu(x) = (\nu_1(x), \ldots, \nu_{m_0}(x)),
$$

the density of the stationary distribution; that is,

$$
|p(x,t,i) - \nu_i(x)| \le K \exp(-\gamma t),
$$

for some $K > 0$ and $\gamma > 0$. The estimate is the well-known spectrum gap condition. This spectrum gap condition helps us to study related asymptotic properties of the system.

Example 1.8 This example is concerned with a switching diffusion process with a state-dependent switching component. We compare the trajectories of a linear stochastic system with drift and diffusion coefficients given by

$$
f(x) = 0.11x \quad \text{and} \quad \sigma(x) = 0.2x \tag{1.33}
$$

and another regime-switching linear system with $\alpha(t) \in \{1, 2\}$ and the drift and diffusion coefficients given by

$$
\begin{aligned}
&f(x,1) = -0.2x, \quad f(x,2) = 0.11x, \\
&\sigma(x,1) = x, \quad \sigma(x,2) = 0.2x,
\end{aligned} \tag{1.34}
$$

respectively. For the switching diffusion, the switching process is continuous-state-dependent with the $Q(x)$ given by

$$
Q(x) = \begin{pmatrix} -5\cos^2 x & 5\cos^2 x \\ 10\cos^2 x & -10\cos^2 x \end{pmatrix}.
$$

What happens if the actual model is the one with switching, but we mismodeled the system so that instead of the switching model, we used a simple linear stochastic differential equation model with drift and diffusion coefficients given by (1.33)? If in fact, the system parameters are really given by (1.34), and if the random environment

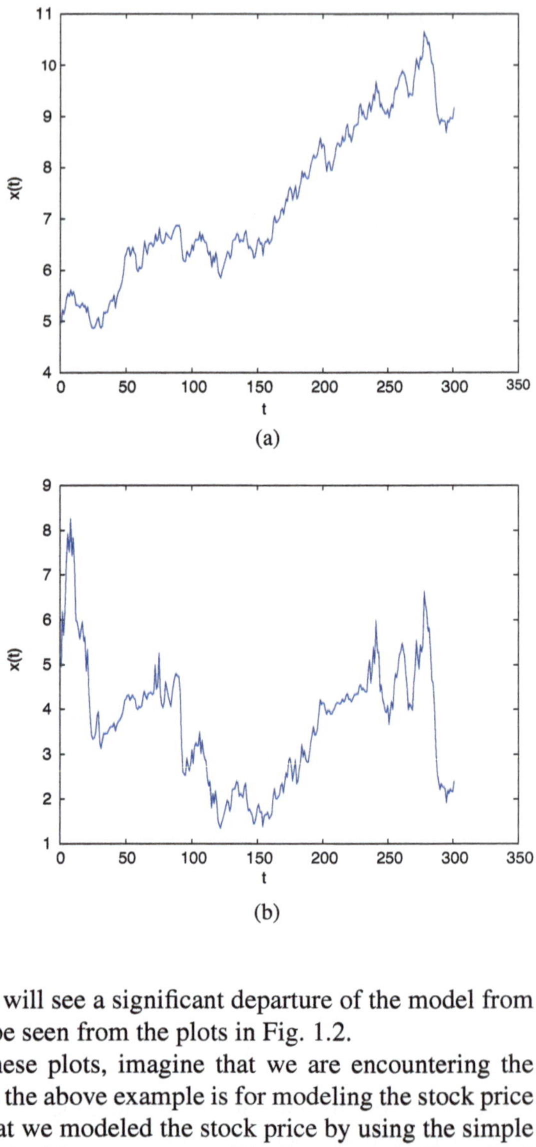

Fig. 1.2 Comparisons of sample paths for a linear diffusion with that of a regime-switching diffusion. (**a**) A sample path of a mismodeled system: A linear stochastic differential equation with constant coefficients. (**b**) A sample path of the true system: A regime-switching stochastic system of differential equations

is included in the model, then we will see a significant departure of the model from that of the true system. This can be seen from the plots in Fig. 1.2.

To get further insight from these plots, imagine that we are encountering the following scenarios. Suppose that the above example is for modeling the stock price of a particular equity. Suppose that we modeled the stock price by using the simple geometric Brownian motion model with a return rate of 11% and a volatility of 0.2 as given in (1.33). However, the market is really subject to random environment perturbation. For the bull market, the rates are as before, but corresponding to the bear market, the return rate and the volatility are given by -20% and 1, respectively. The deviation of the plot in Fig. 1.2 shows that if one uses a simple linear SDE model, one cannot capture the market behavior.

Example 1.9 This example is concerned with a controlled switching diffusion model. We first give a motivation of the study and begin the discussion on Markov decision processes. Consider a real-valued process $X(\cdot) = \{X(t) : t \geq 0\}$ and a feedback control $u(\cdot) = \{u(t) = u(X(t)) : t \geq 0\}$ such that $u(t) \in \Gamma$ for $t \geq 0$ with Γ being a compact subset of an Euclidean space. Γ denotes the control space. A vast literature is concerned with a certain optimal control problem. That is, one aims to find a feedback $u(\cdot)$ so that an appropriate objective function is minimized.

Extending the idea to switching diffusion processes, we consider the following pair of processes $(X(t), \alpha(t)) \in \mathbb{R}^r \times \mathcal{M}$, where $\mathcal{M} = \{1, \ldots, m_0\}$. Let U be the control space or action space, which is a compact set of an Euclidean space. Suppose that $b(\cdot, \cdot, \cdot) : \mathbb{R}^r \times \mathcal{M} \times U \mapsto \mathbb{R}^r$, and $\sigma(\cdot, \cdot, \cdot) : \mathbb{R}^r \times \mathcal{M} \times U \mapsto \mathbb{R}^{r \times r}$ are appropriate functions satisfying certain regularity conditions, and $Q(x) = (q_{ij}(x)) \in \mathbb{R}^{m_0 \times m_0}$ satisfies that for each x, $q_{ij}(x) \geq 0$ for $i \neq j$, $\sum_{j=1}^{m_0} q_{ij}(x) = 0$ for each $i \in \mathcal{M}$. For each $i \in \mathcal{M}$ and suitable smooth function $h(\cdot, i)$, define an operator

$$\begin{aligned}\mathcal{L}h(x, i) = \nabla h'(x, i) b(x, i, u) + \frac{1}{2}\mathrm{tr}[\nabla^2 h(x, i)\sigma(x, i, u)\sigma'(x, i, u)] \\ + \sum_{j=1}^{m_0} q_{ij}(x) h(x, j).\end{aligned} \tag{1.35}$$

Then we call $(X(t), \alpha(t))$ a controlled switching diffusion process. Following the notation of Markov decision processes, we may write the process $(X(t), \alpha(t))$ as $(X(t), \alpha(t)) \sim \mathcal{L}(u)$. Note that owing to the feedback controls used, it is natural to require the matrix-valued function $Q(\cdot)$ to be x-dependent. This model is a generalization of the usual Markov decision processes. When the functions $b(x, i, u) \equiv 0$ and $\sigma(x, i, i) \equiv 0$, the model reduces to a Markov decision process. When the matrix-valued function $Q(x) \equiv 0$, the model reduces to controlled diffusions. Typically, our objective is to select the control $u(\cdot)$ so that a control objective function is achieved.

Example 1.10 Originating from statistical mechanics, mean-field models are concerned with many-body systems with interactions. To overcome the difficulty of interactions due to the many bodies, one of the main ideas is to replace all interactions to any one body with an average or effective interaction. This reduces any multibody problem to an effective one-body problem. Although its main motivation and development are in statistical mechanics, such models have also enjoyed recent applications in, for example, graphical models in artificial intelligence.

If the field or particle exhibits many interactions in the original system, the mean field will be more accurate for such a system. The usefulness, the potential impact on many practical scenarios, and the challenges from both physics and mathematics have attracted much attention in recent years. The work of Dawson [61] presents a detailed study on the cooperative behavior of mean fields.

Owing to the rapid progress in technology, more complicated systems are encountered in applications. In response to such challenges, much effort has been devoted to modeling and analysis for more sophisticated systems. Frequently, there are factors that cannot be described by the traditional models. One of the ideas is to bring regime switching into the formulation, so as to deal with the coexistence of continuous dynamics and discrete events. Although one may use a pure jump process to represent the discrete events, due to the interactions of the many bodies, the discrete process, in fact, is correlated with the diffusive dynamics.

In Xi and Yin [281], the following system is considered. Suppose that $\alpha(t)$ is a right-continuous random process taking values in a finite-state space $\mathcal{M} = \{1, 2, \ldots, m_0\}$. Consider an ℓ-body mean-field model with switching described by the following system of Itô stochastic differential equations. For $i = 1, 2, \ldots, \ell$,

$$\begin{aligned} dX_i(t) = &\big[\gamma(\alpha(t))X_i(t) - X_i^3(t) - \beta(\alpha(t))(X_i(t) - \overline{X}(t))\big]dt \\ &+\sigma_{ii}(X(t), \alpha(t))dw_i(t), \end{aligned} \tag{1.36}$$

where

$$\begin{aligned} \overline{X}(t) &= \frac{1}{\ell}\sum_{j=1}^{\ell} X_j(t), \\ X(t) &= (X_1(t), X_2(t), \ldots, X_\ell(t))', \end{aligned} \tag{1.37}$$

and $\gamma(i) > 0$ and $\beta(i) > 0$ for $i \in \mathcal{M}$. Moreover, the transition rules of $\alpha(t)$ are specified by

$$\mathbb{P}\{\alpha(t+\Delta) = j|\alpha(t) = j, X(t) = x\} = \begin{cases} q_{ij}(x)\Delta + o(\Delta), & \text{if } i \neq j, \\ 1 + q_{ii}(x)\Delta + o(\Delta), & \text{if } i = j, \end{cases} \tag{1.38}$$

which hold uniformly in $\mathbb{R}^\ell$ as $\Delta \downarrow 0$, for $x = (x_1, x_2, \ldots, x_\ell)' \in \mathbb{R}^\ell$ and $\sigma = (\sigma_{ij}) \in \mathbb{R}^\ell \times \mathbb{R}^\ell$. A number of asymptotic properties were obtained in [281] including regularity, Feller properties, and exponential ergodicity. We refer the interested reader to the aforementioned reference for further details.

As shown in the examples above, all the systems considered involve hybrid switching diffusions. To have a better understanding of each of the problems and the properties of the corresponding processes, it is important that we have a thorough understanding of the switching diffusion process. This is our objective in this book.

1.5 Outline of the Book

This chapter serves as a prelude to the book. It gives motivation and presents examples for the switching diffusion processes. After this short introduction, the rest of the book is arranged as follows.

The next three chapters present basic properties such as Feller and strong Feller, recurrence, and ergodicity. Chapter 2 begins with the precise definition of the switching diffusion processes. The existence and uniqueness for solutions of switching diffusions are developed, followed by basic properties such as regularity, weak continuity, Feller and strong Feller properties, and so on. Then continuous and smooth dependence of initial data are presented. The proofs of these results are different from the traditional setup and much more involved due to the coupling resulting from the switching component.

Chapter 3 is concerned with recurrence and positive recurrence. Necessary and sufficient conditions for positive recurrence are given. We show that recurrence and positive recurrence are independent of the open set chosen. Furthermore, we demonstrate that we can work with a fixed discrete component. Based on treatment of elliptic systems of partial differential equations, we obtain criteria of positive recurrence using Lyapunov functions. Then we translate these into conditions on the coefficients of the switching diffusions. We also consider the case under linearization for the continuous component.

Chapter 4 focuses on ergodicity. It is shown that for a positive recurrent switching diffusion, there exists a unique invariant measure. Not only are the existence and uniqueness proved, but the form of the invariance measure is given. It reveals the salient feature of the underlying process owing to the switching. The ergodic measures obtained enable one to carry out control and optimization tasks with replacement of the instantaneous measures by the ergodic measures.

Next, our attention is devoted to numerical solutions of switching diffusions. As their diffusion counterpart, very often closed-form solutions of nonlinear switching diffusions are hard to come by. Thus, numerical approximation becomes a viable and often the only alternative. Developing numerical results is important. Therefore, Chapter 5 is devoted to numerical approximations, in which we present numerical algorithms of the Euler–Maruyama type. We manly concentrate on constant stepsize algorithms. The use of constant stepsize algorithms have been popular and used often by researchers working in the fields of systems theory, control, and signal processing, among others. Using weak convergence methods, we establish the convergence of the numerical approximations. Then we also look at the rate of convergence issue.

To proceed, we focus on stability. Chapters 6 and 7 proceed with the stability analysis. The approach is based on Lyapunov function methods. Chapter 6 studies stability of switching diffusions. The notion is in the sense of stability in probability. First, definitions of stability and instability are given. Then criteria are presented based on Lyapunov functions. In addition, more verifiable conditions on the coefficients of the systems are presented. Necessary and sufficient conditions for pth-moment stability of linear (in the continuous component) systems and linear approximations are provided. It is noted that contrary to common practice, very often nonquadratic Lyapunov functions are easier to use.

Chapter 7 takes up the stability study for fully degenerated systems, namely, switching ordinary differential equations. We treat the case that the diffusion matrix becomes 0, or equivalently, we have switching ordinary differential equations. Based on Lyapunov methods, we obtain sufficient conditions for stability. Similar to the nondegenerate switching diffusions, the results on stability and instability have an "eigenvalue" gap. That is, the stability criteria are in terms of the largest eigenvalue of a certain matrix and the instability is in terms of the smallest eigenvalue of the same matrix. To close up the gap, we introduce a logarithm transformation leading to a necessary and sufficient condition in terms of a one-dimensional function.

Chapter 8 proceeds with the investigation of the invariance principle. In studying deterministic systems represented by ordinary differential equations, the concept of the invariance principle in the sense of LaSalle has come into being (see [182] and [181]). It enables one to obtain far-reaching results compared to the purely equilibrium point analysis alone. Building on this idea, the invariance principle for stochastic counterpart for diffusion processes was considered by Kushner in [174], whereas that for stochastic differential delay equations was treated in [200]. This chapter is focused on switching diffusions using a sample path approach.

Chapter 9 extends the model in another direction, which answers the question "What happens if not all discrete-event states belong to the same recurrent class?" Furthermore, using the methods of perturbed Lyapunov functions, this problem is answered. We divide the state space of the switching component into several irreducible groups. Within each irreducible group, the switching component moves frequently, whereas from one group to another, it moves relatively infrequently. We distinguish the fast and slow motions by introducing a small parameter $\varepsilon > 0$. Aggregating the states in each recurrent group into one superstate, as $\varepsilon \to 0$, we obtain a limit system, whose drift and diffusion coefficients are averaged out with respect to the invariant measure of the switching part. We show that if the limit process is recurrent, so is the original process for sufficiently small $\varepsilon > 0$. Also considered is an application to financial engineering. The equity asset is assumed to follow a Markov regime-switching diffusion, in which both the return rates and the volatility depend on a continuous-time Markov chain. This model is an alternative to the so-called stochastic volatility model of Hull and White [132], which is well-known nowadays. The motivation of our study is similar to the fast mean reverting models treated in [89], in which Fouque, Papanicolaou, and Sircar assumed the volatility to be a fast-varying diffusion process. It was shown that the Black–Scholes pricing formula is a first approximation to the stochastic volatility model when fast mean reversion is observed. We treat an alternative model with regime switching and reach a similar conclusion.

Chapter 10 focuses on models with an additional Lévy jump component. The key here is the utilization of two-time scales. The motivation stems from reduction of complexity. Two different situations are considered. In the first case, the switching component changes an order of magnitude faster than the continuous component. In the second case, the diffusion is fast-varying. In both cases, we obtain appropriation limits using a weak convergence approach.

The next two chapters examine further properties of switching diffusions. Even for continuous-state-dependent switching, our setup in the previous chapters does not involve past dependence. Chapter 11 extends such consideration by allowing the switching to depend on the past information of the continuous states. This extends the model to allow another layer of complex systems.

Chapter 12 is focused on models from mathematical biology. As ecology and evolutionary biology have traditionally been the dominant parts of mathematical biology, there is a vast literature devoted to mathematical models to study ecological systems and population dynamics in recent years. A long-standing and fundamental question is to find minimal conditions (necessary and sufficient) for long-term permanence of a population, or for long-term coexistence of interacting species. Much effort has been placed on studying permanence and extinction of certain species and understanding the traits that enable two species to coexist or lead one species to dominate the habitat of another. It is crucial to determine the fate of an invading species: Will it successfully establish itself or perish in the attempt? Environmental fluctuations make population dynamics inherently stochastic. Hybrid formulation makes the consideration closer to reality, but the analysis is more challenging. The combined effects of biotic interactions and environmental fluctuations are crucial for understanding species richness. Biotic effects can sometimes lead to species extinction, but the addition of environmental randomness might reverse extinction into coexistence. We take a deep dive to study the hybrid stochastic biological and ecological models. Not only is the study important for the specific models given, but it provides much insight to a wider range of models in mathematical biology as well as other related fields.

Finally, for convenience, an appendix including a number of mathematical preliminaries is placed at the end of the book. It serves as a quick reference. Topics are discussed in the appendix including Markov chains, martingales, Gaussian processes, diffusions, jump diffusions, and weak convergence methods. Although detailed developments are often omitted, appropriate references are provided for the reader to facilitate further reading.

1.6 Notes

This book intends to cover a wide range of topics related to switching diffusions. Nevertheless, some topics will still be omitted, for example, computational nonlinear filtering, which is a fundamental problem in control and systems theory. Given the state of a system that is not completely observable, filtering is concerned with state estimation based on noise corrupted observations of the system state. In 1964, Kushner [172] first derived the nonlinear filtering equation (known as the Kushner equation) satisfied by a normalized conditional density using the Itô calculus, which is in fact a probability distribution. Prior to that, Stratonovich worked on similar problem using the Stratonovich integral [265] Subsequently, Duncan [73], Mortensen [217], and Zakai [313] derived the filtering equations for unnormalized conditional densities independently. Nowadays, this unnormalized equation is commonly referred to as the Duncan–Mortensen–Zakai equation. No matter if the equations are normalized or unnormalized, they are all stochastic partial differential equations of infinite dimension. Although the celebrated results of the Kushner equation and the Duncan–Mortensen–Zakai equation settled the matters of nonlinear filtering theoretically, finding the solutions in closed form is difficult or virtually impossible because of the nonlinearity. The systems thus do not have closed-form solution for the filtering problems, which presents a class of challenging problems. The problem persists for more than 60 years. Recently, in lieu of solving the stochastic partial differential equations directly, we used a different alternative, namely, using machine learning and deep neural networks to resolve the problem [243], which provides an effective computational alternative for solving a long-standing problem. For interested reader, we refer to [243] for further details.

Chapter 2
Switching Diffusion

2.1 Introduction

This chapter provides an introduction to switching diffusions. First, the definition of switching diffusion is given. After establishing the existence and uniqueness of solutions to the associated stochastic differential equations, the chapter explores weak continuity, Feller and strong Feller properties. Additionally, the definitions of regularity and criteria ensuring it are discussed. Furthermore, smooth dependence on initial data is demonstrated.

The rest of the chapter is arranged as follows. After this short introductory, Section 2.2 presents the setup for switching processes. In addition, the existence and uniqueness of the solution are established under non-Lipschitz conditions. Section 2.3 is concerned with regularity. Section 2.4 deals with weak continuity of the process $(X(t), \alpha(t))$. Section 2.5 proceeds with Feller properties. Section 2.6 goes one step further to obtain strong Feller properties. Section 2.7 presents smooth dependence properties of solutions of the switching diffusions. Section 2.8 gives remarks on how nonhomogeneous cases in which both the drift and diffusion coefficients depend explicitly on time t can be handled. Finally, Section 2.9 provides additional notes and remarks.

2.2 Switching Diffusions

We work with a probability space $(\Omega, \mathcal{F}, \mathbb{P})$ throughout this book. A family of σ-algebras $\{\mathcal{F}_t\}$, for $t \geq 0$ or $t = 1, 2, \ldots$, is called a filtration if $\mathcal{F}_s \subset \mathcal{F}_t \subset \mathcal{F}$ for all $0 \leq s < t < \infty$. We set $\mathcal{F}_\infty := \sigma(\bigcup_{t\geq 0} \mathcal{F}_t)$. We say that the filtration $\{\mathcal{F}_t\}$ satisfies *the usual conditions* if it is right continuous and $\mathcal{F}_0$ contains all $\mathbb{P}$-null sets in $\mathcal{F}$. A probability space $(\Omega, \mathcal{F}, \mathbb{P})$ together with a filtration $\{\mathcal{F}_t\}$ is said to be a filtered probability space, denoted by $(\Omega, \mathcal{F}, \{\mathcal{F}_t\}, \mathbb{P})$.

H.-D. Nguyen et al., *Hybrid Switching Diffusions*, Probability Theory and Stochastic Modelling 63, https://doi.org/10.1007/978-3-031-93303-5_2

Let $r \geq 1$ be a positive integer and denote $\mathcal{M} = \{1, \ldots, m_0\}$ $(2 \leq m_0 \leq \infty)$. Assume that $b : \mathbb{R}^r \times \mathcal{M} \mapsto \mathbb{R}^r$ and $\sigma : \mathbb{R}^r \times \mathcal{M} \mapsto \mathbb{R}^{r\times r}$ are Borel measurable functions. Let (X, α) be a càdlàg (right continuous with left limits) strong Markov process taking values on $\mathbb{R}^r \times \mathcal{M}$. The first component X satisfies the following stochastic differential equation:

$$dX(t) = b(X(t), \alpha(t))dt + \sigma(X(t), \alpha(t))dw(t), \tag{2.1}$$

where $w(\cdot)$ is an $\mathbb{R}^r$-valued standard Brownian motion. The second component α is a continuous-time random process taking values in the finite or countably infinite set $\mathcal{M}$ such that

$$\mathbb{P}\{\alpha(t+\Delta) = j | \alpha(t) = i, X(s), \alpha(s), s \leq t\} = \begin{cases} q_{ij}(X(t))\Delta + o(\Delta), & \text{if } i \neq j, \\ 1 + q_{ii}(X(t))\Delta + o(\Delta), & \text{if } i = j, \end{cases} \tag{2.2}$$

when $\Delta \downarrow 0$. In (2.2), $Q(x) = (q_{ij}(x)) \in \mathbb{R}^{m_0 \times m_0}$ satisfies the following conditions: For all $i \neq j$, the function $q_{ij} : \mathbb{R}^r \mapsto \mathbb{R}_+$ is Borel measurable and nonnegative, and for all $i \in \mathcal{M}$ and $x \in \mathbb{R}^r$, $q_{ii}(x) = -\sum_{j\neq i} q_{ij}(x)$. Note that $Q(x)$ depends on x and it is called the generator of the process α.

We say the matrix-valued function $Q(x) = (q_{ij}(x)) : \mathbb{R}^r \to \mathbb{R}^{m_0\times m_0}$ satisfies the *q-property*, if $q_{ij}(\cdot)$ is Borel measurable for all $i, j \in \mathcal{M}$; $q_{ij}(x) \geq 0$ for $j \neq i$ and $q_{ii}(x) = -\sum_{j\neq i} q_{ij}(x) > -\infty$ for all $x \in \mathbb{R}^r$. In view of the terminology in [42], we can also say that $Q(x)$ is *totally stable and conservative*. Note that an alternative definition of the q-property can be formulated. Interested readers can find further details in [52].

The evolution of the discrete component $\alpha(\cdot)$ can be represented by a stochastic integral with respect to a Poisson random measure (see, e.g., [100, 262]). Indeed, for $x \in \mathbb{R}^r$, $i \in \mathcal{M}$, let $\Delta_{ij}(x)$, $j \neq i$ be disjoint, left-closed, right-open intervals of the positive half real line, each having length $q_{ij}(x)$. We set $\Delta_{ij}(x) = \emptyset$ if $q_{ij}(x) = 0$, $i \neq j$. A particular way of constructing $\Delta_{ij}(x)$ will be given in (2.29). Define a function $h : \mathbb{R}^r \times \mathcal{M} \times \mathbb{R}_+ \mapsto \mathbb{R}$ by

$$h(x, i, z) = \sum_{j=1}^{m_0} (j - i) I_{\{z\in\Delta_{ij}(x)\}}. \tag{2.3}$$

That is, $h(x, i, z) = j - i$ if $z \in \Delta_{ij}(x)$; otherwise, $h(x, i, z) = 0$. Then (2.2) is equivalent to

$$d\alpha(t) = \int_{\mathbb{R}_+} h(X(t-), \alpha(t-), z)\mathfrak{p}(dt, dz), \tag{2.4}$$

where $\mathfrak{p}(dt, dz)$ is a Poisson random measure with intensity $dt \times \mathfrak{m}(dz)$ and $\mathfrak{m}$ is the Lebesgue measure on $\mathbb{R}$. The Poisson random measure $\mathfrak{p}(\cdot, \cdot)$ is independent of the Brownian motion $w(\cdot)$. Recall that $\mathfrak{p}(dt, dz)$ is a Poisson random measure on $[0, \infty) \times \mathbb{R}_+$ with intensity $dt \times \mathfrak{m}(dz)$ if

(i) $\mathfrak{p} : \Omega \mapsto \mathbf{N}$ is $\mathcal{F}/\mathcal{B}_{\mathbf{N}}$ measurable, where $\mathbf{N}$ is the collection of counting measures on $[0, \infty) \times \mathbb{R}_+$, and $\mathcal{B}_{\mathbf{N}}$ is the smallest σ-algebra on $\mathbf{N}$ with respect to which all $\mathbf{N} \ni \mu \mapsto \mu(B) \in \{0, 1, \dots\} \cup \{\infty\}$, $B \in \mathcal{B}([0, \infty) \times \mathbb{R}_+)$ (the set of Borel measurable sets of $[0, \infty) \times \mathbb{R}_+$) are measurable, and
(ii) for all $0 \le s < t$, $m \in \mathbb{N}$, and disjoint $A_1, A_2, \dots, A_m \in \mathcal{B}(\mathbb{R}_+)$ and $\lambda_i > 0$, we have

$$\mathbb{E}\left[\exp\left[-\sum_{i=1}^{m} \lambda_i \mathfrak{p}((s, t] \times A_i)\right] \Big| \sigma\{\mathfrak{p}((0, r] \times B) : r \le s,\, B \in \mathcal{B}(\mathbb{R})\}\right]$$
$$= \exp\left[-(t - s) \sum_{i=1}^{m} (1 - e^{-\lambda_i}) \mathfrak{m}(A_i)\right], \text{ a.s.}$$

Note that for each $A \in \mathcal{B}(\mathbb{R}_+)$, $\{\mathfrak{p}(t, A), t \ge 0\}$ is a Poisson process with intensity $\mathfrak{m}(A) = \mathbb{E}[\mathfrak{p}(1, A)]$, where $\mathfrak{p}(t, A) := \mathfrak{p}((0, t] \times A)$, $t \ge 0$. We refer to [136] for more discussions on Poisson random measures.

Definition 2.1 We say that a stochastic process $(X, \alpha) \in \mathbb{R}^r \times \mathcal{M}$ on a filtered probability space $(\Omega, \mathcal{F}, \{\mathcal{F}_t\}_{t \ge 0}, \mathbb{P})$ satisfying the usual condition is a *solution to* (2.1) *and* (2.4) *with initial condition* $(X_0, \alpha_0) \in \mathbb{R}^r \times \mathcal{M}$ if:

(i) $\mathbb{P}\{(X(0), \alpha(0)) = (X_0, \alpha_0)\} = 1$,
(ii) there exists an r-dimensional $\{\mathcal{F}_t\}$-Brownian motion $w(\cdot)$ and an independent $\{\mathcal{F}_t\}$-Poisson random measure $\mathfrak{p}(dt, dz)$ on $[0, \infty) \times \mathbb{R}_+$ with intensity $dt \times \mathfrak{m}(dz)$,
(iii) (X, α) has càdlàg sample paths and is adapted to $\{\mathcal{F}_t\}$, and
(iv) for any $t \in [0, \infty)$, we have almost surely

$$\begin{cases} X(t) = X_0 + \int_0^t b(X(s), \alpha(s))ds + \int_0^t \sigma(X(s), \alpha(s))dw(s), \\ \alpha(t) = \alpha_0 + \int_0^t \int_{\mathbb{R}_+} h(X(s-), \alpha(s-), z)\mathfrak{p}(ds, dz). \end{cases}$$

The solution $(X, \alpha) \in \mathbb{R}^r \times \mathcal{M}$ is called a *strong solution* if the filtration $\{\mathcal{F}_t\}$ is generated by the initial condition (X_0, α_0), the Brownian motion $w(\cdot)$, and Poisson random measure $\mathfrak{p}(dt, dz)$, augmented by the $\mathbb{P}$-null sets. In other words, $\{\mathcal{F}_t\}$ is generated as follows. First, we let

$$\mathcal{G}_t := \sigma(X_0, \alpha_0, w(s), \mathfrak{p}((0, s] \times A) : 0 \le s \le t, A \in \mathcal{B}(\mathbb{R}_+)),\ 0 \le t < \infty,$$

and $\mathcal{G}_\infty := \sigma\left(\bigcup_{t \ge 0} \mathcal{G}_t\right)$, as well as the collection of null sets:

$$\mathcal{N} := \{N \subset \Omega : \text{there exists a } G \in \mathcal{G}_\infty \text{ with } N \subset G \text{ and } \mathbb{P}(G) = 0\}.$$

We then create the augmented filtration:

$$\mathcal{F}_t := \sigma(\mathcal{G}_t \cup \mathcal{N}),\ 0 \le t < \infty;\ \ \mathcal{F}_\infty := \sigma\Big(\bigcup_{t\ge 0} \mathcal{F}_t\Big).$$

Sometimes a solution to (2.1) and (2.4) exists only locally and *explodes* in finite time. Therefore, we need to introduce the notion of a *local solution* to (2.1) and (2.4).

Definition 2.2 We say that a stochastic process $(X, \alpha) \in \mathbb{R}^r \times \mathcal{M}$ on a filtered probability space $(\Omega, \mathcal{F}, \{\mathcal{F}_t\}_{t\ge 0}, \mathbb{P})$ satisfying the usual condition is a *solution to* (2.1) *and* (2.4) *up to an explosion time with initial condition* $(X_0, \alpha_0) \in \mathbb{R}^r \times \mathcal{M}$ if

(a) (i), (ii), and (iii) of Definition 2.1 are satisfied,
(b) with $\beta_n := \inf\{t \ge 0 : |X(t)| \vee \alpha(t) \ge n\}$, $n \in \mathbb{N}$, we have almost surely

$$\begin{cases} X(t\wedge\beta_n) = X_0 + \int_0^{t\wedge\beta_n} b(X(s), \alpha(s))ds + \int_0^{t\wedge\beta_n} \sigma(X(s), \alpha(s))dw(s), \\ \alpha(t\wedge\beta_n) = \alpha_0 + \int_0^{t\wedge\beta_n}\int_{\mathbb{R}_+} h(X(s-), \alpha(s-), z)\mathfrak{p}(ds, dz), \end{cases}$$

for all $t \in [0, \infty)$.

We refer to $\beta_\infty := \lim_{n\to\infty} \beta_n$ as the *explosion time* for (X, α) and stipulate that $(X(t), \alpha(t)) = \infty$ for all $t \in [\beta_\infty, \infty)$ if $\beta_\infty < \infty$; in such a case, the process $\{(X(t), \alpha(t)) : 0 \le t < \beta_\infty\}$ is said to be a *local solution* of (2.1) and (2.4). The solution (X, α) is said to be *non-explosive* or a *global solution* if $\beta_\infty = \infty$ a.s.

It will be shown later in Lemma 11.66 that the solution to (2.1) and (2.4) satisfies (2.2) if $q_{ij}(\cdot)$ is a continuous function for each $i, j \in \mathcal{M}$. The two-component process $(X(\cdot), \alpha(\cdot))$ satisfying (2.1)–(2.2) (or, (2.1) and (2.4)) is termed a *switching diffusion* or a *regime-switching diffusion*. Naturally, for the two-component process $(X(t), \alpha(t))$, we call $X(t)$ the continuous component and $\alpha(t)$ the discrete component, in accordance with their sample path properties.

For any solution (X, α) to system (2.1) and (2.4), and any $f : \mathbb{R}^r \times \mathcal{M} \mapsto \mathbb{R}$ so that $f(\cdot, \imath) \in C^2$ for each $\imath \in \mathcal{M}$, the generalized Itô formula [262, Chapter 2] reads

$$f(X(t), \alpha(t)) - f(X(0), \alpha(0)) = \int_0^t \mathcal{L}f(X(s), \alpha(s))ds + M_1(t) + M_2(t), \tag{2.5}$$

where

$$
\begin{aligned}
M_1(t) &= \int_0^t \big\langle \nabla f(X(s), \alpha(s)), \sigma(X(s), \alpha(s)) dw(s) \big\rangle, \\
M_2(t) &= \int_0^t \int_{\mathbb{R}} \big[f(X(s-), \alpha(s-) + h(X(s-), \alpha(s-), z)) \\
&\qquad - f(X(s-), \alpha(s-)) \big] \widetilde{\mathfrak{p}}(ds, dz),
\end{aligned}
$$

are local martingales and

$$
\widetilde{\mathfrak{p}}(ds, dz) = \mathfrak{p}(ds, dz) - ds \times \mathfrak{m}(dz)
$$

is a martingale measure. The operator $\mathcal{L}$ is defined as follows:

$$
\mathcal{L} f(x, \iota) = \mathcal{L}_\iota f(x, \iota) + Q(x) f(x, \cdot)(\iota), \tag{2.6}
$$

in which

$$
\mathcal{L}_\iota f(x, \iota) := \frac{1}{2} \mathrm{tr}\big(a(x, \iota) \nabla^2 f(x, \iota) \big) + \langle b(x, \iota), \nabla f(x, \iota) \rangle, \tag{2.7}
$$

$$
\begin{aligned}
Q(x) f(x, \cdot)(\iota) &:= \sum_{j \in \mathcal{M}} q_{\iota j}(x) [f(x, j) - f(x, \iota)] \\
&= \int_{[0,\infty)} [f(x, \iota + h(x, \iota, z)) - f(x, \iota)] \mathfrak{m}(dz),
\end{aligned} \tag{2.8}
$$

where $a(x, \iota) := \sigma(x, \iota) \sigma'(x, \iota)$, and $\nabla f(x, \iota)$ and $\nabla^2 f(x, \iota)$ denote the gradient and Hessian of $f(x, \iota)$ with respect to x, respectively.

In view of the generalized Itô formula,

$$
M_f(t) = f(X(t), \alpha(t)) - f(X(0), \alpha(0)) - \int_0^t \mathcal{L} f(X(s), \alpha(s)) ds, \, t \geq 0 \tag{2.9}
$$

is a local martingale. If for each $\iota \in \mathcal{M}$, $f(\cdot, \iota) \in C_0^2$ (C^2 functions with compact support), then $M_f(t)$ defined in (2.9) becomes a martingale. Similar to the case of diffusion processes, we can define the notion of the solution of the martingale problem accordingly.

Another consequence of the generalized Itô formula (2.5) is that the Dynkin formula follows. Indeed, let $f : \mathbb{R}^r \times \mathcal{M} \mapsto \mathbb{R}$ so that $f(\cdot, \iota) \in C^2$ for each $\iota \in \mathcal{M}$, and τ_1, τ_2 be bounded stopping times such that $0 \leq \tau_1 \leq \tau_2$ a.s. If $f(X(t), \alpha(t))$ and $\mathcal{L} f(X(t), \alpha(t))$ and so on are bounded on $t \in [\tau_1, \tau_2]$ with probability 1, then

$$
\mathbb{E} f(X(\tau_2), \alpha(\tau_2)) = \mathbb{E} f(X(\tau_1), \alpha(\tau_1)) + \mathbb{E} \int_{\tau_1}^{\tau_2} \mathcal{L} f(X(s), \alpha(s)) ds; \tag{2.10}
$$

see [205] or [262] for details. In what follows, for convenience (with multi-index notation used, for instance) and emphasis on the x-dependence, we often use $D_x f(x, \alpha)$ and $\nabla f(x, \alpha)$ interchangeably to represent the gradient. In addition, for two vectors x and y with appropriate dimensions, we use $x'y$ and $\langle x, y\rangle$ interchangeably to represent their inner product.

For the existence and uniqueness of the strong Markov process (X, α) satisfying system (2.1) and (2.2), we make the following assumptions.

(H2.1) There exists a nondecreasing function $\zeta : [0, \infty) \mapsto [1, \infty)$ that is continuously differentiable and satisfies

$$\int_0^{\infty} \frac{du}{u\zeta(u) + 1} = \infty, \tag{2.11}$$

such that for all $x \in \mathbb{R}^r$ and $k \in \mathcal{M}$,

$$2\langle x, b(x, k)\rangle + |\sigma(x, k)|^2 \le H[|x|^2\zeta(|x|^2) + 1], \tag{2.12}$$

$$q_k(x) := -q_{kk}(x) = \sum_{l \in \mathcal{M}\setminus\{k\}} q_{kl}(x) \le g(k), \tag{2.13}$$

$$\sum_{l \in \mathcal{M}\setminus\{k\}} (f(l) - f(k))q_{kl}(x) \le H(1 + \Phi(x) + f(k)), \tag{2.14}$$

where H is a positive constant, $g : \mathcal{M} \to \mathbb{R}_+$ is a locally bounded function, the function $f : \mathcal{M} \mapsto \mathbb{R}_+$ is nondecreasing satisfying $f(m) \to \infty$ as $m \to \infty$, and

$$\Phi(x) := \exp\left\{\int_0^{|x|^2} \frac{du}{u\zeta(u) + 1}\right\}, \quad x \in \mathbb{R}^r. \tag{2.15}$$

(H2.2) Assume the following conditions hold:

(a) If $r = 1$, then there exist a positive number δ_0 and a nondecreasing and concave function $\varrho : [0, \infty) \mapsto [0, \infty)$ satisfying

$$\int_{0+} \frac{du}{\varrho(u)} = \infty \tag{2.16}$$

such that for all $k \in \mathcal{M}$, $R > 0$, and $x, z \in \mathbb{R}$ with $|x| \vee |z| \le R$ and $|x - z| \le \delta_0$,

$$\operatorname{sign}(x - z)(b(x, k) - b(z, k)) \le \kappa_R \varrho(|x - z|), \tag{2.17}$$

$$|\sigma(x, k) - \sigma(z, k)|^2 \le \kappa_R |x - z|, \tag{2.18}$$

where κ_R is a positive constant and $\operatorname{sign}(a) = 1$ if $a > 0$ and -1 if $a \le 0$. Here in (2.16) and throughout the section, $\int_{0+} \frac{du}{\varrho(u)} := \lim_{\varepsilon \downarrow 0} \int_{\varepsilon}^{1} \frac{du}{\varrho(u)}$.

(b) If $r \geq 2$, then there exist a positive number δ_0 and a nondecreasing and concave function $\varrho : [0, \infty) \mapsto [0, \infty)$ satisfying

$$0 < \varrho(u) \leq (1+u)^2 \varrho(u/(1+u)) \text{ for all } u > 0, \text{ and } \int_{0+} \frac{du}{\varrho(u)} = \infty \qquad (2.19)$$

such that for all $k \in \mathcal{M}$, $R > 0$, and $x, z \in \mathbb{R}^r$ with $|x| \vee |z| \leq R$ and $|x-z| \leq \delta_0$,

$$2\langle x - z, b(x,k) - b(z,k)\rangle + |\sigma(x,k) - \sigma(z,k)|^2 \leq \kappa_R \varrho(|x - z|^2), \qquad (2.20)$$

where κ_R is a positive constant.

(H2.3) There exists a concave function $\gamma : \mathbb{R}_+ \mapsto \mathbb{R}_+$ with $\gamma(0) = 0$ such that for all $x, y \in \mathbb{R}^r$ with $|x| \vee |y| \leq R$, we have

$$\sum_{l \in \mathcal{M} \setminus \{k\}} |q_{kl}(x) - q_{kl}(y)| \leq \kappa_R \gamma(|x - y|), \quad \forall k \in \mathcal{M}. \qquad (2.21)$$

for some positive constant κ_R (which, without loss of generality, can be assumed to be the same positive constant as in that (2.17), (2.18), and (2.20)).

Remark 2.3 We make some comments concerning Assumptions (H2.1), (H2.2), and (H2.3). Assumption (H2.1) places some growth conditions on the coefficients of the system (2.1)–(2.2) so that its solution, if any, cannot explode in finite time a.s. Examples of functions satisfying (2.11) include $\zeta(u) = 1$, $\zeta(u) = \log u$, and $\zeta(u) = \log u \log(\log u)$ for u large. When $\zeta(u) = 1$, (2.12) reduces to the usual linear growth condition. With other choices of ζ, (2.11) allows superlinear growth condition for the coefficients of (2.1) with respect to the variable x for each $k \in \mathcal{M}$. This is motivated by applications such as Lotka–Volterra models [157, 322] and SIR and SIRS epidemic models [63, 220], in which the coefficients have superlinear growth conditions. Conditions (2.13) and (2.14) are imposed so that the α component will not explode in finite time with probability 1; see the proof of Theorem 2.6 and Remark 2.7 for details.

Assumptions (H2.2) and (H2.3) impose certain continuity requirements on the coefficients of the system (2.1) and (2.2) to guarantee the existence and uniqueness of a solution. Examples of functions satisfying (2.16) or (2.19) include $\varrho(u) = u$ and concave and increasing functions such as $\varrho(u) = u \log(1/u)$, $\varrho(u) = u \log(\log(1/u))$, and $\varrho(u) = u \log(1/u) \log(\log(1/u))$ for $u \in (0, \delta)$ with $\delta > 0$ small enough. When $\varrho(u) = u$, Assumption (H2.2) is just the usual local Lipschitz condition. With other choices of continuity modularity, Assumption (H2.2) allows the drift and diffusion coefficients of (2.1) to be non-Lipschitz with respect to the variable x. It is also worth noting that the regularity condition for $Q(\cdot)$ given in Assumption (H2.3) is very mild. These, in turn, present more opportunities for building realistic and flexible mathematical models for a wide range of applications. Indeed, non-Lipschitz coefficients are present in areas such as branching diffusion in biology and the Cox–Ingersoll–Ross model in math finance.

It is also worth pointing out that (2.17), (2.18), and (2.20) of Assumption (H2.2) only require the modulus of continuity to hold in a small neighborhood of the diagonal line $x = z$ in $\mathbb{R}^r \otimes \mathbb{R}^r$ with $|x| \vee |z| \le R$ for each $R > 0$. This is in contrast to those in [186] and adds some subtlety in the proof of pathwise uniqueness for (2.22).

When $r = 1$, Assumption (H2.2) allows the diffusion coefficient $\sigma(\cdot, k)$ to be locally Hölder continuous with exponent $\alpha \in [\frac{1}{2}, 1]$. This is motivated by the pioneering result in [286] for diffusion processes. Such a result was extended to stochastic differential equations with jumps; see, for example, [96, 185] and [186], among others. The continuity assumption (2.17) on the drift coefficient $b(\cdot, k)$ is slightly more general than that in [186]. In particular, (2.17) will be satisfied as long as $b(\cdot, k)$ is decreasing.

Lemma 2.4 *Suppose* (H2.2) *and* (2.12) *hold. Then for each* $k \in \mathcal{M}$*, the stochastic differential equation*

$$X(t) = x + \int_0^t b(X(s), k)ds + \int_0^t \sigma(X(s), k)dw(s) \tag{2.22}$$

has a unique non-explosive strong solution.

Proof The proof is divided into three steps. The first step demonstrates that any solution to (2.22) will not explode in finite time with probability 1 under (2.12). Step 2 establishes the pathwise uniqueness result for (2.22). This is further split into two cases: $r \ge 2$ and $r = 1$. Finally, we establish the existence and uniqueness of a non-explosive strong solution to (2.22) in Step 3.

Step 1. Consider the function Φ defined in (2.15). Note that $\Phi(x) = \phi(|x|^2)$, where $\phi(r) := \exp\left\{\int_0^r \frac{dz}{z\zeta(z)+1}\right\}$ for $r \ge 0$. Then we have

$$\frac{d}{dr}\phi(r) = \frac{\phi(r)}{r\zeta(r)+1} > 0, \text{ and } \frac{d^2}{dr^2}\phi(r) = \phi(r)\frac{1-\zeta(r)-r\zeta'(r)}{(r\zeta(r)+1)^2}.$$

Since $\zeta(r) \ge 1$ and ζ is nondecreasing, it follows that $\phi''(r) \le 0$ and thus ϕ is a concave function. On the other hand, thanks to (2.11), we have $\phi(r) \to \infty$ as $r \to \infty$ and hence $\Phi(x) \to \infty$ as $|x| \to \infty$. Moreover, straightforward computations lead to $D\Phi(x) = 2\frac{d}{dr}\phi(|x|^2)x$ and $D^2\Phi(x) = 2\frac{d}{dr}\phi(|x|^2)I + 4\frac{d^2}{dr^2}\phi(|x|^2)xx'$. Consequently, we can compute

$$\begin{aligned}\mathcal{L}_k\Phi(x) &= 2\frac{d}{dr}\phi(|x|^2)\langle x, b(x)\rangle + \operatorname{tr}\left(\sigma(x)\sigma'(x)\left[\frac{d}{dr}\phi(|x|^2)I + 2\frac{d^2}{dr^2}\phi(|x|^2)xx'\right]\right) \\ &\le \frac{d}{dr}\phi(|x|^2)\big(2\langle x, b(x)\rangle + |\sigma(x)|^2\big) + 2\frac{d^2}{dr^2}\phi(|x|)|\langle x, \sigma(x)\rangle|^2 \\ &\le \frac{\phi(|x|^2)}{|x|^2\zeta(|x|^2)+1}H(|x|^2\zeta(|x|^2)+1) \le H\phi(|x|^2) = H\Phi(x),\end{aligned} \tag{2.23}$$

where we used (2.12) and the fact that $\frac{d^2}{dr^2}\phi(r) \le 0$ to derive the second inequality. Then for any solution X to (2.22), since Φ is positive, thanks to Itô's formula, we see that the process $\{e^{-Ht}\Phi(X(t)), t \ge 0\}$ is a positive supermartingale. Then we can apply the optional sampling theorem to the process $\{e^{-Ht}\Phi(X(t)), t \ge 0\}$ to argue that $\mathbb{P}\{\lim_{n\to\infty} \tau_n = \infty\} = 1$, where $\tau_n := \inf\{t \ge 0 : |X(t)| \ge n\}$. That is, X cannot explode in finite time with probability one and hence the solution X is defined on $[0, \infty)$.

Step 2 (i). In this step, we follow the arguments in the proof of Theorem 3.2 of [186] to show that pathwise uniqueness holds for (2.22) when $r = 1$. First, let $\{a_n\}$ be a strictly decreasing sequence of real numbers satisfying $a_0 = 1$, $\lim_{n\to\infty} a_n = 0$, and $\int_{a_n}^{a_{n-1}} \frac{du}{u} = n$ for each $n \ge 1$. For each $n \ge 1$, let ρ_n be a nonnegative continuous function with support on (a_n, a_{n-1}) so that

$$\int_{a_n}^{a_{n-1}} \rho_n(u)du = 1 \text{ and } \rho_n(u) \le 2(nu)^{-1} \text{ for all } u > 0.$$

For $x \in \mathbb{R}$, define

$$\psi_n(x) = \int_0^{|x|} \int_0^y \rho_n(z)dzdy. \tag{2.24}$$

We can immediately verify that ψ_n is even and twice continuously differentiable, with

$$\frac{d}{du}\psi_n(u) = \text{sign}(u) \int_0^{|u|} \rho_n(z)dz = \text{sign}(u)\left|\frac{d}{du}\psi_n(u)\right|, \tag{2.25}$$

and

$$\left|\frac{d}{du}\psi_n(u)\right| \le 1, \quad 0 \le |u|\frac{d^2}{du^2}\psi_n(u) = |u|\rho_n(|u|) \le \frac{2}{n}, \quad \text{and} \quad \lim_{n\to\infty} \psi_n(u) = |u| \tag{2.26}$$

for $u \in \mathbb{R}$. Furthermore, for each $u > 0$, the sequence $\{\psi_n(u)\}_{n\ge1}$ is nondecreasing. Note also that for each $n \in \mathbb{N}$, ψ_n, $\frac{d}{du}\psi_n$, and $\frac{d^2}{du^2}\psi_n$ all vanish on the interval $(-a_n, a_n)$. Moreover, thanks to (2.18) and the second equation of (2.26), we have

$$\frac{1}{2}\frac{d^2}{du^2}\psi_n(x-z)|\sigma(x,k) - \sigma(z,k)|^2 \le \frac{1}{2} \cdot \frac{2}{n}\kappa_R = \frac{\kappa_R}{n}, \tag{2.27}$$

for all x, z with $|x| \vee |z| \le R$ and $0 < |x - z| \le \delta_0$. On the other hand, for any $x, z \in \mathbb{R}$ with $|x| \vee |z| \le R$ and $|x - z| \le \delta_0$, it follows from (2.17) and (2.25) that

$$\frac{d}{du}\psi_n(x-z)(b(x,k)-b(z,k)) = \text{sign}(x-z)\left|\frac{d}{du}\psi_n(x-z)\right|(b(x,k)-b(z,k))$$
$$\le \kappa_R \varrho(|x-z|). \tag{2.28}$$

Let $\widetilde{X}$ and X be two solutions to (2.22). Denote $\Delta_t := \widetilde{X}(t) - X(t)$ for $t \ge 0$. Assume $|\Delta_0| = |\widetilde{x} - x| < \delta_0$ and define

$$S_{\delta_0} := \inf\{t \ge 0 : |\Delta_t| \ge \delta_0\} = \inf\{t \ge 0 : |\widetilde{X}(t) - X(t)| \ge \delta_0\}.$$

For $R > 0$, let $\tau_R := \inf\{t \ge 0 : |\widetilde{X}(t)| \vee |X(t)| > R\}$. Then $\tau_R \to \infty$ a.s. as $R \to \infty$. Moreover, by Itô's formula, we have

$$\begin{aligned}
&\mathbb{E}[\psi_n(\Delta_{t\wedge S_{\delta_0}\wedge\tau_R})] \\
&= \psi_n(|\Delta_0|) + \mathbb{E}\bigg[\int_0^{t\wedge\tau_R\wedge S_{\delta_0}} \bigg\{\frac{d}{du}\psi_n(\Delta_s)[b(\widetilde{X}(s),k) - b(X(s),k)] \\
&\quad + \frac{1}{2}\frac{d^2}{du^2}\psi_n(\Delta_s)[\sigma(\widetilde{X}(s),k) - \sigma(X(s),k)]^2\bigg\}ds\bigg].
\end{aligned}$$

Furthermore, using (2.27) and (2.28), we obtain

$$\begin{aligned}
\mathbb{E}[\psi_n(\Delta_{t\wedge S_{\delta_0}\wedge\tau_R})] &\le \psi_n(|\Delta_0|) + \mathbb{E}\bigg[\int_0^{t\wedge\tau_R\wedge S_{\delta_0}} \left(\kappa_R\varrho(|\Delta_s|) + \frac{\kappa_R}{n}\right)ds\bigg] \\
&\le \psi_n(|\Delta_0|) + \frac{\kappa_R}{n}t + \int_0^t \kappa_R\varrho\big(\mathbb{E}[|\Delta_{s\wedge\tau_R\wedge S_{\delta_0}}|]\big)ds,
\end{aligned}$$

where the second inequality follows from the concavity of ϱ and Jensen's inequality. Upon passing to the limit as $n \to \infty$, we obtain from the third equation in (2.26) and the monotone convergence theorem that

$$\mathbb{E}[|\Delta_{t\wedge S_{\delta_0}\wedge\tau_R}|] \le |\Delta_0| + \kappa_R \int_0^t \varrho(\mathbb{E}[|\Delta_{s\wedge\tau_R\wedge S_{\delta_0}}|])ds.$$

When $\Delta_0 = 0$, Bihari's inequality then implies that $\mathbb{E}[|\Delta_{t\wedge\tau_R\wedge S_{\delta_0}}|] = 0$. Hence, by Fatou's lemma, we have $\mathbb{E}[|\Delta_{t\wedge S_{\delta_0}}|] = 0$. This implies that $\Delta_{t\wedge S_{\delta_0}} = 0$ a.s.

On the set $\{S_{\delta_0} \le t\}$, we have $|\Delta_{t\wedge S_{\delta_0}}| \ge \delta_0$. Thus, it follows that $0 = \mathbb{E}[|\Delta_{t\wedge S_{\delta_0}}|] \ge \delta_0\mathbb{P}\{S_{\delta_0} \le t\}$. This implies that $\mathbb{P}\{S_{\delta_0} \le t\} = 0$. Consequently, we can compute

$$\begin{aligned}
\mathbb{P}\{|\Delta_t| > 0\} &= \mathbb{P}\{|\Delta_t| > 0, S_{\delta_0} \le t\} + \mathbb{P}\{|\Delta_t| > 0, S_{\delta_0} > t\} \\
&\le \mathbb{P}\{S_{\delta_0} \le t\} + \mathbb{P}\{|\Delta_{t\wedge S_{\delta_0}}| > 0, S_{\delta_0} > t\} \\
&\le 0 + \mathbb{P}\{|\Delta_{t\wedge S_{\delta_0}}| > 0\} = 0.
\end{aligned}$$

This gives $\Delta_t = 0$ a.s. The desired pathwise uniqueness for (2.22) then follows from the fact that $\widetilde{X}$ and X have right continuous sample paths.

Step 2 (ii). We now turn to the case when $r \geq 2$. Let $X(t)$, $\widetilde{X}(t)$, Δ_t, S_{δ_0}, and τ_R be as before. Consider the function $H(r) := \frac{r^2}{1+r^2}$, $r \in \mathbb{R}$. We have $\frac{d}{dr}H(r) = \frac{2r}{(1+r^2)^2}$ and $\frac{d^2}{dr^2}H(r) = \frac{2}{(1+r^2)^2} - \frac{8r^2}{(1+r^2)^3}$. Note that H, $\frac{d}{dr}H$ and $\frac{d^2}{dr^2}H$ are uniformly bounded. Moreover, direct computations yield that for all $x \in \mathbb{R}^r$,

$$DH(|x|) = \frac{2x}{(1+|x|^2)^2}, \quad \text{and} \quad D^2H(|x|) = \frac{2I}{(1+|x|^2)^2} - \frac{8xx'}{(1+|x|^2)^3}.$$

Applying Itô's formula to the process $H(|\Delta_{\cdot\wedge\tau_R\wedge S_{\delta_0}}|)$, we have

$$\begin{aligned}
&\mathbb{E}[H(|\Delta_{t\wedge\tau_R\wedge S_{\delta_0}}|)] - H(|\Delta_0|) \\
&= \mathbb{E}\bigg[\int_0^{t\wedge\tau_R\wedge S_{\delta_0}} \bigg[\frac{\langle 2\Delta_s, b(\widetilde{X}(s)) - b(X(s))\rangle}{(1+|\Delta_s|^2)^2} \\
&\qquad + \frac{1}{2}\mathrm{tr}\bigg((\sigma(\widetilde{X}(s)) - \sigma(X(s)))(\sigma(\widetilde{X}(s)) - \sigma(X(s)))' \\
&\qquad \times \bigg(\frac{2I}{(1+|\Delta_s|^2)^2} - \frac{8\Delta_s\Delta_s'}{(1+|\Delta_s|^2)^3}\bigg)\bigg)\bigg]ds \\
&\leq \mathbb{E}\bigg[\int_0^{t\wedge\tau_R\wedge S_{\delta_0}} \frac{2\langle\Delta_s, b(\widetilde{X}(s)) - b(X(s))\rangle + |\sigma(\widetilde{X}(s)) - \sigma(X(s))|^2}{(1+|\Delta_s|^2)^2}\bigg].
\end{aligned}$$

Then, thanks to (2.20) and the first condition of (2.19), it follows that

$$\begin{aligned}
\mathbb{E}[H(|\Delta_{t\wedge\tau_R\wedge S_{\delta_0}}|)] &\leq H(|\Delta_0|) + \mathbb{E}\bigg[\int_0^{t\wedge\tau_R\wedge S_{\delta_0}} \frac{\kappa_R\varrho(|\Delta_s|^2)}{(1+|\Delta_s|^2)^2}ds\bigg] \\
&\leq H(|\Delta_0|) + \kappa_R\mathbb{E}\bigg[\int_0^{t\wedge\tau_R\wedge S_{\delta_0}} \varrho\bigg(\frac{|\Delta_s|^2}{1+|\Delta_s|^2}\bigg)ds\bigg] \\
&\leq H(|\Delta_0|) + \kappa_R\mathbb{E}\bigg[\int_0^t \varrho(H(|\Delta_{s\wedge\tau_R\wedge S_{\delta_0}}|))ds\bigg] \\
&\leq H(|\Delta_0|) + \kappa_R\int_0^t \varrho(\mathbb{E}[H(|\Delta_{s\wedge\tau_R\wedge S_{\delta_0}}|)])ds.
\end{aligned}$$

When $\widetilde{x} = x$ or $\Delta_0 = 0$, the same arguments as that in the end of the proof of Step 2 (i) reveals that $\mathbb{E}[H(|\Delta_{t\wedge\tau_R\wedge S_{\delta_0}}|)] = 0$. Since $\lim_{R\to\infty}\tau_R = \infty$ a.s. and $0 \leq H(r) \leq 1$ for all $r \geq 0$, the bounded convergence theorem further implies that $\mathbb{E}[H(|\Delta_{t\wedge S_{\delta_0}}|)] = 0$ and hence $\Delta_{t\wedge S_{\delta_0}} = 0$ a.s. As in Step 2 (i), this further implies that $\Delta_t = 0$ a.s. and therefore the pathwise uniqueness result since the sample paths of X and $\widetilde{X}$ are right continuous.

Step 3. Finally, thanks to Theorem IV.2.3 of [136], (2.22) has a weak solution, which further yields the existence and uniqueness of a non-explosive strong solution to (2.22) by the Yamada–Watanabe principle (see, e.g., Section 5.3 of [149]). □

Because of the delicate x-dependence structure in the dynamics of the system (2.1)–(2.2) and that the state space $\mathcal{M}$ for the discrete component α is countably infinite, we need more preparations before we can prove that the system is well-posed. In particular, we need to carefully describe the evolution of the discrete component α so that we can understand the difference of two solutions to (2.1)–(2.2) starting from different initial conditions. Recall that (2.2) is equivalent to (2.4), which relies on the function h of (2.3) and the Poisson random measure $\mathfrak{p}$. Note that the definition of h depends on the intervals $\{\Delta_{ij}(x), i, j \in \mathcal{M}, x \in \mathbb{R}^r\}$. Under certain regularity conditions on $Q(x)$, it can be shown that as long as the intervals $\Delta_{ij}(x)$ for $i \neq j$ are mutually disjoint and have lengths $q_{ij}(x)$, the law of the process $(X(t), \alpha(t))$ satisfying (2.1) and (2.4) is independent of the construction of these intervals. However, certain pathwise comparisons and properties of the solution do depend on the chosen construction. Therefore, we now introduce a specific construction of these intervals that is chosen to align with our objectives.

First, we observe that for any $j \neq i$, condition (2.13) implies that

$$g_{ij} := \sup_{x\in\mathbb{R}^r} q_{ij}(x) \leq \sup_{x\in\mathbb{R}^r} \sum_{k\neq i} q_{ik}(x) \leq g(i) < \infty.$$

We now define the intervals $\Delta_{ij}(x), i \neq j$ as follows:

$$\begin{aligned}
&\Delta_{12}(x) := [0, q_{12}(x)),\\
&\Delta_{21}(x) := [g_{12}, g_{12} + q_{21}(x)), \quad \Delta_{13}(x) := [g_{12} + g_{21}, g_{12} + g_{21} + q_{13}(x)),\\
&\Delta_{31}(x) := [g_{12} + g_{21} + g_{13}, g_{12} + g_{21} + g_{13} + q_{31}(x)),\\
&\Delta_{23}(x) := [g_{12} + g_{21} + g_{13} + g_{31}, g_{12} + g_{21} + g_{13} + g_{31} + q_{23}(x)),\\
&\Delta_{14}(x) := [g_{12} + g_{21} + g_{13} + g_{31} + g_{23}, g_{12} + g_{21} + g_{13} + g_{31} + g_{23} + q_{14}(x)),\\
&\Delta_{41}(x) := [g_{12} + g_{21} + g_{13} + g_{31} + g_{23} + g_{14},\\
&\qquad\qquad g_{12} + g_{21} + g_{13} + g_{31} + g_{23} + g_{14} + q_{41}(x)),\\
&\qquad \vdots
\end{aligned} \tag{2.29}$$

Note that for any $(i, j) \neq (k, l)$ and $x \in \mathbb{R}^r$, we have $\Delta_{ij}(x) \cap \Delta_{kl}(x) = \emptyset$. Moreover, for any $i \neq j \in \mathcal{M}$ and $x \neq \widetilde{x} \in \mathbb{R}^r$, the intervals $\Delta_{ij}(x)$ and $\Delta_{ij}(\widetilde{x})$ start from the same left end point.

Lemma 2.5 *For any bounded and measurable function* $f : \mathcal{M} \times \mathcal{M} \mapsto \mathbb{R}$ *and* $(x, i, \widetilde{x}, j) \in \mathbb{R}^r \times \mathcal{M} \times \mathbb{R}^r \times \mathcal{M}$, *we have*

$$
\begin{aligned}
&\int_{\mathbb{R}_+} [f(i + h(x, i, z), i + h(\widetilde{x}, i, z)) - f(i, i)]\mathfrak{m}(dz) \\
&= \sum_{k \neq i} (q_{ik}(x) - q_{ik}(\widetilde{x}))^+ (f(k, i) - f(i, i)) \\
&\quad + \sum_{k \neq i} (q_{ik}(\widetilde{x}) - q_{ik}(x))^+ (f(i, k) - f(i, i)) \\
&\quad + \sum_{k \neq i} (q_{ik}(x) \wedge q_{ik}(\widetilde{x}))(f(k, k) - f(i, i)),
\end{aligned} \tag{2.30}
$$

and for $i \neq j$,

$$
\begin{aligned}
&\int_{\mathbb{R}_+} [f(i + h(x, i, z), j + h(\widetilde{x}, j, z)) - f(i, j)]\mathfrak{m}(dz) \\
&= \sum_{k \neq i} q_{ik}(x)(f(k, j) - f(i, j)) + \sum_{l \neq j} q_{jl}(\widetilde{x})(f(i, l) - f(i, j)).
\end{aligned} \tag{2.31}
$$

Consequently, for any $i \in \mathcal{M}$ and $x, \widetilde{x} \in \mathbb{R}^r$, we have

$$
\left| \int_{\mathbb{R}_+} I_{\{h(x,i,z) \neq h(\widetilde{x},i,z)\}} \mathfrak{m}(dz) \right| \leq \sum_{k \neq i} |q_{ik}(x) - q_{ik}(\widetilde{x})|. \tag{2.32}
$$

Proof In view of (2.13), all series in (2.30), (2.31), and (2.32) are well-defined. To evaluate the integral $\int_{\mathbb{R}_+} [f(i + h(x, i, z), i + h(\widetilde{x}, i, z)) - f(i, i)]\mathfrak{m}(dz)$, we only need to consider the $z \in \mathbb{R}^+$ for which the integrand $f(i+h(x, i, z), i+h(\widetilde{x}, i, z)) - f(i, i) \neq 0$. Thanks to the construction of $\Delta_{ij}(x)$, we have

$$
\begin{aligned}
&\{z \in \mathbb{R}_+ : f(i + h(x, i, z), i + h(\widetilde{x}, i, z)) - f(i, i) \neq 0\} \\
&\quad \subset \bigcup_{k \neq i} (\Delta_{ik}(x) \setminus \Delta_{ik}(\widetilde{x})) \cup (\Delta_{ik}(\widetilde{x}) \setminus \Delta_{ik}(x)) \cup (\Delta_{ik}(x) \cap \Delta_{ik}(\widetilde{x}));
\end{aligned}
$$

the lengths of these intervals are respectively $(q_{ik}(x) - q_{ik}(\widetilde{x}))^+$, $(q_{ik}(\widetilde{x}) - q_{ik}(x))^+$, and $q_{ik}(x) \wedge q_{ik}(\widetilde{x})$. This gives the identity (2.30) and hence (2.32) by taking $f(i, j) = I_{\{i \neq j\}}$.

When $i \neq j$, as $\Delta_{ik}(x) \cap \Delta_{jl}(\widetilde{x}) = \emptyset$ for all $k \neq i$, $l \neq j$, and $x, \widetilde{x} \in \mathbb{R}^r$, whenever $h(x, i, z) = k - i \neq 0$, we have $h(\widetilde{x}, j, z) = 0$. Likewise, $h(x, i, z) = 0$ whenever $h(\widetilde{x}, j, z) = l - j \neq 0$. Therefore, we have

$$\{z \in \mathbb{R}_+ : f(i + h(x, i, z), j + h(\widetilde{x}, j, z)) - f(i, j) \neq 0\}$$
$$\subset \Big(\bigcup_{k \neq i} \Delta_{ik}(x)\Big) \cup \Big(\bigcup_{l \neq j} \Delta_{jl}(\widetilde{x})\Big).$$

Hence, (2.31) follows. □

We are now ready to present the main result of this section.

Theorem 2.6 *Under Assumptions* (H2.1), (H2.2), *and* (H2.3), *for any* $(x, k) \in \mathbb{R}^r \times \mathcal{M}$, *the system given by* (2.1) *and* (2.2) *has a unique non-explosive strong solution* (X, α) *with initial condition* $(X(0), \alpha(0)) = (x, k)$.

Proof The proof is divided into two steps. First, we show that the system given by (2.1) and (2.2) has a non-explosive solution. The second step then derives the pathwise uniqueness for (2.1) and (2.2). The existence of a unique non-explosive strong solution then follows from the Yamada–Watanabe principle.

Step 1. Let $(\Omega, \mathcal{F}, \{\mathcal{F}_t\}_{t\geq 0}, \mathbb{P})$ be a complete filtered probability space, on which is defined an r-dimensional standard Brownian motion w. In addition, let $\{\xi_n\}$ be a sequence of independent exponential random variables with mean 1 on the probability space $(\Omega, \mathcal{F}, \{\mathcal{F}_t\}_{t\geq 0}, \mathbb{P})$ that is independent of w.

Let $k \in \mathcal{M}$ and consider the stochastic differential equation:

$$X^{(k)}(t) = x + \int_0^t b(X^{(k)}(s), k)ds + \int_0^t \sigma(X^{(k)}(s), k)dw(s). \tag{2.33}$$

Lemma 2.4 guarantees that SDE (2.33) has a unique non-explosive strong solution $X^{(k)}$. As in the proof of Theorem 2.1 of [284], we define

$$\tau_1 = \theta_1 := \inf\left\{t \geq 0 : \int_0^t q_k(X^{(k)}(s))ds > \xi_1\right\}. \tag{2.34}$$

Note that if $\tau_1 = 0$, then for any $\varepsilon > 0$, there exists a $t \in [0, \varepsilon)$ so that $\xi_1 < \int_0^t q_k(X^{(k)}(s))ds \leq g(k)t$, where we used (2.13) to derive the last inequality. Consequently, $\mathbb{P}(\tau_1 = 0) \leq \mathbb{P}(\xi_1 < g(k)t) = 1 - e^{-g(k)t} < 1 - e^{-g(k)\varepsilon}$. Since this is true for any $\varepsilon > 0$, we have $\mathbb{P}(\tau_1 > 0) = 1$. We can now define a process (X, α) on $[0, \tau_1]$ as

$$X(t) = X^{(k)}(t) \text{ for all } t \in [0, \tau_1], \text{ and } \alpha(t) = k \text{ for all } t \in [0, \tau_1).$$

Moreover, we define $\alpha(\tau_1) \in \mathcal{M}$ according to the probability distribution:

$$\mathbb{P}\{\alpha(\tau_1) = l | \mathcal{F}_{\tau_1 -}\} = \frac{q_{kl}(X(\tau_1-))}{q_k(X(\tau_1-))}(1 - \delta_{kl})I_{\{q_k(X(\tau_1-))>0\}} + \delta_{kl}I_{\{q_k(X(\tau_1-))=0\}}, \tag{2.35}$$

for $l \in \mathcal{M}$. In general, having determined (X, α) on $[0, \tau_n]$, we let

$$\theta_{n+1} := \inf\left\{t \geq 0 : \int_0^t q_{\alpha(\tau_n)}(X^{(\alpha(\tau_n))}(s))ds > \xi_{n+1}\right\}, \tag{2.36}$$

where

$$\begin{aligned} X^{(\alpha(\tau_n))}(t) := X(\tau_n) &+ \int_0^t \sigma(X^{(\alpha(\tau_n))}(s), \alpha(\tau_n))dw(s) \\ &+ \int_0^t b(X^{(\alpha(\tau_n))}(s), \alpha(\tau_n))ds. \end{aligned}$$

As before, (2.13) implies that $\mathbb{P}\{\theta_{n+1} > 0\} = 1$. Then we let

$$\tau_{n+1} := \tau_n + \theta_{n+1} \tag{2.37}$$

and define (X, α) on $[\tau_n, \tau_{n+1}]$ by

$$X(t) = X^{(\alpha(\tau_n))}(t - \tau_n) \text{ for } t \in [\tau_n, \tau_{n+1}], \;\; \alpha(t) = \alpha(\tau_n) \text{ for } t \in [\tau_n, \tau_{n+1}), \tag{2.38}$$

and

$$\begin{aligned} \mathbb{P}\left\{\alpha(\tau_{n+1}) = l|\mathcal{F}_{\tau_{n+1}-}\right\} &= \delta_{\alpha(\tau_n),l} I_{\{q_{\alpha(\tau_n)}(X(\tau_{n+1}-))=0\}} \\ &+ \frac{q_{\alpha(\tau_n),l}(X(\tau_{n+1}-))}{q_{\alpha(\tau_n)}(X(\tau_{n+1}-))}(1 - \delta_{\alpha(\tau_n),l}) I_{\{q_{\alpha(\tau_n)}(X(\tau_{n+1}-))>0\}}. \end{aligned} \tag{2.39}$$

As argued in [284], this "interlacing procedure" uniquely determines a solution $(X, \alpha) \in \mathbb{R}^r \times \mathcal{M}$ to (2.1) and (2.2) for all $t \in [0, \tau_\infty)$, where

$$\tau_\infty = \lim_{n\to\infty} \tau_n. \tag{2.40}$$

Since the sequence τ_n is strictly increasing, the limit $\tau_\infty \leq \infty$ exists.

Next, we show that $\tau_\infty = \infty$ a.s. To this end, fix $(X(0), \alpha(0)) = (x, k) \in \mathbb{R}^r \times \mathcal{M}$ as before and for any $m \geq k+1$, denote by $\widetilde{\tau}_m := \inf\{t \geq 0 : \alpha(t) \geq m\}$ the first exit time for the α component from the finite set $\{0, 1, \ldots, m-1\}$. Let $A^c := \{\omega \in \Omega : \tau_\infty > \widetilde{\tau}_m$ for all $m \geq k+1\}$ and $A := \{\omega \in \Omega : \tau_\infty \leq \widetilde{\tau}_{m_0}$ for some $m_0 \geq k+1\}$. Then we have

$$\mathbb{P}\{\tau_\infty = \infty\} = \mathbb{P}\{\tau_\infty = \infty|A^c\}\mathbb{P}(A^c) + \mathbb{P}\{\tau_\infty = \infty|A\}\mathbb{P}(A). \tag{2.41}$$

Let $A_m := \left\{\omega \in \Omega : \tau_\infty \leq \widetilde{\tau}_m\right\}$ for $m \geq k+1$. Then $A = \bigcup_{m=k+1}^\infty A_m$, and $A^c = \bigcap_{m=k+1}^\infty A_m^c$. Also denote $B_{k+1} := A_{k+1}$ and let

$$B_m := A_m \setminus A_{m-1} = \left\{\omega \in \Omega : \widetilde{\tau}_{m-1} < \tau_\infty \le \widetilde{\tau}_m\right\}$$

for $m \ge k+2$. Clearly, $\{B_m\}_{m=k+1}^{\infty}$ is a sequence of disjoint sets, and we have

$$A := \bigcup_{m=k+1}^{\infty} B_m. \tag{2.42}$$

We proceed to show that $\mathbb{P}\{\tau_\infty = \infty, B_m\} = \mathbb{P}(B_m)$ for each m. Note that on the set B_m, $\alpha(\tau_n) \le m$ for all $n = 1, 2, \ldots$ Consequently, using (2.13) in Assumption (H2.1), we have $q_{\alpha(\tau_n)}(X^{(\alpha(\tau_n))}(s)) \le g(m)$ for all n and $s \ge 0$. On the other hand, thanks to the definition of θ_1 in (2.34), for any $\varepsilon > 0$, we have

$$\xi_1 < \int_0^{\theta_1+\varepsilon} q_k(X^{(k)}(s))ds.$$

Consequently, it follows that

$$I_{B_m}\xi_1 \le I_{B_m}\int_0^{\theta_1+\varepsilon} q_k(X^{(k)}(s))ds \le I_{B_m} g(m)(\theta_1+\varepsilon).$$

In the same manner, we have from (2.36) that

$$\xi_n < \int_0^{\theta_n+\varepsilon/2^n} q_{\alpha(\tau_{n-1})}(X^{(\alpha(\tau_{n-1}))}(s))ds,$$

and hence

$$I_{B_m}\xi_n \le I_{B_m}\int_0^{\theta_n+\varepsilon/2^n} q_{\alpha(\tau_{n-1})}(X^{(\alpha(\tau_{n-1}))}(s))ds \le I_{B_m} g(m)(\theta_n+\varepsilon/2^n),$$

for all $n = 1, 2, \ldots$ Summing over these inequalities and noting that $\tau_\infty = \sum_{n=1}^{\infty}\theta_n$, we arrive at

$$I_{B_m}\sum_{n=1}^{\infty}\xi_n \le I_{B_m} g(m)(\tau_\infty + 2\varepsilon), \tag{2.43}$$

which, in turn, implies that $B_m \cap \{\sum_{n=1}^{\infty}\xi_n = \infty\} \subset B_m \cap \{\tau_\infty = \infty\}$. By virtue of Theorem 2.3.2 of [223], we have $\sum_{n=1}^{\infty}\xi_n = \infty$ a.s. Therefore, it follows that

$$\mathbb{P}(B_m) \ge \mathbb{P}(B_m \cap \{\tau_\infty = \infty\}) \ge \mathbb{P}\left(B_m \cap \left\{\sum_{n=1}^{\infty}\xi_n = \infty\right\}\right) = \mathbb{P}(B_m).$$

Consequently, we can use (2.42) to compute

$$\begin{aligned}\mathbb{P}\{\tau_\infty = \infty, A\} &= \mathbb{P}\Big\{\tau_\infty = \infty, \bigcup_{m=k+1}^{\infty} B_m\Big\}\\ &= \sum_{m=k+1}^{\infty} \mathbb{P}\{\tau_\infty = \infty, B_m\} = \sum_{m=k+1}^{\infty} \mathbb{P}(B_m) = \mathbb{P}(A).\end{aligned}$$

This gives

$$\mathbb{P}\{\tau_\infty = \infty | A\} = 1. \tag{2.44}$$

If $\mathbb{P}(A) = 1$ or $\mathbb{P}(A^c) = 0$, then (2.41) and (2.44) imply that $\mathbb{P}\{\tau_\infty = \infty\} = 1$ and the proof is complete. Therefore, it remains to consider the case when $\mathbb{P}(A^c) > 0$. Denote $\widetilde{\tau}_\infty := \lim_{m\to\infty} \widetilde{\tau}_m$. Note that $A^c = \{\tau_\infty \geq \widetilde{\tau}_\infty\}$. Thus, $\mathbb{P}\{\tau_\infty = \infty | A^c\} \geq \mathbb{P}\{\widetilde{\tau}_\infty = \infty | A^c\}$ and hence (2.41) holds if we can show that

$$\mathbb{P}\{\widetilde{\tau}_\infty = \infty | A^c\} = 1. \tag{2.45}$$

Assume on the contrary that (2.45) were false; then there would exist a $T > 0$ such that

$$\delta := \mathbb{P}\{\widetilde{\tau}_\infty \leq T, A^c\} > 0.$$

Let $f : \mathcal{M} \mapsto \mathbb{R}_+$ be as in Assumption (H2.1). Then we have for any $m \geq k+1$,

$$\begin{aligned}f(k) &= \mathbb{E}[e^{-H(T\wedge\tau_\infty\wedge\widetilde{\tau}_m)} f(\alpha(T\wedge\tau_\infty\wedge\widetilde{\tau}_m))]\\ &\quad + \mathbb{E}\Big[\int_0^{T\wedge\tau_\infty\wedge\widetilde{\tau}_m} e^{-Hs}\Big(Hf(\alpha(s)) - \sum_{l\in\mathcal{M}} q_{\alpha(s),l}(X(s))[f(l) - f(\alpha(s))]\Big)ds\Big]\\ &\geq \mathbb{E}[e^{-H(T\wedge\tau_\infty\wedge\widetilde{\tau}_m)} f(\alpha(T\wedge\tau_\infty\wedge\widetilde{\tau}_m))]\\ &\quad + \mathbb{E}\Big[\int_0^{T\wedge\tau_\infty\wedge\widetilde{\tau}_m} e^{-Hs}[Hf(\alpha(s)) - H(1 + f(\alpha(s)) + \Phi(X(s)))]ds\Big]\\ &\geq \mathbb{E}[e^{-H(T\wedge\tau_\infty\wedge\widetilde{\tau}_m)} f(\alpha(T\wedge\tau_\infty\wedge\widetilde{\tau}_m))] - 1 - \mathbb{E}\Big[\int_0^T e^{-Hs} H\Phi(X(s))ds\Big],\end{aligned}$$

where the first inequality above follows from (2.14) in (H2.1). In view of (2.23), we have $\mathcal{L}\Phi(x) \leq H\Phi(x)$ and hence $\{e^{-Ht}\Phi(X(t)), t \geq 0\}$ is a positive supermartingale. Consequently, we have $\mathbb{E}\big[\int_0^T e^{-Hs} H\Phi(X(s))ds\big] \leq HT\Phi(x)$. Substituting this observation into the above displayed equation, it follows that

$$
\begin{aligned}
e^{HT}(1+f(k)+HT\Phi(x)) &\geq \mathbb{E}[f(\alpha(T\wedge\tau_\infty\wedge\widetilde{\tau}_m))] \geq \mathbb{E}[f(\alpha(\widetilde{\tau}_m))I_{\{\widetilde{\tau}_m\leq T\wedge\tau_\infty\}}] \\
&\geq f(m)\mathbb{P}\{\widetilde{\tau}_m\leq T\wedge\tau_\infty\} \geq f(m)\mathbb{P}\{\widetilde{\tau}_m\leq T\wedge\tau_\infty, A^c\} \\
&\geq f(m)\mathbb{P}\{\widetilde{\tau}_\infty\leq T\wedge\tau_\infty, A^c\},
\end{aligned}
\tag{2.46}
$$

where the third inequality follows from the facts that $\alpha(\widetilde{\tau}_m)\geq m$ and that f is nondecreasing, and the last inequality follows from the fact that $\widetilde{\tau}_m\uparrow\widetilde{\tau}_\infty$. Recall that $A^c=\{\tau_\infty\geq\widetilde{\tau}_\infty\}$. Thus,

$$
\begin{aligned}
\mathbb{P}\{\widetilde{\tau}_\infty\leq T\wedge\tau_\infty, A^c\} &= \mathbb{P}\{\widetilde{\tau}_\infty\leq T\wedge\tau_\infty, \widetilde{\tau}_\infty\leq\tau_\infty\} \\
&\geq \mathbb{P}\{\widetilde{\tau}_\infty\leq T, \widetilde{\tau}_\infty\leq\tau_\infty\} = \mathbb{P}\{\widetilde{\tau}_\infty\leq T, A^c\}=\delta>0.
\end{aligned}
$$

Using this observation in (2.46) yields $\infty>e^{HT}(1+f(k)+HT\Phi(x))\geq f(m)\delta\to\infty$ as $m\to\infty$, thanks to the fact that $f(m)\to\infty$ as $m\to\infty$, which is a contradiction. This establishes (2.45) and hence $\mathbb{P}(\tau_\infty=\infty)=1$. In other words, the interlacing procedure uniquely determines a solution $(X,\alpha)=(X^{(x,k)},\alpha^{(x,k)})$ for all $t\in[0,\infty)$.

Thanks to the definition of the event A, we have $A\cap\{\tau_\infty=\infty\}\subset A\cap\{\widetilde{\tau}_\infty=\infty\}\cap\{\tau_\infty=\infty\}\subset A\cap\{\widetilde{\tau}_\infty=\infty\}$. This, together with (2.45) and the fact that $\mathbb{P}(\tau_\infty=\infty)=1$ implies that

$$
\begin{aligned}
\mathbb{P}(\widetilde{\tau}_\infty=\infty) &= \mathbb{P}(A\cap\{\widetilde{\tau}_\infty=\infty\})+\mathbb{P}(A^c\cap\{\widetilde{\tau}_\infty=\infty\}) \\
&\geq \mathbb{P}(A\cap\{\tau_\infty=\infty\})+\mathbb{P}(A^c)=\mathbb{P}(A)+\mathbb{P}(A^c)=1.
\end{aligned}
\tag{2.47}
$$

Next, we show that the solution (X,α) to the system (2.1) and (2.2) is non-explosive a.s. Consider the function $V(x,k):=1+\Phi(x)+f(k)$, where the functions $\Phi:\mathbb{R}^r\mapsto\mathbb{R}_+$ and $f:\mathcal{M}\mapsto\mathbb{R}_+$ are defined in Assumption (H2.1). Note that $V(x,k)\to\infty$ as $|x|\vee k\to\infty$, thanks to Assumption (H2.1). Recall the definition of $\mathcal{L}$ in (2.6). Recall from (2.23) that $\mathcal{L}_k\Phi(x)\leq H\Phi(x)$ for all $x\in\mathbb{R}^r$ and $k\in\mathcal{M}$. Combining this with (2.14), we obtain $\mathcal{L}V(x,k)\leq 2HV(x,k)$. This, together with Itô's formula, shows that the process $\{e^{-2Ht}V(X(t),\alpha(t)), t\geq 0\}$ is a positive supermartingale. Then we can apply the optional sampling theorem to the process $\{e^{-2Ht}V(X(t),\alpha(t)), t\geq 0\}$ to argue that $\mathbb{P}\{\lim_{n\to\infty}\beta_n=\infty\}=1$, where $\beta_n:=\inf\{t\geq 0: |X(t)|\vee\alpha(t)\geq n\}$. This shows that the solution (X,α) has no finite explosion time a.s.

Step 2. Suppose (X,α) and $(\widetilde{X},\widetilde{\alpha})$ are two solutions to (2.1) and (2.2) starting from the same initial condition $(x,k)\in\mathbb{R}^r\times\mathcal{M}$. Then we have

$$
\begin{aligned}
\widetilde{X}(t)-X(t) &= \int_0^t[b(\widetilde{X}(s),\widetilde{\alpha}(s))-b(X(s),\alpha(s))]ds \\
&\quad+\int_0^t[\sigma(\widetilde{X}(s),\widetilde{\alpha}(s))-\sigma(X(s),\alpha(s))]dw(s),
\end{aligned}
$$

and

$$\widetilde{\alpha}(t) - \alpha(t) = \int_0^t \int_{\mathbb{R}_+} [h(\widetilde{X}(s-), \widetilde{\alpha}(s-), z) - h(X(s-), \alpha(s-), z)]\mathfrak{p}(ds, dz).$$

Let $\zeta := \inf\{t \geq 0 : \alpha(t) \neq \widetilde{\alpha}(t)\}$ be the first time when the discrete components differ from each other. Let us also define $\tau_R := \inf\{t \geq 0 : |\widetilde{X}(t)| \vee |X(t)| \vee \widetilde{\alpha}(t) \vee \alpha(t) \geq R\}$ for $R > 0$ and $S_{\delta_0} := \inf\{t \geq 0 : |\widetilde{X}(t) - X(t)| \geq \delta_0\}$. We have $\alpha(t) = \widetilde{\alpha}(t)$ for $t \in [0, \zeta)$. To simplify notation, let us define $\Delta_t := \widetilde{X}(t) - X(t)$. As we have observed in Step 2 (ii) in the proof of Lemma 2.4, $\mathbb{E}[H(|\Delta_{t\wedge\zeta\wedge\tau_R\wedge S_{\delta_0}}|)] = 0$ for $r \geq 2$, where $H(u) := \frac{u^2}{1+u^2}$, $u \geq 0$. When $r = 1$, the proof of Lemma 2.4 reveals that $\mathbb{E}[|\Delta_{t\wedge\zeta\wedge\tau_R\wedge S_{\delta_0}}|] = 0$ and hence $\mathbb{E}[H(|\Delta_{t\wedge\zeta\wedge\tau_R\wedge S_{\delta_0}}|)] = 0$. Note that on the set $\{S_{\delta_0} \leq t \wedge \zeta \wedge \tau_R\}$, $|\Delta_{t\wedge\zeta\wedge\tau_R\wedge S_{\delta_0}}| \geq \delta_0$. Also we can readily check that H is an increasing function. Thus, it follows that

$$\begin{aligned}0 &= \mathbb{E}[H(|\Delta_{t\wedge\zeta\wedge\tau_R\wedge S_{\delta_0}}|)] \geq \mathbb{E}[H(|\Delta_{t\wedge\zeta\wedge\tau_R\wedge S_{\delta_0}}|)I_{\{S_{\delta_0}\leq t\wedge\zeta\wedge\tau_R\}}] \\ &\geq H(\delta_0)\mathbb{P}\{S_{\delta_0} \leq t \wedge \zeta \wedge \tau_R\}.\end{aligned}$$

This implies that $\mathbb{P}\{S_{\delta_0} \leq t \wedge \zeta \wedge \tau_R\} = 0$. Consequently, we have

$$\begin{aligned}&\mathbb{E}[H(|\Delta_{t\wedge\zeta\wedge\tau_R}|)] \\ &= \mathbb{E}[H(|\Delta_{t\wedge\zeta\wedge\tau_R}|)I_{\{S_{\delta_0}\leq t\wedge\zeta\wedge\tau_R\}}] + \mathbb{E}[H(|\Delta_{t\wedge\zeta\wedge\tau_R}|)I_{\{S_{\delta_0}> t\wedge\zeta\wedge\tau_R\}}] \\ &\leq \mathbb{P}\{S_{\delta_0} \leq t \wedge \zeta \wedge \tau_R\} + \mathbb{E}[H(|\Delta_{t\wedge\zeta\wedge\tau_R\wedge S_{\delta_0}}|)I_{\{S_{\delta_0}> t\wedge\zeta\wedge\tau_R\}}] \\ &\leq 0.\end{aligned}$$

It follows that $\mathbb{E}[|\Delta_{t\wedge\zeta\wedge\tau_R}|] = 0$. Recall that the function γ in Assumption (H2.3) is concave with $\gamma(0) = 0$. Thus, we can use Jensen's inequality to see that

$$\mathbb{E}[\gamma(|\Delta_{t\wedge\zeta\wedge\tau_R}|)] \leq \gamma(\mathbb{E}[|\Delta_{t\wedge\zeta\wedge\tau_R}|]) = 0.$$

Note that $\zeta \leq t \wedge \tau_R$ if and only if $\widetilde{\alpha}(t \wedge \zeta \wedge \tau_R) - \alpha(t \wedge \zeta \wedge \tau_R) \neq 0$. Therefore, it follows that

$$\begin{aligned}&\mathbb{P}\{\zeta \leq t \wedge \tau_R\} = \mathbb{E}[I_{\{\widetilde{\alpha}(t\wedge\zeta\wedge\tau_R)-\alpha(t\wedge\zeta\wedge\tau_R)\neq 0\}}] \\ &= \mathbb{E}\bigg[\int_0^{t\wedge\zeta\wedge\tau_R} \int_{\mathbb{R}_+} \big(I_{\{\widetilde{\alpha}(s-)-\alpha(s-)+h(\widetilde{X}(s-),\alpha(s-),z)-h(X(s-),\alpha(s-),z)\neq 0\}} \\ &\qquad\qquad - I_{\{\widetilde{\alpha}(s-)-\alpha(s-)\neq 0\}}\big)\mathfrak{m}(dz)ds\bigg] \\ &= \mathbb{E}\bigg[\int_0^{t\wedge\zeta\wedge\tau_R} \int_{\mathbb{R}_+} I_{\{h(\widetilde{X}(s-),\alpha(s-),z)-h(X(s-),\alpha(s-),z)\neq 0\}}\mathfrak{m}(dz)ds\bigg]\end{aligned}$$

$$\leq \mathbb{E}\left[\int_0^{t\wedge\zeta\wedge\tau_R} \sum_{l\in\mathcal{M},l\neq\alpha(s-)} |q_{\alpha(s-),l}(\widetilde{X}(s-)) - q_{\alpha(s-),l}(X(s-))|ds\right]$$

$$\leq \mathbb{E}\left[\int_0^{t\wedge\zeta\wedge\tau_R} \kappa_R\gamma(|\widetilde{X}(s-)) - X(s-)|ds\right]$$

$$\leq \kappa_R\mathbb{E}\left[\int_0^t \gamma(|\Delta_{s\wedge\zeta\wedge\tau_R}|)ds\right]$$

$$\leq \kappa_R \int_0^t \mathbb{E}[\gamma(|\Delta_{s\wedge\zeta\wedge\tau_R}|)]ds = 0, \tag{2.48}$$

where the first and second inequalities follow from (2.32) and (2.21), respectively. In particular, we have $\mathbb{P}\{\zeta \leq t\wedge\tau_R\} = 0$. Since $\tau_R \to \infty$ a.s. as $R \to \infty$, we have

$$\mathbb{E}[I_{\{\widetilde{\alpha}(t)\neq\alpha(t)\}}] \leq \mathbb{P}\{\zeta \leq t\} = 0. \tag{2.49}$$

Now, we can compute

$$\begin{aligned}
\mathbb{E}[H(|\widetilde{X}(t) - X(t)|)] &= \mathbb{E}[H(|\widetilde{X}(t) - X(t)|)I_{\{\zeta>t\}}] + \mathbb{E}[H(|\widetilde{X}(t) - X(t)|)I_{\{\zeta\leq t\}}] \\
&= \mathbb{E}[H(|\widetilde{X}(t\wedge\zeta) - X(t\wedge\zeta)|)I_{\{\zeta>t\}}] + \mathbb{E}[1\cdot I_{\{\zeta\leq t\}}] \\
&\leq \mathbb{E}[H(|\widetilde{X}(t\wedge\zeta) - X(t\wedge\zeta)|)] + 0 \\
&= 0.
\end{aligned}$$

Thus, $\mathbb{P}\{\widetilde{X}(t) = X(t)\} = 1$. This, together with (2.49), implies that $\mathbb{P}\{(\widetilde{X}(t), \widetilde{\alpha}(t)) = (X(t), \alpha(t))\} = 1$ for all $t \geq 0$. Since the sample paths of (X, α) are right continuous, we obtain the desired pathwise uniqueness result. □

Remark 2.7 Note that in general condition (2.13) alone cannot guarantee that $\tau_\infty = \infty$ a.s. To see this, let us consider a continuous-time Markov chain α with state space $\mathcal{M} = \{1, 2, \ldots, \}$ and Q-matrix given by $Q = (q_{kl})$ such that $-q_{kk} = q_{k,(k+1)^2} = k$ and $q_{kl} = 0$ for all $l \in \mathcal{M}\setminus\{k, (k+1)^2\}$. For this example, (2.13) is satisfied.

Assume $\alpha(0) = 1$; then α will stay in state 1 for an exponential amount of time with mean 1 and then switch to state 4, whose holding time has exponential distribution with mean $\frac{1}{4}$; it next switches to state 25, whose holding time is exponentially distributed with mean $\frac{1}{25}$; and then switches to state 26^2, whose holding time is exponential with mean $\frac{1}{26^2}$; and so on. It is then clear that $\mathbb{P}(\tau_\infty < \infty) = 1$. Of course, we can easily check that condition (2.14) cannot be satisfied for this example.

However, if the upper bound in (2.13) is replaced by a constant $H > 0$, then we have $\tau_\infty = \infty$ a.s. and therefore the proof of Theorem 2.6 can be much simplified. Indeed, with the uniform upper bound, we have $\mathbb{P}\{\theta_k > t\} \geq e^{-Ht}$ for all $k \in \mathcal{M}$ and $t > 0$ and hence

$$\mathbb{P}\{\tau_\infty = \infty\} \ge \mathbb{P}\big\{\{\theta_k > t\} \text{ i.o.}\big\} = \mathbb{P}\Big\{\bigcap_{m=1}^{\infty} \bigcup_{k=m}^{\infty} \{\theta_k > t\}\Big\}$$

$$= \lim_{m\to\infty} \mathbb{P}\Big\{\bigcup_{k=m}^{\infty} \{\theta_k > t\}\Big\} \ge \limsup_{m\to\infty} \mathbb{P}\{\theta_m > t\} \ge e^{-Ht}.$$

Letting $t \downarrow 0$ yields that $\mathbb{P}\{\tau_\infty = \infty\} = 1$. Thus, the "interlacing procedure" in Step 1 of the proof of Theorem 2.6 directly leads to the existence of a solution (X, α) to (2.1) and (2.2) for *all* $t \in [0, \infty)$.

The following result can be obtained from Theorem 2.6 directly; it assumes the usual linear growth and local Lipschitz conditions. These conditions are used extensively in the literature in particular for the usual diffusion processes (without switching).

Corollary 2.8 *Let* $(x, i) \in \mathbb{R}^r \times \mathcal{M}$ *with* $\mathcal{M} = \{1, \ldots, m_0\}$ $(2 \le m_0 \le \infty)$. *Suppose that* $Q(\cdot) : \mathbb{R}^r \mapsto \mathbb{R}^{m_0 \times m_0}$ *is a bounded function satisfying the* q*-property and* (2.21), *that the functions* $b(\cdot, \cdot)$ *and* $\sigma(\cdot, \cdot)$ *satisfy the linear growth condition*

$$|b(x, i)| + |\sigma(x, i)| \le K(1 + |x|), \quad i \in \mathcal{M}, \tag{2.50}$$

as well as the local Lipschitz condition, that is, for every integer $N \ge 1$, *there exists a positive constant* H_N *such that for all* $i \in \mathcal{M}$ *and all* $x, y \in \mathbb{R}^r$ *with* $|x| \vee |y| \le N$,

$$|b(x, i) - b(y, i)| \vee |\sigma(x, i) - \sigma(y, i)| \le H_N |x - y|. \tag{2.51}$$

Then for any given initial data (x, i), *there exists a unique strong solution* $(X(t), \alpha(t))$ *to the system* (2.1) *and* (2.2).

To proceed, we obtain a moment estimate on $X(t)$. This estimate is used frequently in subsequent development. As its diffusion counterpart, the main ingredient of the proof is the use of Gronwall's inequality.

Proposition 2.9 *Assume the linear growth condition* (2.50). *Let* $T > 0$ *be fixed. Then for any positive constant* γ, *we have*

$$\mathbb{E}_{x,i}\Big[\sup_{t \in [0,T]} |X(t)|^\gamma\Big] \le C < \infty, \quad (x, i) \in \mathbb{R}^r \times \mathcal{M}, \tag{2.52}$$

where the constant C *satisfies* $C = C(x, T, \gamma)$.

Proof

Step 1. Because

$$X(t) = x + \int_0^t b(X(s), \alpha(s))ds + \int_0^t \sigma(X(s), \alpha(s))dw(s),$$

we have for any $p \geq 2$ that

$$|X(t)|^p \leq 3^{p-1}\Big[\ |x|^p + \Big|\int_0^t b(X(s), \alpha(s))ds\Big|^p + \Big|\int_0^t \sigma(X(s), \alpha(s))dw(s)\Big|^p\Big].$$

Using Hölder's inequality and the linear growth condition, detailed computations lead to

$$\begin{aligned}\mathbb{E}_{x,i}\Big[\ \sup_{0\leq t\leq T_1}\Big|\int_0^t b(X(s), \alpha(s))ds\Big|^p\Big] &\leq T^{p-1}\int_0^{T_1}|b(X(s), \alpha(s))|^p\, ds\\ &\leq KT^{p-1}\int_0^{T_1}(1+|X(s)|^p)ds\\ &\leq c_1(T,p)\int_0^{T_1}\mathbb{E}_{x,i}\Big[1+\sup_{0\leq u\leq s}|X(u)|^p\Big]ds,\end{aligned}$$

where $0 \leq T_1 \leq T$, and $c_1(T, p)$, $i = 1, 2$ is a positive constant depending only on T, p, and the linear growth constant in (2.50). Similarly, by the Burkholder–Davis–Gundy inequality (see Lemma A.33; see also [205, p. 70]), the Hölder inequality, and the linear growth condition, we compute

$$\begin{aligned}&\mathbb{E}_{x,i}\Big[\sup_{0\leq t\leq T_1}\Big|\int_0^t \sigma(X(s), \alpha(s))dw(s)\Big|^p\Big]\\ &\leq K\mathbb{E}_{x,i}\Big[\int_0^{T_1}|\sigma(X(s), \alpha(s))|^2\, ds\Big]^{p/2}\\ &\leq KT^{p/2-1}\mathbb{E}_{x,i}\int_0^{T_1}|\sigma(X(s), \alpha(s))|^p\, ds\\ &\leq c_2(T,p)\int_0^{T_1}\mathbb{E}_{x,i}\Big[1+\sup_{0\leq u\leq s}|X(u)|^p\Big]ds,\end{aligned}$$

where $0 \leq T_1 \leq T$ and $c_2(T, p)$ is a positive constant depending only on T, p, and the linear growth constant in (2.50). Thus, we have

$$\begin{aligned}&\mathbb{E}_{x,i}\Big[1+\sup_{0\leq t\leq T_1}|X(t)|^p\Big]\\ &\leq c_3(x,p)+c_4(T,p)\int_0^{T_1}\mathbb{E}_{x,i}\Big[1+\sup_{0\leq u\leq s}|X(u)|^p\Big]ds,\end{aligned}$$

where $c_3(x, p) = 1+3^{p-1}|x|^p$, and $c_4(T, p) = 3^{p-1}[c_1(T, p)+c_2(T, p)]$. Hence, Gronwall's inequality (see, e.g., [86]) implies that

$$\begin{aligned}\mathbb{E}_{x,i}\Big[1+\sup_{0\leq t\leq T_1}|X(t)|^p\Big] &\leq c_3(x,p)\exp(c_4(T,p)T_1)\\ &\leq c_3(x,p)\exp(c_4(T,p)T)\\ &:= C(x,T,p).\end{aligned}$$

Thus, we have $\mathbb{E}_{x,i}\left[\sup_{0\le t\le T_1}|X(t)|^p\right]\le C(x,T,p)$. Because $T_1\le T$ is arbitrary, it follows that

$$\mathbb{E}_{x,i}\left[\sup_{0\le t\le T}|X(t)|^p\right]\le C(x,T,p),\quad p\ge 2.$$

Step 2. Note that $\sup_{0\le t\le T}|X(t)|^\gamma=\left(\sup_{0\le t\le T}|X(s)|\right)^\gamma$ for any $\gamma>0$. Thus, we obtain from Hölder's inequality and Step 1 that for any $1\le p<2$,

$$\mathbb{E}_{x,i}\left[\sup_{0\le t\le T}|X(t)|^p\right]\le\left(\mathbb{E}_{x,i}\left[\sup_{0\le t\le T}|X(t)|^{2p}\right]\right)^{1/2}\le C(T,x,p)<\infty.$$

Step 3. Finally, if $0<p<1$, note that

$$|X(t)|^p=|X(t)|^p\,I_{\{|X(t)|\ge 1\}}+|X(t)|^p\,I_{\{|X(t)|<1\}}\le 1+|X(t)|^{1+p}.$$

Hence, it follows from Step 2 that

$$\mathbb{E}_{x,i}\left[\sup_{0\le t\le T}|X(t)|^p\right]\le\mathbb{E}_{x,i}\left[1+\sup_{0\le t\le T}|X(t)|^{1+p}\right]\le C(x,T,p).$$

This completes the proof of the proposition. □

Remark 2.10 By considering x-dependent generator $Q(x)$, our model provides a more realistic formulation allowing the switching component depending on the continuous states. This, in turn, allows the coupling and correlation between $X(t)$ and $\alpha(t)$. It is also worth mentioning that in lieu of a finite-state space, the discrete component α in the system (2.1) and (2.2) can take values from a countably infinite-state space. This is in contrast to many of the existing literature on switching diffusions.

Remark 2.11 In the vast literature on diffusion and switching diffusion processes, the coefficients of the associated stochastic differential equations are usually (locally) Lipschitz. While it is a convenient assumption, it is rather restrictive in many applications. For example, the diffusion coefficients in the Feller branching diffusion and the Cox–Ingersoll–Ross model are only Hölder continuous. We refer to Chapters 12 and 13 of [165] for an introduction to these models. Motivated by these considerations, there has been much efforts devoted to the study of stochastic differential equations with non-Lipschitz coefficients. An incomplete list includes [13, 85, 185, 186, 286], among many others. While there are many works on diffusions and jump diffusions with non-Lipschitz coefficients, the related research on regime-switching diffusions is relatively scarce. Theorem 2.6 establishes the strong existence and uniqueness result for stochastic differential equations associated with switching diffusions, in which the coefficients are non-

Lipschitz and the switching component has countably many states. This result may further facilitate the application of stochastic analysis to real-world problems with inherent complex dynamics.

2.3 Non-explosion

We have established in Theorem 2.6 that if Assumptions (H2.1), (H2.2), and (H2.3) are satisfied, then the system (2.1) and (2.2) has a unique non-explosive strong solution $(X(\cdot), \alpha(\cdot))$, which is defined for all $t \geq 0$. Note that (2.12) of Assumption (H2.1) allows the drift and diffusion coefficients of (2.1) to grow faster than the usual linear growth condition. Likewise, instead of the commonly used boundedness condition, (2.13) and (2.14) place certain growth conditions on the switching rates $q_{ij}(x)$ of (2.2).

This section aims to further relax Assumption (H2.1). We will provide sufficient conditions for non-explosion and explosion in terms of suitable Lyapunov functions. Recall Definition 2.2, where for a (local) solution (X, α) to (2.1) and (2.2), we defined the stopping times

$$\beta_n = \beta_n^{x,k} := \inf\{t \geq 0 : |X(t)| \vee |\alpha(t)| \geq n\}, \quad n \in \mathbb{N}, \tag{2.53}$$

with $(x, k) \in \mathbb{R}^r \times \mathcal{M}$ being the initial condition. Note that the sequence $\{\beta_n\}$ is monotonically increasing and hence has a (finite or infinite) limit. Denote the limit by β_∞. Recall that the switching diffusion process (X, α) with initial data $(X(0), \alpha(0)) = (x, k)$ is *non-explosive* if

$$\beta_\infty = \lim_{n \to \infty} \beta_n = \infty, \quad \text{a.s.} \tag{2.54}$$

In accordance with the terminology in [156, Section 3.4], in what follows, we sometimes refer to non-explosive property as regularity. It is easily seen that the process (X, α) is non-explosive or regular if and only if for any $0 < T < \infty$,

$$\mathbb{P}\left\{ \sup_{0 \leq t \leq T} (|X(t)| \vee \alpha(t)) = \infty \right\} = 0. \tag{2.55}$$

Thus, (2.55) can be used as an alternative definition for non-explosion or regularity. Nevertheless, it is more convenient to use equation (2.54) to delineate non-explosion or regularity for the process (X, α). The following theorem provides a sufficient condition of non-explosion for solutions to (2.1) and (2.2).

Theorem 2.12 *Suppose that for each $k \in \mathcal{M}$, the functions $b(\cdot, k)$ and $\sigma(\cdot, k)$ are locally bounded and $q_k(\cdot)$ satisfies* (2.13) *and there is a nonnegative function $V(\cdot, \cdot) : \mathbb{R}^r \times \mathcal{M} \mapsto \mathbb{R}^+$ that is twice continuously differentiable with respect to $x \in \mathbb{R}^r$ for each $i \in \mathcal{M}$ such that there is a $\gamma_0 > 0$ satisfying*

$$\mathcal{L}V(x,i) \leq \gamma_0 V(x,i), \text{ for all } (x,i) \in \mathbb{R}^r \times \mathcal{M}, \tag{2.56}$$

$$V_R := \inf_{|x| \vee i \geq R} V(x,i) \to \infty \text{ as } R \to \infty. \tag{2.57}$$

Then any solution (X, α) to the system given by (2.1) *and* (2.2) *is non-explosive.*

Proof Let (X, α) be a solution to the system given by (2.1) and (2.2) and define the stopping times $\beta_n, n \in \mathbb{N}$ and β_∞ as in Definition 2.2. It suffices to prove (2.54). Suppose on the contrary that (2.54) were false; then there would exist some $T > 0$ and $\varepsilon > 0$ such that $\mathbb{P}_{x,i}\{\beta_\infty \leq T\} > \varepsilon$. Therefore, we could find some $n_1 \in \mathbb{N}$ such that

$$\mathbb{P}_{x,i}\{\beta_n \leq T\} > \varepsilon, \text{ for all } n \geq n_1. \tag{2.58}$$

Define

$$S(x,i,t) = V(x,i)\exp(-\gamma_0 t), \quad (x,i) \in \mathbb{R}^r \times \mathcal{M}, \text{ and } t \geq 0.$$

Then it satisfies $[(\partial/\partial t) + \mathcal{L}]S(x,i,t) \leq 0$. By virtue of Dynkin's formula, we have

$$\begin{aligned} &\mathbb{E}_{x,i}\left[V(X(\beta_n \wedge T), \alpha(\beta_n \wedge T))\exp(-\gamma_0(\beta_n \wedge T))\right] - V(x,i) \\ &= \mathbb{E}_{x,i}\left[\int_0^{\beta_n \wedge T} \left(\frac{\partial}{\partial t} + \mathcal{L}\right) S(X(u), \alpha(u), u)du\right] \leq 0. \end{aligned}$$

Hence, we have

$$V(x,i) \geq \mathbb{E}_{x,i}\left[V(X(\beta_n \wedge T), \alpha(\beta_n \wedge T))\exp(-\gamma_0(\beta_n \wedge T))\right].$$

Note that $\tau_n \wedge T \leq T$ and V is nonnegative. Thus, we have

$$\begin{aligned} V(x,i)\exp\{\gamma_0 T\} &\geq \mathbb{E}_{x,i}\left[V(X(\beta_n \wedge T), \alpha(\beta_n \wedge T))\right] \\ &\geq \mathbb{E}_{x,i}\left[V(X(\beta_n), \alpha(\beta_n)) I_{\{\beta_n \leq T\}}\right]. \end{aligned}$$

Furthermore, by the definition of β_n and (2.57), we have

$$V(x,i)\exp\{\gamma_0 T\} \geq V_n \mathbb{P}_{x,i}\{\beta_n \leq T\} > \varepsilon V_n \to \infty, \quad \text{as } n \to \infty.$$

This is a contradiction and hence we must have $\beta_\infty = \infty$ a.s. The proof is complete. □

Remark 2.13 In case $\mathcal{M}$ is a finite set, or $q_k(\cdot)$ satisfies (2.13) and (2.14), then $\widetilde{\tau}_\infty = \lim_{n\to\infty} \widetilde{\tau}_n = \infty$ a.s. with $\widetilde{\tau}_n := \inf\{t \geq 0 : \alpha(t) \geq n\}$; see the argument in (2.47). In such a case, the conditions in Theorem 2.12 can be weakened. In particular, (2.57) can be replaced by

$$V_R := \inf_{i \in \mathcal{M}} \inf_{|x| \geq R} V(x,i) \to \infty \text{ as } R \to \infty. \tag{2.59}$$

Indeed, since $\beta_n = \tau_n \wedge \widetilde{\tau}_n$, in which $\tau_n := \inf\{t \geq 0 : |X(t)| \geq n\}$, we can use (2.59), the fact that $\widetilde{\tau}_\infty = \infty$ a.s., and the same arguments as those in the proof of Theorem 2.12 to obtain that $\lim_{n\to\infty} \tau_n = \infty$ a.s. and hence $\beta_\infty = \infty$ a.s. This yields the regularity of the process (X, α).

Similar to the proof of [156, Theorem 3.6], we can also prove the following theorem, which provides a sufficient condition for explosion.

Theorem 2.14 *Assume that for each $k \in \mathcal{M}$, the functions $b(\cdot, k)$ and $\sigma(\cdot, k)$ are locally bounded and $q_k(\cdot)$ satisfies* (2.13) *and* (2.14). *Suppose that Assumptions* (H2.2) *and* (H2.3) *are satisfied and that there exists nonnegative function $V : \mathbb{R}^r \times \mathcal{M} \mapsto \mathbb{R}_+$ and bounded set $E = D \times J \subset \mathbb{R}^r \times M$ satisfying the following:*

(i) $K_1 := \sup_{(x,i)\in E^c} V(x, i) < \infty$,
(ii) *there exists a subset $F \subset E^c$ such that*

$$K_2 := \sup_{(x,i)\in\partial D\times J} V(x, i) < K_3 := \inf_{(x,i)\in F} V(x, i),$$

(iii) *for each $i \in \mathcal{M}$, $V(\cdot, i) \in C^2(D^c)$, and for some positive constant γ_1, V satisfies*

$$\mathcal{L}V(x, i) \geq \gamma_1 V(x, i), \textit{ for all } (x, i) \in E^c.$$

Then for any $(x, k) \in F$, the process $(X^{x,k}(\cdot), \alpha^{x,k}(\cdot))$ explodes in finite time with positive probability.

Proof Fix some $(x, k) \in F$ and for any $n > |x| \vee k$, we define β_n as in (2.53). Let also $\tau_E := \inf\{t \geq 0 : (X(t), \alpha(t)) \in E\}$. Note that the function $S(t, x, i) := e^{-\gamma_1 t} V(x, i)$ satisfies $[\frac{\partial}{\partial t} + \mathcal{L}]S(t, x, i) \geq 0$ for all $(t, x, i) \in [0, \infty) \times E^c$ thanks to condition (iii). Then by virtue of Dynkin's formula, for any $t \geq 0$, we have

$$\mathbb{E}_{x,k}[\exp\{-\gamma_1(t \wedge \beta_n \wedge \tau_E)\}V(X(t \wedge \beta_n \wedge \tau_E), \alpha(t \wedge \beta_n \wedge \tau_E))] - V(x, k) \geq 0.$$

Using conditions (i) and (ii), we can rewrite the above inequality as

$$\begin{aligned} K_3 \leq V(x, k) &\leq \mathbb{E}_{x,k}[e^{-\gamma_1\beta_n} V(X(\beta_n), \alpha(\beta_n))I_{\{\beta_n \leq t\wedge\tau_E\}}] \\ &\quad + \mathbb{E}_{x,k}[e^{-\gamma_1\tau_E} V(X(\tau_E), \alpha(\tau_E))I_{\{\tau_E \leq t\wedge\beta_n\}}] \\ &\quad + \mathbb{E}_{x,k}[e^{-\gamma_1 t} V(X(t), \alpha(t))I_{\{t \leq \beta_n\wedge\tau_E\}}] \\ &\leq K_1\mathbb{E}_{x,k}[e^{-\gamma_1\beta_n}] + K_2 + K_1 e^{-\gamma_1 t}. \end{aligned}$$

Letting $t \to \infty$ and then $n \to \infty$, we have

$$K_3 \leq K_1\mathbb{E}_{x,k}[e^{-\gamma_1\beta_\infty}] + K_2 \text{ or } \mathbb{E}_{x,k}[e^{-\gamma_1\beta_\infty}] \geq \frac{K_3 - K_2}{K_1}.$$

In particular, since $K_2 < K_3$, it holds that $\mathbb{P}_{x,k}\{\beta_\infty < \infty\} > 0$. In other words, the process $(X^{x,k}(\cdot), \alpha^{x,k}(\cdot))$ explodes in finite time with positive probability. □

Remark 2.15 If the Lyapunov function V in Theorem 2.14 is strictly positive and satisfies

$$K_1 := \sup_{(x,i)\in\mathbb{R}^r\times\mathcal{M}} V(x,i) < \infty, \text{ and } \mathcal{L}V(x,i) \ge \gamma_1 V(x,i), \text{ for all } (x,i)\in\mathbb{R}^r\times\mathcal{M},$$

then the process $(X^{x,k}(\cdot), \alpha^{x,k}(\cdot))$ explodes in finite time with positive probability for every initial condition $(x,k) \in \mathbb{R}^r \times \mathcal{M}$.

Indeed, using the same function S and the stopping time β_n as in the proof of Theorem 2.14, we have for any $(x,k) \in \mathbb{R}^r \times \mathcal{M}$ and $t \ge 0$,

$$\begin{aligned}
0 < V(x,k) &\le \mathbb{E}_{x,k}[\exp\{-\gamma_1(t\wedge\beta_n)\}V(X(t\wedge\beta_n), \alpha(t\wedge\beta_n))] \\
&\le K_1(e^{-\gamma_1 t}\mathbb{P}_{x,k}(t \le \beta_n) + \mathbb{E}_{x,k}[e^{-\gamma_1\beta_n} I_{\{\beta_n < t\}}]) \\
&\le K_1(e^{-\gamma_1 t} + \mathbb{P}_{x,k}(\beta_n < t)).
\end{aligned}$$

If $\mathbb{P}_{x,k}(\beta_\infty = \infty) = 1$, then passing to the limit as $n \to \infty$ in the above equation gives us $0 < V(x,k) \le K_1 e^{-\gamma_1 t}$. This is impossible for t sufficiently large.

Remark 2.16 If the transition rate matrix $Q(\cdot)$ is a bounded function satisfying the q-property and (2.21), and if for each $i \in \mathcal{M}$, the coefficients $b(\cdot,i)$ and $\sigma(\cdot,i)$ satisfy both the local Lipschitz condition (2.51) and linear growth condition (2.50), then by Corollary 2.8, there is a unique solution $(X^{x,\alpha}(t), \alpha^{x,\alpha}(t))$ to (2.1)–(2.2) for all $t \ge 0$. The regime-switching diffusion $(X^{x,\alpha}(t), \alpha^{x,\alpha}(t))$ is thus regular. Alternatively, we can use Theorem 2.12 and Remark 2.13 to verify this. In fact, detailed computations show that the function

$$V(x,i) = (|x|^2 + 1)^{r/2}, \quad (x,i) \in \mathbb{R}^r \times \mathcal{M}$$

satisfies all conditions in Remark 2.13. Thus, the desired assertion follows.

In particular, we can verify that $V(\cdot,i)$ is twice continuously differentiable for each $i \in \mathcal{M}$ and that $\mathcal{L}V(x,i) \le \gamma_0 V(x,i)$ for some positive constant γ_0 and all $(x,i) \in \mathbb{R}^r \times \mathcal{M}$. Thus, for any $t > 0$, a similar argument as that in the proof of Theorem 2.12 leads to

$$\mathbb{P}_{x,i}\{\beta_n < t\} \le \frac{V(x,i)e^{\gamma_0 t}}{V_n},$$

where $V_n = (1+n^2)^{r/2}$. Therefore, for any $t > 0$ and $\varepsilon > 0$, there exists an $N \in \mathbb{N}$ such that

$$\mathbb{P}_{x,i}\{\beta_n < t\} < \varepsilon, \text{ for all } n \ge N, \tag{2.60}$$

uniformly for x in any compact set $F \subset \mathbb{R}^r$ and $i \in \mathcal{M}$.

Example 2.17 Consider a real-valued regime-switching diffusion:

$$dX(t) = X(t)\left[(b(\alpha(t)) - a(\alpha(t))X^2(t))dt + \sigma(\alpha(t))dw(t)\right], \qquad (2.61)$$

where $w(\cdot)$ is a one-dimensional standard Brownian motion and $\alpha(\cdot) \in \mathcal{M} = \{1, 2, \ldots, m_0\}$ ($m_0 \leq \infty$) is a switching process with an appropriate bounded generator $Q(x)$ satisfying (2.21). Clearly, the coefficients of (2.61) do not satisfy the linear growth condition if not all $a(i) = 0$ for $i \in \mathcal{M}$. We claim that if $a(i) \geq 0$ for each $i \in \mathcal{M}$, then (2.61) is regular. To see this, we apply Theorem 2.12 and consider

$$V(x, i) := |x|^2, \quad (x, i) \in \mathbb{R} \times \mathcal{M}.$$

Clearly, V is nonnegative and satisfies $\lim_{|x|\to\infty} \inf_{i\in\mathcal{M}} V(x, i) = \infty$, for each $i \in \mathcal{M}$. Thus, by Theorem 2.12, it remains to verify that for all $(x, i) \in \mathbb{R} \times \mathcal{M}$, we have $\mathcal{L}V(x, i) \leq KV(x, i)$ for some positive constant K. To this end, we compute

$$\begin{aligned}\mathcal{L}V(x, i) &= 2x \cdot x[b(i) - a(i)x^2] + \frac{1}{2}2 \cdot x^2\sigma^2(i)\\ &= (2b(i) + \sigma^2(i))x^2 - 2a(i)x^4\\ &\leq KV(x, i),\end{aligned}$$

where $K = \max\left\{2b(i) + \sigma^2(i), i = 1, \ldots, m_0\right\}$. Note that in the above, we used the assumption that $a(i) > 0$ for each $i \in \mathcal{M}$. Hence, it follows from Theorem 2.12 that (2.61) is regular.

For demonstration, we plot a sample path of (2.61) with its coefficients specified as follows. The jump process $\alpha(\cdot) \in \mathcal{M} = \{1, 2\}$ is generated by

$$Q(x) = \begin{pmatrix} -3 - \sin x \cos x & 3 + \sin x \cos x \\ 2 & -2 \end{pmatrix},$$

and

$$\begin{aligned} &b(1) = 3, \quad a(1) = 2, \sigma(1) = 1,\\ &b(2) = 10, a(1) = 1, \sigma(1) = -1.\end{aligned}$$

Figure 2.1 plots a sample path of $(X(t), \alpha(t))$ with initial condition $(x, \alpha) = (2.5, 1)$.

Example 2.18 (Lotka–Volterra Model cont.) We continue the discussion initiated in Example 1.1. Regarding the concept of regularity, we can show that if the ecosystem is self-regulating or competitive, then the population will not explode in finite time almost surely. Recall that the system is competitive if all the values in the community matrix $A(\alpha)$ are nonnegative, that is, $a_{ij}(\alpha) \geq 0$ for all $\alpha \in$

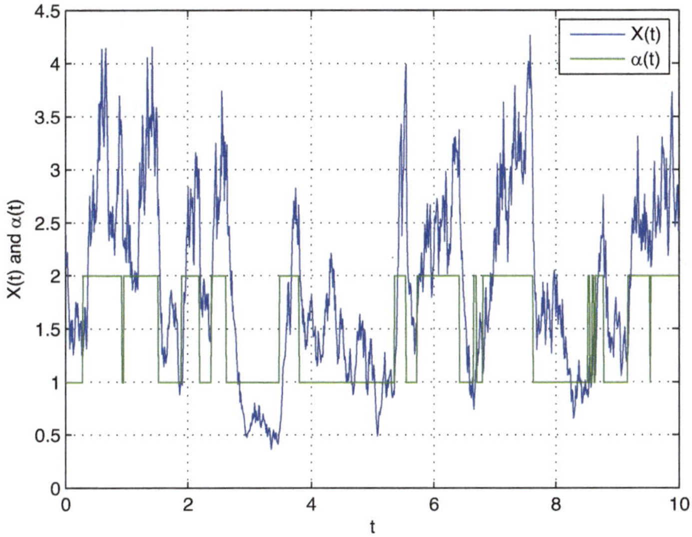

Fig. 2.1 A sample path of switching diffusion (2.61) with initial condition $(x, \alpha) = (2.5, 1)$

$\mathcal{M} = \{1, \ldots, m\}$ and $i, j = 1, 2, \ldots, n$. The competition among the same species is assumed to be strictly positive so that each species grows "logistically" in any environment. To interpret it in another way, members of the same species compete with one another. Take, for instance, bee colonies in a field. They will compete for food strongly with the colonies located near them. We assume that the self-regulating competition within the same species is strictly positive, that is, for each $\alpha \in \mathcal{M} = \{1, 2, \ldots, m\}$ and $i, j = 1, 2, \ldots, n$ with $j \neq i$, $a_{ii}(\alpha) > 0$, and $a_{ij}(\alpha) \geq 0$. It was proved in [322], for any initial conditions $(x(0), \alpha(0)) = (x_0, \alpha) \in \mathbb{R}^n_+ \times \mathcal{M}$, where

$$\mathbb{R}^n_+ = \{(x_1, \ldots, x_n) : x_i > 0, i = 1, \ldots, n\},$$

there is a unique solution $x(t)$ to (1.1) on $t \geq 0$, and the solution will remain in $\mathbb{R}^n_+$ almost surely; that is, $x(t) \in \mathbb{R}^n_+$ for any $t \geq 0$ with probability 1.

Example 2.19 As we observed in Remark 2.3, with $\zeta(u) = \log u$ or $\zeta(u) = \log u \log(\log u)$ for $u > 0$ large, (2.12) in Assumption (H2.1) allows the drift and diffusion coefficient to grow faster than linear with respect to the x variable. Now, we show that if

$$2\langle x, b(x,k)\rangle > |x|^2[\log(|x|^2 + 1) + 1]^{1+\varepsilon}, \text{ and } |\sigma(x,k)| \leq H < \infty, \tag{2.62}$$

for all $(x, k) \in \mathbb{R}^r \times \mathcal{M}$, where H, ε are arbitrary positive constants, then the process $(X^{x,i}(\cdot), \alpha^{x,i}(\cdot))$ explodes in finite time with positive probability for any initial condition $(x, k) \in \mathbb{R}^r \times \mathcal{M}$.

To see this, we consider the function $V(x,i) := \exp\{\arctan((\log(|x|^2+2))^\varepsilon)\}$, which is strictly positive, uniformly bounded, and twice continuously differentiable. Moreover, detailed computations using (2.62) reveal that $\mathcal{L}V(x,i) \geq \gamma_1 V(x,i)$ for all $(x,i) \in \mathbb{R}^r \times \mathcal{M}$, where γ_1 is a positive constant. Therefore, the claimed explosion follows from Theorem 2.14 and Remark 2.15.

2.4 Weak Continuity

Recall that a stochastic process $Y(t)$ with right continuous sample paths is said to be *continuous in probability* at t if for any $\eta > 0$,

$$\lim_{\Delta\to 0} \mathbb{P}(|Y(t+\Delta) - Y(t)| \geq \eta) = 0. \tag{2.63}$$

It is *continuous in mean square* at t if

$$\lim_{\Delta\to 0} \mathbb{E}|Y(t+\Delta) - Y(t)|^2 = 0. \tag{2.64}$$

The process $Y(t)$ is said to be continuous in probability in the interval $[0,T]$ (or in short continuous in probability if $[0,T]$ is clearly understood), if it is continuous in probability at every $t \in [0,T]$. Likewise, it is continuous in mean square in the interval $[0,T]$ if it is continuous in mean square at every $t \in [0,T]$. Clearly mean square continuity implies continuity in probability.

We proceed to obtain a continuity result for the two-component switching diffusion processes $Y(t) = (X(t), \alpha(t))$.

Theorem 2.20 *Suppose Assumptions* (H2.2) *and* (H2.3) *hold. Assume that the coefficients b and σ satisfy the linear growth condition* (2.50) *and that $Q(\cdot)$ satisfies the q-property,* (2.13), *and*

$$\sum_{l\neq k}(l-k)^2 q_{kl}(x) + \Big(\sum_{l\neq k}(l-k)q_{kl}(x)\Big)^2 \leq H(1+|x|^2+k^2),\ \forall (x,k) \in \mathbb{R}^r \times \mathcal{M}, \tag{2.65}$$

where $H > 0$ is a positive constant. Then for any $T \geq 0$, we have

$$\mathbb{E}_{x,k}\Big[\sup_{0\leq t\leq T}(|X(t)|^2 + \alpha(t)^2)\Big] \leq C, \tag{2.66}$$

where $C = C(x,k,T,H)$ is a positive constant. Moreover, the process $Y(t) = (X(t), \alpha(t))$ is mean square continuous and hence continuous in probability.

Proof Under the conditions of the theorem, thanks to Theorem 2.6, for any $(x,k) \in \mathbb{R}^r \times \mathcal{M}$, the system (2.1) and (2.2) has a unique non-explosive strong solution (X,α) with initial condition $(X(0),\alpha(0)) = (x,k)$.

We note that the standard arguments using the linear growth condition (2.50) and the BDG inequality (see, e.g., Lemma A.33 or Theorem IV.48 of [240]) allow us to derive

$$\mathbb{E}_{x,k}\left[\sup_{0\le t\le T}|X(t)|^2\right]\le K_1+K_2\int_0^T\mathbb{E}_{x,k}\left[\sup_{1\le u\le s}|X(u)|^2\right]ds, \tag{2.67}$$

where K_1, K_2 are positive constants depending only on x, H and T.

Using (2.4), we can write

$$\begin{aligned}\alpha(t)=k&+\int_0^t\int_{\mathbb{R}_+}h(X(s-),\alpha(s-),r)\widetilde{\mathfrak{p}}(ds,dr)\\&+\int_0^t\int_{\mathbb{R}_+}h(X(s-),\alpha(s-),r)\mathfrak{m}(dr)ds,\end{aligned}$$

where $\widetilde{\mathfrak{p}}(ds,dr)=\mathfrak{p}(ds,dz)-ds\times\mathfrak{m}(dz)$ is the compensated Poisson random measure. Then we can use the BDG and Hölder's inequalities and (2.65) to compute

$$\begin{aligned}&\mathbb{E}_{x,k}\Big[\sup_{0\le t\le T}\alpha(t)^2\Big]\\
&\le 3k^2+3\mathbb{E}_{x,k}\Big[\sup_{0\le t\le T}\Big(\int_0^t\int_{\mathbb{R}_+}h(X(s-),\alpha(s-),r)\widetilde{\mathfrak{p}}(ds,dr)\Big)^2\Big]\\
&\quad+3\mathbb{E}_{x,k}\Big[\sup_{0\le t\le T}\Big(\int_0^t\int_{\mathbb{R}_+}h(X(s-),\alpha(s-),r)\mathfrak{m}(dr)ds\Big)^2\Big]\\
&\le 3k^2+3\mathbb{E}_{x,k}\Big[\int_0^T\int_{\mathbb{R}_+}h^2(X(s-),\alpha(s-),r)\mathfrak{m}(dr)ds\Big]\\
&\quad+3\mathbb{E}_{x,k}\Big[\Big(\int_0^T\sum_{l\ne\alpha(s-)}(l-\alpha(s-))q_{\alpha(s-),l}(X(s-))ds\Big)^2\Big]\\
&\le 3k^2+3\mathbb{E}_{x,k}\Big[\int_0^T\sum_{l\ne\alpha(s-)}(l-\alpha(s-))^2q_{\alpha(s-),l}(X(s-))ds\Big]\\
&\quad+3\mathbb{E}_{x,k}\Big[T\int_0^T\Big(\sum_{l\ne\alpha(s-)}(l-\alpha(s-))q_{\alpha(s-),l}(X(s-))\Big)^2ds\Big]\\
&\le 3k^2+3H(1+T)\mathbb{E}_{x,k}\Big[\int_0^T(1+|X(s-)|^2+\alpha(s-)^2)ds\Big]\\
&\le 3k^2+3H(1+T)\mathbb{E}_{x,k}\Big[\int_0^T\Big[1+\sup_{0\le u\le s}(|X(u-)|^2+\alpha(u-)^2)\Big]ds\Big].\end{aligned} \tag{2.68}$$

Then (2.66) follows from a combination of (2.67) and (2.68) and Gronwall's inequality.

We next show that the solution (X, α) satisfies

$$\mathbb{E}_{x,k}[|X(t) - X(s)|^2 + |\alpha(t) - \alpha(s)|^2] \le K|t - s|, \quad \forall s, t \in [0, T], \tag{2.69}$$

where K is a positive constant and $T < \infty$. Of course, (2.69) implies the desired mean square continuity and weak continuity of the process (X, α).

Assume without loss of generality that $0 \le s < t \le T < \infty$. The standard computations using the linear growth condition (2.50) and Proposition 2.9 reveal that

$$\mathbb{E}_{x,k}[|X(t) - X(s)|^2] \le C(t - s),$$

for some positive constant $C = C(x, T, K)$, where K is the positive constant given in (2.50).

We next estimate $\mathbb{E}[|\alpha(t)-\alpha(s)|^2]$. Using similar calculations as those for (2.68), we have

$$\begin{aligned}
&\mathbb{E}_{x,k}[|\alpha(t) - \alpha(s)|^2] \\
&\le 2\mathbb{E}_{x,k}\left[\left(\int_s^t \int_{\mathbb{R}_+} |h(X(u-), \alpha(u-), z)|\widetilde{\mathfrak{p}}(du, dz)\right)^2\right] \\
&\quad + 2\mathbb{E}_{x,k}\left[\left(\int_s^t \int_{\mathbb{R}_+} |h(X(u-), \alpha(u-), z)|\mathfrak{m}(dz)du\right)^2\right] \\
&\le 2\mathbb{E}_{x,k}\left[\int_s^t \sum_{l \ne \alpha(u-)} (l - \alpha(u-))^2 q_{\alpha(u-),l}(X(u-)du\right] \\
&\quad + 2\mathbb{E}_{x,k}(t - s)\left[\int_s^t \left(\sum_{l \ne \alpha(u-)} (l - \alpha(u-))q_{\alpha(u-),l}(X(u-))\right)^2 ds\right] \\
&\le 2H(1 + t - s)\mathbb{E}_{x,k}\left[\int_s^t (1 + \alpha(u)^2 + |X(u)|^2)du\right] \\
&\le 2H(1 + t - s)\mathbb{E}_{x,k}\left[\int_s^t (1 + \sup_{0 \le r \le u} [\alpha(r)^2 + |X(r)|^2])du\right] \\
&\le C(t - s),
\end{aligned}$$

where the last inequality follows from (2.66) and C is a positive constant. A combination of the above two displayed equations gives (2.69) and hence completes the proof. □

Remark 2.21 We have derived the moment estimate (2.52) for the X component under the linear growth condition (2.50) in Proposition 2.9. Now, under the additional condition (2.65), (2.66) gives a second-moment estimate for the α component as well, which is interesting in its right. The moment estimate (2.66) in turn helps to establish the mean square continuity and weak continuity for the process (X, α).

Remark 2.22 We make a remark concerning condition (2.65). If $Q(x)$ satisfies (2.13) and the so-called "finite-range switching" condition: For any $k \in \mathcal{M}$, there exists a finite κ_k so that

$$q_{kl}(x) = 0 \text{ for all } (x, l) \in \mathbb{R}^r \times \mathcal{M} \text{ with } |l - k| > \kappa_k,$$

then (2.65) holds.

This in particular includes switching diffusion with birth-and-death type switchings (in which $\kappa_k = 1$) as well as switching diffusions with the α component having a finite state space $\mathcal{M}$ and a bounded generator $Q(x)$.

Another sufficient condition for (2.65) is that $q_{kl}(x)$ decays sufficiently fast for large values of l, uniformly with respect to $x \in \mathbb{R}^r$. For example, if $q_{kl}(x) \leq H(1+|x| + k)2^{-l}$ for all $l \in \mathcal{M}$ large, where H is a positive constant, then (2.65) holds.

In addition, if $q_k(x)$ is uniformly bounded, then (2.65) can be simplified as $\sum_{l \neq k}(l - k)^2 q_{kl}(x) \leq H(1 + |x|^2 + k^2)$ because

$$\Big(\sum_{l \neq k}(l - k) q_{kl}(x)\Big)^2 \leq \|Q\|_\infty \sum_{l \neq k}(l - k)^2 q_{kl}(x), \quad \forall (x, k) \in \mathbb{R}^r \times \mathcal{M}, \tag{2.70}$$

where $\|Q\|_\infty := \sup_{(x,k)\in\mathbb{R}^r \times \mathcal{M}} q_k(x)$. Indeed, (2.70) is trivially true if $q_k(x) = 0$; otherwise, noting that $\frac{q_{kl}(x)}{q_k(x)} \geq 0$ for each $l \neq k$ and $\sum_{l \neq k} \frac{q_{kl}(x)}{q_k(x)} = 1$, we then have

$$\begin{aligned}
\Big(\sum_{l \neq k}(l - k) q_{kl}(x)\Big)^2 &= q_k^2(x)\Big(\sum_{l \neq k}(l - k)\frac{q_{kl}(x)}{q_k(x)}\Big)^2 \\
&\leq q_k^2(x) \sum_{l \neq k} 1^2 \frac{q_{kl}(x)}{q_k(x)} \cdot \sum_{l \neq k}(l - k)^2 \frac{q_{kl}(x)}{q_k(x)} \\
&= q_k(x) \sum_{l \neq k}(l - k)^2 q_{kl}(x),
\end{aligned}$$

establishing (2.70).

2.5 Feller Property

In the theory of Markov processes and their applications, dealing with a Markov process $\xi(t)$ with $\xi(0) = x$, for a suitable function $f(\cdot)$, one often needs to consider the function $u(t,x) = \mathbb{E}_x[f(\xi(t))]$. Following [76], the process $\xi(t)$ is said to be *Feller* if $u(t,\cdot)$ is continuous for any $t \geq 0$ and $\lim_{t\downarrow 0} u(t,x) = f(x)$ for any bounded and continuous function f, and it is said to be *strong Feller* if $u(t,\cdot)$ is continuous for any $t > 0$ and any bounded and measurable function f. This is a natural condition in physical or social modeling. It indicates that a slight perturbation of the initial data should result in a small perturbation in the subsequent movement. For example, letting $X^x(t)$ be a diffusion process of appropriate dimension with $X(0) = x$, $X^x(\cdot)$ satisfies the Feller property or $X^x(\cdot)$ is a Feller process if for any $t \geq 0$ and any bounded and continuous function $g(\cdot)$, $u(t,x) = \mathbb{E}_x[g(X(t))]$ is continuous with respect to x and $\lim_{t\downarrow 0} u(t,x) = g(x)$; see [76, 105] and many references therein for the pioneering work on Feller properties for the diffusions. For switching diffusion processes, do the Feller and the strong Feller properties carry over? This section is concerned with Feller property, whereas the next section deals with strong Feller property.

We show that such properties carry over to the switching diffusions, but the proofs are in fact nontrivial. The difficulty stems from the consideration of x-dependent generator $Q(x)$ for the discrete component $\alpha(t)$. The classical arguments for Feller and strong Feller properties of diffusions will not work here. We need to prove that the function

$$u(t,x,i) = P_t f(x,i) = \mathbb{E}_{x,i}[f(X(t),\alpha(t))] \tag{2.71}$$

is continuous with respect to the initial data (x,i) for any $t \geq 0$ and $\lim_{t\downarrow 0} u(t,x,i) = f(x,i)$ for any bounded and continuous function f. Since the process $(X(\cdot),\alpha(\cdot))$ is càdlàg, it follows from the boundedness and continuity of f that $\lim_{t\downarrow 0} u(t,x,i) = \mathbb{E}_{x,i}[f(X(0),\alpha(0))] = f(x,i)$. Moreover, because $\mathcal{M} = \{1,\ldots,m_0\}$ is a discrete set, it is enough to show that $u(t,x,i)$ is continuous with respect to x for any $t \geq 0$. In this section, we establish the Feller property for switching diffusions under Assumptions (H2.1), (H2.2), and (H2.3). As we commented in Remarks 2.3 and 2.11, these conditions are more general than the commonly used Lipschitz continuity and linear growth conditions in the literature.

Theorem 2.23 *Under Assumptions* (H2.1), (H2.2), *and* (H2.3), *the process* (X,α) *has the Feller property.*

We will use the coupling method [42] to prove Theorem 2.23. To this end, let us first construct a basic coupling operator $\widetilde{\mathcal{L}}$ for $\mathcal{L}$ given in (2.6). For $f(x,i,z,j) \in C_c^2(\mathbb{R}^r \times \mathcal{M} \times \mathbb{R}^r \times \mathcal{M})$, we define

$$\widetilde{\mathcal{L}} f(x,i,z,j) := \big[\widetilde{\Omega}_{\mathrm{d}} + \widetilde{\Omega}_{\mathrm{s}}\big] f(x,i,z,j), \tag{2.72}$$

where $\widetilde{\Omega}_{\mathrm{d}}$ and $\widetilde{\Omega}_{\mathrm{s}}$ are defined as follows. For $x, z \in \mathbb{R}^r$ and $i, j \in \mathcal{M}$, we set $a(x,i) = \sigma(x,i)\sigma'(x,i)$ and

$$a(x,i,z,j) = \begin{pmatrix} a(x,i) & \sigma(x,i)\sigma'(z,j) \\ \sigma(z,j)\sigma'(x,i) & a(z,j) \end{pmatrix}, \quad b(x,i,z,j) = \begin{pmatrix} b(x,i) \\ b(z,j) \end{pmatrix}.$$

Then we define

$$\widetilde{\Omega}_{\mathrm{d}} f(x,i,z,j) := \frac{1}{2}\mathrm{tr}\big(a(x,i,z,j)D^2 f(x,i,z,j)\big) + \langle b(x,i,z,j), Df(x,i,z,j)\rangle, \tag{2.73}$$

where $Df(x,i,z,j) = (D_x f(x,i,z,j), D_z f(x,i,z,j))$ and $D^2 f(x,i,z,j)$ denote the gradient and the Hessian matrix of f with respect to the variables x and z, respectively, and

$$\begin{aligned}\widetilde{\Omega}_{\mathrm{s}} f(x,i,z,j) := &\sum_{l\in\mathcal{M}}[q_{il}(x)-q_{jl}(z)]^+(f(x,l,z,j)-f(x,i,z,j))\\ &+\sum_{l\in\mathcal{M}}[q_{jl}(z)-q_{il}(x)]^+(f(x,i,z,l)-f(x,i,z,j))\\ &+\sum_{l\in\mathcal{M}}[q_{il}(x)\wedge q_{jl}(z)](f(x,l,z,l)-f(x,i,z,j)).\end{aligned} \tag{2.74}$$

For any function $f:\mathbb{R}^r\times\mathbb{R}^r\mapsto\mathbb{R}$, let $\widetilde{f}:\mathbb{R}^r\times\mathcal{M}\times\mathbb{R}^r\times\mathcal{M}\mapsto\mathbb{R}$ be defined by $\widetilde{f}(x,i,z,j) := f(x,z)$. Now, we denote for each $k\in\mathcal{M}$

$$\widetilde{\mathcal{L}}_k f(x,z) := \widetilde{\Omega}_{\mathrm{d}}\widetilde{f}(x,k,z,k), \quad \forall f\in C_c^2(\mathbb{R}^r\times\mathbb{R}^r).$$

We introduce the following notation. Let $(X(\cdot),\alpha(\cdot),\tilde{X}(\cdot),\tilde{\alpha}(\cdot))$ denote the coupling process corresponding to the operator $\widetilde{\mathcal{L}}$ with initial condition (x,k,z,k), in which $\delta_0 > |x-z| > 0$, and δ_0 is the positive constant in Assumption (H2.2). For any $R>0$, let

$$\tau_R := \inf\{t\ge 0 : |\tilde{X}(t)|\vee|X(t)|\vee|\tilde{\alpha}(t)|\vee|\alpha(t)| > R\}. \tag{2.75}$$

In view of Theorem 2.6, $\lim_{R\to\infty}\tau_R=\infty$ a.s. Also denote $\Delta_t=\tilde{X}(t)-X(t)$ and

$$S_{\delta_0} := \inf\{t\ge 0 : |\Delta_t|>\delta_0\} = \inf\{t\ge 0 : |\tilde{X}(t)-X(t)|>\delta_0\}. \tag{2.76}$$

Note that $\alpha(0)=\tilde{\alpha}(0)=k$. We denote by

$$\zeta := \inf\{t\ge 0 : \alpha(t)\ne\tilde{\alpha}(t)\} \tag{2.77}$$

the first time when the switching components α and $\tilde{\alpha}$ differ. We need the following lemma.

Lemma 2.24 *Under Assumption* (H2.2), *the following assertion holds:*

$$\lim_{|\tilde{x}-x|\to 0}\mathbb{E}[|\Delta_{t\wedge\tau_R\wedge S_{\delta_0}\wedge\zeta}|]=0, \quad \forall t\ge 0. \tag{2.78}$$

Proof We prove the lemma by considering the cases $r = 1$ and $r \geq 2$ separately.

Case (i). $d = 1$. Let $\{a_n\}$ be the same sequence defined in the proof of Lemma 2.4. Also let us recall the function $\psi_n(x)$ defined in (2.24). Now, we use (2.17), (2.18), (2.25), and (2.26) to compute

$$\begin{aligned}
&\widetilde{\mathcal{L}}_k \psi_n(x-z) \\
&\quad = \frac{1}{2}\frac{d^2}{dr^2}\psi_n(x-z)|\sigma(x,k)-\sigma(z,k)|^2 + \frac{d}{dr}\psi_n(x-z)(b(x,k)-b(z,k)) \\
&\quad \leq \frac{\kappa_R}{n} + \kappa_R \rho(|x-z|),
\end{aligned} \tag{2.79}$$

for all x, z with $|x| \vee |z| \leq R$ and $0 < |x-z| \leq \delta_0$. Then it follows that

$$\begin{aligned}
\mathbb{E}[\psi_n(\Delta_{t\wedge S_{\delta_0}\wedge\tau_R\wedge\zeta})] &= \mathbb{E}[\psi_n(\tilde{X}(t\wedge S_{\delta_0}\wedge\tau_R\wedge\zeta) - X(t\wedge S_{\delta_0}\wedge\tau_R\wedge\zeta))] \\
&= \psi_n(\tilde{x}-x) + \mathbb{E}\left[\int_0^{t\wedge\tau_R\wedge S_{\delta_0}\wedge\zeta} \widetilde{\mathcal{L}}_k\psi_n(\tilde{X}(s)-X(s))ds\right] \\
&\leq \psi_n(|\Delta_0|) + \mathbb{E}\left[\int_0^{t\wedge\tau_R\wedge S_{\delta_0}\wedge\zeta}\left(\kappa_R\rho(|\Delta_s|) + \frac{\kappa_R}{n}\right)ds\right] \\
&\leq \psi_n(|\Delta_0|) + \frac{\kappa_R}{n}t + \kappa_R\int_0^t \rho\big(\mathbb{E}[|\Delta_{s\wedge\tau_R\wedge S_{\delta_0}\wedge\zeta}|]\big)ds,
\end{aligned}$$

where the first inequality follows from (2.79) and the second inequality follows from the concavity of ρ and Jensen's inequality. Then we use the monotone convergence theorem and (2.26) to derive

$$\mathbb{E}[|\Delta_{t\wedge\tau_R\wedge S_{\delta_0}\wedge\zeta}|] \leq |\Delta_0| + \kappa_R\int_0^t \rho(\mathbb{E}[|\Delta_{s\wedge\tau_R\wedge S_{\delta_0}\wedge\zeta}|])ds.$$

Let $u(t) := \mathbb{E}[|\Delta_{t\wedge\tau_R\wedge S_{\delta_0}\wedge\zeta}|]$. Then u satisfies

$$0 \leq u(t) \leq v(t) := |\Delta_0| + \kappa_R\int_0^t \rho(u(s))ds.$$

Define the function $\Gamma(r) := \int_1^r \frac{ds}{\rho(s)}$ for $r > 0$. Thanks to (2.16), we can verify that Γ is nondecreasing and satisfies $\Gamma(r) > -\infty$ for all $r > 0$ and $\lim_{r\to 0}\Gamma(r) = -\infty$. Then we have

$$\begin{aligned}
\Gamma(u(t)) \leq \Gamma(v(t)) &= \Gamma(|\Delta_0|) + \int_0^t \frac{d}{dr}\Gamma(v(s))v'(s)ds \\
&= \Gamma(|\Delta_0|) + \kappa_R\int_0^t \frac{\rho(u(s))}{\rho(v(s))}ds \leq \Gamma(|\Delta_0|) + \kappa_R\int_0^t 1ds = \Gamma(|\Delta_0|) + \kappa_R t,
\end{aligned}$$

where we use the assumption that ρ is nondecreasing to obtain the last inequality. Taking the limit $|\Delta_0| = |\tilde{x} - x| \to 0$, we have $\Gamma(u(t)) \to -\infty$ since $\lim_{r\to 0}\Gamma(r) = -\infty$. Moreover, since $\Gamma(r) > -\infty$ for all $r > 0$, we must have $\lim_{|\tilde{x}-x|\to 0} u(t) = 0$. This gives (2.78) as desired.

Case (ii). $d \geq 2$. Consider the function $f(x, z) := |x - z|^2$. Then

$$\begin{aligned}\widetilde{\mathcal{L}}_k f(x, z) &= 2\langle x - z, b(x, k) - b(z, k)\rangle + |\sigma(x, k) - \sigma(z, k)|^2\\ &\leq \kappa_R \rho(|x - z|^2),\end{aligned}$$

for all $x, z \in \mathbb{R}^r$ with $|x| \vee |z| \leq R$ and $|x - z| \leq \delta_0$. Consequently,

$$\begin{aligned}\mathbb{E}\big[|\Delta_{t\wedge\tau_R\wedge S_{\delta_0}\wedge\zeta}|^2\big] &= \mathbb{E}[f(\tilde{X}(t\wedge\tau_R\wedge S_{\delta_0}\wedge\zeta), X(t\wedge\tau_R\wedge S_{\delta_0}\wedge\zeta))]\\ &= f(\tilde{x}, x) + \mathbb{E}\bigg[\int_0^{t\wedge\tau_R\wedge S_{\delta_0}\wedge\zeta} \widetilde{\mathcal{L}}_k f(\tilde{X}(s), X(s))ds\bigg]\\ &\leq |\Delta_0| + \mathbb{E}\bigg[\int_0^{t\wedge\tau_R\wedge S_{\delta_0}\wedge\zeta} \kappa_R \rho(|\tilde{X}(s) - X(s)|^2)ds\bigg]\\ &\leq |\Delta_0| + \kappa_R \int_0^t \rho(\mathbb{E}[|\Delta_{s\wedge\tau_R\wedge S_{\delta_0}\wedge\zeta}|^2])ds,\end{aligned}$$

where the last inequality follows from the concavity of ρ and Jensen's inequality. Using the same argument as that in Case (i), we can show that

$$\lim_{|\tilde{x}-x|\to 0} \mathbb{E}[|\Delta_{t\wedge\tau_R\wedge S_{\delta_0}\wedge\zeta}|^2] = 0,$$

which, together with Hölder's inequality, leads to (2.78). Combining the two cases completes the proof. □

Proof of Theorem 2.23 We need to show that for each $(x, k) \in \mathbb{R}^r \times \mathbb{S}$ and each $f \in C_b(\mathbb{R}^r \times \mathcal{M})$, the limit $(P_t f)(\tilde{x}, \tilde{k}) \to (P_t f)(x, k)$ as $(\tilde{x}, \tilde{k}) \to (x, k)$ holds for all $t \geq 0$. Since $\mathcal{M} = \{1, 2, ...\}$ has a discrete topology, it is enough to consider only $(\tilde{x}, k) \to (x, k)$. First, observe that

$$\begin{aligned}&|(P_t f)(\tilde{x}, \tilde{k}) - (P_t f)(x, k)|\\ &= |\mathbb{E}[f(\tilde{X}(t), \tilde{\alpha}(t))] - \mathbb{E}[f(X(t), \alpha(t))]|\\ &\leq |\mathbb{E}[f(\tilde{X}(t), \tilde{\alpha}(t))] - \mathbb{E}[f(\tilde{X}(t), \alpha(t))]|\\ &\quad + |\mathbb{E}[f(\tilde{X}(t), \alpha(t))] - \mathbb{E}[f(X(t), \alpha(t))]|\\ &= |\mathbb{E}[(f(\tilde{X}(t), \tilde{\alpha}(t)) - f(\tilde{X}(t), \alpha(t)))1_{\{\zeta\leq t\}}]|\\ &\quad + |\mathbb{E}[(f(\tilde{X}(t), \tilde{\alpha}(t)) - f(\tilde{X}(t), \alpha(t)))1_{\{\zeta>t\}}]|\end{aligned}$$

$$+ |\mathbb{E}[f(\tilde{X}(t), \alpha(t))] - \mathbb{E}[f(X(t), \alpha(t))]|$$
$$\le 2||f||_\infty \mathbb{P}\{\zeta \le t\} + |\mathbb{E}[f(\tilde{X}(t), \alpha(t))] - \mathbb{E}[f(X(t), \alpha(t))]|. \tag{2.80}$$

We proceed to show that both terms on the right-hand side of (2.80) converge to 0 as $\tilde{x} \to x$.

Consider the function $\Xi(x, k, z, l) := 1_{\{k \ne l\}}$. It follows directly from the definition that

$$\widetilde{\mathcal{L}}\Xi(x, k, z, l) = \widetilde{\Omega}_s \Xi(x, k, z, l) \le 0, \text{ if } k \ne l.$$

When $k = l$, we have from (2.21) that

$$\begin{aligned}
\widetilde{\mathcal{L}}\Xi(x, k, z, l) &= \widetilde{\Omega}_s \Xi(x, k, z, k) \\
&= \sum_{i \in \mathcal{M}} [q_{ki}(x) - q_{ki}(z)]^+ (I_{\{i \ne k\}} - I_{\{k \ne k\}}) \\
&\quad + \sum_{i \in \mathcal{M}} [q_{ki}(z) - q_{ki}(x)]^+ (I_{\{i \ne k\}} - I_{\{k \ne k\}}) \\
&\le \sum_{i \in \mathcal{M}, i \ne k} |q_{ki}(x) - q_{ki}(z)| \le \kappa_R \gamma(|x - z|).
\end{aligned}$$

Hence,

$$\widetilde{\mathcal{L}}\Xi(x, k, z, l) \le \kappa_R \gamma(|x - z|) \tag{2.81}$$

for all $k, l \in \mathcal{M}$ and $x, z \in \mathbb{R}^r$ with $|x| \vee |z| \le R$.

Note that $\zeta \le t \wedge \tau_R \wedge S_{\delta_0}$ if and only if $\tilde{\alpha}(t \wedge \tau_R \wedge S_{\delta_0} \wedge \zeta) \ne \alpha(t \wedge \tau_R \wedge S_{\delta_0} \wedge \zeta)$. In addition, $\alpha(s) = \widetilde{\alpha}(s)$ for all $s \in [0, t \wedge \tau_R \wedge S_{\delta_0} \wedge \zeta)$. Thus, we can use (2.81) to compute

$$\begin{aligned}
&\mathbb{P}\{\zeta \le t \wedge \tau_R \wedge S_{\delta_0}\} \\
&\quad = \mathbb{E}[\Xi(\tilde{X}(t \wedge \tau_R \wedge S_{\delta_0} \wedge \zeta), \tilde{\alpha}(t \wedge \tau_R \wedge S_{\delta_0} \wedge \zeta), \\
&\qquad\qquad X(t \wedge \tau_R \wedge S_{\delta_0} \wedge \zeta), \alpha(t \wedge \tau_R \wedge S_{\delta_0} \wedge \zeta))] \\
&\quad = \Xi(\tilde{x}, k, x, k) + \mathbb{E}\bigg[\int_0^{t \wedge \tau_R \wedge S_{\delta_0} \wedge \zeta} \widetilde{\mathcal{L}}\Xi(\tilde{X}(s), \tilde{\alpha}(s), X(s), \alpha(s)) ds\bigg] \\
&\quad \le \kappa_R \mathbb{E}\bigg[\int_0^{t \wedge \tau_R \wedge S_{\delta_0} \wedge \zeta} \gamma(|\Delta_s|) ds\bigg] \le \kappa_R \int_0^t \mathbb{E}\big[\gamma(|\Delta_{s \wedge \tau_R \wedge S_{\delta_0} \wedge \zeta}|)\big] ds \\
&\quad \le \kappa_R \int_0^t \gamma\big(\mathbb{E}[|\Delta_{s \wedge \tau_R \wedge S_{\delta_0} \wedge \zeta}|]\big) ds,
\end{aligned}$$

where the last inequality follows from the concavity of γ. Since $\gamma(0) = 0$ and $|\Delta_{s\wedge\tau_R\wedge S_{\delta_0}\wedge\zeta}| \le \delta_0$ for all $s \ge 0$, (2.78) and the bounded convergence theorem imply

$$\lim_{|\widetilde{x}-x|\to 0} \mathbb{P}\{\zeta \le t \wedge \tau_R \wedge S_{\delta_0}\} = 0. \tag{2.82}$$

Note also that on the set $\{S_{\delta_0} \le t \wedge \zeta \wedge \tau_R\}$, we have $\delta_0 \le |\Delta_{S_{\delta_0}\wedge t\wedge\zeta\wedge\tau_R}|$. This implies

$$\delta_0 \mathbb{P}\{S_{\delta_0} \le t \wedge \zeta \wedge \tau_R\} \le \mathbb{E}[|\Delta_{t\wedge S_{\delta_0}\wedge\zeta\wedge\tau_R}| I_{\{S_{\delta_0}\le t\wedge\zeta\wedge\tau_R\}}] \le \mathbb{E}[|\Delta_{t\wedge S_{\delta_0}\wedge\zeta\wedge\tau_R}|].$$

Therefore, it follows from (2.78) that

$$\lim_{|\widetilde{x}-x|\to 0} \mathbb{P}\{S_{\delta_0} \le t \wedge \zeta \wedge \tau_R\} = 0. \tag{2.83}$$

Fix two arbitrary positive numbers ε and $\hat{\varepsilon}$. Then (2.78) also implies that

$$\lim_{|\widetilde{x}-x|\to 0} \mathbb{P}\{|\Delta_{t\wedge S_{\delta_0}\wedge\tau_R\wedge\zeta}| > \varepsilon\} \le \lim_{|\widetilde{x}-x|\to 0} \varepsilon^{-1}\mathbb{E}\big[|\Delta_{t\wedge S_{\delta_0}\wedge\tau_R\wedge\zeta}|\big] = 0. \tag{2.84}$$

Since $\lim_{R\to\infty} \tau_R = \infty$ a.s., we can choose R sufficiently large so that

$$\mathbb{P}\{\tau_R < t\} < \hat{\varepsilon}. \tag{2.85}$$

Then

$$\begin{aligned}
&\mathbb{P}\{|\Delta_t| > \varepsilon\} \\
&= \mathbb{P}\{|\Delta_t| > \varepsilon, \tau_R < t\} + \mathbb{P}\{|\Delta_t| > \varepsilon, \tau_R \ge t, \zeta \le t \wedge S_{\delta_0} \wedge \tau_R\} \\
&\quad + \mathbb{P}\{|\Delta_t| > \varepsilon, \tau_R \ge t, \zeta > t \wedge S_{\delta_0} \wedge \tau_R, S_{\delta_0} \le t \wedge \tau_R \wedge \zeta\} \\
&\quad + \mathbb{P}\{|\Delta_t| > \varepsilon, \tau_R \ge t, \zeta > t \wedge S_{\delta_0} \wedge \tau_R, S_{\delta_0} > t \wedge \tau_R \wedge \zeta\} \\
&\le \hat{\varepsilon} + \mathbb{P}\{\zeta \le t \wedge S_{\delta_0} \wedge \tau_R\} \\
&\quad + \mathbb{P}\{S_{\delta_0} \le t \wedge \tau_R \wedge \zeta\} + \mathbb{P}\{|\Delta_t| > \varepsilon, t \le S_{\delta_0} \wedge \tau_R \wedge \zeta\} \\
&\le \hat{\varepsilon} + \mathbb{P}\{\zeta \le t \wedge S_{\delta_0} \wedge \tau_R\} + \mathbb{P}\{S_{\delta_0} \le t \wedge \tau_R \wedge \zeta\} + \mathbb{P}\{|\Delta_{t\wedge S_{\delta_0}\wedge\tau_R\wedge\zeta}| > \varepsilon\}.
\end{aligned}$$

Plugging (2.82)–(2.84) into the above equation, we have

$$\lim_{|\widetilde{x}-x|\to 0} \mathbb{P}\{|\Delta_t| > \varepsilon\} \le \hat{\varepsilon}.$$

Since $\hat{\varepsilon}$ is arbitrary, we conclude that $\lim_{|\widetilde{x}-x|\to 0} \mathbb{P}\{|\Delta_t| > \varepsilon\} = 0$. In other words, $\widetilde{X}(t) \to X(t)$ in probability as $\widetilde{x} \to x$, with the metric d on $\mathbb{R}^r \times \mathcal{M}$ defined by $d((x,i),(y,j)) := |x-y| + 1_{\{i\ne j\}}$, we see immediately that $(\widetilde{X}(t), \Lambda(t)) \to$

$(X(t), \Lambda(t))$ in probability as $\tilde{x} \to x$. Because the function f is continuous, we also have $f(\tilde{X}(t), \Lambda(t)) \to f(X(t), \Lambda(t))$ in probability as $\tilde{x} \to x$. Then the bounded convergence theorem implies

$$|\mathbb{E}[f(\tilde{X}(t), \alpha(t))] - \mathbb{E}[f(X(t), \alpha(t))]| \to 0 \quad \text{as} \quad \tilde{x} \to x. \tag{2.86}$$

Next, we show that $\lim_{\tilde{x}\to x} \mathbb{P}\{\zeta \le t\} = 0$ holds. Thanks to (2.85), we can compute

$$\begin{aligned}
&\mathbb{P}\{\zeta \le t\} \\
&= \mathbb{P}\{\zeta \le t, \tau_R < t\} + \mathbb{P}\{\zeta \le t, \tau_R \ge t\} \\
&\le \mathbb{P}\{\tau_R < t\} + \mathbb{P}\{\zeta \le t, \tau_R \ge t, S_{\delta_0} \le t \wedge \zeta\} + \mathbb{P}\{\zeta \le t, \tau_R \ge t, S_{\delta_0} > t \wedge \zeta\} \\
&\le \hat{\varepsilon} + \mathbb{P}\{S_{\delta_0} \le t \wedge \zeta \wedge \tau_R\} + \mathbb{P}\{\zeta \le t \wedge \tau_R \wedge S_{\delta_0}\}.
\end{aligned}$$

It then follows from (2.82) and (2.83) that $\lim_{|\tilde{x}-x|\to 0} \mathbb{P}\{\zeta \le t\} \le \hat{\varepsilon}$. Again since $\hat{\varepsilon}$ is arbitrary, we have

$$\lim_{|\tilde{x}-x|\to 0} \mathbb{P}\{\zeta \le t\} = 0. \tag{2.87}$$

Finally, we plug (2.86) and (2.87) into (2.80) to complete the proof. □

Corollary 2.25 *Assume that the conditions of Theorem* 2.23 *hold. Then the process* $(X(t), \alpha(t))$ *is strong Markov.*

Proof This follows directly from [94, Theorem 2.2.4] and Theorem 2.23. □

2.6 Strong Feller Property

This section establishes the strong Feller property for the process (X, α); namely, for any bounded and measurable function f, the function $P_t f$ defined in (2.71) is continuous for any $t > 0$. To this end, we assume the following condition holds throughout the section.

(H2.4) For every $k \in \mathcal{M}$, the following assertions hold:

(i) For every $R > 0$, there exists a constant $\lambda_R > 0$ such that

$$\langle \xi, a(x, k)\xi \rangle \ge \lambda_R |\xi|^2, \qquad \xi \in \mathbb{R}^r, \tag{2.88}$$

for all $x \in \mathbb{R}^r$ with $|x| \le R$, where $a(x, k) := \sigma(x, k)\sigma'(x, k)$.

(ii) There exist a positive constant δ_0 and a nonnegative function $g \in C(0, 1)$ satisfying

$$\int_0^1 g(r)dr < \infty, \tag{2.89}$$

such that for each $R > 0$, there exists a constant $\kappa_R > 0$ so that either (a) or (b) below holds:

(a) If $r = 1$, then

$$2\langle x - z, b(x, k) - b(z, k)\rangle \le 2\kappa_R |x - z| g(|x - z|), \tag{2.90}$$

for all $x, z \in \mathbb{R}$ with $|x| \vee |z| \le R$ and $|x - z| \le \delta_0$.

(b) If $r \ge 2$, then

$$\begin{aligned} &|\sigma_{\lambda_R}(x, k) - \sigma_{\lambda_R}(z, k)|^2 \\ &\qquad + 2\langle x - z, b(x, k) - b(z, k)\rangle \le 2\kappa_R |x - z| g(|x - z|), \end{aligned} \tag{2.91}$$

for all $x, z \in \mathbb{R}^r$ with $|x| \vee |z| \le R$ and $|x - z| \le \delta_0$, where σ_{λ_R} is the unique symmetric nonnegative definite matrix-valued function such that $\sigma^2_{\lambda_R}(x, k) = a(x, k) - \lambda_R I$.

Theorem 2.26 *Suppose that Assumptions* (H2.1)–(H2.4) *hold. Then the process* (X, α) *has strong Feller property.*

As in Section 2.5, we will use the coupling method [42] to prove Theorem 2.26. To this end, we first define

$$\widehat{a}(x, i, z, j) := \begin{pmatrix} a(x, i) & \hat{g}(x, i, z, j) \\ \hat{g}'(x, i, z, j) & a(z, j) \end{pmatrix} \quad \text{and} \quad b(x, i, z, j) := \begin{pmatrix} b(x, i) \\ b(z, j) \end{pmatrix}$$

where

$$\hat{g}(x, i, z, j) := \lambda_R(I - 2u(x, z)u'(x, z)) + \sigma_{\lambda_R}(x, i)\sigma'_{\lambda_R}(z, j),$$

and $u(x, z) := \frac{x-z}{|x-z|}$. Define the coupling operator $\widehat{\mathcal{L}}$ for $\mathcal{L}$ of (2.6) as follows:

$$\widehat{\mathcal{L}} f(x, i, z, j) := [\widehat{\Omega}_{\rm d} + \widetilde{\Omega}_{\rm s}] f(x, i, z, j), \quad f \in C_c^2(\mathbb{R}^r \times \mathcal{M} \times \mathbb{R}^r \times \mathcal{M}), \tag{2.92}$$

where

$$\widehat{\Omega}_{\rm d} f(x, i, z, j) = \frac{1}{2}\text{tr}\left(\widehat{a}(x, i, z, j) D^2 f(x, i, z, j)\right) + \langle b(x, i, z, j), Df(x, i, z, j)\rangle, \tag{2.93}$$

and $\widetilde{\Omega}_{\rm s}$ is defined as in (2.74). In addition, as in Section 2.5, for each $k \in \mathcal{M}$ and any $F \in C_c^2(\mathbb{R}^r \times \mathbb{R}^r)$, we write $f(x, k, z, k) := F(x, z)$ and denote

$$\widehat{\mathcal{L}}_k F(x, z) := \widehat{\Omega}_{\rm d} f(x, k, z, k). \tag{2.94}$$

Note that $\widehat{\mathcal{L}}_k$ is a coupling operator for $\mathcal{L}_k$ defined in (2.7) and $\widehat{\mathcal{L}}_k F(x,z) = \widehat{\mathcal{L}} F(x,z)$.

Furthermore, to facilitate future presentations, we introduce the following notation. For any $x, z \in \mathbb{R}^r$ and $i, j \in \mathbb{S}$, we let

$$\begin{aligned}
A(x,i,z,j) &:= a(x,i) + a(z,j) - 2\hat{g}(x,i,z,j),\\
\bar{A}(x,i,z,j) &:= \frac{1}{|x-z|^2}\langle x-z, A(x,i,z,j)(x-z)\rangle,\\
B(x,i,z,j) &:= \langle x-z, b(x,i) - b(z,j)\rangle.
\end{aligned}$$

Lemma 2.27 *For all $x, z \in \mathbb{R}^r$ and $i, j \in \mathcal{M}$, we have the following:*

(i) $\widehat{a}(x,i,z,j)$ *is symmetric and positive definite.*
(ii) $\operatorname{tr} A(x,i,z,j) = |\sigma_{\lambda_R}(x,i) - \sigma_{\lambda_R}(z,j)|^2 + 4\lambda_R$.
(iii) $\bar{A}(x,i,z,j) \ge 4\lambda_R$.

Proof The proof involves elementary and straightforward computations; similar computations can be found in [43] and [238]. We shall omit the details here. □

Now, let $\phi \in C^2([0,\infty))$. As in [43], for each $k \in \mathcal{M}$ and all $x, z \in \mathbb{R}^r$ with $x \neq z$, we can verify that

$$\begin{aligned}
\widehat{\mathcal{L}}_k \phi(|x-z|) &= \frac{1}{2}\frac{d^2}{dr^2}\phi(|x-z|)\bar{A}(x,k,z,k) \\
&\quad + \frac{\frac{d}{dr}\phi(|x-z|)}{2|x-z|}[\operatorname{tr} A(x,k,z,k) - \bar{A}(x,k,z,k) + 2B(x,k,z,k)].
\end{aligned} \tag{2.95}$$

Motivated by [238], for any $R > 0$, we consider the function G_R defined by

$$G_R(\rho) := \int_0^\rho \exp\left\{-\int_0^s \frac{\kappa_R}{2\lambda_R} g(w)dw\right\}\int_s^1 \exp\left\{\int_0^v \frac{\kappa_R}{2\lambda_R} g(u)du\right\}dvds,$$

for $\rho \in [0,1]$, where the function g and the constants κ_R, λ_R are given in Assumption (H2.4). Since $g \ge 0$, we see that

$$\begin{aligned}
&\frac{d}{d\rho}G_R(\rho) = e^{-\int_0^\rho \frac{\kappa_R}{2\lambda_R} g(w)dw}\int_\rho^1 e^{\int_0^v \frac{\kappa_R}{2\lambda_R} g(u)du}dv \ge 0, \quad \text{and}\\
&\frac{d^2}{d\rho^2}G_R(\rho) = -1 - \frac{\kappa_R}{2\lambda_R} g(\rho)\frac{d}{d\rho}G(\rho) \le 0.
\end{aligned} \tag{2.96}$$

Note also that G_R is concave and $\lim_{\rho\to 0} G_R(\rho) = 0$. Since $\frac{d}{d\rho}G_R(0) \ge 1$ and $G_R(0) = 0$, there exists a constant $\theta \in (0,1)$ so that

$$\rho \le G_R(\rho) \quad \text{for all } \rho \in [0,\theta]. \tag{2.97}$$

Lemma 2.28 *Suppose that Assumption* (H2.4) *holds. Then for any $R > 0$ and $k \in \mathcal{M}$, we have*

$$\widehat{\mathfrak{L}}_k G_R(|x-z|) \le -2\lambda_R \tag{2.98}$$

for all $x, z \in \mathbb{R}^r$ with $|z| \vee |x| \le R$ and $0 < |x-z| \le \delta_0$.

Proof Fix any $R > 0$ and $k \in \mathcal{M}$. For notational simplicity, we write $G = G_R$ throughout the proof. Let us first prove the lemma for the case when $r \ge 2$. In view of (2.95), it follows from (2.96) and Lemma 2.27 that

$$\begin{aligned}
\widehat{\mathfrak{L}}_k G(|x-z|) &= \frac{\frac{d^2}{d\rho^2} G(|x-z|)}{2} \bar{A}(x,k,z,k) \\
&\quad + \frac{\frac{d}{d\rho} G(|x-z|)}{2|x-z|} [\operatorname{tr} A(x,k,z,k) - \bar{A}(x,k,z,k) + 2B(x,k,z,k)] \\
&\le \frac{\frac{d^2}{d\rho^2} G(|x-z|)}{2} 4\lambda_R + \frac{\frac{d}{d\rho} G(|x-z|)}{2|x-z|} [|\sigma_{\lambda_R}(x,k) - \sigma_{\lambda_R}(z,k)|^2 + 2B(x,k,z,k)] \\
&= 2\lambda_R \left(-1 - \frac{\kappa_R}{2\lambda_R} g(|x-z|) \frac{d}{d\rho} G(|x-z|) \right) \\
&\quad + \frac{\frac{d}{d\rho} G(|x-z|)}{2|x-z|} [|\sigma_{\lambda_R}(x,k) - \sigma_{\lambda_R}(z,k)|^2 + 2B(x,k,z,k)] \\
&= -2\lambda_R + \frac{d}{d\rho} G(|x-z|) \\
&\quad \times \left(-\kappa_R g(|x-z|) + \frac{|\sigma_{\lambda_R}(x,k) - \sigma_{\lambda_R}(z,k)|^2 + 2B(x,k,z,k)}{2|x-z|} \right) \\
&\le -2\lambda_R,
\end{aligned} \tag{2.99}$$

where the last inequality follows from (2.91). This gives (2.98) when $r \ge 2$.

Finally, we look at the case when $r = 1$. First, we notice that

$$\bar{A}(x,i,z,j) = \operatorname{tr} A(x,i,z,j) = (\sigma_{\lambda_R}(x,i) - \sigma_{\lambda_R}(z,j))^2 + 4\lambda_R.$$

Using this observation and similar calculations as those in (2.99), we obtain

$$\begin{aligned}
\widehat{\mathfrak{L}}_k G(|x-z|) &\le -2\lambda_R + \left(-\kappa_R g(|x-z|) + \frac{2B(x,k,z,k)}{2|x-z|} \right) G'(|x-z|) \\
&\le -2\lambda_R,
\end{aligned}$$

where we used (2.90) to derive the last inequality. The proof is complete. □

Throughout the rest of the section, we use the following notation. For any $x, \tilde{x} \in \mathbb{R}^r$ and $k \in \mathcal{M}$, denote by $(X(\cdot), \alpha(\cdot), \tilde{X}(\cdot), \tilde{\alpha}(\cdot))$ the process corresponding to the coupling operator $\widehat{\mathcal{L}}$ with initial condition $(x, k, \tilde{x}, k)$. As in Section 2.5, denote $\Delta_t := \tilde{X}(t) - X(t)$ for $t \geq 0$. Let τ_R, S_{δ_0}, and ζ be defined as in (2.75), (2.76), and (2.77), respectively. In addition, for each $n \in \mathbb{N}$, we define

$$T_n := \inf\left\{t \geq 0 : |X(t) - \tilde{X}(t)| < \frac{1}{n}\right\}. \tag{2.100}$$

Then $\lim_{n\to\infty} T_n = T$, where

$$T := \inf\{t \geq 0 : X(t) = \tilde{X}(t)\}. \tag{2.101}$$

Lemma 2.29 *Suppose Assumption* (H2.4) *holds. Then the following assertions hold for every $t \geq 0$:*

$$\lim_{|\tilde{x}-x|\to 0} \mathbb{E}[G(|\Delta_{t\wedge\tau_R\wedge S_{\bar{\delta}}\wedge\zeta}|)] = 0, \text{ and} \tag{2.102}$$

$$\lim_{|\tilde{x}-x|\to 0} \mathbb{E}[|\Delta_{t\wedge\tau_R\wedge S_{\bar{\delta}}\wedge\zeta}|] = 0, \tag{2.103}$$

where $\bar{\delta} := \delta_0 \wedge \theta$, δ_0 is the constant given in Assumption (H2.4) *(ii), and $\theta \in (0, 1)$ is the constant given in* (2.97).

Proof Assume without loss of generality that $\bar{\delta} \geq |x - \tilde{x}| > 0$. As in the proof of Lemma 2.28, we write $G = G_R$ throughout the proof. We apply Itô's formula to the process $G(|\tilde{X}(\cdot) - X(\cdot)|) = G(|\Delta_\cdot|)$:

$$\begin{aligned}\mathbb{E}[G(|\Delta_{t\wedge\tau_R\wedge S_{\bar{\delta}}\wedge\zeta}|)] &= G(|\Delta_0|) + \mathbb{E}\left[\int_0^{t\wedge\tau_R\wedge S_{\bar{\delta}}\wedge\zeta} \widehat{\mathcal{L}}G(|\Delta_s|)ds\right] \\ &\leq G(|\Delta_0|) - 2\lambda_R\mathbb{E}[t\wedge\tau_R\wedge S_{\bar{\delta}}\wedge\zeta],\end{aligned}$$

where the last inequality follows from (2.98). Hence,

$$\mathbb{E}[G(|\Delta_{t\wedge\tau_R\wedge S_{\bar{\delta}}\wedge\zeta}|)] + 2\lambda_R\mathbb{E}[t\wedge\tau_R\wedge S_{\bar{\delta}}\wedge\zeta] \leq G(|\Delta_0|) = G(|x - \tilde{x}|).$$

Since $\lim_{r\to 0} G(r) = 0$, (2.102) follows. Since $|\Delta_{t\wedge\tau_R\wedge S_{\bar{\delta}}\wedge\zeta}| \leq \bar{\delta} \leq \theta$, it follows from (2.97) that $|\Delta_{t\wedge\tau_R\wedge S_{\bar{\delta}}\wedge\zeta}| \leq G(|\Delta_{t\wedge\tau_R\wedge S_{\bar{\delta}}\wedge\zeta}|)$ and therefore (2.103) follows as well. □

Lemma 2.30 *Suppose that Assumptions* (H2.3) *and* (H2.4) *hold. Then*

$$\lim_{|\tilde{x}-x|\to 0} \mathbb{P}\{t < T\} = 0 \tag{2.104}$$

holds for every $t \geq 0$.

Proof We may assume without loss of generality that $\bar{\delta} \geq |x - \tilde{x}| > \frac{1}{n_0} > 0$ for some $n_0 \in \mathbb{N}$. Let $\varepsilon > 0$ and choose a sufficiently large R so that $|x| \vee |\tilde{x}| < R$ and $\mathbb{P}\{\tau_R \leq t\} < \varepsilon$. Write $G = G_R$ as before. For each $n \geq n_0$, define T_n and T as in (2.100) and (2.101), respectively.

We first observe that

$$\begin{aligned}
\mathbb{P}\{t < T\} &= \mathbb{P}\{t < T, \tau_R < t\} + \mathbb{P}\{t < T, \tau_R \geq t\} \\
&\leq \mathbb{P}\{\tau_R < t\} + \mathbb{P}\{t < T, \tau_R \geq t, S_{\bar{\delta}} < t\} + \mathbb{P}\{t < T, \tau_R \geq t, S_{\bar{\delta}} \geq t\} \\
&\leq \varepsilon + \mathbb{P}\{S_{\bar{\delta}} \leq t \wedge T \wedge \tau_R\} + \mathbb{P}\{t \leq T \wedge \tau_R \wedge S_{\bar{\delta}}\} \\
&= \varepsilon + \mathbb{P}\{S_{\bar{\delta}} \leq t \wedge T \wedge \tau_R, S_{\bar{\delta}} \leq \zeta\} + \mathbb{P}\{S_{\bar{\delta}} \leq t \wedge T \wedge \tau_R, S_{\bar{\delta}} > \zeta\} \\
&\quad + \mathbb{P}\{t \leq T \wedge \tau_R \wedge S_{\bar{\delta}}, t < \zeta\} + \mathbb{P}\{t \leq T \wedge \tau_R \wedge S_{\bar{\delta}}, t \geq \zeta\} \\
&\leq \varepsilon + \mathbb{P}\{S_{\bar{\delta}} \leq t \wedge T \wedge \tau_R \wedge \zeta\} + \mathbb{P}\{\zeta < S_{\bar{\delta}} \wedge t \wedge T \wedge \tau_R\} \\
&\quad + \mathbb{P}\{t \leq T \wedge \tau_R \wedge S_{\bar{\delta}} \wedge \zeta\} + \mathbb{P}\{\zeta \leq t\} \\
&\leq \varepsilon + \mathbb{P}\{S_{\bar{\delta}} \leq t \wedge T \wedge \tau_R \wedge \zeta\} \\
&\quad + \mathbb{P}\{\zeta \leq t\} + \mathbb{P}\{t \leq T \wedge \tau_R \wedge S_{\bar{\delta}} \wedge \zeta\} + \mathbb{P}\{\zeta \leq t\} \\
&= \varepsilon + \mathbb{P}\{S_{\bar{\delta}} \leq t \wedge T \wedge \tau_R \wedge \zeta\} + \mathbb{P}\{t \leq T \wedge \tau_R \wedge S_{\bar{\delta}} \wedge \zeta\} + 2\mathbb{P}\{\zeta \leq t\} \\
&\leq \varepsilon + \mathbb{P}\{S_{\bar{\delta}} \leq T \wedge \tau_R \wedge \zeta\} + \frac{\mathbb{E}[T \wedge \tau_R \wedge S_{\bar{\delta}} \wedge \zeta]}{t} + 2\mathbb{P}\{\zeta \leq t\}.
\end{aligned} \tag{2.105}$$

Note that on the set $\{S_{\bar{\delta}} \leq T_n \wedge \tau_R \wedge \zeta\}$, we have $\bar{\delta} \leq |\Delta_{S_{\bar{\delta}} \wedge T_n \wedge \tau_R \wedge \zeta}|$. Since G is increasing, $0 < G(\bar{\delta}) \leq G(|\Delta_{S_{\bar{\delta}} \wedge T_n \wedge \tau_R \wedge \zeta}|)$. Thus,

$$\begin{aligned}
G(\bar{\delta})\mathbb{P}\{S_{\bar{\delta}} \leq T_n \wedge \tau_R \wedge \zeta\} &\leq \mathbb{E}[G(|\Delta_{S_{\bar{\delta}} \wedge T_n \wedge \tau_R \wedge \zeta}|) I_{\{S_{\bar{\delta}} \leq T_n \wedge \tau_R \wedge \zeta\}}] \\
&\leq \mathbb{E}[G(|\Delta_{S_{\bar{\delta}} \wedge T_n \wedge \tau_R \wedge \zeta}|)] \\
&= G(|x - \tilde{x}|) + \mathbb{E}\left[\int_0^{S_{\bar{\delta}} \wedge T_n \wedge \tau_R \wedge \zeta} \widehat{\mathfrak{L}}_k G(|\Delta_s|) ds\right] \\
&\leq G(|x - \tilde{x}|) - 2\lambda_R \mathbb{E}[T_n \wedge \tau_R \wedge S_{\bar{\delta}} \wedge \zeta],
\end{aligned}$$

where the last inequality follows from (2.98). So

$$G(\bar{\delta})\mathbb{P}\{S_{\bar{\delta}} \leq T_n \wedge \tau_R \wedge \zeta\} + 2\lambda_R \mathbb{E}[T_n \wedge \tau_R \wedge S_{\bar{\delta}} \wedge \zeta] \leq G(|x - \tilde{x}|).$$

Passing to the limit as $n \to \infty$, we obtain

$$G(\bar{\delta})\mathbb{P}\{S_{\bar{\delta}} \leq T \wedge \tau_R \wedge \zeta\} + 2\lambda_R \mathbb{E}[T \wedge \tau_R \wedge S_{\bar{\delta}} \wedge \zeta] \leq G(|x - \tilde{x}|).$$

Then, in view of (2.105), we have

$$\begin{aligned}\mathbb{P}\{t < T\} &\le \varepsilon + \mathbb{P}\{S_{\bar{\delta}} \le T \wedge \tau_R \wedge \zeta\} + \frac{\mathbb{E}[T \wedge \tau_R \wedge S_{\bar{\delta}} \wedge \zeta]}{t} + 2\mathbb{P}\{\zeta \le t\}\\ &\le \varepsilon + \frac{G(|x-\tilde{x}|)}{G(\bar{\delta})} + \frac{G(|x-\tilde{x}|)}{2t\lambda_R} + 2\mathbb{P}\{\zeta \le t\}.\end{aligned}$$

From (2.87) and $\lim_{|\tilde{x}-x|\to 0} G(|x-\tilde{x}|) = 0$, we obtain $\lim_{|\tilde{x}-x|\to 0} \mathbb{P}\{t < T\} \le \varepsilon$. Since ε was arbitrary, we obtain (2.104). □

Now, we are ready to present the proof of Theorem 2.26.

Proof of Theorem 2.26 Given $x \in \mathbb{R}^r$ and $k \in \mathcal{M}$. We want to show that for every $f \in \mathcal{B}_b(\mathbb{R}^r \times \mathcal{M})$ (the set of bounded and Borel measurable functions), the limit $(P_t f)(\tilde{x}, \tilde{k}) \to (P_t f)(x, k)$ as $(\tilde{x}, \tilde{k}) \to (x, k)$ holds for all $t > 0$. As in the proof of Theorem 2.23, we only need to consider the case when $\tilde{k} = k$.

For any given $\varepsilon > 0$, we can choose a sufficiently large R so that $\mathbb{P}\{\tau_R \le t\} < \varepsilon$. Let $\tilde{x} \in \mathbb{R}^r$ be such that $\bar{\delta} \ge |x - \tilde{x}| > 0$, where $\bar{\delta} := \delta_0 \wedge \theta$. Denote the coupling process corresponding to the coupling operator $\widehat{\mathcal{L}}$ defined in (2.92) with initial condition $(x, k, \tilde{x}, k)$ by $(X(t), \alpha(t), \tilde{X}(t), \tilde{\alpha}(t))$. Denote by

$$\tilde{T} := \inf\{t \ge 0 : (X(t), \alpha(t)) = (\tilde{X}(t), \tilde{\alpha}(t))\} \tag{2.106}$$

the coupling time of $(X(t), \alpha(t))$ and $(\tilde{X}(t), \tilde{\alpha}(t))$. Recall the stopping time T defined in (2.101). We make the following observations: (i) $T \le \tilde{T}$. (ii) $T < \zeta$ implies $T = \tilde{T}$. We then have

$$\begin{aligned}I_{\{t<\tilde{T}\}} &= I_{\{t<T\}} + I_{\{T\le t<\tilde{T}\}} = I_{\{t<T\}} + I_{\{T\le t<\tilde{T},\zeta\le t\}} + I_{\{T\le t<\tilde{T},\zeta>t\}}\\ &\le I_{\{t<T\}} + I_{\{\zeta\le t\}} + I_{\{T\le t,t<\zeta,t<\tilde{T}\}} \le I_{\{t<T\}} + I_{\{\zeta\le t\}} + I_{\{T<\zeta,t<\tilde{T}\}}\\ &\le I_{\{t<T\}} + I_{\{\zeta\le t\}} + I_{\{T=\tilde{T},t<\tilde{T}\}} \le I_{\{t<T\}} + I_{\{\zeta\le t\}} + I_{\{t<T\}}\\ &= 2\cdot I_{\{t<T\}} + I_{\{\zeta\le t\}}.\end{aligned}$$

Then it follows that

$$\begin{aligned}|(P_t f)(\tilde{x}, \tilde{k}) - (P_t f)(x, k)| &= |\mathbb{E}[f(\tilde{X}(t), \tilde{\alpha}(t))] - \mathbb{E}[f(X(t), \alpha(t))]|\\ &\le \mathbb{E}[|f(\tilde{X}(t), \tilde{\alpha}(t)) - f(X(t), \alpha(t))| I_{\{t<\tilde{T}\}}]\\ &\quad + \mathbb{E}[|f(\tilde{X}(t), \tilde{\alpha}(t)) - f(X(t), \alpha(t))| I_{\{t\ge\tilde{T}\}}]\\ &= \mathbb{E}[|f(\tilde{X}(t), \tilde{\alpha}(t)) - f(X(t), \alpha(t))| I_{\{t<\tilde{T}\}}]\\ &\le 2||f||_\infty \mathbb{E}[I_{\{t<\tilde{T}\}}]\\ &\le 2||f||_\infty \mathbb{E}[2\cdot I_{\{t<T\}} + I_{\{\zeta\le t\}}]\\ &= 4||f||_\infty \mathbb{P}\{t < T\} + 2||f||_\infty \mathbb{P}\{\zeta \le t\}.\end{aligned}$$

A combination of (2.87) and (2.104) then gives

$$\lim_{|\tilde{x}-x|\to 0} |(P_t f)(\tilde{x}, \tilde{k}) - (P_t f)(x, k)| = 0.$$

This establishes the strong Feller property and concludes the proof. □

Remark 2.31 We note that Assumptions (H2.1)–(H2.4) improve significantly over those in the literature such as [257, 284], which require Lipschitz condition for the coefficients of the associated stochastic differential equations. In contrast, (2.90) and (2.91) place very mild conditions on the coefficients. It allows us to treat, for example, the case of Hölder continuous coefficients by taking $g(u) = u^{-p}$ for $0 \le p < 1$. For the case when $r = 1$, only the drift coefficient is required to satisfy the regularity conditions. Moreover, a careful examination of the proof of Theorem 2.26 reveals that if a weak solution to (2.1) and (2.2) exists and is unique in the sense of probability law, then we only need Assumptions (H2.3) and (H2.4) to establish strong Feller property for the process (X, α).

A slight modification of the standard argument in [76, Theorem 13.5] leads to the following theorem. We omit the detailed proof here.

Theorem 2.32 *Suppose Assumptions* (H2.2), (H2.3), *and* (H2.4) *hold. Let* $U := D \times J \subset \mathbb{R}^r \times \mathcal{M}$, *where* $D \subset \mathbb{R}^r$ *is nonempty, open, and bounded with sufficiently smooth boundary* ∂D. *Then the process* $(X(\cdot \wedge \tau_U), \alpha(\cdot \wedge \tau_U))$ *possesses the Feller property, where* τ_U *is the first exit time from* U.

We end this section with a brief discussion of strong Feller processes and $\mathcal{L}$-harmonic functions. More properties of $\mathcal{L}$-harmonic functions are treated in Chapter 3. As in [100], for any $U = D \times J$, where $D \subset \mathbb{R}^r$ is a nonempty domain, and $J \subset \mathcal{M}$, a Borel measurable function $u : U \mapsto \mathbb{R}$ is said to be *$\mathcal{L}$-harmonic* in U if u is bounded in compact subsets of U and that for all $(x, i) \in U$ and any $V = \widetilde{D} \times \widetilde{J}$ with $\widetilde{D} \subset\subset D$ being a neighborhood of x and $i \in \widetilde{J} \subset J$, we have

$$u(x, i) = \mathbb{E}_{x,i}[u(X(\tau_V), \alpha(\tau_V))],$$

where τ_V denotes the first exit time of the process $(X(t), \alpha(t))$ from V, and $\widetilde{D} \subset\subset D$ means $\widetilde{D} \subset \overline{\widetilde{D}} \subset D$ and $\overline{\widetilde{D}}$ is compact, with $\overline{\widetilde{D}} = \widetilde{D} \cup \partial\widetilde{D}$ denoting the closure of D. It is now obvious that if the process $(X(t), \alpha(t))$ is strong Feller, then any $\mathcal{L}$-harmonic function is continuous. Therefore, we have the following.

Proposition 2.33 *Suppose Assumptions* (H2.2), (H2.3), *and* (H2.4) *hold. Then any $\mathcal{L}$-harmonic function is continuous in its domain.*

2.7 Continuity and Smooth Dependence on the Initial Data x

Well-posedness is an important concept in dynamic systems. It appears in ordinary differential equations and partial differential equations, together with initial and/or boundary data. They are time-honored phenomena, which naturally carry over to stochastic differential equations as well as stochastic differential equations with random switching. A problem for the associated switching diffusion is well posed if there is a unique solution for the initial value problem and the solution continuously depends on the initial data.

In Section 2.5, we established the Feller property for the switching diffusion process (X, α) under Assumptions (H2.1), (H2.2), and (H2.3). The Feller property, in particular, indicates that (X, α) depends on the initial data continuously. In the subsequent development, we also need to use smooth dependence on the initial data, which is even more difficult to obtain. In this section, we devote our attention to this smoothness property. Again we need to use the notion of multi-index. Recall that a vector $\beta = (\beta_1, \dots, \beta_r)$ with nonnegative integer components is referred to as a multi-index. Put

$$|\beta| = \beta_1 + \dots + \beta_r,$$

and define D_x^β as

$$D_x^\beta = \frac{\partial^\beta}{\partial x^\beta} = \frac{\partial^{|\beta|}}{\partial x_1^{\beta_1} \cdots \partial x_r^{\beta_r}}. \tag{2.107}$$

Next, we define the L^p differentiability of a smooth random function $\Phi(x_1, \dots, x_r, t)$. Its partial derivative in probability with respect to x_i is defined as a random variable $\Psi(x_1, \dots, x_r, t)$ such that

$$\frac{\Phi(x_1, \dots, x_i + h, \dots, x_r, t) - \Phi(x_1, \dots, x_i, \dots, x_r, t)}{h} \to \Psi(x_1, \dots, x_r, t)$$

in probability as $h \to 0$. If for some $p > 0$,

$$\mathbb{E}\left|\frac{\Phi(x_1, \dots, x_i + h, \dots, x_r, t) - \Phi(x_1, \dots, x_i, \dots, x_r, t)}{h} - \Psi(x_1, \dots, x_r, t)\right|^p \to 0$$

as $h \to 0$, we say that $\Phi(x_1, \dots, x_r, t)$ has partial derivative with respect to x_i in L^p. We proceed to obtain the smoothness of the switching diffusion with respect to the initial data in the L^p sense. The main result of this section is the following theorem.

Theorem 2.34 *Assume that $b(\cdot, i)$ and $\sigma(\cdot, i)$ are Lipschitz for each $i \in \mathcal{M}$ so that there exists a positive constant H so that*

$$|b(x, i) - b(y, i)| \vee |\sigma(x, i) - \sigma(y, i)| \le H|x - y|, \quad \textit{for all } x, y \in \mathbb{R}^r \textit{ and } i \in \mathcal{M}, \tag{2.108}$$

and that for each $i \in \mathcal{M}$, $b(\cdot, i)$ and $\sigma(\cdot, i)$ have continuous partial derivatives with respect to the variable x up to the second order satisfying

$$|D_x^\beta b(x,i)| + |D_x^\beta \sigma(x,i)| \le K_0(1+|x|^{\gamma_0}), \tag{2.109}$$

where K_0 and γ_0 are positive constants and β is a multi-index with $|\beta| \le 2$. Let $(X^{x,\alpha}(t), \alpha^{x,\alpha}(t))$ be the solution to the system given by (2.1) *and* (2.2) *with initial data $(X(0), \alpha(0)) = (x, \alpha)$.*

(a) *Suppose that $Q(x) = (q_{ij}(x))$ satisfies* (2.13) *and* (2.21), *then $X^{x,\alpha}(t)$ is twice continuously differentiable with respect to x in probability.*
(b) *Suppose that $Q(x)$ satisfies* (2.13) *and is locally Hölder continuous so that there exist constants $\lambda \in (0, 1]$, $K_1 > 0$, and $\gamma_1 > 0$ such that*

$$\sum_{j\neq k} |q_{kj}(x) - q_{kj}(y)| \le K_1(1+|x|^{\gamma_1}+|y|^{\gamma_1})|x-y|^\lambda, \ \forall k \in \mathcal{M} \text{ and } x, y \in \mathbb{R}^r. \tag{2.110}$$

Then $X^{x,\alpha}(t)$ is twice continuously differentiable in L^p with respect to x for any $0 < p < \lambda$.

Proof For ease of presentation and without loss of generality, we prove the result when $X(t)$ is one dimensional. Consider the joint process $(X(\cdot), \alpha(\cdot), \widetilde{X}(\cdot), \widetilde{\alpha}(\cdot))$ given by

$$\begin{cases} dX(t) = b(X(t), \alpha(t))dt + \sigma(X(t), \alpha(t))dw(t), \\ d\alpha(t) = \int_{\mathbb{R}_+} h(X(t-), \alpha(t-), z)\mathfrak{p}(dt, dz), \\ d\widetilde{X}(t) = b(\widetilde{X}(t), \widetilde{\alpha}(t))dt + \sigma(\widetilde{X}(t), \widetilde{\alpha}(t))dw(t), \\ d\widetilde{\alpha}(t) = \int_{\mathbb{R}_+} h(\widetilde{X}(t-), \widetilde{\alpha}(t-), z)\mathfrak{p}(dt, dz), \end{cases} \tag{2.111}$$

with initial condition $(X(0), \alpha(0), \widetilde{X}(0), \widetilde{\alpha}(0)) = (x, \alpha, \widetilde{x}, \alpha) \in \mathbb{R}^r \times \mathcal{M} \times \mathbb{R}^r \times \mathcal{M}$, in which $h := \widetilde{x} - x \neq 0$. Fix an arbitrary $T > 0$. We need to show that $Z^h(T) := \frac{\widetilde{X}(T) - X(T)}{h}$ converges in probability in part (a) and in the L^p sense for any $0 < p < \lambda$ in part (b) as $h \to 0$.

Denote $\zeta = \zeta^h := \inf\{t \ge 0 : \alpha(t) \neq \widetilde{\alpha}(t)\}$. Since $\alpha(t)$ and $\widetilde{\alpha}(t)$ are the same up to ζ, and $b(\cdot, i)$ and $\sigma(\cdot, i)$ are Lipschitz continuous for each $i \in \mathcal{M}$, the moment estimate (2.52) holds. The standard arguments using the Lipschitz condition and Gronwall's inequality (see, e.g., [205, Lemma 3.3]) reveal that

$$\mathbb{E}\left[\sup_{0\le t\le T\wedge\zeta} |X(t) - \widetilde{X}(t)|^2\right] \le K|x - \widetilde{x}|^2, \tag{2.112}$$

where $K = K(T)$ is a positive constant. This, in particular, implies that

$$\sup_{t\in[0,T\wedge\zeta]} |X(t) - \widetilde{X}(t)| \to 0 \text{ in probability as } h \to 0. \tag{2.113}$$

In addition, in view of the Lipschitz condition (2.108) and (2.32) in Lemma 2.5, we can use the same arguments as those for (2.87) to show that $\lim_{h\to 0}\mathbb{P}\{\zeta \le t\} = 0$ for any $t \ge 0$.

Put $Z^h(t) := \frac{\widetilde{X}(t)-X(t)}{h}$ for $t \ge 0$. Then we have

$$Z^h(t\wedge\zeta) = 1 + \frac{1}{h}\int_0^{t\wedge\zeta}\Big[b(\widetilde{X}(s),\alpha(s)) - b(X(s),\alpha(s))\Big]ds + \frac{1}{h}\int_0^{t\wedge\zeta}\Big[\sigma(\widetilde{X}(s),\alpha(s)) - \sigma(X(s),\alpha(s))\Big]dw(s).$$

Now, we evaluate the drift:

$$\begin{aligned}
&\frac{1}{h}\int_0^{t\wedge\zeta}\big[b(\widetilde{X}(s),\alpha(s)) - b(X(s),\alpha(s))\big]\,ds \\
&= \frac{1}{h}\int_0^t I_{\{s\le\zeta\}}\big[b(\widetilde{X}(s),\alpha(s)) - b(X(s),\alpha(s))\big]\,ds \\
&= \frac{1}{h}\int_0^t I_{\{s\le\zeta\}}\left(\int_0^1 \frac{d}{dv}b(X(s)+v(\widetilde{X}(s)-X(s)),\alpha(s))dv\right)ds \qquad (2.114)\\
&= \int_0^t Z^h(s)I_{\{s\le\zeta\}}\int_0^1 b_x(X(s)+v(\widetilde{X}(s)-X(s)),\alpha(s))dvds. \\
&= \int_0^t Z^h(s\wedge\zeta)I_{\{s\le\zeta\}}\int_0^1 b_x(X(s)+v(\widetilde{X}(s)-X(s)),\alpha(s))dvds.
\end{aligned}$$

In view of (2.114) and a similar evaluation for the diffusion part, we have that $U_h(t) = Z^h(t\wedge\zeta)$ satisfies the equation:

$$U_h(t) = 1 + \int_0^t A_h(s)U_h(s)ds + \int_0^t B_h(s)U_h(s)dw(s),$$

where

$$A_h(s) := I_{\{s\le\zeta\}}\int_0^1 b_x(X(s)+v(\widetilde{X}(s)-X(s)),\alpha(s))dv,$$

$$B_h(s) := I_{\{s\le\zeta\}}\int_0^1 \sigma_x(X(s)+v(\widetilde{X}(s)-X(s)),\alpha(s))dv.$$

By (2.112), (2.52), and (2.87), we have

$$A_h(s) \to b_x(X(s),\alpha(s)) \text{ and } B_h(s) \to \sigma_x(X(s),\alpha(s)) \text{ in probability}$$

as $h \to 0$. In view of [94, Theorem 5.5.3],

$$\mathbb{E}\Big[\Big|\frac{\widetilde{X}(T\wedge\zeta)-X(T\wedge\zeta)}{h}-\xi(T)\Big|^2\Big]=\mathbb{E}\Big[\big|U_h(T)-\xi(T)\big|^2\Big]\to 0 \text{ as } h\to 0, \tag{2.115}$$

where $\xi(t)$ is the solution to

$$\xi(t)=1+\int_0^t b_x(X(s),\alpha(s))\xi(s)ds+\int_0^t \sigma_x(X(s),\alpha(s))\xi(s)dw(s). \tag{2.116}$$

(a) Note that

$$Z^h(T)=U_h(T)+I_{\{\zeta\le T\}}\frac{\widetilde{X}(T)-\widetilde{X}(T\wedge\zeta)-X(T)+X(T\wedge\zeta)}{h}. \tag{2.117}$$

In light of (2.87) and noting $I_{\{\zeta\le T\}}$ equals 0 or 1, for any $\delta>0$, we have

$$\mathbb{P}\left\{I_{\{\zeta\le T\}}\left|\frac{\widetilde{X}(T)-\widetilde{X}(T\wedge\zeta)-X(T)+X(T\wedge\zeta)}{h}\right|>\frac{\delta}{2}\right\}\le\mathbb{P}\{\zeta\le T\}\to 0$$

as $h\to 0$. This, together with (2.115) and (2.117), implies that

$$\begin{aligned}\mathbb{P}\{\big|Z^h(T)-\xi(T)\big|>\delta\}&\le\mathbb{P}\Big\{|U_h(T)-\xi(T)|>\frac{\delta}{2}\Big\}\\&\quad+\mathbb{P}\left\{I_{\{\zeta\le T\}}\left|\frac{\widetilde{X}(T)-\widetilde{X}(T\wedge\zeta)-X(T)+X(T\wedge\zeta)}{h}\right|>\frac{\delta}{2}\right\}\\&\to 0 \text{ as } h\to 0.\end{aligned}$$

This shows that $Z^h(T)$ converges in probability to $\xi(T)$ and hence $X(T)$ is differentiable in probability.

(b) Using the same calculations as those in (2.48), we have

$$\mathbb{P}\{\zeta\le T\}\le\mathbb{E}\Big[\int^{T\wedge\zeta}\rho(X(t),\widetilde{X}(t),\alpha(t))dt\Big],$$

where $\rho(x,\widetilde{x},k):=\sum_{j\in\mathcal{M},j\ne k}|q_{kj}(x)-q_{kj}(\widetilde{x})|$. Therefore, if (2.110) is satisfied, then we can use Hölder's inequality, (2.110), (2.112), and (2.52) to derive

$$\begin{aligned}&\mathbb{P}\{\zeta\le T\}\\&\le\mathbb{E}\int_0^{T\wedge\zeta}K_1(1+|X(t)|^{\gamma_1}+|\widetilde{X}(t)|^{\gamma_1})|X(t)-\widetilde{X}(t)|^{\lambda}dt\end{aligned}$$

$$
\begin{aligned}
&\le K_1 T \mathbb{E}\Big[\sup_{t\le T\wedge\zeta}(1+|X(t)|^{\gamma_1}+|\widetilde{X}(t)|^{\gamma_1})|X(t)-\widetilde{X}(t)|^{\lambda}\Big] \\
&\le K_1 T\Big(\mathbb{E}\Big[\sup_{t\le T\wedge\zeta}(1+|X(t)|^{\gamma_1}+|\widetilde{X}(t)|^{\gamma_1})^{\frac{2}{2-\lambda}}\Big]\Big)^{\frac{2-\lambda}{2}} \\
&\quad\times\Big(\mathbb{E}\Big[\sup_{t\le T\wedge\zeta}|X(t)-\widetilde{X}(t)|^2\Big]\Big)^{\frac{\lambda}{2}} \le K_\lambda |h|^{\lambda},
\end{aligned}
\tag{2.118}
$$

where $K_\lambda > 0$ is a positive constant. For any $0 < p < \lambda$, let $\theta = \frac{\lambda-p}{2}$. It follows from Hölder's inequality, (2.52), and (2.118) that

$$
\begin{aligned}
&\mathbb{E}\Big[I_{\{\zeta<T\}}\Big|\widetilde{X}(T)-\widetilde{X}(T\wedge\zeta)-X(T)+X(T\wedge\zeta)\Big|^p\Big] \\
&\le 2^p\mathbb{E}\Big[I_{\{\zeta<T\}}\Big(\sup_{t\in[0,T]}\{|X(t)|+|\widetilde{X}(t)|\}\Big)^p\Big] \\
&\le 2^p(\mathbb{E}[I_{\{\zeta<T\}}])^{\frac{p+\theta}{\lambda}}\Big(\mathbb{E}\Big[\sup_{t\in[0,T]}\{|X(t)|+|\widetilde{X}(t)|\}^{\frac{p\lambda}{\theta}}\Big]\Big)^{\frac{\theta}{\lambda}} \\
&\le K_2|h|^{p+\theta},
\end{aligned}
\tag{2.119}
$$

where K_2 is a positive constant. It then follows from (2.119) that

$$
\mathbb{E}\Big[I_{\{\zeta<T\}}\Big|\frac{\widetilde{X}(T)-\widetilde{X}(T\wedge\zeta)-X(T)+X(T\wedge\zeta)}{h}\Big|^p\Big] \to 0 \text{ as } h\to 0.
\tag{2.120}
$$

As a result of (2.115), (2.117), and (2.120), we obtain that

$$
\mathbb{E}\Big[\Big|\frac{\widetilde{X}(T)-X(T)}{h}-\xi(T)\Big|^p\Big] \to 0 \text{ as } h\to 0.
\tag{2.121}
$$

Thus, $X(t)$ is differentiable in L^p for $p < \lambda$ if (2.110) is satisfied. Likewise, it can be shown that $X(t)$ is twice differentiable in probability under conditions (2.13) and (2.21) and in L^p for $p < \lambda$ under conditions (2.13) and (2.110). □

2.8 A Remark on Nonhomogeneous Markov Processes

Throughout the book, for notational simplicity, we have decided to mainly concern ourselves with time-homogeneous switching diffusion processes. As a consequence, the drift and diffusion coefficients are all independent of t.

The discussion throughout the book can be extended to nonhomogeneous Markov processes. The setup can be changed as follows. Suppose that $b(\cdot,\cdot,\cdot) : \mathbb{R} \times \mathbb{R}^r \times \mathcal{M} \mapsto \mathbb{R}^r$ and that $\sigma(\cdot,\cdot,\cdot) : \mathbb{R} \times \mathbb{R}^r \times \mathcal{M} \mapsto \mathbb{R}^{r\times d}$. Let $w(\cdot)$ be an $\mathbb{R}^d$-valued standard Brownian motion defined in the filtered probability space $(\Omega, \mathcal{F}, \{\mathcal{F}_t\}, \mathbb{P})$, and $\alpha(\cdot)$ be a pure jump process whose generator is given by $Q(x)$ as before such that for $f(\cdot,\cdot,\cdot) : \mathbb{R} \times \mathbb{R}^r \times \mathcal{M}$,

$$Q(x)f(t,x,\cdot)(\imath) = \sum_{\jmath\in\mathcal{M}} q_{\imath\jmath}(x)(f(t,x,\jmath) - f(t,x,\imath)), \quad \text{for each } \imath \in \mathcal{M}. \tag{2.122}$$

In lieu of (2.1), we may consider the two-component process $(X(\cdot), \alpha(\cdot))$ satisfying

$$X(t) = X(0) + \int_0^t b(s, X(s), \alpha(s))ds + \int_0^t \sigma(s, X(s), \alpha(s))dw(s), \tag{2.123}$$

and as $\Delta \to 0$,

$$\begin{aligned} &\mathbb{P}\{\alpha(t+\Delta) = \jmath | X(t) = x, \alpha(t) = \imath, X(s), \alpha(s), s \le t\} \\ &\quad = q_{\imath\jmath}(x)\Delta + o(\Delta), \quad \text{for } \imath \ne \jmath. \end{aligned} \tag{2.124}$$

It follows that the associated operator can be defined as follows: For each $\iota \in \mathcal{M}$, $f(\cdot,\cdot,\iota) \in C^{1,2}$, where $C^{1,2}$ denotes the class of functions that are continuously differentiable with respect to t and twice continuously differential with respect to x, we define

$$\begin{aligned} \mathcal{L}f(t,x,\iota) = &\sum_{i=1}^r b_i(t,x,\iota)\frac{\partial f(t,x,\iota)}{\partial x_i} + \frac{1}{2}\sum_{i,j=1}^r a_{ij}(t,x,\iota)\frac{\partial^2 f(t,x,\iota)}{\partial x_i \partial x_j} \\ &+ Q(x)f(t,x,\cdot)(\iota). \end{aligned} \tag{2.125}$$

Now, the generalized Itô lemma (2.5) is changed to

$$\begin{aligned} &f(t, X(t), \alpha(t)) - f(0, X(0), \alpha(0)) \\ &= \int_0^t \left(\frac{\partial}{\partial s} + \mathcal{L}\right) f(s, X(s), \alpha(s))ds + M_1(t) + M_2(t), \end{aligned} \tag{2.126}$$

where

$$\begin{aligned} M_1(t) &= \int_0^t \big\langle \nabla f(s, X(s), \alpha(s)), \sigma(s, X(s), \alpha(s))dw(s)\big\rangle, \\ M_2(t) &= \int_0^t \int_{\mathbb{R}} \big[f(s, X(s), \alpha(s-) + h(X(s), \alpha(s-), z)) \\ &\qquad\qquad - f(s, X(s), \alpha(s-))\big]\widetilde{\mathfrak{p}}(ds, dz). \end{aligned}$$

In view of the generalized Itô formula, for any $f(\cdot,\cdot,\imath) \in C^{1,2}$ with $\imath \in \mathcal{M}$, and τ_1, τ_2 being bounded stopping times such that $0 \le \tau_1 \le \tau_2$ a.s., if $f(t, X(t), \alpha(t))$ and $\mathcal{L}f(t, X(t), \alpha(t))$ etc. are bounded on $t \in [\tau_1, \tau_2]$ with probability 1, then Dynkin's formula becomes

$$\mathbb{E}f(\tau_2, X(\tau_2), \alpha(\tau_2)) = \mathbb{E}f(\tau_1, X(\tau_1), \alpha(\tau_1)) + \mathbb{E}\int_{\tau_1}^{\tau_2} \mathcal{L}f(s, X(s), \alpha(s))ds. \tag{2.127}$$

Moreover, for each $f(\cdot,\cdot,\imath) \in C_b^{1,2}$ or $f(\cdot,\cdot,\imath) \in C_0^{1,2}$,

$$\begin{aligned} M_f(t) &= f(t, X(t), \alpha(t)) - f(0, X(0), \alpha(0)) \\ &\quad - \int_0^t \left(\frac{\partial}{\partial s} + \mathcal{L}\right) f(s, X(s), \alpha(s))ds \end{aligned}$$

is a martingale. With the setup changed to nonhomogeneous Markov processes, most of the subsequent results will carry over. Nevertheless, modifications are necessary to take into account the added complexity due to nonhomogeneous processes. In order to present the main results and ideas without much notational complication, we confine ourselves to the homogeneous switching diffusions throughout the book.

2.9 Notes

The connection between generators of Markov processes and martingales is explained in Ethier and Kurtz [81]. An account of piecewise-deterministic processes is in Davis [60]. Results on basic probability theory can be found in Chow and Teicher [51] and Chung [53]. For an introduction to the theory of stochastic processes, we refer to Gihman and Skorohod [103], Khasminskii [156], and Liptser and Shiryayev [190], among others. More detailed discussions related to martingales and diffusions are in Elliott [80], and an in-depth study of stochastic differential equations and diffusion processes is contained in Ikeda and Watanabe [136]. In Section 2.7, we dealt with smoothness properties of solutions of stochastic differential equations with x-dependent switching. It is interesting to note that even the time-honored concept of well-posedness cannot be easily obtained. Once the switching is x-dependent, the difficulty rises considerably.

Chapter 3
Recurrence

3.1 Introduction

This chapter is concerned with recurrence of switching diffusion processes. Because practical systems in applications are often in operation for a relatively long time, it is of foremost importance to understand the systems' asymptotic behavior. By asymptotic behavior, we mean the properties of the underlying processes in a neighborhood of "∞" and in a neighborhood of an equilibrium point. Properties concerning a neighborhood of ∞ are treated in this chapter, whereas stability of an equilibrium point is dealt with in Chapter 6.

Dealing with dynamic systems, one often wishes to examine if the underlying system is sensitive to perturbations. In accordance with LaSalle and Lefschetz [181], a deterministic system $\dot{x} = h(t, x)$ satisfying appropriate conditions is Lagrange stable if the solutions are ultimately uniformly bounded. When diffusions are used, one may wish to add probability modifications such as "almost surely" or "in probability" to the aforementioned uniform boundedness. Nevertheless, for instance, if almost sure boundedness is used, it will exclude many systems due to the presence of Brownian motion. Thus, as pointed out by Wonham [278], such boundedness is inappropriate; an alternative notion of stability in a certain weak sense should be used. In lieu of requiring the system to be bounded, one aims to find conditions under which the systems return to a prescribed compact region in finite time. This chapter focuses on weak-sense stability for switching diffusions. We define recurrence, positive recurrence, and null recurrence; we also develop Lyapunov-function-based criteria together with more easily verifiable conditions on the coefficients of the processes for positive recurrence as well as nonrecurrence and null recurrence. Despite the growing interest in treating regime-switching systems, the results regarding such issues as recurrence and positive recurrence (or weak stochastic stability as coined by Wonham [278]) are still scarce. These are not simple extensions of their diffusion counterpart. Due to the coupling and interactions, elliptic systems instead of a single elliptic equation must be treated. Moreover, even

H.-D. Nguyen et al., *Hybrid Switching Diffusions*, Probability Theory and Stochastic Modelling 63, https://doi.org/10.1007/978-3-031-93303-5_3

though the classical approaches such as Lyapunov function methods and Dynkin's formula are still applicable for switching diffusions, the analysis is more delicate than the diffusion counterparts and requires careful handling of the discrete-event component $\alpha(\cdot)$.

The rest of the chapter is arranged as follows. In Section 3.2, in addition to introducing certain notations, we also provide definitions of recurrence, transience, positive recurrence, and null recurrence, as well as some preliminary results. Section 3.3 focuses on recurrence and transience. Section 3.4 proceeds with the study of positive and null recurrence. We present results of necessary and sufficient conditions for recurrence using Lyapunov functions. We also consider the case under "linearization" for the continuous component. Section 3.5 is devoted to a number of examples as applications of the general results. Section 3.6, containing the proofs of several technical lemmas, is provided to facilitate reading. Discussions and further remarks are made in Section 3.7.

3.2 Formulation and Preliminaries

3.2.1 *Problem Setup*

We work with $(\Omega, \mathcal{F}, \{\mathcal{F}_t\}_{t\geq 0}, \mathbb{P})$, a complete probability space with a filtration $\{\mathcal{F}_t\}_{t\geq 0}$ satisfying the usual condition (i.e., it is right continuous with $\mathcal{F}_0$ containing all $\mathbb{P}$-null sets). Let $x \in \mathbb{R}^r$, $\mathcal{M} = \{1, \ldots, m_0\}$, and $Q(x) = (q_{ij}(x))$ an $m_0 \times m_0$ matrix depending on x satisfying that for any $x \in \mathbb{R}^r$, $q_{ij}(x) \geq 0$ for $i \neq j$ and $\sum_{j=1}^{m_0} q_{ij}(x) = 0$. For any twice continuously differentiable function $f(\cdot, i)$, $i \in \mathcal{M}$, define $\mathcal{L}$ by

$$\begin{aligned}
\mathcal{L}f(x,i) &= \frac{1}{2}\sum_{j,k=1}^{r} a_{jk}(x,i)\frac{\partial^2 f(x,i)}{\partial x_j \partial x_k} + \sum_{j=1}^{r} b_j(x,i)\frac{\partial f(x,i)}{\partial x_j} + Q(x)f(x,\cdot)(i) \\
&= \frac{1}{2}\text{tr}(a(x,i)\nabla^2 f(x,i)) + b'(x,i)\nabla f(x,i) + Q(x)f(x,\cdot)(i),
\end{aligned} \tag{3.1}$$

where $\nabla f(\cdot, i)$ and $\nabla^2 f(\cdot, i)$ denote the gradient and Hessian of $f(\cdot, i)$, respectively, $b'(x,i)\nabla f(x,i)$ denotes the usual inner product on $\mathbb{R}^r$ with z' being the transpose of z for $z \in \mathbb{R}^{\iota_1 \times \iota_2}$ with $\iota_1,\ \iota_2 \geq 1$, and

$$Q(x)f(x,\cdot)(i) = \sum_{j\in\mathcal{M}} q_{ij}(x)(f(x,j) - f(x,i)), \quad i \in \mathcal{M}. \tag{3.2}$$

Consider a Markov process $Y(t) = (X(t), \alpha(t))$, whose associated operator is given by $\mathcal{L}$. Note that $Y(t)$ has two components, an r-dimensional continuous component $X(t)$ and a discrete component $\alpha(t)$ taking value in $\mathcal{M} = \{1, \ldots, m_0\}$. In this chapter, unless otherwise specified, we use $m_0 < \infty$ as our derivations often rely on results on *finite* systems of partial differential equations.

Recall that the process $Y(t) = (X(t), \alpha(t))$ may be described by the following pair of equations:

$$\begin{aligned} &dX(t) = b(X(t), \alpha(t))dt + \sigma(X(t), \alpha(t))dw(t), \\ &X(0) = x, \quad \alpha(0) = \alpha, \end{aligned} \tag{3.3}$$

and

$$\begin{aligned} &\mathbb{P}\{\alpha(t+\Delta) = j | \alpha(t) = i, X(s), \alpha(s), s \le t\} \\ &\qquad = \begin{cases} q_{ij}(X(t))\Delta + o(\Delta), & \text{if } i \ne j, \\ 1 + q_{ii}(X(t))\Delta + o(\Delta), & \text{if } i = j, \end{cases} \end{aligned} \tag{3.4}$$

when $\Delta \downarrow 0$, where $w(t)$ is a d-dimensional standard Brownian motion, $b(\cdot, \cdot) : \mathbb{R}^r \times \mathcal{M} \mapsto \mathbb{R}^r$, and $\sigma(\cdot, \cdot) : \mathbb{R}^r \times \mathcal{M} \mapsto \mathbb{R}^{r \times d}$ satisfying $\sigma(x, i)\sigma'(x, i) = a(x, i)$. Note that (3.3) depicts the system dynamics and (3.4) delineates the probability structure of the jump process. Note that if $\alpha(\cdot)$ is a continuous-time Markov chain independent of the Brownian motion $w(\cdot)$ and $Q(x) = Q$ or $Q(x) = Q(t)$ (independent of x), then (3.3) together with the generator Q or $Q(t)$ is sufficient to characterize the underlying process. As long as there is an x-dependence, (3.4) is needed in delineating the dynamics of the switching diffusion.

In this chapter, our study is carried out with the use of the operator $\mathcal{L}$ given in (3.1). Throughout the chapter, we assume that both $b(\cdot, i)$ and $\sigma(\cdot, i)$ satisfy the usual local Lipschitz and linear growth conditions for each $i \in \mathcal{M}$ and that $Q(\cdot)$ is bounded and continuous. Thanks to Theorem 2.6, under these conditions, the system given by (3.3)–(3.4) has a unique strong solution. In what follows, denote the solution of (3.3)–(3.4) by $(X^{x,\alpha}(t), \alpha^{x,\alpha}(t))$ when we need to emphasize the dependence on initial data. To study recurrence and ergodicity of the process $Y(t) = (X(t), \alpha(t))$, we further assume that the following condition holds throughout the chapter.

(H3.1) The operator $\mathcal{L}$ satisfies the following conditions:

(a) For each $i \in \mathcal{M}$, $a(x, i) = (a_{jk}(x, i))$ is symmetric and satisfies

$$\kappa_1 |\xi|^2 \le \xi' a(x, i) \xi \le \kappa_1^{-1} |\xi|^2, \quad \text{for all } \xi \in \mathbb{R}^r, \tag{3.5}$$

with some constant $\kappa_1 \in (0, 1]$ for all $x \in \mathbb{R}^r$.

(b) $Q(\cdot) : \mathbb{R}^r \mapsto \mathbb{R}^{m_0 \times m_0}$ is a bounded and continuous function. Moreover, $Q(x)$ is irreducible for each $x \in \mathbb{R}^r$.

3.2.2 Definitions of Recurrence and Positive Recurrence

This subsection is devoted to the definitions of recurrence, positive recurrence, and null recurrence. First, we introduce the following notation and conventions. For any $D \subset \mathbb{R}^r$, $J \subset \mathcal{M}$, and $U = D \times J \subset \mathbb{R}^r \times \mathcal{M}$, denote

$$\begin{aligned} \tau_U &:= \inf\{t \geq 0 : (X(t), \alpha(t)) \notin U\}, \\ \sigma_U &:= \inf\{t \geq 0 : (X(t), \alpha(t)) \in U\}. \end{aligned} \tag{3.6}$$

In particular, if $U = D \times \mathcal{M}$ is a "cylinder," we set

$$\begin{aligned} \tau_D &:= \inf\{t \geq 0 : X(t) \notin D\}, \\ \sigma_D &:= \inf\{t \geq 0 : X(t) \in D\}. \end{aligned} \tag{3.7}$$

Definition 3.1 Recurrence and positive and null recurrence are defined as follows.

- *Recurrence and Transience.* For $U := D \times J$, where $J \subset \mathcal{M}$ and $D \subset \mathbb{R}^r$ is an open set with compact closure, let

$$\sigma_U^{x,\alpha} = \inf\{t : (X^{x,\alpha}(t), \alpha^{x,\alpha}(t)) \in U\}.$$

  A regular process $(X^{x,\alpha}(\cdot), \alpha^{x,\alpha}(\cdot))$ is *recurrent* with respect to U if $\mathbb{P}\{\sigma_U^{x,\alpha} < \infty\} = 1$ for any $(x, \alpha) \in D^c \times \mathcal{M}$, where D^c denotes the complement of D; otherwise, the process is *transient* with respect to U.
- *Positive Recurrence and Null Recurrence.* A recurrent process with finite mean recurrence time for some set $U = D \times J$, where $J \subset \mathcal{M}$ and $D \subset \mathbb{R}^r$ is a bounded open set with compact closure, is said to be *positive recurrent* with respect to U; otherwise, the process is *null recurrent* with respect to U.

3.2.3 Preparatory Results

We first prove the following theorem, which asserts that under Assumption (H3.1), the process $Y(t) = (X(t), \alpha(t))$ will exit every bounded "cylinder" with finite mean exit time.

Theorem 3.2 *Let $D \subset \mathbb{R}^r$ be a nonempty open set with compact closure $\overline{D}$. Let $\tau_D := \inf\{t \geq 0 : X(t) \notin D\}$. Then*

$$\mathbb{E}_{x,i}\tau_D < \infty, \quad \textit{for any} \quad (x, i) \in D \times \mathcal{M}. \tag{3.8}$$

Proof First, note that from the uniform ellipticity (3.5), we have

$$\kappa_1 \le a_{11}(x,i) \le \kappa_1^{-1}, \quad \text{for any} \quad (x,i) \in D \times \mathcal{M}. \tag{3.9}$$

For each $i \in \mathcal{M}$, consider

$$W(x,i) = k - (x_1 + \beta)^c,$$

where the constants k, c (with $c \ge 2$), and β are to be specified, and $x_1 = e_1' x$ is the first component of x with $e_1 = (1, 0, \ldots, 0)'$ being the standard unit vector. Direct computation leads to

$$\mathcal{L}W(x,i) = -c(x_1+\beta)^{c-2}\left[b_1(x,i)(x_1+\beta) + \frac{c-1}{2}a_{11}(x,i)\right].$$

Set

$$c = \frac{2}{\kappa_1}\left(\sup_{(x,i)\in\overline{D}\times\mathcal{M}} |b_1(x,i)(x_1+\beta)| + 1\right) + 1.$$

Then we have from (3.9) that

$$\begin{aligned}&\frac{c-1}{2}a_{11}(x,i) + b_1(x,i)(x_1+\beta)\\ &\quad \ge \frac{c-1}{2}\kappa_1 - \sup_{(x,i)\in\overline{D}\times\mathcal{M}} |b_1(x,i)(x_1+\beta)| \ge 1.\end{aligned}$$

Meanwhile, since $x \in D \subset \overline{D}$ and $\overline{D}$ is compact, we can choose β such that $1 \le x_1 + \beta \le M$ for all $x \in D$, where M is some positive constant. Thus, we have $(x_1+\beta)^{c-2} \ge 1^{c-2} = 1$. Finally, we choose k large enough so that $W(x,i) = k - (x_1+\beta)^c > 0$ for all $(x,i) \in D \times \mathcal{M}$. Therefore, $W(\cdot,\cdot)$ is a Lyapunov function satisfying

$$\mathcal{L}W(x,i) \le -c, \qquad \text{for all} \quad (x,i) \in D \times \mathcal{M}. \tag{3.10}$$

Now let $\tau_D(t) = t \wedge \tau_D := \min\{t, \tau_D\}$. Then we have from Dynkin's formula and (3.10) that

$$\begin{aligned}&\mathbb{E}_{x,i} W(X(\tau_D(t)), \alpha(\tau_D(t))) - W(x,i)\\ &\quad = \mathbb{E}_{x,i}\int_0^{\tau_D(t)} \mathcal{L}W(X(u),\alpha(u))du \le -c\mathbb{E}_{x,i}\tau_D(t).\end{aligned}$$

Because the function $W(\cdot,\cdot)$ is nonnegative, we have

$$\mathbb{E}_{x,i}\tau_D(t) \le \frac{1}{c}W(x,i). \tag{3.11}$$

Because $\mathbb{E}_{x,i}\tau_D(t) = \mathbb{E}_{x,i}\tau_D I_{\{\tau_D \le t\}} + \mathbb{E}_{x,i} t I_{\{\tau_D > t\}}$, we have from (3.11) that

$$t\mathbb{P}_{x,i}[\tau_D > t] \le \frac{1}{c} W(x,i).$$

Letting $t \to \infty$, we obtain

$$\mathbb{P}_{x,i}[\tau_D = \infty] = 0 \text{ or } \mathbb{P}_{x,i}[\tau_D < \infty] = 1.$$

This yields that $\tau_D(t) \to \tau_D$ almost surely (a.s.) $\mathbb{P}_{x,i}$ as $t \to \infty$. Now, applying Fatou's lemma in (3.11), as $t \to \infty$, we obtain

$$\mathbb{E}_{x,i}\tau_D \le \frac{1}{c} W(x,i) \le \frac{k}{c} < \infty, \tag{3.12}$$

as desired. □

Remark 3.3 A closer examination of the proof shows that the conclusion of Theorem 3.2 remains valid if we replace the uniform ellipticity condition (3.5) by a weaker condition: There exist some $\iota = 1, 2, \ldots, r$ and positive constant κ such that

$$a_{\iota\iota}(x,i) \ge \kappa \text{ for any } (x,i) \in D \times \mathcal{M}. \tag{3.13}$$

Let us recall the definition of $\mathcal{L}$-harmonic functions. For any $U = D \times J$, where $D \subset \mathbb{R}^r$ is a nonempty domain, and $J \subset \mathcal{M}$, a Borel measurable function $u : U \mapsto \mathbb{R}$ is said to be *$\mathcal{L}$-harmonic* in U if u is bounded in compact subsets of U and that for all $(x,i) \in U$ and any $V = \widetilde{D} \times \widetilde{J}$ with $\widetilde{D} \subset\subset D$ being a neighborhood of x and $i \in \widetilde{J} \subset J$, we have

$$u(x,i) = \mathbb{E}_{x,i} u(X(\tau_V), \alpha(\tau_V)),$$

where τ_V denotes the first exit time of the process $(X(t), \alpha(t))$ from V and $\widetilde{D} \subset\subset D$ means $\widetilde{D} \subset \overline{\widetilde{D}} \subset D$ and $\overline{\widetilde{D}}$ is compact, with $\overline{\widetilde{D}} = \widetilde{D} \cup \partial\widetilde{D}$ denoting the closure of $\widetilde{D}$.

Lemma 3.4 *For any $U = D \times J \subset \mathbb{R}^r \times \mathcal{M}$, where $D \subset \mathbb{R}^r$ is a nonempty domain, the functions*

$$f(x,i) = \mathbb{P}_{x,i}\{\tau_U < \infty\} \textit{ and } g(x,i) = \mathbb{E}_{x,i}\phi(X(\tau_U), \alpha(\tau_U))$$

are $\mathcal{L}$-harmonic in U, where ϕ is any bounded and Borel measurable function on $\partial D \times \mathcal{M}$.

Proof Fix any $(x,i) \in U$. Consider any $V = \widetilde{D} \times \widetilde{J} \subset U$ such that $x \in \widetilde{D} \subset\subset D$ and $i \in \widetilde{J} \subset J$. Note that $\tau_V \le \tau_U$. Then it follows from the strong Markov property that

$$\begin{aligned} f(x,i) &= \mathbb{E}_{x,i}[I_{\{\tau_U<\infty\}}] = \mathbb{E}_{x,i}\left[\mathbb{E}_{x,i}[I_{\{\tau_U<\infty\}}]\middle|\mathcal{F}_{\tau_V}\right] \\ &= \mathbb{E}_{x,i}\left[\mathbb{E}_{X(\tau_V),\alpha(\tau_V)}[I_{\{\tau_U<\infty\}}]\right] = \mathbb{E}_{x,i} f(X(\tau_V),\alpha(\tau_V)). \end{aligned}$$

This shows that f is $\mathcal{L}$-harmonic in U. A very similar argument shows that g is also $\mathcal{L}$-harmonic in U. □

Following the well-known arguments in [76, Vol. II, Chapter 13], we obtain the following two lemmas. (Note that Lemma 3.5 was also proved in [100, Lemma 4.3], and in [46], when the operator $\mathcal{L}$ is in divergence form.)

Lemma 3.5 *Suppose Assumption* (H3.1) *holds. Let* $U = D \times \mathcal{M} \subset \mathbb{R}^r \times \mathcal{M}$, *where* $D \subset \mathbb{R}^r$ *is a nonempty domain. Suppose* $f : U \mapsto \mathbb{R}$ *is Borel measurable and bounded on compact subsets of* U. *Then*

$$\mathcal{L}f(x,i) = 0 \ \textit{for any} \ (x,i) \in U \tag{3.14}$$

if and only if f *is* $\mathcal{L}$*-harmonic in* U. *Moreover, assume that* ∂D *is sufficiently smooth,* $\overline{D}$ *is compact, and* $\varphi(\cdot, i)$ *is an arbitrary continuous function on* ∂D *for any* $i \in \mathcal{M}$. *Then*

$$u(x,i) := \mathbb{E}_{x,i}\varphi(X(\tau_U),\alpha(\tau_U)) \tag{3.15}$$

is the unique solution of the differential equation (3.14) *with boundary condition*

$$\lim_{x\to x_0, x\in D} u(x,i) = \varphi(x_0,i) \ \textit{for any} \ (x_0,i) \in \partial D \times \mathcal{M}. \tag{3.16}$$

Proof We divide the proof into several steps.
Step 1. Assume (3.14). Let $(x,i) \in V \subset U$, V, and τ_V be as in the proof of Lemma 3.4 and $t > 0$. Then Dynkin's formula and (3.14) lead to

$$\begin{aligned} \mathbb{E}_{x,i} f(X(\tau_V \wedge t),\alpha(\tau_V \wedge t)) &= f(x,i) + \mathbb{E}_{x,i}\int_0^{\tau_V\wedge t} \mathcal{L}f(X(s),\alpha(s))ds \\ &= f(x,i). \end{aligned}$$

Note that Theorem 3.2 implies that $\mathbb{P}_{x,i}\{\tau_V < \infty\} = 1$. Letting $t \to \infty$, we obtain by virtue of the bounded convergence theorem that

$$f(x,i) = \mathbb{E}_{x,i} f(X(\tau_V),\alpha(\tau_V)).$$

This shows that f is $\mathcal{L}$-harmonic in U.
Step 2. Assume that f is $\mathcal{L}$-harmonic in U. Note that by virtue of Proposition 2.33, f is continuous in U. Consider $V = \widetilde{D} \times \mathcal{M} \subset U$ with $\widetilde{D} \subset\subset D$ and $\partial\widetilde{D}$ sufficiently smooth. Then by virtue of [79], the boundary value problem

$$\begin{cases} \mathcal{L}\widetilde{f}(x,i) = 0, & (x,i) \in V, \\ \widetilde{f}(x,i) = f(x,i), & (x,i) \in \partial\widetilde{D} \times \mathcal{M}, \end{cases} \tag{3.17}$$

has a unique classical solution. We show that $\widetilde{f}$ agrees with f in V. In fact, for any $(x,i) \in V$, the same argument as in that Step 1 shows that

$$\widetilde{f}(x,i) = \mathbb{E}_{x,i}\widetilde{f}(X(\tau_V), \alpha(\tau_V)).$$

But the boundary condition in (3.17) implies that

$$\mathbb{E}_{x,i}\widetilde{f}(X(\tau_V), \alpha(\tau_V)) = \mathbb{E}_{x,i} f(X(\tau_V), \alpha(\tau_V)).$$

Therefore, the assumption that f is $\mathcal{L}$-harmonic in U further leads to

$$\widetilde{f}(x,i) = \mathbb{E}_{x,i} f(X(\tau_V), \alpha(\tau_V)) = f(x,i).$$

This shows that $f(\cdot,i) \in C^2(D)$ for each $i \in \mathcal{M}$ and that f satisfies the differential equation (3.14).
Step 3. Now assume ∂D is sufficiently smooth, $\overline{D}$ is compact, and $\varphi(\cdot,i)$ is an arbitrary continuous function on ∂D for any $i \in \mathcal{M}$. We show that the function u defined in (3.15) is the unique solution of (3.14) with boundary condition (3.16). Indeed, it follows from Lemma 3.4 that u is $\mathcal{L}$-harmonic in U. Then we have from Step 2 that u satisfies the differential equation (3.14). Finally, the boundary condition (3.16) is satisfied by the assumptions that ∂D is sufficiently smooth, that $\overline{D}$ is compact, and that ϕ is continuous. This completes the proof of the lemma. □

Using a similar argument, we can prove the following lemma.

Lemma 3.6 *Let $U = D \times \mathcal{M} \subset \mathbb{R}^r \times \mathcal{M}$, where $D \subset \mathbb{R}^r$ is a nonempty open set with compact closure. Suppose that $g(\cdot,i) \in C_b(\overline{D})$ and $f(\cdot,\cdot) : \overline{D} \times \mathcal{M} \mapsto \mathbb{R}$. Then f solves the boundary value problem:*

$$\begin{cases} \mathcal{L}f(x,i) = -g(x,i), & (x,i) \in D \times \mathcal{M} \\ f(x,i) = 0, & (x,i) \in \partial D \times \mathcal{M} \end{cases}$$

if and only if

$$f(x,i) = \mathbb{E}_{x,i}\int_0^{\tau_U} g(X(t), \alpha(t))dt, \quad \textit{for all } (x,i) \in D \times \mathcal{M}.$$

Using Lemmas 3.5 and 3.6, we proceed to prove that if the process $Y(t) = (X(t), \alpha(t))$ is recurrent (resp., positive recurrent) with respect to some "cylinder" $D \times \mathcal{M} \subset \mathbb{R}^r \times \mathcal{M}$, then it is recurrent (resp., positive recurrent) with respect to any "cylinder" $E \times \mathcal{M} \subset \mathbb{R}^r \times \mathcal{M}$, where D is any nonempty domain in $\mathbb{R}^r$ with compact closure. These results are proved in the following two lemmas. To preserve the flow of presentation, the proofs are postponed to Section 3.6.

Lemma 3.7 *Let $D \subset \mathbb{R}^r$ be a nonempty open set with compact closure. Suppose that*

$$\mathbb{P}_{x,i}\{\sigma_D < \infty\} = 1 \text{ for any } (x, i) \in D^c \times \mathcal{M}. \tag{3.18}$$

Then for any nonempty open set $E \subset \mathbb{R}^r$, we have

$$\mathbb{P}_{x,i}\{\sigma_E < \infty\} = 1 \text{ for any } (x, i) \in E^c \times \mathcal{M}.$$

Lemma 3.8 *Let $D \subset \mathbb{R}^r$ be a nonempty open set with compact closure. Suppose that*

$$\mathbb{E}_{x,i}\sigma_D < \infty \quad \text{for any } (x, i) \in D^c \times \mathcal{M}. \tag{3.19}$$

Then for any nonempty open set $E \subset \mathbb{R}^r$, we have

$$\mathbb{E}_{x,i}\sigma_E < \infty \quad \text{for any } (x, i) \in E^c \times \mathcal{M}.$$

The following lemma shows that if the process $Y(t) = (X(t), \alpha(t))$ reaches the "cylinder" $D \times \mathcal{M}$ in finite time a.s. $\mathbb{P}_{x,i}$, then it will visit the set $D \times \{\ell\}$ in finite time a.s. $\mathbb{P}_{x,i}$ for any $\ell \in \mathcal{M}$. Its proof together with the proof of Lemma 3.10 is placed in Section 3.6 as well.

Lemma 3.9 *Let $D \subset \mathbb{R}^r$ be a nonempty open set with compact closure satisfying*

$$\mathbb{P}_{y,j}\{\sigma_D < \infty\} = 1 \quad \text{for any } (y, j) \in D^c \times \mathcal{M}. \tag{3.20}$$

Then for any $(x, i) \in \mathbb{R}^r \times \mathcal{M}$,

$$\mathbb{P}_{x,i}\{\sigma_{D\times\{\ell\}} < \infty\} = 1 \quad \text{for any } \ell \in \mathcal{M}. \tag{3.21}$$

With Lemma 3.9, we can now prove that if the process $Y(t) = (X(t), \alpha(t))$ is positive recurrent with respect to some "cylinder" $D \times \mathcal{M}$, then it is positive recurrent with respect to the set $D \times \{\ell\} \subset \mathbb{R}^r \times \mathcal{M}$.

Lemma 3.10 *Let $D \subset \mathbb{R}^r$ be a nonempty open set with compact closure satisfying*

$$\mathbb{E}_{y,j}\sigma_D < \infty \quad \text{for any } (y, j) \in D^c \times \mathcal{M}. \tag{3.22}$$

Then for any $(x, i) \in \mathbb{R}^r \times \mathcal{M}$,

$$\mathbb{E}_{x,i}\sigma_{D\times\{\ell\}} < \infty \quad \text{for any } \ell \in \mathcal{M}. \tag{3.23}$$

Remark 3.11 By virtue of Lemmas 3.7–3.10, under Assumption (H3.1), the process $Y(t) = (X(t), \alpha(t))$ is recurrent (resp., positive recurrent) with respect to some "cylinder" $D \times \mathcal{M}$ if and only if it is recurrent (resp., positive recurrent) with respect to the product set $D \times \{\ell\} \subset \mathbb{R}^r \times \mathcal{M}$ for any $\ell \in \mathcal{M}$. Also we have proved that the properties of recurrence and positive recurrence are independent of the choice of the set D. We summarize these in the following theorem.

Theorem 3.12 *Suppose that Assumption* (H3.1) *holds. Then the following assertions hold:*

- *The process $(X(t), \alpha(t))$ is recurrent (resp., positive recurrent) with respect to $D \times \mathcal{M}$ if and only if it is recurrent (resp., positive recurrent) with respect to $D \times \{\ell\}$, where $D \subset \mathbb{R}^r$ is a nonempty open set with compact closure and $\ell \in \mathcal{M}$.*
- *If the process $(X(t), \alpha(t))$ is recurrent (resp., positive recurrent) with respect to some $U = D \times \mathcal{M}$, where $D \subset \mathbb{R}^r$ is a nonempty open set with compact closure, then it is recurrent (resp., positive recurrent) with respect to any $\widetilde{U} = \widetilde{D} \times \mathcal{M}$, where $\widetilde{D} \subset \mathbb{R}^r$ is any nonempty open set.*

Remark 3.13 In view of Theorem 3.12, we make the following remarks.

- Recurrence is a property independent of the region chosen; henceforth, a process $(X(t), \alpha(t))$ with the associated generator $\mathcal{L}$ satisfying Assumption (H3.1) is said to be *recurrent*, if it is recurrent with respect to some $U = D \times \{\ell\}$, where $D \subset \mathbb{R}^r$ is a nonempty bounded open set and $\ell \in \mathcal{M}$; otherwise, it is said to be *transient*.
- Henceforth, we call a recurrent process $(X(t), \alpha(t))$ *positive recurrent* if it is positive recurrent with respect to some bounded domain $U = D \times \{\ell\} \subset \mathbb{R}^r \times \mathcal{M}$; otherwise, we have a *null recurrent* process.

3.3 Recurrence and Transience

3.3.1 Recurrence

To study the recurrence of the process $(X(t), \alpha(t))$, we first present the following criterion based upon the existence of certain Lyapunov functions.

Theorem 3.14 *Assume that there exists a nonempty bounded open set $D \subset \mathbb{R}^r$ such that there exists $V(\cdot, \cdot) : D^c \times \mathcal{M} \mapsto \mathbb{R}^+$ satisfying*

$$\begin{aligned} &V_n := \inf_{|x| \geq n, i \in \mathcal{M}} V(x, i) \to \infty, \quad \text{as } n \to \infty, \\ &\mathcal{L}V(x, i) \leq 0, \quad \text{for all } (x, i) \in D^c \times \mathcal{M}. \end{aligned} \tag{3.24}$$

Then the process $(X(t), \alpha(t))$ is recurrent.

Proof Fix any $(x,\alpha) \in D^c \times \mathcal{M}$. Define $\sigma_D = \sigma_D^{x,\alpha} := \inf\{t \geq 0 : X^{x,\alpha}(t) \in D\}$ and $\sigma_D^{(n)}(t) := \sigma_D \wedge t \wedge \beta_n$, where $\beta_n = \inf\{t : |X^{(n)}(t)| \geq n\}$. Then it follows from Dynkin's formula that

$$\begin{aligned} \mathbb{E}V(X(\sigma_D^{(n)}(t)), \alpha(\sigma_D^{(n)}(t))) &= V(x,\alpha) + \mathbb{E}\int_0^{\sigma_D^{(n)}(t)} \mathcal{L}V(X(u),\alpha(u))du \\ &\leq V(x,\alpha). \end{aligned}$$

Note that as $t \to \infty$, $\sigma_D \wedge t \wedge \beta_n \to \sigma_D \wedge \beta_n$ a.s. Hence, Fatou's lemma implies that

$$\mathbb{E}V(X(\sigma_D \wedge \beta_n), \alpha(\sigma_D \wedge \beta_n)) \leq V(x,\alpha).$$

Then we have

$$\begin{aligned} V(x,\alpha) &\geq \mathbb{E}V(X(\sigma_D \wedge \beta_n), \alpha(\sigma_D \wedge \beta_n)) \\ &\geq \mathbb{E}V(X(\beta_n), \alpha(\beta_n))I_{\{\beta_n < \sigma_D\}} \geq V_n \mathbb{P}\{\beta_n < \sigma_D\}. \end{aligned}$$

Hence, it follows from (3.24) that as $n \to \infty$,

$$\mathbb{P}\{\beta_n < \sigma_D\} \leq \frac{V(x,\alpha)}{V_n} \to 0.$$

Note that $\mathbb{P}\{\sigma_D = \infty\} \leq \mathbb{P}\{\beta_n < \sigma_D\}$. Thus, we have $\mathbb{P}\{\sigma_D < \infty\} = 1$. □

The above result is based on a Lyapunov function argument. However, constructing Lyapunov functions is generally difficult. It would be nice if we could place certain conditions on the coefficients of processes. The following theorem is an attempt in this direction.

Theorem 3.15 *Either one of the following conditions implies that the process $(X(t),\alpha(t))$ is recurrent:*

(i) *There exist constants $\gamma > 0$ and $c_i \in \mathbb{R}$ with $i \in \mathcal{M}$ such that for $(x,i) \in \{x \in \mathbb{R}^r : |x| \geq 1\} \times \mathcal{M}$,*

$$\frac{x'b(x,i)}{|x|^2} + \frac{\operatorname{tr}(a(x,i))}{2|x|^2} + (\gamma - 2)\frac{x'a(x,i)x}{2|x|^4} - \frac{1}{k - \gamma c_i}\sum_{j=1}^{m_0} q_{ij}(x)c_j \leq 0, \tag{3.25}$$

where k is a positive constant sufficiently large so that $k - \gamma c_i > 0$ for each $i \in \mathcal{M}$.

(ii) *There exist a positive constant γ and symmetric and positive definite matrices P_i for $i \in \mathcal{M}$ such that for $(x,i) \in \{x \in \mathbb{R}^r : |x| \geq 1\} \times \mathcal{M}$,*

$$\frac{x'P_ib(x,i)}{x'P_ix}+\frac{\operatorname{tr}(\sigma'(x,i)P_i\sigma(x,i))}{2x'P_ix} +(\gamma-2)\frac{|\sigma(x,i)'P_ix|^2}{|x'P_ix|^2}+\sum_{j=1}^{m_0}q_{ij}(x)\frac{|x'P_jx|^{\gamma/2}}{|x'P_ix|^{\gamma/2}}\le 0. \tag{3.26}$$

Proof

(i) Consider the Lyapunov function $V(x,i)=(k-\gamma c_i)|x|^\gamma$, $(x,i)\in\mathbb{R}^r\times\mathcal{M}$. Then direct computation shows that for $x\neq 0$, we have

$$\begin{aligned}\nabla V(x,i)&=(k-\gamma c_i)\gamma|x|^{\gamma-2}x,\\ \nabla^2V(x,i)&=(k-\gamma c_i)\gamma\left[|x|^{\gamma-2}I+(\gamma-2)|x|^{\gamma-4}xx'\right].\end{aligned}$$

Hence, it follows that

$$\begin{aligned}\mathcal{L}V(x,i)=(k-\gamma c_i)\gamma|x|^\gamma\Bigg\{&\frac{x'b(x,i)}{|x|^2}+\frac{\operatorname{tr}(a(x,i))}{|x|^2}-(\gamma-2)\frac{x'a(x,i)x}{|x|^4}\\ &-\frac{1}{k-\gamma c_i}\sum_{j=1}^{m_0}q_{ij}(x)c_j\Bigg\}.\end{aligned}$$

Therefore, Theorem 3.14 implies the desired conclusion.
Assertion (ii) can be established using a similar argument as in (i) by considering

$$W(x,i)=(x'P_ix)^{\gamma/2}\ \text{ for }\ (x,i)\in\left\{x\in\mathbb{R}^r:|x|\ge 1\right\}\times\mathcal{M}$$

and verifying that

$$\begin{aligned}\nabla W(x,i)&=\gamma(x'P_ix)^{(\gamma-2)/2}P_ix,\\ \nabla^2W(x,i)&=\gamma(x'P_ix)^{(\gamma-2)/2}P_i+\gamma(\gamma-2)(x'P_ix)^{(\gamma-4)/2}P_ixx'P_i.\end{aligned}$$

The details are omitted for brevity. □

Lemma 3.16 *If there exists some $(x_0,\ell)\in\mathbb{R}^r\times\mathcal{M}$ such that for any $\varepsilon>0$,*

$$\mathbb{P}_{x_0,\ell}\left\{(X(t_n),\alpha(t_n))\in B(x_0,\varepsilon)\times\{\ell\},\ \textit{for a sequence } t_n\uparrow\infty\right\}=1, \tag{3.27}$$

then for any $U:=D\times\{j\}\subset\mathbb{R}^r\times\mathcal{M}$, where $D\subset\mathbb{R}^r$ is a nonempty bounded domain and $j\in\mathcal{M}$, we have

$$\mathbb{P}_{x_0,\ell}\{\sigma_U<\infty\}=1.$$

In particular, if (3.27) *is true for any* $(x, i) \in \mathbb{R}^r \times \mathcal{M}$, *then the process* $(X(t), \alpha(t))$ *is recurrent.*

Proof It is enough to consider two cases, namely, when $x_0 \notin D$ and when $x_0 \in D$ with $j \neq \ell$. For the first case, the proof in Lemma 3.7 can be adopted to show that $\mathbb{P}_{x_0,\ell}\{\sigma_D < \infty\} = 1$. Then by virtue of Lemma 3.9, it follows that $\mathbb{P}_{x_0,\ell}\left\{\sigma_{D\times\{j\}} < \infty\right\} = 1$. The second case follows from a slight modification of the argument in the proof of Lemma 3.9. □

Lemma 3.17 *If the process* $(X(t), \alpha(t))$ *is recurrent, then for any* $(x, i) \in \mathbb{R}^r \times \mathcal{M}$ *and* $\varepsilon > 0$, *we have*

$$\mathbb{P}_{x,i}\left\{(X(t_n), \alpha(t_n)) \in B(x, \varepsilon) \times \{i\}, \text{ for a sequence } t_n \uparrow \infty\right\} = 1. \tag{3.28}$$

Proof Denote $B = B(x, \varepsilon)$, $B_1 = B(x, \varepsilon/2)$, and $B_2 = B(x, 2\varepsilon)$. Define a sequence of stopping times by

$$\eta_1 := \inf\{t \geq 0 : X(t) \notin B_2\};$$

and for $n = 1, 2, \ldots,$

$$\begin{aligned} \eta_{2n} &:= \inf\{t \geq \eta_{2n-1} : (X(t), \alpha(t)) \in B_1 \times \{i\}\}, \\ \eta_{2n+1} &:= \inf\{t \geq \eta_{2n} : X(t) \notin B_2\}. \end{aligned}$$

Note that the process $(X(t), \alpha(t))$ is recurrent, in particular, with respect to $B_1 \times \{i\}$. This, together with Theorem 3.2, implies that $\eta_n < \infty$ a.s. $\mathbb{P}_{x,i}$. Thus, (3.28) follows. □

Combining Lemmas 3.16 and 3.17, we obtain the following theorem.

Theorem 3.18 *The process* $(X(t), \alpha(t))$ *is recurrent if and only if every point* $(x, i) \in \mathbb{R}^r \times \mathcal{M}$ *is recurrent in the sense that for any* $\varepsilon > 0$,

$$\mathbb{P}_{x,i}\left\{(X(t_n), \alpha(t_n)) \in B(x, \varepsilon) \times \{i\}, \text{ for a sequence } t_n \uparrow \infty\right\} = 1.$$

Theorem 3.18 enables us to provide another criterion for recurrence in terms of mean sojourn time. This is motivated by the results in [189]. To this end, for any $U := D \times J \subset \mathbb{R}^r \times \mathcal{M}$ and any $\lambda \geq 0$, define

$$R_\lambda(x, i, U) := \mathbb{E}_{x,i}\left[\int_0^\infty e^{-\lambda t} I_U(X(t), \alpha(t))dt\right]. \tag{3.29}$$

Note that $R_0(x, i, U)$ denotes the mean sojourn time of the process $(X^{x,i}(\cdot), \alpha^{x,i}(\cdot))$ in the domain U. We state a proposition below, whose proof, being similar to a result in [189], is relegated to Section 3.6.

Proposition 3.19 *Suppose Assumption* (H3.1) *holds. If for some point* $(x_0, \ell) \in \mathbb{R}^r \times \mathcal{M}$ *and any* $\rho > 0$,

$$R_0(x_0, \ell, B(x_0, \rho) \times \{\ell\}) = \infty, \tag{3.30}$$

then (x_0, ℓ) *is a recurrent point; that is, for every* $\rho > 0$,

$$\mathbb{P}_{x_0,\ell}\left\{(X(t_n), \alpha(t_n)) \in B(x_0, \rho) \times \{\ell\},\ \text{for a sequence } t_n \uparrow \infty\right\} = 1. \tag{3.31}$$

In particular, if (3.30) is true for any $(x, i) \in \mathbb{R}^r \times \mathcal{M}$*, then the process* $(X(t), \alpha(t))$ *is recurrent.*

3.3.2 Transience

We first argue that if the process $(X(t), \alpha(t))$ is transient, then the norm of the continuous component $|X(t)| \to \infty$ a.s. as $t \to \infty$ and vice versa. Then we provide two criteria for transience in terms of mean sojourn time and Lyapunov functions, respectively. If the coefficients $b(x, i)$ and $\sigma(x, i)$ for $i \in \mathcal{M}$ are linearizable in x, we also obtain easily verifiable conditions for transience.

Theorem 3.20 *The process* $(X(t), \alpha(t))$ *is transient if and only if*

$$\lim_{t\to\infty} |X(t)| = \infty \ \text{a.s. } \mathbb{P}_{x,\alpha} \ \text{for any } (x, \alpha) \in \mathbb{R}^r \times \mathcal{M}.$$

The proof of Theorem 3.20 follows the argument of Bhattacharya [27]. We next obtain a criterion for transience under certain conditions. To preserve the flow of the presentation, the proofs of both Theorem 3.20 and Proposition 3.21 are placed in Section 3.6.

Proposition 3.21 *Suppose Assumption* (H3.1) *holds. If for some* $U := D \times J \subset \mathbb{R}^r \times \mathcal{M}$ *containing the point* (x_0, ℓ)*, where D is a nonempty open and bounded set,* $R_0(x_0, \ell, U) < \infty$*, then the process* $(X(t), \alpha(t))$ *is transient.*

Remark 3.22 It follows from Propositions 3.19 and 3.21 that the process $(X(t), \alpha(t))$ is recurrent if and only if for every $(x, i) \in \mathbb{R}^r \times \mathcal{M}$ and every $\rho > 0$, we have

$$R_0(x, i, B(x, \rho) \times \{i\}) = \infty.$$

Next, we obtain a sufficient condition for transience in terms of the existence of a Lyapunov function.

Theorem 3.23 *Assume there exists a nonempty bounded domain* $D \subset \mathbb{R}^r$ *and a function* $V(\cdot, \cdot) : D^c \times \mathcal{M} \mapsto \mathbb{R}$ *satisfying*

$$\begin{aligned}&\sup_{(x,i)\in\partial D\times\mathcal{M}} V(x,i)\le 0,\\&\mathcal{L}V(x,i)\ge 0 \textit{ for any } (x,i)\in D^c\times\mathcal{M},\\&\sup_{(x,i)\in D^c\times\mathcal{M}} V(x,i)\le M<\infty,\\&V(y,\ell)>0 \textit{ for some } (y,\ell)\in D^c\times\mathcal{M}.\end{aligned} \tag{3.32}$$

Then the process $(X(t),\alpha(t))$ is either transient or not regular.

Proof Assuming the process $(X(t),\alpha(t))$ is regular, we need to show that it is transient. Fix $(y,\ell)\in D^c\times\mathcal{M}$ with $V(y,\ell)>0$. Define the stopping times $\sigma_D=\sigma_D^{y,\ell}$, $\beta_n=\beta_n^{y,\ell}$, and $\beta_n\wedge\sigma_D\wedge t$ as in the proof of Theorem 3.14, with $n>(n_0\vee|y|)$, where n_0 is an integer such that $D\subset\{x:|x|<n_0\}$ and $\beta_n^{y,\ell}=\inf\{t:|X^{y,\ell}(t)|=n\}$. By virtue of Dynkin's formula and (3.32),

$$\begin{aligned}&\mathbb{E}V(X(\beta_n\wedge\sigma_D\wedge t),\alpha(\beta_n\wedge\sigma_D\wedge t))-V(y,\ell)\\&=\mathbb{E}\int_0^{\beta_n\wedge\sigma_D\wedge t}\mathcal{L}V(X(u),\alpha(u))du\ge 0.\end{aligned}$$

Therefore, we have from (3.32) that

$$\begin{aligned}V(y,\ell)&\le\mathbb{E}V(X(\beta_n\wedge\sigma_D\wedge t),\alpha(\beta_n\wedge\sigma_D\wedge t))\\&=\mathbb{E}V(X(\sigma_D),\alpha(\sigma_D))I_{\{\sigma_D\le\beta_n\wedge t\}}\\&\quad+\mathbb{E}V(X(\beta_n\wedge t),\alpha(\beta_n\wedge t))I_{\{\sigma_D>\beta_n\wedge t\}}\\&\le M\mathbb{P}\{\sigma_D>\beta_n\wedge t\}.\end{aligned}$$

Let $A_n:=\{\omega\in\Omega:\sigma_D(\omega)>\beta_n(\omega)\wedge t\}$. Then

$$\mathbb{P}(A_n)\ge\frac{V(y,\ell)}{M},\quad\text{for any } n\ge n_0.$$

Note that $\beta_n\le\beta_{n+1}$ implies $A_n\supset A_{n+1}$. This, together with the regularity, yields that

$$\bigcap_{n=n_0}^{\infty}A_n=\lim_{n\to\infty}A_n=\{\sigma_D>t\}.$$

Therefore, $\mathbb{P}\{\sigma_D>t\}\ge\frac{V(y,\ell)}{M}$. Finally, by letting $t\to\infty$, we obtain that

$$\mathbb{P}\{\sigma_D=\infty\}>0.$$

Thus the process $(X(t),\alpha(t))$ is transient. This completes the proof. □

To proceed, we focus on linearizable (in the continuous component) systems. In addition to Assumption (H3.1), we also assume that the following condition holds.

(H3.2) For each $i \in \mathcal{M}$, there exist $b(i), \sigma_j(i) \in \mathbb{R}^{r\times r}$, $j = 1, 2, \ldots, d$, and $\widehat{Q} = (\widehat{q}_{ij})$, a generator of a continuous-time Markov chain $\widehat{\alpha}(t)$ such that as $|x| \to \infty$,

$$\begin{aligned} &\frac{b(x,i)}{|x|} = b(i)\frac{x}{|x|} + o(1),\\ &\frac{\sigma(x,i)}{|x|} = (\sigma_1(i)x, \sigma_2(i)x, \ldots, \sigma_d(i)x)\frac{1}{|x|} + o(1),\\ &Q(x) = \widehat{Q} + o(1), \end{aligned} \tag{3.33}$$

where $o(1) \to 0$ as $|x| \to \infty$. Moreover, $\widehat{\alpha}(t)$ is irreducible with stationary distribution denoted by $\pi = (\pi_1, \pi_2, \ldots, \pi_{m_0}) \in \mathbb{R}^{1\times m_0}$.

An application of Theorem 3.23 leads to the following easily verifiable condition for transience under Assumptions (H3.1) and (H3.2).

Theorem 3.24 *Suppose Assumptions* (H3.1) *and* (H3.2). *If for each* $i \in \mathcal{M}$,

$$\lambda_{\min}\Big(b(i) + b'(i) + \sum_{j=1}^{d}\sigma_j(i)\sigma_j'(i)\Big) - \frac{1}{2}\sum_{j=1}^{d}\Big[\rho(\sigma_j(i) + \sigma_j'(i))\Big]^2 > 0, \tag{3.34}$$

then the process $(X(t), \alpha(t))$ *is transient.*

Proof Let $D = \{x \in \mathbb{R}^r : |x| < k\}$, where k is a sufficiently large positive number. Consider $W(x,i) = k^\beta - |x|^\beta$, $(x,i) \in D^c \times \mathcal{M}$, where $\beta < 0$ is a sufficiently small constant. Then we have $W(x,i) = 0$ for all $(x,i) \in \partial D \times \mathcal{M}$ and $k^\beta \geq W(x,i) > 0$ for all $(x,i) \in D^c \times \mathcal{M}$. Thus, all conditions in (3.32) will be satisfied if we can verify that $\mathcal{L}W(x,i) \geq 0$ for all $(x,i) \in D^c \times \mathcal{M}$.

To this end, we compute for $x \neq 0$ that

$$\begin{aligned} &\nabla W(x,i) = -\beta|x|^{\beta-2}x,\\ &\nabla^2 W(x,i) = -\beta\Big[|x|^{\beta-2}I + (\beta-2)|x|^{\beta-4}xx'\Big]. \end{aligned}$$

Hence, it follows from (3.33) that for all $(x,i) \in D^c \times \mathcal{M}$, we have

$$\begin{aligned} \mathcal{L}W(x,i) = -\beta|x|^\beta\bigg\{&\frac{x'b(i)x}{|x|^2} + \frac{1}{2}\sum_{j=1}^{d}\bigg(\frac{x'\sigma_j'(i)\sigma_j(i)x}{|x|^2}\\ &+ (\beta-2)\frac{(x'\sigma_j'(i)x)^2}{|x|^4}\bigg) + o(1)\bigg\}. \end{aligned} \tag{3.35}$$

Note that

$$\frac{x'b(i)x}{|x|^2} + \frac{1}{2}\sum_{j=1}^{d}\frac{x'\sigma_j'(i)\sigma_j(i)x}{|x|^2} = \frac{x'\left(b'(i)+b(i)+\sum_{j=1}^{d}\sigma_j'(i)\sigma_j(i)\right)x}{2|x|^2}$$
$$\geq \frac{1}{2}\lambda_{\min}\left(b(i)+b'(i)+\sum_{j=1}^{d}\sigma_j'(i)\sigma_j(i)\right). \tag{3.36}$$

Meanwhile, for any symmetric matrix A with real eigenvalues $\lambda_1 \geq \lambda_2 \geq \cdots \geq \lambda_n$, using the transformation $x = Uy$ with U being a real orthogonal matrix satisfying $U'AU = \text{diag}(\lambda_1, \lambda_2, \ldots, \lambda_n)$ (see [107, Theorem 8.1.1]), we have

$$|x'Ax| = |\lambda_1 y_1^2 + \lambda_2 y_2^2 + \cdots + \lambda_n y_n^2| \leq \rho(A)|y|^2 = \rho(A)|x|^2, \tag{3.37}$$

where $\rho(A) := \max\{|\lambda_1|, \ldots, |\lambda_n|\}$ is the spectral radius of A. Therefore, it follows from (3.35)–(3.37) that for all $(x, i) \in D^c \times \mathcal{M}$,

$$\begin{aligned}\mathcal{L}W(x,i) &\geq -\frac{1}{2}\beta|x|^\beta\Bigg\{\lambda_{\min}\left(b(i)+b'(i)+\sum_{j=1}^{d}\sigma_j'(i)\sigma_j(i)\right)\\ &\quad -\frac{1}{2}\sum_{j=1}^{d}\left[\rho(\sigma_j'(i)+\sigma_j(i))\right]^2 + o(1) + O(\beta)\Bigg\}\\ &\geq 0,\end{aligned}$$

where in the last step, we used the fact that $\beta < 0$ and condition (3.34). Hence, the second equation in (3.32) of Theorem 3.23 is valid. Thus, Theorem 3.23 implies that the process $(X(t), \alpha(t))$ is not recurrent. □

3.4 Positive and Null Recurrence

This section takes up the positive recurrence issue. It entails the use of appropriate Lyapunov functions. By virtue of Theorem 3.12, the process $Y(t) = (X(t), \alpha(t))$ is recurrent (resp., positive recurrent) with respect to some "cylinder" $D \times \mathcal{M}$ if and only if it is recurrent (resp., positive recurrent) with respect to $D \times \{\ell\}$, where $D \subset \mathbb{R}^r$ is a nonempty open set with compact closure and $\ell \in \mathcal{M}$. Moreover, the properties of recurrence or positive recurrence do not depend on the choice of the open set $D \subset \mathbb{R}^r$ or $\ell \in \mathcal{M}$. The result in Example 3.35 is quite interesting, which shows that the combination of a transient diffusion and a positive recurrent diffusion can be a positive recurrent switching diffusion.

3.4.1 General Criteria for Positive Recurrence

Theorem 3.25 *A necessary and sufficient condition for positive recurrence with respect to a domain* $U = D \times \{\ell\} \subset \mathbb{R}^r \times \mathcal{M}$ *is: For each* $i \in \mathcal{M}$*, there exists a nonnegative function* $V(\cdot, i) : D^c \mapsto \mathbb{R}$ *such that* $V(\cdot, i)$ *is twice continuously differentiable and that*

$$\mathcal{L}V(x, i) = -1, \quad (x, i) \in D^c \times \mathcal{M}. \tag{3.38}$$

Let $u(x, i) = \mathbb{E}_{x,i}\sigma_D$*. Then* $u(x, i)$ *is the smallest positive solution of*

$$\begin{cases} \mathcal{L}u(x, i) = -1, & \text{if } (x, i) \in D^c \times \mathcal{M}, \\ u(x, i) = 0, & \text{if } (x, i) \in \partial D \times \mathcal{M}, \end{cases} \tag{3.39}$$

where ∂D *denotes the boundary of* D.

Proof The proof is divided into three steps.
Step 1: Show that the process $Y(t) = (X(t), \alpha(t))$ is positive recurrent if there exists a nonnegative function $V(\cdot, \cdot)$ satisfying the conditions of the theorem. Choose n_0 to be a positive integer sufficiently large so that $D \subset \{|x| < n_0\}$. Fix any $(x, i) \in D^c \times \mathcal{M}$. For any $t > 0$ and $n \in \mathbb{N}$ with $n > n_0$, we define

$$\sigma_D^{(n)}(t) = \sigma_D \wedge t \wedge \beta_n,$$

where $\beta_n = \inf\{t : |X^{(n)}(t)| \ge n\}$ is defined as in (2.53) and σ_D is the first entrance time $X(t)$ to D. That is, $\sigma_D = \inf\{t \ge 0 : X(t) \in D\}$. Now Dynkin's formula and Eq. (3.38) imply that

$$\begin{aligned} &\mathbb{E}_{x,i} V(X(\sigma_D^{(n)}(t)), \alpha(\sigma_D^{(n)}(t))) - V(x, i) \\ &= \mathbb{E}_{x,i} \int_0^{\sigma_D^{(n)}(t)} \mathcal{L}V(X(s), \alpha(s))ds = -\mathbb{E}_{x,i}\sigma_D^{(n)}(t). \end{aligned}$$

Since the function V is nonnegative, we have $\mathbb{E}_{x,i}\sigma_D^{(n)}(t) \le V(x, i)$. Meanwhile, because the process $Y(t) = (X(t), \alpha(t))$ is regular, it follows that $\sigma_D^{(n)}(t) \to \sigma_D(t)$ a.s. as $n \to \infty$, where $\sigma_D(t) = \sigma_D \wedge t$. By virtue of Fatou's lemma, we obtain

$$\mathbb{E}_{x,i}\sigma_D(t) \le V(x, i). \tag{3.40}$$

Now the argument after Eq. (3.11) in the proof of Theorem 3.2 yields that $\mathbb{E}_{x,i}\sigma_D \le V(x, i) < \infty$. Then Lemma 3.10 implies that $\mathbb{E}_{x,i}\sigma_U = \mathbb{E}_{x,i}\sigma_{D\times\{\ell\}} < \infty$. Since $(x, i) \in D^c \times \mathcal{M}$ is arbitrary, we conclude that $Y(t)$ is positive recurrent with respect to U.

Step 2: Show that $u(x,i) := \mathbb{E}_{x,i}\sigma_D$ is the smallest positive solution of (3.39). To this end, let n_0 be defined as in Step 1. For $n \ge n_0$, set $\sigma_D^{(n)} = \sigma_D \wedge \beta_n$. Clearly, we have $\sigma_D^{(n)} \le \sigma_D^{(n+1)}$ for all $n \ge n_0$. Then the regularity of the process $Y(t)$ implies that $\sigma_D^{(n)} \nearrow \sigma_D$ a.s. as $n \to \infty$. Hence, the monotone convergence theorem implies that as $n \to \infty$,

$$\mathbb{E}_{x,i}\sigma_D^{(n)} \nearrow \mathbb{E}_{x,i}\sigma_D. \tag{3.41}$$

Note that $\mathbb{E}_{x,i}\sigma_D < \infty$ from Step 1. Meanwhile, Lemma 3.6 implies that the function $u_n(x,i) = \mathbb{E}_{x,i}\sigma_D^{(n)}$ solves the boundary value problem

$$\begin{aligned} &\mathcal{L}u_n(x,i) = -1, \\ &u_n(x,i)|_{x\in\partial D} = 0, \\ &u_n(x,i)|_{|x|=n} = 0, \quad i \in \mathcal{M}. \end{aligned} \tag{3.42}$$

Thus, the function $v_n(x,i) := u_{n+1}(x,i) - u_n(x,i)$ is $\mathcal{L}$-harmonic in the domain $(D^c \cap \{|x| < n\}) \times \mathcal{M}$. Since $\sigma_D^{(n)} \le \sigma_D^{(n+1)}$, it follows that $\mathbb{E}_{x,i}\sigma_D^{(n)} \le \mathbb{E}_{x,i}\sigma_D^{(n+1)}$ and hence $v_n(x,i) \ge 0$. Now, (3.41) implies that

$$u(x,i) = u_{n_0}(x,i) + \sum_{k=n_0}^{\infty} v_k(x,i). \tag{3.43}$$

Using Harnack's inequality for $\mathcal{L}$-elliptic systems of equations (see [7, 47] and also [276] for general references on elliptic systems), it can be shown by a slight modification of the well-known arguments (see, e.g., [104, pp. 21–22]) that the sum of a convergent series of positive $\mathcal{L}$-harmonic functions is also an $\mathcal{L}$-harmonic function. Hence, we conclude that $u(x,i)$ is twice continuously differentiable and satisfies Eq. (3.39). To verify that $u(x,i)$ is the smallest positive solution of (3.39), let $w(x,i)$ be any positive solution of (3.39). Note that $u_n(x,i) = \mathbb{E}_{x,i}\sigma_D^{(n)}$ satisfies the boundary conditions

$$\begin{aligned} &u_n(x,i)|_{x\in\partial D} = 0, \\ &u_n(x,i)|_{|x|=n} = 0, \quad i \in \mathcal{M}. \end{aligned}$$

Then the functions $u_n(x,i) - w(x,i)$ for $i \in \mathcal{M}$ are $\mathcal{L}$-harmonic and satisfy $u_n(x,i) - w(x,i) = 0$ for $(x,i) \in \partial D \times \mathcal{M}$ and $u_n(x,i) - w(x,i) < 0$ for $(x,i) \in \{|x| = n\} \times \mathcal{M}$. Hence, it follows from the maximum principle for $\mathcal{L}$-elliptic system of equations [239, p. 192] that $u_n(x,i) \le w(x,i)$ in $(D^c \cap \{|x| < n\}) \times \mathcal{M}$ for all $n \ge n_0$. Letting $n \to \infty$, we obtain $u(x,i) \le w(x,i)$, as desired.

Step 3: Show that there exists a nonnegative function V satisfying the conditions of the theorem if the process $Y(t) = (X(t), \alpha(t))$ is positive recurrent with respect to the domain $U = D \times \{\ell\}$. Then $\mathbb{E}_{x,i}\sigma_D < \infty$ for all $(x,i) \in D^c \times \mathcal{M}$ and consequently Eq. (3.43) and Harnack's inequality for $\mathcal{L}$-elliptic system of equations

[7, 47] imply that the bounded monotone increasing sequence $u_n(x, i)$ converges uniformly on every compact subset of $D^c \times \mathcal{M}$. Moreover, its limit $u(x, i)$ satisfies the equation $\mathcal{L}u(x, i) = -1$ for each $i \in \mathcal{M}$. Therefore, the function $V(x, i) := u(x, i)$ satisfies Eq. (3.38). This completes the proof of the theorem. □

Theorem 3.26 *A necessary and sufficient condition for positive recurrence with respect to a domain $U = D \times \{\ell\} \subset \mathbb{R}^r \times \mathcal{M}$ is: For each $i \in \mathcal{M}$, there exists a nonnegative function $V(\cdot, i) : D^c \mapsto \mathbb{R}$ such that $V(\cdot, i)$ is twice continuously differentiable and that for some $\gamma > 0$,*

$$\mathcal{L}V(x, i) \le -\gamma, \quad (x, i) \in D^c \times \mathcal{M}. \tag{3.44}$$

Proof Necessity: This part follows immediately from the necessity of Theorem 3.25 with $\gamma = -1$.

Sufficiency: Suppose that there exists a nonnegative function V satisfying the conditions of the theorem. Define the stopping time $\sigma_D^{(n)}(t) = \sigma_D \wedge t \wedge \beta_n$ as in the proof of Theorem 3.25. Now Dynkin's formula and Eq. (3.44) imply that for any $(x, i) \in D^c \times \mathcal{M}$,

$$\begin{aligned}
&\mathbb{E}_{x,i} V(X(\sigma_D^{(n)}(t)), \alpha(\sigma_D^{(n)}(t))) - V(x, i) \\
&\quad = \mathbb{E}_{x,i} \int_0^{\sigma_D^{(n)}(t)} \mathcal{L}V(X(s), \alpha(s))ds \le -\gamma \mathbb{E}_{x,i} \sigma_D^{(n)}(t).
\end{aligned}$$

Hence, we have by the nonnegativity of the function V that $\mathbb{E}_{x,i}\sigma_D^{(n)}(t) \le V(x, i)/\gamma$. Meanwhile, the regularity of the process $Y(t) = (X(t), \alpha(t))$ implies that $\sigma_D^{(n)}(t) \to \sigma_D(t)$ a.s. as $n \to \infty$, where $\sigma_D(t) = \sigma_D \wedge t$. Therefore, Fatou's lemma leads to $\mathbb{E}_{x,i}\sigma_D(t) \le V(x, i)/\gamma$. Moreover, from the proof of Theorem 3.25, $\sigma_D(t) \to \sigma_D$ a.s. as $t \to \infty$. Thus, we obtain by applying Fatou's lemma again that

$$\mathbb{E}_{x,i}\sigma_D \le \frac{1}{\gamma} V(x, i).$$

Then Lemma 3.10 implies that $\mathbb{E}_{x,i}\sigma_U = \mathbb{E}_{x,i}\sigma_{D\times\{\ell\}} < \infty$. Because $(x, i) \in D^c \times \mathcal{M}$ is arbitrary, we conclude that $Y(t)$ is positive recurrent with respect to U. This completes the proof of the theorem. □

Example 3.27 Let us continue our discussion from Example 1.1. Consider (1.1) and assume that for each $\alpha \in \mathcal{M} = \{1, 2, \ldots, m_0\}$ and $i, j = 1, 2, \ldots, n$ with $j \ne i$, $a_{ii}(\alpha) > 0$, and $a_{ij}(\alpha) \ge 0$. Consider the Lyapunov function

$$V(x, \alpha) = \sum_{i=1}^{n} (x_i - 1 - \log x_i), \quad (x, \alpha) \in \mathbb{R}_+^n \times \mathcal{M}.$$

One readily verifies that V is nonnegative. Moreover, direct computations lead to

$$\begin{aligned}\mathcal{L}V(x,\alpha) &= \sum_{i=1}^{n}\left[(x_i-1)\left(r_i(\alpha)-\sum_{j=1}^{n}a_{ij}(\alpha)x_j\right)+\frac{1}{2}\sigma_i^2(\alpha)\right]\\ &\le \sum_{i=1}^{n}\left[-a_{ii}(\alpha)x_i^2+\left(r_i(\alpha)+\sum_{j=1}^{n}a_{ji}(\alpha)\right)x_i-b_i(\alpha)\right].\end{aligned}$$

Note that for each $i=1,\dots,n$ and $\alpha\in\mathcal{M}$, we have

$$\begin{aligned}&-a_{ii}(\alpha)x_i^2+\left(r_i(\alpha)+\sum_{j=1}^{n}a_{ji}(\alpha)\right)x_i-b_i(\alpha)\\ &\le \frac{\left(r_i(\alpha)+\sum_{j=1}^{n}a_{ji}(\alpha)\right)^2}{4a_{ii}(\alpha)}-b_i(\alpha) := \widehat{K}_i(\alpha).\end{aligned}$$

Denote $K_i(\alpha) := \widehat{K}_i(\alpha)\vee 0$ for each $i=1,\dots,n$ and $\alpha=1,\dots,m_0$. We can further find a number $\rho_i(\alpha)>0$ sufficiently large so that

$$-a_{ii}(\alpha)x_i^2+\left(r_i(\alpha)+\sum_{j=1}^{n}a_{ji}(\alpha)\right)x_i-b_i(\alpha)\le -\sum_{j=1}^{n}K_j(\alpha)-1,$$

for any $x_i>\rho_i(\alpha)$. Denote $\rho := \max\{\rho_i(\alpha), i=1,\dots,n,\alpha=1,\dots,m_0\}$ and $E_\rho := \{x\in\mathbb{R}_+^n : 0<x_i<\rho, i=1,2,\dots,n\}$. Then for any $(x,\alpha)\in(\mathbb{R}_+^n\setminus\overline{E}_\rho)\times\mathcal{M}$, there exists at least one index, say, $\ell\in\{1,\dots,n\}$ such that $x_\ell>\rho$. Then it follows that for any $(x,\alpha)\in(\mathbb{R}_+^n\setminus\overline{E}_\rho)\times\mathcal{M}$

$$\begin{aligned}\mathcal{L}V(x,\alpha) &\le -a_{\ell\ell}(\alpha)x_\ell^2+\left(r_\ell(\alpha)+\sum_{j=1}^{n}a_{j\ell}(\alpha)\right)x_\ell-b_\ell(\alpha)\\ &\quad+\sum_{i\ne\ell}\left[-a_{ii}(\alpha)x_i^2+\left(r_i(\alpha)+\sum_{j=1}^{n}a_{ji}(\alpha)\right)x_i-b_i(\alpha)\right]\\ &\le -\sum_{j=1}^{n}K_j(\alpha)-1+\sum_{i\ne\ell}K_i(\alpha)\le -1.\end{aligned}$$

Therefore, the solution $x(t)$ to (1.1) is positive recurrent with respect to the domain E_ρ.

3.4.2 Path Excursions

Applications of the positive recurrence criteria enable us to establish path excursions of the underlying processes. Suppose that $Y(t)=(X(t),\alpha(t))$ is positive recurrent and that the Lyapunov functions $V(x, i)$ (with $i \in \mathcal{M}$) are given in Theorem 3.26, so is the set D. Let D_0 be a bounded open set with compact closure satisfying $D \subset D_0$, and τ_1 be the first exit time of $(X(t), \alpha(t))$ from $D_0 \times \mathcal{M}$; that is, $\tau_1 = \min\{t > 0 : X(t) \notin D_0\}$. Define $\sigma_1 = \min\{t > \tau_1 : X(t) \in D_0\}$. We can obtain

$$
\begin{aligned}
&\mathbb{P}(\sup_{\tau_1 \le t \le \sigma_1} V(X(t), \alpha(t)) \ge \gamma) \le \frac{\mathbb{E}V(X(\tau_1), \alpha(\tau_1))}{\gamma}, \quad \text{for } \gamma > 0, \\
&\mathbb{E}(\sigma_1 - \tau_1) \le \frac{\mathbb{E}V(X(\tau_1), \alpha(\tau_1))}{\gamma},
\end{aligned} \tag{3.45}
$$

where γ is as given in Theorem 3.26. The idea can be continued in the following way for $k \ge 1$. Define

$$
\tau_{k+1} = \min\{t > \sigma_k : X(t) \notin D_0\}
$$

and

$$
\sigma_{k+1} = \min\{t \ge \tau_{k+1} : X(t) \in D_0\}.
$$

Then we can obtain estimates of $\mathbb{E}[\sigma_{k+1} - \tau_{k+1}]$, which gives us the expected difference of $(k+1)$st return time and exit time.

3.4.3 Positive Recurrence Under Linearization

This subsection is devoted to positive recurrence of regime-switching diffusions under linearization with respect to the continuous component. By linearizable systems, we mean such systems that are linearizable with respect to the continuous component. These systems are important inasmuch as linearization is widely used in many applications because it is much easier to deal with linear systems.

With Theorem 3.26 at hand, we proceed to study positive recurrence of linearizable (in the continuous component) systems as described in Assumption (H3.2). We are ready to present an easily verifiable condition for positive recurrence.

Theorem 3.28 *Suppose Assumptions* (H3.1) *and* (H3.2) *hold. If*

$$
\sum_{i=1}^{m_0} \pi_i \lambda_{\max}\left(b(i) + b'(i) + \sum_{j=1}^{d} \sigma_j(i)\sigma_j'(i) \right) < 0, \tag{3.46}
$$

where $\pi = (\pi_1, \ldots, \pi_{m_0}) \in \mathbb{R}^{1 \times m_0}$ *is as in Assumption* (H3.2), *then the process* $(X(t), \alpha(t))$ *is positive recurrent.*

Proof For notational simplicity, define the column vector $\mu = (\mu_1, \ldots, \mu_{m_0})' \in \mathbb{R}^{m_0}$ with

$$\mu_i = \frac{1}{2}\lambda_{\max}\Big(b(i) + b'(i) + \sum_{j=1}^{d} \sigma_j(i)\sigma_j'(i)\Big).$$

Let

$$\beta := -\pi\mu = -\frac{1}{2}\sum_{i=1}^{m_0} \pi_i \Big(b(i) + b'(i) + \sum_{j=1}^{d} \sigma_j(i)\sigma_j'(i)\Big).$$

Note that $\beta > 0$ by (3.46). Because

$$\pi(\mu + \beta\mathbb{1}) = \pi\mu + \beta \cdot \pi\mathbb{1} = -\beta + \beta = 0,$$

Assumption (H3.2) and Lemma A.12 yield that the equation

$$\widehat{Q}c = \mu + \beta\mathbb{1}$$

has a solution $c = (c_1, c_2, \ldots, c_{m_0})' \in \mathbb{R}^{m_0}$. Thus, we have

$$\mu_i - \sum_{j=1}^{m_0} \widehat{q}_{ij} c_j = -\beta, \quad i \in \mathcal{M}. \tag{3.47}$$

For each $i \in \mathcal{M}$, consider the Lyapunov function

$$V(x, i) = (1 - \gamma c_i)|x|^{\gamma},$$

where $0 < \gamma < 1$ is sufficiently small so that $1 - \gamma c_i > 0$ for each $i \in \mathcal{M}$. It is readily seen that for each $i \in \mathcal{M}$, $V(\cdot, i)$ is continuous and nonnegative and has continuous second-order partial derivatives with respect to x in any deleted neighborhood of 0. Detailed calculations reveal that for $x \neq 0$, we have

$$\begin{aligned}
&\nabla V(x, i) = (1 - \gamma c_i)\gamma |x|^{\gamma-2} x,\\
&\nabla^2 V(x, i) = (1 - \gamma c_i)\gamma \left[|x|^{\gamma-2} I + (\gamma - 2)|x|^{\gamma-4} x x'\right].
\end{aligned}$$

Meanwhile, it follows from (3.33) that

$$\frac{a(x, i)}{|x|^2} = \frac{\sigma(x, i)\sigma'(x, i)}{|x|^2} = \sum_{j=1}^{d} \sigma_j(i) x x' \sigma_j'(i) \frac{1}{|x|^2} + o(1),$$

where $o(1) \to 0$ as $|x| \to \infty$. Therefore, we have that

$$
\begin{aligned}
\mathcal{L}V(x,i) = & \frac{\gamma}{2}(1-\gamma c_i)\sum_{j=1}^{d}\Big(x'\sigma_j'(i)|x|^{\gamma-2}I\sigma_j(i)x \\
& + x'\sigma_j'(i)(\gamma-2)|x|^{\gamma-4}xx'\sigma_j(i)x\Big) \\
& + (1-\gamma c_i)\gamma|x|^{\gamma-2}x'b(i)x + o(|x|^{\gamma}) - \sum_{j\neq i} q_{ij}(x)|x|^{\gamma}\gamma(c_j-c_i) \\
= & \gamma(1-\gamma c_i)|x|^{\gamma}\left\{\frac{1}{2}\sum_{j=1}^{d}\left(\frac{x'\sigma_j'(i)\sigma_j(i)x}{|x|^2} + (\gamma-2)\frac{(x'\sigma_j'(i)x)^2}{|x|^4}\right)\right. \\
& \left. + \frac{x'b(i)x}{|x|^2} - \sum_{j\neq i} q_{ij}(x)\frac{c_j-c_i}{1-\gamma c_i} + o(1)\right\}.
\end{aligned}
\tag{3.48}
$$

Note that

$$
\frac{x'b(i)x}{|x|^2} + \frac{1}{2}\sum_{j=1}^{d}\frac{x'\sigma_j'(i)\sigma_j(i)x}{|x|^2} \le \frac{1}{2}\lambda_{\max}\left(b(i)+b'(i)+\sum_{j=1}^{d}\sigma_j'(i)\sigma_j(i)\right) = \mu_i.
\tag{3.49}
$$

Next, using Assumption (H3.2),

$$
\begin{aligned}
\sum_{j\neq i} q_{ij}(x)\frac{c_j-c_i}{1-\gamma c_i} & = \sum_{j=1}^{m_0} q_{ij}(x)c_j + \sum_{j\neq i} q_{ij}(x)\frac{c_i(c_j-c_i)}{1-\gamma c_i}\gamma \\
& = \sum_{j=1}^{m_0}\widehat{q}_{ij}c_j + O(\gamma) + o(1),
\end{aligned}
\tag{3.50}
$$

where $O(\gamma)/\gamma$ is bounded as $\gamma \to 0$. Hence, it follows from (3.48)–(3.50) that when $|x| > R$ with R sufficiently large and $0 < \gamma < 1$ sufficiently small, we have

$$
\mathcal{L}V(x,i) \le \gamma(1-\gamma c_i)|x|^{\gamma}\left\{\mu_i - \sum_{j=1}^{m_0}\widehat{q}_{ij}c_j + o(1) + O(\gamma)\right\}.
$$

Furthermore, by virtue of (3.47), we have

$$
\mathcal{L}V(x,i) \le \gamma(1-\gamma c_i)|x|^{\gamma}\left(-\beta + o(1) + O(\gamma)\right) \le -K < 0,
$$

for any $(x,i) \in \mathbb{R}^r \times \mathcal{M}$ with $|x| > R$, where K is a positive constant. Therefore, we conclude from Theorem 3.26 that the process $(X(t), \alpha(t))$ is positive recurrent. □

The above result further specializes to the following corollary.

Corollary 3.29 *Suppose that the continuous component $X(t)$ is one-dimensional and that as $|x| \to \infty$,*

$$\begin{aligned} \frac{b(x,i)}{|x|} &= b_i \frac{x}{|x|} + o(1), \\ \frac{\sigma(x,i)}{|x|} &= \sigma_i \frac{x}{|x|} + o(1), \end{aligned} \tag{3.51}$$

for some constants b_i, σ_i, $i \in \mathcal{M}$. If

$$\pi b - \frac{1}{2}\pi\sigma^2 := \sum_{i=1}^{m_0} \pi_i \left(b_i - \frac{\sigma_i^2}{2} \right) < 0, \tag{3.52}$$

where $\pi = (\pi_1, \ldots, \pi_{m_0}) \in \mathbb{R}^{1\times m_0}$ is as in Assumption (H3.2)*, then the process $(X(t), \alpha(t))$ is positive recurrent.*

Next, we develop a sufficient condition for nonpositive recurrence.

Theorem 3.30 *Suppose that there exists a nonempty bounded domain D such that there exist functions $V(x,i)$ and $W(x,i)$ defined on $D^c \times \mathcal{M}$ satisfying the following:*

(a) *$V(x,i) \geq 0$ for all $(x,i) \in D^c \times \mathcal{M}$, and for some positive constant k,*

$$0 \leq \mathcal{L}V(x,i) \leq k \ \text{ for all } \ (x,i) \in D^c \times \mathcal{M}.$$

(b) *$W(x,i) \leq 0$ for all $(x,i) \in \partial D \times \mathcal{M}$, and*

$$\mathcal{L}W(x,i) \geq 0 \ \text{ for all } \ (x,i) \in D^c \times \mathcal{M}.$$

(c) *For an increasing sequence of bounded domains $E_n \supset D$ with boundaries Γ_n,*

$$\frac{\inf_{(x,i)\in\Gamma_n\times\mathcal{M}} V(x,i)}{\sup_{(x,i)\in\Gamma_n\times\mathcal{M}} W(x,i)} = R_n \to \infty \ \text{ as } \ n \to \infty. \tag{3.53}$$

If there exists some $(x,\alpha) \in D^c \times \mathcal{M}$ satisfying $W(x,\alpha) > 0$, then the process $(X(t), \alpha(t))$ is not positive recurrent. That is, $\mathbb{E}\sigma_D = \infty$ for all $(x,\alpha) \in D^c \times \mathcal{M}$ such that $W(x,\alpha) > 0$, where

$$\sigma_D = \sigma_D^{x,\alpha} = \inf\left\{t \geq 0 : X^{x,\alpha}(t) \in D\right\}.$$

Proof Consider the function $V - R_n W$ in $(D^c \cap E_n) \times \mathcal{M}$. Then from conditions (a), (b), and (c), we have

$$\begin{aligned}
&\mathcal{L}(V - R_n W)(x, i) \le k,\\
&(V - R_n W)(x, i)|_{x\in\partial D} \ge 0,\\
&(V - R_n W)(x, i)|_{x\in\Gamma_n} \ge 0.
\end{aligned}$$

Fix any $(x, \alpha) \in D^c \times \mathcal{M}$ with $W(x, \alpha) > 0$. Then we have $\mathbb{P}\{\sigma_D \wedge \beta_n < \infty\} = 1$ for any $n > (n_0 \vee |x|)$, where n_0 is a sufficiently large integer such that $D \subset \{x \in \mathbb{R}^r : |x| < n_0\}$ and $\beta_n := \inf\{t \ge 0 : X(t) \notin E_n\}$. Hence, Dynkin's formula implies that

$$\begin{aligned}
&\mathbb{E}(V - R_n W)(X(\sigma_D \wedge \beta_n), \alpha(\sigma_D \wedge \beta_n)) - (V - R_n W)(x, \alpha)\\
&= \mathbb{E}\int_0^{\sigma_D\wedge\beta_n} \mathcal{L}(V - R_n W)(X(u), \alpha(u))du\\
&\le k\mathbb{E}[\sigma_D \wedge \beta_n].
\end{aligned}$$

That is,

$$\mathbb{E}\sigma_D \wedge \beta_n \ge \frac{1}{k}[R_n W(x, \alpha) - V(x, \alpha)].$$

Therefore, we have from (3.53) that

$$\mathbb{E}\sigma_D \wedge \beta_n \to \infty, \quad \text{as } n \to \infty.$$

Meanwhile, note that $\sigma_D \ge \sigma_D \wedge \beta_n$. We have $\mathbb{E}\sigma_D = \infty$. Therefore, the process $(X(t), \alpha(t))$ is not positive recurrent by virtue of Theorem 3.12. □

3.4.4 Null Recurrence

Null recurrence is more complex. Even for diffusions alone, the available results are scarce. For simplicity here, we assume that the continuous component X of the system (X, α) has no drift term.

Theorem 3.31 *Consider a real-valued regime-switching diffusion process*

$$\begin{cases} dX(t) = \sigma(X(t), \alpha(t))dw(t)\\ \mathbb{P}\{\alpha(t + \Delta) = j|\alpha(t) = i, X(s), \alpha(s), s \le t\} = q_{ij}(X(t))\Delta + o(\Delta). \end{cases} \tag{3.54}$$

If for some constants $0 \le \beta \le 1$, $k_1 > 0$, *and* $k_2 > 0$,

$$\sigma^2(x, i) \le k_1|x|^{1-\beta}, \ \textit{for all } (x, i) \in \{x \in \mathbb{R} : |x| \ge k_2\} \times \mathcal{M}, \tag{3.55}$$

then the process $(X(t), \alpha(t))$ *defined by* (3.54) *is null recurrent.*

Proof In fact, by Theorem 3.12, it suffices to prove that the process is null recurrent with respect to $D \times \{\ell\}$, where $D = (-k_2, k_2)$ and $\ell \in \mathcal{M}$. We now consider the functions:

$$V(x, i) = \begin{cases} |x|^{1+\beta}, & \text{if } \beta > 0, \\ |x|(\ln|x| - 1) + K, & \text{if } \beta = 0, \end{cases}$$

where $(x, i) \in D^c \times \mathcal{M}$ and K is a sufficiently large positive constant, and

$$W(x, i) = |x| - k_2 \text{ for } (x, i) \in D^c \times \mathcal{M}.$$

Then detailed computations show that the nonnegative function W satisfies (3.24) in Theorem 3.14. This implies that the process is recurrent with respect to $D \times \{\ell\}$.

Similarly, by virtue of (3.55), detailed computations show that conditions (a)–(c) of Theorem 3.30 are satisfied with the functions V and W as chosen above. It thus follows from Theorem 3.30 that the process $(X(t), \alpha(t))$ is not positive recurrent with respect to $D \times \{\ell\}$. The details are omitted. Therefore, the process defined by (3.54) is null recurrent. □

Remark 3.32 If for some positive constants ϱ, k_1, and k_2, we have

$$\sigma^2(x, i) \geq k_1|x|^{1+\varrho}, \quad \text{for all } (x, i) \in \{x \in \mathbb{R} : |x| \geq k_2\} \times \mathcal{M}, \tag{3.56}$$

then the process $(X(t), \alpha(t))$ defined by (3.54) is positive recurrent.

To verify this, we choose $1 > \varsigma > 0$ such that $\varsigma + \varrho > 1$ and define $V(x, i) = |x|^{\varsigma}$ for $(x, i) \in \{x \in \mathbb{R} : |x| \geq k_2\} \times \mathcal{M}$. Then we can show

$$\mathcal{L}V(x, i) = \frac{1}{2}\varsigma(\varsigma - 1)|x|^{\varsigma-2}\sigma^2(x, i) \leq \frac{1}{2}k_1\varsigma(\varsigma - 1)k_2^{\varsigma+\varrho-1} < 0,$$

where $(x, i) \in \{x \in \mathbb{R} : |x| \geq k_2\} \times \mathcal{M}$. Therefore, it follows from Theorem 3.26 that the process $(X(t), \alpha(t))$ is positive recurrent.

3.5 Examples

In this section, we provide several examples to illustrate the results obtained thus far.

Example 3.33 Suppose that for each $x \in \mathbb{R}^r$ and each $i \in \mathcal{M}$, there exist positive constants c and γ such that for all x with $|x| \geq c$,

$$b'(x, i)\frac{x}{|x|} < -\gamma. \tag{3.57}$$

That is, the drifts are pointed inward. Then the process $Y(t) = (X(t), \alpha(t))$ is positive recurrent.

First, note that (3.5) implies that for all x with $|x| \geq r/(\gamma\kappa_1)$, we have

$$\operatorname{tr}(a(x,i)) = \sum_{j=1}^{r} e_j' a(x,i) e_j \leq \sum_{j=1}^{r} \kappa_1^{-1} \leq \gamma |x|.$$

By Theorem 3.12, it is enough to prove that the process $Y(t) = (X(t), \alpha(t))$ is positive recurrent with respect to the domain $U := \{|x| < \varrho\} \times \{\ell\}$ for some $\ell \in \mathcal{M}$, where $\varrho := \max\{c, r/(\gamma\kappa_1)\}$. To this end, consider the function

$$V(x,i) = \frac{1}{2} x' x, \quad \text{for each } i \in \mathcal{M} \text{ and for all } |x| \geq \varrho.$$

For each $i \in \mathcal{M}$, $\nabla V(\cdot, i) = x$ and $\nabla^2 V(\cdot, i) = I$, where I is the $r \times r$ identity matrix. Thus, by the definition of $\mathcal{L}$, we have for all $(x,i) \in \{|x| \geq \varrho\} \times \mathcal{M}$ that

$$\mathcal{L}V(x,i) = \frac{1}{2}\operatorname{tr}(a(x,i)) + b'(x,i)\frac{x}{|x|}|x| < \frac{1}{2}\gamma|x| - \gamma|x| = -\frac{1}{2}\gamma|x| \leq -\frac{1}{2}\kappa\varrho.$$

Then the conclusion follows from Theorem 3.26 immediately.

Remark 3.34 Suppose that the continuous component $X(t)$ of the process $Y(t) = (X(t), \alpha(t))$ is one-dimensional and that there exist constants $c_0 > 0$ and $c_1 > 0$ such that for each $i \in \mathcal{M}$,

$$b(x,i) \begin{cases} < -c_1, & \text{for } x > c_0, \\ > c_1, & \text{for } x < -c_0. \end{cases} \tag{3.58}$$

Then the process $Y(t) = (X(t), \alpha(t))$ is positive recurrent. In fact, the conclusion follows immediately if we observe that (3.58) satisfies (3.57). Alternatively, we can verify this directly by defining the Lyapunov function $V(x,i) = |x|$ for each $i \in \mathcal{M}$.

Example 3.35 To illustrate the utility of Theorem 3.26, consider a real-valued process

$$dX(t) = b(X(t), \alpha(t))dt + \sigma(X(t), \alpha(t))dw(t), \tag{3.59}$$

where $\alpha(t)$ is a two-state random jump process, with x-dependent generator

$$Q(x) = \begin{pmatrix} -\frac{1}{3} - \frac{1}{4}\cos x & \frac{1}{3} + \frac{1}{4}\cos x \\ \frac{7}{3} + \frac{1}{2}\sin x & -\frac{7}{3} - \frac{1}{2}\sin x \end{pmatrix},$$

and

$$b(x,1) = -x, \quad \sigma(x,1) = 1, \quad b(x,2) = x, \quad \sigma(x,2) = 1.$$

Thus, (3.59) can be regarded as the result of the following two diffusions:

$$dX(t) = -X(t)dt + dw(t), \tag{3.60}$$

and

$$dX(t) = X(t)dt + dw(t), \tag{3.61}$$

switching back and forth from one to the other according to the movement of $\alpha(t)$.

Note that (3.60) is positive recurrent, whereas (3.61) is a transient diffusion process. But the switching diffusion (3.59) is positive recurrent. We verify these as follows. Consider the Lyapunov function $V(x,1) = |x|$. Let $\mathcal{L}_1$ be the operator associated with (3.60). Then we have for all $|x| \geq 1$, $\mathcal{L}_1 V(x,1) = -x \operatorname{sign}(x) = -|x| \leq -1 < 0$. It follows from [156, Theorem 3.9] that (3.60) is positive recurrent. Recall that the real-valued diffusion process

$$dX(t) = b(X(t))dt + \sigma(X(t))dw(t)$$

with $\sigma(x) \neq 0$ for all $x \in \mathbb{R}$ is recurrent if and only if

$$\int_0^x \exp\left\{-2\int_0^u \frac{b(z)}{\sigma^2(z)}dz\right\} du \to \pm\infty \tag{3.62}$$

as $x \to \pm\infty$; see [156, Example 3.10]. Direct computation shows that (3.61) fails to satisfy this condition and hence is transient.

Next, we use Theorem 3.26 to demonstrate that the switching diffusion (3.59) is positive recurrent for appropriate Q. Consider the Lyapunov functions

$$V(x,1) = |x|, \quad V(x,2) = \frac{7}{3}|x|.$$

We have

$$\begin{aligned}
\mathcal{L}V(x,1) &= -x \cdot \operatorname{sign} x + \left(\frac{1}{3} + \frac{1}{4}\cos x\right)\left(\frac{7}{3} - 1\right)|x| \leq -\frac{2}{9}|x| \leq -\frac{2}{9}, \\
\mathcal{L}V(x,2) &= x \cdot \frac{7}{3}\operatorname{sign} x + \left(\frac{7}{3} + \frac{1}{2}\sin x\right)\left(1 - \frac{7}{3}\right)|x| \leq -\frac{1}{9}|x| \leq -\frac{1}{9},
\end{aligned}$$

for all $|x| \geq 1$. Thus, the switching diffusion (3.59) is positive recurrent by Theorem 3.26.

Example 3.36 To illustrate the result of Corollary 3.29, we consider a real-valued process given by

$$\begin{cases} dX(t) = b(X(t), \alpha(t))dt + \sigma(X(t), \alpha(t))dw(t) \\ \mathbb{P}\{\alpha(t+\Delta) = j|\alpha(t) = i, X(s), \alpha(s), s \le t\} = q_{ij}(X(t))\Delta + o(\Delta), \end{cases} \tag{3.63}$$

with the following specifications. The jump component $\alpha(t)$ has three states and is generated by

$$Q(x) = \begin{pmatrix} -\dfrac{2+|x|}{1+x^2} - \dfrac{1+3x^2}{2+x^2} & \dfrac{2+|x|}{1+x^2} & \dfrac{1+3x^2}{2+x^2} \\ 1 - \dfrac{\sin x}{1+x^2} & \dfrac{\sin x}{1+x^2} - \dfrac{\cos x}{2+|x|} - 3 & 2 + \dfrac{\cos x}{2+|x|} \\ \dfrac{1+\cos^2 x}{2+x^2} & 2 & -\dfrac{\cos^2 x + 1}{2+x^2} - 2 \end{pmatrix},$$

and the drift and diffusion coefficients are given by

$$\begin{aligned} &b(x,1) = 3x - 1,\ b(x,2) = \frac{3}{2}x + 1,\ b(x,3) = -x + \frac{x}{1+|x|}, \\ &\sigma(x,1) = \sqrt{3+x^2},\ \sigma(x,2) = \sqrt{2 - \sin x + 2x^2},\ \sigma(x,3) = 3 + \sqrt{4+x^2}. \end{aligned}$$

Hence, associated with (3.63), there are three diffusions:

$$dX(t) = (3X(t) - 1)dt + \sqrt{3 + X^2(t)}dw(t), \tag{3.64}$$

$$dX(t) = \left(\frac{3}{2}X(t) + 1\right)dt + \sqrt{2 - \sin(X(t)) + 2X^2(t)}dw(t), \tag{3.65}$$

$$dX(t) = \left(\frac{X(t)}{1+|X(t)|} - X(t)\right)dt + \left(3 + \sqrt{4 + X^2(t)}\right)dw(t), \tag{3.66}$$

switching back and forth from one to another according to the movement of the jump component $\alpha(t)$.

Note that (3.66) is positive recurrent, whereas (3.64) and (3.65) are transient diffusions. But due to the stabilization effect of $\alpha(t)$, the switching diffusion (3.63) is positive recurrent. We verify these as follows. Consider the Lyapunov function $V(x) = |x|^\gamma$ with $0 < \gamma < 1$ sufficiently small; let $\mathcal{L}_1$ be the operator associated with the third equation in (3.64). Detailed computation shows that for all $|x| \ge 1$, we have $\mathcal{L}_1 V(x) \le -\frac{1}{2}\gamma < 0$. Thus, it follows from [156, Theorem 3.9] that (3.66) is positive recurrent. Detailed computations show that (3.64) and (3.65) fail to satisfy (3.62) and hence these diffusions are transient.

Next, we use Corollary 3.29 to demonstrate that the switching diffusion (3.63) is positive recurrent. In fact, it is readily seen that as $x \to \infty$, the constants $b(i)$ and $\sigma^2(i)$, $i = 1, 2, 3$, as in (3.51) are given by

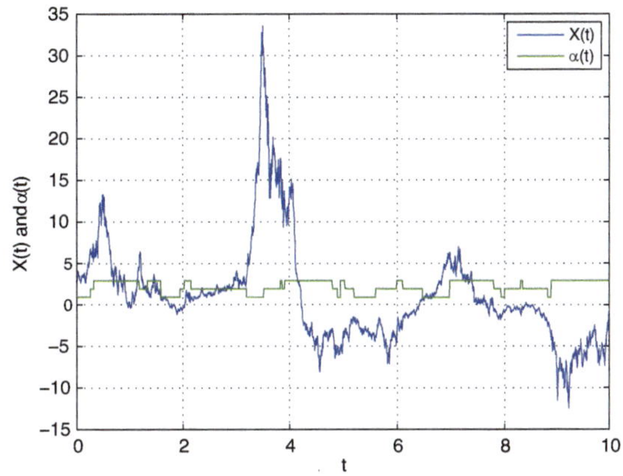

Fig. 3.1 Sample path of switching diffusion (3.63) with initial condition $(x, \alpha) = (3, 1)$

$$\begin{aligned} b(1) = 3, \quad b(2) = \frac{3}{2}, \quad b(3) = -1, \\ \sigma^2(1) = 1, \, \sigma^2(2) = 2, \, \sigma^2(3) = 1. \end{aligned} \tag{3.67}$$

In addition, as $|x| \to \infty$, $Q(x)$ tends to

$$\widehat{Q} = \begin{pmatrix} -3 & 0 & 3 \\ 1 & -3 & 2 \\ 0 & 2 & -2 \end{pmatrix}.$$

The stationary distribution π associated with $\widetilde{Q}$ is

$$\pi = \left(\frac{2}{17}, \frac{6}{17}, \frac{9}{17} \right). \tag{3.68}$$

Thus, by virtue of (3.67) and (3.68), we have

$$\sum_{i=1}^{3} \pi_i \left(b(i) - \frac{\sigma^2(i)}{2} \right) = -\frac{23}{34} < 0.$$

Therefore Corollary 3.29 implies that (3.63) is positive recurrent; see the sample path demonstrated in Fig. 3.1.

Example 3.37 Consider a two-dimensional (in the continuous component) regime-switching diffusion:

$$\begin{cases} dX(t) = b(X(t), \alpha(t))dt + \sigma(X(t), \alpha(t))dw(t) \\ \mathbb{P}\{\alpha(t + \Delta) = j | \alpha(t) = i, X(s), \alpha(s), s \le t\} = q_{ij}(X(t))\Delta + o(\Delta) \end{cases} \tag{3.69}$$

with $(X(t), \alpha(t)) \in \mathbb{R}^2 \times \{1, 2\}$. The discrete component $\alpha(t)$ is generated by

$$Q(x_1, x_2) = \begin{pmatrix} -3 + \dfrac{2(1+\cos(x_1^2))}{3 + x_1^2 + x_2^2} & 3 - \dfrac{2(1+\cos(x_1^2))}{3 + x_1^2 + x_2^2} \\ 1 + \dfrac{1}{\sqrt{1 + x_1^2 + x_2^2}} & -1 - \dfrac{1}{\sqrt{1 + x_1^2 + x_2^2}} \end{pmatrix},$$

and

$$\begin{aligned} b(x_1, x_2, 1) &= \begin{pmatrix} -x_1 + 2x_2 \\ 2x_2 \end{pmatrix}, \\ \sigma(x_1, x_2, 1) &= \begin{pmatrix} 3 + \sin x_1 + \sqrt{2 + x_2^2} & 0 \\ 0 & 2 - \cos x_2 + \sqrt{1 + x_1^2} \end{pmatrix}, \\ b(x_1, x_2, 2) &= \begin{pmatrix} -3x_1 - x_2 \\ x_1 - 2x_2 \end{pmatrix}, \\ \sigma(x_1, x_2, 2) &= \begin{pmatrix} 1 & 1 \\ 0 & 10 + \sqrt{3 + x_1^2} \end{pmatrix}. \end{aligned}$$

Associated with the regime-switching diffusion (3.69), there are two diffusions

$$dX(t) = b(X(t), 1)dt + \sigma(X(t), 1)dw(t), \tag{3.70}$$

and

$$dX(t) = b(X(t), 2)dt + \sigma(X(t), 2)dw(t), \tag{3.71}$$

switching back and forth from one to another according to the movement of the jump component $\alpha(t)$, where $w(t) = (w_1(t), w_2(t))'$ is a two-dimensional standard Brownian motion. By selecting appropriate Lyapunov functions as in [156, Examples 3.10 & 3.11], or using the criteria in [27], we can verify that (3.70) is transient and (3.71) is positive recurrent.

Next, we use Theorem 3.28 to show that the switching diffusion (3.69) is positive recurrent owing to the presence of the stabilizing effect of the discrete component $\alpha(t)$. Note that the matrices $b(i)$ and $\sigma_j(i)$, $i, j = 1, 2$, and $\widehat{Q}$ in Assumption (H3.2) are

$$\begin{aligned} &b(1) = \begin{pmatrix} -1 & 2 \\ 0 & 2 \end{pmatrix}, \quad b(2) = \begin{pmatrix} -3 & -1 \\ 1 & -2 \end{pmatrix}, \quad \sigma_1(2) = 0, \\ &\sigma_1(1) = \sigma_2(2) = \sigma_2'(1) = \begin{pmatrix} 0 & 1 \\ 0 & 0 \end{pmatrix}, \quad \text{and } \widehat{Q} = \begin{pmatrix} -3 & 3 \\ 1 & -1 \end{pmatrix}. \end{aligned}$$

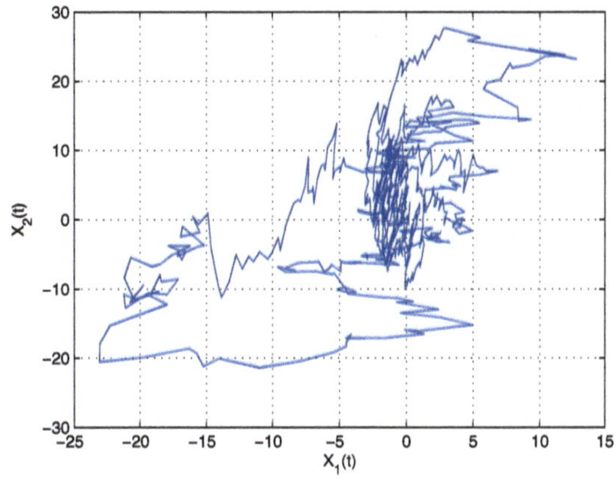

Fig. 3.2 Phase portrait of switching diffusion (3.69) with initial condition (x, α), where $x = [2.5, 2.5]'$ and $\alpha = 1$

Thus, the stationary distribution associated with $\widehat{Q}$ is $\pi = (0.25, 0.75)$, and

$$\begin{aligned}
\lambda_{\max}(b(1) + b'(1) + \sigma_1(1)\sigma_1'(1) + \sigma_2(1)\sigma_2'(1)) &= 5.6056,\\
\lambda_{\max}(b(2) + b'(2) + \sigma_1(2)\sigma_1'(2) + \sigma_2(2)\sigma_2'(2)) &= -4.
\end{aligned}$$

This yields that

$$\sum_{i=1}^{2} \pi_i \lambda_{\max}(b(i) + b'(i) + \sigma_1(i)\sigma_1'(i) + \sigma_2(i)\sigma_2'(i)) = -1.5986 < 0.$$

Therefore, we conclude from Theorem 3.28 that the switching diffusion (3.69) is positive recurrent. For comparison, we begin by considering a sample path of the switching diffusion. Next, treating t as a parameter and eliminating it from the two components x_1 and x_2, we plot a curve of x_2 versus x_1 in the "phase space." Borrowing the terminology from ordinary differential equations, and abusing the terminology slightly, we still call such plots phase portraits henceforth. The phase portrait in Fig. 3.2 confirms our findings. For comparison and better visualization, we also present the componentwise sample paths in Fig. 3.3(a) and (b).

3.6 Proofs of Several Results

This section serves as an appendix of this chapter. It provides several proofs of some technical estimates.

Proof of Lemma 3.7 It suffices to prove the lemma when $E \cup \partial E \subset D$ and ∂E is sufficiently smooth. Fix any $(x, i) \in E^c \times \mathcal{M}$. Let $G \subset \mathbb{R}^r$ be an open and bounded set with sufficiently smooth boundary such that $D \cup \partial D \subset G$. Without

Fig. 3.3 Componentwise sample path of switching diffusion (3.69) with initial condition (x, α), where $x = [2.5, 2.5]'$ and $\alpha = 1$. (**a**) Sample path $(X_1(t), \alpha(t))$ of (3.69). (**b**) Sample path $(X_2(t), \alpha(t))$ of (3.69)

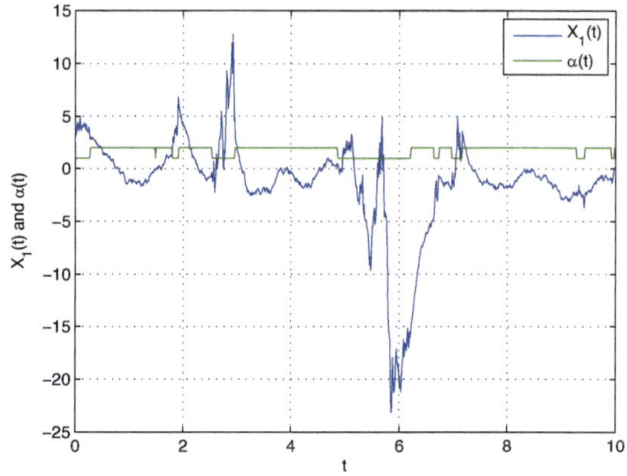

(a) Sample path $(X_1(t), \alpha(t))$ of (3.69).

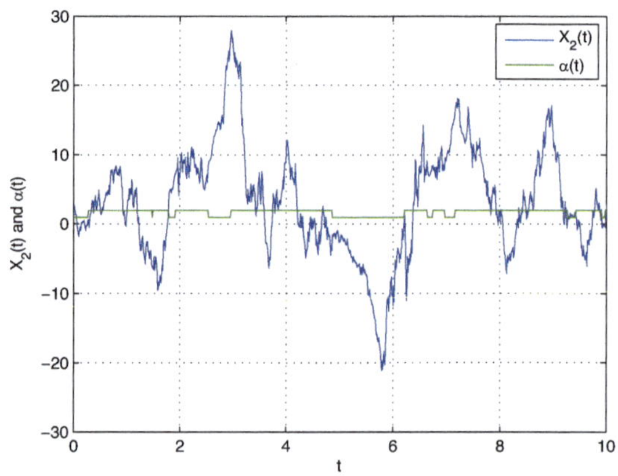

(b) Sample path $(X_2(t), \alpha(t))$ of (3.69).

loss of generality, we may further assume that $(x, i) \in G \times \mathcal{M}$. Define a sequence of stopping times by

$$\varsigma_1 := \inf\{t \geq 0 : X(t) \in \partial G\}, \tag{3.72}$$

and for $n = 1, 2, \ldots$,

$$\begin{aligned} \varsigma_{2n} &:= \inf\{t \geq \varsigma_{2n-1} : X(t) \in \partial D\}, \\ \varsigma_{2n+1} &:= \inf\{t \geq \varsigma_{2n} : X(t) \in \partial G\}. \end{aligned} \tag{3.73}$$

Let us also define $\varsigma_0 := 0$ for convenience. It follows from (3.18) and Theorem 3.2 that $\varsigma_n < \infty$ a.s. $\mathbb{P}_{x,i}$ for $n = 1, 2, \ldots$ Let $H := G \setminus \overline{E}$ and define $u(x, i) := \mathbb{P}_{x,i}\{X(\tau_H) \in \partial E\}$. Note that $u(x, j)|_{x \in \partial E} = 1$ and $u(x, j)|_{x \in \partial G} = 0$ for all $j \in$

$\mathcal{M}$. Therefore, it follows that

$$\begin{aligned} u(x,i) &= \sum_{j=1}^{m_0} \int_{\partial E} \mathbb{P}_{x,i}\left\{(X(\tau_H), \alpha(\tau_H)) \in (dy \times \{j\})\right\} u(y,j) \\ &\quad + \sum_{j=1}^{m_0} \int_{\partial G} \mathbb{P}_{x,i}\left\{(X(\tau_H), \alpha(\tau_H)) \in (dy \times \{j\})\right\} u(y,j) \\ &= \mathbb{E}_{x,i} u(X(\tau_H), \alpha(\tau_H)). \end{aligned}$$

Thus, $u(x,i) \geq 0$ is $\mathcal{L}$-harmonic in $H \times \mathcal{M}$ by Lemma 3.5. Moreover, u is not identically zero since $u(x,i) = 1$ for $(x,i) \in \partial E \times \mathcal{M}$. Therefore, the maximum principle for $\mathcal{L}$-harmonic functions [100] implies that

$$\inf_{(x,i) \in K \times \mathcal{M}} u(x,i) \geq \delta_1 > 0, \tag{3.74}$$

where K is some compact subset of H containing x and ∂D. Define

$$A_0 := \{X(t) \in \partial E, \quad \text{for some } \ t \in [0, \varsigma_1)\}, \tag{3.75}$$

and for $n = 1, 2, \ldots,$

$$A_n := \{X(t) \in \partial E, \ \text{for some } t \in [\varsigma_{2n}, \varsigma_{2n+1})\}. \tag{3.76}$$

Note that the event $A_0^c \subset \{X(\tau_H) = X(\varsigma_1) \in \partial G\}$. Hence, we have from (3.74) that

$$\mathbb{P}_{x,i}(A_0^c) \leq \mathbb{P}_{x,i}(X(\tau_H) \in \partial G) = 1 - u(x,i) \leq 1 - \delta_1.$$

Then it follows from the strong Markov property and (3.74) that

$$\mathbb{P}_{x,i}\left\{\bigcap_{k=0}^{n} A_k^c\right\} \leq (1-\delta_1)^{n+1}. \tag{3.77}$$

Thus, we have

$$\begin{aligned} \mathbb{P}_{x,i}\{\sigma_E = \infty\} &= \mathbb{P}_{x,i}\{X(t) \notin \partial E, \ \text{for any } t \geq 0\} \leq \lim_{n\to\infty} \mathbb{P}_{x,i}\left\{\bigcap_{k=0}^{n} A_k^c\right\} \\ &\leq \lim_{n\to\infty} (1-\delta_1)^{n+1} = 0. \end{aligned}$$

It follows that $\mathbb{P}_{x,i}\{\sigma_E < \infty\} = 1$ as desired. □

Proof of Lemma 3.8 As in Lemma 3.7, it is enough to prove the lemma when $E \cup \partial E \subset D$ and ∂E is sufficiently smooth. Fix any $(x,i) \in E^c \times \mathcal{M}$. Let $G \subset \mathbb{R}^r$ be an open and bounded set with sufficiently smooth boundary such that $D \cup \partial D \subset G$. As in the proof of Lemma 3.7, we may further assume that $(x, i) \in G \times \mathcal{M}$. Define the stopping times $\varsigma_0, \varsigma_1, \varsigma_2, \ldots$ and the events $A_0, A_1, A_2, \ldots$ as in (3.72), (3.73), (3.75), and (3.76) in the proof of Lemma 3.7. It follows from (3.19) and Lemma 3.7 that $\mathbb{P}_{x,i}\{\sigma_E < \infty\} = 1$. Note that for $n = 1, 2, \ldots$,

$$\{\varsigma_{2n} \le \sigma_E < \varsigma_{2n+2}\} = \{\varsigma_{2n} \le \sigma_E < \varsigma_{2n+1}\} \subset \{\varsigma_{2n} \le \sigma_E\} \subset \bigcap_{k=0}^{n-1} A_k^c.$$

Hence, it follows from (3.77) that

$$\mathbb{P}_{x,i}\{\varsigma_{2n} \le \sigma_E\} \le \mathbb{P}_{x,i}\left\{\bigcap_{k=0}^{n-1} A_k^c\right\} \le (1 - \delta_1)^n, \quad n = 1, 2, \ldots.$$

Therefore, we have

$$\begin{aligned}
\mathbb{E}_{x,i}[\sigma_E] &= \sum_{n=0}^{\infty} \mathbb{E}_{x,i}[\sigma_E I_{\{\varsigma_{2n} \le \sigma_E < \varsigma_{2n+1}\}}] \\
&\le \sum_{n=0}^{\infty} \mathbb{E}_{x,i}[\varsigma_{2n+1} I_{\{\varsigma_{2n} \le \sigma_E\}}] \\
&= \sum_{n=0}^{\infty} \mathbb{E}_{x,i}\big[\mathbb{E}_{x,i}[\varsigma_{2n+1} I_{\{\varsigma_{2n} \le \sigma_E\}} | \mathcal{F}_{\varsigma_{2n}}]\big] \\
&= \sum_{n=0}^{\infty} \mathbb{E}_{x,i}\big[I_{\{\varsigma_{2n} \le \sigma_E\}} \mathbb{E}_{x,i}[\varsigma_{2n+1} | \mathcal{F}_{\varsigma_{2n}}]\big] \\
&= \sum_{n=0}^{\infty} \mathbb{E}_{x,i}\big[I_{\{\varsigma_{2n} \le \sigma_E\}} \mathbb{E}_{X(\varsigma_{2n}),\alpha(\varsigma_{2n})}[\varsigma_{2n+1}]\big],
\end{aligned} \tag{3.78}$$

where the last step follows from the strong Markov property. We now prove that $\mathbb{E}_{X(\varsigma_{2n}),\alpha(\varsigma_{2n})}[\varsigma_{2n+1}]$ is uniformly bounded. Indeed, for any $n = 0, 1, \ldots$, we have

$$\mathbb{E}_{X(\varsigma_{2n}),\alpha(\varsigma_{2n})}[\varsigma_{2n+1}] \le \mathbb{E}_{x,i}[\tau_G] \vee \sup_{(y,j) \in \partial D \times \mathcal{M}} \mathbb{E}_{y,j}[\tau_G],$$

where $\tau_G := \inf\{t \ge 0 : X(t) \notin G\}$. Thanks to Theorem 3.2, $\mathbb{E}_{y,j}[\tau_G] < \infty$ for any $(y, j) \in G \times \mathcal{M}$. Since G is a bounded domain, the proof of Theorem 3.2 implies that $\mathbb{E}_{y,j}[\tau_G]$ is uniformly bounded above by a positive constant M for any $(y, j) \in G \times \mathcal{M}$; see (3.12). Plugging this observation into (3.78), we then obtain

$$
\begin{aligned}
\mathbb{E}_{x,i}[\sigma_E] &\le \sum_{n=0}^{\infty} \mathbb{E}_{x,i}\big[I_{\{\varsigma_{2n} \le \sigma_E\}} M\big] \\
&\le M \sum_{n=0}^{\infty} \mathbb{P}_{x,i}\{\varsigma_{2n} \le \sigma_E\} \\
&\le M \sum_{n=0}^{\infty} (1-\delta_1)^n < \infty.
\end{aligned}
\tag{3.79}
$$

This completes the proof of the lemma. □

Proof of Lemma 3.9 Fix any $\ell \in \mathcal{M}$. It suffices to prove (3.21) when $(x, i) \in D \times (\mathcal{M} \setminus \{\ell\})$ because the process $Y(t) = (X(t), \alpha(t))$, starting from $(y, j) \in D^c \times \mathcal{M}$, will reach $D \times \mathcal{M}$ in finite time a.s. $\mathbb{P}_{y,j}$ by (3.20). Choose $\varepsilon > 0$ sufficiently small such that $B \subset \overline{B} \subset B_1 \subset \overline{B}_1 \subset D$, where

$$
B = B(x, \varepsilon) = \{y \in \mathbb{R}^r : |y - x| < \varepsilon\}, \quad \text{and} \quad B_1 = B(x, 2\varepsilon). \tag{3.80}
$$

Redefine $\varsigma_0 := 0$,

$$
\varsigma_1 := \inf\{t \ge 0 : X(t) \in \partial B\}, \tag{3.81}
$$

and for $n = 1, 2, \ldots$,

$$
\begin{aligned}
\varsigma_{2n} &:= \inf\{t \ge \varsigma_{2n-1} : X(t) \in \partial B_1\}, \\
\varsigma_{2n+1} &:= \inf\{t \ge \varsigma_{2n} : X(t) \in \partial B\}.
\end{aligned}
\tag{3.82}
$$

Note that Eq. (3.20), Theorem 3.2, and Lemma 3.7 imply that $\varsigma_n < \infty$ a.s. $\mathbb{P}_{x,i}$. Set

$$
u(x, i) := \mathbb{P}_{x,i}\left\{\sigma_{\overline{B}\times\{\ell\}} < \tau_{B_1}\right\}.
$$

As in the proof of Lemma 3.7, we can verify that $u(x, i)$ is $\mathcal{L}$-harmonic in $B_1 \times \mathcal{M}$. Moreover, u is not identically zero, because $u(x, \ell)|_{x \in \partial B} = 1$. Therefore, the maximum principle [100] implies that

$$
\inf_{(x,i)\in\overline{B}\times\mathcal{M}} u(x, i) \ge \delta_2 > 0. \tag{3.83}
$$

Redefine

$$
A_0 := \{\alpha(t) = \ell, \ \text{for some } t \in [0, \varsigma_2)\}, \tag{3.84}
$$

and for $n = 1, 2, \ldots$,

$$
A_n := \{\alpha(t) = \ell, \ \text{for some } t \in [\varsigma_{2n+1}, \varsigma_{2n+2})\}. \tag{3.85}
$$

Using almost the same argument as in the proof of Lemma 3.7, we obtain that

$$\mathbb{P}_{x,i}(A_0^c) \le 1 - \delta_2, \quad \text{and} \quad \mathbb{P}_{x,i}\left\{\bigcap_{k=0}^{n} A_k^c\right\} \le (1-\delta_2)^{n+1}. \tag{3.86}$$

Thus, we have

$$\begin{aligned}
&\mathbb{P}_{x,i}\{(X(t),\alpha(t)) \notin D \times \{\ell\}, \text{ for any } t \ge 0\} \\
&\le \mathbb{P}_{x,i}\left\{(X(t),\alpha(t)) \notin \overline{B_1} \times \{\ell\}, \text{ for any } t \ge 0\right\} \\
&\le \lim_{n\to\infty} \mathbb{P}_{x,i}\left\{\bigcap_{k=0}^{n} A_k^c\right\} \le \lim_{n\to\infty}(1-\delta_2)^{n+1} = 0.
\end{aligned}$$

As a result,

$$\mathbb{P}_{x,i}\{\sigma_{D\times\{\ell\}} = \infty\} = \mathbb{P}_{x,i}\{(X(t),\alpha(t)) \notin D \times \{\ell\}, \text{ for any } t \ge 0\} = 0,$$

or $\mathbb{P}_{x,i}\{\sigma_{D\times\{\ell\}} < \infty\} = 1$. This completes the proof of the lemma. □

Proof of Lemma 3.10 Fix any $\ell \in \mathcal{M}$. As in Lemma 3.9, it is enough to prove (3.23) when $(x,i) \in D \times (\mathcal{M} \setminus \{\ell\})$. Let the balls B and B_1, stopping times $\varsigma_0, \varsigma_1, \varsigma_2, \ldots$, and events $A_0, A_1, A_2, \ldots$ be as in (3.80)–(3.82), (3.84), and (3.85) in the proof of Lemma 3.9. It follows from Eq. (3.22) and Lemma 3.9 that

$$\mathbb{P}_{x,i}\{\sigma_{D\times\{\ell\}} < \infty\} = 1.$$

Observe that

$$\{\varsigma_{2n} \le \sigma_{D\times\{\ell\}} < \varsigma_{2n+2}\} \subset \{\varsigma_{2n} \le \sigma_{D\times\{\ell\}}\} \subset \bigcap_{k=0}^{n-1} A_k^c.$$

Hence we have from (3.86) that

$$\mathbb{P}_{x,i}\{\varsigma_{2n} \le \sigma_{D\times\{\ell\}} < \varsigma_{2n+2}\} \le \mathbb{P}_{x,i}\{\varsigma_{2n} \le \sigma_{D\times\{\ell\}}\} \le (1-\delta_2)^n.$$

On the other hand, using similar arguments as those in (3.78) in the proof of Lemma 3.8, we have

$$\begin{aligned}
\mathbb{E}_{x,i}[\sigma_{D\times\{\ell\}}] &= \sum_{n=0}^{\infty} \mathbb{E}_{x,i}[\sigma_{D\times\{\ell\}} I_{\{\varsigma_{2n} \le \sigma_{D\times\{\ell\}} < \varsigma_{2n+2}\}}] \\
&\le \sum_{n=0}^{\infty} \mathbb{E}_{x,i}[\varsigma_{2n+2} I_{\{\varsigma_{2n} \le \sigma_{D\times\{\ell\}}\}}] \\
&= \sum_{n=0}^{\infty} \mathbb{E}_{x,i}\left[I_{\{\varsigma_{2n} \le \sigma_{D\times\{\ell\}}\}} \mathbb{E}_{X(\varsigma_{2n}),\alpha(\varsigma_{2n})}[\varsigma_{2n+2}]\right].
\end{aligned} \tag{3.87}$$

Next, we observe

$$\begin{aligned}
&\mathbb{E}_{X(\varsigma_{2n}),\alpha(\varsigma_{2n})}[\varsigma_{2n+2}] \\
&\quad = \mathbb{E}_{X(\varsigma_{2n}),\alpha(\varsigma_{2n})}[\varsigma_{2n+1} + \varsigma_{2n+2} - \varsigma_{2n+1}] \\
&\quad = \mathbb{E}_{X(\varsigma_{2n}),\alpha(\varsigma_{2n})}[\varsigma_{2n+1}] + \mathbb{E}_{X(\varsigma_{2n}),\alpha(\varsigma_{2n})}[\mathbb{E}_{X(\varsigma_{2n+1}),\alpha(\varsigma_{2n+1})}[\varsigma_{2n+2} - \varsigma_{2n+1}]] \\
&\quad \le \sup_{(y,j)\in\partial B_1\times\mathcal{M}} \mathbb{E}_{y,j}[\sigma_B] + \sup_{(y,j)\in\partial B\times\mathcal{M}} \mathbb{E}_{y,j}[\tau_{B_1}],
\end{aligned}$$

where $\sigma_B := \inf\{t \ge 0 : X(t) \in B\}$ and $\tau_{B_1} := \inf\{t \ge 0 : X(t) \notin B_1\}$. Recall that we have observed in the end of the proof of Lemma 3.8 that $\sup_{(y,j)\in\partial B\times\mathcal{M}} \mathbb{E}_{y,j}[\tau_{B_1}]$ is finite. In addition, in view of (3.79), $\sup_{(y,j)\in\partial B_1\times\mathcal{M}} \mathbb{E}_{y,j}[\sigma_B]$ is also finite. Therefore, $\mathbb{E}_{X(\varsigma_{2n}),\alpha(\varsigma_{2n})}[\varsigma_{2n+2}]$ is uniformly bounded above by a positive constant, say K. Using this observation in (3.87), we have

$$\mathbb{E}_{x,i}[\sigma_{D\times\{\ell\}}] \le \sum_{n=0}^{\infty} \mathbb{P}_{x,i}\{\varsigma_{2n} \le \sigma_{D\times\{\ell\}}\}K \le \sum_{n=0}^{\infty}(1-\delta_2)^n K < \infty.$$

The proof of the lemma is thus complete. □

Proof of Proposition 3.19 In view of Theorem 3.18, it is enough to prove the first assertion only. Denote $B = B(x_0, \rho)$ and $U = B \times \{\ell\}$. Define for any $T > 0$ and any $0 < \varepsilon < \rho$ a sequence of stopping times by $\varsigma_0 := 0$, and for $n \ge 1$,

$$\varsigma_n := \inf\{t > \varsigma_{n-1} + T : (X(t), \alpha(t)) \in B(x_0, \varepsilon) \times \{\ell\}\}.$$

(We use the convention that $\inf\emptyset = \infty$.) Then by virtue of the strong Markov property, we can show

$$\sup_{x\in B(x_0,\varepsilon)} \mathbb{P}_{x,\ell}\{\varsigma_n < \infty\} \le \Big(\sup_{x\in B(x_0,\varepsilon)} \mathbb{P}_{x,\ell}\{\varsigma_1 < \infty\}\Big)^n. \tag{3.88}$$

Hence, we have

$$\begin{aligned}
R_0(x_0, \ell, U) &= \sum_{n=0}^{\infty} \mathbb{E}_{x_0,\ell}\left[I_{\{\varsigma_n<\infty\}} \int_{\varsigma_n}^{\varsigma_{n+1}} I_U(X(t), \alpha(t))dt\right] \\
&= \sum_{n=0}^{\infty} \mathbb{E}_{x_0,\ell}\left[I_{\{\varsigma_n<\infty\}} \mathbb{E}_{X(\varsigma_n),\alpha(\varsigma_n)}\left(\int_0^{\varsigma_1} I_U(X(t), \alpha(t))dt\right)\right].
\end{aligned}$$

Note that

$$\int_0^{\varsigma_1} I_U(X(t), \alpha(t))dt = \int_0^{T} I_U(X(t), \alpha(t))dt + \int_T^{\varsigma_1} I_U(X(t), \alpha(t))dt \le T.$$

It follows from (3.88) that

$$R_0(x_0, \ell, U) \leq T \sum_{n=0}^{\infty} \mathbb{P}_{x_0,\ell} \{\varsigma_n < \infty\}$$
$$\leq T \sum_{n=0}^{\infty} \left(\sup_{x \in B(x_0,\ell)} \mathbb{P}_{x,\ell} \{\varsigma_1 < \infty\} \right)^n .$$

But $R_0(x_0, \ell, U) = \infty$; thus,

$$\sup_{x \in B(x_0,\varepsilon)} \mathbb{P}_{x,\ell} \left\{ \sigma^T_{B(x_0,\varepsilon)\times\{\ell\}} < \infty \right\} = \sup_{x \in B(x_0,\ell)} \mathbb{P}_{x,\ell} \{\varsigma_1 < \infty\} = 1,$$

where $\sigma^T_{B(x_0,\varepsilon)\times\{\ell\}} := \varsigma_1 = \inf\{t > T : (X(t), \alpha(t)) \in B(x_0, \varepsilon) \times \{\ell\}\}$. Because $B(x_0, \varepsilon) \subset B$,

$$\mathbb{P}_{x,\ell} \left\{ \sigma^T_{B(x_0,\varepsilon)\times\{\ell\}} < \infty \right\} \leq \mathbb{P}_{x,\ell} \left\{ \sigma^T_{B\times\{\ell\}} < \infty \right\},$$

and $\sup_{x \in B(x_0,\varepsilon)} \mathbb{P}_{x,\ell} \left\{ \sigma^T_{B\times\{\ell\}} < \infty \right\} = 1$. Finally, because $(X(t), \alpha(t))$ is strong Feller, we obtain by letting $\varepsilon \to 0$ that

$$\mathbb{P}_{x_0,\ell} \left\{ \sigma^T_{B\times\{\ell\}} < \infty \right\} = 1$$

for any $T > 0$. Thus, (3.31) follows. This completes the proof of the proposition.

□

Proof of Theorem 3.20 Sufficiency. This is clear by a contradiction argument.

Necessity. Assume the process $(X(t), \alpha(t))$ is transient. Fix any $(x, \alpha) \in \mathbb{R}^r \times \mathcal{M}$. Let $\rho \in \mathbb{R}$ be sufficiently large such that $\rho > 1 \vee |x|$. The process $(X(t), \alpha(t))$ is transient; therefore, it is transient with respect to the "cylinder" $B(0, \rho) \times \mathcal{M}$. Thus, there exists some $(y_0, j_0) \in (\mathbb{R}^r \setminus B(0, \rho)) \times \mathcal{M}$ such that

$$\mathbb{P}_{y_0, j_0} \left\{ \sigma_{B(0,\rho)} < \infty \right\} < 1. \tag{3.89}$$

Assume $|y_0| = r_0 > \rho$. Then by virtue of Lemma 3.4, the function $(y, j) \mapsto \mathbb{P}_{y,j} \left\{ \sigma_{B(0,\rho)} < \infty \right\}$ is $\mathcal{L}$-harmonic. Hence, it follows from the maximum principle for $\mathcal{L}$-harmonic functions [100] and (3.89) that

$$\sup_{|y|=r_0, j \in \mathcal{M}} \mathbb{P}_{y,j} \left\{ \sigma_{B(0,\rho)} < \infty \right\} = \delta < 1. \tag{3.90}$$

Then using the standard argument (see [27, Theorem 3.2]), we can show that

$$\mathbb{P}_{x,\alpha} \left\{ \liminf_{t\to\infty} |X(t)| > \rho - 1 \right\} = 1.$$

Because this is true for all $\rho > 0$,

$$\mathbb{P}_{x,\alpha}\{|X(t)| \to \infty \text{ as } t \to \infty\} = 1,$$

as desired. □

Proof of Proposition 3.21 By virtue of Theorem 3.18, it is enough to show that the point (x_0, ℓ) is transient in the sense that there exists some $\varepsilon_0 > 0$ and a finite time $T_0 > 0$ such that

$$\mathbb{P}_{x_0,\ell}\{(X(t), \alpha(t)) \notin B(x_0, \varepsilon_0) \times \{\ell\}, \text{ for all } t \geq T_0\} = 1. \tag{3.91}$$

Clearly, $R_0(x_0, \ell, U) = \mathbb{E}_{x_0,\ell} \int_0^\infty I_U(X(t), \alpha(t))dt > 0$. Thus, there exists some $t > 0$ such that

$$\mathbb{E}_{x_0,\ell} \int_0^t I_U(X(t), \alpha(t))dt > 0.$$

Because the process $(X(t), \alpha(t))$ is strong Feller by virtue of Theorem 2.26, we conclude that there exists a neighborhood $E \subset D$ of x_0 such that

$$\inf_{y \in E} \mathbb{E}_{y,\ell} \int_0^t I_U(X(t), \alpha(t))dt > 0.$$

Hence, we have

$$\delta := \inf_{x \in E} R_0(x, \ell, U) > 0. \tag{3.92}$$

For each $T > 0$, define

$$\sigma^T_{E\times\{\ell\}} = \inf\{t > T : (X(t), \alpha(t)) \in E \times \{\ell\}\}. \tag{3.93}$$

Then it follows from the strong Markov property that

$$\begin{aligned}
R_0(x_0, \ell, U) &\geq \mathbb{E}_{x_0,\ell} \int_T^\infty I_U(X(t), \alpha(t))dt \\
&\geq \mathbb{E}_{x_0,\ell}\left[I_{\left\{\sigma^T_{E\times\{\ell\}} < \infty\right\}} \int_{\sigma^T_{E\times\{\ell\}}}^\infty I_U(X(t), \alpha(t))dt \right] \\
&= \mathbb{E}_{x_0,\ell}\left[I_{\left\{\sigma^T_{E\times\{\ell\}} < \infty\right\}} R_0(X(\sigma^T_{E\times\{\ell\}}), \alpha(\sigma^T_{E\times\{\ell\}}), U) \right] \\
&\geq \inf_{x\in E} R_0(x, \ell, U) \mathbb{P}_{x_0,\ell}\left\{\sigma^T_{E\times\{\ell\}} < \infty\right\}.
\end{aligned}$$

Hence, we have from (3.92) that

$$\mathbb{P}_{x_0,\ell}\left\{\sigma^T_{E\times\{\ell\}} < \infty\right\} \le \frac{1}{\delta}\mathbb{E}_{x_0,\ell}\int_T^\infty I_U(X(t),\alpha(t))dt.$$

For any $\varepsilon > 0$, in view of the assumption that $R_0(x_0,\ell,U) < \infty$, we can choose some $\widetilde{T}_0 > 0$ such that

$$\mathbb{E}_{x_0,\ell}\int_{\widetilde{T}_0}^\infty I_U(X(t),\alpha(t))dt < \delta\varepsilon$$

and hence

$$\mathbb{P}_{x_0,\ell}\left\{\sigma^{\widetilde{T}_0}_{E\times\{\ell\}} < \infty\right\} < \varepsilon.$$

Note that for any $0 < T_1 < T_2$, $\{\sigma^{T_1}_{E\times\{\ell\}} < \infty\} \supset \{\sigma^{T_2}_{E\times\{\ell\}} < \infty\}$. Thus,

$$\mathbb{P}_{x_0,\ell}\left\{\bigcap_{n=\widetilde{T}_0}^\infty \{\sigma^n_{E\times\{\ell\}} < \infty\}\right\} = \lim_{n\to\infty}\mathbb{P}_{x_0,\ell}\left\{\sigma^T_{E\times\{\ell\}} < \infty\right\} \le \varepsilon.$$

This implies that there exists some $T_0 \ge \widetilde{T}_0$ such that

$$\mathbb{P}_{x_0,\ell}\{(X(t),\alpha(t)) \notin E\times\{\ell\},\ \text{for all } t \ge T_0\} \ge 1-\varepsilon$$

and hence (3.91) follows. □

3.7 Notes

Under general conditions, necessary and sufficient conditions for recurrence, nonrecurrence, and positive recurrence have been studied in this chapter. We refer the reader to Chapter 2 and also Skorohod [262] for related stochastic differential equations involving Poisson measures describing the evolution of the switching processes. In our formulation, the switching process depicts a random environment that has right-continuous sample paths and that cannot be described by a stochastic differential equation of Itô type. Consequently, both continuous dynamics (diffusions) and discrete events (jumps) coexist yielding hybrid dynamic systems, which provide a more realistic formulation for many applications. The discrete events are frequently used to provide more realistic models and to capture random evolutions. For instance, the switching may be used to describe stochastic volatility resulting from market modes and interest rates, as well as other economic factors in modeling financial markets, to enhance the versatility in risk management practice, to better understand ruin probability in insurance, and to carry out dividend optimization tasks.

Regime-switching diffusions have received much attention in recent years. For instance, optimal controls of switching diffusions were studied in [22] using a martingale problem formulation; jump-linear systems were treated in [143]; stability of semi-linear stochastic differential equations with Markovian switching was considered in [11]; ergodic control problems of switching diffusions were studied in [100]; stability of stochastic differential equations with Markovian switching was dealt with in [199, 236, 311]; asymptotic expansions for solutions of integrodifferential equations for transition densities of singularly perturbed switching-diffusion processes were developed in [138]; switching diffusions were used for stock liquidation models in [314]. For some recent applications of hybrid systems in communication networks, air traffic management, control problems, and so on, we refer the reader to [121, 122, 208, 237, 268] and references therein.

In [11, 199, 311, 314], $Q(x) = Q$, a constant matrix. In such cases, $\alpha(\cdot)$ is a continuous-time Markov chain. Moreover, it is assumed that the Markov chain $\alpha(\cdot)$ is independent of Brownian motion. In our formulation, x-dependent $Q(x)$ is considered, and as a result, the transition rates of the discrete event $\alpha(\cdot)$ depend on the continuous dynamic $X(\cdot)$, as depicted in (3.4). Although the pair $(X(\cdot), \alpha(\cdot))$ is a Markov process, for x-dependent $Q(x)$, the process $\alpha(\cdot)$ itself is not a continuous-time Markov chain and it is not independent of the underlying Brownian motion in general. Such formulation enables us to describe complex systems and their inherent uncertainty and randomness in the environment. However, it adds much difficulty in analysis. Our formulation is motivated by the fact that in many applications, the discrete event and continuous dynamic are intertwined. Hence it is useful to relax the independence assumption of the discrete-event process and Brownian motion.

As seen in this chapter, the study of switching diffusions is connected with systems of partial differential equations. The works [1, 78, 93, 97, 104, 178, 187, 276] and references therein provide a systematic treatment of partial differential equations and systems of partial differential equations, which are handy to use.

Chapter 4
Ergodicity

4.1 Introduction

Continuing with the study of basic properties of switching diffusion processes, this chapter is concerned with ergodicity. Many applications in control and optimization require minimizing an expected cost of certain objective functions. Treating average cost per unit time problems, we often wish to "replace" the time-dependent instantaneous measure by a steady-state (or ergodic) measure. Significant effort has been devoted to approximating such expected values by replacing the instantaneous measures with stationary measures when the time horizon is long enough. To justify such a replacement, ergodicity is needed. Thus, we face the following questions: Do the systems possess an ergodic property? Under what conditions do the systems have the desired ergodicity? For diffusion processes, we refer the readers to, for example, [21, 176], among others, for the study of ergodic control problems. In what follows, we study ergodicity and reveal the main features of the ergodic measures. We carry out our study on ergodicity by constructing cycles and using induced discrete-time Markov chains.

We consider the two-component process $Y(t) = (X(t), \alpha(t))$ as in Chapter 3. Let $w(t)$ be a d-dimensional standard Brownian motion, $b(\cdot, \cdot) : \mathbb{R}^r \times \mathcal{M} \mapsto \mathbb{R}^r$, and $\sigma(\cdot, \cdot) : \mathbb{R}^r \times \mathcal{M} \mapsto \mathbb{R}^{r\times d}$ satisfying $\sigma(x, i)\sigma'(x, i) = a(x, i)$. For $t \geq 0$, let $X(t) \in \mathbb{R}^r$ and $\alpha(t) \in \mathcal{M}$ be such that

$$\begin{aligned} &dX(t) = b(X(t), \alpha(t))dt + \sigma(X(t), \alpha(t))dw(t), \\ &X(0) = x, \quad \alpha(0) = \alpha, \end{aligned} \tag{4.1}$$

and

$$\mathbb{P}\{\alpha(t + \Delta) = j|\alpha(t) = i, X(s), \alpha(s), s \leq t\}$$

H.-D. Nguyen et al., *Hybrid Switching Diffusions*, Probability Theory and Stochastic Modelling 63, https://doi.org/10.1007/978-3-031-93303-5_4

$$= \begin{cases} q_{ij}(X(t))\Delta + o(\Delta), & \text{if } i \neq j, \\ 1 + q_{ii}(X(t))\Delta + o(\Delta), & \text{if } i = j, \end{cases} \tag{4.2}$$

where the function $Q(x) = (q_{ij}(x)) : \mathbb{R}^r \to \mathbb{R}^{m_0 \times m_0}$ satisfies the *q-property*; see Section 2.2 for more details.

The rest of the chapter is arranged as follows. Section 4.2 begins with the discussion of ergodicity. The analysis is carried out by using cycles and induced Markov chains in discrete time. Then the desired result is obtained together with the representation of the stationary density. By establishing the irreducibility of the process (X, α), Section 4.3 derives an alternative set of sufficient conditions for the existence and uniqueness of a stationary distribution. Section 4.4 takes up the issue of making a switching diffusion process ergodic by means of feedback controls. Section 4.5 discusses some ramifications, and Section 4.6 obtains asymptotic normality when the continuous component belongs to a compact set. Section 4.7 presents some further remarks.

4.2 Ergodicity

In this section, we study the ergodic properties of the process $Y(t) = (X(t), \alpha(t))$ under the assumption that the process is positive recurrent with respect to some bounded domain $U = E \times \{\ell\}$, where $E \subset \mathbb{R}^r$ and $\ell \in \mathcal{M}$ are fixed throughout this section. We also assume that the boundary ∂E of E is sufficiently smooth. Let the operator $\mathcal{L}$ satisfy Assumption (H3.1). Then it follows from Theorem 3.12 that the process is positive recurrent with respect to any nonempty open set.

Let $D \subset \mathbb{R}^r$ be a bounded open set with sufficiently smooth boundary ∂D such that $E \cup \partial E \subset D$. Let $\varsigma_0 = 0$ and define the stopping times $\varsigma_1, \varsigma_2, \ldots$ inductively as follows: ς_{2n+1} is the first time after ς_{2n} at which the process $Y(t) = (X(t), \alpha(t))$ reaches the set $\partial E \times \{\ell\}$ and ς_{2n+2} is the first time after ς_{2n+1} at which the process reaches the set $\partial D \times \{\ell\}$. Now, we can divide an arbitrary sample path of the process $Y(t) = (X(t), \alpha(t))$ into cycles:

$$[\varsigma_0, \varsigma_2), [\varsigma_2, \varsigma_4), \ldots, [\varsigma_{2n}, \varsigma_{2n+2}), \ldots \tag{4.3}$$

Figure 4.1 presents a demonstration of such cycles when the discrete component $\alpha(\cdot)$ has three states.

The process $Y(t) = (X(t), \alpha(t))$ is positive recurrent with respect to $E \times \{\ell\}$ and hence positive recurrent with respect to $D \times \{\ell\}$ by Theorem 3.12. It follows that all the stopping times $\varsigma_0 < \varsigma_1 < \varsigma_2 < \varsigma_3 < \varsigma_4 < \cdots$ are finite almost surely (a.s.). Because the process $Y(t) = (X(t), \alpha(t))$ is positive recurrent, we may assume without loss of generality that $Y(0) = (X(0), \alpha(0)) = (x, \ell) \in \partial D \times \{\ell\}$. It follows from the strong Markov property of the process $Y(t) = (X(t), \alpha(t))$ that the sequence $\{Y_n\}$ is a Markov chain on $\partial D \times \{\ell\}$, where $Y_n = Y(\varsigma_{2n}) = (X_n, \ell)$,

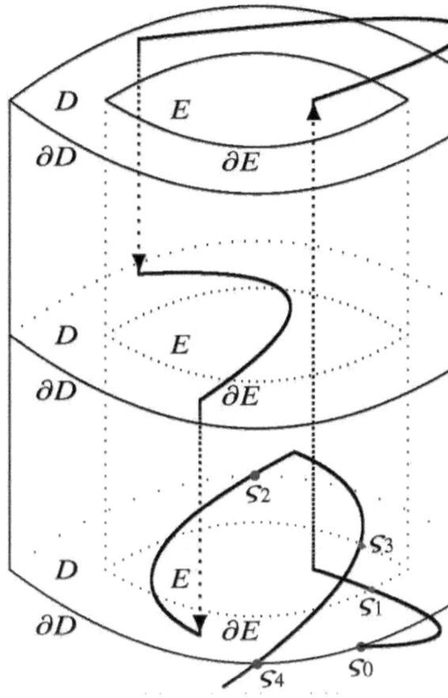

Fig. 4.1 A sample path of the process $Y(t) = (X(t), \alpha(t))$ when $m_0 = 3$

$n = 0, 1, \ldots$ Let $\widetilde{P}(x, A)$ denote the one-step transition probabilities of this Markov chain; that is,

$$\widetilde{P}(x, A) = \mathbb{P}(Y_1 \in (A \times \{\ell\}) \mid Y_0 = (x, \ell))$$

for any $x \in \partial D$ and $A \in \mathcal{B}(\partial D)$, where $\mathcal{B}(\partial D)$ denotes the collection of Borel measurable sets on ∂D. Note that the process $Y(t) = (X(t), \alpha(t))$, starting from (x, ℓ), may jump many times before it reaches the set (A, ℓ); see [262] for more details. Denote by $\widetilde{P}^{(n)}(x, A)$ the n-step transition probability of the Markov chain for any $n \geq 1$. For any Borel measurable function $f : \mathbb{R}^r \mapsto \mathbb{R}$, set

$$\mathbb{E}_x f(X_1) := \mathbb{E}_{x,\ell} f(X_1) = \int_{\partial D} f(y)\widetilde{P}(x, dy). \tag{4.4}$$

Throughout this section, we write $\mathbb{E}_x$ in lieu of $\mathbb{E}_{x,\ell}$ for simplicity. We show that the process $Y(t) = (X(t), \alpha(t))$ possesses a unique stationary distribution. To this end, we need the following lemma.

Lemma 4.1 *The Markov chain $Y_i = (X_i, \ell)$ has a unique stationary distribution $m(\cdot)$ such that the n-step transition matrix $\widetilde{P}^{(n)}(x, A)$ satisfies*

$$\left|\widetilde{P}^{(n)}(x, A) - m(A)\right| < \lambda^n, \quad \textit{for any } A \in \mathcal{B}(\partial D), \tag{4.5}$$

for some constant $0 < \lambda < 1$.

Proof Note that

$$\begin{aligned}\widetilde{P}(x, A) &= \mathbb{P}\{Y_1 \in (A \times \{\ell\}) | Y_0 = (x, \ell)\}\\ &= \int_{\partial E} \mathbb{P}_{x,\ell}\{(X(\varsigma_1), \alpha(\varsigma_1)) \in (dy \times \{\ell\})\} \mathbb{P}_{y,\ell}\{(X(\varsigma_2), \alpha(\varsigma_2)) \in (A \times \{\ell\})\}.\end{aligned}$$

Using the harmonic measure defined and Lemmas 2.2 and 2.3 in Chen and Zhao [47], relating the kernel and surface area (similar to the solution for the diffusion process without switching in the form of double-layer potential given in [156, Remark 3.10]) and the harmonic measure, we can finish the proof of this lemma analogously to that of [156, Lemma 4.6]. The details are omitted. □

Remark 4.2 Note that

$$(X^{s,X(s),\alpha(s)}(t), \alpha^{s,X(s),\alpha(s)}(t)) = (X^{0,X(0),\alpha(0)}(t+s), \alpha^{0,X(0),\alpha(0)}(t+s)), \tag{4.6}$$

where $(X^{0,X(0),\alpha(0)}(\cdot), \alpha^{0,X(0),\alpha(0)}(\cdot))$ denotes the switching diffusion process started from $(X(0), \alpha(0))$ at time 0, and similarly, $(X^{s,X(s),\alpha(s)}(\cdot), \alpha^{s,X(s),\alpha(s)}(\cdot))$ is the switching diffusion that started from $(X(s), \alpha(s))$ at time s. When no confusion arises, we simply write

$$(X(u), \alpha(u)) = (X^{0,X(0),\alpha(0)}(u), \alpha^{0,X(0),\alpha(0)}(u)).$$

Let τ be an $\mathcal{F}_t$ stopping time with $\mathbb{E}_{x,i}\tau < \infty$ and let $f : \mathbb{R}^r \times \mathcal{M} \mapsto \mathbb{R}$ be a Borel measurable function. Then it follows from the Markov property that for any $t \ge 0$

$$\mathbb{E}_{x,i} \int_0^\tau f(X(s+t), \alpha(s+t))ds = \mathbb{E}_{x,i} \int_0^\tau \mathbb{E}_{X(s),\alpha(s)} f(X(t), \alpha(t))ds. \tag{4.7}$$

Now, we can construct the stationary distribution of the process $Y(t) = (X(t), \alpha(t))$ explicitly.

Theorem 4.3 *The positive recurrent process* $Y(t) = (X(t), \alpha(t))$ *has a stationary distribution* $\nu(\cdot, \cdot) \in \mathcal{P}(\mathbb{R}^r \times \mathcal{M})$.

Proof Recall the cycles defined in (4.3). Let $A \in \mathcal{B}(\mathbb{R}^r)$ and $i \in \mathcal{M}$. Denote by $\tau^{A\times\{i\}}$ the time spent by the path of $Y(t) = (X(t), \alpha(t))$ in the set $(A \times \{i\})$ during the first cycle. Set

$$\widehat{\nu}(A, i) := \int_{\partial D} m(dx) \mathbb{E}_x \tau^{A\times\{i\}}, \tag{4.8}$$

where $m(\cdot)$ is the stationary distribution of $Y_i = (X_i, \ell)$, whose existence is guaranteed by Lemma 4.1. It is easy to verify that $\widehat{\nu}(\cdot, \cdot)$ is a positive measure defined on $\mathcal{B}(\mathbb{R}^r) \times \mathcal{M}$. Thus, for any bounded Borel measurable function $g(\cdot) : \mathbb{R}^r \mapsto \mathbb{R}$, it follows from (4.4) and Fubini's theorem that

$$\int_{\partial D} \mathbb{E}_x g(X_1) m(dx) = \int_{\partial D} m(dx) \int_{\partial D} g(y) \widetilde{P}(x, dy) = \int_{\partial D} g(y) m(dy). \tag{4.9}$$

Now, we claim that for any bounded and continuous function $f(\cdot, \cdot)$,

$$\sum_{j=1}^{m_0} \int_{\mathbb{R}^r} f(y, j) \widehat{\nu}(dy, j) = \int_{\partial D} m(dx) \mathbb{E}_x \int_0^{\varsigma_2} f(X(t), \alpha(t)) dt \tag{4.10}$$

holds. In fact, if f is an indicator function, that is, $f(y, j) = I_{A \times \{i\}}(y, j)$ for some $A \in \mathcal{B}(\mathbb{R}^r)$ and $i \in \mathcal{M}$, then from (4.8),

$$\begin{aligned} \sum_{j=1}^{m_0} \int_{\mathbb{R}^r} I_{A \times \{i\}}(y, j) \widehat{\nu}(dy, j) = \widehat{\nu}(A, i) &= \int_{\partial D} m(dx) \mathbb{E}_x \tau^{A \times \{i\}} \\ &= \int_{\partial D} m(dx) \mathbb{E}_x \int_0^{\varsigma_2} I_{A \times \{i\}}(X(t), \alpha(t)) dt. \end{aligned}$$

Similarly, (4.10) holds for f being a simple function of the form $f(y, j) = \sum_{p=1}^{n} c_p I_{U_p}(y, j)$, where U_p is a Borel subset of $\mathbb{R}^r \times \mathcal{M}$. Finally, if f is a bounded and continuous function, (4.10) follows by approximating f by simple functions. It follows from (4.10), (4.6), and (4.7) that

$$\begin{aligned} &\sum_{i=1}^{m_0} \int_{\mathbb{R}^r} \mathbb{E}_{x,i} f(X(t), \alpha(t)) \widehat{\nu}(dx, i) \\ &\quad = \int_{\partial D} m(dx) \mathbb{E}_x \int_0^{\varsigma_2} \mathbb{E}_{X(s), \alpha(s)} f(X(t), \alpha(t)) ds \\ &\quad = \int_{\partial D} m(dx) \mathbb{E}_x \int_0^{\varsigma_2} f(X(t+s), \alpha(t+s)) ds \\ &\quad = \int_{\partial D} m(dx) \mathbb{E}_x \int_0^{\varsigma_2} f(X(u), \alpha(u)) du \\ &\qquad + \int_{\partial D} m(dx) \mathbb{E}_x \int_{\varsigma_2}^{t+\varsigma_2} f(X(u), \alpha(u)) du \\ &\qquad - \int_{\partial D} m(dx) \mathbb{E}_x \int_0^{t} f(X(u), \alpha(u)) du. \end{aligned}$$

Applying (4.9) with

$$g(x) = \mathbb{E}_x \int_{\varsigma_2}^{\varsigma_2 + t} f(X(u), \alpha(u)) du,$$

we obtain

$$\begin{aligned}
&\int_{\partial D} m(dx)\mathbb{E}_x \int_{\varsigma_2}^{\varsigma_2+t} f(X(u), \alpha(u))du \\
&\quad = \int_{\partial D} m(dx)\mathbb{E}_x \mathbb{E}_{X_1,\ell} \int_{\varsigma_2}^{\varsigma_2+t} f(X(u), \alpha(u)) \\
&\quad = \int_{\partial D} m(dx)\mathbb{E}_x \int_0^t f(X(u), \alpha(u))du.
\end{aligned}$$

Note that in the above deduction, we used the strong Markov property. Therefore, the above two equations and (4.10) yield that

$$\sum_{i=1}^{m_0} \int_{\mathbb{R}^r} \mathbb{E}_{x,i} f(X(t), \alpha(t))\widehat{\nu}(dx, i) = \sum_{i=1}^{m_0} \int_{\mathbb{R}^r} f(x, i)\widehat{\nu}(dx, i).$$

Since the above equation holds for any $t \geq 0$ and any $f \in C_b(\mathbb{R}^r \times \mathcal{M})$, it follows that $\widehat{\nu}$ is an invariant measure for the process (X, α). Thus, the normalized measure

$$\nu(A, i) = \frac{\widehat{\nu}(A, i)}{\sum_{j=1}^{m_0} \widehat{\nu}(\mathbb{R}^r, j)}, \quad i \in \mathcal{M}, \tag{4.11}$$

defines the desired stationary distribution. The theorem thus follows. □

Theorem 4.4 *Let $f(\cdot, \cdot) : \mathbb{R}^r \times \mathcal{M} \mapsto \mathbb{R}$ be a Borel measurable function such that*

$$\sum_{i=1}^{m_0} \int_{\mathbb{R}^r} |f(x, i)|\nu(dx, i) < \infty. \tag{4.12}$$

Then

$$\mathbb{P}_{x,i}\left(\lim_{T\to\infty} \frac{1}{T} \int_0^T f(X(t), \alpha(t))dt = \overline{f}\right) = 1, \tag{4.13}$$

for any $(x, i) \in \mathbb{R}^r \times \mathcal{M}$, where

$$\overline{f} = \sum_{i=1}^{m_0} \int_{\mathbb{R}^r} f(x, i)\nu(dx, i). \tag{4.14}$$

Proof We first prove (4.13) if the initial distribution is the stationary distribution of the Markov chain $Y_i = (X_i, \alpha_i)$; that is,

$$\mathbb{P}\{(X(0), \alpha(0)) \in (A \times \{\ell\})\} = m(A) \tag{4.15}$$

for any $A \in \mathcal{B}(\partial D)$. Consider the sequence of random variables

$$\eta_n = \int_{\varsigma_{2n}}^{\varsigma_{2n+2}} f(X(t), \alpha(t))dt, \qquad n = 0, 1, 2, \ldots \tag{4.16}$$

Then it follows from Eq. (4.15) that $\{\eta_n\}$ is a strictly stationary sequence. Also from Eqs. (4.8) and (4.10), we have

$$\mathbb{E}\eta_n = \sum_{i=1}^{m_0} \int_{\mathbb{R}^r} f(x, i)\widehat{\nu}(dx, i), \tag{4.17}$$

for all $n = 0, 1, 2, \ldots$ Meanwhile, Eq. (4.5) implies that the sequence η_n is metrically transitive. Let $\upsilon(T)$ denote the number of cycles completed up to time T. That is,

$$\upsilon(T) := \max\left\{n \in \mathbb{N} : \sum_{k=1}^{n}(\varsigma_{2k} - \varsigma_{2k-2}) \le T\right\}.$$

Then we can decompose $\int_0^T f(X(t), \alpha(t))dt$ into

$$\int_0^T f(X(t), \alpha(t))dt = \sum_{n=0}^{\upsilon(T)} \eta_n + \int_{\varsigma_{2\upsilon(T)}}^{T} f(X(t), \alpha(t))dt, \tag{4.18}$$

with η_n given in (4.16). We may assume without loss of generality that $f(x, i) \ge 0$ (for the general case, we can write $f(x, i)$ as a difference of two nonnegative functions). Then it follows from (4.18) that

$$\sum_{n=0}^{\upsilon(T)} \eta_n \le \int_0^T f(X(t), \alpha(t))dt \le \sum_{n=0}^{\upsilon(T)+1} \eta_n.$$

Inasmuch as the sequence $\{\eta_n\}$ is stationary and metrically transitive, the law of large numbers for such sequences implies that

$$\mathbb{P}\left\{\frac{1}{n}\sum_{k=0}^{n} \eta_k \to \sum_{i=1}^{m_0} \int_{\mathbb{R}^r} f(x, i)\widehat{\nu}(dx, i), \text{ as } n \to \infty\right\} = 1. \tag{4.19}$$

In particular, if $f(x, i) \equiv 1$, then the above equation reduces to

$$\mathbb{P}\left\{\frac{\varsigma_{2n+2}}{n} \to \sum_{i=1}^{m_0} \widehat{\nu}(\mathbb{R}^r, i), \text{ as } n \to \infty\right\} = 1. \tag{4.20}$$

Note that the positive recurrence of the process $Y(t) = (X(t), \alpha(t))$ implies that $\upsilon(T) \to \infty$ as $T \to \infty$. Clearly, $\upsilon(T)/(\upsilon(T)+1) \to 1$ almost surely as $T \to \infty$. Thus, it follows from (4.20) that as $T \to \infty$,

$$\frac{\varsigma_{2\upsilon(T)}}{\varsigma_{2\upsilon(T)+2}} = \frac{\dfrac{\varsigma_{2\upsilon(T)}}{\upsilon(T)}}{\dfrac{\varsigma_{2\upsilon(T)+2}}{\upsilon(T)+1}} \frac{\upsilon(T)}{\upsilon(T)+1} \to 1 \text{ a.s.} \tag{4.21}$$

Meanwhile, since $\varsigma_{2\upsilon(T)} \le T \le \varsigma_{2\upsilon(T)+2}$, we have

$$\frac{\varsigma_{2\upsilon(T)}}{\varsigma_{2\upsilon(T)+2}} \le \frac{\varsigma_{2\upsilon(T)}}{T} \le \frac{\varsigma_{2\upsilon(T)}}{\varsigma_{2\upsilon(T)}} = 1.$$

Therefore, we have from (4.21) that

$$\frac{\varsigma_{2\upsilon(T)}}{T} \to 1 \text{ a.s. as } T \to \infty. \tag{4.22}$$

Moreover, (4.20) implies that

$$\frac{\upsilon(T)}{\varsigma_{2\upsilon(T)}} \to \frac{1}{\sum_{i=1}^{m_0} \widehat{\nu}(\mathbb{R}^r, i)} \text{ a.s. as } T \to \infty. \tag{4.23}$$

Now, using Eqs. (4.19), (4.22), and (4.23), we obtain

$$\begin{aligned}\mathbb{P}\Bigg\{\frac{1}{T}\int_0^T f(X(t),\alpha(t))dt &= \frac{\int_0^T f(X(t),\alpha(t))dt}{\upsilon(T)} \frac{\upsilon(T)}{\varsigma_{2\upsilon(T)}} \frac{\varsigma_{2\upsilon(T)}}{T} \\ &\overset{T\to\infty}{\longrightarrow} \sum_{i=1}^{m_0} \int_{\mathbb{R}^r} f(x,i)\nu(dx,i)\Bigg\} = 1.\end{aligned}$$

Then it follows that (4.13) holds for a.e. $x \in \partial D$ with respect to the measure m (and hence also with respect to the surface measure on ∂D). This proves (4.13) if the initial distribution is (4.15).

Let $(x,i) \in \mathbb{R}^r \times \mathcal{M}$. Because the process $Y(t) = (X(t), \alpha(t))$ is positive recurrent with respect to the domain $D \times \{\ell\}$, we have

$$\begin{aligned}&\mathbb{P}_{x,i}\Big[\lim_{T\to\infty} \frac{1}{T}\int_0^T f(X(t),\alpha(t))dt = \overline{f}\Big] \\ &= \mathbb{P}_{x,i}\Big[\lim_{T\to\infty} \frac{1}{T}\int_{\varsigma_2}^T f(X(t),\alpha(t))dt = \overline{f}\Big].\end{aligned}$$

Using the result from the previous paragraph, we can further write the latter as

$$\begin{aligned}
&\mathbb{P}_{x,i}\Big[\lim_{T\to\infty}\frac{1}{T}\int_{\varsigma_2}^{T} f(X(t),\alpha(t))dt = \overline{f}\Big]\\
&= \int_{\partial D} \mathbb{P}_{x,i}[(X(\varsigma_2),\alpha(\varsigma_2))\in(dy,\ell)]\mathbb{P}_{y,\ell}\Big[\lim_{T\to\infty}\frac{1}{T}\int_0^T f(X(t),\alpha(t))dt = \overline{f}\Big]\\
&= \int_{\partial D} \mathbb{P}_{x,i}\big[(X(\varsigma_2),\alpha(\varsigma_2))\in(dy,\ell)\big] = 1.
\end{aligned}$$

Therefore, (4.13) holds for all $(x,i)\in\mathbb{R}^r\times\mathcal{M}$. This completes the proof of the theorem. □

As a consequence of Theorem 4.4, we obtain the following corollary.

Corollary 4.5 *Let the assumptions of Theorem* 4.4 *be satisfied and let* $u(t,x,i)$ *be the solution of the Cauchy problem:*

$$\begin{cases} \dfrac{\partial u(t,x,i)}{\partial t} = \mathcal{L}u(t,x,i), & t>0,\ (x,i)\in\mathbb{R}^r\times\mathcal{M},\\ u(0,x,i)=f(x,i), & (x,i)\in\mathbb{R}^r\times\mathcal{M}, \end{cases} \tag{4.24}$$

where $f(\cdot,i)\in C_b(\mathbb{R}^r)$ *for each* $i\in\mathcal{M}$. *Then as* $T\to\infty$,

$$\frac{1}{T}\int_0^T u(t,x,i)dt \to \sum_{i=1}^{m_0}\int_{\mathbb{R}^r} f(x,i)\nu(dx,i). \tag{4.25}$$

Proof Thanks to the Feynman–Kac formula, u admits the stochastic representation $u(t,x,i)=\mathbb{E}_{x,i}f(X(t),\alpha(t))$; see, for example, Lemma 3.7 of [323]. Thus, we have

$$\frac{1}{T}\int_0^T u(t,x,i)dt = \mathbb{E}_{x,i}\left(\frac{1}{T}\int_0^T f(X(t),\alpha(t))dt\right). \tag{4.26}$$

Meanwhile, (4.13) implies that

$$\frac{1}{T}\int_0^T f(X(t),\alpha(t))dt \overset{T\to\infty}{\longrightarrow} \sum_{i=1}^{m_0}\int_{\mathbb{R}^r} f(x,i)\nu(dx,i) \text{ a.s.}$$

with respect to the probability $\mathbb{P}_{x,i}$. Then (4.25) follows from the dominated convergence theorem. □

4.3 Irreducibility and Uniqueness of Stationary Distribution

We proved in Theorem 4.3 that a positive recurrent switching diffusion (X, α) possesses a stationary distribution ν. Moreover, Theorem 4.4 implies that the stationary distribution ν is unique. These results, however, rest on the assumption that the process (X, α) is positive recurrent, which in turn requires the existence of a suitable Lyapunov function (see Theorem 3.26). The aim of this section is to present an alternative set of conditions in Proposition 4.10 for ensuring the existence and uniqueness of stationary distributions for switching diffusions. Note that these conditions do not require that the state space $\mathcal{M}$ to be finite and hence $m_0 \leq \infty$ throughout this section. To this end, we first present sufficient conditions for irreducibility, which together with the strong Feller property derived in Theorem 2.26, implies that (X, α) can have at most one stationary distribution.

Let us recall the notion of irreducibility. Denote the transition probability of the process (X, α) by

$$\begin{aligned} P(t, (x, k), B \times \{l\}) :&= P_t I_{B\times\{l\}}(x, k) \\ &= \mathbb{P}\{(X(t), \Lambda(t)) \in B \times \{l\} | (X(0), \Lambda(0)) = (x, k)\}, \end{aligned}$$

for $B \in \mathcal{B}(\mathbb{R}^r)$ and $l \in \mathcal{M}$; recall the semigroup P_t was defined in (2.71). The process (X, α) or the semigroup P_t is said to be *irreducible* if for any $t > 0$ and $(x, k) \in \mathbb{R}^r \times \mathcal{M}$, we have

$$P(t, (x, k), B \times \{l\}) > 0$$

for all $l \in \mathcal{M}$ and all nonempty open set $B \in \mathcal{B}(\mathbb{R}^r)$.

(H4.1) There exist positive constants λ and κ so that for any $x \in \mathbb{R}^r$ and $k \in \mathcal{M}$, we have

$$2\langle x, b(x, k)\rangle^+ + |\sigma(x, k)|^2 \leq \kappa(|x|^2 + 1), \tag{4.27}$$

and

$$\langle \xi, a(x, k)\xi\rangle \geq \lambda|\xi|^2, \qquad \xi \in \mathbb{R}^r. \tag{4.28}$$

(H4.2) There exists a positive constant κ_0 such that

$$0 \leq q_{kl}(x) \leq \kappa_0 l 3^{-l} \tag{4.29}$$

for all $x \in \mathbb{R}^r$ and $k \neq l \in \mathcal{M}$. In addition, for any $k, l \in \mathcal{M}$, there exist $k_0, k_1, ..., k_n \in \mathcal{M}$ with $k_i \neq k_{i+1}, k_0 = k$, and $k_n = l$ such that the set $\{x \in \mathbb{R}^r : q_{k_i k_{i+1}}(x) > 0\}$ has positive Lebesgue measure for all $i = 0, 1, \ldots, n-1$.

The main result of this section is the following theorem.

Theorem 4.6 *Under Assumptions* (H2.2), (H2.3), (H2.4) *(ii),* (H4.1), *and* (H4.2), *the process* (X, α) *is irreducible.*

Remark 4.7 As we mentioned in Remarks 2.3 and 2.31, Assumptions (H2.2), (H2.3), and (H2.4) (ii) place very mild regularity condition for the coefficients of the stochastic differential equations given by (4.1) and (4.2). The uniform ellipticity condition (4.28) is a common assumption in the literature for deriving irreducibility of (jump) diffusions; see, for example, [235, 244] and others. In addition, since we are dealing with a two-component process (X, α), one can expect that the q-matrix $Q(x)$ must satisfy some sort of irreducibility condition so that (X, α) is irreducible; Assumption (H4.2) is therefore in force. Finally, the linear growth condition (4.27) as well as (4.29) is imposed to facilitate our technical analyses.

To prove Theorem 4.6, we first consider the SDE:

$$X^{(k)}(t) = x + \int_0^t b(X^{(k)}(s), k)ds + \int_0^t \sigma(X^{(k)}(s), k)dW(s), \ x \in \mathbb{R}^r, k \in \mathcal{M}. \tag{4.30}$$

Thanks to Assumptions (H2.1), (H2.2), (H2.4) (ii), and (H4.1), for each $k \in \mathbb{S}$ and $x \in \mathbb{R}^r$, (4.30) has a unique non-explosive strong solution $X^{(k)}$ with initial condition x; see Lemma 2.4. Denote the transition probability of the process $X^{(k)}$ by

$$P^{(k)}(t, x, B) := \mathbb{P}\{X^{(k)}(t) \in B | X^{(k)}(0) = x\}, \quad B \in \mathcal{B}(\mathbb{R}^r).$$

The corresponding semigroup $P_t^{(k)}$ is said to be *irreducible* if $P^{(k)}(t, x, B) > 0$ for all $t > 0$, $x \in \mathbb{R}^r$, and all nonempty open set $B \subset \mathbb{R}^r$. We next kill the process $X^{(k)}$ with killing rate $q_k(\cdot)$ and denote the killed process by $\tilde{X}^{(k)}$, that is, we define

$$\tilde{X}^{(k)}(t) = \begin{cases} X^{(k)}(t) & \text{if } t < \tau, \\ \partial & \text{if } t \geq \tau, \end{cases}$$

where $\tau := \inf\{t \geq 0 : \alpha(t) \neq \alpha(0)\}$ and ∂ is a cemetery point added to $\mathbb{R}^r$. Then the semigroup of the killed process $\tilde{X}^{(k)}$ is given by

$$\begin{aligned} \tilde{P}_t^{(k)} f(x) : &= \mathbb{E}_x[f(\tilde{X}^{(k)}(t))] \\ &= \mathbb{E}\bigg[f(X^{(k)}(t)) \exp\bigg\{ \int_0^t q_{kk}(X^{(k)}(s))ds \bigg\} \bigg| X^{(k)}(0) = x \bigg], \end{aligned}$$

where $f \in \mathfrak{B}_b(\mathbb{R}^r)$. We also denote its sub-transition probability by $\tilde{P}^{(k)}(t, x, B) := \tilde{P}_t^{(k)} I_B(x) = \mathbb{E}_x[I_B(\tilde{X}^{(k)}(t))]$, where $B \in \mathcal{B}(\mathbb{R}^r)$.

Lemma 4.8 *Under Assumptions* (H2.2), (H2.4) *(ii), and* (H4.1), *the semigroup* $P_t^{(k)}$ *is strong Feller and irreducible.*

Proof The strong Feller property of $P_t^{(k)}$ follows directly from Theorem 2.26. We now prove that $P_t^{(k)}$ is irreducible. The standard arguments reveal that for every $T > 0$, there exists a constant $K := K(T, x) > 0$ so that

$$\mathbb{E}[|X^{(k)}(t)|^2] \le K \tag{4.31}$$

for all $t \in [0, T]$. Consider the function F given by

$$F(u) := \int_0^{\frac{u}{1+u}} e^{-\int_0^s g(w)dw} ds, \qquad u \in [0, \infty), \tag{4.32}$$

where g is the function given in Assumption (H2.4)(ii). Since $g \ge 0$, we see that

$$0 \le F(u) \le \frac{u}{1+u} \le 1, \tag{4.33}$$

$$0 \le F'(u) = \frac{1}{(1+u)^2} e^{-\int_0^{\frac{u}{1+u}} g(w)dw} \le \frac{1}{(1+u)^2} \le 1, \tag{4.34}$$

and

$$0 \ge F''(u) = -\left[\frac{2}{1+u} + \frac{g(\frac{u}{1+u})}{(1+u)^2}\right] F'(u). \tag{4.35}$$

In addition, for any $x \in \mathbb{R}^r$, we have

$$\nabla F(|x|^2) = 2F'(|x|^2)x, \quad \nabla^2 F(|x|^2) = 4F''(|x|^2)xx^T + 2F'(|x|^2)I.$$

Let $T > 0$, $\rho > 0$ and $x, a \in \mathbb{R}^r$ be arbitrary but fixed. We will show that

$$P^{(k)}(T, x, B(a; \rho)) = \mathbb{P}\{|X^{(k)}(T) - a| < \rho | X^{(k)}(0) = x\} > 0$$

or equivalently $\mathbb{P}\{|X^{(k)}(T) - a| \ge \rho | X^{(k)}(0) = x\} < 1$. Let us choose some $t_0 \in (0, T)$. For any $n \in \mathbb{N}$, we set $X_n^{(k)}(t_0) := X^{(k)}(t_0) I_{\{|X^{(k)}(t_0)| \le n\}}$. Since $\lim_{\rho \to 0} F(\rho) = 0$ and $0 \le F \le 1$, the bounded convergence theorem implies that

$$\lim_{n \to \infty} \mathbb{E}[F(|X_n^{(k)}(t_0) - X^{(k)}(t_0)|^2)] = 0. \tag{4.36}$$

For $t \in [t_0, T]$, define

$$J^n(t) := \frac{T-t}{T-t_0} X_n^{(k)}(t_0) + \frac{t-t_0}{T-t_0} a, \text{ and } h^n(t) := \frac{a - X_n^{(k)}(t_0)}{T-t_0} - b(J^n(t), k).$$

Observe that $J^n(t_0) = X_n^{(k)}(t_0)$ and $J^n(T) = a$. In addition, J^n satisfies the following SDE:

$$J^n(t) = X_n^{(k)}(t_0) + \int_{t_0}^t b(J^n(s), k)ds + \int_{t_0}^t h^n(s)ds, \quad t \in [t_0, T].$$

Consider the SDE:

$$Y(t) = X^{(k)}(t_0) + \int_{t_0}^t [b(Y(s), k) + h^n(s)]ds + \int_{t_0}^t \sigma(Y(s), k)dW(s), \tag{4.37}$$

for $t \in [t_0, T]$. Also denote $\Delta_t := Y(t) - J^n(t)$ for $t \in [t_0, T]$. Note that $\Delta_{t_0} = X^{(k)}(t_0) - X_n^{(k)}(t_0)$ and $\Delta_T = Y(T) - a$. In addition, Δ_t satisfies the SDE:

$$\Delta_t = \Delta_{t_0} + \int_{t_0}^t [b(Y(s), k) - b(J^n(s), k)]ds + \int_{t_0}^t \sigma(Y(s), k)dW(s).$$

Consequently, the generator of the process Δ_t is given by

$$\begin{aligned}\mathbb{L}f(x) :&= \frac{1}{2}\mathrm{tr}\big(\sigma(Y(s), k)\sigma(Y(s), k)^T \nabla^2 f(x)\big) \\ &\quad + \langle b(Y(s), k) - b(J^n(s), k), \nabla f(x)\rangle,\end{aligned}$$

where $f \in C_c^2(\mathbb{R}^r)$. We can now compute

$$\begin{aligned}\mathbb{L}F(|\Delta_s|^2) &= \frac{1}{2}\mathrm{tr}\left(\sigma(Y(s), k)\sigma(Y(s), k)^T \left[4F''(|\Delta_s|^2)\Delta_s\Delta_s^T + 2F'(|\Delta_s|^2)I\right]\right) \\ &\quad + \langle b(Y(s), k) - b(J^n(s), k), 2F'(|\Delta_s|)\Delta_s\rangle \\ &= 2F''(|\Delta_s|^2)|\Delta_s^T\sigma(Y(s), k)|^2 + F'(|\Delta_s|^2)|\sigma(Y(s), k)|^2 \\ &\quad + 2F'(|\Delta_s|^2)\langle b(Y(s), k) - b(J^n(s), k), \Delta_s\rangle \\ &\le F'(|\Delta_s|^2)\left[|\sigma(Y(s), k)|^2 + 2\langle b(Y(s), k) - b(J^n(s), k), \Delta_s\rangle\right],\end{aligned}$$

where the inequality follows from (4.35). On the other hand, when $|Y(s)| \le R$, $|J^n(s)| \le R$, and $|\Delta_s| \le \delta_0$, we can use (2.91) and (4.27) to obtain

$$\begin{aligned}&|\sigma(Y(s), k)|^2 + 2\langle b(Y(s), k) - b(J^n(s), k), \Delta_s\rangle \\ &\qquad \le \kappa(|Y(s)|^2 + 1) + 2\kappa_R|\Delta_s|g(|\Delta_s|) \le K_R + \kappa|Y(s)|^2,\end{aligned}$$

where $K_R = \kappa + 2\kappa_R \max_{r\in[0,\delta_0]}\{rg(r)\} < \infty$. Then it follows that

$$\mathbb{L}F(|\Delta_s|^2) \le F'(|\Delta_s|^2)[K_R + \kappa|Y(s)|^2] \le K_R + \kappa|Y(s)|^2.$$

In view of (4.27) and (4.31), we can use the standard arguments to show that $\mathbb{E}[\sup_{t_0\le s\le T}|Y(s)|^2] \le K$, where K is a positive constant independent of t_0. For any $R > 0$, we define $\tau_R := \inf\{t \ge t_0 : |Y(t)| \vee |J^n(t)| > R\} \wedge T$ and $S_{\delta_0} := \inf\{t \ge t_0 : |Y(t) - J^n(t)| \ge \delta_0\} \wedge T$. Then we can compute

$$\begin{aligned}
\mathbb{E}[F(|\Delta_{T\wedge\tau_R\wedge S_{\delta_0}}|^2)] &= \mathbb{E}[F(|\Delta_{t_0}|^2)] + \mathbb{E}\left[\int_{t_0}^{T\wedge\tau_R\wedge S_{\delta_0}} \mathbb{L}F(|\Delta_s|^2)ds\right] \\
&\le \mathbb{E}[F(|\Delta_{t_0}|^2)] + \mathbb{E}\left[\int_{t_0}^{T\wedge\tau_R\wedge S_{\delta_0}} (K_R + \kappa|Y(s)|^2)ds\right] \\
&\le \mathbb{E}[F(|\Delta_{t_0}|^2)] + K_R(T - t_0) + \mathbb{E}\left[\int_{t_0}^{T} \kappa|Y(s)|^2 ds\right] \\
&\le \mathbb{E}[F(|\Delta_{t_0}|^2)] + (K_R + \kappa K)(T - t_0). \qquad (4.38)
\end{aligned}$$

Next, we show that

$$\mathbb{E}[F(|\Delta_T|^2)] \le \frac{1}{F(\delta_0^2)}\mathbb{E}[F(|\Delta_{T\wedge S_{\delta_0}}|^2)]. \qquad (4.39)$$

To this end, we note that $|\Delta_{T\wedge S_{\delta_0}\wedge\tau_R}| \ge \delta_0$ on the set $\{S_{\delta_0} < T \wedge \tau_R\}$. Since F is increasing, we have $F(\delta_0^2) \le F(|\Delta_{T\wedge S_{\delta_0}}|^2)$. This together with the fact that $0 \le F \le 1$ give the following

$$\begin{aligned}
&\frac{\mathbb{E}[F(|\Delta_{T\wedge\tau_R\wedge S_{\delta_0}}|^2)]}{F(\delta_0^2)} - \mathbb{E}[F(|\Delta_{T\wedge\tau_R}|^2)] \\
&= \frac{\mathbb{E}[F(|\Delta_{T\wedge\tau_R\wedge S_{\delta_0}}|^2)(I_{\{T\wedge\tau_R\le S_{\delta_0}\}} + I_{\{T\wedge\tau_R > S_{\delta_0}\}})]}{F(\delta_0^2)} - \mathbb{E}[F(|\Delta_{T\wedge\tau_R}|^2)] \\
&\ge \frac{\mathbb{E}[F(|\Delta_{T\wedge\tau_R}|^2)I_{\{T\wedge\tau_R\le S_{\delta_0}\}}] + F(\delta_0^2)\mathbb{P}\{T\wedge\tau_R > S_{\delta_0}\}}{F(\delta_0^2)} - \mathbb{E}[F(|\Delta_{T\wedge\tau_R}|^2)] \\
&\ge \mathbb{P}\{T\wedge\tau_R > S_{\delta_0}\} + \mathbb{E}[F(|\Delta_{T\wedge\tau_R}|^2)I_{\{T\wedge\tau_R\le S_{\delta_0}\}}] - \mathbb{E}[F(|\Delta_{T\wedge\tau_R}|^2)] \\
&= \mathbb{P}\{T\wedge\tau_R > S_{\delta_0}\} - \mathbb{E}[F(|\Delta_{T\wedge\tau_R}|^2)I_{\{T\wedge\tau_R > S_{\delta_0}\}}] \\
&\ge \mathbb{P}\{T\wedge\tau_R > S_{\delta_0}\} - \mathbb{E}[1\cdot I_{\{T\wedge\tau_R > S_{\delta_0}\}}] = 0.
\end{aligned}$$

Consequently, we have $\mathbb{E}[F(|\Delta_{T\wedge\tau_R}|^2)] \le \frac{\mathbb{E}[F(|\Delta_{T\wedge\tau_R\wedge S_{\delta_0}}|^2)]}{F(\delta_0^2)}$. Since $\lim_{R\to\infty}\tau_R = T$ a.s. and $0 \le F \le 1$, the bounded convergence theorem gives (4.39).

Recall that Y satisfies the stochastic differential equation (4.37) for $t \in [t_0, T]$. For $t \in [0, t_0]$, we define $Y(t) := X^{(k)}(t)$ and $X^{(k)}(t)$ is the weak solution to (4.30) with initial condition x. Then the process Y satisfies the SDE:

$$Y(t) = x + \int_0^t [b(Y(s), k) + h^n(s)I_{\{s>t_0\}}]ds + \int_0^t \sigma(Y(s), k)dW(s)$$

for $t \in [0, T]$. Next, we set $H(t) := I_{\{t>t_0\}}\sigma^{-1}(Y(t), k)h^n(t)$, and

$$M(t) := \exp\left\{ \int_0^t \langle H(s), dW(s)\rangle - \frac{1}{2}\int_0^t |H(s)|^2 ds \right\}.$$

As argued in [244], it follows from (4.28) that $|H(t)|^2$ is bounded and hence M is a martingale under $\mathbb{P}$ by Novikov's criteria. Moreover, $\mathbb{E}[M(T)] = 1$. Define

$$\mathbb{Q}(B) := \mathbb{E}[M(T)I_B], \quad B \in \mathcal{F}_T \quad \text{and} \quad \tilde{W}(t) := W(t) + \int_0^t H(s)ds, \quad t \in [0, T].$$

It follows from the Girsanov theorem [149, Theorem 3.5.1] that $\mathbb{Q}$ is a probability measure and $\tilde{W}$ is a $\mathbb{Q}$-Brownian motion. Furthermore, under the measure $\mathbb{Q}$, Y solves the SDE:

$$Y(t) = x + \int_0^t b(Y(s), k)ds + \int_0^t \sigma(Y(s), k)d\tilde{W}(s)$$

for $t \in [0, T]$. By the uniqueness in law of the solution to the SDE, we have that the law of $\{X^{(k)}(t) : t \in [0, T]\}$ under $\mathbb{P}$ is the same as the law of $\{Y(t) : t \in [0, T]\}$ under $\mathbb{Q}$. In particular, we have $\mathbb{P}\{|X^{(k)}(T) - a| \geq \rho | X^{(k)}(0) = x\} = \mathbb{Q}\{|Y(T) - a| \geq \rho | Y(0) = x\}$. Since $\mathbb{P}$ and $\mathbb{Q}$ are equivalent, the desired assertion $\mathbb{P}\{|X^{(k)}(T) - a| \geq \rho | X^{(k)}(0) = x\} = \mathbb{Q}\{|Y(T) - a| \geq \rho | Y(0) = x\} < 1$ will follow if we can show that $\mathbb{P}\{|Y(T) - a| \geq \rho | Y(0) = x\} < 1$. To this end, for any $\varepsilon > 0$, we first choose an $R > 0$ sufficiently large so that $\mathbb{P}\{\tau_R < T\} < \varepsilon$. Since F is bounded and increasing, we can use (4.39) and (4.38) to compute

$$\begin{aligned}
&\mathbb{P}\{|Y(T) - a| \geq \rho | Y(0) = x\} \\
&\quad = \mathbb{P}\{|Y(T) - a|^2 \geq \rho^2 | Y(0) = x\} \\
&\quad \leq \mathbb{P}\{F(|Y(T) - a|^2) \geq F(\rho^2) | Y(0) = x\} \\
&\quad \leq \frac{\mathbb{E}[F(|Y(T) - a|^2)]}{F(\rho^2)} = \frac{\mathbb{E}[F(|\Delta_T|^2)]}{F(\rho^2)} \leq \frac{\mathbb{E}[F(|\Delta_{T\wedge S_{\delta_0}}|^2)]}{F(\rho^2)F(\delta_0^2)} \\
&\quad = \frac{\mathbb{E}[F(|\Delta_{T\wedge S_{\delta_0}\wedge\tau_R}|^2)I_{\{\tau_R \geq T\wedge S_{\delta_0}\}}] + \mathbb{E}[F(|\Delta_{T\wedge S_{\delta_0}}|^2)I_{\{\tau_R < T\wedge S_{\delta_0}\}}]}{F(\rho^2)F(\delta_0^2)} \\
&\quad \leq \frac{\mathbb{E}[F(|\Delta_{t_0}|^2)] + (K_R + \kappa K)(T - t_0) + \mathbb{P}\{\tau_R < T\}}{F(\rho^2)F(\delta_0^2)} \\
&\quad \leq \frac{\mathbb{E}[F(|\Delta_{t_0}|^2)] + (K_R + \kappa K)(T - t_0) + \varepsilon}{F(\rho^2)F(\delta_0^2)}.
\end{aligned}$$

Thanks to (4.36), we have $\mathbb{E}[F(|\Delta_{t_0}|^2)] \to 0$ as $n \to \infty$. Hence, we can choose n sufficiently large and t_0 sufficiently close to T to make the last term less than 1 as desired. This concludes the proof. □

We are now ready to present the proof of Theorem 4.6.

Proof of Theorem 4.6 Given $t > 0$ and $(x,k) \in \mathbb{R}^r \times \mathcal{M}$. We want to show that $P(t,(x,k), B \times \{l\}) > 0$ for all $l \in \mathcal{M}$ and all $B \in \mathcal{B}(\mathbb{R}^r)$ with positive Lebesgue measure. Under Assumption (H4.2) and from Lemma 4.8, as in the proof of Theorem 4.8 of [283], we can write

$$\begin{aligned}
&P(t,(x,k), B \times \{l\}) = \delta_{kl}\tilde{P}^{(k)}(t,x,B)\\
&\quad + \sum_{m=1}^{\infty} \int \cdots \int_{0<t_1<\cdots<t_m<t} \sum_{\substack{l_0,l_1,l_2,\ldots,l_m \in \mathcal{M}\\ l_i \neq l_{i+1}, l_0=k, l_m=l}} \int_{\mathbb{R}^r} \cdots \int_{\mathbb{R}^r} \tilde{P}^{(l_0)}(t_1,x,dy_1) q_{l_0 l_1}(y_1)\\
&\qquad \times \tilde{P}^{(l_1)}(t_2-t_1,y_1,dy_2)\cdots q_{l_{m-1}l_m}(y_m)\tilde{P}^{(l_m)}(t-t_m,y_m,B)dt_1dt_2\cdots dt_m,
\end{aligned}$$

where δ_{kl} is the Kronecker symbol. From Assumption (H4.2) (ii), we know that the set $\{y \in \mathbb{R}^r : q_{l_i l_{i+1}}(y) > 0\}$ has positive Lebesgue measure. Then it suffices to show that $\tilde{P}^{(k)}(s,y,B) > 0$ for all $k \in \mathcal{M}$, $s > 0$, and $B \in \mathcal{B}(\mathbb{R}^r)$. We calculate

$$\begin{aligned}
\tilde{P}^{(k)}(s,y,B) &= \mathbb{P}_y\{\tilde{X}^{(k)}(s) \in B\} = \mathbb{E}_y\left[I_B(X^{(k)}(s))\exp\left(\int_0^s q_{kk}(X^{(k)}(r))dr\right)\right]\\
&\geq \mathbb{E}_y\left[I_B(X^{(k)}(s))e^{-M}\right] = e^{-M}P^{(k)}(s,y,B).
\end{aligned}$$

From Lemma 4.8, the semigroup associated with the process $X^{(k)}$ is irreducible and therefore $P^{(k)}(s,y,B) > 0$. This completes the proof. □

Proposition 4.9 *Suppose that Assumptions* (H2.1), (H2.2), *and* (H2.3) *hold. In addition, assume there exist constants $c, d > 0$, a compact subset $C \subset \mathbb{R}^r$, a compact subset $N \subset \mathcal{M}$, a measurable function $f : \mathbb{R}^r \times \mathcal{M} \to [1,\infty)$, and a twice continuously differentiable function $V : \mathbb{R}^r \times \mathcal{M} \to [0,\infty)$ such that*

$$\mathcal{L}V(x,k) \leq -cf(x,k) + dI_{C\times N}(x,k), \quad \forall (x,k) \in \mathbb{R}^r \times \mathcal{M}. \tag{4.40}$$

Then the process (X,α) has an invariant probability measure π.

Proof Since the proof is very similar to those in [280, 285], we shall only give the sketch here. We first use (4.40) and Itô's formula to derive

$$ct \leq V(x,k) + d\int_0^t P(s,(x,k), C \times N)ds, \qquad \forall t > 0,$$

which, in turn, implies that

$$\liminf_{t\longrightarrow\infty} \frac{1}{t} \int_0^t P(s, (x, k), C \times N)ds \geq \frac{c}{d} > 0. \tag{4.41}$$

We have shown in Theorem 2.23 that the process (X, α) is Feller under Assumptions (H2.1), (H2.2), and (H2.3). Then, in view of [88] (see also the proof of Theorem 4.5 of [212]), (4.41) implies that an invariant measure π exists. □

Proposition 4.10 *Suppose that Assumptions* (H2.2), (H2.3), (H2.4) *(ii),* (H4.1), *and* (H4.2) *hold. If there exists a twice continuously differentiable function* $V : \mathbb{R}^r \times \mathcal{M} \to [0, \infty)$ *such that* (4.40) *holds, then the process* (X, α) *has a unique invariant measure.*

Proof Since (4.27) and (4.29) imply the growth conditions for the coefficients of the system (4.1)–(4.2) given in Assumption (H2.1), the existence of an invariant measure follows directly from Proposition 4.9. For the uniqueness, we note that the process (X, α) is strong Feller and irreducible by Theorems 2.26 and 4.6, respectively. Then by Cerrai [39] and also [108], the process (X, α) can admit at most one invariant measure. This completes the proof. □

4.4 Feedback Controls for Weak Stabilization

Many applications in control and optimization require minimizing an expected cost of a certain objective function. For example, one often wishes to minimize an average cost per unit time cost function. The computation could be difficult, complicated, and time consuming. Significant effort has been devoted to approximating such expected values by replacing the instantaneous measure with stationary measures when the time horizon is long enough. To justify such a replacement, ergodicity is needed. As we have proved in Section 4.2, a positive recurrent regime-switching diffusion possesses an ergodic measure. One question of both theoretical and practical interest is: If a switching diffusion is not ergodic, can we design suitable controls so that the controlled regime-switching diffusion becomes positive recurrent and hence ergodic? Because positive recurrence was named as weak stability by Wonham [278], the problem of interest can be restated as: Can we find suitable controls to stabilize (in the weak sense) the regime-switching diffusion system?

In this section, our goal is to design suitable controls so that the resulting regime-switching diffusion is positive recurrent and hence ergodic. Consider the regime-switching diffusion (4.1) with the transition of the jump process specified by (4.2) and initial condition

$$X(0) = x, \quad \alpha(0) = \alpha.$$

Consider also its controlled system

$$dX(t) = b(X(t), \alpha(t))dt + B(\alpha(t))u(X(t), \alpha(t))dt + \sigma(X(t), \alpha(t))dw(t), \tag{4.42}$$

where $B(i) \in \mathbb{R}^{r\times r}, i \in \mathcal{M}$ are constant matrices and $u(\cdot, \cdot) : \mathbb{R}^r \times \mathcal{M} \mapsto \mathbb{R}^r$ denotes the feedback control to be identified.

It is well-known that the regime-switching diffusion (4.1) can be regarded as the following m_0 sub-diffusions:

$$dX(t) = b(X(t), i)dt + \sigma(X(t), i)dw(t), \quad i \in \mathcal{M}, \tag{4.43}$$

coupled by the discrete-event component $\alpha(t)$ according to the transition laws specified in (4.2). Often the system is observable only when it operates in some modes but not all. Accordingly, it is natural to decompose the discrete state space $\mathcal{M}$ into two disjoint subsets $\mathcal{M}_1$ and $\mathcal{M}_2$, namely, $\mathcal{M} = \mathcal{M}_1 \cup \mathcal{M}_2$, where for each mode $i \in \mathcal{M}_2$, the process (4.43) cannot be stabilized by feedback control, but it can be stabilized for each $i \in \mathcal{M}_1$. Thus, we consider feedback control of the following form:

$$u(X(t), \alpha(t)) = -L(\alpha(t))X(t),$$

where for each $i \in \mathcal{M}$, $L(i) \in \mathbb{R}^{r\times r}$ is a constant matrix. Moreover, if $i \in \mathcal{M}_2$, $L(i) = 0$. Thus, (4.42) can be rewritten as

$$dX(t) = [b(X(t), \alpha(t)) - B(\alpha(t))L(\alpha(t))X(t)]\, dt + \sigma(X(t), \alpha(t))dw(t). \tag{4.44}$$

Suppose Assumptions (H3.1) and (H3.2) hold. In other words, we assume that the nonlinear system is locally linearizable in a neighborhood of ∞. Then we have the following theorem by applying Theorem 3.26 to (4.44).

Theorem 4.11 *If for each $i \in \mathcal{M}_1$, there exists a constant matrix $L(i) \in \mathbb{R}^{r\times r}$ such that*

$$\begin{aligned}\sum_{i\in\mathcal{M}_1} \pi_i \lambda_{\max}\Big(b(i) - B(i)L(i) + b'(i) - L'(i)B'(i) + \sum_{j=1}^{d} \sigma_j(i)\sigma_j'(i)\Big) \\ + \sum_{i\in\mathcal{M}_2} \pi_i \lambda_{\max}\Big(b(i) + b'(i) + \sum_{j=1}^{d} \sigma_j(i)\sigma_j'(i)\Big) < 0,\end{aligned} \tag{4.45}$$

then the resulting controlled regime-switching system (4.42) *is weakly stabilizable. That is, the controlled regime-switching diffusion is positive recurrent.*

Theorem 4.11 ensures that under simple conditions, there are many choices for the matrices $L(i)$ with $i \in \mathcal{M}_1$ in order to make the regime-switching diffusion (4.1) be positive recurrent. Take, for example,

$$L(i) = \theta_i I, \quad i \in \mathcal{M}_1,$$

where I is the $r \times r$ identity matrix and θ_i are nonnegative constants to be determined. Hence, it follows that for $i \in \mathcal{M}_1$, we have

$$\begin{aligned}
&\lambda_{\max}\left(b(i) - B(i)L(i) + b'(i) - L'(i)B'(i) + \sum_{j=1}^{d} \sigma_j(i)\sigma_j'(i)\right) \\
&\le \lambda_{\max}\left(b(i) + b'(i) + \sum_{j=1}^{d} \sigma_j(i)\sigma_j'(i)\right) + \lambda_{\max}(-B(i)L(i) - L'(i)B'(i)) \\
&= \lambda_{\max}\left(b(i) + b'(i) + \sum_{j=1}^{d} \sigma_j(i)\sigma_j'(i)\right) - \theta_i \lambda_{\min}(B(i) + B'(i)).
\end{aligned} \tag{4.46}$$

Assume that for some $\iota \in \mathcal{M}_1$ such that the symmetric matrix $B(\iota)+B'(\iota)$ is positive definite. Hence, $\lambda_{\min}(B(\iota) + B'(\iota)) > 0$. Then if $\theta_\iota > 0$ is sufficiently large, and $\theta_i = 0$ for $i \neq \iota$, we have from (4.46) that the left-hand side of (4.45) is less than 0. Thus, the resulting controlled regime-switching diffusion (4.42) is positive recurrent. We summarize the discussion in the following theorem.

Theorem 4.12 *If for some $\iota \in \mathcal{M}_1$, the symmetric matrix $B(\iota) + B'(\iota)$ is positive definite, then there exists a feedback control $u(\cdot, \cdot)$ such that the controlled regime-switching diffusion* (4.42) *is positive recurrent.*

Example 4.13 In this example, we apply Theorems 4.11 and 4.12 to stabilize (in the weak sense) a switching diffusion. Consider a two-dimensional (in the continuous component) regime-switching diffusion:

$$\begin{cases} dX(t) = b(X(t), \alpha(t))dt + \sigma(X(t), \alpha(t))dw(t) \\ \mathbb{P}\{\alpha(t+\Delta) = j | \alpha(t) = i, X(s), \alpha(s), s \le t\} = q_{ij}(X(t))\Delta + o(\Delta), \end{cases} \tag{4.47}$$

and its controlled system

$$\begin{cases} dX(t) = b(X(t), \alpha(t)) + B(\alpha(t))u(X(t), \alpha(t))dt + \sigma(X(t), \alpha(t))dw(t) \\ \mathbb{P}\{\alpha(t+\Delta) = j | \alpha(t) = i, X(s), \alpha(s), s \le t\} = q_{ij}(X(t))\Delta + o(\Delta), \end{cases} \tag{4.48}$$

respectively. Suppose that

$$\begin{aligned}
b(x_1, x_2, 1) &= \begin{pmatrix} x_1 + x_2 \\ 2x_2 \end{pmatrix}, \quad \sigma(x_1, x_2, 1) = \begin{pmatrix} \sqrt{3 + 2x_2^2} & 5 \\ 0 & \sqrt{4 + x_2^2} - \sin(x_1 x_2) \end{pmatrix}, \\
b(x_1, x_2, 2) &= \begin{pmatrix} 2x_1 + x_2 \\ -x_1 + 3x_2 \end{pmatrix}, \quad \sigma(x_1, x_2, 2) = \begin{pmatrix} -2 + \sqrt{1 + x_2^2} & 0 \\ -3 & \sqrt{2 + x_1^2} \end{pmatrix},
\end{aligned}$$

and

$$B(1) = \begin{pmatrix} -2 & 3 \\ 1 & -1 \end{pmatrix}, \quad B(2) = \begin{pmatrix} 2 & -3 \\ 1 & 3 \end{pmatrix}.$$

The generator of $\alpha(\cdot)$ is given by

$$Q(x) = \begin{pmatrix} -2 - \dfrac{\sin x_1 \cos(x_1^2) + \sin(x_2^2)}{1 + \sqrt{x_1^2 + x_2^2}} & 2 + \dfrac{\sin x_1 \cos(x_1^2) + \sin(x_2^2)}{1 + \sqrt{x_1^2 + x_2^2}} \\ 1 - \dfrac{2x_2 - \cos x_1}{3 + x_1^2 + x_2^2} & -1 + \dfrac{2x_2 - \cos x_1}{3 + x_1^2 + x_2^2} \end{pmatrix}.$$

Thus, associated with the regime-switching diffusion (4.47), there are two diffusions:

$$dX(t) = b(X(t), 1)dt + \sigma(X(t), 1)dw(t), \tag{4.49}$$

and

$$dX(t) = b(X(t), 2)dt + \sigma(X(t), 2)dw(t) \tag{4.50}$$

switching back and forth from one to another according to the movement of the discrete component $\alpha(t)$, where $w(t) = (w_1(t), w_2(t))'$ is a two-dimensional standard Brownian motion. Assume that the system is observable when the discrete component $\alpha(\cdot)$ is in state 2. Detailed calculations using the methods in [156, Section 3.8] or [27] allow us to verify that both (4.49) and (4.50) are transient diffusions; see also the phase portraits in Fig. 4.2(a) and (b).

Next, we use Theorem 3.24 to verify that the switching diffusion (4.47) is transient. To this end, we compute the matrices $b(i)$, $\sigma_j(i)$ for $i, j = 1, 2$ and $\widehat{Q}$ as in Assumption (H3.2):

$$b(1) = \begin{pmatrix} 1 & 1 \\ 0 & 2 \end{pmatrix}, \quad \sigma_1(1) = \begin{pmatrix} 0 & \sqrt{2} \\ 0 & 0 \end{pmatrix}, \quad \sigma_2(1) = \begin{pmatrix} 0 & 0 \\ 0 & 1 \end{pmatrix},$$

$$b(2) = \begin{pmatrix} 2 & 1 \\ -1 & 3 \end{pmatrix}, \quad \sigma_1(2) = \begin{pmatrix} 0 & 1 \\ 0 & 0 \end{pmatrix}, \quad \sigma_2(2) = \begin{pmatrix} 0 & 0 \\ 1 & 0 \end{pmatrix},$$

and

$$\widehat{Q} = \begin{pmatrix} -2 & 2 \\ 1 & -1 \end{pmatrix}.$$

Thus, the stationary distribution is $\pi = (1/3,\ 2/3)$ and

Fig. 4.2 Phase portraits of transient diffusions in (4.49) and (4.50) with initial condition x, where $x = [-5, 3]'$. (**a**) Phase portrait of (4.49). (**b**) Phase portrait of (4.50)

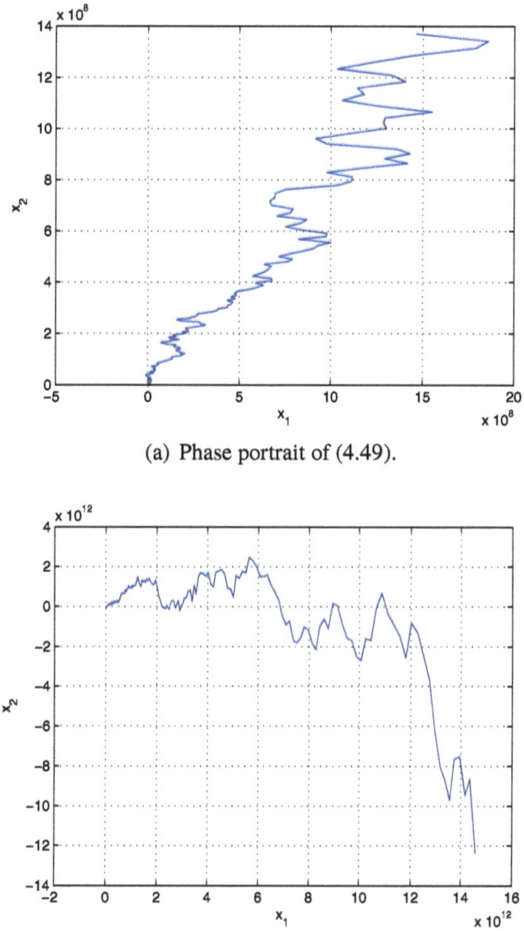

(a) Phase portrait of (4.49).

(b) Phase portrait of (4.50).

$$\lambda_{\min}(b(1) + b'(1) + \sigma_1(1)\sigma_1'(1) + \sigma_2(1)\sigma_2'(1)) - \frac{1}{2}\sum_{j=1}^{2}[\rho(\sigma_j(1) + \sigma_j'(1))]^2 = 0.382,$$

$$\lambda_{\min}(b(2) + b'(2) + \sigma_1(2)\sigma_1'(2) + \sigma_2(2)\sigma_2'(2)) - \frac{1}{2}\sum_{j=1}^{2}[\rho(\sigma_j(2) + \sigma_j'(2))]^2 = 4.$$

Hence, Theorem 3.24 implies that the switching diffusion (4.47) is transient. Figure 4.3(a) confirms our analysis.

By our assumption, the switching diffusion is observable when the discrete component $\alpha(\cdot)$ is in state 2. Note that

$$B(2) + B'(2) = \begin{pmatrix} 4 & -2 \\ -2 & 6 \end{pmatrix}$$

is symmetric and positive definite. Therefore, it follows from Theorem 4.12 that (4.48) is stabilizable in the weak sense. In fact, as we discussed in Section 4.4, if we take $\theta_1 = 0$ and $\theta_2 = -4$, then direct computation leads to

$$\begin{aligned}&\pi_2\lambda_{\max}\big(b(2) + b'(2) - 4B(2) - 4B'(2) + \sigma_1(2)\sigma_1'(2) + \sigma_2(2)\sigma_2'(2)\big)\\ &+\pi_1\lambda_{\max}\big(b(1) + b'(1) + \sigma_1(1)\sigma_1'(1) + \sigma_2(1)\sigma_2'(1)\big) = -1.7647 < 0.\end{aligned}$$

Therefore, the controlled switching diffusion (4.48) is positive recurrent under the feedback control law

$$u(x, 1) = 0 \text{ and } u(x, 2) = -4x. \tag{4.51}$$

Figure 4.3(b) demonstrates the phase portrait of (4.48) under the feedback control (4.51). We also demonstrate the componentwise sample path of (4.48) under the feedback control (4.51) in Fig. 4.4(a) and (b).

4.5 Ramifications

4.5.1 Remark on a Tightness Result

Under positive recurrence, we may obtain tightness (or boundedness in the sense of in probability) of the underlying process. Suppose that $(X(t), \alpha(t))$ is positive recurrent. We can use the result in Chapter 3 about path excursion (3.45) to prove that for any compact set B, the set

$$\bigcup_{x\in B}\{(X(t), \alpha(t)) : t \geq 0,\ X(0) = x,\ \alpha(0) = \alpha\}$$

is tight (or bounded in probability). The idea is along the line of the diffusion counterpart; see [175].

Suppose that for each $i \in \mathcal{M}$, there is a Lyapunov function $V(\cdot, i)$ such that $\min_x V(x, i) = 0$, $V(x, i) \to \infty$ as $|x| \to \infty$. Let $a_1 > a_0 > 0$ and $X(0) = x$ and $\alpha(0) = \alpha$. Using the argument in Chapter 3, because recurrence is independent of the chosen set, we can work with a fixed $\ell \in \mathcal{M}$ and consider the sets

$$\begin{aligned}B_{a_0} &= \{x \in \mathbb{R}^r : V(x, \ell) \leq a_0\},\\ B_{a_1} &= \{x \in \mathbb{R}^r : V(x, \ell) \leq a_1\}.\end{aligned}$$

Fig. 4.3 Phase portraits of switching diffusion (4.47) and its controlled system (4.48) under feedback control law (4.51) with initial condition (x, α), where $x = [-5, 3]'$ and $\alpha = 2$. (**a**) Phase portrait of (4.47). (**b**) Phase portrait of (4.48)

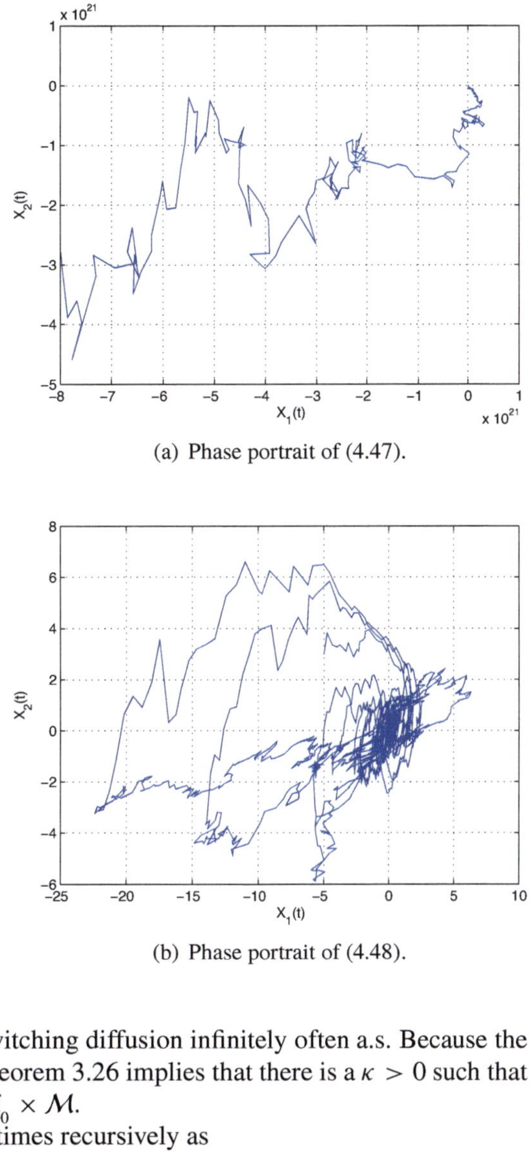

(a) Phase portrait of (4.47).

(b) Phase portrait of (4.48).

Then B_{a_0} will be visited by the switching diffusion infinitely often a.s. Because the switching is positive recurrent, Theorem 3.26 implies that there is a $\kappa > 0$ such that $\mathcal{L}V(x, i) \leq -\kappa$ for all $(x, i) \in B_{a_0}^c \times \mathcal{M}$.

Define a sequence of stopping times recursively as

$$\begin{aligned}
\tau_1 &= \min\{t : X(t) \in B_{a_0}\}, \\
\tau_{2n} &= \min\{t > \tau_{2n-1} : X(t) \in \partial B_{a_1}\}, \\
\tau_{2n+1} &= \min\{t > \tau_{2n} : X(t) \in \partial B_{a_0}\}.
\end{aligned}$$

Using the argument as in (3.45), we can obtain

Fig. 4.4 Componentwise sample path of (4.48) under feedback control law (4.51) with initial condition (x, α), where $x = [-5, 3]'$ and $\alpha = 2$. (**a**) Sample path $(X_1(t), \alpha(t))$ of (4.48). (**b**) Sample path $(X_2(t), \alpha(t))$ of (4.48)

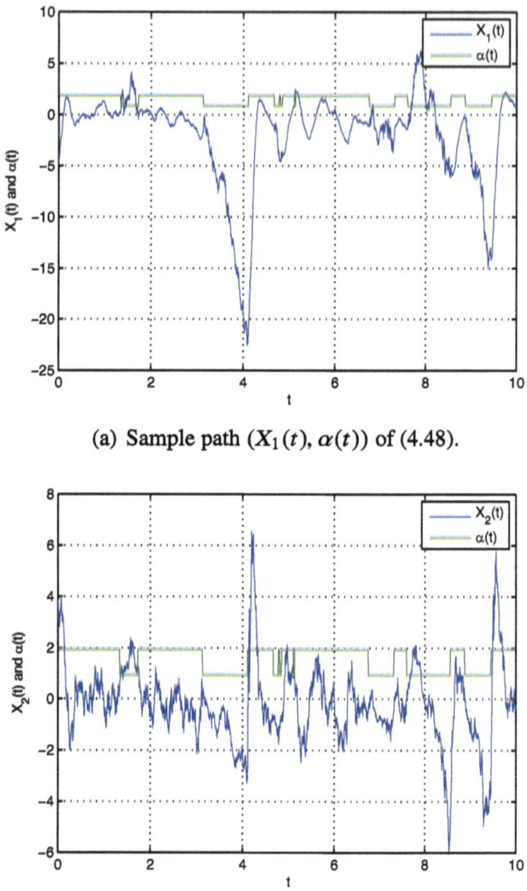

(a) Sample path $(X_1(t), \alpha(t))$ of (4.48).

(b) Sample path $(X_2(t), \alpha(t))$ of (4.48).

$$\mathbb{E}_{\tau_{2n}}(\tau_{2n+1} - \tau_{2n}) I_{\{\tau_{2n}<\infty\}} \leq \frac{a_1}{\kappa} I_{\{\tau_{2n}<\infty\}},$$
$$I_{\{\tau_n<\infty\}} \mathbb{P}_{\tau_n} \left(\sup_{\tau_n \leq t \leq \tau_{n+1}} V(X(t), \alpha(t)) \geq a \right) \to 0 \text{ as } a \to \infty,$$

uniformly in $n \geq 1$. Here and hereafter, $\mathbb{E}_\tau[\cdot] := \mathbb{E}[\cdot|\mathcal{F}_\tau]$ and $\mathbb{P}_\tau(\cdot) := \mathbb{P}(\cdot|\mathcal{F}_\tau)$ for a stopping time τ. Then there exist $\Delta_i > 0$ for $i = 0, 1$ satisfying $I_{\{\tau_{2n-1}<\infty\}} \mathbb{P}_{\tau_{2n-1}}(\tau_{2n} - \tau_{2n-1} \geq \Delta_0) \geq \Delta_1 I_{\{\tau_{2n-1}<\infty\}}$. Working with the estimates up to now and using the argument as that in [175, pp. 147–148], we can show that for any $\Delta > 0$ and compact set B, there is an a_Δ such that $\mathbb{P}_{x,\alpha}(V(X(t), \alpha(t)) \geq a_\Delta) \leq \Delta$ for any $(x, \alpha) \in B \times \mathcal{M}$ and $t \geq 0$. The condition $V(x, i) \to \infty$ as $|x| \to \infty$ then implies the desired tightness claim.

4.5.2 Occupation Measures

To illustrate another utility of Theorem 4.4, take $f(x,i) = I_{B\times J}(x,i)$, as the indicator function of the set $B \times J$, where $B \subset \mathbb{R}^r$ and $J \subset \mathcal{M}$. Then Theorem 4.4 becomes a result regarding the occupation measure. In fact, we have

$$\frac{1}{T}\int_0^T I_{B\times J}(X(t),\alpha(t))dt \to \sum_{i\in J}\int_B \nu(dx,i) = \nu(B\times J) \text{ a.s. as } T\to\infty.$$

4.5.3 Stochastic Approximation

Consider a parameter optimization problem. We wish to find θ_*, a vector-valued parameter so that the cost function

$$J(\theta) = \lim_{T\to\infty}\mathbb{E}\frac{1}{T}\int_0^T \widehat{J}(\theta, Y(t))dt$$

is minimized, where $Y(t)$ is a positive recurrent switching diffusion as considered in this chapter and for each θ, $\widehat{J}(\theta,\cdot,\cdot)$ satisfies the conditions of Theorem 4.4. For simplicity, we assume that the gradient of $\widehat{J}(\cdot,x,i)$ with respect to θ is available for each x and each $i \in \mathcal{M}$. Then we consider a constant stepsize recursive algorithm:

$$\theta_{n+1} = \theta_n - \varepsilon\frac{1}{T}\int_{nT}^{nT+T} \nabla\widehat{J}(\theta_n, Y(t))dt,$$

or a decreasing stepsize algorithm:

$$\theta_{n+1} = \theta_n - \varepsilon_n\frac{1}{T}\int_{nT}^{nT+T} \nabla\widehat{J}(\theta_n, Y(t))dt,$$

where $\varepsilon > 0$ and $\varepsilon_n \to 0$ as $n \to \infty$ and $\sum_n \varepsilon_n = \infty$. Modifications and variants are possible. For example, additional measurement noise may be included, and the gradient of $\widehat{J}(\cdot)$ may be replaced by its gradient estimates. The motivation for such algorithms stems from optimization of average cost per unit time problems arising from parameter estimations in switching systems of SDEs, manufacturing systems, and queueing networks; see related work in [177, Chapter 9] and [300]. The ergodicity of the switching diffusion is crucial in the study of the asymptotic behavior of the algorithms.

4.6 Asymptotic Distribution

Based on ergodicity of hybrid diffusions, this section is concerned with a centered and scaled sequence. It begins with a lemma on the convergence of the transition density to the invariant density, followed by the verification of a couple of lemmas. Then a result concerning the correlation function is presented. Finally, an asymptotic distribution result is established. Here, we work with a compact state space for the x component. To be more specific, we assume that the state space is a torus $\mathbb{S}$ in $\mathbb{R}^r$. More general compact manifolds can also be considered with essentially the same argument.

The following lemma can be obtained in the same spirit as that of [138]. We omit the detailed argument and refer the reader to the reference.

Lemma 4.14 *Denote by $p(y, j; t, x, i)$ the transition density of the process $(X(t), \alpha(t))$ with*

$$\begin{aligned}\mathbb{P}_{y,j}(X(t) \in S, \alpha(t) = i) &= \mathbb{P}(X(t) \in S, \alpha(t) = i \big| X(0) = y, \alpha(0) = j)) \\ &= \int_S p(y, j; t, x, i) dx,\end{aligned}$$

where $S \in \mathcal{B}(\mathbb{S})$. Then $p(y, j; t, x, i)$ converges exponentially fast to $\mu(\cdot, \cdot)$. That is, for some $C > 0$ and $\kappa_3 > 0$,

$$|p(y, j; t, x, i) - \mu(x, i)| \le C \exp(-\kappa_3 t) \ \text{for any} \ (y, j),\ (x, i) \in \mathbb{S} \times \mathcal{M}, \tag{4.52}$$

where $\mu(x, i)$ is the invariant density as in [138], which is the solution of the system of Kolmogorov–Fokker–Planck equations:

$$\begin{aligned}&\mathcal{L}^* g(x, i) = 0, \ \text{for each} \ i \in \mathcal{M}, \\ &\sum_{i=1}^{m_0} \int_{\mathbb{S}} g(x, i) dx = 1,\end{aligned} \tag{4.53}$$

where $\mathcal{L}^*$ is the adjoint operator of $\mathcal{L}$ defined in (3.1).

Lemma 4.15 *Let the assumptions of Lemma* 4.14 *be satisfied. Suppose that $f(\cdot, \cdot) : \mathbb{S} \times \mathcal{M} \mapsto \mathbb{R}$ is a Borel measurable function. Then for any $0 \le s \le t$, we have*

$$\left|\mathbb{E}_{X(s),\alpha(s)} f(X(t), \alpha(t)) - \mathbb{E} f(X(t), \alpha(t))\right| \le C \exp(-\kappa_3 (t - s)), \tag{4.54}$$

for some positive constant C and κ_3 given by (4.52)*, where $\mathbb{E} f(X(t), \alpha(t))$ is the expectation with respect to the stationary measure:*

$$\mathbb{E} f(X(t), \alpha(t)) = \sum_{i=1}^{m_0} \int_{\mathbb{S}} f(x, i) \mu(x, i) dx \overset{\text{def}}{=} f_{av}. \tag{4.55}$$

Proof Note that

$$f(X(t), \alpha(t)) = \sum_{i=1}^{m_0} f(X(t), i) I_{\{\alpha(t)=i\}}.$$

Inasmuch as we work with $\mathbb{S}$, a torus in $\mathbb{R}^r$,

$$\sum_{i=1}^{m_0} \int_{\mathbb{S}} |f(x, i)| dx < \infty.$$

Moreover,

$$\begin{aligned}
\mathbb{E}_{X(s),\alpha(s)} f(X(t), \alpha(t)) &= \sum_{i=1}^{m_0} \mathbb{E}_{X(s),\alpha(s)} f(X(t), i) I_{\{\alpha(t)=i\}} \\
&= \sum_{i=1}^{m_0} \int_{\mathbb{S}} f(x, i) \mathbb{P}_{X(s),\alpha(s)}(X(t) \in dx, \alpha(t) = i) \\
&= \sum_{i=1}^{m_0} \int_{\mathbb{S}} f(x, i) p(X(s), \alpha(s); t - s, x, i) dx.
\end{aligned}$$

Thus, in view of (4.55), we have from (4.52) that

$$\begin{aligned}
&\left|\mathbb{E}_{X(s),\alpha(s)} f(X(t), \alpha(t)) - \mathbb{E} f(X(t), \alpha(t))\right| \\
&= \left|\sum_{i=1}^{m_0} \int_{\mathbb{S}} f(x, i)[p(X(s), \alpha(s), t - s, x, i) - \mu(x, i)] dx\right| \\
&\le C \exp(-\kappa_3(t - s)).
\end{aligned}$$

Hence, the lemma follows. □

Using Lemma 4.15, we obtain another lemma.

Lemma 4.16 *Assume the conditions of Lemma* 4.15. *Then the following assertion holds:*

$$|\mathbb{E}[f(X(t), \alpha(t)) - f_{av}][f(X(s), \alpha(s)) - f_{av}]| \le C \exp(-\kappa_3 |t - s|) \tag{4.56}$$

for some positive constant C.

Proof Assume without loss of generality that $s \le t$. Note that from (4.55), $\mathbb{E} f(X(u), \alpha(u)) - f_{av} = 0$ for any $u \ge 0$. Hence, we have

$$\begin{aligned}
&|\mathbb{E}[f(X(t),\alpha(t)) - f_{av}][f(X(s),\alpha(s)) - f_{av}]| \\
&= \Big|\mathbb{E}[f(X(t),\alpha(t)) - f_{av}][f(X(s),\alpha(s)) - f_{av}] \\
&\quad -\mathbb{E}[f(X(t),\alpha(t)) - f_{av}]\mathbb{E}[f(X(s),\alpha(s)) - f_{av}]\Big| \\
&= \Big|\mathbb{E}\big\{[f(X(s),\alpha(s)) - f_{av}](\mathbb{E}_{X(s),\alpha(s)}[f(X(t),\alpha(t)) - f_{av}] \\
&\quad -\mathbb{E}[f(X(t),\alpha(t)) - f_{av}])\big\}\Big| \\
&\le C\mathbb{E}^{1/2}|f(X(s),\alpha(s)) - f_{av}|^2 \\
&\quad \times\mathbb{E}^{1/2}\Big|\mathbb{E}_{X(s),\alpha(s)}[f(X(t),\alpha(t)) - f_{av}] - \mathbb{E}[f(X(t),\alpha(t)) - f_{av}]\Big|^2 \\
&\le C\exp(-\kappa_3(t-s))
\end{aligned}$$

by virtue of Lemma 4.15. In the above, from the next to the last line to the last line, we used C as a generic positive constant whose values may be different for different appearances. The proof of the lemma is concluded. □

In what follows, denote

$$\rho(t-s) = \mathbb{E}[f(X(t),\alpha(t)) - f_{av}][f(X(s),\alpha(s)) - f_{av}], \quad 0 \le s \le t \le T. \tag{4.57}$$

This is the covariance function. The process is time homogeneous; thus, it is a function of the time difference only.

Lemma 4.17 *Assume the conditions of Lemma 4.15 hold. Then*

$$\int_0^\infty |\rho(t)|dt < \infty. \tag{4.58}$$

Proof In view of (4.57), we have from Lemma 4.16 that

$$\begin{aligned}
|\rho(t)| &= |\rho(s+t-s)| \\
&= |\mathbb{E}[f(X(s+t),\alpha(s+t)) - f_{av}][f(X(s),\alpha(s)) - f_{av}]| \\
&\le C\exp(-\kappa_3 t).
\end{aligned}$$

Then Eq. (4.58) follows immediately. □

Theorem 4.18 *Let the conditions of Lemma* 4.15 *be satisfied. Then*

$$\frac{1}{\sqrt{T}}\int_0^T [f(X(t),\alpha(t)) - f_{av}]dt \ \text{ converges in distribution to } \ N(0,\sigma_{av}^2), \tag{4.59}$$

a normal random variable with mean 0 and variance

$$\sigma_{av}^2 = 2\int_0^\infty \rho(t)dt.$$

Proof Define

$$\zeta(T) = \int_0^T [f(X(t), \alpha(t)) - f_{av}]dt, \text{ and } \xi(T) = \frac{1}{\sqrt{T}}\zeta(T). \tag{4.60}$$

Note that

$$\begin{aligned}
\mathbb{E}\xi(T) &= \mathbb{E}\frac{1}{\sqrt{T}}\int_0^T [f(X(t), \alpha(t)) - f_{av}]dt \\
&= \frac{1}{\sqrt{T}}\int_0^T \mathbb{E}[f(X(t), \alpha(t)) - f_{av}]dt = 0.
\end{aligned}$$

To calculate the asymptotic variance, note that

$$\begin{aligned}
&\mathbb{E}\left[\frac{1}{\sqrt{T}}\int_0^T [f(X(t), \alpha(t)) - f_{av}]dt\right]^2 \\
&= \mathbb{E}\frac{1}{T}\int_0^T\int_0^T [f(X(t), \alpha(t)) - f_{av}][f(X(s), \alpha(s)) - f_{av}]dsdt \\
&= 2\mathbb{E}\frac{1}{T}\int_0^T\int_0^t [f(X(t), \alpha(t)) - f_{av}][f(X(s), \alpha(s)) - f_{av}]dsdt.
\end{aligned} \tag{4.61}$$

Equation (4.57) and changing variables lead to

$$\begin{aligned}
&\mathbb{E}\frac{1}{T}\int_0^T\int_0^t [f(X(t), \alpha(t)) - f_{av}][f(X(s), \alpha(s)) - f_{av}]dsdt \\
&= \frac{1}{T}\int_0^T\int_0^t \rho(t-s)dsdt \\
&= \frac{1}{T}\int_0^T\int_u^T \rho(u)dtdu \\
&= \int_0^T \left(1 - \frac{u}{T}\right)\rho(u)du.
\end{aligned} \tag{4.62}$$

Choose some $0 < \Delta < 1$. Then by virtue of Lemma 4.17, it follows that as $T \to \infty$,

$$\begin{aligned}
\int_0^T \frac{u}{T}\rho(u)du &= \int_0^{T^\Delta} \frac{u}{T}\rho(u)du + \int_{T^\Delta}^T \frac{u}{T}\rho(u)du \\
&\le \frac{T^\Delta}{T}\int_0^{T^\Delta} \rho(u)du + \frac{T}{T}\int_{T^\Delta}^T \rho(u)du \\
&\to 0.
\end{aligned}$$

Combining the above, we arrive at

$$
\begin{aligned}
&\mathbb{E}\frac{1}{T}\int_0^T\int_0^t [f(X(t),\alpha(t)) - f_{av}][f(X(s),\alpha(s)) - f_{av}]dsdt \\
&\quad \to \int_0^\infty \rho(t)dt \text{ as } T \to \infty.
\end{aligned}
$$

Therefore, we have

$$
\mathbb{E}[\xi(T)]^2 \to \sigma_{av}^2 = 2\int_0^\infty \rho(t)dt \text{ as } T \to \infty. \tag{4.63}
$$

The desired result thus follows. □

4.7 Notes

Many applications of diffusion processes without switching require using invariant distributions of the underlying processes. One of them is the formulation of two-time-scale diffusions. Dealing with diffusions having fast and slow components, under the framework of diffusion approximation, it has been shown that in the limit, the slow component is averaged out with respect to the invariant measure of the fast component; see Khasminskii [114] for a weak convergence limit result and Papanicolaou, Stroock, and Varadhan [230] for using a martingale problem formulation. In any event, a crucial step in these references is the use of the invariance measure. In this chapter, in contrast to the diffusion counterpart, we treat regime-switching diffusions, where in addition to the continuous component, there are discrete events. Our interest has been devoted to obtaining ergodicity.

For regime-switching diffusions, asymptotic stability for the density of the two-state random process $(X(t), \alpha(t))$ was established in [236]; asymptotic stability in distribution (or the convergence to the stationary measures of the switching diffusions) for the process $(X(t), \alpha(t))$ was obtained in [11, 311]. Here, we addressed ergodicity for $(X(t), \alpha(t))$ under different conditions from those in [11, 236, 311].

Taking into consideration of the many applications in which discrete events and continuous dynamics are intertwined and the discrete-event process depends on the continuous state, we allow the discrete component $\alpha(\cdot)$ to have x-dependent generator $Q(x)$. Another highlight is that we obtain the explicit representation of the invariant measure of the process $(X(t), \alpha(t))$ by considering certain cylinder sets and by defining cycles appropriately. As a byproduct, we demonstrate a strong law of large numbers type theorem for positive recurrent regime-switching diffusions. It reveals that positive recurrence and ergodicity of switching diffusions are equivalent.

In this chapter, we first developed ergodicity for positive recurrent switching diffusions. Focusing on a compact space, we then obtained asymptotic distributions as a consequence of the ergodicity of the process. The asymptotic distributions are important in treating limit ergodic control problems as well as in applications of Markov chain Monte Carlo. A crucial step in the proof is the verification of the centered and scaled process being a mixing process with exponential mixing rate.

A number of important problems remain open. Obtaining large deviation-type bounds is a worthwhile undertaking, which will have an important impact on studying the associated control and optimization problems.

Chapter 5
Numerical Approximation

5.1 Introduction

As is the case for deterministic dynamic systems or diffusions, closed-form solutions for switching diffusions are often difficult to obtain, and numerical approximation is frequently a viable or possibly the only alternative. Being extremely important, numerical methods have drawn much attention. To date, a number of works (e.g., [205–207, 296]) have focused on numerical approximations where the switching process is independent of the continuous component and is modeled by a continuous-time Markov chain. Nevertheless, it is necessary to be able to handle the coupling and dependence of the continuous states and discrete events. This chapter is devoted to numerical approximation methods for switching diffusions whose switching component is x-dependent. Section 5.2 presents the setup of the problem together with the numerical algorithms. Section 5.3 establishes the convergence of the numerical algorithms. Section 5.4 studies rates of convergence of the algorithms. Section 5.5 proceeds with a couple of examples. Section 5.6 provides two remarks on algorithms for Markovian switching and algorithms with random truncations. Finally, Section 5.7 concludes the chapter.

5.2 Formulation and Algorithms

5.2.1 Setup and Assumptions

As in Chapter 2, we consider the switching diffusion $(X, \alpha) \in \mathbb{R}^r \times \mathcal{M}$ (here $\mathcal{M} = \{1, \ldots, m_0\}$ with $m_0 \leq 2 \leq \infty$):

H.-D. Nguyen et al., *Hybrid Switching Diffusions*, Probability Theory and Stochastic Modelling 63, https://doi.org/10.1007/978-3-031-93303-5_5

$$\begin{aligned} dX(t) &= b(X(t), \alpha(t))dt + \sigma(X(t), \alpha(t))dw(t), \\ X(0) &= X_0, \quad \alpha(0) = \alpha_0, \end{aligned} \tag{5.1}$$

and

$$\mathbb{P}(\alpha(t+\Delta) = j | \alpha(t) = i, X(s), \alpha(s), s \le t) = q_{ij}(X(t))\Delta + o(\Delta), \quad i \ne j, \tag{5.2}$$

where $w(\cdot)$ is an r-dimensional standard Brownian motion, $b(\cdot,\cdot) : \mathbb{R}^r \times \mathcal{M} \mapsto \mathbb{R}^r$, $\sigma(\cdot,\cdot) : \mathbb{R}^r \times \mathcal{M} \mapsto \mathbb{R}^{r\times r}$ are appropriate Borel measurable functions satisfying certain regularity conditions, and $Q(x) = (q_{ij}(x)) \in \mathbb{R}^{m_0 \times m_0}$ satisfies that for each x, $q_{ij}(x) \ge 0$ for $i \ne j$, $\sum_{j=1}^{m_0} q_{ij}(x) = 0$ for each $i \in \mathcal{M}$. There is an associated operator for the switching diffusion process defined as follows. For each $i \in \mathcal{M}$ and suitable smooth function $h(\cdot, i)$, define the operator:

$$\begin{aligned} \mathcal{L}h(x,i) = {} & \nabla h'(x,i) b(x,i) + \frac{1}{2}\mathrm{tr}[\nabla^2 h(x,i)\sigma(x,i)\sigma'(x,i)] \\ & + \sum_{j\in\mathcal{M}} q_{ij}(x)[h(x,j) - h(x,i)], \end{aligned} \tag{5.3}$$

where $\nabla h(x,i)$ and $\nabla^2 h(x,i)$ denote the gradient and Hessian matrix of h with respect to the x variable, respectively.

In this chapter, our aim is to develop numerical approximation schemes for solving (5.1)–(5.2). Note that in our setup, $Q(x)$, the generator of the switching process $\alpha(t)$, is state dependent. It is the x-dependence of $Q(x)$ that makes the analysis much more difficult. One of the main difficulties is that due to the continuous-state dependence, $\alpha(t)$ and $X(t)$ are dependent; $\alpha(t)$ is a Markov chain only for a fixed x but is otherwise non-Markovian. The essence of our approach is to treat the pair of processes $(X(t), \alpha(t))$ jointly since the two-component process is Markovian. Nevertheless, much care needs to be exercised in handling the mixture distributions. To proceed, we use the following conditions:

(H5.1) The function $Q(\cdot)$ is bounded, continuous, and satisfies (2.21) for some concave function $\gamma : \mathbb{R}_+ \mapsto \mathbb{R}_+$ with $\gamma(0) = 0$.

(H5.2) The functions $b(\cdot,\cdot)$ and $\sigma(\cdot,\cdot)$ satisfy for some $K_0 > 0$

$$|b(x,\alpha)| + |\sigma(x,\alpha)| \le K_0(1 + |x|),$$

and

$$|b(x,\alpha) - b(y,\alpha)| + |\sigma(x,\alpha) - \sigma(y,\alpha)| \le K_0|x-y|$$

for all $x, y \in \mathbb{R}^r$ and $\alpha \in \mathcal{M}$.

Recall from Theorem 2.6 that under (H5.1) and (H5.2), the system (5.1)–(5.2) has a unique nonexplosive strong solution.

We first construct Euler's scheme (also called the Euler-Maruyama scheme) with a constant stepsize for approximating solutions of switching diffusions. It should be mentioned in particular, our analysis differs from the usual approach. To obtain convergence of the algorithms, we first obtain weak convergence of the algorithm by means of martingale problem formulation. This is particularly suited for x-dependent switching processes because the convergence result would be much more difficult to obtain otherwise. As a demonstration, we provide numerical experiments to delineate sample path properties of the approximating solutions. Further discussion of uniform convergence in the sense of L^2 and the associated rates of convergence are derived.

5.2.2 Numerical Algorithms

To approximate (5.1), choosing a sequence of independent and identically distributed random variables $\{\xi_n\}$ with mean 0 and finite variance, we propose the following algorithm:

$$X_{n+1} = X_n + \varepsilon b(X_n, \alpha_n) + \sqrt{\varepsilon}\sigma(X_n, \alpha_n)\xi_n, \quad n \geq 0, \tag{5.4}$$

where $\varepsilon > 0$ is small. We proceed to describe the terms involved above. First, we would like to have α_n as a discrete-time stochastic process that approximates $\alpha(t)$ in an appropriate sense. It is natural when $X_{n-1} = x$ that α_n has a transition probability matrix $\exp(Q(x)\varepsilon)$, which may be approximated further by $I + \varepsilon Q(x) + O(\varepsilon^2)$ by virtue of the boundedness and the continuity of $Q(\cdot)$. Based on this observation, in what follows, we discard the $O(\varepsilon^2)$ term and simply use $I + \varepsilon Q(x)$ for the transition matrix for α_n when $X_{n-1} = x$. Since $Q(\cdot)$ is bounded thanks to Assumption (H5.1), $1 + \varepsilon q_{ii}(x) \geq 0$ for all $(x, i) \in \mathbb{R}^r \times \mathcal{M}$ as long as $\varepsilon > 0$ is sufficiently small. The sequence $\{\xi_n\}$ of independent and identically distributed random variables with 0 mean and finite variance is used to approximate the Brownian motion w.

We put what has been said above into the following assumption.

(H5.3) The following conditions hold:

(a) $\{\xi_n\}$ is a sequence of independent and identically distributed random variables with $\mathbb{E}\xi_n = 0$, $\mathbb{E}|\xi_n|^p < \infty$ for some $p \geq 2$, and $\mathbb{E}\xi_n\xi_n' = I$.

(b) Put $\mathcal{G}_n := \{X_k, \alpha_k : k \leq n\}$. Then, $\{\alpha_n\}$ is a discrete-time stochastic process taking value in $\mathcal{M}$ and satisfying

$$\mathbb{P}\{\alpha_{n+1} = j | \alpha_n = i, X_n = x, \mathcal{G}_n\} = P_{ij}(x), \quad \forall i, j \in \mathcal{M}, \ x \in \mathbb{R}^r,$$

where $P(x) = (P_{ij}(x))$ is defined to be $P(x) := I + \varepsilon Q(x)$.

Remark 5.1 Note that for each n, ξ_n is independent of $\mathcal{G}_n$. As mentioned before, the boundedness of $Q(x)$ implies that $P(x) = I + \varepsilon Q(x)$ is truly a transition matrix for all $x \in \mathbb{R}^r$ and ε sufficiently small. In addition, Assumption (H5.3) (b) says that when $X_n = x$ in (5.4), $\{\alpha_{n+1}\}$ has the transition matrix $P(x)$.

One remarkable feature of the algorithm (5.4) is that it is easily implementable. In lieu of discretizing a Brownian motion, we generate a sequence of independent and identically distributed (normal) random variables to approximate the Brownian motion. This facilitates the computational task. In addition, instead of using transition matrix $\exp(\varepsilon Q(x))$ for a fixed x, we use another fold of approximation $I + \varepsilon Q(x)$ based on a truncated Taylor series. All of these stem from consideration of numerical computation and Monte Carlo implementation. For simplicity, we can choose ξ_n to be Gaussian. In fact, any sequence with mean 0 and finite second moment can be used. Moreover, correlated sequence $\{\xi_n\}$ can be used. However, from a Monte Carlo perspective, there seems to be no strong reason that we need to use correlated sequences.

5.3 Convergence of the Algorithm

5.3.1 Moment Estimates

Throughout this chapter, we assume the stepsize $0 < \varepsilon < 1$. Note that by T/ε, we mean the integer part of T/ε, that is, $\lfloor T/\varepsilon \rfloor$. However, for simplicity, we do not use the floor function notation in what follows and retain this notation only if it is necessary. We first obtain an estimate on the pth moment of $\{X_n\}$. This is stated as follows.

Lemma 5.2 *Under Assumptions* (H5.1)–(H5.3), *for any fixed* $p \geq 2$ *and* $T > 0$

$$\sup_{0 \leq n \leq T/\varepsilon} \mathbb{E}|X_n|^p \leq (|X_0|^p + KT) \exp(KT) < \infty. \tag{5.5}$$

Proof Define $U(x) = |x|^p$ and use $\mathbb{E}_n$ to denote the conditional expectation with respect to the σ-algebra $\mathcal{G}_n$, where $\mathcal{G}_n$ is given in Assumption (H5.3). Note that

$$\begin{aligned}&\mathbb{E}_n \sigma(X_n, \alpha_n)\xi_n = \sigma(X_n, \alpha_n)\mathbb{E}_n \xi_n = 0 \text{ and}\\ &\mathbb{E}_n |\sigma(X_n, \alpha_n)|^2 |\xi_n|^2 = |\sigma(X_n, \alpha_n)|^2 \mathbb{E}_n |\xi_n|^2 \leq K|\sigma(X_n, \alpha_n)|^2,\end{aligned}$$

where K is a generic positive constant. Thus

$$\begin{aligned}\mathbb{E}_n U(X_{n+1}) - U(X_n) &= \mathbb{E}_n (\nabla U(X_n))'[X_{n+1} - X_n]\\ &\quad + \frac{1}{2}\mathbb{E}_n (X_{n+1} - X_n)' \nabla^2 U(X_n^+)(X_{n+1} - X_n),\end{aligned} \tag{5.6}$$

where ∇U and $\nabla^2 U$ denote the gradient and Hessian of U with respect to to x, and X_n^+ denotes a vector on the line segment joining X_n and X_{n+1}. Using (5.4), we may write

$$X_n^+ = (1-\theta)X_n + \theta X_{n+1} = X_n + \varepsilon\theta b(X_n, \alpha_n) + \sqrt{\varepsilon}\theta\sigma(X_n, \alpha_n)\xi_n,$$

for some $\theta \in [0, 1]$. In addition, using (5.4) and Assumption (H5.3), we can compute

$$\begin{aligned}\mathbb{E}_n(\nabla U(X_n))'[X_{n+1} - X_n] &\le K\mathbb{E}_n[|X_n|^{p-1}[\varepsilon|b(X_n, \alpha_n)| + \sqrt{\varepsilon}|\sigma(X_n, \alpha_n)||\xi_n|]] \\ &\le K\varepsilon(1 + |X_n|^p).\end{aligned}$$

On the other hand, since $|\nabla^2 U(x)| \le K|x|^{p-2}$, using the expression for X_n^+, we have $|\nabla^2 U(X_n^+)| \le K(1 + |X_n|^{p-2} + |X_n|^{p-2}|\xi_n|^{p-2})$. Then, it follows that

$$\begin{aligned}&\mathbb{E}_n(X_{n+1} - X_n)'\nabla^2 U(X_n^+)(X_{n+1} - X_n) \\ &\le K\varepsilon\mathbb{E}_n[(1 + |X_n|^{p-2} + |X_n|^{p-2}|\xi_n|^{p-2}) \cdot (\varepsilon(1 + |X_n|^2) + (1 + |X_n|^2)|\xi_n|^2)] \\ &\le K\varepsilon\mathbb{E}_n[1 + |X_n|^p(1 + |\xi_n|^2 + |\xi_n|^{p-2} + |\xi_n|^p)] \\ &\le K\varepsilon(1 + |X_n|^p).\end{aligned}$$

Plugging the previous two displayed equations into (5.6) gives $\mathbb{E}_n U(X_{n+1}) - U(X_n) \le K\varepsilon(1 + |X_n|^p)$. Recall that $U(X_n) = |X_n|^p$, we obtain

$$\mathbb{E}_n|X_{n+1}|^p \le |X_n|^p + K\varepsilon + K\varepsilon|X_n|^p.$$

Taking the expectation on both sides and iterating on the resulting recursion, we obtain

$$\mathbb{E}|X_{n+1}|^p \le |X_0|^p + K\varepsilon n + K\varepsilon\sum_{k=0}^{n}\mathbb{E}|X_k|^p.$$

An application of discrete Gronwall's inequality yields that

$$\mathbb{E}|X_{n+1}|^p \le (|X_0|^p + KT)\exp(KT)$$

as desired. □

Remark 5.3 In view of the estimate above, $\{X_n : 0 \le n \le T/\varepsilon\}$ is tight in $\mathbb{R}^r$ by means of the well-known Tchebyshev inequality. That is, for each $\eta > 0$, there is a K_η satisfying $K_\eta > \sqrt{(1/\eta)}$ such that

$$\mathbb{P}(|X_n| > K_\eta) \le \frac{\sup_{0\le n\le T/\varepsilon}\mathbb{E}|X_n|^2}{K_\eta^2} \le K\eta.$$

This indicates that the sequence of iterates is "mass preserving" or no probability is lost.

To proceed, we take continuous-time interpolations defined by

$$X^\varepsilon(t) = X_n,\ \alpha^\varepsilon(t) = \alpha_n, \quad \text{for } t \in [n\varepsilon, n\varepsilon + \varepsilon). \tag{5.7}$$

We will show that $X^\varepsilon(\cdot)$ and $\alpha^\varepsilon(\cdot)$ are tight in suitable function spaces. Denote by $\mathcal{F}_t^\varepsilon$ the σ-algebra generated by $\{X^\varepsilon(u), \alpha^\varepsilon(u) : u \le t\}$. Write the conditional expectation with respect to $\mathcal{F}_t^\varepsilon$ as $\mathbb{E}_t^\varepsilon$. Mimic the proof in Lemma 5.2, we can also obtain that for any $t > 0$ and $0 < s < \infty$

$$\sup_{t/\varepsilon \le n \le (t+s)/\varepsilon} \mathbb{E}_t^\varepsilon |X_n|^2 < \infty. \tag{5.8}$$

Lemma 5.4 *Suppose Assumptions* (H5.1)–(H5.3). *Define*

$$\begin{aligned} &\chi_n = (I_{\{\alpha_n=1\}}, \ldots, I_{\{\alpha_n=m_0\}}) \in \mathbb{R}^{1\times m_0}, \quad \textit{and} \\ &\chi^\varepsilon(t) = \chi_n, \quad \textit{for } t \in [\varepsilon n, \varepsilon n + \varepsilon). \end{aligned}$$

Then, for any $t, s > 0$

$$\mathbb{E}_t^\varepsilon[\chi^\varepsilon(t+s) - \chi^\varepsilon(t)] = O(s). \tag{5.9}$$

Proof Write $\lfloor \frac{t}{\varepsilon} \rfloor = m$ and $\lfloor \frac{t+s}{\varepsilon} \rfloor = n$. If $n = m$, then $\chi^\varepsilon(t+s) = \chi^\varepsilon(t)$, and hence, (5.9) holds true trivially. We now focus on the case when $n > m$. Suppose $k \ge m$ and $\alpha_k = i \in \mathcal{M}$. Then, for any $j \in \mathcal{M}$, using Assumption (H5.3)(b), we can write

$$\begin{aligned} \mathbb{E}[I_{\{\alpha_{k+1}=j\}} - I_{\{\alpha_k=j\}} | \mathcal{G}_k] &= I_{\{j\ne i\}}(1-0)\varepsilon q_{ij}(X_k) + I_{\{j=i\}}(0-1)\sum_{l\ne i} \varepsilon q_{il}(X_k) \\ &= I_{\{j\ne i\}}\varepsilon q_{ij}(X_k) + I_{\{j=i\}}\varepsilon q_{ii}(X_k). \end{aligned}$$

This is true for any $i \in \mathcal{M}$, and therefore, we have

$$\mathbb{E}[\chi_{k+1} - \chi_k | \mathcal{G}_k] = \varepsilon \chi_k Q(X_k). \tag{5.10}$$

Then, it follows that

$$\begin{aligned} \mathbb{E}_t^\varepsilon[\chi^\varepsilon(t+s) - \chi^\varepsilon(t)] &= \sum_{k=m}^{n-1} \mathbb{E}_t^\varepsilon[\chi_{k+1} - \chi_k] = \sum_{k=m}^{n-1} \mathbb{E}_t^\varepsilon[\mathbb{E}[\chi_{k+1} - \chi_k | \mathcal{G}_k]] \\ &= \sum_{k=m}^{n-1} \mathbb{E}_t^\varepsilon[\varepsilon \chi_k Q(X_k)]. \end{aligned}$$

Furthermore, since $Q(x)$ is uniformly bounded thanks to Assumption (H5.1) and $|\chi_k| = 1$, it follows that

$$\mathbb{E}_t^\varepsilon[|\chi^\varepsilon(t+s) - \chi^\varepsilon(t)|] \le \sum_{k=m}^{n-1} \mathbb{E}_t^\varepsilon[K\varepsilon] = K(n-m)\varepsilon = O(s),$$

where $K := \sup_{x\in\mathbb{R}^r} |Q(x)| < \infty$. This implies (5.9) and completes the proof. □

Lemma 5.5 *Under the conditions of Lemma* 5.4, $\{\alpha^\varepsilon(\cdot)\}$ *is tight in* $D([0,\infty) : \mathcal{M})$, *the space of* $\mathcal{M}$*-valued càdlàg functions endowed with the Skorohod topology.*

Proof Let $t, s \ge 0$. Note that $|\chi^\varepsilon(t)|^2 = \chi^\varepsilon(t)\chi^{\varepsilon,\prime}(t) = 1$. Then, we can use Lemma 5.4 to compute

$$\begin{aligned}
&\mathbb{E}_t^\varepsilon|\chi^\varepsilon(t+s) - \chi^\varepsilon(t)|^2 \\
&= \mathbb{E}_t^\varepsilon[\chi^\varepsilon(t+s)\chi^{\varepsilon,\prime}(t+s) - 2\chi^\varepsilon(t+s)\chi^{\varepsilon,\prime}(t) + \chi^\varepsilon(t)\chi^{\varepsilon,\prime}(t)] \\
&= 2 - 2\mathbb{E}_t^\varepsilon[\chi^\varepsilon(t+s)\chi^{\varepsilon,\prime}(t)] \\
&= 2 - 2\mathbb{E}_t^\varepsilon[(\chi^\varepsilon(t) + O(s))\chi^{\varepsilon,\prime}(t)] \\
&= O(s).
\end{aligned}$$

Therefore, we have

$$\lim_{s\to 0}\limsup_{\varepsilon\to 0} \mathbb{E}[\mathbb{E}_t^\varepsilon|\chi^\varepsilon(t+s) - \chi^\varepsilon(t)|^2] = 0.$$

The tightness criterion in [175, p. 47] yields that $\{\chi^\varepsilon(\cdot)\}$ is tight. Consequently, $\{\alpha^\varepsilon(\cdot)\} = \chi^\varepsilon(\cdot)(1, 2, \dots)'$ is tight as well. □

Lemma 5.6 *Assume that the conditions of Lemma 5.5 are satisfied. Then,* $\{X^\varepsilon(\cdot)\}$ *is tight in* $D([0,\infty) : \mathbb{R}^r)$, *the space of* $\mathbb{R}^r$*-valued càdlàg functions endowed with the Skorohod topology.*

Proof For any $\eta > 0$, $t \ge 0$, $0 \le s \le \eta$, denote $i := \lfloor \frac{t}{\varepsilon} \rfloor$ and $j := \lfloor \frac{t+s}{\varepsilon} \rfloor$. If $j = i$, then $\mathbb{E}_t^\varepsilon|X^\varepsilon(t+s) - X^\varepsilon(t)|^2 = 0$ thanks to (5.7); otherwise, we can compute

$$\begin{aligned}
&\mathbb{E}_t^\varepsilon|X^\varepsilon(t+s) - X^\varepsilon(t)|^2 \\
&= \mathbb{E}_t^\varepsilon\left|\varepsilon\sum_{k=i}^{j-1} b(X_k,\alpha_k) + \sqrt{\varepsilon}\sum_{k=i}^{j-1}\sigma(X_k,\alpha_k)\xi_k\right|^2 \\
&\le 2\varepsilon^2(j-i)\mathbb{E}_t^\varepsilon\sum_{k=i}^{j-1}|b(X_k,\alpha_k)|^2 + 2\varepsilon\mathbb{E}_t^\varepsilon\sum_{k=i}^{j-1}|\sigma(X_k,\alpha_k)|^2|\xi_k|^2 \\
&\quad + 4\varepsilon\mathbb{E}_t^\varepsilon\sum_{p<q}\sigma(X_p,\alpha_p)\xi_p\cdot\sigma(X_q,\alpha_q)\xi_q
\end{aligned}$$

$$
\begin{aligned}
&\leq K\varepsilon^2(j-i)\mathbb{E}_t^\varepsilon \sum_{k=i}^{j-1}(1+|X_k|^2) + K\varepsilon\mathbb{E}_t^\varepsilon \sum_{k=i}^{j-1}(1+|X_k|^2)|\xi_k|^2 \\
&\quad + 4\varepsilon \sum_{p<q} \mathbb{E}_t^\varepsilon[\sigma(X_p,\alpha_p)\xi_p \cdot \sigma(X_q,\alpha_q)\mathbb{E}_{q\varepsilon}^\varepsilon[\xi_q]] \\
&= K\varepsilon^2(j-i)\mathbb{E}_t^\varepsilon \sum_{k=i}^{j-1}(1+|X_k|^2) + K\varepsilon\mathbb{E}_t^\varepsilon \sum_{k=i}^{j-1}(1+|X_k|^2)|\xi_k|^2,
\end{aligned}
$$

where we used Assumption (H5.3) to derive $\mathbb{E}_{q\varepsilon}^\varepsilon[\xi_q] = \mathbb{E}[\xi_q] = 0$ and hence the last step. Then

$$
\begin{aligned}
&\mathbb{E}_t^\varepsilon |X^\varepsilon(t+s) - X^\varepsilon(t)|^2 \\
&\leq K\varepsilon^2(j-i)\sum_{k=i}^{j-1}(1+\mathbb{E}_t^\varepsilon|X_k|^2) + K\varepsilon \sum_{k=i}^{j-1}\mathbb{E}_t^\varepsilon|\sigma(X_k,\alpha_k)|^2\mathbb{E}_t^\varepsilon|\xi_k|^2 \\
&\leq K\varepsilon^2(j-i)\sum_{k=i}^{j-1}\left(1+\sup_{i\leq k\leq j-1}\mathbb{E}_t^\varepsilon|X_k|^2\right) + K\varepsilon\sum_{k=i}^{j-1}\left(1+\sup_{i\leq k\leq j-1}\mathbb{E}_t^\varepsilon|X_k|^2\right) \\
&\leq O\left(\frac{t+s}{\varepsilon} - \frac{t}{\varepsilon}\right) O(\varepsilon) = O(s).
\end{aligned}
\tag{5.11}
$$

In the above, we have used (5.8) to ensure that $\sup_{i\leq k\leq j-1}\mathbb{E}_t^\varepsilon|X_k|^2 < \infty$. Therefore, (5.11) leads to

$$
\lim_{\eta\to 0}\limsup_{\varepsilon\to 0}\mathbb{E}\mathbb{E}_t^\varepsilon|X^\varepsilon(t+s) - X^\varepsilon(t)|^2 = 0.
$$

On the other hand, Lemma 5.2 together with the definition given in (5.7) implies that $\sup_{\varepsilon>0}\sup_{0\leq t\leq T}\mathbb{E}[|X^\varepsilon(t)|^2] \leq K < \infty$, where K is a constant independent of ε. Therefore, for each $t \in [0, T]$

$$
\sup_{\varepsilon>0}\mathbb{P}\{|X^\varepsilon(t)| > R\} \leq \frac{\sup_{\varepsilon>0}\mathbb{E}|X^\varepsilon(t)|^2}{R^2} \to 0
$$

as $R \to \infty$. The tightness of $\{X^\varepsilon(\cdot)\}$ then follows from the Corollary to Theorem 13.2 of [28] or Theorem 3.7.2 of [81]. □

Combining Lemmas 5.4–5.6, we obtain the following tightness result.

Lemma 5.7 *Under Assumptions* (H5.1)–(H5.3), $\{X^\varepsilon(\cdot), \alpha^\varepsilon(\cdot)\}$ *is tight in* $D([0,\infty) : \mathbb{R}^r \times \mathcal{M})$.

5.3.2 Weak Convergence

The main result of this section is the following theorem.

Theorem 5.8 *Under Assumptions* (H5.1)–(H5.3), $(X^\varepsilon(\cdot), \alpha^\varepsilon(\cdot))$ *converges weakly to* $(X(\cdot), \alpha(\cdot))$, *which is a solution to* (5.1)–(5.2).

Proof Because $(X^\varepsilon(\cdot), \alpha^\varepsilon(\cdot))$ is tight, by Prohorov's theorem (see Theorem A.20 of this book and also [81, 177]), we may select a convergent subsequence. For simplicity, we still denote the subsequence by $(X^\varepsilon(\cdot), \alpha^\varepsilon(\cdot))$ with the limit denoted by $(\widetilde{X}(\cdot), \widetilde{\alpha}(\cdot))$.

By the Skorohod representation theorem (see Theorem A.21 and also [81, 177]), without loss of generality and without changing notation, we may assume that $(X^\varepsilon(\cdot), \alpha^\varepsilon(\cdot))$ converges to $(\widetilde{X}(\cdot), \widetilde{\alpha}(\cdot))$ w.p.1, and the convergence is uniform on each bounded interval. We proceed to characterize the limit process.

Step 1: We first work with the marginal of the switching component and characterize the limit of $\alpha^\varepsilon(\cdot)$. The a.s. convergence of $\alpha^\varepsilon(\cdot)$ to $\widetilde{\alpha}(\cdot)$ yields that $\chi^\varepsilon(\cdot)$ converges to $\widetilde{\chi}(\cdot)$ a.s., where $\chi^\varepsilon(\cdot)$ is defined in the statement of Lemma 5.4, and $\widetilde{\chi}(\cdot)$ is defined similarly using $\widetilde{\alpha}(\cdot)$. For each $t > 0$ and $s > 0$, each positive integer κ, each $0 \le t_\iota \le t$ with $\iota \le \kappa$, and each bounded and continuous function $\rho_\iota(\cdot, i)$ for each $i \in \mathcal{M}$

$$\mathbb{E}\prod_{\iota=1}^{\kappa}\rho_\iota(X^\varepsilon(t_\iota), \alpha^\varepsilon(t_\iota))\left[\chi^\varepsilon(t+s) - \chi^\varepsilon(t) - \sum_{k=t/\varepsilon}^{(t+s)/\varepsilon-1}(\chi_{k+1} - \chi_k)\right] = 0. \tag{5.12}$$

The convergence of $(X^\varepsilon(\cdot), \alpha^\varepsilon(\cdot))$ to $(\widetilde{X}(\cdot), \widetilde{\alpha}(\cdot))$ and the bounded convergence theorem imply that

$$\begin{aligned}
&\lim_{\varepsilon\to 0}\mathbb{E}\prod_{\iota=1}^{\kappa}\rho_\iota(X^\varepsilon(t_\iota), \alpha^\varepsilon(t_\iota))[\chi^\varepsilon(t+s) - \chi^\varepsilon(t)] \\
&= \mathbb{E}\prod_{\iota=1}^{\kappa}\rho_\iota(\widetilde{X}(t_\iota), \widetilde{\alpha}(t_\iota))[\chi(t+s) - \chi(t)].
\end{aligned} \tag{5.13}$$

Denote $m := \lfloor \frac{t}{\varepsilon} \rfloor$ and $n := \lfloor \frac{t+s}{\varepsilon} \rfloor$. Using (5.10), we can compute

$$\begin{aligned}
&\mathbb{E}\prod_{\iota=1}^{\kappa}\rho_\iota(X^\varepsilon(t_\iota), \alpha^\varepsilon(t_\iota))\sum_{k=m}^{n-1}(\chi_{k+1} - \chi_k) \\
&= \sum_{k=m}^{n-1}\mathbb{E}\left[\prod_{\iota=1}^{\kappa}\rho_\iota(X^\varepsilon(t_\iota), \alpha^\varepsilon(t_\iota))\mathbb{E}[(\chi_{k+1} - \chi_k)|\mathcal{G}_k]\right]
\end{aligned}$$

$$
\begin{aligned}
&= \sum_{k=m}^{n-1} \mathbb{E}\left[\prod_{\iota=1}^{\kappa} \rho_\iota(X^\varepsilon(t_\iota), \alpha^\varepsilon(t_\iota))\varepsilon \chi_k Q(X_k)\right] \\
&= \mathbb{E}\left[\prod_{\iota=1}^{\kappa} \rho_\iota(X^\varepsilon(t_\iota), \alpha^\varepsilon(t_\iota)) \sum_{k=m}^{n-1} \varepsilon \chi^\varepsilon(k\varepsilon) Q(X^\varepsilon(k\varepsilon))\right] \\
&= \mathbb{E}\left[\prod_{\iota=1}^{\kappa} \rho_\iota(X^\varepsilon(t_\iota), \alpha^\varepsilon(t_\iota)) \int_t^{t+s} \chi^\varepsilon(u) Q(X^\varepsilon(u)) du\right] + O(\varepsilon) \\
&\to \mathbb{E}\left[\prod_{\iota=1}^{\kappa} \rho_\iota(\widetilde{X}(t_\iota), \widetilde{\alpha}(t_\iota)) \int_t^{t+s} \widetilde{\chi}(u) Q(\widetilde{X}(u)) du\right],
\end{aligned}
$$

where the last step follows from the convergence of $(X^\varepsilon(\cdot), \alpha^\varepsilon(\cdot))$ to $(\widetilde{X}(\cdot), \widetilde{\alpha}(\cdot))$, the continuity of the functions ρ_ι and Q, and the bounded convergence theorem. Plugging this equation and (5.13) into (5.12) yields

$$
\mathbb{E}\prod_{\iota=1}^{\kappa} \rho_\iota(\widetilde{X}(t_\iota), \widetilde{\alpha}(t_\iota)) \left[\chi(t+s) - \chi(t) - \int_t^{t+s} \chi(u) Q(\widetilde{X}(u)) du\right] = 0. \tag{5.14}
$$

Therefore, the limit process $\widetilde{\alpha}(\cdot)$ has a generator $Q(\widetilde{X}(\cdot))$.

Step 2: For t, s, κ, t_ι, and $\rho_\iota(\cdot, \cdot)$ as before, and for all functions $h : \mathbb{R}^r \times \mathcal{M} \mapsto \mathbb{R}$ so that $h(\cdot, i) \in C_c^2(\mathbb{R}^r)$ for each $i \in \mathcal{M}$, we need to show that

$$
\begin{aligned}
\mathbb{E}\prod_{\iota=1}^{\kappa} \rho_\iota(\widetilde{X}(t_\iota), \widetilde{\alpha}(t_\iota)) \Big[& h(\widetilde{X}(t+s), \widetilde{\alpha}(t+s)) \\
& - h(\widetilde{X}(t), \widetilde{\alpha}(t)) - \int_t^{t+s} \mathcal{L}h(\widetilde{X}(u), \widetilde{\alpha}(u)) du\Big] = 0.
\end{aligned} \tag{5.15}
$$

This yields that

$$
h(\widetilde{X}(t), \widetilde{\alpha}(t)) - \int_0^t \mathcal{L}h(\widetilde{X}(u), \widetilde{\alpha}(u)) du, \quad t \geq 0
$$

is a continuous-time martingale, which in turn implies that $(\widetilde{X}(\cdot), \widetilde{\alpha}(\cdot))$ is a solution of the martingale problem with operator $\mathcal{L}$ defined in (5.3). To this end, we first observe that

$$
\begin{aligned}
\mathbb{E}\prod_{\iota=1}^{\kappa} \rho_\iota(X^\varepsilon(t_\iota), \alpha^\varepsilon(t_\iota)) \Big[& h(X^\varepsilon(t+s), \alpha^\varepsilon(t+s)) \\
& - h(X^\varepsilon(t), \alpha^\varepsilon(t)) - \sum_{k=m}^{n-1} [h(X_{k+1}, \alpha_{k+1}) - h(X_k, \alpha_k)]\Big] = 0,
\end{aligned} \tag{5.16}
$$

where as before we denote $m = \lfloor \frac{t}{\varepsilon} \rfloor$ and $n = \lfloor \frac{t+s}{\varepsilon} \rfloor$.

Since $(X^\varepsilon(\cdot), \alpha^\varepsilon(\cdot))$ converges to $(\widetilde{X}(\cdot), \widetilde{\alpha}(\cdot))$ w.p.1, by virtue of the bounded convergence theorem, it is readily seen that

$$\begin{aligned} &\mathbb{E}\prod_{\iota=1}^{\kappa} \rho_\iota(X^\varepsilon(t_\iota), \alpha^\varepsilon(t_\iota))[h(X^\varepsilon(t+s), \alpha^\varepsilon(t+s)) - h(X^\varepsilon(t), \alpha^\varepsilon(t))] \\ &\quad \to \mathbb{E}\prod_{\iota=1}^{\kappa} \rho_\iota(\widetilde{X}(t_\iota), \widetilde{\alpha}(t_\iota))[h(\widetilde{X}(t+s), \widetilde{\alpha}(t+s)) - h(\widetilde{X}(t), \widetilde{\alpha}(t))] \end{aligned} \tag{5.17}$$

as $\varepsilon \to 0$.

On the other hand, direct calculation shows that

$$\begin{aligned} &\sum_{k=m}^{n-1}[h(X_{k+1}, \alpha_{k+1}) - h(X_k, \alpha_k)] \\ &= \sum_{k=m}^{n-1}[h(X_{k+1}, \alpha_k) - h(X_k, \alpha_k)] + \sum_{k=m}^{n-1}[h(X_{k+1}, \alpha_{k+1}) - h(X_{k+1}, \alpha_k)] \\ &= \sum_{k=m}^{n-1}\Big[\nabla h(X_k, \alpha_k)'(X_{k+1} - X_k) \\ &\quad + \frac{1}{2}\mathrm{tr}(\nabla^2 h(X_k, \alpha_k)(X_{k+1} - X_k)(X_{k+1} - X_k)')\Big] \\ &\quad + \sum_{k=m}^{n-1} o(|X_{k+1} - X_k|^2) + \sum_{k=m}^{n-1}[h(X_{k+1}, \alpha_{k+1}) - h(X_{k+1}, \alpha_k)]I_{\{\alpha_{k+1} \neq \alpha_k\}} \\ &= \sum_{k=m}^{n-1} \varepsilon\Big[\nabla h(X_k, \alpha_k)'b(X_k, \alpha_k) + \frac{1}{2}\mathrm{tr}(\nabla^2 h(X_k, \alpha_k)\sigma\sigma'(X_k, \alpha_k))|\xi_k|^2\Big] \\ &\quad + \sum_{k=m}^{n-1}\Big[\sqrt{\varepsilon}\nabla h(X_k, \alpha_k)'\sigma(X_k, \alpha_k)\xi_k + \frac{\varepsilon^2}{2}\mathrm{tr}(\nabla^2 h(X_k, \alpha_k)b(X_k, \alpha_k)b'(X_k, \alpha_k))\Big] \\ &\quad + \sum_{k=m}^{n-1} o(|X_{k+1} - X_k|^2) + \sum_{k=m}^{n-1}[h(X_{k+1}, \alpha_{k+1}) - h(X_{k+1}, \alpha_k)]I_{\{\alpha_{k+1} \neq \alpha_k\}} \end{aligned}$$

Hence, it follows that

$$\mathbb{E}\prod_{\iota=1}^{\kappa} \rho_\iota(X^\varepsilon(t_\iota), \alpha^\varepsilon(t_\iota)) \sum_{k=m}^{n-1}[h(X_{k+1}, \alpha_{k+1}) - h(X_k, \alpha_k)] = (\mathrm{I}) + (\mathrm{II}) + (\mathrm{III}), \tag{5.18}$$

where

$$(\mathrm{I}) = \mathbb{E}\prod_{\iota=1}^{\kappa}\rho_\iota(X^\varepsilon(t_\iota),\alpha^\varepsilon(t_\iota))\sum_{k=m}^{n-1}\varepsilon\Big[\nabla h(X_k,\alpha_k)'b(X_k,\alpha_k)$$
$$+\frac{1}{2}\mathrm{tr}(\nabla^2 h(X_k,\alpha_k)\sigma\sigma'(X_k,\alpha_k))|\xi_k|^2\Big],$$

$$(\mathrm{II}) = \mathbb{E}\prod_{\iota=1}^{\kappa}\rho_\iota(X^\varepsilon(t_\iota),\alpha^\varepsilon(t_\iota))\sum_{k=m}^{n-1}[h(X_{k+1},\alpha_{k+1})-h(X_{k+1},\alpha_k)]I_{\{\alpha_{k+1}\neq\alpha_k\}},$$

and

$$(\mathrm{III}) = \mathbb{E}\prod_{\iota=1}^{\kappa}\rho_\iota(X^\varepsilon(t_\iota),\alpha^\varepsilon(t_\iota))\sum_{k=m}^{n-1}\Big[\sqrt{\varepsilon}\nabla h(X_k,\alpha_k)'\sigma(X_k,\alpha_k)\xi_k$$
$$+\frac{\varepsilon^2}{2}\mathrm{tr}(\nabla^2 h(X_k,\alpha_k)b(X_k,\alpha_k)b'(X_k,\alpha_k))+o(|X_{k+1}-X_k|^2)\Big].$$

Let us analyze these terms separately. First, using Assumption (H5.3) and the bounded convergence theorem, we can compute

$$\begin{aligned}(\mathrm{I}) &= \sum_{k=m}^{n-1}\mathbb{E}\Big[\prod_{\iota=1}^{\kappa}\rho_\iota(X^\varepsilon(t_\iota),\alpha^\varepsilon(t_\iota))\varepsilon\Big[\nabla h(X_k,\alpha_k)'b(X_k,\alpha_k)\\
&\qquad+\frac{1}{2}\mathrm{tr}(\nabla^2 h(X_k,\alpha_k)\sigma\sigma'(X_k,\alpha_k))|\xi_k|^2\Big]\\
&= \sum_{k=m}^{n-1}\mathbb{E}\Big[\prod_{\iota=1}^{\kappa}\rho_\iota(X^\varepsilon(t_\iota),\alpha^\varepsilon(t_\iota))\varepsilon\Big[\nabla h(X_k,\alpha_k)'b(X_k,\alpha_k)\\
&\qquad+\frac{1}{2}\mathrm{tr}(\nabla^2 h(X_k,\alpha_k)\sigma\sigma'(X_k,\alpha_k))\mathbb{E}[|\xi_k|^2|\mathcal{G}_k]\Big]\\
&= \mathbb{E}\prod_{\iota=1}^{\kappa}\rho_\iota(X^\varepsilon(t_\iota),\alpha^\varepsilon(t_\iota))\sum_{k=m}^{n-1}\varepsilon\Big[\nabla h(X_k,\alpha_k)'b(X_k,\alpha_k)\\
&\qquad+\frac{1}{2}\mathrm{tr}(\nabla^2 h(X_k,\alpha_k)\sigma\sigma'(X_k,\alpha_k))\Big]\\
&= \mathbb{E}\prod_{\iota=1}^{\kappa}\rho_\iota(X^\varepsilon(t_\iota),\alpha^\varepsilon(t_\iota))\int_t^{t+s}\Big[\nabla h(X^\varepsilon(u),\alpha^\varepsilon(u))'b(X^\varepsilon(u),\alpha^\varepsilon(u))\\
&\qquad+\frac{1}{2}\mathrm{tr}(\nabla^2 h(X^\varepsilon(u),\alpha^\varepsilon(u))\sigma\sigma'(X^\varepsilon(u),\alpha^\varepsilon(u)))\Big]du+O(\varepsilon)\end{aligned}$$

$$\to \mathbb{E}\prod_{\iota=1}^{\kappa}\rho_\iota(\widetilde{X}(t_\iota),\widetilde{\alpha}(t_\iota))\int_t^{t+s}\Big[\nabla h(\widetilde{X}(u),\widetilde{\alpha}(u))'b(\widetilde{X}(u),\widetilde{\alpha}(u))$$
$$+\frac{1}{2}\mathrm{tr}(\nabla^2 h(\widetilde{X}(u),\widetilde{\alpha}(u))\sigma\sigma'(\widetilde{X}(u),\widetilde{\alpha}(u)))\Big]du \tag{5.19}$$

as $\varepsilon \to 0$. Next, we rewrite (II) as

$$\begin{aligned}(\mathrm{II}) &= \mathbb{E}\prod_{\iota=1}^{\kappa}\rho_\iota(X^\varepsilon(t_\iota),\alpha^\varepsilon(t_\iota))\sum_{k=m}^{n-1}[h(X_k,\alpha_{k+1})-h(X_k,\alpha_k)]I_{\{\alpha_{k+1}\neq\alpha_k\}}\\ &\quad+\mathbb{E}\prod_{\iota=1}^{\kappa}\rho_\iota(X^\varepsilon(t_\iota),\alpha^\varepsilon(t_\iota))\sum_{k=m}^{n-1}[h(X_{k+1},\alpha_{k+1})-h(X_k,\alpha_{k+1})\\ &\qquad\qquad -h(X_{k+1},\alpha_k)+h(X_k,\alpha_k)]I_{\{\alpha_{k+1}\neq\alpha_k\}}\\ &:= (\mathrm{IIa})+(\mathrm{IIb}).\end{aligned}$$

Again, we use Assumption (H5.3) and the bounded convergence theorem to compute

$$\begin{aligned}(\mathrm{IIa}) &= \sum_{k=m}^{n-1}\mathbb{E}\left[\prod_{\iota=1}^{\kappa}\rho_\iota(X^\varepsilon(t_\iota),\alpha^\varepsilon(t_\iota))\mathbb{E}[[h(X_k,\alpha_{k+1})-h(X_k,\alpha_k)]I_{\{\alpha_{k+1}\neq\alpha_k\}}|\mathcal{G}_k]\right]\\ &= \sum_{k=m}^{n-1}\mathbb{E}\left[\prod_{\iota=1}^{\kappa}\rho_\iota(X^\varepsilon(t_\iota),\alpha^\varepsilon(t_\iota))\sum_{j\neq\alpha_k}[h(X_k,j)-h(X_k,\alpha_k)]\varepsilon q_{\alpha_k,j}(X_k)\right]\\ &= \mathbb{E}\left[\prod_{\iota=1}^{\kappa}\rho_\iota(X^\varepsilon(t_\iota),\alpha^\varepsilon(t_\iota))\int_t^{t+s}Q(X^\varepsilon(u)h(X^\varepsilon(u),\cdot)(\alpha^\varepsilon(u))du\right]+O(\varepsilon)\\ &\to \mathbb{E}\left[\prod_{\iota=1}^{\kappa}\rho_\iota(\widetilde{X}(t_\iota),\widetilde{\alpha}(t_\iota))\int_t^{t+s}Q(\widetilde{X}(u)h(\widetilde{X}(u),\cdot)(\widetilde{\alpha}(u))du\right],\end{aligned} \tag{5.20}$$

as $\varepsilon \to 0$, where

$$Q(x)h(x,\cdot)(i) := \sum_{j\neq I}q_{ij}(x)[h(x,j)-h(x,i)],\quad \forall (x,i)\in\mathbb{R}^r\times\mathcal{M}.$$

On the other hand, recalling that the functions ρ_ι, $\iota = 1,\ldots\kappa$ are bounded and that $h(\cdot,i)\in C_c^2(\mathbb{R}^r)$ for each $i\in\mathcal{M}$ and using the Hölder inequality, we have

$$
\begin{aligned}
\text{(IIb)} &\le K\mathbb{E}\Big[\sum_{k=m}^{n-1} |X_{k+1} - X_k| I_{\{\alpha_{k+1}\neq\alpha_k\}}\Big] \\
&= K\sum_{k=m}^{n-1} \mathbb{E}\big[[\varepsilon(1+|X_k|) + \sqrt{\varepsilon}(1+|X_k|)|\xi_k|] I_{\{\alpha_{k+1}\neq\alpha_k\}}\big] \\
&\le K\sum_{k=m}^{n-1} \mathbb{E}^{1/3}[(\varepsilon(1+|X_k|) + \sqrt{\varepsilon}(1+|X_k|)|\xi_k|)^3]\mathbb{E}^{2/3}[I_{\{\alpha_{k+1}\neq\alpha_k\}}] \\
&\le K\sum_{k=m}^{n-1} \varepsilon^{1/2}\cdot\varepsilon^{2/3} = O(\varepsilon^{1/6}) \to 0
\end{aligned} \tag{5.21}
$$

as $\varepsilon \to 0$, where we used Lemma 5.2 and the fact that $\mathbb{P}\{\alpha_{k+1} \neq \alpha_k\} = O(\varepsilon)$ (since $Q(\cdot)$ is bounded) to derive the last inequality. Finally, similar calculations as before reveal that

$$
\begin{aligned}
\text{(III)} &= \mathbb{E}\prod_{\iota=1}^{\kappa} \rho_\iota(X^\varepsilon(t_\iota), \alpha^\varepsilon(t_\iota)) \sum_{k=m}^{n-1} \Big[\frac{\varepsilon^2}{2}\text{tr}(\nabla^2 h(X_k, \alpha_k) b(X_k, \alpha_k) b'(X_k, \alpha_k)) \\
&\qquad + o(|X_{k+1} - X_k|^2)\Big] \\
&= o(\varepsilon) \to 0, \ \text{as } \varepsilon \to 0.
\end{aligned} \tag{5.22}
$$

Now, plugging (5.19)–(5.22) into (5.18) leads to

$$
\begin{aligned}
&\mathbb{E}\prod_{\iota=1}^{\kappa} \rho_\iota(X^\varepsilon(t_\iota), \alpha^\varepsilon(t_\iota)) \sum_{k=m}^{n-1} [h(X_{k+1}, \alpha_{k+1}) - h(X_k, \alpha_k)] \\
&\quad \to \mathbb{E}\prod_{\iota=1}^{\kappa} \rho_\iota(\widetilde{X}(t_\iota), \widetilde{\alpha}(t_\iota)) \int_t^{t+s} \Big[\nabla h(\widetilde{X}(u), \widetilde{\alpha}(u))' b(\widetilde{X}(u), \widetilde{\alpha}(u)) \\
&\qquad + \frac{1}{2}\text{tr}(\nabla^2 h(\widetilde{X}(u), \widetilde{\alpha}(u)) \sigma\sigma'(\widetilde{X}(u), \widetilde{\alpha}(u))) + Q(\widetilde{X}(u) h(\widetilde{X}(u), \cdot)(\widetilde{\alpha}(u))\Big] du \\
&\quad = \mathbb{E}\prod_{\iota=1}^{\kappa} \rho_\iota(\widetilde{X}(t_\iota), \widetilde{\alpha}(t_\iota)) \int_t^{t+s} \mathcal{L}h(\widetilde{X}(u), \widetilde{\alpha}(u)) du, \ \text{ as } \varepsilon \to 0.
\end{aligned} \tag{5.23}
$$

Finally, passing to the limit as $\varepsilon \to 0$ in (5.16) and using (5.17) and (5.23) imply the desired equation (5.15).

Step 3: Combining Steps 1 and 2, we conclude that $(\widetilde{X}(\cdot), \widetilde{\alpha}(\cdot))$, the weak limit of $(X^\varepsilon(\cdot), \alpha^\varepsilon(\cdot))$, is a solution of the martingale problem with operator $\mathcal{L}$ defined in (5.3). Because of the equivalence of solutions to the martingale problem and its corresponding stochastic differential equation (see Theorem A.25), it follows that $(\widetilde{X}(\cdot), \widetilde{\alpha}(\cdot))$ is a weak solution to the system (5.1)–(5.2). Recall from Theorem 2.6 that under (H5.1) and (H5.2), pathwise uniqueness holds for the system (5.1)–(5.2). Therefore, it follows from the Yamada-Watanabe principle (see, e.g., Section 5.3 of [149]) that $(\widetilde{X}(\cdot), \widetilde{\alpha}(\cdot))$ is the unique nonexplosive strong solution to (5.1)–(5.2). □

5.4 Convergence Rates

Because we are dealing with numerical algorithms, it is desirable that we obtain estimation error bounds on the rates of convergence. This section takes up the rates of convergence issue. In Kloeden and Platen [166, p. 323], the rate of convergence was defined as follows. For a finite time $T > 0$, if there exists a positive constant K independent of ε such that $\mathbb{E}|\widetilde{X}^{\varepsilon}(T) - X(T)| \le K\varepsilon^{\gamma}$ for some $\gamma > 0$, then the (piecewise linear) approximation $\widetilde{X}^{\varepsilon}$ is said to converge strongly to X with order γ. In Mao and Yuan [205], continuous approximation was considered. The convergence rate concentrates on the error bounds of uniform convergence of the form $\mathbb{E}\sup_{0\le t\le T} |\overline{X}^{\varepsilon}(t) - X(t)|^2$, where $\overline{X}^{\varepsilon}(t)$ is the continuous approximation. Such an approach is now commonly adopted; the mode of convergence is referred to as strong convergence and the corresponding rate of convergence is referred to as strong convergence rate. Using this approach, we consider the strong convergence together with the convergence rate in this section.

To begin, we assume the following condition holds.

(H5.4) Assume $Q(x)$ is Lipschitz continuous so that for some positive constant K_1, we have

$$\sum_{l\neq k} |q_{kl}(x) - q_{kl}(\widetilde{x})| \le K_1|x - \widetilde{x}|, \quad \forall k \in \mathcal{M}, x, \widetilde{x} \in \mathbb{R}^r. \tag{5.24}$$

In the spirit of the definition of continuous approximation as done in [124, 258, 297], we consider the following interpolation process of (5.4) given by

$$\begin{cases} \overline{X}^{\varepsilon}(t) = X_0 + \int_0^t b(\overline{X}^{\varepsilon}(s_\varepsilon), \overline{\alpha}^{\varepsilon}(s_\varepsilon))ds + \int_0^t \sigma(\overline{X}^{\varepsilon}(s_\varepsilon), \overline{\alpha}^{\varepsilon}(s_\varepsilon))dw(s), \\ \overline{\alpha}^{\varepsilon}(t) = \alpha_0 + \int_0^t \int_{\mathbb{R}_+} h(\overline{X}^{\varepsilon}(s_\varepsilon), \overline{\alpha}^{\varepsilon}(s-), z)\mathfrak{p}(ds, dz), \end{cases} \tag{5.25}$$

where $s_\varepsilon := \lfloor \frac{s}{\varepsilon} \rfloor \varepsilon$, $(X_0, \alpha_0) \in \mathbb{R}^r \times \mathcal{M}$ is the initial condition of the system (5.1)-(5.2), h is the function defined in (2.3), and $\mathfrak{p}$ is the Poisson random measure defined in Section 2.2. Note that (5.25) uniquely determines the process $(\overline{X}^{\varepsilon}(t), \overline{\alpha}^{\varepsilon}(t))$ first on $[0, \varepsilon]$ and then on $[\varepsilon, 2\varepsilon]$, $[2\varepsilon, 3\varepsilon]$ and so on. In addition, we have

Lemma 5.9 *Suppose Assumptions* (H5.1) *and* (H5.2) *hold, then for any* $T > 0$ *and* $p \ge 2$, *we have*

$$\mathbb{E}\Big[\sup_{0\le t\le T}|\overline{X}^{\varepsilon}(t)|^p\Big]\le K, \tag{5.26}$$

and hence

$$\max_{1\le k\le\lfloor\frac{T}{\varepsilon}\rfloor}\mathbb{E}\big[|\overline{X}^{\varepsilon}(k\varepsilon)|^p\big]\le K,$$

where $K = K(T, p, \mathbb{E}[|X_0|^p], K_0)$ *is a positive constant independent of* ε. *In addition, we have*

$$\sup_{s\in[0,T]}\mathbb{E}[|\overline{X}^{\varepsilon}(s)-\overline{X}^{\varepsilon}(s_\varepsilon)|^2]\le K\varepsilon, \tag{5.27}$$

$$\sup_{s\in[0,T]}\mathbb{P}\{\overline{\alpha}^{\varepsilon}(s)\ne\overline{\alpha}^{\varepsilon}(s_\varepsilon)|\mathcal{F}_{s_\varepsilon}\}\le K\varepsilon. \tag{5.28}$$

Proof The proof for (5.26) is standard; similar arguments can be found in the proof of Theorem 2.20 or Proposition 3.5 in [323]. Next, we note that $s - s_\varepsilon \in [0, \varepsilon)$ for any $s \in [0, T]$. Hence, it follows from the linear growth condition and (5.26) that

$$\begin{aligned}
&\mathbb{E}[|\overline{X}^{\varepsilon}(s)-\overline{X}^{\varepsilon}(s_\varepsilon)|^2]\\
&=\mathbb{E}[|b(\overline{X}^{\varepsilon}(s_\varepsilon),\overline{\alpha}^{\varepsilon}(s_\varepsilon))(s-s_\varepsilon)+\sigma(\overline{X}^{\varepsilon}(s_\varepsilon),\overline{\alpha}^{\varepsilon}(s_\varepsilon))(w(s)-w(s_\varepsilon))|^2]\\
&\le K\mathbb{E}[(1+|\overline{X}^{\varepsilon}(s_\varepsilon)|^2)(\varepsilon+w(s)-w(s_\varepsilon))|^2]\le K\varepsilon.
\end{aligned}$$

This establishes (5.27).

It remains to establish (5.28). To this end, we fix an arbitrary $s \in (0, T]$. Since $\overline{\alpha}^{\varepsilon}$ is a pure jump process whose evolution is described by the second equation of (5.25), for any $f \in \mathcal{B}_b(\mathcal{M}; \mathbb{R})$, the set of bounded and Borel measurable functions from $\mathcal{M}$ to $\mathbb{R}$, we can write

$$\begin{aligned}
&f(\overline{\alpha}^{\varepsilon}(s))-f(\overline{\alpha}^{\varepsilon}(s_\varepsilon))=\sum_{u\in(s_\varepsilon,s]}[f(\overline{\alpha}^{\varepsilon}(u))-f(\overline{\alpha}^{\varepsilon}(u-))]\\
&=\int_{s_\varepsilon}^{s}\int_{\mathbb{R}_+}[f(\overline{\alpha}^{\varepsilon}(u-)+h(\overline{X}^{\varepsilon}(u_\varepsilon),\overline{\alpha}^{\varepsilon}(u-),z))-f(\overline{\alpha}^{\varepsilon}(u-))]\mathfrak{p}(du,dz)\\
&=\int_{s_\varepsilon}^{s}\int_{\mathbb{R}_+}[f(\overline{\alpha}^{\varepsilon}(u-)+h(\overline{X}^{\varepsilon}(u_\varepsilon),\overline{\alpha}^{\varepsilon}(u-),z))-f(\overline{\alpha}^{\varepsilon}(u-))]\mathfrak{m}(dz)du+M(s)\\
&=\int_{s_\varepsilon}^{s}\sum_{j\ne\overline{\alpha}^{\varepsilon}(u-)}q_{\overline{\alpha}^{\varepsilon}(u-),j}(\overline{X}^{\varepsilon}(u_\varepsilon))[f(j)-f(\overline{\alpha}^{\varepsilon}(u-))]du+M(s),
\end{aligned}$$

where

$$M(s) = \int_{s_\varepsilon}^{s} \int_{\mathbb{R}_+} [f(\overline{\alpha}^\varepsilon(u-) + h(\overline{X}^\varepsilon(u_\varepsilon), \overline{\alpha}^\varepsilon(u-), z)) - f(\overline{\alpha}^\varepsilon(u-))]\widetilde{\mathfrak{p}}(ds, dz).$$

In particular, using the function $f(i) := I_{\{i \neq \overline{\alpha}^\varepsilon(s_\varepsilon)\}}, i \in \mathcal{M}$ and noting that $\sup_{i\in\mathcal{M}, x\in\mathbb{R}^r} q_i(x) := \sup_{i\in\mathcal{M}, x\in\mathbb{R}^r} \sum_{l\neq i} q_{il}(x) < K < \infty$ thanks to Assumption (H5.1), we have

$$\begin{aligned} &\mathbb{P}\{\overline{\alpha}^\varepsilon(s) \neq \overline{\alpha}^\varepsilon(s_\varepsilon)|\mathcal{F}_{s_\varepsilon}\} = \mathbb{E}[I_{\{\overline{\alpha}^\varepsilon(s)\neq\overline{\alpha}^\varepsilon(s_\varepsilon)\}}|\mathcal{F}_{s_\varepsilon}] \\ &= \mathbb{E}\bigg[\int_{s_\varepsilon}^{s} \sum_{j\neq\overline{\alpha}^\varepsilon(u-)} q_{\overline{\alpha}^\varepsilon(u-),j}(\overline{X}^\varepsilon(r_\varepsilon))[f(j) - f(\overline{\alpha}^\varepsilon(u-))]dr|\mathcal{F}_{s_\varepsilon}\bigg] \\ &\le \int_{s_\varepsilon}^{s} 2K dr \le 2K\varepsilon. \end{aligned}$$

This establishes (5.28) and completes the proof. □

We are now ready to present the main result of this section:

Theorem 5.10 *Suppose Assumptions* (H5.1)–(H5.4) *hold. Then, we have the following rate of convergence estimate:*

$$\mathbb{E}\bigg[\sup_{0\le t\le T} |\overline{X}^\varepsilon(t) - X(t)|^2\bigg] = O(\varepsilon^\gamma) \text{ as } \varepsilon \to 0, \tag{5.29}$$

where $\gamma \in (0, \frac{1}{2})$.

Proof For the processes α and $\overline{\alpha}^\varepsilon$, we have $\alpha(0) = \overline{\alpha}^\varepsilon(0) = \alpha_0 \in \mathcal{M}$. Thus, $\mathbb{P}\{\zeta^\varepsilon > 0\} = 1$, where $\zeta^\varepsilon := \inf\{t \ge 0 : \alpha(t) \neq \overline{\alpha}^\varepsilon(t)\}$. Furthermore, we can write

$$\begin{aligned} \mathbb{E}\bigg[\sup_{0\le t\le T} |\overline{X}^\varepsilon(t) - X(t)|^2\bigg] &= \mathbb{E}\bigg[\sup_{0\le t\le T} |\overline{X}^\varepsilon(t) - X(t)|^2 (I_{\{\zeta^\varepsilon > T\}} + I_{\{\zeta^\varepsilon \le T\}})\bigg] \\ &\le \mathbb{E}\bigg[\sup_{0\le t\le T\wedge\zeta^\varepsilon} |\overline{X}^\varepsilon(t) - X(t)|^2\bigg] + \mathbb{E}\bigg[I_{\{\zeta^\varepsilon\le T\}} \sup_{0\le t\le T} |\overline{X}^\varepsilon(t) - X(t)|^2\bigg]. \end{aligned} \tag{5.30}$$

We will establish appropriate bounds for these two terms in the rest of the proof.

We first fix an arbitrary $t \in (0, T]$ and compute

$$\begin{aligned} &\mathbb{E}\bigg[\sup_{0\le u\le t\wedge\zeta^\varepsilon} |\overline{X}^\varepsilon(u) - X(u)|^2\bigg] \\ &\le 2\mathbb{E} \sup_{0\le u\le t\wedge\zeta^\varepsilon} \bigg|\int_0^u [b(\overline{X}^\varepsilon(s_\varepsilon), \overline{\alpha}^\varepsilon(s_\varepsilon)) - b(X(s), \alpha(s))]ds\bigg|^2 \end{aligned}$$

$$
\begin{aligned}
&+ 2\mathbb{E} \sup_{0\le u\le t\wedge\zeta^\varepsilon} \left| \int_0^u [\sigma(\overline{X}^\varepsilon(s_\varepsilon), \overline{\alpha}^\varepsilon(s_\varepsilon)) - \sigma(X(s), \alpha(s))] dw(s) \right|^2 \\
&\le K\mathbb{E} \int_0^{t\wedge\zeta^\varepsilon} |b(\overline{X}^\varepsilon(s_\varepsilon), \overline{\alpha}^\varepsilon(s_\varepsilon)) - b(X(s), \alpha(s))|^2 ds \\
&+ K\mathbb{E} \int_0^{t\wedge\zeta^\varepsilon} |\sigma(\overline{X}^\varepsilon(s_\varepsilon), \overline{\alpha}^\varepsilon(s_\varepsilon)) - \sigma(X(s), \alpha(s))|^2 ds,
\end{aligned} \tag{5.31}
$$

where $K > 0$ is a generic constant independent of ε. Note that in (5.31), the first inequality is obtained from the elementary inequality $(a + b)^2 \le 2(a^2 + b^2)$ for two real numbers a and b. The term on the fourth line of (5.31) is an application of Hölder's inequality to the term on the second line, and the last line of (5.31) is a consequence of the well-known Doob martingale inequality (see (A.19) in the appendix) applied to the term on the third line.

To proceed, we treat the drift and diffusion terms separately. By virtue of the triangle inequality

$$
\begin{aligned}
&\mathbb{E} \int_0^{t\wedge\zeta^\varepsilon} |b(\overline{X}^\varepsilon(s_\varepsilon), \overline{\alpha}^\varepsilon(s_\varepsilon)) - b(X(s), \alpha(s))|^2 ds \\
&\quad \le 2\mathbb{E} \int_0^{t\wedge\zeta^\varepsilon} |b(\overline{X}^\varepsilon(s_\varepsilon), \overline{\alpha}^\varepsilon(s_\varepsilon)) - b(\overline{X}^\varepsilon(s), \overline{\alpha}^\varepsilon(s))|^2 ds \\
&\quad\quad + 2\mathbb{E} \int_0^{t\wedge\zeta^\varepsilon} |b(\overline{X}^\varepsilon(s), \overline{\alpha}^\varepsilon(s)) - b(X(s), \alpha(s))|^2 ds.
\end{aligned} \tag{5.32}
$$

Using the Lipschitz continuity of $b(\cdot, \alpha)$ and (5.27), we can compute

$$
\begin{aligned}
&\mathbb{E} \int_0^{t\wedge\zeta^\varepsilon} |b(\overline{X}^\varepsilon(s_\varepsilon), \overline{\alpha}^\varepsilon(s_\varepsilon)) - b(\overline{X}^\varepsilon(s), \overline{\alpha}^\varepsilon(s))|^2 ds \\
&\le 2\mathbb{E} \int_0^t |b(\overline{X}^\varepsilon(s_\varepsilon), \overline{\alpha}^\varepsilon(s_\varepsilon)) - b(\overline{X}^\varepsilon(s_\varepsilon), \overline{\alpha}^\varepsilon(s))|^2 ds \\
&\quad + 2\mathbb{E} \int_0^t |b(\overline{X}^\varepsilon(s), \overline{\alpha}^\varepsilon(s)) - b(\overline{X}^\varepsilon(s_\varepsilon), \overline{\alpha}^\varepsilon(s))|^2 ds \\
&\le K\mathbb{E} \int_0^t (1 + |\overline{X}^\varepsilon(s_\varepsilon)|^2) I_{\{\overline{\alpha}^\varepsilon(s_\varepsilon)\ne\overline{\alpha}^\varepsilon(s)\}} ds + K\mathbb{E} \int_0^t |\overline{X}^\varepsilon(s) - \overline{X}^\varepsilon(s_\varepsilon)|^2 ds \\
&\le K\mathbb{E} \int_0^t (1 + |\overline{X}^\varepsilon(s_\varepsilon)|^2) \mathbb{E}[I_{\{\overline{\alpha}^\varepsilon(s_\varepsilon)\ne\overline{\alpha}^\varepsilon(s)\}} | \mathcal{F}_{s_\varepsilon}] ds + K\varepsilon \\
&\le K\varepsilon,
\end{aligned} \tag{5.33}
$$

where the last inequality follows from (5.26) and (5.28).

We next concentrate on the last term of (5.32). Since $\overline{\alpha}^{\varepsilon}(s) = \alpha(s)$ for $s \le \zeta^{\varepsilon}$, in view of (5.27) together with the Lipschitz continuity of $b(\cdot, \alpha)$, we have

$$\begin{aligned}
&\mathbb{E}\int_0^{t\wedge\zeta^{\varepsilon}} |b(\overline{X}^{\varepsilon}(s), \overline{\alpha}^{\varepsilon}(s)) - b(X(s), \alpha(s))|^2 ds \\
&\le K\mathbb{E}\int_0^{t\wedge\zeta^{\varepsilon}} |\overline{X}^{\varepsilon}(s) - X(s)|^2 ds \\
&\le K\mathbb{E}\int_0^{t} \sup_{0\le u\le s\wedge\zeta^{\varepsilon}} |\overline{X}^{\varepsilon}(u) - X(u)|^2 ds \\
&= K\int_0^{t} \mathbb{E}\Big[\sup_{0\le u\le s\wedge\zeta^{\varepsilon}} |\overline{X}^{\varepsilon}(u) - X(u)|^2\Big] ds.
\end{aligned} \tag{5.34}$$

We now plugging (5.33) and (5.34) into (5.32) to obtain

$$\begin{aligned}
&\mathbb{E}\int_0^{t\wedge\zeta^{\varepsilon}} |b(\overline{X}^{\varepsilon}(s_{\varepsilon}), \overline{\alpha}^{\varepsilon}(s_{\varepsilon})) - b(X(s), \alpha(s))|^2 ds \\
&\le K\int_0^{t} \mathbb{E}\Big[\sup_{0\le u\le s\wedge\zeta^{\varepsilon}} |\overline{X}^{\varepsilon}(u) - X(u)|^2\Big] ds + K\varepsilon.
\end{aligned}$$

Similar calculations reveal that

$$\begin{aligned}
&\mathbb{E}\int_0^{t\wedge\zeta^{\varepsilon}} |\sigma(\overline{X}^{\varepsilon}(s_{\varepsilon}), \overline{\alpha}^{\varepsilon}(s_{\varepsilon})) - \sigma(X(s), \alpha(s))|^2 ds \\
&\le K\int_0^{t} \mathbb{E}\Big[\sup_{0\le u\le s\wedge\zeta^{\varepsilon}} |\overline{X}^{\varepsilon}(u) - X(u)|^2\Big] ds + K\varepsilon.
\end{aligned}$$

Using these two displayed equations in (5.31), we have

$$\mathbb{E}\Big[\sup_{0\le u\le t\wedge\zeta^{\varepsilon}} |\overline{X}^{\varepsilon}(u) - X(u)|^2\Big] \le K\int_0^{t} \mathbb{E}\Big[\sup_{0\le r\le s\wedge\zeta^{\varepsilon}} |\overline{X}^{\varepsilon}(r) - X(r)|^2\Big] ds + K\varepsilon.$$

An application of Gronwall's inequality then leads to

$$\mathbb{E}\Big[\sup_{0\le u\le t\wedge\zeta^{\varepsilon}} |\overline{X}^{\varepsilon}(u) - X(u)|^2\Big] \le e^{Kt}K\varepsilon \le K\varepsilon, \tag{5.35}$$

where K is independent of $t \in (0, T]$ and $\varepsilon > 0$. In particular, (5.35) holds for $t = T$.

We now concentrate on the second term of (5.30). To this end, we first estimate $\mathbb{P}\{\zeta^{\varepsilon} \le T\}$. Thanks to the SDEs for α and $\overline{\alpha}^{\varepsilon}$, we have

$$\overline{\alpha}^{\varepsilon}(t) - \alpha(t) = \int_0^t \int_{\mathbb{R}_+} [h(\overline{X}^{\varepsilon}(s_{\varepsilon}), \overline{\alpha}^{\varepsilon}(s-), z) - h(X(s), \alpha(s-), z)]\mathfrak{p}(ds, dz).$$

Consequently, and noting the fact that $\overline{\alpha}^{\varepsilon}(t) = \alpha(t)$ for all $t < \zeta^{\varepsilon}$, we can use (2.32) and (5.24) in the first two inequalities below to compute

$$\begin{aligned}
\mathbb{P}\{\zeta^{\varepsilon} \le T\} &= \mathbb{E}[I_{\{\overline{\alpha}^{\varepsilon}(T\wedge\zeta^{\varepsilon}) \neq \alpha(T\wedge\zeta^{\varepsilon})\}}] \\
&= \mathbb{E}\bigg[\int_0^{T\wedge\zeta^{\varepsilon}} \int_{\mathbb{R}_+} I_{\{h(\overline{X}^{\varepsilon}(s_{\varepsilon}),\overline{\alpha}^{\varepsilon}(s-),z) - h(X(s),\alpha(s-),z) \neq 0\}} \mathfrak{p}(ds, dz)\bigg] \\
&= \mathbb{E}\bigg[\int_0^{T\wedge\zeta^{\varepsilon}} \int_{\mathbb{R}_+} I_{\{h(\overline{X}^{\varepsilon}(s_{\varepsilon}),\overline{\alpha}^{\varepsilon}(s-),z) - h(X(s),\alpha(s-),z) \neq 0\}} \mathfrak{m}(dz)ds\bigg] \\
&\le \mathbb{E}\bigg[\int_0^{T\wedge\zeta^{\varepsilon}} \sum_{l \neq \alpha(s-)} |q_{\alpha(s-),l}(\overline{X}^{\varepsilon}(s_{\varepsilon})) - q_{\alpha(s-),l}(X(s))|ds\bigg] \\
&\le \mathbb{E}\bigg[\int_0^{T\wedge\zeta^{\varepsilon}} K|\overline{X}^{\varepsilon}(s_{\varepsilon}) - X(s)|ds\bigg] \\
&\le K\mathbb{E}\bigg[\int_0^{T\wedge\zeta^{\varepsilon}} (|\overline{X}^{\varepsilon}(s) - X(s)| + |\overline{X}^{\varepsilon}(s_{\varepsilon}) - \overline{X}^{\varepsilon}(s)|)ds\bigg] \\
&\le K\int_0^T \bigg[\bigg(\mathbb{E}\bigg[\sup_{0\le u\le s\wedge\zeta^{\varepsilon}} |\overline{X}^{\varepsilon}(u) - X(u)|^2\bigg]\bigg)^{\frac{1}{2}} \\
&\quad + \mathbb{E}[|\overline{X}^{\varepsilon}(s_{\varepsilon}) - \overline{X}^{\varepsilon}(s)|]\bigg]ds \\
&\le K\int_0^T \Big(K\varepsilon^{\frac{1}{2}} + K\varepsilon^{\frac{1}{2}}\Big)ds \\
&= K\varepsilon^{\frac{1}{2}},
\end{aligned} \tag{5.36}$$

where the last inequality follows from (5.35) and (5.27). Then, we can use the Hölder inequality and (5.26) to derive

$$\begin{aligned}
&\mathbb{E}\bigg[I_{\{\zeta^{\varepsilon}\le T\}} \sup_{0\le t\le T} |\overline{X}^{\varepsilon}(t) - X(t)|^2\bigg] \\
&\quad \le (\mathbb{P}\{\zeta^{\varepsilon} \le T\})^{\frac{1}{p}} \bigg(\mathbb{E}\bigg[\sup_{0\le t\le T} |\overline{X}^{\varepsilon}(t) - X(t)|^{2q}\bigg]\bigg)^{\frac{1}{q}} \\
&\quad \le K\varepsilon^{\frac{1}{2p}},
\end{aligned} \tag{5.37}$$

where $p, q > 1$ satisfy $\frac{1}{p} + \frac{1}{q} = 1$.

Finally, by substituting (5.35) and (5.37) into (5.30), we obtain

$$\mathbb{E}\left[\sup_{0\le t\le T} |\overline{X}^{\varepsilon}(t) - X(t)|^2\right] \le K\varepsilon + K\varepsilon^{\frac{1}{2p}} \le K\varepsilon^{\frac{1}{2p}}.$$

In particular, for any $\gamma \in (0, \frac{1}{2})$, we can select $p = \frac{1}{2\gamma}$ to derive (5.29) and therefore completes the proof. □

Remark 5.11 Consider the discrete distance $\rho(i, j) := \delta_{ij}$ on $\mathcal{M}$, where $\delta_{ij} = 1$ if $i \ne j$ and 0 if $i = j$. Also define $d((x, i), (y, j)) := |x - y| + \rho(i, j)$ for $(x, i), (y, j) \in \mathbb{R}^r \times \mathcal{M}$. One can verify that $d(\cdot, \cdot)$ is a distance function on $\mathbb{R}^r \times \mathcal{M}$. If the conditions of Theorem 5.10 hold true, then (5.36) implies that

$$\mathbb{E}\left[\sup_{0\le t\le T} \rho(\alpha(t), \overline{\alpha}^{\varepsilon}(t))\right] \le \mathbb{P}\{\zeta^{\varepsilon} \le T\} \le K\varepsilon^{\frac{1}{2}}.$$

This together with (5.29) leads to

$$\mathbb{E}\left[\sup_{0\le t\le T} d((X(t), \alpha(t)), (\overline{X}^{\varepsilon}(t), \overline{\alpha}^{\varepsilon}(t)))\right] = O(\varepsilon^{\gamma})$$

in which $\gamma \in (0, \frac{1}{4})$. This gives the convergence rate for the numerical scheme $(\overline{X}^{\varepsilon}, \overline{\alpha}^{\varepsilon})$ of (5.25) to the true solution (X, α) of the system (5.1)–(5.2).

Corollary 5.12 *In addition to the conditions of Theorem 5.10, if the coefficients b and σ are bounded, then*

$$\mathbb{E}\left[\sup_{0\le t\le T} |\overline{X}^{\varepsilon}(t) - X(t)|^2\right] = O(\varepsilon^{\frac{1}{2}}).$$

Proof In lieu of (5.31), we have for any $t \in (0, T]$

$$\mathbb{E}\left[\sup_{0\le u\le t} |\overline{X}^{\varepsilon}(u) - X(u)|^2\right] \le K\mathbb{E}\int_0^t |b(\overline{X}^{\varepsilon}(s_{\varepsilon}), \overline{\alpha}^{\varepsilon}(s_{\varepsilon})) - b(X(s), \alpha(s))|^2 ds$$
$$+ K\mathbb{E}\int_0^t |\sigma(\overline{X}^{\varepsilon}(s_{\varepsilon}), \overline{\alpha}^{\varepsilon}(s_{\varepsilon})) - \sigma(X(s), \alpha(s))|^2 ds. \tag{5.38}$$

Furthermore, using similar calculations as those in the proof of Theorem 5.10, we can compute

$$
\begin{aligned}
&\mathbb{E}\int_0^t |b(\overline{X}^\varepsilon(s_\varepsilon), \overline{\alpha}^\varepsilon(s_\varepsilon)) - b(X(s), \alpha(s))|^2 ds \\
&\le 2\mathbb{E}\int_0^t |b(\overline{X}^\varepsilon(s_\varepsilon), \alpha(s)) - b(X(s), \alpha(s))|^2 ds \\
&\quad + 2\mathbb{E}\int_0^t |b(\overline{X}^\varepsilon(s_\varepsilon), \overline{\alpha}^\varepsilon(s_\varepsilon)) - b(\overline{X}^\varepsilon(s_\varepsilon), \alpha(s))|^2 ds \\
&\le 2K\mathbb{E}\int_0^t |\overline{X}^\varepsilon(s_\varepsilon) - X(s)|^2 ds \\
&\quad + 2\mathbb{E}\int_0^t |b(\overline{X}^\varepsilon(s_\varepsilon), \overline{\alpha}^\varepsilon(s_\varepsilon)) - b(\overline{X}^\varepsilon(s_\varepsilon), \alpha(s))|^2 I_{\{\overline{\alpha}^\varepsilon(s_\varepsilon)\ne\alpha(s)\}} ds \\
&\le K\mathbb{E}\int_0^t \Big(\sup_{0\le u\le s} |\overline{X}^\varepsilon(s) - X(s)|^2 + |\overline{X}^\varepsilon(s_\varepsilon) - \overline{X}^\varepsilon(s)|^2\Big) ds \\
&\quad + \mathbb{E}\int_0^t K(I_{\{\overline{\alpha}^\varepsilon(s_\varepsilon)\ne\overline{\alpha}^\varepsilon(s)\}} + I_{\{\overline{\alpha}^\varepsilon(s)\ne\alpha(s)\}}) ds \\
&\le K\int_0^t \mathbb{E}\Big[\sup_{0\le u\le s} |\overline{X}^\varepsilon(s) - X(s)|^2\Big] ds + K\varepsilon + K\mathbb{E}\int_0^t (K\varepsilon + \mathbb{P}\{\zeta^\varepsilon \le s\}) ds \\
&\le K\int_0^t \mathbb{E}\Big[\sup_{0\le u\le s} |\overline{X}^\varepsilon(u) - X(u)|^2\Big] ds + K\varepsilon + K\varepsilon^{\frac{1}{2}},
\end{aligned}
$$

where the third inequality above follows from the boundedness of b, the fourth inequality follows from (5.27) and (5.28), and the last inequality follows from (5.36) and the fact that $\mathbb{P}\{\zeta^\varepsilon \le s\} \le \mathbb{P}\{\zeta^\varepsilon \le T\}$. In the above, K is a generic constant independent of ε and t. Likewise

$$
\begin{aligned}
&\mathbb{E}\int_0^t |\sigma(\overline{X}^\varepsilon(s_\varepsilon), \overline{\alpha}^\varepsilon(s_\varepsilon)) - \sigma(X(s), \alpha(s))|^2 ds \\
&\quad \le K\int_0^t \mathbb{E}\Big[\sup_{0\le u\le s} |\overline{X}^\varepsilon(u) - X(u)|^2\Big] ds + K\varepsilon + K\varepsilon^{\frac{1}{2}}.
\end{aligned}
$$

Using these equations in (5.38), it follows that

$$
\mathbb{E}\Big[\sup_{0\le u\le t} |\overline{X}^\varepsilon(u) - X(u)|^2\Big] \le K\int_0^t \mathbb{E}\Big[\sup_{0\le u\le s} |\overline{X}^\varepsilon(u) - X(u)|^2\Big] ds + K\varepsilon^{\frac{1}{2}}.
$$

An application of Gronwall's inequality then leads to

$$
\mathbb{E}\Big[\sup_{0\le u\le t} |\overline{X}^\varepsilon(u) - X(u)|^2\Big] \le K\varepsilon^{\frac{1}{2}}, \quad \forall t \in (0, T].
$$

The proof is complete. □

Corollary 5.13 *In addition to the conditions of Theorem 5.10, if* $\sigma(x,i) = \sigma$, *a constant matrix, then*

$$\mathbb{E}\Big[\sup_{0\le t\le T} |\overline{X}^{\varepsilon}(t) - X(t)|\Big] = O(\varepsilon^{\gamma}), \quad \gamma \in \Big(0, \frac{1}{2}\Big).$$

If, in addition, b is bounded, then $\mathbb{E}\big[\sup_{0\le t\le T} |\overline{X}^{\varepsilon}(t) - X(t)|\big] = O(\varepsilon^{\frac{1}{2}})$.

Proof When $\sigma(x,i)$ is a constant matrix, in lieu of (5.31), we can write

$$\begin{aligned}
&\mathbb{E}\Big[\sup_{0\le t\le T} |\overline{X}^{\varepsilon}(t) - X(t)|\Big] \\
&\le K\mathbb{E}\int_0^T |b(\overline{X}^{\varepsilon}(s_\varepsilon), \overline{\alpha}^{\varepsilon}(s_\varepsilon)) - b(X(s), \alpha(s))|ds \\
&\le \mathbb{E}\int_0^T |b(\overline{X}^{\varepsilon}(s_\varepsilon), \alpha(s)) - b(X(s), \alpha(s))|ds \\
&\quad + \mathbb{E}\int_0^T |b(\overline{X}^{\varepsilon}(s_\varepsilon), \overline{\alpha}^{\varepsilon}(s_\varepsilon)) - b(\overline{X}^{\varepsilon}(s_\varepsilon), \alpha(s))|ds \\
&\le K\mathbb{E}\int_0^T |\overline{X}^{\varepsilon}(s_\varepsilon) - X(s)|ds + K\mathbb{E}\int_0^T (1 + |\overline{X}^{\varepsilon}(s_\varepsilon)|) I_{\{\alpha^{\varepsilon}(s_\varepsilon) \ne \alpha(s)\}} ds.
\end{aligned}$$

The rest of the proof is similar to those in the proofs of Theorem 5.10 and Corollary 5.12. We shall omit the details for brevity. □

5.5 Examples

Here, we present two examples for demonstration. It would be ideal if we could compare the numerical solutions using our algorithms with the analytic solutions. Unfortunately, due to the complexity of the x-dependent switching process, closed-form solutions are not available. We are thus contended with the numerical solutions. In both examples, we use the state-dependent generator $Q(x)$ given by

$$Q(x) = \begin{pmatrix} -5\cos^2 x & 5\cos^2 x \\ 10\cos^2 x & -10\cos^2 x \end{pmatrix}.$$

Example 5.14 Consider (5.1) that is linear in the continuous state, which is known as a jump linear system. Suppose that the drift is given by

$$b(x,i) = A(i)x \ \text{ with } \ A(1) = -3.3 \ \text{ and } \ A(2) = -2.7,$$

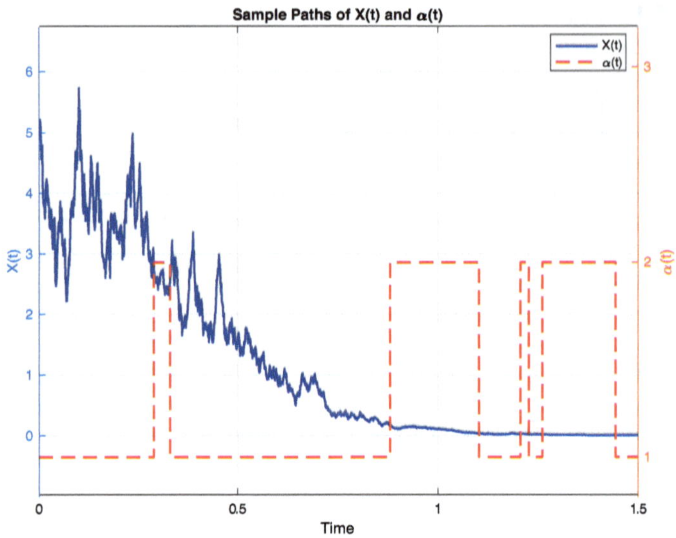

Fig. 5.1 A sample path of a numerical approximation to a jump linear system

and the diffusion is specified as

$$\sigma(x, 1) = 2x \text{ and } \sigma(x, 2) = x.$$

We use the constant stepsize algorithm (5.4). Specify the initial conditions as $X_0 = 5$ and $\alpha_0 = 1$, and use the constant stepsize 0.001. A sample path of the computed iterations is depicted in Fig. 5.1.

Example 5.15 This example is concerned with a switching diffusion system with nonlinear drift with respect to the continuous-state variable. For each $i \in \{1, 2\}$, the drifts are nonlinear functions:

$$b(x, 1) = -(2 + \sin x) \text{ and } b(x, 2) = -(1 + \sin x \cos x).$$

The diffusion terms are given as the follows:

$$\sigma(x, 1) = 0.5x, \quad \sigma(x, 2) = 0.2x.$$

Using the same initial data and stepsize as in Example 5.14, the calculation is carried out. The computational result is displayed in Fig. 5.2.

For these numerical examples, we have tried different stepsizes. They all produced similar sample path behavior as displayed above. For numerical experiment purposes, we have also tested different functions $b(\cdot, \cdot)$ and $\sigma(\cdot, \cdot)$.

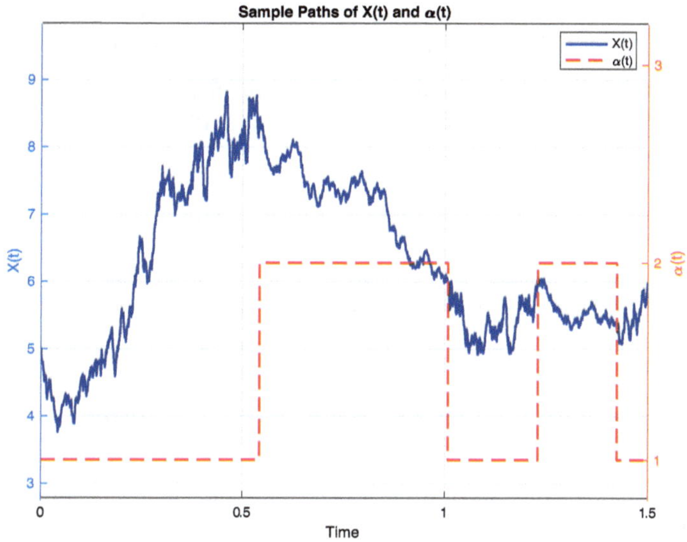

Fig. 5.2 A sample path of switching diffusion with nonlinear drifts

5.6 Two Remarks

5.6.1 Remarks on Markovian Switching Diffusions

In the previous results presented, it is seen that the rate of convergence is much slower than the convergence results of numerical algorithms for approximation of diffusion processes. Suppose that $y(t)$ is a diffusion process and $Y^\varepsilon(t)$ is its piecewise constant interpolation and $\overline{Y}^\varepsilon(t)$ is its continuous interpolation. Then, it is well known that

$$\mathbb{E} \sup_{t\in[0,T]} |y(t) - \overline{Y}^\varepsilon(t)|^2 = O(\varepsilon).$$

We now consider a Markovian switching diffusion $(X(t), \alpha(t))$ as in (5.1), where $\alpha(t)$ is a continuous-time Markov chain independent of the Brownian motion. We fix some $\varepsilon > 0$ and define $\alpha_n := \alpha(n\varepsilon)$ for each $n = 0, 1, \ldots$ Also let $\{X_n\}$ be the sequence generated by the Euler scheme (similarly defined as in (5.4)). We then define the piecewise constant interpolation $X^\varepsilon(\cdot)$ and the continuous interpolation $\overline{X}^\varepsilon(\cdot)$ as

$$\begin{aligned} &X^\varepsilon(t) = X_n\ \alpha^\varepsilon(t) = \alpha_n, \ \text{for } t \in [n\varepsilon, n\varepsilon + \varepsilon), \\ &\overline{X}^\varepsilon(t) = X_0 + \int_0^t b(X^\varepsilon(s), \alpha^\varepsilon(s))ds + \int_0^t \sigma(X^\varepsilon(s), \alpha^\varepsilon(s))dw(s). \end{aligned} \tag{5.39}$$

We obtain the convergence rate result as follows. As can be seen that weaker conditions are used, tighter estimation error bounds can be obtained.

Theorem 5.16 *Consider the numerical algorithm* (5.4). *Suppose* (H5.1), (H5.2), *and* (H5.3) (a) *are satisfied. Then, for* $\overline{X}^{\varepsilon}(t)$ *defined in* (5.39), *we have the following uniform convergence rate:*

$$\mathbb{E} \sup_{t\in[0,T]} |\overline{X}^{\varepsilon}(t) - X(t)|^2 = O(\varepsilon). \tag{5.40}$$

For a proof of the result, we refer the reader to the book [205] for details.

5.6.2 Remarks on a Class of Truncation Algorithms

We give a remark in this section on a class of projection/truncation algorithms. Consider (5.1), in which $\alpha(\cdot)$ is a continuous-time Markov chain independent of the Brownian motion $w(\cdot)$ with state space $\mathcal{M} = \{1, \dots, m_0\}$, a finite set. Suppose that the coefficients of the SDEs are locally Lipschitz and there is a nonnegative function $V : \mathbb{R}^r \times \mathcal{M} \to \mathbb{R}_+$ satisfying $V(\cdot, i) \in C^2$, and for some $K > 0$

$$\begin{aligned} &\mathcal{L}V(x,i) \le KV(x,i) \ \text{ for each } \ i \in \mathcal{M}, \\ &V_R = \inf_{|x|>R} V(x,i) \to \infty \ \text{ as } \ R \to \infty \ \text{ for each } \ i \in \mathcal{M}. \end{aligned} \tag{5.41}$$

The growth condition (5.41) is often referred to as the Khasminskii condition. Our primary concern is the coefficients of (5.1) (especially the drift) with respect to the variable x grow faster than linear. A straightforward implementation of the Euler-Maruyama numerical scheme may lead to numerical finite explosion times (see [221, 222, 242, 306]) even if the solutions of the corresponding switching stochastic differential equations are well behaved. In contrast to the previous work, in lieu of modifying the drift and/or diffusion coefficients, we generate randomly increasing truncation regions.

Our approach is inspired by a class of stochastic approximation/optimization algorithms. The essence is to generate randomly growing truncation bounds. It stems from stochastic approximation/optimization algorithms for solving root-finding or optimization problems under noisy observations. In such problems, it is often required to restrict the growth rate of the functions involved. As a remedy, some truncation or projection algorithms are often used to push the estimates to remain in certain bounded regions [177]. In [41], Chen and Zhu proposed to generate a randomly increasing sequence of truncation bounds in lieu of a fixed truncation bound. When the iterates are outside the truncation region, the scheme forces the iterate to return to a fixed point within the bounding regions. Such an idea was subsequently extended in Andradóttir [4] by replacing the fixed point of return with a sequence of sets if the iterate ever exits the bounding regions. One of the advantages is that neither the growth rate of the function nor the prior knowledge of the root or optimum needs to be specified. Motivated by Andradóttir [4] and Chen and Zhu [41], to prevent the unwanted excursion for numerical solutions of the SDEs, we proposed a class of new algorithms using a sequence of randomly varying truncation bounds. In reference to [306], and in contrast to [4, 41], in which the iterates are eventually bounded, our random-varying truncation bounds enable the iterates to be well behaved (no numerical finite explosion time).

We present an explicit numerical scheme with randomly generated truncations to approximate the solution of (5.1), which was initially proposed and developed in [242]. Given a step size $\varepsilon > 0$, and finite time horizon $[0, T]$ for some $T > 0$, we set $N = \lfloor \frac{T}{\varepsilon} \rfloor$. For the discretization, we consider a sequence of Brownian increments $\{\xi_n\}$ defined by

$$\xi_n = \frac{w(\varepsilon(n+1)) - w(\varepsilon n)}{\sqrt{\varepsilon}} = \frac{\Delta w(\varepsilon n)}{\sqrt{\varepsilon}}. \tag{5.42}$$

It is readily seen that $\mathbb{E}\xi_n = 0$ and $\mathbb{E}\xi_n\xi_n' = I$, the r-dimensional identity matrix. For a fixed $\varepsilon > 0$, we let $\{\alpha_n := \alpha(n\varepsilon), n = 0, 1, \ldots\}$ be the skeleton of the continuous-time Markov chain $\alpha(\cdot)$. Also, let $\{s(n)\}$ be an increasing sequence of positive real numbers such that $s(n) \to \infty$ as $n \to \infty$. In the computation, we choose $s(0)$ to be sufficiently large. Our numerical algorithm is

$$\begin{cases} X_0 = x_0, \, \rho(0) = 0, \\ \widetilde{X}_{n+1} = X_n + \varepsilon b(X_n, \alpha_n) + \sqrt{\varepsilon}\sigma(X_n, \alpha_n)\xi_n, \\ \rho(n+1) = \rho(n) + I_{\{|\widetilde{X}_{n+1}| > s(\rho(n))\}}, \\ X_{n+1} = \widetilde{X}_{n+1} I_{\{|\widetilde{X}_{n+1}| \le s(\rho(n))\}} + \pi_n(\widetilde{X}_{n+1}) I_{\{|\widetilde{X}_{n+1}| > s(\rho(n))\}}, \end{cases} \tag{5.43}$$

where $\pi_n(x)$ is the projection operator acting on x so that $\pi_n(x)$ takes x to the boundary of $O_n := B(0, s(\rho(n)))$, the ball with radius $s(\rho(n))$ centered at 0. Define

$$q_n := \left[\pi_n(\widetilde{X}_{n+1}) - \widetilde{X}_{n+1}\right] I_{\{|\widetilde{X}_{n+1}| > s(\rho(n))\}}. \tag{5.44}$$

Note that

$$I_{\{|\widetilde{X}_{n+1}| \le s(\rho(n))\}} = 1 - I_{\{|\widetilde{X}_{n+1}| > s(\rho(n))\}}.$$

Then, the numerical scheme (5.43) can be rewritten as the following compact form:

$$X_{n+1} = X_n + \varepsilon b(X_n, \alpha_n) + \sqrt{\varepsilon}\sigma(X_n, \alpha_n)\xi_n + q_n. \tag{5.45}$$

The rational is that at time n, if the iterate is within the ball O_n, we keep the iterate as is. If it ever exits O_n, we project it back to the boundary of O_n. As we update the approximation to the solution of the stochastic differential equations, we also update the truncation region O_n. In addition, the truncation bounds of O_n grow and are generated randomly due to the randomness of $\rho(n)$ defined in (5.43).

One of the features of (5.43) is that it can be easily implemented by adding a logic statement in coding. The drift function need not be specified. In addition, it performs better than a straightforward use of the Euler-Maruyama scheme, for which the iterates could blow up at finite time.

Because $\{s(n)\}$ is an appropriate increasing sequence of positive real numbers and $\{\rho(n)\}$ is a monotone increasing sequence with probability one (w.p.1), $\{s(\rho(n))\}$ is a monotone increasing sequence w.p.1. Note that either $\{s(\rho(n))\}$ has a finite limit w.p.1 or diverges w.p.1. It can be shown that it cannot have a finite limit. Note that this is a distinctive feature compared with the stochastic approximation problems studied in [4, 41]. In the aforementioned references, the iterates will be eventually confined to a bounded region. Because of the Brownian motion, the w.p.1 boundedness no longer holds. Nevertheless, our truncation algorithm has the advantage of suppressing the faster than linear growth of the stochastic differential equation and still ensure no finite explosion time. In this section, with the truncation algorithm presented, we state the convergence result below.

Theorem 5.17 *Assume that the drift and diffusion coefficients $b(\cdot, i)$ and $\sigma(\cdot, i)$ are locally Lipschitz for each $i \in \mathcal{M}$ and that the Khasminskii condition* (5.41) *holds. Define the piecewise-constant interpolation process as $x^{\varepsilon}(t) = X_n$ for $t \in [n\varepsilon, n\varepsilon + \varepsilon)$. Then, $x^{\varepsilon}(\cdot)$ converges weakly to $x(\cdot)$, which is a solution of* (5.1).

For a proof of the convergence result, we refer the reader to [242, Theorem 3.4 and Corollary 3.6]. Without assuming the the global Lipschitz conditions, it seems that strong convergence rate will be virtually impossible to obtain for our algorithm. As an alternative, a scaled and interpolated sequence that converges weakly to a diffusion limit was obtained in the aforementioned reference. The scaling factor together with the stationary covariance gives us the desired convergence rate; see [242, Section 4] for the detailed development.

5.7 Notes

Numerical methods for stochastic differential equations have been studied extensively, for example, in [166, 213] among others. A comprehensive study of the early results for solving diffusion equations is contained in Kloeden and Platen [166]. Accelerated rates of convergence are given in Milstein [213]. As a natural extension, further results are considered in Milstein and Tretyakov [214], among others. Numerical solutions for stochastic differential equations modulated by Markovian switching have also been well studied; recent progress for switching diffusions is contained in [205] and references therein. Recently, we have examined Milstein-type algorithms for stochastic differential equations with Markovian switching [222]. It is confirmed that the scheme also produces optimal or accelerated rates of convergence.

Although numerical methods for Markov modulated switching diffusions have been considered by many researchers, less is known for processes with continuous-state-dependent switching. In fact, the study of switching diffusions with state-dependent switching is still in its infancy. There are certain difficulties. For example, the usual Picard iterations cannot be used. When Markovian regime-switching processes are treated, with the given generator of the switching process, we can pre-generate the switching process throughout the iterations. However, in the continuous state-dependent switching case, since the generator of the x-dependent switching processes is different in every step, we can no longer pre-generate the switching process without interacting with the continuous-state process and use the Lipschitz condition directly.

This chapter extends the results on Yin, Mao, Yuan, and Cao [297]. The approach uses local analysis and weak convergence methods. It relies on the solutions of associated martingale problems. The approach is different from the usual techniques developed in the literature for numerical solutions of stochastic differential equations to date. The idea and the techniques used are interesting in their own right. Rates of convergence are then ascertained. Using the results established in this chapter dealing with x-dependent switching combined with the techniques provided in [222], one may proceed to the study Milstein-type algorithms to establish accelerated rates of convergence.

Chapter 6
Stability

6.1 Introduction

Continuing our effort of studying positive recurrence and ergodicity of switching diffusion processes in Chapters 3 and 4, this chapter focuses on stability issues. For some of the recent progress in stability analysis, we refer the reader to [92, 199, 236, 310, 311] and references therein. To treat dynamic systems arising in science and engineering, linearization techniques are used most often. Nevertheless, nonlinear systems and their linearized counterpart may or may not share similar asymptotic behavior. A problem of great interest is: If a linear system is stable, what can we say about the associated nonlinear systems? This chapter provides a systematic approach to treat such problems for switching diffusions. Our main machineries are Lyapunov function methods.

The rest of the chapter is arranged as follows. Section 6.2 begins with the formulation of the problem together with an auxiliary result to be used in our stability analysis. Stability and instability in probability is treated in Section 6.3. Section 6.4 presents p-stability and exponential p-stability results. Sufficient conditions in terms of the existence of appropriate Lyapunov functions for almost sure exponential stability are established in Section 6.5. Easily verifiable conditions for stability and instability of linearized systems are provided in Section 6.6. To demonstrate our results, we provide several examples in Section 6.7. Further remarks are made in Section 6.8 to conclude this chapter.

6.2 Formulation and Auxiliary Results

Recall that we use z' to denote the transpose of $z \in \mathbb{R}^{\ell_1 \times \ell_2}$ with $\ell_i \geq 1$, and $i = 1, 2$, whereas $\mathbb{R}^{\ell \times 1}$ is simply written as $\mathbb{R}^{\ell}$; $\mathbb{1} = (1, 1, \dots, 1)' \in \mathbb{R}^{m_0}$ is a column vector with all entries being 1; $|x|$ is the Euclidean norm for a row or column

H.-D. Nguyen et al., *Hybrid Switching Diffusions*, Probability Theory and Stochastic Modelling 63, https://doi.org/10.1007/978-3-031-93303-5_6

vector x; I denotes the identity matrix. For a matrix A, its trace norm is denoted by $|A| = \sqrt{\mathrm{tr}(A'A)}$. If B is a set, its indicator function is denoted by $I_B(\cdot)$.

Suppose that $(X(t), \alpha(t))$ is a two-component Markov process such that $X(\cdot)$ is a continuous component taking values in $\mathbb{R}^r$ and $\alpha(\cdot)$ is a jump process taking values in a finite set $\mathcal{M} = \{1, 2, \ldots, m_0\}$. For simplicity of presentation, we assume throughout the chapter that $m_0 < \infty$, even though most of the results in this chapter can be easily generalized to the case when $m_0 = \infty$. The process $(X(t), \alpha(t))$ has a generator $\mathcal{L}$ given as follows. For any twice continuously differentiable function $g(\cdot, i)$, $i \in \mathcal{M}$,

$$\begin{aligned}\mathcal{L}g(x,i) &= \frac{1}{2}\sum_{j,k=1}^{r} a_{jk}(x,i)\frac{\partial^2 g(x,i)}{\partial x_j \partial x_k} + \sum_{j=1}^{r} b_j(x,i)\frac{\partial g(x,i)}{\partial x_j} + Q(x)g(x,\cdot)(i)\\ &= \frac{1}{2}\mathrm{tr}(a(x,i)\nabla^2 g(x,i)) + b'(x,i)\nabla g(x,i) + Q(x)g(x,\cdot)(i),\end{aligned} \tag{6.1}$$

where $x \in \mathbb{R}^r$, $\nabla g(\cdot, i)$, and $\nabla^2 g(\cdot, i)$ denote the gradient and Hessian of $g(\cdot, i)$, respectively, and $Q(x) = (q_{ij}(x))$ is an $m_0 \times m_0$ matrix depending on x satisfying $q_{ij}(x) \geq 0$ for $i \neq j$ and $\sum_{j \in \mathcal{M}} q_{ij}(x) = 0$ for each $i \in \mathcal{M}$, and

$$Q(x)g(x,\cdot)(i) = \sum_{j\in\mathcal{M}} q_{ij}(x)g(x,j) = \sum_{j\in\mathcal{M}} q_{ij}(x)(g(x,j) - g(x,i)), \quad i \in \mathcal{M}.$$

The process $(X(t), \alpha(t))$ can be described by

$$\begin{aligned}&dX(t) = b(X(t),\alpha(t))dt + \sigma(X(t),\alpha(t))dw(t),\\ &X(0) = x, \quad \alpha(0) = \alpha,\end{aligned} \tag{6.2}$$

and for $i \neq j$

$$\begin{aligned}\mathbb{P}\{\alpha(t+\Delta t) = j | \alpha(t) = i, (X(s),\alpha(s)), s \leq t\}\\ = q_{ij}(X(t))\Delta t + o(\Delta t),\end{aligned} \tag{6.3}$$

where $w(t)$ is a d-dimensional standard Brownian motion, $b(\cdot,\cdot) : \mathbb{R}^r \times \mathcal{M} \mapsto \mathbb{R}^r$, and $\sigma(\cdot,\cdot) : \mathbb{R}^r \times \mathcal{M} \mapsto \mathbb{R}^{r\times d}$ satisfies $\sigma(x,i)\sigma'(x,i) = a(x,i)$.

To proceed, we need conditions on the smoothness and growth of the functions involved and the condition that 0 is the only equilibrium point of the random dynamic system. Hence, we assume that the following conditions are valid throughout this chapter.

(H6.1) The matrix-valued function $Q(\cdot)$ is bounded, continuous, and satisfies (2.21) for some concave function $\gamma : \mathbb{R}_+ \mapsto \mathbb{R}_+$ with $\gamma(0) = 0$.

(H6.2) $b(0,\alpha) = \sigma(0,\alpha) = 0$ for each $\alpha \in \mathcal{M}$. Moreover, assume that $\sigma(x,\alpha)$ vanishes only at $x = 0$ for each $\alpha \in \mathcal{M}$.

(H6.3) There exists a constant $K_0 > 0$ such that for each $\alpha \in \mathcal{M}$ and for any $x, y \in \mathbb{R}^r$

$$|b(x, \alpha) - b(y, \alpha)| + |\sigma(x, \alpha) - \sigma(y, \alpha)| \le K_0|x - y|. \tag{6.4}$$

Under these conditions, the system given by (6.2) and (6.3) has a unique solution; see Chapter 2 for more details. In what follows, a process starting from (x, α) is denoted by $(X^{x,\alpha}(t), \alpha^{x,\alpha}(t))$ if the emphasis on initial condition is desired; if the context is clear, we simply write $(X(t), \alpha(t))$.

To study stability of the equilibrium point $x = 0$, we first present the following "nonzero" property, which asserts that almost all the sample paths of any solution of the system given by (6.2) and (6.3) starting from a nonzero state will never reach the origin. For diffusion processes, such a result was established in [156, Lemma 5.3]; for Markovian regime-switching processes, similar results were obtained in [199, Lemma 2.1]. In what follows, we give a proof for switching diffusions with continuous-state-dependent switching processes. Throughout the rest of the chapter, by a deleted neighborhood of the origin, we mean a neighborhood of the origin with the point 0 removed. The result is useful, as it provides us with flexibility for choices of Lyapunov functions and enables us to build Lyapunov functions in a deleted neighborhood of the origin.

Lemma 6.1 *Under Assumptions* (H6.1)–(H6.3), *we have*

$$\mathbb{P}\{X^{x,\alpha}(t) \ne 0, t \ge 0\} = 1, \quad \textit{for any } x \ne 0,\ \alpha \in \mathcal{M}, \tag{6.5}$$

and for any $\beta \in \mathbb{R}$ *and* $t > 0$

$$\mathbb{E}\left[\left|X^{x,\alpha}(t)\right|^{\beta}\right] \le |x|^{\beta} e^{Kt}, \quad x \ne 0,\ \alpha \in \mathcal{M}, \tag{6.6}$$

where K *is a constant depending only on* β, m_0, *and the Lipschitz constant* K_0 *in* (6.4).

Proof For $x \ne 0$, for each $i \in \mathcal{M}$, define $V(x, i) = |x|^{\beta}$ for any $\beta \in \mathbb{R} \setminus \{0\}$. For any $\Delta > 0$ and $|x| > \Delta$

$$\begin{aligned}
\nabla V(x, i) &= \beta|x|^{\beta-2}x,\\
\nabla^2 V(x, i) &= \beta|x|^{\beta-4}\left(|x|^2 I + (\beta - 2)xx'\right).
\end{aligned}$$

Then, it follows that

$$\begin{aligned}
\mathcal{L}V(x, i) = {}& \langle \beta\, |x|^{\beta-2} x, b(x, i)\rangle \\
& + \frac{1}{2}\mathrm{tr}\left[\sigma(x, i)\sigma'(x, i)\beta|x|^{\beta-4}\left(|x|^2 I + (\beta - 2)xx'\right)\right].
\end{aligned}$$

Moreover, using Assumptions (H6.2) and (H6.3)

$$|\mathcal{L}V(x,i)| \le K\,|x|^{\beta}, \quad \text{for any } \ (x,i) \in \mathbb{R}^r \times \mathcal{M} \text{ with } x \neq 0. \tag{6.7}$$

Let τ_Δ be the first exit time from $\{x \in \mathbb{R}^r : |x| > \Delta\}$. Denote $(X(t), \alpha(t)) = (X^{x,\alpha}(t), \alpha^{x,\alpha}(t))$. Now, applying generalized Itô's formula (see (2.5)) to V, we obtain

$$\begin{aligned}|X(\tau_\Delta \wedge t)|^{\beta} = |x|^{\beta} + \int_0^{\tau_\Delta \wedge t} \mathcal{L}\,|X(u)|^{\beta}\, du \\ + \int_0^{\tau_\Delta \wedge t} \beta\,|X(u)|^{\beta-2}\, X'(u)\sigma(X(u),\alpha(u))dw(u).\end{aligned} \tag{6.8}$$

Note that by virtue of Assumptions (H6.2) and (H6.3)

$$\mathbb{E}\int_0^{\tau_\Delta \wedge t} \Big|\beta\,|X(u)|^{\beta-2}\, X'(u)\sigma(X(u),\alpha(u))\Big|^2 du \le \mathbb{E}\int_0^{\tau_\Delta \wedge t} K\,|X(u)|^{2\beta}\, du.$$

Thus, if $\beta > 0$, (2.52) implies that

$$\mathbb{E}\int_0^{\tau_\Delta \wedge t} K\,|X(u)|^{2\beta}\, du \le \mathbb{E}\int_0^{t} |X(u)|^{2\beta}\, du \le KCt < \infty,$$

where $C = C(x,t,\beta) > 0$. On the other hand, if $\beta < 0$, then by the definition of τ_Δ, we have

$$\mathbb{E}\int_0^{\tau_\Delta \wedge t} |X(u)|^{2\beta}\, du \le \mathbb{E}\int_0^{\tau_\Delta \wedge t} K\Delta^{2\beta} du \le K\Delta^{2\beta} t < \infty.$$

Therefore, we have verified that the stochastic integral in (6.8) is a martingale with mean 0. Hence, by taking expectations on both sides of (6.8), and taking into account (6.7), it follows that

$$\begin{aligned}\mathbb{E}|X(\tau_\Delta \wedge t)|^{\beta} &\le |x|^{\beta} + K\mathbb{E}\int_0^{\tau_\Delta \wedge t} |X(u)|^{\beta} du \\ &\le |x|^{\beta} + K\mathbb{E}\int_0^{t} |X(u \wedge \tau_\Delta)|^{\beta} du.\end{aligned}$$

In the above, we have used the fact $\tau_\Delta \wedge u = u$ for $u \le \tau_\Delta \wedge t$. An application of Gronwall's inequality implies that for any $\beta \neq 0$

$$\mathbb{E}|X(\tau_\Delta \wedge t)|^{\beta} \le |x|^{\beta} \exp(Kt). \tag{6.9}$$

In particular, with $\beta = -1$, we have

$$\mathbb{E}\Big[|X(\tau_\Delta \wedge t)|^{-1}\Big] \le |x|^{-1} \exp(Kt).$$

By Chebyshev's inequality, for any $\Delta > 0$

$$\begin{aligned}\mathbb{P}(\tau_\Delta < t) &= \mathbb{P}(|X(\tau_\Delta \wedge t)| \le \Delta) = \mathbb{P}\Big(|X(\tau_\Delta \wedge t)|^{-1} \ge \Delta^{-1}\Big) \\ &\le \Delta\Big[\mathbb{E}\,|X(\tau_\Delta \wedge t)|^{-1}\Big] \le \Delta\Big[|x|^{-1}\exp(Kt)\Big].\end{aligned}$$

Letting $\Delta \to 0$, we have $\mathbb{P}(\tau_\Delta < t) \to 0$ as $\Delta \to 0$. Therefore, (6.6) follows by letting $\Delta \to 0$ in (6.9) and Fatou's lemma.

Finally, assuming (6.5) were false, then there would exist some $(x_0, \alpha_0) \in \mathbb{R}^r \times \mathcal{M}$ and $T > 0$ such that

$$\mathbb{P}\left\{X^{x_0,\alpha_0}(T) = 0\right\} > 0.$$

Then, we would have

$$\mathbb{E}\Big[\big|X^{x_0,\alpha_0}(T)\big|^{-1}\Big] = \infty,$$

which would contradict with (6.6) that we just proved. This completes the proof of the lemma. □

Remark 6.2 In view of (6.5), we can work with functions $V(\cdot, i)$, $i \in \mathcal{M}$ that are twice continuously differentiable and are defined on a deleted neighborhood of 0 in what follows.

To proceed, we present an auxiliary result, namely, the solvability of a system of deterministic equations. Suppose that Q, an $m_0 \times m_0$ constant matrix, is the generator of a continuous-time Markov chain $r(t)$ and that Q is irreducible.

Remark 6.3 In the above, by the irreducibility, we mean that the system of equations

$$\begin{cases}\nu Q = 0\\ \nu \mathbb{1} = 1,\end{cases}$$

has a unique solution such that $\nu = (\nu_1, \ldots, \nu_{m_0})$ satisfies $\nu_i > 0$; see Definition A.7 and discussions there for further details.

Note that if Q is irreducible, the rank of Q is $m_0 - 1$. Denote by $\mathcal{R}(Q)$ and $\mathcal{N}(Q)$ the range and the null space of Q, respectively. It follows that $\mathcal{N}(Q)$ is one-dimensional spanned by $\mathbb{1}$ (i.e., $\mathcal{N}(Q) = \text{span}\{\mathbb{1}\}$). As a consequence, the Markov chain $r(t)$ is ergodic; see, for example, [56]. In what follows, denote the associated stationary distribution by

$$\nu = (\nu_1, \nu_2, \ldots, \nu_{m_0}) \in \mathbb{R}^{1\times m_0}. \tag{6.10}$$

We are interested in solving a linear system of equations:

$$Qc = \eta, \tag{6.11}$$

where $Q \in \mathbb{R}^{m_0 \times m_0}$ and $\eta \in \mathbb{R}^{m_0}$ are given and $c \in \mathbb{R}^{m_0}$ is an unknown vector. Note that (6.11) is a Poisson equation. The properties of solutions of (6.11) are provided in Lemma A.12. Basically, it indicates that under the irreducibility of Q, (6.11) has a solution if and only if $\nu\eta = 0$. Moreover, suppose that c_1 and c_2 are two solutions of (6.11). Then, $c_1 - c_2 = \alpha_0 \mathbb{1}$ for some $\alpha_0 \in \mathbb{R}$.

We end this section with the following definition on various notions of stability of the equilibrium point $x = 0$ of the system given by (6.2) and (6.3). These are in line with the counterparts for diffusion processes in [156].

Definition 6.4 The equilibrium point $x = 0$ of the system given by (6.2) and (6.3) is said to be

(i) *stable in probability*, if for any $\varepsilon > 0$ and any $\alpha \in \mathcal{M}$

$$\lim_{x \to 0} \mathbb{P}\{\sup_{t \ge 0} |X^{x,\alpha}(t)| > \varepsilon\} = 0,$$

and $x = 0$ is said to be *unstable in probability* if it is not stable in probability,
(ii) *asymptotically stable in probability*, if it is stable in probability and satisfies

$$\lim_{x \to 0} \mathbb{P}\left\{\lim_{t \to \infty} X^{x,\alpha}(t) = 0\right\} = 1, \text{ for any } \alpha \in \mathcal{M},$$

(iii) *p-stable (for $p > 0$)*, if

$$\lim_{\delta \to 0} \sup_{|x| \le \delta, \alpha \in \mathcal{M}, t \ge 0} \mathbb{E}[|X^{x,\alpha}(t)|^p] = 0,$$

(iv) *asymptotically p-stable*, if it is p-stable and satisfies $\mathbb{E}[|X^{x,\alpha}(t)|^p] \to 0$ as $t \to \infty$ for any $(x, \alpha) \in \mathbb{R}^r \times \mathcal{M}$,
(v) *exponentially p-stable*, if there exists a $\delta > 0$ independent of $(x, \alpha) \in \mathbb{R}^n \times \mathcal{M}$ such that

$$\limsup_{t \to \infty} \frac{1}{t} \log \mathbb{E}[|X^{x,\alpha}(t)|^p] \le -\delta,$$

(vi) *almost surely exponentially stable* if there exists a $\delta > 0$ independent of $(x, \alpha) \in \mathbb{R}^r \times \mathcal{M}$ such that

$$\limsup_{t \to \infty} \frac{1}{t} \log |X^{x,\alpha}(t)| \le -\delta \text{ a.s.}$$

6.3 Stability and Instability in Probability

Using similar arguments as those of [156, Theorems 5.3, 5.5, and 5.6], we establish the following three lemmas. The statements are given together with their proofs.

Lemma 6.5 *Let $D \subset \mathbb{R}^r$ be a neighborhood of* 0. *Suppose that for each $i \in \mathcal{M}$, there exists a nonnegative function $V(\cdot, i) : D \mapsto \mathbb{R}$ such that:*

(i) *$V(\cdot, i)$ is continuous in D and vanishes only at $x = 0$,*
(ii) *$V(\cdot, i)$ is twice continuously differentiable in $D - \{0\}$ and satisfies*

$$\mathcal{L}V(x, i) \le 0, \quad \textit{for all} \;\; x \in D - \{0\}.$$

Then, the equilibrium point $x = 0$ is stable in probability.

Proof Let $\varsigma > 0$ be such that the ball $B_\varsigma = \{x \in \mathbb{R}^r : |x| < \varsigma\}$ and its boundary $\partial B_\varsigma = \{x \in \mathbb{R}^r : |x| = \varsigma\}$ are contained in D. Set

$$V_\varsigma := \inf\left\{V(y, j) : y \in D \setminus B_\varsigma, j \in \mathcal{M}\right\}.$$

Then, $V_\varsigma > 0$ by assumption (i). Next, by virtue of assumption (ii) and Dynkin's formula, we have

$$\begin{aligned}\mathbb{E}_{x,i} V(X(t \wedge \tau_\varsigma), \alpha(t \wedge \tau_\varsigma)) &= V(x, i) + \mathbb{E}_{x,i} \int_0^{\tau_\varsigma \wedge t} \mathcal{L}V(X(s), \alpha(s))ds \\ &\le V(x, i),\end{aligned}$$

where $(x, i) \in B_\varsigma \times \mathcal{M}$ and τ_ς is the first exit time from B_ς, that is, $\tau_\varsigma := \inf\{t \ge 0 : |X(t)| \ge \varsigma\}$. Because V is nonnegative, we further have

$$V_\varsigma \mathbb{P}\left\{\tau_\varsigma \le t\right\} \le \mathbb{E}_{x,i}\left[V(X(\tau_\varsigma), \alpha(\tau_\varsigma)) I_{\{\tau_\varsigma \le t\}}\right] \le V(x, i).$$

Note that $\tau_\varsigma \le t$ if and only if $\sup_{0 \le u \le t} |X(u)| > \varsigma$. Therefore, it follows that

$$\mathbb{P}_{x,i}\left\{\sup_{0 \le u \le t} |X(u)| > \varsigma\right\} \le \frac{V(x, i)}{V_\varsigma}.$$

Letting $t \to \infty$, we obtain

$$\mathbb{P}_{x,i}\left\{\sup_{0 \le t} |X(t)| > \varsigma\right\} \le \frac{V(x, i)}{V_\varsigma}.$$

Finally, the desired conclusion follows from the assumptions that $V(0, i) = 0$ and $V(\cdot, i)$ is continuous for each $i \in \mathcal{M}$. □

Introduce the notation

$$\tau_{\varepsilon,r_0}^{x,\alpha} := \inf\{t \geq 0 : |X^{x,\alpha}(t)| = \varepsilon \text{ or } |X^{x,\alpha}(t)| = r_0\}, \tag{6.12}$$

for any $0 < \varepsilon < r_0$ and any $(x, \alpha) \in \mathbb{R}^r \times \mathcal{M}$ with $\varepsilon < |x| < r_0$.

Lemma 6.6 *Assume the conditions of Lemma 6.5. If for any sufficiently small* $0 < \varepsilon < r_0$ *and any* $(x, \alpha) \in \mathbb{R}^r \times \mathcal{M}$ *with* $\varepsilon < |x| < r_0$, *we have*

$$\mathbb{P}\{\tau_{\varepsilon,r_0}^{x,\alpha} < \infty\} = 1, \tag{6.13}$$

then the equilibrium point $x = 0$ *is asymptotically stable in probability.*

Proof We divide the proof into several steps.

Step 1. Let $\varsigma > 0$ and $(x, i) \in B_\varsigma \times \mathcal{M}$, and define τ_ς as in the proof of Lemma 6.5. For any $t \geq 0$, let $g(t) := V(X^{x,i}(\tau_\varsigma \wedge t), \alpha^{x,i}(\tau_\varsigma \wedge t))$. Then, as in the proof of Lemma 6.5, the assumption that $\mathcal{L}V(y, j) \leq 0$ for all $(y, j) \in (D - \{0\}) \times \mathcal{M}$ implies that g is a nonnegative supermartingale. Therefore, the martingale convergence theorem implies that

$$\lim_{t\to\infty} g(t) = g(\infty) \quad \text{exists a.s.} \tag{6.14}$$

Step 2. By virtue of Lemma 6.5, the equilibrium point $x = 0$ is stable in probability. Therefore, for any $\varepsilon > 0$, there exists a $\delta > 0$ (we may further assume that $\delta < \varsigma$) such that

$$\mathbb{P}_{y,j}\left\{\tau_\varsigma < \infty\right\} < \varepsilon/2, \quad \text{for all } (y, j) \in B_\delta \times \mathcal{M}. \tag{6.15}$$

Now, choose an arbitrary point $(x, \alpha) \in B_\delta \times \mathcal{M}$. Then, both (6.14) and (6.15) hold. Hence, it follows from (6.13) and (6.15) that for any $\rho > 0$ with $\rho < |x|$, we have

$$\mathbb{P}_{x,\alpha}\left\{\tau_\rho < \infty\right\} \geq \mathbb{P}_{x,\alpha}\left\{\tau_{\rho,\varsigma} < \infty\right\} - \mathbb{P}_{x,\alpha}\left\{\tau_\varsigma < \infty\right\} \geq 1 - \varepsilon/2.$$

This implies that $\mathbb{P}_{x,\alpha}\left\{\inf_{t\geq 0} |X(t)| \leq \rho\right\} \geq 1 - \varepsilon/2$. Since $\rho > 0$ can be arbitrarily small

$$\mathbb{P}_{x,\alpha}\left\{\inf_{t\geq 0} |X(t)| = 0\right\} \geq 1 - \varepsilon/2.$$

Now, let

$$A := \left\{\omega \in \Omega : \tau_\varsigma(\omega) = \infty, \inf_{t\geq 0} |X(t,\omega)| = 0\right\}.$$

Then

$$\mathbb{P}_{x,\alpha}(A) = \mathbb{P}\Big\{ \inf_{t\geq 0} |X(t)| = 0 \Big\} - \mathbb{P}\Big\{ \inf_{t\geq 0} |X(t)| = 0, \, \tau_\varsigma < \infty \Big\}$$
$$\geq 1 - \frac{\varepsilon}{2} - \frac{\varepsilon}{2} = 1 - \varepsilon.$$

Step 3. We claim that for almost all $\omega \in A$, we have

$$\liminf_{t\to\infty} \left| X^{x,\alpha}(t,\omega) \right| = 0.$$

If the claim were false, there would exist a $B \subset A$ with $\mathbb{P}_{x,\alpha}(B) > 0$ such that for all $\omega \in B$, we would have

$$\liminf_{t\to\infty} \left| X^{x,\alpha}(t,\omega) \right| \geq \theta > 0.$$

Then, for any $\omega \in B$, there exists a $T = T(\omega) > 0$ such that $|X^{x,\alpha}(t,\omega)| \geq \theta$ for all $t \geq T$. Therefore, for any $\omega \in B$ and $n \in \mathbb{N}$ sufficiently large, $\tau_{1/n}(\omega) \leq T$, where $\tau_{1/n}(\omega) := \inf\{t \geq 0 : |X^{x,\alpha}(t,\omega)| \leq 1/n\}$. Hence, it follows that

$$\lim_{n\to\infty} \tau_{1/n}(\omega) \leq T(\omega) < \infty.$$

Then, we would have

$$\mathbb{P}_{x,\alpha}\left\{ \lim_{n\to\infty} \tau_{1/n} < \infty \right\} \geq \mathbb{P}_{x,\alpha}(B) > 0.$$

But this would lead to a contradiction because by virtue of Lemma 6.1, the equilibrium point 0 is inaccessible with probability 1. Thus, it follows that

$$\mathbb{P}_{x,\alpha}\left\{ \lim_{n\to\infty} \tau_{1/n} = \infty \right\} = 1, \quad \text{or } \mathbb{P}_{x,\alpha}\left\{ \lim_{n\to\infty} \tau_{1/n} < \infty \right\} = 0.$$

Hence, the claim is verified.

Step 4. Since $V(\cdot, i)$ is continuous and $V(0,i) = 0$ for each $i \in \mathcal{M}$, we have from Step 3 that

$$\liminf_{t\to\infty} V(X^{x,\alpha}(t), \alpha^{x,\alpha}(t)) = 0, \quad \text{for almost all } \omega \in A. \tag{6.16}$$

Now, by virtue of (6.14) and the definition of A, we have

$$\begin{aligned}\lim_{t\to\infty} g(t) &= \lim_{t\to\infty} V(X^{x,\alpha}(\tau_\varsigma \wedge t), \alpha^{x,\alpha}(\tau_\varsigma \wedge t)) \\ &= \lim_{t\to\infty} V(X^{x,\alpha}(t), \alpha^{x,\alpha}(t)) = g(\infty) \text{ a.s.}\end{aligned} \tag{6.17}$$

Thus, it follows from (6.16) and (6.17) that

$$\lim_{t\to\infty} V(X^{x,\alpha}(t), \alpha^{x,\alpha}(t)) = 0, \quad \text{for almost all } \omega \in A.$$

But $V(\cdot, i)$ vanishes only at $x = 0$ for each $i \in \mathcal{M}$, so $\lim_{t\to\infty} X^{x,\alpha}(t) = 0$ on A. Thus, we have

$$\mathbb{P}_{x,\alpha}\left\{\lim_{t\to\infty} X(t) = 0\right\} \geq \mathbb{P}_{x,\alpha}(A) \geq 1 - \varepsilon.$$

Note that (x, α) is an arbitrary point in $B_\delta \times \mathcal{M}$. Thus, for any $\varepsilon > 0$, there exists a $\delta > 0$ such that

$$\mathbb{P}_{x,\alpha}\left\{\lim_{t\to\infty} X(t) = 0\right\} \geq 1 - \varepsilon, \quad \text{for all } (x, \alpha) \in B_\delta \times \mathcal{M}.$$

That is, the equilibrium point $x = 0$ is asymptotically stable in probability as desired. □

Lemma 6.7 *Let $D \subset \mathbb{R}^r$ be a neighborhood of 0. Assume that the conditions of Lemma 6.6 hold and that for each $i \in \mathcal{M}$, there exists a nonnegative function $V(\cdot, i) : D \mapsto \mathbb{R}$ such that $V(\cdot, i)$ is twice continuously differentiable in every deleted neighborhood of 0, and*

$$\mathcal{L}V(x, i) \leq 0 \text{ for all } x \in D - \{0\}; \tag{6.18}$$

$$\lim_{|x|\to 0} V(x, i) = \infty, \text{ for each } i \in \mathcal{M}. \tag{6.19}$$

Then, the equilibrium point $x = 0$ is unstable in probability if (6.13) *holds.*

Proof Let $\varsigma > 0$ and $(x, \alpha) \in B_\varsigma \times \mathcal{M}$, and define τ_ς as in the proof of Lemma 6.5. Let also $0 < \varepsilon < |x|$ and define $\tau_\varepsilon := \inf\{t \geq 0 : |X^{x,\alpha}(t)| \leq \varepsilon\}$. Then for any $t > 0$, we have

$$\begin{aligned}
&\mathbb{E}_{x,\alpha} V(X(t \wedge \tau_{\varepsilon,\varsigma}), \alpha(t \wedge \tau_{\varepsilon,\varsigma})) \\
&= V(x, \alpha) + \mathbb{E}_{x,\alpha} \int_0^{t\wedge\tau_{\varepsilon,\varsigma}} \mathcal{L}V(X(s), \alpha(s))ds \\
&\leq V(x, \alpha).
\end{aligned} \tag{6.20}$$

Letting $t \to \infty$ in (6.20), we obtain by virtue of Fatou's lemma and (6.13) that

$$\mathbb{E}_{x,\alpha} V(X(\tau_\varepsilon \wedge \tau_\varsigma), \alpha(\tau_\varepsilon \wedge \tau_\varsigma)) \leq V(x, \alpha).$$

Furthermore, since V is nonnegative, we have

$$\begin{aligned} V(x,\alpha) &\ge \mathbb{E}_{x,\alpha}\left[V(X(\tau_\varepsilon),\alpha(\tau_\varepsilon))I_{\{\tau_\varepsilon<\tau_\varsigma\}}\right] \\ &\ge \inf\{V(y,j): |y|=\varepsilon,\ j\in\mathcal{M}\}\,\mathbb{P}_{x,\alpha}\left\{\tau_\varepsilon<\tau_\varsigma\right\} \\ &= V_\varepsilon \mathbb{P}_{x,\alpha}\left\{\sup_{0\le t\le\tau_\varepsilon}|X(t)|<\varsigma\right\}, \end{aligned}$$

where $V_\varepsilon = \inf\{V(y,j): |y|=\varepsilon,\ j\in\mathcal{M}\}$. By Lemma 6.1, the equilibrium point 0 is inaccessible with probability 1 and hence $\tau_\varepsilon\to\infty$ a.s. as $\varepsilon\to 0$. Also, it follows from (6.19) that $V_\varepsilon\to\infty$ as $\varepsilon\to 0$. Therefore, it follows that as $\varepsilon\to 0$, we have

$$\mathbb{P}_{x,\alpha}\left\{\sup_{t\ge 0}|X(t)|<\varsigma\right\}=0.$$

This shows that the equilibrium point $x=0$ is unstable in probability. □

Remark 6.8 Note that (6.13) is an essential assumption in Lemmas 6.6 and 6.7. Here, we present two sufficient conditions for which (6.13) holds.

(i) Let $N\subset\mathbb{R}^r$ be a neighborhood of 0. Assume that for each $i\in\mathcal{M}$, there exists a nonnegative function $V(\cdot,i): N\mapsto\mathbb{R}$ such that $V(\cdot,i)$ is twice continuously differentiable in every deleted neighborhood of 0 and that for any sufficiently small $0<\varepsilon<r_0$ there is a positive constant $\kappa=\kappa(\varepsilon)$ such that

$$\mathcal{L}V(x,i)\le-\kappa,\quad \text{for all } x\in N \text{ with } \varepsilon<|x|<r_0. \tag{6.21}$$

Then, (6.13) holds.

(ii) If for any sufficiently small $0<\varepsilon<r_0$, there exist some $\iota=1,2,\ldots,r$ and some constant $\kappa=\kappa(\varepsilon)>0$ such that

$$a_{\iota\iota}(x,i)\ge\kappa,\ \text{for all } (x,i)\in(\{x:\varepsilon<|x|<r_0\})\times\mathcal{M}, \tag{6.22}$$

then (6.13) holds.

Assertion (i) can be established using almost the same proof as that for [156, Corollary 5.1]. Also, (ii) follows by observing that if (6.22) is satisfied, then we can construct some Lyapunov function $V(\cdot,\cdot)$ satisfying (6.21); see the proof of Theorem 3.2 and Remark 3.3.

6.4 Exponential p-Stability

This section is concerned with exponential stability of the equilibrium point $x = 0$ for the system given by (6.2) and (6.3).

6.4.1 Auxiliary Results

Concerning the exponential p-stability of the equilibrium point $x = 0$ of the system given by (6.2) and (6.3), sufficient conditions in terms of the existence of certain Lyapunov functions being homogeneous of degree p were obtained in [199]. In what follows, we first show that the function u defined in (6.24) is sufficiently smooth and satisfies a Kolmogorov-type backward equation. Then we present a necessary and sufficient condition for exponential p-stability.

Note that in Theorem 6.10, we do not assume that the operator $\mathcal{L}$ is uniformly parabolic. In other words, the operator $\mathcal{L}$ may be degenerate. Nevertheless, we prove that the function $u(t, x, i)$ defined in (6.24) is a *classical* solution to the initial value problem (6.47)–(6.48).

Theorem 6.9 *Assume that for each* $i \in \mathcal{M}$, *the coefficients of the operator* $\mathcal{L}$ *defined in* (6.1) *satisfy Assumption* (H6.3) *and that* $b(\cdot, i) \in C^2, \sigma(\cdot, i), q_{ij}(\cdot) \in C^2$. *Suppose that* $\phi(\cdot, i) \in C^2$ *and that*

$$\left|\nabla_x^\beta b(x, i)\right| + \left|\nabla_x^\beta \sigma(x, i)\right| + \left|\nabla_x^\beta \phi(x, i)\right| \le K(1 + |x|^\gamma), \quad i \in \mathcal{M}, \tag{6.23}$$

where K *and* γ *are positive constants,* β *is a multi-index with* $|\beta| \le 2$, *and* ∇_x^β *is the partial differentiation operator as defined in* (2.107). *Assume further that* $\sup_{(x,i)\in\mathbb{R}^r\times\mathcal{M}} q_i(x) < M$ *with* $q_i(x) = \sum_{j\neq i} q_{ij}(x)$ *for each* $i \in \mathcal{M}$ *and that* $|D^\beta q_{ij}(\cdot)|$ *are Lipschitz and bounded uniformly by some positive constant* M *for all* $|\beta| \le 2$. *Then, the function*

$$u(t, x, i) := \mathbb{E}_{x,i}[\phi(X(t), \alpha(t))] = \mathbb{E}[\phi(X^{x,i}(t), \alpha^{x,i}(t))] \tag{6.24}$$

is twice continuously differentiable with respect to the variable x *and satisfies*

$$\left|\nabla_x^\beta u(t, x, i)\right| \le K(1 + |x|^\gamma),$$

where $t \in [0, T]$, $x \in \mathbb{R}^r$, *and* $i \in \mathcal{M}$.

Proof For notational simplicity, we work out the details for the one-dimensional case. Denote by $(\partial/\partial x)\phi(\cdot, i)$ for each $i \in \mathcal{M}$ and $(d/dx)q_{ij}(\cdot)$ the derivatives of $\phi(\cdot, i)$ and $q_{ij}(\cdot)$ with respect to x, respectively. Let $\chi(t)$ be the continuous-time Markov chain with state space $\mathcal{M}$ and generator $\widehat{Q} = (\widehat{q}_{ij})$ in which $\hat{q}_{ij} = 1$ if $i \neq j$ and $\hat{q}_{ii} = -m_0 + 1$. Let $\theta_n^{i_0}$ be the n-th jump time of $\chi(t)$ given that $\chi(0) = i_0$. Let

$Z^{x_0,i_0}(t)$ be the solution with initial value $Z^{x_0,i_0}(0) = x_0$ to

$$dZ(t) = b(Z(t), \chi(t))dt + \sigma(Z(t), \chi(t))dw(t). \tag{6.25}$$

Let $\tau_n^{x_0,i_0}$ be the n-th jump time of $\alpha^{x_0,i_0}(t)$. By a change of measure (see [79]), we have

$$\begin{aligned}
&\mathbb{E}\Big[\phi(X^{x_0,i_0}(T), \alpha^{x_0,i_0}(T))I_{\{\tau_n^{x_0,i_0} \le T < \tau_{n+1}^{x_0,i_0}\}} \prod_{k=1}^{n} I_{\{\alpha^{x_0,i_0}(\tau_k^{x_0,i_0})=i_k\}}\Big] \\
&= \exp((m_0-1)T)\mathbb{E}\Big[\phi(Z^{x_0,i_0}(T), i_n)I_{\{\theta_n^{i_0} \le T < \theta_{n+1}^{i_0}\}}\Big(\prod_{k=1}^{n} I_{\{\chi(\theta_k^{i_0})=i_k\}}\Big) \\
&\quad \times \exp\Big\{-\int_0^T q_{\chi(s)}(Z^{x_0,i_0}(s))ds\Big\}\prod_{k=0}^{n-1}\Big(q_{i_k i_{k+1}}(Z^{x_0,i_0}(\theta_{k+1}^{i_0}))\Big)\Big].
\end{aligned} \tag{6.26}$$

The change of measure formula (6.26) allows us to write the quotient $\frac{1}{\Delta}(u(T, x + \Delta, i) - u(T, x, i))$ in terms of the expectation of functionals of the process (Z, χ) and then show that it converges to a well-defined limit as $\Delta \to 0$. Let us now fix x and $\alpha = i_0$ and denote by $(X(t), \alpha(t))$ and $(\widetilde{X}(t), \widetilde{\alpha}(t))$ the switching-diffusion processes satisfying (6.2) and (6.3) with initial conditions (x, α) and $(\widetilde{x}, \alpha)$, respectively, where $\widetilde{x} = x + \Delta, \alpha = i_0$. Likewise, let $Z(t)$, $\widetilde{Z}(t)$ be the solutions to (6.25) with initial conditions $x, \widetilde{x}$ respectively.

Fix an arbitrary $T > 0$. By standard arguments (e.g., [205, Lemma 3.3] and [321]), we have

$$\begin{cases}
\mathbb{E}\sup\limits_{t\in[0,T]} |\widetilde{Z}(t) - Z(t)|^4 \le K\Delta^4 \text{ for some } K > 0, \\
\mathbb{E}\sup\limits_{t\in[0,T]} |Z(t)|^m \le K_m(1 + |x|^m) \text{ for any } m > 0.
\end{cases} \tag{6.27}$$

Furthermore, adapting the proofs of [94, Theorems 5.5.2, 5.5.3] with a slight modification by replacing fixed times with stopping times, we can show that

$$\mu_\Delta := \sup_{\{\vartheta\in\mathcal{T}\}} \frac{\mathbb{E}|\widetilde{Z}(\vartheta) - Z(\vartheta) - \eta(\vartheta)\Delta|^2}{\Delta^2} \to 0 \text{ as } \Delta \to 0, \tag{6.28}$$

where $\eta(t)$ is the derivative of $Z(t)$ with respect to the initial condition x which exists in L^2-sense and $\mathcal{T}$ is the set of all stopping times that are bounded above by T almost surely. Note that $\eta(t)$ satisfies

$$\eta(t) = 1 + \int_0^t \eta(s)b_x(Z(s), \chi(s))ds + \int_0^t \eta(s)\sigma_x(Z(s), \chi(s))dw(s).$$

Using the above Eq. (6.27), the Burkholder-Davis-Gundy inequality, Gronwall's inequality and standard arguments in [94, Theorem 5.2.2], we can show that

$$\mathbb{E}\sup_{t\in[0,T]}|\eta(t)|^2 \le C(T,x), \tag{6.29}$$

where $C(T,x)$ is a constant depending on T and x.

By Taylor's expansion, it can be easily derived that

$$\begin{aligned}
&\Big|\exp\Big\{-\int_0^T q_{\chi(s)}(\widetilde{Z}(s))ds\Big\} - \exp\Big\{-\int_0^T q_{\chi(s)}(Z(s))ds\Big\} \\
&\quad + \Big(\int_0^T (\widetilde{Z}(s)-Z(s))\frac{d}{dz}q_{\chi(s)}(Z(s))ds\Big)\exp\Big\{-\int_0^T q_{\chi(s)}(Z(s))ds\Big\}\Big| \\
&\le M\int_0^T |\widetilde{Z}(s)-Z(s)|^2 ds.
\end{aligned} \tag{6.30}$$

In view of (6.30), we have

$$\begin{aligned}
&\Big|\frac{1}{\Delta}\Big(\exp\Big\{-\int_0^T q_{\chi(s)}(\widetilde{Z}(s))ds\Big\} - \exp\Big\{-\int_0^T q_{\chi(s)}(Z(s))ds\Big\}\Big) \\
&\quad + \Big(\int_0^T \eta(s)\frac{d}{dz}q_{\chi(s)}(Z(s))ds\Big)\exp\Big\{-\int_0^T q_{\chi(s)}(Z(s))ds\Big\}\Big| \\
&\le \frac{1}{\Delta}\Big|\exp\Big\{-\int_0^T q_{\chi(s)}(\widetilde{Z}(s))ds\Big\} - \exp\Big\{-\int_0^T q_{\chi(s)}(Z(s))ds\Big\} \\
&\quad + \Big(\int_0^T (\widetilde{Z}(s)-Z(s))\frac{d}{dz}q_{\chi(s)}(Z(s))ds\Big)\exp\Big\{-\int_0^T q_{\chi(s)}(Z(s))ds\Big\}\Big| \\
&\quad + \Big|\Big(\int_0^T \Big[\eta(s) - \frac{\widetilde{Z}(s)-Z(s)}{\Delta}\Big]\frac{d}{dz}q_{\chi(s)}(Z(s))ds\Big) \\
&\qquad \exp\Big\{-\int_0^T q_{\chi(s)}(Z(s))ds\Big\}\Big| \\
&\le \frac{M}{\Delta}\int_0^T \Big(|\widetilde{Z}(s)-Z(s)|^2 + |\widetilde{Z}(s)-Z(s)-\Delta\eta(s)|\Big)ds.
\end{aligned} \tag{6.31}$$

We have from Taylor's expansion

$$\begin{aligned}
&\Big|\frac{q_{i_k i_{k+1}}(\widetilde{Z}(\theta_{k+1})) - q_{i_k i_{k+1}}(Z(\theta_{k+1}))}{\Delta} - \eta(\theta_{k+1})\frac{d}{dz}q_{i_k i_{k+1}}(Z(\theta_{k+1}))\Big| \\
&\le \frac{1}{\Delta}\Big|q_{i_k i_{k+1}}(\widetilde{Z}(\theta_{k+1})) - q_{i_k i_{k+1}}(Z(\theta_{k+1}))
\end{aligned}$$

$$
\begin{aligned}
&\qquad - \Big(\widetilde{Z}(\theta_{k+1}) - Z(\theta_{k+1})\Big)\frac{d}{dz}q_{i_k i_{k+1}}(Z(\theta_{k+1}))\Big| \\
&+ \frac{1}{\Delta}\Big|\widetilde{Z}(\theta_{k+1}) - Z(\theta_{k+1}) - \Delta\eta(\theta_{k+1})\Big|\,\Big|\frac{d}{dz}q_{i_k i_{k+1}}(Z(\theta_{k+1}))\Big| \\
\le{}& \frac{M}{\Delta}\Big(|\widetilde{Z}(\theta_{k+1}) - Z(\theta_{k+1})|^2 + |\widetilde{Z}(\theta_{k+1}) - Z(\theta_{k+1}) - \Delta\eta(\theta_{k+1})|\Big). \qquad (6.32)
\end{aligned}
$$

Similar to [94, (5.22) on p.123], we have

$$
\psi_\Delta := \left|\frac{\phi(\widetilde{Z}(T),\chi(T)) - \phi(Z(T),\chi(T))}{\Delta} - \eta(T)\frac{\partial}{\partial z}\phi(Z(T),\chi(T))\right| \to 0 \qquad (6.33)
$$

in $L^2(\Omega)$ as $\Delta \to 0$.

To simplify notation, we drop the superscript of $\theta_n^{i_0}$ the n-th jump time of $\chi(t)$, and for each $n \in \mathbb{N}$, let

$$
A_n := \{(i_1,\ldots,i_{n+1}) : i_k \in \mathcal{M}, i_k \neq i_{k-1}, k = 1,\ldots,n+1\}.
$$

For each $\mathbf{I} = (i_1,\ldots,i_{n+1}) \in A_n$, let $f_{\mathbf{I}} := I_{\{\theta_n \le T < \theta_{n+1}\}}\Big(\prod_{k=1}^n I_{\{\chi(\theta_k)=i_k\}}\Big)$. Also, denote

$$
\begin{aligned}
g_{\mathbf{I}} :={}& \eta(T)\frac{\partial}{\partial z}\phi(Z(T),i_n)f_{\mathbf{I}}\exp\Big\{-\int_0^T q_{\chi(s)}(Z(s))ds\Big\}\prod_{k=0}^{n-1} q_{i_k i_{k+1}}(Z(\theta_{k+1})) \\
&+ \sum_{j=0}^{n-1}\phi(Z(T),i_n)f_{\mathbf{I}}\exp\Big\{-\int_0^T q_{\chi(s)}(Z(s))ds\Big\}\eta(\theta_{j+1})\frac{d}{dz}q_{i_j i_{j+1}}(Z(\theta_{j+1})) \\
&\times \prod_{k=0,k\neq j}^{n-1} q_{i_k i_{k+1}}(Z(\theta_{k+1})) \\
&- \phi(Z(T),i_n)f_{\mathbf{I}}\exp\Big\{-\int_0^T q_{\chi(s)}(Z(s))ds\Big\}\Big(\int_0^T \eta(s)\frac{d}{dz}q_{\chi(s)}(Z(s))ds\Big) \\
&\times \prod_{k=0}^{n-1} q_{i_k i_{k+1}}(Z(\theta_{k+1}))
\end{aligned}
$$

and $\hat{\xi}_n(T) := \sum_{\mathbf{I}\in A_n} g_{\mathbf{I}}$. We compute

$$
\begin{aligned}
&\bigg|\frac{1}{\Delta}\Big[\phi(\widetilde{Z}(T),i_n)f_{\mathbf{I}}\exp\Big\{-\int_0^T q_{\chi(s)}(\widetilde{Z}(s))ds\Big\}\prod_{k=0}^{n-1} q_{i_k i_{k+1}}(\widetilde{Z}(\theta_{k+1})) \\
&\qquad - \phi(Z(T),i_n)f_{\mathbf{I}}\exp\Big\{-\int_0^T q_{\chi(s)}(Z(s))ds\Big\}\prod_{k=0}^{n-1} q_{i_k i_{k+1}}(Z(\theta_{k+1}))\Big] - g_{\mathbf{I}}\bigg|
\end{aligned}
$$

$$\leq Cf_{\mathbf{I}}M^n\bigg[\psi_\Delta + (1+|\widetilde{Z}(T)|^\gamma + |Z(T)|^\gamma)$$
$$\times \int_0^T \frac{|\widetilde{Z}(s)-Z(s)|^2 + |\widetilde{Z}(s)-Z(s)-\Delta\eta(s)|}{\Delta}ds$$
$$+ (1+|\widetilde{Z}(T)|^\gamma + |Z(T)|^\gamma)$$
$$\times \sum_{j=0}^{n} \frac{|\widetilde{Z}(\theta_{j+1})-Z(\theta_{j+1})|^2 + |\widetilde{Z}(\theta_{j+1})-Z(\theta_{j+1})-\Delta\eta(\theta_{j+1})|}{\Delta}\bigg]. \tag{6.34}$$

Note that $\sum_{\mathbf{I}\in A_n} f_{\mathbf{I}} = I_{\{\theta_n\leq T<\theta_{n+1}\}}$. It then follows from (6.34) that for each $n \in \mathbb{N}$

$$\mathbb{E}\bigg|\sum_{\mathbf{I}\in A_n}\frac{1}{\Delta}\Big[\phi(\widetilde{Z}(T),i_n)f_{\mathbf{I}}\exp\Big\{-\int_0^T q_{\chi(s)}(\widetilde{Z}(s))ds\Big\}\prod_{k=0}^{n-1} q_{i_k i_{k+1}}(\widetilde{Z}(\theta_{k+1}))$$
$$-\phi(Z(T),i_n)f_{\mathbf{I}}\exp\Big\{-\int_0^T q_{\chi(s)}(Z(s))ds\Big\}\prod_{k=0}^{n-1} q_{i_k i_{k+1}}(Z(\theta_{k+1}))\Big] - \hat{\zeta}_n(T)\bigg|$$
$$\leq CM^n\mathbb{E}[I_{\{\theta_n\leq T<\theta_{n+1}\}}\psi_\Delta] + \frac{C}{\Delta}M^n\mathbb{E}\bigg[I_{\{\theta_n\leq T<\theta_{n+1}\}}(1+|\widetilde{Z}(T)|^\gamma + |Z(T)|^\gamma)$$
$$\times \int_0^T \Big(|\widetilde{Z}(s)-Z(s)|^2 + |\widetilde{Z}(s)-Z(s)-\Delta\eta(s)|\Big)ds\bigg]$$
$$+\frac{C}{\Delta}M^n\mathbb{E}\bigg[I_{\{\theta_n\leq T<\theta_{n+1}\}}(1+|\widetilde{Z}(T)|^\gamma + |Z(T)|^\gamma)$$
$$\times \sum_{j=0}^{n}\Big(|\widetilde{Z}(\theta_{j+1})-Z(\theta_{j+1})|^2 + |\widetilde{Z}(\theta_{j+1})-Z(\theta_{j+1})-\Delta\eta(\theta_{j+1})|\Big)\bigg]. \tag{6.35}$$

By Hölder's inequality and the fact that

$$\mathbb{P}\{\theta_n \leq T < \theta_{n+1}\} = \frac{\exp(-m_0T)(m_0T)^n}{n!}, \tag{6.36}$$

we can obtain the following estimate:

$$\mathbb{E}I_{\{\theta_n\leq T<\theta_{n+1}\}}\psi_\Delta \leq \Big(\mathbb{P}\{\theta_n \leq T < \theta_{n+1}\}\mathbb{E}\psi_\Delta^2\Big)^{\frac{1}{2}}$$
$$\leq \left(\frac{\exp(-m_0T)(m_0T)^n}{n!}\right)^{\frac{1}{2}}(\mathbb{E}\psi_\Delta^2)^{\frac{1}{2}}. \tag{6.37}$$

Similarly, with the aid of Hölder's and Minkowski's inequalities, we have

$$\begin{aligned}
&\frac{1}{\Delta}\mathbb{E}\bigg[I_{\{\theta_n \le T < \theta_{n+1}\}}(1 + |\widetilde{Z}(T)|^\gamma + |Z(T)|^\gamma) \\
&\quad \times \int_0^T \Big(|\widetilde{Z}(s) - Z(s)|^2 + |\widetilde{Z}(s) - Z(s) - \Delta\eta(s)| \Big) ds \bigg] \\
&\le \Big(\mathbb{P}\{\theta_n \le T < \theta_{n+1}\} \Big)^{\frac{1}{4}} \Big(1 + (\mathbb{E}|\widetilde{Z}(T)|^{4\gamma})^{\frac{1}{4}} + (\mathbb{E}|Z(T)|^{4\gamma})^{\frac{1}{4}} \Big) \\
&\quad \times \frac{1}{\Delta}\bigg[\int_0^T \Big(\mathbb{E}|\widetilde{Z}(s) - Z(s)|^4 \Big)^{\frac{1}{2}} ds + \int_0^T \Big(\mathbb{E}|\widetilde{Z}(s) - Z(s) - \Delta\eta(s)|^2 \Big)^{\frac{1}{2}} ds \bigg) \\
&\le K \left(\frac{\exp(-m_0 T)(m_0 T)^n}{n!} \right)^{\frac{1}{4}} \Big(\Delta + \mu_\Delta^{\frac{1}{2}} \Big), \qquad (6.38)
\end{aligned}$$

where the last inequality follows from (6.27) and (6.28). Likewise

$$\begin{aligned}
&\frac{1}{\Delta}\mathbb{E}\bigg[I_{\{\theta_n \le T < \theta_{n+1}\}}(1 + |\widetilde{Z}(T)|^\gamma + |Z(T)|^\gamma) \\
&\qquad \times \sum_{j=0}^{n} \Big(|\widetilde{Z}(\theta_{j+1}) - Z(\theta_{j+1})|^2 + |\widetilde{Z}(\theta_{j+1}) - Z(\theta_{j+1}) - \Delta\eta(\theta_{j+1})| \Big) \bigg] \\
&\quad \le nK \left(\frac{\exp(-m_0 T)(m_0 T)^n}{n!} \right)^{\frac{1}{4}} \Big(\Delta + \mu_\Delta^{\frac{1}{2}} \Big). \qquad (6.39)
\end{aligned}$$

We now plugging (6.37), (6.38), and (6.39) into (6.35) to obtain

$$\begin{aligned}
&\mathbb{E}\bigg| \frac{1}{\Delta} \sum_{\mathbf{I} \in A_n} \bigg[\phi(\widetilde{Z}(T), i_n) f_{\mathbf{I}} \exp\Big\{ -\int_0^T q_{\chi(s)}(\widetilde{Z}(s)) ds \Big\} \prod_{k=0}^{n-1} q_{i_k i_{k+1}}(\widetilde{Z}(\theta_{k+1})) \\
&\qquad - \phi(Z(T), i_n) f_{\mathbf{I}} \exp\Big\{ -\int_0^T q_{\chi(s)}(Z(s)) ds \Big\} \prod_{k=0}^{n-1} q_{i_k i_{k+1}}(Z(\theta_{k+1})) \bigg] - \hat{\xi}_n(T) \bigg| \\
&\quad \le (n+1)\widetilde{K} M^n \left(\frac{\exp(-m_0 T)(m_0 T)^n}{n!} \right)^{\frac{1}{4}} \Big(\mu_\Delta^{\frac{1}{2}} + \big(\mathbb{E}\psi_\Delta^2 \big)^{\frac{1}{2}} + \Delta \Big) \qquad (6.40)
\end{aligned}$$

for some $\widetilde{K}$ independent of n. Note that

$$\sum_{n=0}^{\infty} (n+1)\widetilde{K} M^n \left(\frac{\exp(-m_0 T)(m_0 T)^n}{n!} \right)^{\frac{1}{4}} < \infty. \qquad (6.41)$$

Moreover, by virtue of (6.28) and (6.33), we have

$$\lim_{\Delta\to 0}\left(\mu_{\Delta}^{\frac{1}{2}} + (\mathbb{E}\psi_{\Delta}^{2})^{\frac{1}{2}} + \Delta\right) = 0.$$

Thus

$$\begin{aligned}
&\mathbb{E}\sum_{n=0}^{\infty}\Bigg|\sum_{\mathbf{I}\in A_n}\frac{1}{\Delta}\Bigg[\phi(\widetilde{Z}(T), i_n) f_{\mathbf{I}} \exp\left\{-\int_0^T q_{\chi(s)}(\widetilde{Z}(s))ds\right\}\prod_{k=0}^{n-1} q_{i_k i_{k+1}}(\widetilde{Z}(\theta_{k+1}))\\
&-\phi(Z(T), i_n) f_{\mathbf{I}} \exp\left\{-\int_0^T q_{\chi(s)}(Z(s))ds\right\}\prod_{k=0}^{n-1} q_{i_k i_{k+1}}(Z(\theta_{k+1}))\Bigg] - \hat{\xi}_n(T)\Bigg|\\
&\to 0
\end{aligned} \tag{6.42}$$

as $\Delta \to 0$.

Thanks to the uniform boundedness of $q_i(x)$ and $|D^{\beta} q_{ij}(x)|$ and (6.23), we have

$$|\hat{\xi}_n(T)| \le K M^n I_{\{\theta_n \le t < \theta_{n+1}\}}(1+T)(n+1)\left(1 + \sup_{0\le s\le T}|Z(s)|^{\gamma}\right)\sup_{0\le s\le T}|\eta(s)|. \tag{6.43}$$

Then, similar to (6.40) and (6.41), we can use Holder's inequality, (6.27), (6.29), and (6.36) to show that

$$\mathbb{E}\sum_{n=0}^{\infty}|\hat{\xi}_n(T)| \le \hat{C}(T, x) < \infty. \tag{6.44}$$

As a result of (6.42) and (6.44), we deduce

$$\begin{aligned}
&\frac{1}{\Delta}[u(T, x+\Delta, i) - u(T, x, i)]\\
&\quad = \frac{1}{\Delta}\mathbb{E}\Big[\phi(\widetilde{X}(T), \widetilde{\alpha}(T)) - \phi(X(T), \alpha(T))\Big]\\
&\quad = \frac{1}{\Delta}\mathbb{E}\sum_{n=0}^{\infty}\sum_{\mathbf{I}\in A_n}\Bigg[\phi(\widetilde{Z}(T), i_n) f_{\mathbf{I}} \exp\left\{-\int_0^T q_{\chi(s)}(\widetilde{Z}(s))ds\right\}\prod_{k=0}^{n-1} q_{i_k i_{k+1}}(\widetilde{Z}(\theta_{k+1}))\\
&\qquad\qquad -\phi(Z(T), i_n) f_{\mathbf{I}} \exp\left\{-\int_0^T q_{\chi(s)}(Z(s))ds\right\}\prod_{k=0}^{n-1} q_{i_k i_{k+1}}(Z(\theta_{k+1}))\Bigg]\\
&\quad \to \mathbb{E}\sum_{n=0}^{\infty}\hat{\xi}_n(T), \qquad \text{as } \Delta \to 0.
\end{aligned} \tag{6.45}$$

We have therefore proved that $u(T, x, i)$ is differentiable with respect to x with

$$\frac{\partial}{\partial x} u(T, x, i) = \mathbb{E}_{x,i} \sum_{n=0}^{\infty} \hat{\zeta}_n(T). \tag{6.46}$$

With Taylor's expansion up to the second order and using the same method, we can show that $u(t, x, i)$ is twice continuously differentiable with respect to x. □

Theorem 6.10 *Assume the conditions of Theorem 6.9. Then, the function u defined in (6.24) is continuously differentiable with respect to the variable t. Moreover, u satisfies the system of Kolmogorov backward equations:*

$$\frac{\partial u(t, x, i)}{\partial t} = \mathcal{L}u(t, x, i), \quad (t, x, i) \in (0, T] \times \mathbb{R}^r \times \mathcal{M}, \tag{6.47}$$

with initial condition

$$\lim_{t \downarrow 0} u(t, x, i) = \phi(x, i), \quad (x, i) \in \mathbb{R}^r \times \mathcal{M}, \tag{6.48}$$

where $\mathcal{L}u(t, x, i)$ in (6.47) is to be interpreted as $\mathcal{L}$ applied to the function $(x, i) \mapsto u(t, x, i)$.

Proof Since the process $(X(t), \alpha(t))$ is càdlàg, the initial condition (6.48) follows from the continuity of ϕ. We divide the rest of the proof into several steps.

Step 1. For fixed $(x, i) \in \mathbb{R}^r \times \mathcal{M}$, $u(t, x, i)$ is absolutely continuous with respect to $t \in [0, T]$. In fact, for any $0 \le s \le t \le T$, we have from Dynkin's formula that

$$\begin{aligned}
u(t, x, i) - u(s, x, i) &= \mathbb{E}_{x,i}\phi(X(t), \alpha(t)) - \mathbb{E}_{x,i}\phi(X(s), \alpha(s)) \\
&= \mathbb{E}_{x,i}\left[\mathbb{E}_{x,i}[(\phi(X(t), \alpha(t)) - \phi(X(s), \alpha(s)))|\mathcal{F}_s]\right] \\
&= \mathbb{E}_{x,i}\mathbb{E}_{x,i}\left[\int_s^t \mathcal{L}\phi(X(v), \alpha(v))dv|\mathcal{F}_s\right] \\
&\le \mathbb{E}_{x,i}\int_s^t \mathbb{E}_{x,i}\left[|\mathcal{L}\phi(X(v), \alpha(v))|\,|\mathcal{F}_s\right]dv.
\end{aligned}$$

Using (6.23), for some positive constants K and γ_0, we have

$$|\mathcal{L}\phi(x, i)| \le K(1 + |x|^{\gamma_0}) \text{ for all } (x, i) \in \mathbb{R}^r \times \mathcal{M}.$$

Hence, it follows from Proposition 2.9 that

$$\mathbb{E}_{x,i}\left[|\mathcal{L}\phi(X(v), \alpha(v))|\,|\mathcal{F}_s\right] \le K\mathbb{E}_{x,i}\left[(1 + |X(v)|^{\gamma_0})|\mathcal{F}_s\right] \le C,$$

where C is independent of t, s, or v. Thus, we have

$$|u(t, x, i) - u(s, x, i)| \le C\,|t - s|.$$

Thus, u is absolutely continuous with respect to t, $(\partial/\partial t)u(t,x,i)$ exists a.s. on $[0,T]$, and we have

$$u(t,x,i) = u(0,x,i) + \int_0^t \frac{\partial u(v,x,i)}{\partial v} dv. \tag{6.49}$$

Step 2. For any $h > 0$, we have from the strong Markov property that

$$\begin{aligned} u(t+h,x,i) &= \mathbb{E}_{x,i}\phi(X(t+h),\alpha(t+h)) \\ &= \mathbb{E}_{x,i}\left[\mathbb{E}_{x,i}[\phi(X(t+h),\alpha(t+h))|\mathcal{F}_h]\right] \\ &= \mathbb{E}_{x,i}\left[\mathbb{E}_{X(h),\alpha(h)}\phi(X(t+h),\alpha(t+h))\right] \\ &= \mathbb{E}_{x,i}u(t,X(h),\alpha(h)). \end{aligned} \tag{6.50}$$

Now, let $g(x,i) := u(t,x,i)$. Then, Theorem 6.9 implies that $g(\cdot,i) \in C^2$ for each $i \in \mathcal{M}$ and for some $K > 0$ and $\gamma_0 > 0$

$$\left|D_x^\beta g(x,i)\right| \le K(1+|x|^{\gamma_0}), \quad i \in \mathcal{M}.$$

Thus, it follows from Dynkin's formula that

$$\mathbb{E}_{x,i}g(X(h),\alpha(h)) - g(x,i) = \mathbb{E}_{x,i}\int_0^h \mathcal{L}g(X(v),\alpha(v))dv.$$

Using the same argument as in the proof of [94, Theorem 5.6.1], we can show that

$$\frac{1}{h}\mathbb{E}_{x,i}\int_0^h \mathcal{L}g(X(v),\alpha(v))dv \to \mathcal{L}g(x,i) \text{ as } h \downarrow 0. \tag{6.51}$$

Therefore

$$\lim_{h\downarrow 0}\frac{\mathbb{E}_{x,i}g(X(h),\alpha(h)) - g(x,i)}{h} = \mathcal{L}g(x,i).$$

But by the definition of g, we have from (6.50) that

$$\lim_{h\downarrow 0}\frac{u(t+h,x,i) - u(t,x,i)}{h} = \mathcal{L}g(x,i) = \mathcal{L}u(t,x,i). \tag{6.52}$$

Thus, a combination of (6.49) and (6.52) leads to

$$u(t,x,i) = u(0,x,i) + \int_0^t \mathcal{L}u(s,x,i)ds. \tag{6.53}$$

Step 3. We claim that $\mathcal{L}u(t, x, i)$ is continuous with respect to the variable t. Note that

$$\mathcal{L}u(t, x, i) = b(x, i)\frac{\partial u(t, x, i)}{\partial x} + \frac{1}{2}\sigma^2(x, i)\frac{\partial^2 u(t, x, i)}{\partial x^2} + \sum_{j=1}^{m_0} q_{ij}(x)u(t, x, j).$$

The claim is verified if we can show $(\partial/\partial x)u(t, x, i)$ and $(\partial^2/\partial x^2)u(t, x, i)$ are continuous with respect to t, since Step 1 above establishes that $u(t, x, i)$ is continuous with respect to t. To this end, we fix some $t_0 \geq 0$ and then let $T > t_0 + 1$. For any $t \in [0, T]$, recall (6.46) and the expression $\hat{\zeta}_n(t)$ in the proof of Theorem 6.9. Similar to the observations in (6.43) and (6.44), we have the estimates

$$|\hat{\zeta}_n(t)| \leq KM^n I_{\{\theta_n \leq t < \theta_{n+1}\}}(1 + T)(n + 1)\left(1 + \sup_{0\leq s\leq T} |Z(s)|^\gamma\right) \sup_{0\leq s\leq T} |\eta(s)|;$$

and

$$\mathbb{E}_{x,i} \sum_{n=0}^{\infty} KM^n I_{\{\theta_n \leq t < \theta_{n+1}\}}(1+T)(n+1)\left(1+ \sup_{0\leq s\leq T} |Z(s)|^\gamma\right) \sup_{0\leq s\leq T} |\eta(s)| < \infty.$$

Therefore, it follows from the dominated convergence theorem and the mean square continuity of the processes (Z, χ) and η that

$$\lim_{t\to t_0} \frac{\partial}{\partial x}u(t, x, i) = \lim_{t\to t_0} \mathbb{E}_{x,i} \sum_{n=0}^{\infty} \hat{\zeta}_n(t) = \mathbb{E}_{x,i} \sum_{n=0}^{\infty} \hat{\zeta}_n(t_0) = \frac{\partial}{\partial x}u(t_0, x, i).$$

This reveals that $(\partial/\partial x)u(t, x, i)$ is continuous with respect to the variable t. Similarly, we can show that $(\partial^2/\partial x^2)u(t, x, i)$ is also continuous with respect to the variable t. Therefore, $\mathcal{L}u(t, x, i)$ is continuous with respect to the variable t.

Step 4. Finally, by virtue of (6.53) and Step 3 above, we conclude that $(\partial/\partial t)u(t, x, i)$ exists everywhere for $t \in (0, T]$ and that

$$\frac{\partial u(t, x, i)}{\partial t} = \mathcal{L}u(t, x, i).$$

This finishes the proof of the theorem. □

The following corollary follows from Theorems 6.9 and 6.10 immediately.

Corollary 6.11 *Let* $(X^{x,\alpha}(t), \alpha^{x,\alpha}(t))$ *be the solution to the system given by* (6.2) *and* (6.3) *with initial data* $X(0) = x$, $\alpha(0) = \alpha$. *Assume that for each* $i \in \mathcal{M}$, $b(\cdot, i)$ *and* $\sigma(\cdot, i)$ *have continuous partial derivatives with respect to the variable* x *up to the second order and* $b(0, i) = \sigma(0, i) = 0$. *Then, for a fixed* $T > 0$ *and for any* $p > 0$, *the function* $u(t, x, i) : [0, T] \times \mathbb{R}^r \times \mathcal{M} \mapsto \mathbb{R}$ *defined by*

$$u(t,x,i) := \mathbb{E}\left|X^{x,i}(t)\right|^p \tag{6.54}$$

is twice continuously differentiable with respect to x *(except possibly at* $x = 0$*) and satisfies the system of parabolic equations together with the initial data:*

$$\begin{cases} \dfrac{\partial u(t,x,i)}{\partial t} = \mathcal{L}u(t,x,i), \quad (t,x,i) \in (0,T) \times \mathbb{R}^n \times \mathcal{M}, \\ u(0,x,i) = |x|^p, \quad (x,i) \in \mathbb{R}^n \times \mathcal{M}. \end{cases} \tag{6.55}$$

6.4.2 Necessary and Sufficient Conditions for Exponential p-Stability

Theorem 6.12 *Assume that for each* $i \in \mathcal{M}$, $b(\cdot,i)$ *and* $\sigma(\cdot,i)$ *have continuous partial derivatives with respect to the variable* x *up to the second oder and* $b(0,i) = \sigma(0,i) = 0$*. Then the equilibrium point 0 is exponentially* p*-stable if and only if there exists a function* $V : \mathbb{R}^r \times \mathcal{M} \mapsto \mathbb{R}$ *such that for some positive constants* k_1, k_2, k_3*, and* k_4

$$k_1|x|^p \le V(x,i) \le k_2|x|^p, \quad x \in N, \tag{6.56}$$

$$\mathcal{L}V(x,i) \le -k_3|x|^p \quad \text{for all} \quad x \in N - \{0\}, \tag{6.57}$$

for all $1 \le j,k \le r$, $x \in N - \{0\}$*, where* N *is a neighborhood of* 0.

Proof The proof of sufficiency follows from the same argument as in the proof of [199, Theorem 3.1]. The reminder of the proof is devoted to establishing necessity. For each $i \in \mathcal{M}$, consider the function:

$$V(x,i) = \int_0^T \mathbb{E}|X^{x,i}(t)|^p dt.$$

It follows from Corollary 6.11 that for each $i \in \mathcal{M}$, the function $V(\cdot,i) \in C^2(\mathbb{R}^r \setminus \{0\})$. Furthermore, since the equilibrium point 0 is exponentially p-stable, there is a $\beta > 0$ such that

$$V(x,i) \le \int_0^T K|x|^p \exp(-\beta u)du \le \frac{K}{\beta}|x|^p.$$

On the other hand, using Assumptions (H6.2) and (H6.3), we can readily verify that $|\mathcal{L}|x|^p| \le K|x|^p$ for some positive constant K. Therefore, an application of Itô's lemma to $g(x) = |x|^p$ implies that

$$\begin{aligned}\mathbb{E}|X^{x,i}(T)|^p - |x|^p &= \mathbb{E}\int_0^T \mathcal{L}|X^{x,i}(u)|^p du \\ &\geq -K\int_0^T \mathbb{E}|X^{x,i}(u)|^p du = -KV(x,i).\end{aligned}$$

Again, thanks to the definition of exponential p-stability, we can choose T large enough so that $\mathbb{E}|X^{x,i}(T)|^p \leq \frac{1}{2}|x|^p$. As a result, we have $V(x,i) \geq |x|^p/(2K)$ and hence (6.56).

To verify (6.57), we use the continuities of $\frac{\partial}{\partial x_j}u(t,x,i)$ and $\frac{\partial^2}{\partial x_j \partial x_k}u(t,x,i)$, $j,k = 1,2,\ldots,r$, where $u(t,x,i) := \mathbb{E}|X^{x,i}(t)|^p$, and (6.55), to derive

$$\begin{aligned}\mathcal{L}V(x,i) = \int_0^T \mathcal{L}u(t,x,i)dt = \int_0^T \frac{\partial}{\partial t}u(t,x,i)dt = u(T,x,i) - u(0,x,i) \\ < -\frac{1}{2}|x|^p,\end{aligned}$$

where the last inequality follows from the definition of exponential p-stability. Thus, the proof is complete. □

We end this section with the following results on linear systems. Assume that the evolution (6.2) is replaced by

$$dX(t) = b(\alpha(t))X(t)dt + \sum_{j=1}^{d} \sigma_j(\alpha(t))X(t)dw_j(t), \tag{6.58}$$

where $b(i)$ and $\sigma_j(i)$ are $r \times r$ constant matrices and $w_j(t)$ are independent one-dimensional standard Brownian motions for $i = 1,2,\ldots,m_0$, and $j = 1,2,\ldots,d$. Then, we have the following two theorems.

Theorem 6.13 *The equilibrium point $x = 0$ of system* (6.58) *together with* (6.3) *is exponentially p-stable if and only if for each $i \in \mathcal{M}$, there is a function $V(\cdot,i) : \mathbb{R}^r \mapsto \mathbb{R}$ satisfying equations* (6.56) *and* (6.57) *for some constants $k_i > 0$, $i = 1,2,3$.*

Proof The proof of sufficiency was contained in [199]. The necessity follows from Theorem 6.12 because the coefficients of (6.58) and (6.3) satisfy the conditions of Theorem 6.12. We omit the details for brevity. □

Theorem 6.14 *Let $Q(x) \equiv Q$ be a constant matrix. Assume also that the Markov chain $\alpha(t)$ is independent of the Brownian motion:*

$$w(t) = (w_1(t), w_2(t), \ldots, w_d(t))'$$

(or equivalently, $\alpha(0)$ is independent of the Brownian motion $w(\cdot)$). If the equilibrium point $x = 0$ of the system given by (6.3) *and* (6.58) *is stable in probability, then it is p-stable for sufficiently small $p > 0$.*

Proof The proof follows from a crucial observation. In this case, because (6.58) is linear in $X(t)$, $X^{\lambda x,\alpha}(t) = \lambda X^{x,\alpha}(t)$. Using a similar argument as that for [156, Lemma 6.2], we can conclude the proof; a few details are omitted. □

6.5 Almost Sure Exponential Stability

This section is devoted to establishing sufficient conditions in terms of the existence of appropriate Lyapunov functions for a.s. exponential stability of the equilibrium point of the system (6.2) and (6.3).

Theorem 6.15 *Suppose Assumptions* (H6.1)–(H6.3) *hold. Let* $V : \mathbb{R}^r \times \mathcal{M} \mapsto \mathbb{R}^+$ *be such that* $V(\cdot, i) \in C^2(\mathbb{R}^r)$ *for each* $i \in \mathcal{M}$. *Suppose there exist* $p > 0$, $c_1(i) > 0$, $c_2(i) \in \mathbb{R}$, *and nonnegative constants* $c_3(i)$ *and* $c_4(i)$ *such that for all* $x \neq 0$ *and* $i \in \mathcal{M}$:

(i) $c_1(i)|x|^p \leq V(x,i)$
(ii) $\mathcal{L}V(x,i) \leq c_2(i)V(x,i)$
(iii) $|\nabla V(x,i)\cdot\sigma(x,i)|^2 \geq c_3(i)V(x,i)^2$
(iv) $\sum_{j\in\mathcal{M}} q_{ij}(x)\left(\log V(x,j) - \frac{V(x,j)}{V(x,i)}\right) \leq -c_4(i)$

Then

$$\limsup_{t\to\infty} \frac{1}{t}\log|X(t)| \leq \frac{1}{p}\max_{i\in\mathcal{M}}\left\{c_2(i) - 0.5c_3(i) - c_4(i)\right\} =: \delta \ \ a.s. \tag{6.59}$$

In particular, if $\delta < 0$, *then the equilibrium point of the system* (6.2) *and* (6.3) *is a.s. exponentially stable.*

Remark 6.16 It is worth pointing out that conditions (i)–(iii) are standard and similar to those in Theorem 3.1 of [6]. Condition (iv) is needed so that we can control the fluctuations of $\frac{1}{t}\log|X(t)|$ due to the presence of regime switching. This condition is quite natural because of the following observation. Using the elementary inequality $\log y \leq y - 1$ for $y > 0$, we derive

$$\log V(x,j) - \frac{V(x,j)}{V(x,i)} \leq \log V(x,i) - 1.$$

Then, it follows that for every $i \in \mathcal{M}$

$$\begin{aligned}
&\sum_{j\in\mathcal{M}} q_{ij}(x)\left(\log V(x,j) - \frac{V(x,j)}{V(x,i)}\right) \\
&\quad = \sum_{j\neq i} q_{ij}(x)\left(\log V(x,j) - \frac{V(x,j)}{V(x,i)}\right) + q_{ii}(x)(\log V(x,i) - 1) \\
&\quad \leq \sum_{j\neq i} q_{ij}(x)\left(\log V(x,i) - 1\right) + q_{ii}(x)(\log V(x,i) - 1) = 0.
\end{aligned}$$

In light of this observation, the inclusion of a nonpositive constant $-c_4(i)$ in condition (iv) is reasonable. This constant may vary depending on $i \in \mathcal{M}$. In fact, the constants $c_1(i), \ldots, c_4(i)$ in Conditions (i)–(iv) may all depend on $i \in \mathcal{M}$; this allows for some extra flexibility for the selection of the Lyapunov function V and, more importantly, the sufficient condition for a.s. exponential stability in Theorem 6.15 and Corollary 6.17.

Proof of Theorem 6.15 Recall that thanks to Lemma 6.2, for every $(x_0, \alpha_0) \in \mathbb{R}^r_0 \times \mathcal{M}$, $X(t) := X^{x_0,\alpha_0}(t) \neq 0$ for all $t \geq 0$ a.s. Let $U(x, i) = \log V(x, i)$ for $(x, i) \in \mathbb{R}^r_0 \times \mathcal{M}$. Since

$$\nabla U(x,i) = \frac{\nabla V(x,i)}{V(x,i)} \text{ and } \nabla^2 U(x,i) = \frac{\nabla^2 V(x,i)}{V(x,i)} - \frac{\nabla V(x,i)\nabla V(x,i)'}{V^2(x,i)},$$

we have

$$\begin{aligned}
&\mathcal{L}U(x,i)\\
&= \frac{\langle \nabla V(x,i), b(x,i)\rangle}{V(x,i)} + \frac{1}{2V(x,i)}\mathrm{tr}\big(\sigma\sigma'(x,i)\nabla^2 V(x,i)\big)\\
&\quad - \frac{|\langle \nabla V(x,i), \sigma(x,i)\rangle|^2}{2V^2(x,i)} + \sum_{j\in\mathcal{M}} q_{ij}(x)\log V(x,j)\\
&= \frac{\mathcal{L}V(x,i)}{V(x,i)} - \frac{|\langle \nabla V(x,i), \sigma(x,i)\rangle|^2}{2V^2(x,i)} + \sum_{j\in\mathcal{M}} q_{ij}(x)\left(\log V(x,j) - \frac{V(x,j)}{V(x,i)}\right).
\end{aligned} \tag{6.60}$$

Now, we apply Itô's formula to the process $U(X(t), \alpha(t))$:

$$U(X(t),\alpha(t)) = U(x_0,\alpha_0) + \int_0^t \mathcal{L}U(X(s),\alpha(s))ds + M(t), \tag{6.61}$$

where $M(t) = M_1(t) + M_2(t)$, and

$$\begin{aligned}
M_1(t) &= \int_0^t \frac{\langle \nabla V(X(s),\alpha(s)), \sigma(X(s),\alpha(s))\rangle}{V(X(s),\alpha(s))} dW(s),\\
M_2(t) &= \int_0^t \int_{\mathbb{R}} \log \frac{V(X(s-),\alpha(s-) + h(X(s-),\alpha(s-),y))}{V(X(s-),\alpha(s-))}\widetilde{\mathfrak{p}}(ds,dy).
\end{aligned}$$

By the exponential martingale inequality [5, Theorem 5.2.9], for any $k \in \mathbb{N}$ and $\theta \in (0, 1)$, we have

$$\begin{aligned}
\mathbb{P}\bigg\{\sup_{0\le t\le k}\bigg[M(t) - \frac{\theta}{2}\int_0^t \frac{|\langle \nabla V(X(s),\alpha(s)), \sigma(X(s),\alpha(s))\rangle|^2}{|V(X(s),\alpha(s))|^2}ds - f_\theta(t)\bigg] > \theta\sqrt{k}\bigg\}\\
\le e^{-\theta^2\sqrt{k}},
\end{aligned}$$

where

$$f_\theta(t) = \frac{1}{\theta} \int_0^t \int_{\mathbb{R}} \Bigg[\bigg(\frac{V(X(s-), \alpha(s-) + h(X(s-), \alpha(s-), y))}{V(X(s-), \alpha(s-))} \bigg)^\theta - 1 \\ - \theta \log \frac{V(X(s-), \alpha(s-) + h(X(s-), \alpha(s-), y))}{V(X(s-), \alpha(s-))} \Bigg] \mathfrak{m}(dy) ds.$$

We can verify that $\sum_k e^{-\theta^2 \sqrt{k}} < \infty$. Therefore, the Borel-Cantelli lemma implies that there exists an $\Omega_0 \subset \Omega$ with $\mathbb{P}(\Omega_0) = 1$ such that for every $\omega \in \Omega_0$, there exists an integer $k_0 = k_0(\omega)$ so that for all $k \geq k_0$ and $0 \leq t \leq k$, we have

$$M(t) \leq \frac{\theta}{2} \int_0^t \frac{|\langle \nabla V(X(s), \alpha(s)), \sigma(X(s), \alpha(s)) \rangle|^2}{|V(X(s), \alpha(s))|^2} ds + \theta \sqrt{k} + f_\theta(t). \tag{6.62}$$

Now putting (6.62) and (6.60) into (6.61), it follows that for all $\omega \in \Omega_0$ and $0 \leq t \leq k$, we have

$$\begin{aligned} &U(X(t), \alpha(t)) - U(x_0, \alpha_0) \\ &\leq \int_0^t \frac{\mathcal{L}V(X(s), \alpha(s))}{V(X(s), \alpha(s))} ds - \frac{1-\theta}{2} \int_0^t \frac{|\langle \nabla V(X(s), \alpha(s)), \sigma(X(s), \alpha(s)) \rangle|^2}{|V(X(s), \alpha(s))|^2} ds \\ &\quad + \theta \sqrt{k} + f_\theta(t) + \int_0^t \sum_{j \in \mathcal{M}} q_{\alpha(s), j}(X(s)) \bigg(\log V(X(s), j) - \frac{V(X(s), j)}{V(X(s), \alpha(s))} \bigg) ds \\ &\leq \int_0^t \Big[c_2(\alpha(s)) - \frac{1-\theta}{2} c_3(\alpha(s)) - c_4(\alpha(s)) \Big] ds + \theta \sqrt{k} + f_\theta(t). \end{aligned} \tag{6.63}$$

Next, we argue that for any $t \geq 0$, $f_\theta(t) \to 0$ as $\theta \downarrow 0$. To this end, we first use the elementary inequality $e^a \geq a + 1$ for $a \in \mathbb{R}$ to obtain

$$\frac{1}{\theta} \Bigg[\bigg(\frac{V(x, i + h(x, i, z), i)}{V(x, i)} \bigg)^\theta - 1 - \theta \log \frac{V(x, i + h(x, i, z))}{V(x, i)} \Bigg] \geq 0$$

for any $(x, i, z) \in \mathbb{R}^r \times \mathcal{M} \times \mathbb{R}$. Next, the inequality $a^\theta \leq 1 + \theta(a-1)$ for $0 < \theta \leq 1$ and $a > 0$ leads to

$$\begin{aligned} &\frac{1}{\theta} \Bigg[\bigg(\frac{V(x, i + h(x, i, z))}{V(x, i)} \bigg)^\theta - 1 - \theta \log \frac{V(x, i + h(x, i, z))}{V(x, i)} \Bigg] \\ &\quad \leq \frac{1}{\theta} \Bigg[1 + \theta \bigg(\frac{V(x, i + h(x, i, z))}{V(x, i)} - 1 \bigg) - 1 - \theta \log \frac{V(x, i + h(x, i, z))}{V(x, i)} \Bigg] \\ &\quad = \frac{V(x, i + h(x, i, z))}{V(x, i)} - 1 - \log \frac{V(x, i + h(x, i, z))}{V(x, i)}; \end{aligned}$$

note that the last expression in the above equation is nonnegative thanks to the inequality $a - 1 - \log a \geq 0$ for $a > 0$. On the other hand, using the assumption that $Q(\cdot)$ is bounded, we can readily verify that

$$\int_0^t \int_{\mathbb{R}} \Bigg[\frac{V(X(s-), \alpha(s-) + h(X(s-), \alpha(s-), y))}{V(X(s-), \alpha(s-))} - 1 \\ - \log \frac{V(X(s-), \alpha(s-) + h(X(s-), \alpha(s-), y))}{V(X(s-), \alpha(s-))} \Bigg] \mathfrak{m}(dy)ds < \infty \text{ a.s.}$$

In addition, we can verify that $\lim_{\theta \downarrow 0}[\frac{1}{\theta}(a^\theta - 1) - \log a] = 0$ for $a > 0$. Then, the dominated convergence theorem leads to

$$\lim_{\theta \downarrow 0} f_\theta(t) = \int_0^t \int_{\mathbb{R}} \lim_{\theta \downarrow 0} \frac{1}{\theta} \Bigg[\left(\frac{V(X(s-), \alpha(s-) + h(X(s-), \alpha(s-), z))}{V(X(s-), \alpha(s-))} \right)^\theta - 1 \\ - \theta \log \frac{V(X(s-), \alpha(s-) + h(X(s-), \alpha(s-), z))}{V(X(s-), \alpha(s-))} \Bigg] \mathfrak{m}(dz)ds = 0.$$

Now, passing to the limit as $\theta \downarrow 0$ in (6.63) leads to

$$U(X(t), \alpha(t)) - U(x_0, \alpha_0) \leq \int_0^t \big[c_2(\alpha(s)) - 0.5c_3(\alpha(s)) - c_4(\alpha(s))\big]ds \tag{6.64}$$

for all $\omega \in \Omega_0$, $k \geq k_0 = k_0(\omega)$ and $0 \leq t \leq k$. Recall that $U(x, i) = \log V(x, i)$. Then, inserting condition (i) into (6.64) yields that for almost all $\omega \in \Omega$, $k \geq k_0$, and $k - 1 \leq t \leq k$, we have

$$\begin{aligned} \frac{1}{t}[p \log |X(t)| + \log c_1(\alpha(t))] &\leq \frac{1}{t} \log V(X(t), \alpha(t)) \\ &\leq \frac{1}{t} \int_0^t \big[c_2(\alpha(s)) - 0.5c_3(\alpha(s)) - c_4(\alpha(s))\big]ds + \frac{\log V(x_0, \alpha_0)}{t} \\ &\leq \frac{1}{t} \int_0^t \big[c_2(\alpha(s)) - 0.5c_3(\alpha(s)) - c_4(\alpha(s))\big]ds + \frac{\log V(x_0, \alpha_0)}{k-1} \\ &\leq \max_{i \in \mathcal{M}} \big\{c_2(i) - 0.5c_3(i) - c_4(i)\big\} + \frac{\log V(x_0, \alpha_0)}{k-1}; \end{aligned} \tag{6.65}$$

the last inequality yields (6.59) by letting $t \to \infty$. □

Corollary 6.17 *In addition to the conditions of Theorem 6.15, suppose also that the discrete component α in (6.2) and (6.3) is an irreducible continuous-time Markov chain with an invariant distribution $\pi = (\pi_i, i \in \mathcal{M})$, then (6.59) can be strengthened to*

$$\limsup_{t \to \infty} \frac{1}{t} \log |X(t)| \leq \frac{1}{p} \sum_{i \in \mathcal{M}} \pi_i [c_2(i) - 0.5c_3(i) - c_4(i)] \ \textit{a.s.} \tag{6.66}$$

Proof This follows from applying the ergodic theorem of continuous-time Markov chain (see, e.g., [223, Theorem 3.8.1]) to the right-hand side of (6.65):

$$\lim_{t\to\infty}\frac{1}{t}\int_0^t \big[c_2(\alpha(s)) - 0.5c_3(\alpha(s)) - c_4(\alpha(s))\big]ds$$
$$= \sum_{i=1}^{m} \pi_i[c_2(i) - 0.5\, c_3(i) - c_4(i)] \text{ a.s.}$$

Then, (6.66) follows directly. □

Theorem 6.18 *Suppose Assumptions* (H6.1)–(H6.3) *hold. Assume that there exists a function* $V : \mathbb{R}^r \times \mathcal{M} \mapsto \mathbb{R}^+$ *such that* $V(\cdot, i) \in C^2(\mathbb{R}^r)$ *for each* $i \in \mathcal{M}$ *satisfying:*

(i) $c_1|x|^p \le V(x,i) \le c_2|x|^p$
(ii) $\mathcal{L}V(x,i) \le -c_3 V(x,i)$

for all $(x,i) \in \mathbb{R}^r \times \mathcal{M}$, *where* p, c_1, c_2, *and* c_3 *are positive constants. Let* $(X(0), \alpha(0)) = (x,i) \in \mathbb{R}^r \times \mathcal{M}$. *Then, we have*

(a) $\mathbb{E}[|X(t)|^p] \le \frac{c_2}{c_1}|x|^p e^{-c_3 t}$. *In particular, the equilibrium point of* (6.2) *and* (6.3) *is exponentially p-stable with Lyapunov exponent being less than or equal to* $-c_3$,
(b) there exists an almost surely finite and positive random variable Ξ *such that*

$$|X(t)|^p \le \frac{\Xi}{c_1} e^{-c_3 t} \quad \textit{for all } t \ge 0 \textit{ a.s.} \tag{6.67}$$

In particular, the equilibrium point of (6.2) *and* (6.3) *is almost sure exponentially stable with Lyapunov exponent less than or equal to* $-\frac{c_3}{p}$.

Proof The proof of part (a) is very similar to that of Theorem 3.1 in [199]; see also Theorem 4.1 in [6]. For brevity, we omit the details here.

Part (b) is motivated by Khasminskii [156, Theorem 5.11]. For any $t \ge 0$ and $(x,i) \in \mathbb{R}^r \times \mathcal{M}$, we consider the function $f(t,x,i) := e^{c_3 t}V(x,i)$. Condition (i) and Theorem 2.20 imply that

$$\mathbb{E}[f(t, X(t), \alpha(t))] = \mathbb{E}[e^{c_3 t}V(X(t), \alpha(t))] \le c_2 e^{c_3 t}\mathbb{E}[|X(t)|^p] < \infty.$$

On the other hand, thanks to Itô's formula, we have for all $0 \le s < t$

$$\begin{aligned} &f(t, X(t), \alpha(t)) \\ &= f(s, X(s), \alpha(s)) + \int_s^t e^{c_3 r}(c_3 + \mathcal{L})V(X(r), \alpha(r))dr + M(s,t) \\ &\le f(s, X(s), \alpha(s)) + M(s,t), \end{aligned}$$

where we used condition (ii) to obtain the inequality, and

$$
\begin{aligned}
M(s,t) &:= \int_s^t e^{c_3 r} \nabla V(X(r),\alpha(r)) \cdot \sigma(X(r),\alpha(r)) dW(r) \\
&\quad + \int_s^t \int_{\mathbb{R}_+} e^{c_3 r} \big[V(X(r-),\alpha(r-) + h(X(r-),\alpha(r-),z)) \\
&\qquad\qquad - V(X(r-),\alpha(r-)) \big] \widetilde{\mathfrak{p}}(dr,dz).
\end{aligned}
$$

Let $\tau_n := \inf\{t \ge 0 : |X(t)| \ge n\}$. Then, we have $\lim_{n\to\infty} \tau_n = \infty$ a.s. and

$$
\mathbb{E}[f(t\wedge\tau_n, X(t\wedge\tau_n), \alpha(t\wedge\tau_n))|\mathcal{F}_s] \le f(s\wedge\tau_n, X(s\wedge\tau_n), \alpha(s\wedge\tau_n)) \text{ a.s.}
$$

Passing to the limit as $n \to \infty$, and noting that f is positive, we obtain from Fatou's lemma that $\mathbb{E}[f(t, X(t), \alpha(t))|\mathcal{F}_s] \le f(s, X(s), \alpha(s))$ a.s. Therefore, it follows that the process $\{f(t, X(t), \alpha(t)), t \ge 0\}$ is a positive supermartingale. The martingale convergence theorem (see, e.g., Theorem 3.15 and Problem 3.16 in [149]) then implies that $f(t, X(t), \alpha(t))$ converges a.s. to a finite limit as $t \to \infty$. Consequently, there exists an a.s. finite and positive random variable Ξ such that

$$
\sup_{t\ge 0}\{e^{c_3 t} V(X(t),\alpha(t))\} = \sup_{t\ge 0}\{f(t, X(t),\alpha(t))\} \le \Xi < \infty, \text{ a.s.}
$$

Furthermore, it follows from condition (i) that $|X(t)|^p \le \frac{1}{c_1} V(X(t),\alpha(t))$. Putting this observation into the above displayed equation yields (6.67). □

6.6 Stability and Instability of Linearized Systems

This section provides criteria for stability and instability. To proceed, we put an assumption.

(H6.4) For each $i \in \mathcal{M}$, there exist $b(i), \sigma_j(i) \in \mathbb{R}^{r\times r}$, $j = 1, 2, \ldots, d$, and a generator of a continuous-time Markov chain $\widehat{Q} = (\widehat{q}_{ij})$ such that as $x \to 0$

$$
\begin{aligned}
&b(x,i) = b(i)x + o(|x|), \\
&\sigma(x,i) = (\sigma_1(i)x, \sigma_2(i)x, \ldots, \sigma_d(i)x) + o(|x|), \\
&Q(x) = \widehat{Q} + o(1).
\end{aligned}
\tag{6.68}
$$

Moreover, $\widehat{Q}$ is irreducible and $\widehat{\alpha}(t)$ is a Markov chain with generator $\widehat{Q}$.

Remark 6.19 Note that Assumption (H6.4) is rather natural. It is equivalent to $Q(x)$ being continuous at $x = 0$ and $b(x, i)$ and $\sigma(x, i)$ continuously differentiable at $x = 0$. It follows from Assumption (H6.4) that $\widehat{\alpha}(t)$ is an ergodic Markov chain. Denote the stationary distribution of $\widehat{\alpha}(t)$ by $\pi = (\pi_1, \pi_2, \ldots, \pi_{m_0}) \in \mathbb{R}^{1\times m_0}$.

Remark 6.20 For any square matrix $A \in \mathbb{R}^{r\times r}$, A can be decomposed to the sum of a symmetric matrix A_1 and an antisymmetric matrix A_2. In fact, $A_1 = (A+A')/2$ and $A_2 = (A - A')/2$. Moreover, the quadratic form satisfies

$$x'Ax = x'A_1x = x'\frac{A + A'}{2}x. \tag{6.69}$$

This observation is used in what follows.

Theorem 6.21 *Suppose Assumption* (H6.4) *holds. Then, the equilibrium point* $x = 0$ *of system given by* (6.2) *and* (6.3) *is asymptotically stable in probability if*

$$\sum_{i=1}^{m_0} \pi_i \lambda_{\max}\left(b(i) + b'(i) + \sum_{j=1}^{d} \sigma_j'(i)\sigma_j(i)\right) < 0, \tag{6.70}$$

and is unstable in probability if

$$\sum_{i=1}^{m_0} \pi_i \left(\lambda_{\min}\left(b(i) + b'(i) + \sum_{j=1}^{d} \sigma_j'(i)\sigma_j(i)\right) - \frac{1}{2}\sum_{j=1}^{d}\left[\rho(\sigma_j(i) + \sigma_j'(i))\right]^2\right) > 0. \tag{6.71}$$

Proof

(a) We first prove that the equilibrium point $x = 0$ of the system given by (6.2) and (6.3) is asymptotically stable in probability if (6.70) holds. For notational simplicity, define the column vector:

$$\mu = (\mu_1, \mu_2, \ldots, \mu_{m_0})' \in \mathbb{R}^{m_0}$$

with

$$\mu_i = \lambda_{\max}\left(b(i) + b'(i) + \sum_{j=1}^{d} \sigma_j'(i)\sigma_j(i)\right).$$

Also, let $\beta := -\pi\mu$. Note that $\beta > 0$ by (6.70). It follows from Assumption (H6.4) and Lemma A.12 that the equation

$$\widehat{Q}c = \mu + \beta\mathbb{1}$$

has a solution $c = (c_1, c_2, \ldots, c_{m_0})' \in \mathbb{R}^{m_0}$. Thus, we have

$$\mu_i - \sum_{j=1}^{m_0} \widehat{q}_{ij}c_j = -\beta, \quad i \in \mathcal{M}. \tag{6.72}$$

For each $i \in \mathcal{M}$, consider the Lyapunov function:

$$V(x,i) = (1-\gamma c_i)|x|^\gamma,$$

where $0 < \gamma < 1$ is sufficiently small so that $1-\gamma c_i > 0$ for each $i \in \mathcal{M}$. It is readily seen that for each $i \in \mathcal{M}$, $V(\cdot, i)$ is continuous and nonnegative and vanishes only at $x = 0$. Detailed calculations reveal that for $x \neq 0$, we have

$$\begin{aligned}
\nabla V(x,i) &= (1-\gamma c_i)\gamma |x|^{\gamma-2}x,\\
\nabla^2 V(x,i) &= (1-\gamma c_i)\gamma \left[|x|^{\gamma-2}I + (\gamma-2)|x|^{\gamma-4}xx'\right].
\end{aligned}$$

In addition, it follows from (6.68) that

$$a(x,i) = \sigma(x,i)\sigma'(x,i) = \sum_{j=1}^{d} \sigma_j(i)xx'\sigma_j'(i) + o(|x|^2).$$

Note that for any matrix $A \in \mathbb{R}^{r\times r}$, we have $\operatorname{tr}(\sigma_j(i)xx'\sigma_j'(i)A) = x'\sigma_j'(i)A\sigma_j(i)x$. Therefore, it follows that

$$\begin{aligned}
\mathcal{L}V(x,i) &= \frac{1}{2}\operatorname{tr}\left[\left(\sum_{j=1}^{d}\sigma_j(i)xx'\sigma_j'(i) + o(|x|^2)\right)\nabla^2 V(x,i)\right]\\
&\quad + \nabla V'(x,i)(b(i)x + o(|x|)) - \sum_{j\neq i} q_{ij}(x)|x|^\gamma \gamma(c_j - c_i)\\
&= \gamma(1-\gamma c_i)|x|^\gamma \left\{\frac{1}{2}\sum_{j=1}^{d}\left(\frac{x'\sigma_j'(i)\sigma_j(i)x}{|x|^2} + (\gamma-2)\frac{(x'\sigma_j'(i)x)^2}{|x|^4}\right)\right.\\
&\quad \left. + \frac{x'b(i)x}{|x|^2} - \sum_{j\neq i} q_{ij}(x)\frac{c_j - c_i}{1-\gamma c_i} + o(1)\right\}.
\end{aligned} \tag{6.73}$$

By virtue of Remark 6.20, we obtain

$$\begin{aligned}
\frac{x'b(i)x}{|x|^2} + \frac{1}{2}\sum_{j=1}^{d}\frac{x'\sigma_j'(i)\sigma_j(i)x}{|x|^2} &\le \frac{1}{2}\lambda_{\max}\left(b(i) + b'(i) + \sum_{j=1}^{d}\sigma_j'(i)\sigma_j(i)\right)\\
&= \mu_i.
\end{aligned} \tag{6.74}$$

Next, it follows from Assumption (H6.4) that when $|x|$ and γ are sufficiently small

$$\begin{aligned}\sum_{j\neq i} q_{ij}(x)\frac{c_j - c_i}{1-\gamma c_i} &= \sum_{j=1}^{m_0} q_{ij}(x)c_j + \sum_{j\neq i} q_{ij}(x)\frac{c_i(c_j - c_i)}{1-\gamma c_i}\gamma \\ &= \sum_{j=1}^{m_0} \widehat{q}_{ij}c_j + O(\gamma) + o(1),\end{aligned} \tag{6.75}$$

where $o(1) \to 0$ as $|x| \to 0$. Hence, it follows from (6.73) and (6.75) that when $|x| < r_0$ with r_0 and $0 < \gamma < 1$ sufficiently small, we have

$$\mathcal{L}V(x,i) \leq \gamma(1-\gamma c_i)|x|^\gamma \left\{ \mu_i - \sum_{j=1}^{m_0} \widehat{q}_{ij}c_j + o(1) + O(\gamma) \right\}.$$

Furthermore, by virtue of (6.72), we have

$$\mathcal{L}V(x,i) \leq \gamma(1-\gamma c_i)|x|^\gamma \left(-\beta + o(1) + O(\gamma)\right) \leq -\kappa(\varepsilon) < 0,$$

for any $(x,i) \in N \times \mathcal{M}$ with $\varepsilon < |x| < r_0$, where $N \subset \mathbb{R}^r$ is a small neighborhood of 0 and $\kappa(\varepsilon)$ is a positive constant. Therefore, we conclude from Lemma 6.6 and Remark 6.8 that the equilibrium point $x = 0$ is asymptotically stable in probability.

(b) Now, we prove that the equilibrium point $x = 0$ is unstable in probability if (6.71) holds. Define the column vector $\theta = (\theta_1, \theta_2, \ldots, \theta_{m_0})' \in \mathbb{R}^{m_0}$ by

$$\theta_i := \frac{1}{2}\lambda_{\min}\left(b(i) + b'(i) + \sum_{j=1}^{d} \sigma_j'(i)\sigma_j(i)\right) - \frac{1}{4}\sum_{j=1}^{d}\left[\rho(\sigma_j(i) + \sigma_j'(i))\right]^2,$$

and set

$$\delta := -\pi\theta = \sum_{i=1}^{m_0} \pi_i\theta_i < 0.$$

As in part (a), Assumption (H6.4), the definition of δ, and Lemma A.12 imply that the equation $\widehat{Q}c = \theta + \delta\mathbb{1}$ has a solution $c = (c_1, c_2, \ldots, c_{m_0})' \in \mathbb{R}^{m_0}$ and

$$\theta_i - \sum_{j=1}^{d} \widehat{q}_{ij}c_j = -\delta > 0, \quad i \in \mathcal{M}. \tag{6.76}$$

For $i \in \mathcal{M}$, consider the Lyapunov function:

$$V(x,i) = (1-\gamma c_i)|x|^\gamma,$$

where $-1 < \gamma < 0$ is sufficiently small so that $1 - \gamma c_i > 0$ for each $i \in \mathcal{M}$. It is readily seen that the nonnegative function $V(\cdot, i), i \in \mathcal{M}$ satisfies (6.19). Similar to the arguments in part (a), Remark 6.20 implies that

$$\frac{x'b(i)x}{|x|^2} + \frac{\sum_{j=1}^{d} x'\sigma_j'(i)\sigma_j(i)x}{2|x|^2} \geq \frac{1}{2}\lambda_{\min}\Big(b(i) + b'(i) + \sum_{j=1}^{d} \sigma_j'(i)\sigma_j(i)\Big).$$

Note that for any symmetric matrix A with real eigenvalues $\lambda_1 \geq \lambda_2 \geq \cdots \geq \lambda_n$, using the transformation $x = Uy$, where U is a real orthogonal matrix such that $U'AU = \mathrm{diag}(\lambda_1, \lambda_2, \ldots, \lambda_n)$ (see [107, Theorem 8.1.1]), we have

$$|x'Ax| = |\lambda_1 y_1^2 + \lambda_2 y_2^2 + \cdots + \lambda_n y_n^2| \leq \rho_A |y|^2 = \rho_A |x|^2. \tag{6.77}$$

Thus, by applying (6.77) to the matrix $\sigma_j'(i) + \sigma_j(i)$, we obtain that

$$\frac{(x'\sigma_j'(i)x)^2}{|x|^4} = \frac{(x'(\sigma_j'(i) + \sigma_j(i))x)^2}{4|x|^4} \leq \frac{1}{4}\Big[\rho(\sigma_j(i) + \sigma_j'(i))\Big]^2.$$

Therefore, detailed computations as in part (a) (taking into account the extra term involving $\frac{1}{4}[\rho(\sigma_j(i) + \sigma_j'(i))]^2$) show that for any sufficiently small $0 < \varepsilon < r$, we have

$$\mathcal{L}V(x, i) \leq -\kappa(\varepsilon) < 0, \quad \text{for any } (x, i) \in N \times \mathcal{M} \text{ with } \varepsilon < |x| < r_0,$$

where $N \subset \mathbb{R}^r$ is a small neighborhood of 0 and $\kappa(\varepsilon)$ is a positive constant. Therefore, Lemma 6.7 and Remark 6.8 imply that the equilibrium point $x = 0$ is unstable in probability. The proof of the theorem is concluded. □

Remark 6.22 Suppose that for all $i \in \mathcal{M}$ and $j = 1, 2, \ldots, d$, the matrices $(\sigma_j'(i) + \sigma_j(i))$ are nonnegative definite. Then, we have

$$\rho(\sigma_j(i) + \sigma_j'(i)) = \lambda_{\max}(\sigma_j(i) + \sigma_j'(i)) \geq \lambda_{\min}(\sigma_j(i) + \sigma_j'(i)) \geq 0.$$

Consequently, a close examination of the proof of Theorem 6.21 reveals that the conditions (6.70) and (6.71) can be replaced by

$$\sum_{i=1}^{m_0} \pi_i \Big(\lambda_{\max}\Big(b(i) + b'(i) + \sum_{j=1}^{d} \sigma_j(i)\sigma_j'(i)\Big) - \frac{1}{2}\sum_{j=1}^{d}\Big[\lambda_{\min}(\sigma_j(i) + \sigma_j'(i))\Big]^2\Big)$$
$$< 0, \tag{6.78}$$

and

$$\sum_{i=1}^{m_0} \pi_i \left(\lambda_{\min} \left(b(i) + b'(i) + \sum_{j=1}^{d} \sigma_j(i)\sigma_j'(i) \right) - \frac{1}{2} \sum_{j=1}^{d} \left[\lambda_{\max}(\sigma_j(i) + \sigma_j'(i)) \right]^2 \right)$$
$$> 0, \tag{6.79}$$

respectively. In a sense, the above two inequalities, in particular (6.78), strengthen the corresponding results in Theorem 6.21.

Theorem 6.21 gives sufficient conditions in terms of the maximum and minimum eigenvalues of the matrices for stability and instability of the equilibrium point $x = 0$. Because there is a "gap" between the maximum and minimum eigenvalues, a natural question arises: Can we obtain the necessary and sufficient conditions for stability? If the component $X(t)$ is one dimensional, we have the following result.

We replace the first and second equations of (6.68) in Assumption (H6.4) by

$$\begin{aligned} b(x, i) &= b_i x + o(x), \\ \sigma(x, i) &= \sigma_i x + o(|x|), \end{aligned} \tag{6.80}$$

where $x \in \mathbb{R}$ and b_i and σ_i^2 are real constants with $\sigma_i^2 \geq 0, i \in \mathcal{M}$. Then, we immediately have the following corollary from Theorem 6.21 and Remark 6.22.

Corollary 6.23 *Let Assumption* (H6.4) *and* (6.80) *be valid. Then, the equilibrium point $x = 0$ is asymptotically stable in probability if*

$$\sum_{i=1}^{m_0} \pi_i \left(b_i - \frac{\sigma_i^2}{2} \right) < 0,$$

and is unstable in probability if

$$\sum_{i=1}^{m_0} \pi_i \left(b_i - \frac{\sigma_i^2}{2} \right) > 0.$$

Remark 6.24 As can be seen from Corollary 6.23, if the continuous component of the system is one dimensional, we obtain a necessary and sufficient condition for stability. One question of particular interest is: Will we be able to obtain a similar condition for a multidimensional counterpart. For linear systems of stochastic differential equations with constant coefficients without switching, such a condition was obtained in Khasminskii [156, Theorems 6.11 & 6.12]. The main ingredient is the use of the transformations $y = x/|x|$ and $\ln |x|$. The result is a sharp necessary and sufficient condition. In Mao et al. [204], inspired by the approach of [156], Markov-modulated regime-switching diffusions were considered, and necessary and sufficient conditions were obtained for exponential stability. The main ingredient is the use of a logarithm transformation technique leading to the derivation of the so-called Lyapunov exponent. Such an approach can be adopted to treat switching diffusions with state-dependent switching with no essential difficulty. Because in Chapter 7, we will also examine a related problem for switched ordinary differential equations (a completely degenerate switching diffusion with the absence of the diffusion terms), we will not dwell on it here.

We end this section with some discussions on one-dimensional linear systems:

$$dX(t) = a(\alpha(t))X(t)dt + b(\alpha(t))X(t)dW(t), \tag{6.81}$$

where α is an irreducible continuous-time Markov chain generated by $Q = (q_{ij})$ with an invariant distribution $\pi = (\pi_i, i \in \mathcal{M})$. Suppose for each $i \in \mathcal{M}$, $a_i = a(i)$, $b_i = b(i)$ are real numbers. Furthermore, let $\mathcal{P}(\mathcal{M})$ be the set of probability measures on the state space $\mathcal{M}$; then under the irreducibility and ergodicity assumptions, the empirical measure of the continuous-time Markov chain $\alpha(\cdot)$ satisfies the large deviations principle with the rate function:

$$\Im(\mu) := -\inf_{u_1,\dots,u_{m_0}>0} \sum_{i,j\in\mathcal{M}} \frac{\mu_i q_{ij} u_j}{u_i}, \tag{6.82}$$

where $\mu = (\mu_1, \dots, \mu_{m_0}) \in \mathcal{P}(\mathcal{M})$; we refer to [64] for details. It is known that $\Im(\mu)$ is lower semicontinuous and $\Im(\mu) = 0$ if and only if $\mu = \pi$. In addition, by virtue of [325], if $a = (a_1, \dots, a_{m_0})' \in \mathbb{R}^{m_0}$, then we have

$$\Upsilon(a) := \lim_{t\to\infty} \frac{1}{t} \log\left(\mathbb{E}\left[\exp\left\{\int_0^t a(\alpha(s))ds\right\}\right]\right) = \sup_{\mu\in\mathcal{P}(\mathcal{M})} \left\{\sum_{i\in\mathcal{M}} a_i\mu_i - \Im(\mu)\right\}.$$

Note that $\sum_{i\in\mathcal{M}} a_i\pi_i \le \Upsilon(a) \le \max_{i\in\mathcal{M}} a_i$.

Proposition 6.25 *The solution to* (6.81) *satisfies the following property:*

(i) The equilibrium point of (6.81) *is almost surely exponentially stable if and only if $\delta < 0$, where*

$$\lim_{t\to+\infty} \frac{1}{t} \log|X(t)| = \delta := \sum_{i=1}^{m_0} \pi_i \left[a_i - \frac{1}{2}b_i^2\right] \quad a.s.$$

(ii) Let $p > 0$ and $f(i) = f_p(i) = pa(i) + \frac{1}{2}p(p-1)b^2(i)$ for each $i \in \mathcal{M}$. Then, we have

$$\lim_{t\to\infty} \frac{1}{t} \log(\mathbb{E}[|x(t)|^p]) = \Upsilon(f) := \sup_{\mu\in\mathcal{P}(\mathcal{M})} \left\{\sum_{i\in\mathcal{M}} f(i)\mu_i - \Im(\mu)\right\}.$$

Therefore, the trivial solution of (6.81) *is pth moment exponentially stable if and only if* $\Upsilon(f) < 0$.

Proof The proof of this proposition can be found in [205, Chapter 5]; see also [40] and [325] for the case of switching jump diffusions. □

6.7 Examples

Example 6.26 To illustrate Theorem 6.21 and Corollary 6.23, we consider a real-valued process given by

$$\begin{cases} dX(t) = b(X(t), \alpha(t))dt + \sigma(X(t), \alpha(t))dw(t) \\ \mathbb{P}\{\alpha(t+\Delta) = j|\alpha(t) = i, X(s), \alpha(s), s \le t\} = q_{ij}(X(t))\Delta + o(\Delta), \end{cases} \tag{6.83}$$

for $j \neq i$, where the jump process $\alpha(t)$ has three states and is generated by

$$Q(x) = \begin{bmatrix} -3 - \sin x \cos x + \sin x^2 & 1 + \sin x \cos x & 2 - \sin x^2 \\ 2 & -2 - \dfrac{x^2}{2+x^2} & \dfrac{x^2}{2+x^2} \\ 4 - \sin x & \sin^2 x & -4 + \sin x - \sin^2 x \end{bmatrix},$$

and the drift and diffusion coefficients are given by

$$\begin{aligned} &b(x,1) = x - x\sin x,\ b(x,2) = x - x\sin x^2,\quad b(x,3) = 4x + x\sin x, \\ &\sigma(x,1) = -\frac{3x}{1+x^2},\quad \sigma(x,2) = x + \frac{1}{3}x\sin x,\ \sigma(x,3) = x - \frac{1}{2}x\sin^2 x. \end{aligned}$$

Associated with (6.83), there are three diffusions:

$$dX(t) = (X(t) - X(t)\sin X(t))dt - \frac{3X(t)}{1+X^2(t)}dw(t), \tag{6.84}$$

$$dX(t) = \left(X(t) - X(t)\sin X^2(t)\right)dt + \left(X(t) + \frac{1}{3}X(t)\sin X(t)\right)dw(t), \tag{6.85}$$

$$dX(t) = (4X(t) + X(t)\sin X(t))\,dt + \left(X(t) - \frac{1}{2}X(t)\sin^2 X(t)\right)dw(t), \tag{6.86}$$

switching from one to another according to the movement of the jump process $\alpha(t)$. It is readily seen that as $x \to 0$, the constants $b_i, \sigma_i^2, i = 1, 2, 3$, as in (6.80) are given by

$$
\begin{aligned}
&b_1 = 1,\ b_2 = 1,\ b_3 = 4 \\
&\sigma_1^2 = 9,\ \sigma_2^2 = 1,\ \sigma_3^2 = 1.
\end{aligned}
\tag{6.87}
$$

Also, as $x \to 0$, $Q(x)$ tends to

$$
\widehat{Q} = \begin{pmatrix} -3 & 1 & 2 \\ 2 & -2 & 0 \\ 4 & 0 & -4 \end{pmatrix}.
$$

The matrix is irreducible. By solving the system of equations $\pi \widehat{Q} = 0$ and $\mathbb{1}\pi = 1$, we obtain the stationary distribution π associated with $\widehat{Q}$:

$$
\pi = (0.5,\ 0.25,\ 0.25)\,. \tag{6.88}
$$

Finally, by virtue of (6.87) and (6.88), we examine that

$$
\sum_{i=1}^{3} \pi_i \left(b_i - \frac{\sigma_i^2}{2} \right) = -0.75 < 0.
$$

Therefore, we conclude from Corollary 6.23 that the equilibrium point $x = 0$ of (6.83) is asymptotically stable in probability.

This example is interesting and provides insight. It was proven in [156, Example 7.1] that a one-dimensional nonlinear diffusion is stable if and only if its linear approximation is stable. Hence, we can check that the equilibrium point $x = 0$ of (6.84) is stable in probability, whereas (6.85) and (6.86) are unstable in probability. Therefore, the jump process $\alpha(t)$ could be considered as a stabilization factor. Note that similar examples were demonstrated in [199] under the assumptions that the jump component $\alpha(\cdot)$ is generated by a constant matrix Q and that the Markov chain $\alpha(\cdot)$ is independent of the Brownian motion $w(\cdot)$. Note also Examples 4.1 and 4.2 in [92] are also concerned with stability of switching systems. Their result indicates that if the switching takes place sufficiently fast, the system will be stable even if the individual mode may be unstable. Essentially, it is related to singularly perturbed systems. Due to the fast variation, there is a limit system that is an average with respect to the stationary distribution of the Markov chain and that is stable. Then, if the rate of switching is fast enough, the original system will also be stable. Such an idea was also used in an earlier paper [29].

To illustrate, we plot a sample path of (6.83) in Fig. 6.1. For comparison, we also demonstrate the sample paths of (6.84), (6.85), and (6.86) (without switching) in Figs. 6.2, 6.3, and 6.4, respectively.

Example 6.27 (Lot-Volterra Model) This is a continuation of the discussion of the Lot-Volterra model given in Example 1.1. The notation and problem setup are as in

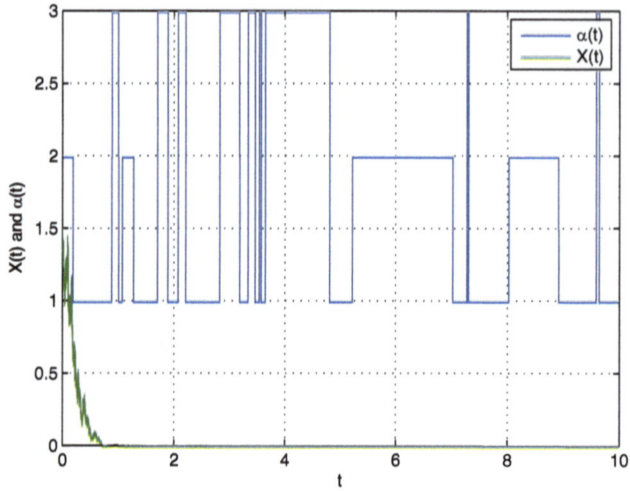

Fig. 6.1 A sample path of (6.83) with initial condition $(x, \alpha) = (1.5, 1)$

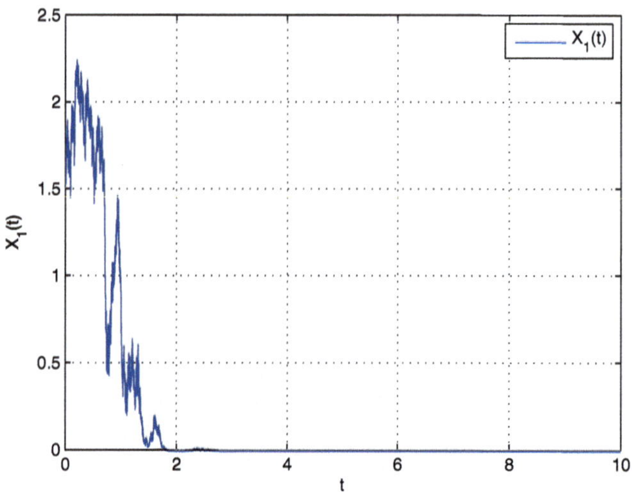

Fig. 6.2 A sample path of (6.84) with initial condition $x = 1.5$

Example 1.1. Motivated by the works [70] and [203], define $V(t, x, \alpha) = e^t \log(|x|)$ for $(t, x, \alpha) \in [0, \infty) \times \mathbb{R}^n_+ \times \mathcal{M}$. It follows from Itô's lemma that

$$e^t \log(|x(t)|) - \log(|x(0)|) = \int_0^t e^s \sum_{i=1}^n \frac{x_i^2(s)}{|x(s)|^2} \Bigg[r_i(\alpha(s)) - \sum_{j=1}^n a_{ij}(\alpha(s)) x_j(s)$$

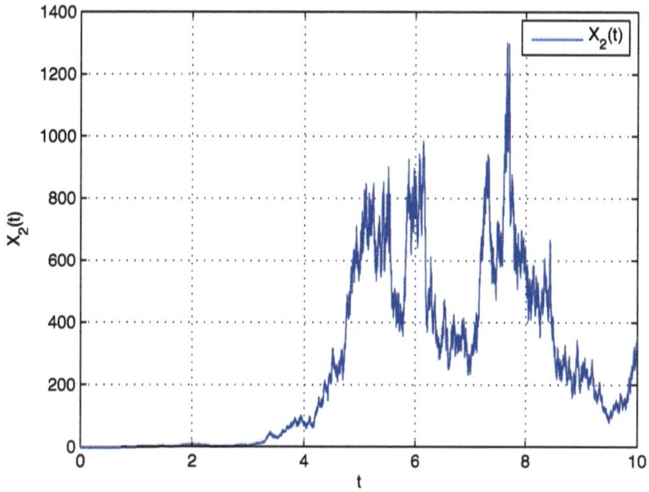

Fig. 6.3 A sample path of (6.85) with initial condition $x = 1.5$

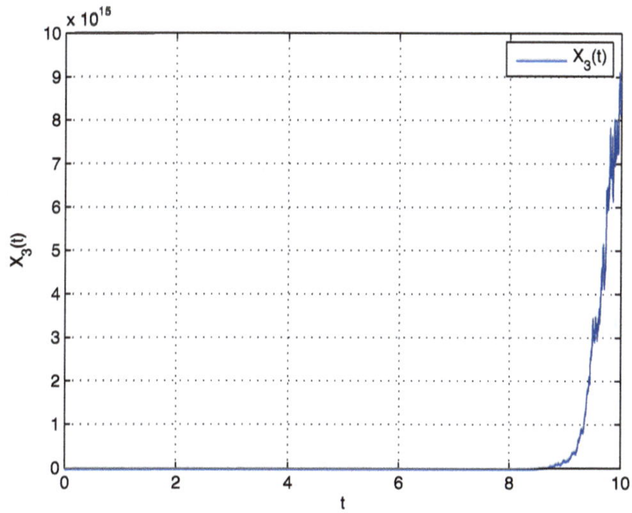

Fig. 6.4 A sample path of (6.86) with initial condition $x = 1.5$

$$+\frac{1}{2}\left(1-\frac{2x_i^2(s)}{|x(s)|^2}\right)\sigma_i^2(\alpha(s))\Big]ds$$

$$+\int_0^t e^s \log(|x(s)|)ds + \int_0^t e^s \sum_{i=1}^n \frac{x_i^2(s)}{|x(s)|^2}\sigma_i(\alpha(s))dw_i(s).$$

Denote

$$M_i(t) = \int_0^t e^s \frac{x_i^2(s)}{|x(s)|^2} \sigma_i(\alpha(s)) dw_i(s),$$

whose quadratic variation is

$$\langle M_i(t), M_i(t) \rangle = \int_0^t \frac{e^{2s}}{|x(s)|^4} x_i^4(s) \sigma_i^2(\alpha(s)) ds.$$

By virtue of the exponential martingale inequality [94], for any positive constants T, δ, and β, we have

$$\mathbb{P}\left\{ \sup_{0 \le t \le T} [M_i(t) - \frac{\delta}{2} \langle M_i(t), M_i(t) \rangle] > \beta \right\} \le e^{-\delta\beta}.$$

Choose $T = k\gamma$, $\delta = n\varepsilon e^{-k\delta}$, and $\beta = (\theta e^{k\delta} \log(k))/(\varepsilon n)$, where $k \in \mathbb{N}$, $0 < \varepsilon < 1$, $\theta > 1$, and $\gamma > 0$ in the above equation. Then, it follows that

$$\mathbb{P}\left\{ \sup_{0 \le t \le k\gamma} [M_i(t) - \frac{n\varepsilon e^{-k\gamma}}{2} \langle M_i(t), M_i(t) \rangle] > \frac{\theta e^{k\delta} \log k}{\varepsilon n} \right\} \le k^{-\theta}.$$

Because $\sum_{k=1}^{\infty} k^{-\theta} < \infty$, it follows from the Borel-Cantelli lemma that there exists some $\Omega_i \subset \Omega$ with $\mathbb{P}(\Omega_i) = 1$ such that for any $\omega \in \Omega_i$, an integer $k_i = k_i(\omega)$ such that for any $k > k_i$, we have

$$M_i(t) \le \frac{n\varepsilon e^{-k\gamma}}{2} \langle M_i(t), M_i(t) \rangle + \frac{\theta e^{k\gamma} \log k}{\varepsilon n} \quad \text{for all } 0 \le t \le k\gamma.$$

Now, let $\Omega_0 := \bigcap_{i=1}^n \Omega_i$. Then, $\mathbb{P}(\Omega_0) = 1$. Moreover, for any $\omega \in \Omega_0$, let

$$k_0(\omega) := \max\{k_i(\omega), i = 1, 2, \ldots, n\}.$$

Then, for any $\omega \in \Omega_0$ and any $k \ge k_0(\omega)$, we have

$$\begin{aligned}
&\sum_{i=1}^n \int_0^t e^s \frac{x_i^2(s)}{|x(s)|^2} \sigma_i(\alpha(s)) dw_i(s) \\
&= \sum_{i=1}^n M_i(t) \le \frac{n\varepsilon e^{-k\gamma}}{2} \sum_{i=1}^n \langle M_i(t), M_i(t) \rangle + \frac{\theta e^{k\gamma} \log k}{\varepsilon},
\end{aligned}$$

where $0 \le t \le k\gamma$. Then, it follows that

$$\begin{aligned}
&e^t \log(|x(t)|) - \log(|x(0)|) \\
&\le \int_0^t e^s \sum_{i=1}^n \frac{x_i^2(s)}{|x(s)|^2} \Big[r_i(\alpha(s)) - \sum_{j=1}^n a_{ij}(\alpha(s))x_j(s) \\
&+ \frac{1}{2}\left(1 - \frac{2x_i^2(s)}{|x(s)|^2}\right) \sigma_i^2(\alpha(s)) \Big] ds \\
&+ \int_0^t e^s \log(|x(s)|)ds + \int_0^t \frac{n\varepsilon e^{-k\gamma}}{2} e^{2s} \sum_{i=1}^n \frac{x_i^4(s)}{|x(s)|^4}\sigma_i^2(\alpha(s))ds + \frac{\theta e^{k\gamma}\log k}{\varepsilon} \\
&\le \int_0^t e^s \Big\{ \log(|x(s)|) + \sum_{i=1}^n \frac{x_i^2(s)}{|x(s)|^2}\Big[b_i(\alpha(s)) + \sigma_i^2(\alpha(s)) - \sum_{j=1}^n a_{ij}(\alpha(s))x_j(s) \Big] \\
&+ \sum_{i=1}^n \left(\frac{\varepsilon n e^{-k\gamma}}{2} e^s - 1 \right) \frac{x_i^4(s)\sigma_i^2(\alpha(s))}{|x(s)|^4} \Big\} + \frac{\theta e^{k\gamma}\log k}{\varepsilon}.
\end{aligned}$$

Note that for any $t \in [0, k\gamma]$, $s \in [0, t]$, and $(x, \alpha) \in \mathbb{R}_+^n \times \mathcal{M}$, we have

$$\begin{aligned}
&\log(|x|) + \sum_{i=1}^n \Big[\frac{x_i^2}{|x|^2}\Big[b_i(\alpha) + \sigma_i^2(\alpha) - \sum_{j=1}^n a_{ij}(\alpha)x_j \Big] \\
&+ \left(\frac{\varepsilon n e^{s-k\gamma}}{2} - 1 \right) \frac{x_i^4 \sigma_i^2(\alpha)}{|x|^4} \Big] \\
&\le \log(|x|) + \kappa - \frac{1}{|x|^2}\beta \sum_{i=1}^n x_i^3 + K \\
&\le \log(|x|) + \kappa - \frac{\beta}{\sqrt{n}}|x| + K \le K.
\end{aligned}$$

Hence, it follows that for all $0 \le t \le k\gamma$ with $k \ge k_0(\omega)$, we have

$$\begin{aligned}
e^t \log(|x(t)|) - \log(|x(0)|) &\le \int_0^t K e^s ds + \frac{\theta e^{k\gamma}\log k}{\varepsilon} \\
&= K(e^t - 1) + \frac{\theta e^{k\gamma}\log k}{\varepsilon}.
\end{aligned}$$

Thus, for $(k-1)\gamma \le t \le k\gamma$, we have

$$\begin{aligned}
\log(|x(t)|) &\le e^{-t}\log(|x(0)|) + K(1 - e^{-t}) + \frac{\theta e^{k\gamma}\log k}{\varepsilon e^{(k-1)\gamma}} \\
&= e^{-t}\log(|x(0)|) + K(1 - e^{-t}) + \frac{\theta e^{\gamma}\log k}{\varepsilon},
\end{aligned}$$

and hence, it follows that

$$\frac{\log(|x(t)|)}{\log t} \le \frac{\log(|x(0)|)}{e^t \log t} + \frac{K(1-e^{-t})}{\log t} + \frac{\theta e^\gamma \log k}{\varepsilon \log((k-1)\gamma)}.$$

Now, let $k \to \infty$ (and so $t \to \infty$) and we obtain

$$\limsup_{t\to\infty} \frac{\log(|x(t)|)}{\log t} \le \frac{\theta e^\gamma}{\varepsilon}.$$

Finally, by sending $\gamma \downarrow 0$, $\varepsilon \uparrow 1$, and $\theta \downarrow 1$, we have

$$\limsup_{t\to\infty} \frac{\log(|x(t)|)}{\log t} \le 1,$$

as desired. Then, it follows that

$$\limsup_{T\to\infty} \frac{\log|x(T)|}{T} = \limsup_{T\to\infty} \frac{\log|x(T)|}{\log T} \limsup_{T\to\infty} \frac{\log T}{T} \le \limsup_{T\to\infty} \frac{\log T}{T} = 0,$$

for all $\omega \in \Omega_0$. In particular, we have

$$\limsup_{T\to\infty} \frac{\log|x(T)|}{T} \le 0 \text{ a.s.} \tag{6.89}$$

6.8 Notes

The framework of this chapter is derived from the paper of Khasminskii, Zhu, and Yin [164]. Some new results such as Theorems 6.9 and 6.10 are taken from Nguyen, Yin, and Zhu [219], where we derived systems of Kolmogorov backward equations without assuming nondegeneracy of the diffusion part. The results on almost sure exponential stability in Section 6.5 are based on [40]. These results are interesting in their own right.

For brevity, we are not trying to cover every angle of the stability analysis. For example, sufficient conditions for exponential q-instability can be derived similar to that of [156], so the verbatim proof for the sufficiency is omitted. Nevertheless, necessary conditions for stability and stability under linearization are provided. Finally, we note that using linearization to infer the stability of the associated nonlinear systems should be of interest for many people owing to its wide range of applications.

Chapter 7
Stability of Switching ODEs

7.1 Introduction

The main motivational forces for this chapter are the work of Davis [60] on piecewise deterministic systems and the work of Kac and Krasovskii [146] on stability of randomly switched systems. In recent years, growing attention has been drawn to deterministic dynamic systems formulated as differential equations modulated by a random switching process. This is because of the increasing demands for modeling large-scale and complex systems, designing optimal controls, and carrying out optimization tasks. In this chapter, we consider stability of such hybrid systems modulated by a random-switching process, which are "equivalent" to a number of ordinary differential equations coupled by a switching or jump process.

In this chapter, for random-switching systems, we first obtain sufficient conditions for stability and instability. Our approach leads to a necessary and sufficient condition for systems whose continuous component is one dimensional. For multidimensional systems, our conditions involve the use of minimal and maximal eigenvalues of appropriate matrices. The difference of maximal and minimal eigenvalues results in a gap for stability and instability. To close this gap, we introduce a logarithm transformation leading to the continuous component taking values in the unit sphere. This, in turn, enables us to obtain necessary and sufficient conditions for stability. The essence is the utilization of the so-called Lyapunov exponent.

Because the systems we are interested in have continuous components (representing continuous dynamics) as well as discrete components (representing discrete events), their asymptotic behavior can be quite different from a single differential equation. As noted, a random-switching differential system may be considered as several differential equations coupled by a switching process. We show that even though some of the individual equations are not stable, the entire switching system may still be stable as long as certain conditions are satisfied.

H.-D. Nguyen et al., *Hybrid Switching Diffusions*, Probability Theory and Stochastic Modelling 63, https://doi.org/10.1007/978-3-031-93303-5_7

For random switching systems that are linear in their continuous component, suppose that corresponding to different discrete states, some of the associated differential equations are stable and the others are not. If the switching component is ergodic, we show that as long as the stable part of the differential equations dominates the rest (in an appropriate sense), the coupled hybrid system will be stable. For nonlinear differential equations, the well-known Hartman-Grobman theorem (see [234, Section 2.8]) provides an important result concerning the local qualitative behavior. It says that near a hyperbolic equilibrium point x_0, the nonlinear system $\dot{x} = f(x)$ has the same qualitative structure as that of the linear system $\dot{x} = \nabla f(x_0)x$, although the topological equivalence may not hold for a non-hyperbolic equilibrium point (e.g., a center). Treating hybrid systems, consider the differential equations $\dot{x}(t) = f(x(t), \alpha(t))$ and $\dot{x}(t) = \nabla f(x_0, \alpha(t))x(t)$ for $\alpha(t)$ belonging in a finite set. We show although some of the linear equations have centers, as long as the spectrum of the coefficients of the differential equation corresponding to the stable node dominates that of the centers, the overall system will still be topologically equivalent to the linear (in continuous component) system. To reveal the salient features, we present several examples and display the corresponding phase portraits. The results are quite revealing.

The rest of the chapter is arranged as follows. Section 7.2 begins with the formulation of the random-switching systems and provides definitions of stability, instability, and asymptotical stability of the equilibrium point of the random-switching hybrid systems and gives some preliminary results. For the purpose of our asymptotic analysis, we also present sufficient conditions for stability, instability, and asymptotical stability. Easily verifiable conditions for stability and instability of the systems are provided in Section 7.3. Section 7.4 presents a sharper condition for systems that are linear in the continuous state variable. Discussions on the Lyapunov exponent are given in Section 7.5. To demonstrate our results, we provide several examples in Section 7.6. Finally, we conclude the chapter with further remarks in Section 7.7.

7.2 Formulation and Preliminary Results

7.2.1 Problem Setup

Throughout the chapter, we use z' to denote the transpose of $z \in \mathbb{R}^{\ell_1 \times \ell_2}$ with $\ell_i \geq 1$, whereas $\mathbb{R}^{\ell \times 1}$ is simply written as $\mathbb{R}^{\ell}$; $\mathbb{1} = (1, 1, \ldots, 1)' \in \mathbb{R}^{m_0}$ is a column vector with all entries being 1; the Euclidean norm for a row or column vector x is denoted by $|x|$. As usual, I denotes the identity matrix. For a matrix A, its trace norm is denoted by $|A| = \sqrt{\text{tr}(A'A)}$. When B is a set, $I_B(\cdot)$ denotes the indicator function of B. For $A \in \mathbb{R}^{r \times r}$ being a symmetric matrix, we use $\lambda_{\max}(A)$ and $\lambda_{\min}(A)$ to denote the maximum and minimum eigenvalues of A, respectively.

Consider the system with random switching:

$$\dot{X}(t) = f(X(t), \alpha(t)), \quad X(0) = x \in \mathbb{R}^r, \quad \alpha(0) = \alpha \in \mathcal{M}, \tag{7.1}$$

where $X(t)$ is the continuous state, $f(\cdot,\cdot) : \mathbb{R}^r \times \mathcal{M} \mapsto \mathbb{R}^r$, and $\alpha(\cdot)$ is a switching process taking value in a finite state space $\mathcal{M} = \{1, 2, \dots, m_0\}$ with generator $Q(x) = q_{ij}(x)$ satisfying $q_{ij}(x) \geq 0$ for $j \neq i$ and $\sum_{j\in\mathcal{M}} q_{ij}(x) = 0$ for all $x \in \mathbb{R}^r$ and $i \in \mathcal{M}$. The evolution of the switching component is described by

$$\begin{aligned} &\mathbb{P}\{\alpha(t+\Delta) = j|\alpha(t) = i, (X(s), \alpha(s)), s \leq t\} \\ &\quad = q_{ij}(X(t))\Delta + o(\Delta), \quad i \neq j. \end{aligned} \tag{7.2}$$

Note that in our formulation, x-dependent $Q(x)$ is considered, whereas in [11, 146, 199, 311], the constant generator Q was used.

Associated with the process $(X(t), \alpha(t))$ defined by (7.1)–(7.2), there is an operator $\mathcal{L}$ defined as follows. For each $i \in \mathcal{M}$ and any $g(\cdot, i) \in C^1(\mathbb{R}^r)$

$$\mathcal{L}g(x,i) = f'(x,i)\nabla g(x,i) + Q(x)g(x,\cdot)(i), \tag{7.3}$$

where $\nabla g(x,i)$ denotes the gradient (with respect to the variable x) of $g(x,i)$, and

$$Q(x)g(x,\cdot)(i) = \sum_{j\in\mathcal{M}} q_{ij}(x)g(x,j) \ \text{ for each } \ i \in \mathcal{M}. \tag{7.4}$$

For further references on the associated operator (or generator) of the hybrid system (7.1)–(7.2), we refer the reader to Chapter 2 of this book; see also [146] and [262].

To proceed, we need conditions regarding the smoothness and growth of the functions involved and the condition that 0 is an equilibrium point of the dynamic system. We assume the following hypotheses throughout the chapter:

(H7.1) The matrix-valued function $Q(\cdot)$ is bounded and continuous, and satisfies (2.21) for some concave function $\gamma : \mathbb{R}_+ \mapsto \mathbb{R}_+$ with $\gamma(0) = 0$.

(H7.2) For each $\alpha \in \mathcal{M}$, $f(\cdot, \alpha)$ is locally Lipschitz continuous and satisfies $f(0, \alpha) = 0$.

(H7.3) There exists a constant $K_0 > 0$ such that for each $\alpha \in \mathcal{M}$

$$|f(x,\alpha)| \leq K_0(1 + |x|), \quad \text{for all } x \in \mathbb{R}^r. \tag{7.5}$$

It is well known that under these conditions, system (7.1)–(7.2) has a unique solution; see Chapter 2. In what follows, a process starting from (x, α) is denoted by $Y^{x,\alpha}(t) = (X^{x,\alpha}(t), \alpha^{x,\alpha}(t))$ to emphasize the dependence on the initial condition. If the context is clear, we simply write $Y(t) = (X(t), \alpha(t))$.

7.2.2 Preliminary Results

In this subsection, we first recall the definitions of stability, instability, asymptotic stability, and exponential p-stability. Then, we present some preparatory results of stability and instability in terms of Lyapunov functions.

Definition 7.1 ([146]) The equilibrium point $x = 0$ of system (7.1)–(7.2) is said to be

(i) *stable in probability*, if for any $\alpha = 1, \ldots, m_0$ and $r_0 > 0$

$$\lim_{x \to 0} \mathbb{P}\{\sup_{t \geq 0} |X^{x,\alpha}(t)| > r_0\} = 0,$$

and it is said to be *unstable in probability* if it is not stable in probability

(ii) *asymptotically stable in probability*, if it is stable in probability and

$$\lim_{x \to 0} \mathbb{P}\{\lim_{t \to \infty} X^{x,\alpha}(t) = 0\} = 1, \text{ for each } \alpha = 1, \ldots, m_0,$$

(iii) *exponentially p-stable*, if for some positive constants K and γ

$$\mathbb{E}|X^{x,\alpha}(t)|^p \leq K|x|^p \exp\{-\gamma t\}, \text{ for any } (x, \alpha) \in \mathbb{R}^r \times \mathcal{M}.$$

The definitions above should be compared to Definition 6.4. They are of the same spirit although now we have a completely degenerate case with the diffusion matrix being identically 0. To study stability of the equilibrium point $x = 0$, we first observe that almost all trajectories of the system (7.1)–(7.2) starting from a nonzero state will never reach the origin with probability one.

Proposition 7.2 *Let Assumptions* (H7.1)–(H7.3) *be satisfied. Then*

$$\mathbb{P}\{X^{x,\alpha}(t) \neq 0, t \geq 0\} = 1, \quad \textit{for any } (x, \alpha) \in \mathbb{R}^r \times \mathcal{M} \textit{ with } x \neq 0. \tag{7.6}$$

Proof This proposition can be proved using a slight modification of the argument in Lemma 6.1. The details are omitted for brevity. □

In view of (7.6), we can work with functions $V(\cdot, i), i \in \mathcal{M}$, which are continuously differentiable in a deleted neighborhood of 0 in what follows. This turns out to be quite convenient. Another immediate consequence of (7.6) is the following L^p estimate for the solution of the system (7.1)–(7.2). The result is interesting in its own right.

Theorem 7.3 *Let Assumptions* (H7.1)–(H7.3) *be satisfied. Then, for any* $p \geq 1$ *and any* $(x, \alpha) \in \mathbb{R}^r \times \mathcal{M}$ *with* $x \neq 0$, *we have*

$$\mathbb{E}|X^{x,\alpha}(t)|^p \leq \left(|x|^p + \frac{1}{2}\right) \exp(2pK_0 t). \tag{7.7}$$

Proof For each $\alpha \in \mathcal{M}$, the function $V(x,\alpha) = |x|^p$ is continuously differentiable in the domain $|x| > \delta$ for any $\delta > 0$. Let τ_δ be the first exit time of the process $X^{x,\alpha}(\cdot)$ from the set $\{x \in \mathbb{R}^r : |x| > \delta\} \times \mathcal{M}$. That is, $\tau_\delta := \inf\{t \geq 0 : |X^{x,\alpha}(t)| \leq \delta\}$. For any $t > 0$, set $\tau_\delta(t) := \min\{\tau_\delta, t\}$. Because $V(x,i)$ is independent of i, $\sum_{j\in\mathcal{M}} q_{ij}(x)V(x,j) = 0$. Then, it follows from the Cauchy-Schwartz inequality and (7.5) that

$$\begin{aligned}
&\mathbb{E}\left|X^{x,\alpha}(\tau_\delta(t))\right|^p \\
&= |x|^p + \mathbb{E}\int_0^{\tau_\delta(t)} p\left|X^{x,\alpha}(s)\right|^{p-2}\langle X^{x,\alpha}(s), f(X^{x,\alpha}(s), \alpha^{x,\alpha}(s))\rangle ds \\
&\leq |x|^p + pK_0\mathbb{E}\int_0^{\tau_\delta(t)} \left|X^{x,\alpha}(s)\right|^{p-1}(1 + \left|X^{x,\alpha}(s)\right|)ds \\
&\leq |x|^p + pK_0\mathbb{E}\int_0^{\tau_\delta(t)} (2\left|X^{x,\alpha}(s)\right|^p + \left|X^{x,\alpha}(s)\right|^{p-1} I_{\{\delta\leq|X^{x,\alpha}(s)|<1\}})ds \\
&\leq |x|^p + 2pK_0\mathbb{E}\int_0^{\tau_\delta(t)} \left(\left|X^{x,\alpha}(s)\right|^p + \frac{1}{2}\right)ds.
\end{aligned}$$

Note that for all $s \leq \tau_\delta(t)$, we have $s = \tau_\delta(s)$. Hence, we have

$$\begin{aligned}
\mathbb{E}\,|X^{x,\alpha}(\tau_\delta(t))|^p &\leq |x|^p + 2pK_0\mathbb{E}\int_0^{\tau_\delta(t)} \left(\left|X^{x,\alpha}(\tau_\delta(s))\right|^p + \frac{1}{2}\right)ds \\
&\leq |x|^p + 2pK_0\int_0^t \left(\mathbb{E}\left|X^{x,\alpha}(\tau_\delta(s))\right|^p + \frac{1}{2}\right)ds.
\end{aligned}$$

Applying Gronwall's inequality to $[\mathbb{E}\,|X^{x,\alpha}(\tau_\delta(t))|^p + (1/2)]$ leads to

$$\mathbb{E}\left|X^{x,\alpha}(\tau_\delta(t))\right|^p + \frac{1}{2} \leq \left(|x|^p + \frac{1}{2}\right)\exp(2pK_0t), \tag{7.8}$$

or equivalently,

$$\mathbb{E}\,|X^{x,\alpha}(\tau_\delta(t))|^p \leq \left(|x|^p + \frac{1}{2}\right)\exp(2pK_0t) - \frac{1}{2} \leq \left(|x|^p + \frac{1}{2}\right)\exp(2pK_0t). \tag{7.9}$$

Note that we have from (7.6) that

$$\tau_\delta(t) \to t \text{ as } \delta \to 0 \text{ with probability 1 for any } t > 0.$$

Finally, letting $\delta \to 0$ in (7.9), by Fatou's lemma, we obtain (7.7). □

Remark 7.4 If condition (7.5) is replaced by

$$|f(x,\alpha)| \leq K|x|, \quad \text{for all } (x,\alpha) \in \mathbb{R}^r \times \mathcal{M}, \tag{7.10}$$

where K is some positive constant, then the conclusion Theorem 7.3 can be strengthened to the following. For any $\beta \in \mathbb{R}$ and any $(x, \alpha) \in \mathbb{R}^r \times \mathcal{M}$ with $x \neq 0$, we have

$$\mathbb{E}|X^{x,\alpha}(t)|^\beta \leq |x|^\beta e^{\rho t}, \tag{7.11}$$

where ρ is a constant depending only on β and the constant K given in (7.10).

In fact, by virtue of (7.10), we obtain by a slight modification of the argument in the proof of Theorem 7.3 that

$$\mathbb{E}|X^{x,\alpha}(\tau_\delta(t))|^\beta \leq |x|^\beta e^{\rho t}.$$

Then, similar to the proof of Theorem 7.3, (7.11) follows from (7.6) and Fatou's lemma.

We finally note that if $f(x, \alpha)$ is Lipschitzian with a global Lipschitz constant L_0, and $f(0, \alpha) = 0$, then (7.10) is verified.

Next, concerning stability and asymptotical stability of the equilibrium point $x = 0$ of the system (7.1)–(7.2), we have the following results.

Proposition 7.5 *Let $D \subset \mathbb{R}^r$ be a neighborhood of 0. Suppose that for each $i \in \mathcal{M}$, there exists a nonnegative function $V(\cdot, i) : D \mapsto \mathbb{R}$ such that*

(i) *$V(\cdot, i)$ is continuous in D and vanishes only at $x = 0$,*
(ii) *$V(\cdot, i)$ is continuously differentiable in $D \setminus \{0\}$ and satisfies*

$$\mathcal{L}V(x, i) \leq 0, \quad \text{for all } x \in D \setminus \{0\}. \tag{7.12}$$

Then, the equilibrium point $x = 0$ is stable in probability.

Proof The proof is almost identical to that of Lemma 6.5 and is thus omitted. □

Proposition 7.6 *Assume the conditions of Proposition 7.5. Suppose also that for each $i \in \mathcal{M}$, the function $V(\cdot, i)$ satisfies*

$$\mathcal{L}V(x, i) \leq -\kappa(\varrho) < 0, \quad \text{for all } x \in D \setminus \{x \in \mathbb{R}^r : |x| > \varrho\}, \tag{7.13}$$

where $\varrho > 0$ and $\kappa(\varrho)$ is a positive constant. Then, the equilibrium point $x = 0$ is asymptotically stable in probability.

Proof By virtue of Proposition 7.5, the equilibrium point $x = 0$ is stable in probability. It remains to show that

$$\lim_{x \to 0} \mathbb{P}_{x,i} \left\{ \lim_{t \to \infty} X(t) = 0 \right\} = 1.$$

The equilibrium point $x = 0$ is stable in probability; therefore, it follows that for any $\varepsilon > 0$ and $r_0 > 0$, there exists some $\delta > 0$ (without loss of generality, we may assume that $\delta < r_0$) such that

$$\mathbb{P}_{x,i}\left\{\sup_{t\geq 0}|X(t)| < r_0\right\} \geq 1 - \frac{\varepsilon}{2}, \text{ for any } (x,i) \in B_\delta \times \mathcal{M}, \tag{7.14}$$

where $B_\delta := \{x \in \mathbb{R}^r : |x| < \delta\}$. Now, fix some $(x, \alpha) \in (B_\delta \setminus \{0\}) \times \mathcal{M}$ and let $\varrho_1 > \varrho > 0$ be arbitrary satisfying $\varrho_1 < |x|$. Define

$$\tau_\varrho := \inf\{t \geq 0 : |X(t)| \leq \varrho\}, \quad \text{and} \quad \tau_{r_0} := \inf\{t \geq 0 : |X(t)| \geq r_0\}.$$

Then, it follows that for any $t > 0$

$$\begin{aligned}&\mathbb{E}_{x,\alpha}V(X(t\wedge\tau_\varrho\wedge\tau_{r_0}), \alpha(t\wedge\tau_\varrho\wedge\tau_{r_0})) - V(x,\alpha)\\ &\leq \mathbb{E}_{x,\alpha}\int_0^{t\wedge\tau_\varrho\wedge\tau_{r_0}} \mathcal{L}V(X(s),\alpha(s))ds \leq -\kappa(\varrho)\mathbb{E}_{x,\alpha}[t\wedge\tau_\varrho\wedge\tau_{r_0}].\end{aligned}$$

Because V is nonnegative, we have $\mathbb{E}_{x,\alpha}[t\wedge\tau_\varrho\wedge\tau_{r_0}] \leq V(x,\alpha)/(\kappa(\varrho))$ and hence $t\mathbb{P}_{x,\alpha}\left\{\tau_\varrho\wedge\tau_{r_0} > t\right\} \leq V(x,\alpha)/(\kappa(\varrho))$. Letting $t \to \infty$, we obtain

$$\mathbb{P}_{x,\alpha}\left\{\tau_\varrho\wedge\tau_{r_0} = \infty\right\} = 0 \text{ or } \mathbb{P}_{x,\alpha}\left\{\tau_\varrho\wedge\tau_{r_0} < \infty\right\} = 1. \tag{7.15}$$

Note that (7.15) allows us to complete the proof using a similar argument as that for Lemma 6.6. However, we present a different proof here. Using (7.14), we have $\mathbb{P}_{x,\alpha}\left\{\tau_{r_0} < \infty\right\} \leq \varepsilon/2$ and hence

$$\mathbb{P}_{x,\alpha}\left\{\tau_\varrho < \infty\right\} \geq \mathbb{P}_{x,\alpha}\left\{\tau_\varrho\wedge\tau_{r_0} < \infty\right\} - \mathbb{P}_{x,\alpha}\left\{\tau_{r_0} < \infty\right\} \geq 1 - \frac{\varepsilon}{2}. \tag{7.16}$$

Now, let $\tau_{\varrho_1} := \inf\left\{t \geq \tau_\varrho : |X(t)| \geq \varrho_1\right\}$. We use the convention that $\inf\emptyset = \infty$. Then, for any $t > 0$, we have

$$\begin{aligned}&\mathbb{E}_{x,\alpha}V(X(t\wedge\tau_{\varrho_1}), \alpha(t\wedge\tau_{\varrho_1}))\\ &= \mathbb{E}_{x,\alpha}V(X(t\wedge\tau_\varrho), \alpha(t\wedge\tau_\varrho)) + \mathbb{E}_{x,\alpha}\int_{t\wedge\tau_\varrho}^{t\wedge\tau_{\varrho_1}} \mathcal{L}V(X(s),\alpha(s))ds\\ &\leq \mathbb{E}_{x,\alpha}V(X(t\wedge\tau_\varrho), \alpha(t\wedge\tau_\varrho)).\end{aligned} \tag{7.17}$$

Note that $\tau_\varrho \leq \tau_{\varrho_1}$ by definition and hence $\tau_\varrho \geq t$ implies that $\tau_{\varrho_1} \geq t$ and thus $I_{\{\tau_\varrho \geq t\}} \leq I_{\{\tau_{\varrho_1} \geq t\}}$. Therefore, it follows that

$$
\begin{aligned}
&\mathbb{E}_{x,\alpha}\big[I_{\{\tau_{\varrho_1}<t\}}V(X(\tau_{\varrho_1}\wedge t),\alpha(\tau_{\varrho_1}\wedge t))\big]\\
&\quad=\mathbb{E}_{x,\alpha}\big[V(X(\tau_{\varrho_1}\wedge t),\alpha(\tau_{\varrho_1}\wedge t))\big]-\mathbb{E}_{x,\alpha}\big[I_{\{\tau_{\varrho_1}\geq t\}}V(X(\tau_{\varrho_1}\wedge t),\alpha(\tau_{\varrho_1}\wedge t))\big]\\
&\quad\leq\mathbb{E}_{x,\alpha}\big[V(X(\tau_{\varrho_1}\wedge t),\alpha(\tau_{\varrho_1}\wedge t))\big]-\mathbb{E}_{x,\alpha}\big[I_{\{\tau_{\varrho}\geq t\}}V(X(\tau_{\varrho_1}\wedge t),\alpha(\tau_{\varrho_1}\wedge t))\big]\\
&\quad\leq\mathbb{E}_{x,\alpha}\big[V(X(\tau_{\varrho}\wedge t),\alpha(\tau_{\varrho}\wedge t))\big]-\mathbb{E}_{x,\alpha}\big[I_{\{\tau_{\varrho}\geq t\}}V(X(\tau_{\varrho}\wedge t),\alpha(\tau_{\varrho}\wedge t))\big]\\
&\quad=\mathbb{E}_{x,\alpha}\big[I_{\{\tau_{\varrho}<t\}}V(X(\tau_{\varrho}\wedge t),\alpha(\tau_{\varrho}\wedge t))\big]\\
&\quad\leq\widehat{V}_{\varrho},
\end{aligned}
\tag{7.18}
$$

where $\widehat{V}_{\varrho} := \sup\{V(y,j) : |y| = \varrho,\, j \in \mathcal{M}\}$; note that we used (7.17) to derive the second inequality. Furthermore, using (7.18), we can derive

$$
\begin{aligned}
\widehat{V}_{\varrho} &\geq \mathbb{E}_{x,\alpha}\big[I_{\{\tau_{\varrho_1}<t\}}V(X(\tau_{\varrho_1}\wedge t),\alpha(\tau_{\varrho_1}\wedge t))\big]=\mathbb{E}_{x,\alpha}\big[I_{\{\tau_{\varrho_1}<t\}}V(X(\tau_{\varrho_1}),\alpha(\tau_{\varrho_1}))\big]\\
&\geq V_{\varrho_1}\mathbb{P}_{x,\alpha}\left\{\tau_{\varrho_1}<t\right\},
\end{aligned}
$$

where $V_{\varrho_1} := \inf\{V(y,j) : |y| = \varrho_1,\, j \in \mathcal{M}\}$. Because for each $i \in \mathcal{M}$, $V(\cdot,i)$ vanishes only at $x = 0$, $V_{\varrho_1} > 0$. Since V is continuous, for any $\varrho_1 > 0$, we can choose ϱ sufficiently small so that

$$
\mathbb{P}_{x,\alpha}\left\{\tau_{\varrho_1}<t\right\}\leq\frac{\widehat{V}_{\varrho}}{V_{\varrho_1}}\leq\frac{\varepsilon}{2}.
$$

Letting $t \to \infty$, we obtain

$$
\mathbb{P}_{x,\alpha}\left\{\tau_{\varrho_1}<\infty\right\}\leq\frac{\varepsilon}{2}. \tag{7.19}
$$

Finally, it follows from (7.16) and (7.19) that

$$
\begin{aligned}
\mathbb{P}_{x,\alpha}\left\{\tau_{\varrho}<\infty,\tau_{\varrho_1}=\infty\right\}&\geq\mathbb{P}_{x,\alpha}\left\{\tau_{\varrho}<\infty\right\}-\mathbb{P}_{x,\alpha}\left\{\tau_{\varrho_1}<\infty\right\}\\
&\geq 1-\frac{\varepsilon}{2}-\frac{\varepsilon}{2}=1-\varepsilon.
\end{aligned}
$$

This implies that

$$
\mathbb{P}_{x,\alpha}\left\{\limsup_{t\to\infty}|X(t)|\leq\varrho_1\right\}\geq 1-\varepsilon.
$$

Because $\varrho_1 > 0$ can be chosen to be arbitrarily small, we have

$$
\mathbb{P}_{x,\alpha}\left\{\lim_{t\to\infty}X(t)=0\right\}\geq 1-\varepsilon.
$$

This finishes the proof of the proposition. □

Proposition 7.7 *Let $D \subset \mathbb{R}^r$ be a neighborhood of 0. Assume that for each $i \in \mathcal{M}$, there exists a nonnegative function $V(\cdot, i) : D \mapsto \mathbb{R}$ such that $V(\cdot, i)$ is continuously differentiable in every deleted neighborhood of 0,*

$$\mathcal{L}V(x, i) \leq -\kappa(\varepsilon) < 0, \quad \textit{for all } x \in D \setminus \{x \in \mathbb{R}^r : |x| \leq \varepsilon\}, \tag{7.20}$$

where $\varepsilon > 0$ and $\kappa(\varepsilon)$ is a positive constant, and

$$\lim_{|x| \to 0} V(x, i) = \infty, \quad \textit{for each } i \in \mathcal{M}. \tag{7.21}$$

Then, the equilibrium point $x = 0$ is unstable in probability.

Proof We omit the proof as it is almost identical to that of Lemma 6.7. □

7.3 Stability and Instability: Sufficient Conditions

In the previous section, we obtained sufficient conditions for stability, instability, and asymptotic stability, using a Lyapunov function argument. Because the results are based on the existence of Lyapunov functions, to apply them, it is necessary to find appropriate Lyapunov functions. Nevertheless, finding suitable Lyapunov functions is more often than not a very challenging task. In many applications, it is often more convenient to be able to analyze the stability through conditions on the coefficients of the corresponding stochastic differential equations. Thus, in this section, we continue our study by providing easily verifiable conditions on the coefficients of the system (7.1)–(7.2).

Note that if the generator Q is irreducible, then all the states of the Markov chain belong to the same ergodic class. For multiple ergodic class cases, one may use the idea of two-timescale formulation and singular perturbation methods as in [303] and [307]. To proceed, we assume the following condition holds throughout the rest of the section.

(H7.4) For each $i \in \mathcal{M}$, there exist $A_i \in \mathbb{R}^{r \times r}$ and $\widehat{Q} = (\widehat{q}_{ij}) \in \mathbb{R}^{m_0 \times m_0}$, a generator of a continuous-time Markov chain $\widehat{\alpha}(t)$ such that as $x \to 0$

$$\begin{aligned} f(x, i) &= A_i x + o(|x|), \\ Q(x) &= \widehat{Q} + o(1). \end{aligned} \tag{7.22}$$

Moreover, $\widehat{Q}$ is irreducible. Denote the unique stationary distribution of the associated Markov chain $\widehat{\alpha}(t)$ by

$$\pi = (\pi_1, \pi_2, \dots, \pi_{m_0}) \in \mathbb{R}^{1 \times m_0}.$$

Theorem 7.8 *Let Assumptions* (H7.1)–(H7.4) *be satisfied. Then, the following assertions hold.*

(i) *If there exists a symmetric and positive definite matrix G such that*

$$\sum_{i=1}^{m_0} \pi_i \lambda_{\max}(GA_iG^{-1} + G^{-1}A_i'G) < 0, \tag{7.23}$$

then the equilibrium point $x = 0$ *of the system* (7.1)–(7.2) *is asymptotically stable in probability.*

(ii) *If there exists a symmetric and positive definite matrix G such that*

$$\sum_{i=1}^{m_0} \pi_i \lambda_{\min}(GA_iG^{-1} + G^{-1}A_i'G) > 0, \tag{7.24}$$

then the equilibrium point $x = 0$ *of the system* (7.1)–(7.2) *is unstable in probability.*

Proof

(a) We first prove that the equilibrium point $x = 0$ of system (7.1)–(7.2) is asymptotically stable in probability if (7.23) holds for some symmetric and positive definite matrix G. For notational simplicity, define the column vector $\mu = (\mu_1, \mu_2, \ldots, \mu_{m_0})' \in \mathbb{R}^{m_0}$ with

$$\mu_i = \frac{1}{2}\lambda_{\max}(GA_iG^{-1} + G^{-1}A_i'G),$$

where G is as in (7.23). Also, let

$$\beta := -\pi\mu = -\sum_{i=1}^{m_0} \pi_i \mu_i .$$

Note that $\beta > 0$ by (7.23). By virtue of Assumption (H7.4) and Lemma A.12, the equation

$$\widehat{Q}c = \mu + \beta \mathbb{1}$$

has a solution $c = (c_1, c_2, \ldots, c_{m_0})' \in \mathbb{R}^{m_0}$. Thus, we have

$$\mu_i - \sum_{j=1}^{m_0} \widehat{q}_{ij} c_j = -\beta, \quad i \in \mathcal{M}. \tag{7.25}$$

For each $i \in \mathcal{M}$, consider the Lyapunov function:

$$V(x,i) = (1-\gamma c_i)(x'G^2x)^{\gamma/2},$$

where $0 < \gamma < 1$ is sufficiently small so that $1 - \gamma c_i > 0$ for each $i \in \mathcal{M}$. It is readily seen that for each $i \in \mathcal{M}$, $V(\cdot, i)$ is continuous and nonnegative and vanishes only at $x = 0$. In addition, since $\gamma > 0$ and $1 - \gamma c_i > 0$, we have

$$\lim_{|x|\to\infty} V(x,i) \geq \lim_{|x|\to\infty} (1-\gamma c_i)(\lambda_{\min}(G^2))^{\gamma/2}|x|^{\gamma} = \infty. \tag{7.26}$$

Detailed calculation reveals that for $x \neq 0$, we have

$$\nabla V(x,i) = (1-\gamma c_i)\gamma(x'G^2x)^{\gamma/2-1}G^2x.$$

It yields that for $x \neq 0$,

$$\begin{aligned}\mathcal{L}V(x,i) &= (1-\gamma c_i)\gamma(x'G^2x)^{\gamma/2-1}x'G^2(A_ix + o(|x|)) \\ &\quad - \sum_{j\neq i} q_{ij}(x)(x'G^2x)^{\gamma/2}\gamma(c_j - c_i) \\ &= (1-\gamma c_i)\gamma(x'G^2x)^{\gamma/2}\Big[\frac{x'G^2A_ix}{x'G^2x} + o(1) - \sum_{j\neq i} q_{ij}(x)\frac{c_j - c_i}{1-\gamma c_i}\Big].\end{aligned} \tag{7.27}$$

It follows from Assumption (H7.4) that for sufficiently small $|x|$

$$\begin{aligned}\sum_{j\neq i} q_{ij}(x)\frac{c_j - c_i}{1-\gamma c_i} &= \sum_{j=1}^{m_0} q_{ij}(x)c_j + \sum_{j\neq i} q_{ij}(x)\frac{c_i(c_j - c_i)}{1-\gamma c_i}\gamma \\ &= \sum_{j=1}^{m_0} \widehat{q}_{ij}c_j + O(\gamma) + o(1),\end{aligned} \tag{7.28}$$

where $o(1) \to 0$ as $|x| \to 0$ and $O(\gamma) \to 0$ as $\gamma \to 0$. Meanwhile, using the transformation $y = Gx$, we have

$$\begin{aligned}\frac{x'G^2A_ix}{x'G^2x} &= \frac{x'(G^2A_i + A_i'G^2)x}{2x'G^2x} = \frac{y'G^{-1}(G^2A_i + A_i'G^2)G^{-1}y}{2y'y} \\ &\leq \frac{1}{2}\lambda_{\max}(G^{-1}(G^2A_i + A_i'G^2)G^{-1}) = \frac{1}{2}\lambda_{\max}(GA_iG^{-1} + G^{-1}A_i'G) = \mu_i.\end{aligned} \tag{7.29}$$

Moreover, note that

$$(\lambda_{\min}(G^2))^{\gamma/2}|x|^{\gamma} \leq (x'G^2x)^{\gamma/2} \leq (\lambda_{\max}(G^2))^{\gamma/2}|x|^{\gamma}. \tag{7.30}$$

When $|x| < \delta$ with δ and $0 < \gamma < 1$ sufficiently small, (7.27)–(7.30) lead to

$$\mathcal{L}V(x,i) \le \gamma(1-\gamma c_i)(\lambda_{\min}(G^2))^{\gamma/2}|x|^\gamma \left\{ \mu_i - \sum_{j=1}^{m_0} \widehat{q}_{ij} c_j + o(1) + O(\gamma) \right\}.$$

Furthermore, by virtue of (7.25), we have

$$\begin{aligned}\mathcal{L}V(x,i) &\le \gamma(1-\gamma c_i)(\lambda_{\min}(G^2))^{\gamma/2}|x|^\gamma \left(-\beta + o(1) + O(\gamma)\right) \\ &\le -\kappa(\varepsilon) < 0,\end{aligned}$$

for any $(x,i) \in N_0 \times \mathcal{M}$ with $|x| > \varepsilon$, where $N_0 \subset \mathbb{R}^r$ is a small neighborhood of 0 and $\kappa(\varepsilon)$ is a positive constant. Therefore, we conclude from Proposition 7.6 that the equilibrium point $x = 0$ is asymptotically stable in probability.

(b) Now, we prove that the equilibrium point $x = 0$ is unstable in probability if (7.24) holds for some symmetric and positive definite matrix G. Define the column vector $\theta = (\theta_1, \theta_2, \ldots, \theta_{m_0})' \in \mathbb{R}^{m_0}$ by

$$\theta_i := \frac{1}{2}\lambda_{\min}(GA_iG^{-1} + G^{-1}A_i'G),$$

and set $\delta = -\pi\theta$. Note that $\delta = -\sum_{i=1}^{m_0} \pi_i \theta_i < 0$. As in part (a), Assumption (H7.4), the definition of δ, and Lemma A.12 imply that the equation $\widehat{Q}c = \theta + \delta\mathbb{1}$ has a solution $c = (c_1, c_2, \ldots, c_{m_0})' \in \mathbb{R}^{m_0}$ and

$$\theta_i - \sum_{j=1}^{m_0} \widehat{q}_{ij} c_j = -\delta > 0, \quad i \in \mathcal{M}. \tag{7.31}$$

For $i \in \mathcal{M}$, consider the Lyapunov function:

$$V(x,i) = (1-\gamma c_i)(x'G^2x)^{\gamma/2},$$

where $-1 < \gamma < 0$ is sufficiently small so that $1 - \gamma c_i > 0$ for each $i \in \mathcal{M}$. Similar to the argument in (7.26), we can verify that $V(\cdot, i)$ satisfies (7.21) for each $i \in \mathcal{M}$. Detailed computations as in part (a) show that for any sufficiently small $0 < \varepsilon < r_0$,

$$\mathcal{L}V(x,i) \le -\kappa(\varepsilon) < 0 \ \text{ for any } (x,i) \in N_0 \times \mathcal{M} \text{ with } |x| > \varepsilon,$$

where $N_0 \subset \mathbb{R}^r$ is a small neighborhood of 0 and $\kappa(\varepsilon)$ is a positive constant. Therefore, Proposition 7.7 implies that the equilibrium point $x = 0$ is unstable in probability. This completes the proof of the theorem.

□

Corollary 7.9 *Suppose Assumptions* (H7.1)–(H7.4). *Then, the following assertions hold.*

(i) *The equilibrium point* $x = 0$ *is asymptotically stable in probability if*

$$\sum_{i=1}^{m_0} \pi_i \lambda_{\max}(A_i + A_i') < 0. \tag{7.32}$$

(ii) *The equilibrium point* $x = 0$ *is unstable in probability if*

$$\sum_{i=1}^{m_0} \pi_i \lambda_{\min}(A_i + A_i') > 0. \tag{7.33}$$

Proof This corollary follows from Theorem 7.8 immediately by choosing the symmetric and positive definite matrix G in (7.23) and (7.24) to be the identity matrix I. □

Theorem 7.8 and Corollary 7.9 give sufficient conditions in terms of the maximum and minimum eigenvalues of the matrices for stability and instability of the equilibrium point $x = 0$. Because there is a "gap" between the maximum and minimum eigenvalues, a natural question arises: Can we obtain a necessary and sufficient condition for stability. If the component $X(t)$ is one dimensional, we have the following result from Theorem 7.8, which is a necessary and sufficient condition.

Corollary 7.10 *Suppose Assumptions* (H7.1)–(H7.4). *Let the continuous component* $X(t)$ *of the hybrid process* $(X(t), \alpha(t))$ *given by* (7.1) *and* (7.2) *be one dimensional. Then, the equilibrium point* $x = 0$ *is asymptotically stable in probability if*

$$\sum_{i=1}^{m_0} \pi_i A_i < 0,$$

and is unstable in probability if

$$\sum_{i=1}^{m_0} \pi_i A_i > 0.$$

7.4 A Sharper Result

This section deals with systems that are linear in the continuous state variable x. The system we are interested in is given by

$$\dot{X}(t) = A(\alpha(t))X(t), \tag{7.34}$$

where $A(i) = A_i \in \mathbb{R}^{r\times r}$, $X(t) \in \mathbb{R}^r$, and $\alpha(t)$ is a continuous-time Markov chain with a generator Q independent of t. Assume moreover that the Markov chain $\alpha(t)$ is irreducible. Denote the corresponding stationary distribution by $\pi = (\pi_1, \pi_2, \ldots, \pi_{m_0})$. Our main objective is to obtain a necessary and sufficient condition or, in other words, to close the gap due to the presence of minimal and maximal eigenvalues as we mentioned in the previous section. Denote the solution of (7.34) by $X(t)$ with initial condition $(X(0), \alpha(0))$. Assume that $X(0) \neq 0$. Then, according to Corollary 7.2, $X(t) \neq 0$ for any $t \geq 0$ with probability 1. As a result, $Y(t) = X(t)/|X(t)|$ is well defined and takes values on the unit sphere

$$\mathbb{S}_r = \{y \in \mathbb{R}^r : |y| = 1\}.$$

Define $z(t) = \ln |X(t)|$. It is readily seen that

$$\dot{z}(t) = \frac{X'(t)\dot{X}(t)}{|X(t)|^2}. \tag{7.35}$$

In what follows, we deal with a case that the Markov chain is fast varying. It acts as a "noise," whereas the $X(t)$ is slowly varying. In the end, the noise is averaged out and replaced by its stationary distribution. To put this in a mathematical form, we suppose that there is a small parameter $\varepsilon > 0$ and

$$Q = Q^\varepsilon = \frac{Q_0}{\varepsilon},$$

where Q_0 is the generator of an irreducible Markov chain. Note that in this case, $\alpha(t)$ should really be written as $\alpha^\varepsilon(t)$. In what follows, we adopt this notation.

Using (7.35), for any $T > 0$, we have

$$\begin{aligned}
\frac{z(T) - z(0)}{T} &= \frac{1}{T}\int_0^T \frac{X'(t)A(\alpha^\varepsilon(t))X(t)}{|X(t)|^2}dt \\
&= \frac{1}{T}\sum_{i=1}^{m_0}\int_0^T \frac{X'(t)A_iX(t)}{|X(t)|^2}[I_{\{\alpha^\varepsilon(t)=i\}} - \pi_i]dt \\
&\quad + \frac{1}{T}\sum_{i=1}^{m_0}\int_0^T \frac{X'(t)A_iX(t)}{|X(t)|^2}\pi_i dt.
\end{aligned} \tag{7.36}$$

For arbitrary $T > 0$, we partition the interval $[0, T]$ by use of $T^{-\Delta} > 0$ where $\Delta > 0$ is a parameter. Then, we choose ε as a function of $T^{-\Delta}$. Such a choice will lead to the desired result. To proceed, choose a real number $\Delta > 1/2$. Denote $\delta_\Delta = T^{-\Delta}$ and $N = \lfloor T/\delta_\Delta \rfloor$, where $\lfloor y \rfloor$ is the usual floor function notation for a real number y. Note that $N = O(T^{1+\Delta})$ and $N\delta_\Delta = O(T)$. Next, let $0 = t_0 < t_1 < \cdots < t_N = T$ be a partition of $[0, T]$ such that $t_k = k\delta_\Delta$ for $k = 0, 1, \ldots, N$. For notational simplicity, for $i = 1, \ldots, m_0$, denote

$$\zeta_i(t) = \frac{X'(t)A_iX(t)}{|X(t)|^2},$$

$$\widetilde{\zeta}_i(t) = \begin{cases} \zeta_i(0), & \text{if } 0 \le t < t_2, \\ \zeta_i(t_{k-1}), & \text{if } t_k \le t < t_{k+1},\ k = 2, \ldots, N-1, \\ \zeta_i(t_{N-1}), & \text{if } t_N \le t \le T. \end{cases}$$

$$I_i^\varepsilon(t) = I_{\{\alpha^\varepsilon(t)=i\}} - \pi_i.$$

Note that in the above, $\widetilde{\zeta}_i(t)$ is a piecewise constant approximation of $\zeta_i(t)$ with the interpolation intervals $[t_k, t_{k+1})$, $k = 0, 1, \ldots, N$.

Lemma 7.11 *Suppose that Q_0 is irreducible and that*

$$\varepsilon = o(T^{-\Delta}) \text{ as } T \to \infty. \tag{7.37}$$

Then

$$\frac{1}{T}\sum_{i=1}^{m_0}\int_0^T \frac{X'(t)A_iX(t)}{|X(t)|^2}[I_{\{\alpha^\varepsilon(t)=i\}} - \pi_i]dt \to 0 \text{ in probability as } T \to \infty.$$

Proof The assertion of Lemma 7.11 will follow immediately if we can show that for each $i \in \mathcal{M}$,

$$\mathbb{E}\Big|\frac{1}{T}\int_0^T \zeta_i(t)I_i^\varepsilon(t)dt\Big|^2 \to 0 \text{ as } T \to \infty.$$

Using the triangle inequality and the Cauchy-Schwarz inequality, we can verify that for each $i \in \mathcal{M}$

$$\begin{aligned} &\mathbb{E}\Big|\frac{1}{T}\int_0^T \zeta_i(t)I_i^\varepsilon(t)dt\Big|^2 \\ &\le \frac{2}{T^2}\mathbb{E}\Big|\int_0^T [\zeta_i(t) - \widetilde{\zeta}_i(t)]I_i^\varepsilon(t)dt\Big|^2 + \frac{2}{T^2}\mathbb{E}\Big|\int_0^T \widetilde{\zeta}_i(t)I_i^\varepsilon(t)dt\Big|^2 \\ &\le \frac{K}{T}\int_0^T \mathbb{E}[\zeta_i(t) - \widetilde{\zeta}_i(t)]^2dt + \frac{2}{T^2}\mathbb{E}\Big|\int_0^T \widetilde{\zeta}_i(t)I_i^\varepsilon(t)dt\Big|^2. \end{aligned} \tag{7.38}$$

In the above and hereafter, K is used as a generic positive constant, whose values may change for different appearances. Clearly, for each $i \in \mathcal{M}$, $\zeta_i(\cdot)$ is uniformly bounded by $|A_i|$, the norm of A_i. Thus

$$\sup_{0\le t\le T} \mathbb{E}|\zeta_i(t)|^2 < \infty.$$

Because $X(t)/|X(t)|$ takes values on the unit sphere, it is readily verified that

$$\sup_{0\le t\le T} \left| \frac{d}{dt} \frac{X'(t)A_iX(t)}{|X(t)|^2} \right| \le K, \tag{7.39}$$

where K is independent of T. As a result, $\zeta_i(\cdot)$ is Lipschitz continuous, uniformly on $[0, T]$. Consequently

$$\begin{aligned} \int_0^T \mathbb{E}[\zeta_i(t) - \widetilde{\zeta}_i(t)]^2 dt &= \sum_{k=0}^{N-1} \int_{t_k}^{t_{k+1}} \mathbb{E}[\zeta_i(t) - \zeta_i(t_{k-1})]^2 dt \\ &\le K \sum_{k=0}^{N-1} \int_{t_k}^{t_{k+1}} [t - t_{k-1}]^2 dt \\ &= O(N\delta_\Delta^3) = O(T\delta_\Delta^2) \to 0 \text{ as } T \to \infty, \end{aligned} \tag{7.40}$$

by the choice of δ_Δ.

By means of (7.38) and (7.40), it remains to show that the last term of (7.38) converges to 0 as $T \to \infty$. For any $0 \le t \le T$, define

$$h_i(t) = \mathbb{E}\Big[\int_0^t \widetilde{\zeta}_i(s) I_i^\varepsilon(s) ds \Big]^2. \tag{7.41}$$

Then

$$\frac{dh_i(t)}{dt} = 2 \int_0^t \mathbb{E}[\widetilde{\zeta}_i(s) I_i^\varepsilon(s) \widetilde{\zeta}_i(t) I_i^\varepsilon(t)] ds. \tag{7.42}$$

For $0 \le t \le t_2$

$$\begin{aligned} &\left| \int_0^t \mathbb{E}[\widetilde{\zeta}_i(s) I_i^\varepsilon(s) \widetilde{\zeta}_i(t) I_i^\varepsilon(t)] ds \right| \\ &\le \int_0^{t_2} \mathbb{E}^{1/2} |\widetilde{\zeta}_i(s)|^2 \mathbb{E}^{1/2} |\widetilde{\zeta}_i(t)|^2 ds \le K t_2 = O(T^{-\Delta}). \end{aligned}$$

For $t_k \le t < t_{k+1}$ with $k = 2, \ldots, N$, we have

$$\frac{dh_i(t)}{dt} = 2 \left(\int_0^{t_{k-1}} + \int_{t_{k-1}}^t \right) \mathbb{E}\left[\widetilde{\zeta}_i(s) I_i^\varepsilon(s) \widetilde{\zeta}_i(t) I_i^\varepsilon(t) \right] ds.$$

Furthermore

$$\left|\int_{t_{k-1}}^{t} \mathbb{E}[\widetilde{\zeta}_i(s)I_i^{\varepsilon}(s)\widetilde{\zeta}_i(t)I_i^{\varepsilon}(t)]ds\right| \le \int_{t_{k-1}}^{t} \mathbb{E}^{1/2}|\widetilde{\zeta}_i(s)|^2\mathbb{E}^{1/2}|\widetilde{\zeta}_i(t)|^2 ds$$
$$\le K(t-t_{k-1}) = O(T^{-\Delta}).$$

It follows that

$$\frac{dh_i(t)}{dt} = 2\int_0^{t_{k-1}} \mathbb{E}[\widetilde{\zeta}_i(s)I_i^{\varepsilon}(s)\widetilde{\zeta}_i(t)I_i^{\varepsilon}(t)]ds + O(T^{-\Delta}). \tag{7.43}$$

For $s \le t_{k-1} \le t < t_{k+1}$, by use of (7.37),

$$\begin{aligned}
\left|\mathbb{E}[\widetilde{\zeta}_i(s)I_i^{\varepsilon}(s)\widetilde{\zeta}_i(t)I_i^{\varepsilon}(t)]\right| &= |\mathbb{E}[\widetilde{\zeta}_i(s)I_i^{\varepsilon}(s)\mathbb{E}(\widetilde{\zeta}_i(t)I_i^{\varepsilon}(t)|\mathcal{F}_{t_{k-1}})]| \\
&= |\mathbb{E}[\widetilde{\zeta}_i(s)I_i^{\varepsilon}(s)\zeta_i(t_{k-1})\mathbb{E}(I_i^{\varepsilon}(t)|\mathcal{F}_{t_{k-1}})]| \\
&\le \mathbb{E}^{1/2}|\widetilde{\zeta}_i(s)|^2\mathbb{E}^{1/2}|\zeta_i(t_{k-1})|^2 O(\varepsilon + \exp(-(t-t_{k-1})/\varepsilon)) \\
&= O(\varepsilon + \exp(-(t_k - t_{k-1})/\varepsilon)) \\
&= O(\varepsilon + \exp(-T^{-\Delta}/\varepsilon)).
\end{aligned} \tag{7.44}$$

The next to the last inequality follows from the asymptotic expansions obtained in, for example, [303, Lemma 5.1].

Using (7.43) and (7.44), we obtain that

$$\sup_{0\le t\le T}\left|\frac{dh_i(t)}{dt}\right| \le O(T)O\Big(\varepsilon + \exp\Big(-\frac{T^{-\Delta}}{\varepsilon}\Big)\Big) \quad \text{for each } i \in \mathcal{M}.$$

Because $h_i(0) = 0$, we conclude

$$|h_i(T)| \le O(T^2)O\Big(\varepsilon + \exp\Big(-\frac{T^{-\Delta}}{\varepsilon}\Big)\Big).$$

Thus, $h_i(T)/T^2 \to 0$ as $T \to \infty$ as long as $\varepsilon = o(T^{-\Delta})$. The lemma then follows. □

By virtue of Lemma 7.11, we need only consider the last term in (7.36). To this end, for each $i \in \mathcal{M}$, there is a $\lambda_i \in \mathbb{R}$ such that

$$\begin{aligned}
\frac{1}{T}\int_0^T \frac{X'(t)A_iX(t)}{|X(t)|^2}dt &= \frac{1}{T}\int_0^T Y'(t)A_iY(t)dt \\
&\to \lambda_i \ \text{ in probability as } \ T \to \infty,
\end{aligned} \tag{7.45}$$

where the existence of the limit follows from a slight modification of [110, p. 344, Theorem 6]. The number λ_i above is precisely the average of $Y'(t)A_iY(t)$, known as the mean value of $Y'(\cdot)A_iY(\cdot)$ and denoted by $M[Y'A_iY]$ in the literature of almost periodic functions [110, Appendix]. Define

$$\lambda = \sum_{i=1}^{m_0} \pi_i \lambda_i. \tag{7.46}$$

It follows from Lemma 7.11, Eqs. (7.36), (7.45), and (7.46) that

$$\frac{1}{T} \ln \frac{|X(T)|}{|X(0)|} = \frac{1}{T} \int_0^T \frac{X'(t) A(\alpha^\varepsilon(t)) X(t)}{|X(t)|^2} dt \to \lambda \text{ as } T \to \infty. \tag{7.47}$$

To proceed, note that (7.47) can be rewritten as

$$\frac{1}{T} \ln \frac{|X(T)|}{|X(0)|} = \lambda + o(1), \tag{7.48}$$

where $o(1) \to 0$ in probability as $T \to \infty$. If $\lambda < 0$, denote $\lambda = -\lambda_0$ with $\lambda_0 > 0$. We can make $-\lambda_0 + o(1) \le -\lambda_1$ where $0 < \lambda_1 < \lambda_0$. Thus, it follows from (7.48) that

$$|X(T)| \le |X(0)| \exp(-\lambda_1 T) \to 0 \text{ in probability as } T \to \infty.$$

Likewise, if $\lambda > 0$, we can find $0 < \lambda_2 < \lambda$ such that $\lambda + o(1) \ge \lambda_2$, which in turn, implies that

$$|X(T)| \ge |X(0)| \exp(\lambda_2 T) \to \infty \text{ in probability as } T \to \infty. \tag{7.49}$$

We summarize the result obtained thus far in the following theorem.

Theorem 7.12 *The equilibrium point $x = 0$ of* (7.34) *is asymptotically stable in probability if $\lambda < 0$ and asymptotically unstable in probability in the sense of* (7.49) *if $\lambda > 0$.*

Note that the results in Theorem 7.12 present a dichotomy for the property of the equilibrium point $x = 0$ according to $\lambda < 0$ or $\lambda > 0$. One naturally asks the question: What can we say about the equilibrium point $x = 0$ if $\lambda = 0$. To answer this question, we need the following lemma.

Lemma 7.13 *If the equilibrium point $x = 0$ of* (7.34) *is asymptotically stable in probability, then it is exponentially p-stable for all sufficiently small positive p.*

Proof This lemma can be established using a similar argument as that of [156, Theorem 6.5]; we omit the details here for brevity. □

With Lemma 7.13 at our hand, we are now able to describe the local qualitative behavior of the equilibrium point $x = 0$ when $\lambda = 0$.

Theorem 7.14 *Suppose that $\lambda = 0$. Then, the equilibrium point $x = 0$ is neither asymptotically stable nor asymptotically unstable in probability.*

Proof Assume first that the equilibrium point $x = 0$ is asymptotically stable in probability. Then, Lemma 7.13 implies that it is exponentially p-stable for all sufficiently small positive p. Now, applying Jensen's inequality to the convex function $\varphi(x) = e^x$, from equation (7.36) and Definition 7.1 (iii), we obtain that

$$\begin{aligned}
&|X(0)|^p \exp\left\{p\mathbb{E}\int_0^T \frac{X'(t)A(\alpha^\varepsilon(t))X(t)}{|X(t)|^2}dt\right\} \\
&= |X(0)|^p \exp\{p\mathbb{E}(\ln|X(T)| - \ln|X(0)|)\} \\
&\leq \mathbb{E}|X(T)|^p \leq K|X(0)|^p e^{-kT}.
\end{aligned}$$

Consequently, with positive probability

$$\lim_{T\to\infty}\frac{1}{T}\int_0^T \frac{X'(t)A(\alpha^\varepsilon(t))X(t)}{|X(t)|^2}dt < 0.$$

According to (7.47), this contradicts the assumption that $\lambda = 0$. Thus, the equilibrium point $x = 0$ is not asymptotically stable. Similarly, we can prove that the equilibrium point $x = 0$ is not asymptotically unstable in probability if $\lambda = 0$.

□

7.5 Remarks on Lyapunov Exponent

This section is divided into two parts. In the first part, we consider a more general case with the process given by (7.34), where $\alpha(t)$ is a continuous-time Markov chain whose generator Q does not involve a small parameter $\varepsilon > 0$. Here, the result is presented by the use of the Lyapunov exponent and the stationary distributions. Then, in the second part, we consider the problems of finding the stationary density of the random process.

7.5.1 Stability Under General Setup

To date, a method used frequently for analyzing stability of stochastic systems is based on the work set forth in Khasminskii's book [156]. It is particularly effective in treating linear stochastic differential equations, using transformation techniques. Such an approach depends on the calculation of a limit quantity later commonly referred to as the Lyapunov exponent.

Assume that the Markov chain $\alpha(t)$ is irreducible. Our main objective is to obtain a necessary and sufficient condition for stability. It follows that $Y(t) = X(t)/|X(t)|$

is well defined and takes values on the unit sphere $\mathbb{S}_r = \{y \in \mathbb{R}^r : |y| = 1\}$. Differentiating $Y(t)$ with respect to t and expressing the result in terms of y again, we obtain

$$\dot{Y}(t) = -[Y'(t)A(\alpha(t))y(t)]Y(t) + A(\alpha(t))Y(t). \tag{7.50}$$

Defining a process $z(t) = \ln |X(t)|$ as before, we obtain

$$\dot{z}(t) = Y'(t)A(\alpha(t))Y(t). \tag{7.51}$$

Thus

$$\ln |X(t)| = \ln |X(0)| + \int_0^t Y'(u)A(\alpha(u))Y(u)du. \tag{7.52}$$

In view of the discussion in Chapter 4 and the well-known results for Markov processes, denote the transition density of $(Y(t), \alpha(t))$ by $p(y, i, t)$ for $i \in \mathcal{M}$. (Note that in general, we would use $p(y_0, i_0; y, i, t)$ for each $i \in \mathcal{M}$ to denote the transition probability density function, where (y_0, i_0) denotes the initial position of the process $(Y(t), \alpha(t))$. Nevertheless, in view of (7.34), the process is time homogeneous. Using the convention in Markov processes, we can simply write the transition density by $p(y, i, t)$.) Then, $p(y, i, t)$ satisfies the system of forward equations:

$$\begin{aligned}\frac{\partial p(y, i, t)}{\partial t} &= \mathcal{L}^* p(y, i, t)\\ &= -\sum_{l=1}^{r} \frac{\partial}{\partial y_l}((A(i)y)_l p(y, i, t)) + Q'p(y, \cdot, t)(i), \quad i \in \mathcal{M},\end{aligned}$$

where $(A(i)y)_l$ denotes the lth component of $A(i)y$ and Q' is the transpose of Q. Moreover, for any $\Gamma \subset \mathbb{R}^r$ and $i \in \mathcal{M}$,

$$\mathbb{P}(y(t) \in \Gamma, \alpha(t) = i) = \int_\Gamma p(y, i, t)dx.$$

Note that the process $(Y(t), \alpha(t))$ is on the compact set $\mathbb{S}_r \times \mathcal{M}$. The existence of the invariant density of $(Y(t), \alpha(t))$ is thus guaranteed owing to the compactness. We further assume the uniqueness of the invariant density and denote it by $(\mu(y, i) : i = 1, \ldots, m_0)$. Then, $\mu(y, i)$ satisfies

$$\mathcal{L}^* \mu(y, i) = 0, \quad \sum_{i=1}^{m_0} \int_{\mathbb{R}^r} \mu(y, i)dy = 1.$$

Under this condition, as in the proof similar to [138, Theorem 1], it can be shown that there is a $\kappa_0 > 0$ such that

$$|p(y, i, t) - \mu(y, i)| \le K \exp(-\kappa_0 t). \tag{7.53}$$

Similar to (7.36), we obtain

$$\frac{z(T) - z(0)}{T} = \frac{1}{T} \int_0^T Y'(t) A(\alpha(t)) Y(t) dt = \frac{1}{T} \int_0^T Y'(t) A(\alpha(t)) Y(t) dt.$$

Redefine λ as

$$\lambda = \sum_{i=1}^{m_0} \int_{\mathbb{R}^r} y' A(i) y \mu(y, i) dy,$$

where $\mu(y, i)$ is the stationary density of $(Y(t), \alpha(t))$. Then, we have

$$\begin{aligned} &\mathbb{E}\left|\frac{1}{T} \int_0^T Y'(t) A(\alpha(t)) Y(t) dt - \lambda\right|^2 \\ &= \left|\mathbb{E}\frac{1}{T^2} \int_0^T \int_0^T \left(Y'(t) A(\alpha(t)) Y(t) - \lambda\right)' \left(Y'(s) A(\alpha(s)) Y(s) - \lambda\right) dt ds\right|. \end{aligned} \tag{7.54}$$

By virtue of (7.53), for $t \ge s$

$$\left|\mathbb{E}\left(Y'(t) A(\alpha(t)) Y(t) - \lambda\right)' \left(Y'(s) A(\alpha(s)) Y(s) - \lambda\right)\right| \le K \exp(-\kappa_0 (t - s)).$$

This together with (7.54) then yields that

$$\mathbb{E}\left|\frac{1}{T} \int_0^T Y'(t) A(\alpha(t)) Y(t) dt - \lambda\right|^2 \le \frac{K}{T} \to 0 \text{ as } T \to \infty.$$

Because $z(0)/T \to 0$ as $T \to \infty$, we conclude that $z(T)/T \to \lambda$ in probability as $T \to \infty$ as desired. The rest of the arguments are the same as in the previous section. In fact, we can also obtain the limit in the sense of convergence w.p.1. That is,

$$\lim_{t \to \infty} \frac{\ln |X(t)|}{t} = \lambda = \sum_{i=1}^{m_0} \int_{\mathbb{S}_r} y' A(i) y \mu(y, i) dy \text{ w.p.1.} \tag{7.55}$$

The limit λ is precisely the Lyapunov exponent.

7.5.2 Invariant Density

The last section was devoted to calculation of the Lyapunov exponent. In obtaining the Lyapunov exponent, it is crucial to find the associated stationary density or invariant density. For ordinary differential equations with Markovian random switching, when the continuous component is two dimensional, using Lyapunov exponents and the framework of [156], stability analysis was carried out in [20, 135, 195], and interesting results were obtained for special classes of problems such as certain forms of Markov modulated harmonic oscillators. Nevertheless, general results on sufficient conditions guaranteeing the existence and uniqueness of invariant density are still scarce.

In this section, we discuss the problem of finding the associated invariant densities of $(X(t), \alpha(t))$. In the literature, considerable attention has been devoted to the case that $X(t) \in \mathbb{R}^2$; see [20, 135, 195]. First, second-order differential equations with random switching are frequently used in mechanical systems to describe oscillatory motions, as well as in mathematical physics to study problems involving Schrödinger equations with random potential. Linearization of the regime-switching Liénard equations (in particular, Van der Pol equations) also leads to switching systems that are linear in the x component. In addition, for two-dimensional systems, polar coordinate transformation can be used to facilitate the study. To give better insight and for simplicity, we let $\alpha(t) \in \mathcal{M} = \{1, 2\}$. Conditions for existence and uniqueness of solutions are provided. The equations satisfied by the invariant density are nonlinear and complex. It appears that closed-form solutions are difficult to obtain.

Assume that the Markov chain $\alpha(t)$ is irreducible. To proceed, for simplicity, we consider the case that $x \in \mathbb{R}^2$ and

$$A(i) = \begin{pmatrix} a_{11}(i) & a_{12}(i) \\ a_{21}(i) & a_{22}(i) \end{pmatrix} \in \mathbb{R}^{2\times 2}, \quad \mathcal{M} = \{1, 2\},$$

and the generator

$$Q = \begin{pmatrix} -q_1 & q_1 \\ q_2 & -q_2 \end{pmatrix} \in \mathbb{R}^{2\times 2}.$$

For $x = (x_1, x_2)' \in \mathbb{R}^2$, it is more convenient to use coordinate transformation. Introduce the polar coordinate system by defining $\varphi = \varphi(x) = \tan^{-1}(x_2/x_1)$. Conditioning on $\alpha(t) = i$, with the use of variable φ, we obtain

$$\begin{aligned} \gamma(\varphi, i) = \frac{d\varphi}{dt} &= -[\sin\varphi\cos\varphi a_{11}(i) + \sin^2\varphi a_{12}(i)] \\ &\quad +[\cos^2\varphi a_{21}(i) + a_{22}(i)\cos\varphi\sin\varphi]. \end{aligned} \tag{7.56}$$

To proceed, for each $i \in \mathcal{M}$, consider a Lyapunov function $V(\cdot, i) : [0, 2\pi] \mapsto \mathbb{R}$, where $\mathbb{S}$ is the unit circle. It can be verified that

$$\mathcal{L}V(\varphi, i) = \frac{\partial V}{\partial \varphi}\gamma(\varphi, i) + QV(\varphi, \cdot)(i), \quad i \in \mathcal{M}. \tag{7.57}$$

We slightly change the notation and write the vector-valued function $\mu(\varphi) = (\mu_1(\varphi), \mu_2(\varphi))' \in \mathbb{R}^{2\times 1}$ in what follows. This invariant density then is a solution of the system of forward equations:

$$\begin{cases} -\dfrac{\partial}{\partial \varphi}\begin{pmatrix}\gamma_1(\varphi)\mu_1(\varphi)\\ \gamma_2(\varphi)\mu_2(\varphi)\end{pmatrix} + Q'\begin{pmatrix}\mu_1(\varphi)\\ \mu_2(\varphi)\end{pmatrix} = 0, \\ \displaystyle\sum_{i=1}^{2}\int_0^{2\pi} \mu_i(\varphi)d\varphi = 1, \\ \mu_i(\cdot) \text{ is } 2\pi \text{ periodic for } i \in \mathcal{M}. \end{cases} \tag{7.58}$$

Recall that Q' denotes the transpose of Q. To ensure the existence of the solution of (7.58), we propose the following conditions.

(H7.5) For each $i \in \mathcal{M}$

$$[a_{22}(i) - a_{11}(i)]^2 + 4a_{12}(i)a_{21}(i) < 0. \tag{7.59}$$

Next, we proceed with solving the system of equations (7.58). Suppress the argument when there is no confusion. Then, the first part of (7.58) leads to

$$\begin{cases} -\dfrac{d}{d\varphi}(\gamma_1\mu_1) - q_1\mu_1 + q_2\mu_2 = 0, \\ -\dfrac{d}{d\varphi}(\gamma_2\mu_2) + q_1\mu_1 - q_2\mu_2 = 0. \end{cases} \tag{7.60}$$

Thus, we have

$$\frac{d}{d\varphi}(\gamma_1\mu_1 + \gamma_2\mu_2) = 0. \tag{7.61}$$

As a result, $\gamma_1(\varphi)\mu_1(\varphi) + \gamma_2(\varphi)\mu_2(\varphi) = c$ for all $\varphi \in [0, 2\pi]$, where c is a constant.

Let $\overline{\gamma}_i(\varphi) = \gamma_i(\varphi)\mu_i(\varphi)$ for $i = 1, 2$. Then, (7.60) and (7.61) can be rewritten as

$$\begin{cases} \dfrac{d\overline{\gamma_1}}{d\varphi} + q_1\dfrac{\overline{\gamma_1}}{\gamma_1} - q_2\dfrac{\overline{\gamma_2}}{\gamma_2} = 0 \\ \dfrac{d\overline{\gamma_2}}{d\varphi} - q_1\dfrac{\overline{\gamma_1}}{\gamma_1} + q_2\dfrac{\overline{\gamma_2}}{\gamma_2} = 0 \\ \overline{\gamma_1} + \overline{\gamma_2} = c. \end{cases} \tag{7.62}$$

Denote

$$h(\varphi) = \exp\left(\int_0^{\varphi} (\frac{q_1}{\gamma_1(s)} + \frac{q_2}{\gamma_2(s)})ds\right).$$

By solving the above system of linear differential equations, we have

$$\begin{cases} \overline{\gamma_1} = \dfrac{\displaystyle\int_0^{\varphi} q_2 c \frac{h(s)}{\gamma_2(s)} ds + \overline{c}}{h(\varphi)} \\ \overline{\gamma_2} = \dfrac{\displaystyle\int_0^{\varphi} q_1 c \frac{h(s)}{\gamma_1(s)} ds + c - \overline{c}}{h(\varphi)}, \end{cases} \tag{7.63}$$

where $\overline{c} = \overline{\gamma}_1(0)$. Note that in deriving the second equation in (7.62), we have used $\overline{\gamma}_2(0) = c - \overline{\gamma}_1(0)$. Thus

$$\begin{cases} \mu_1(\varphi) = \dfrac{\displaystyle\int_0^{\varphi} q_2 c \frac{h(s)}{\gamma_2(s)} ds + \overline{c}}{h(\varphi)\gamma_1(\varphi)} \\ \mu_2(\varphi) = \dfrac{\displaystyle\int_0^{\varphi} q_1 c \frac{h(s)}{\gamma_1(s)} ds + c - \overline{c}}{h(\varphi)\gamma_2(\varphi)}. \end{cases} \tag{7.64}$$

We also know that $\mu_i(\varphi)$ must be a 2π-periodic function of φ for $i \in \mathcal{M}$, so $\mu_1(0) = \mu_1(2\pi)$. We thus have a system of linear equations for c and $\overline{c}$, namely,

$$D\begin{pmatrix} c \\ \overline{c} \end{pmatrix} = \begin{pmatrix} 0 \\ 1 \end{pmatrix}, \tag{7.65}$$

where $D = (d_{ij}) \in \mathbb{R}^{2\times 2}$, with

$$\begin{aligned} d_{11} &= \int_0^{2\pi} \frac{q_2 h(s)}{\gamma_2(s)} ds, \\ d_{12} &= 1 - h(2\pi), \\ d_{21} &= \int_0^{2\pi} \frac{\gamma_2(\varphi)\displaystyle\int_0^{\varphi} q_2 \frac{h(s)}{\gamma_2(s)} ds + \gamma_1(\varphi)\displaystyle\int_0^{\varphi} q_1 \frac{h(s)}{\gamma_1(s)} ds + \gamma_1(\varphi)}{h(\varphi)\gamma_1(\varphi)\gamma_2(\varphi)} d\varphi, \\ d_{22} &= \int_0^{2\pi} \frac{\gamma_2(\varphi) - \gamma_1(\varphi)}{h(\varphi)\gamma_1(\varphi)\gamma_2(\varphi)} d\varphi. \end{aligned}$$

(H7.6) The matrix D is nonsingular or, equivalently, $\det(D) \neq 0$.

The uniqueness of (7.65) leads to the uniqueness of $\mu(y)$. Thus, we arrive at the following conclusion: Under Assumptions (H7.5) and (H7.6), (7.58) has a unique solution.

7.6 Examples

For simplicity, by a slightly abuse of notation, we call a system that is linear with respect to the continuous state a linear system or hybrid linear system. To illustrate, we provide several examples in this section. In addition, we demonstrate results that are associated with the well-known Hartman-Grobman theorem for random switching systems.

Example 7.15 Consider a system (7.34) (linear in the variable x) with the following specifications. The Markov chain $\alpha(t)$ has two states and is generated by

$$Q = \begin{pmatrix} -3 & 3 \\ 1 & -1 \end{pmatrix},$$

and

$$A_1 = A(1) = \begin{pmatrix} -1 & 2 \\ 0 & 2 \end{pmatrix}, \quad A_2 = A(2) = \begin{pmatrix} -3 & -1 \\ 1 & -2 \end{pmatrix}.$$

Thus, associated with the hybrid system (7.34), there are two ordinary differential equations:

$$\dot{X}(t) = A_1 X(t), \tag{7.66}$$

and

$$\dot{X}(t) = A_2 X(t), \tag{7.67}$$

switching back and forth from one to another according to the movement of the jump process $\alpha(t)$. It is readily seen that the eigenvalues of A_1 are -1 and 2. Hence, the motion of (7.66) is unstable. Similarly, by computing the eigenvalues of A_2, the motion of (7.67) is asymptotically stable.

Next, we use Corollary 7.9 to show that the hybrid system (7.34) is asymptotically stable owing to the presence of the Markov chain $\alpha(t)$. That is, the Markov chain becomes a stabilizing factor. The stationary distribution of the Markov chain $\alpha(t)$ is $\pi = (0.25, \ 0.75)$, which is obtained by solving the system of equations $\pi Q = 0$ and $\pi \mathbb{1} = 1$. The maximum eigenvalues of $A_1 + A_1'$ and $A_2 + A_2'$ are

$$\lambda_{\max}(A_1 + A_1') = 4.6056, \quad \lambda_{\max}(A_2 + A_2') = -4,$$

respectively. This yields that

$$\pi_1 \lambda_{\max}(A_1 + A_1') + \pi_2 \lambda_{\max}(A_2 + A_2') = -1.8486 < 0.$$

Therefore, we conclude from Corollary 7.9 that the hybrid system (7.34) is asymptotically stable. The phase portrait Fig. 7.1a confirms our findings. It is interesting to note the dynamic movements and the interactions of the continuous and discrete components. To see the difference between hybrid systems and ordinary differential equations, we also present the phase portraits of (7.66) and (7.67) in Fig. 7.1b.

Example 7.16 Consider a system (linear in the x variable)

$$\dot{X}(t) = A(\alpha(t))X(t) \tag{7.68}$$

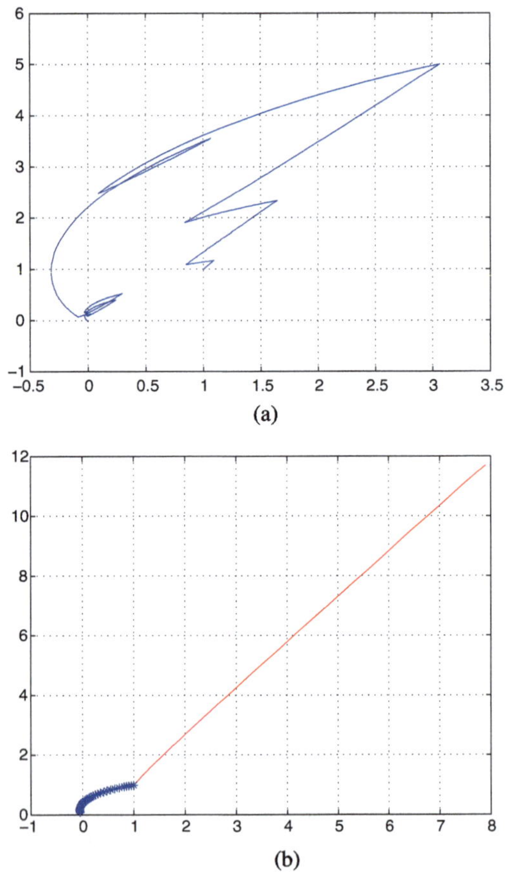

Fig. 7.1 Comparisons of switching system and the associated ordinary differential equations. (**a**) Hybrid linear system: Phase portrait of (7.34) with initial condition $(x, \alpha) = ([1, 1]', 1)$. (**b**) Phase portraits of (7.66) (solid line) and (7.67) (starred line) with the same initial condition $x = [1, 1]'$

with the following specifications. Here, we consider an x-dependent generator $Q(x)$. The discrete event process $\alpha(t)$ has two states and is generated by

$$\begin{aligned}Q(x) &= Q(x_1, x_2)\\ &= \begin{pmatrix} -2 - \sin x_1 \cos x_1^2 - \sin x_2^2 & 2 + \sin x_1 \cos(x_1^2) + \sin(x_2^2) \\ 1 - 0.5\sin(x_1 x_2) & -1 + 0.5\sin(x_1 x_2) \end{pmatrix},\end{aligned}$$

and

$$A_1 = A(1) = \begin{pmatrix} 0 & 0 \\ 3 & 0 \end{pmatrix}, \qquad A_2 = A(2) = \begin{pmatrix} -1 & 2 \\ -2 & -1 \end{pmatrix}.$$

Note that the distinct feature of this example compared with the last one is that the Q matrix is x dependent, which satisfies the approximation condition posed in Section 7.3. Associated with the hybrid system (7.34), there are two ordinary differential equations:

$$\dot{X}(t) = A_1 X(t), \tag{7.69}$$

and

$$\dot{X}(t) = A_2 X(t) \tag{7.70}$$

switching from one to another according to the movement of the jump process $\alpha(t)$. Solving (7.69), we obtain

$$x_1(t) = c_1, \quad x_2(t) = 3c_1 t + c_2$$

for some constants c_1 and c_2. Hence, the motion of (7.69) is unstable if the initial point (c_1, c_2) is not in the y-axis. In addition, by computing the eigenvalues of A_2, we obtain that the motion of (7.70) is asymptotically stable.

By virtue of Corollary 7.9, the hybrid system (7.68) is asymptotically stable due to the stabilizing jump process $\alpha(t)$. We first note that as $x \to 0$

$$Q(x) \to \widehat{Q} = \begin{pmatrix} -2 & 2 \\ 1 & -1 \end{pmatrix},$$

and hence, the stationary distribution of the Markov chain $\widehat{\alpha}(t)$ is $\pi = (1/3,\ 2/3)$. The maximum eigenvalues of $A_1 + A_1'$ and $A_2 + A_2'$ are

$$\lambda_{\max}(A_1 + A_1') = 3, \quad \lambda_{\max}(A_2 + A_2') = -2,$$

respectively. It follows that

$$\pi_1\lambda_{\max}(A_1 + A_1') + \pi_2\lambda_{\max}(A_2 + A_2') = -1/3 < 0.$$

Therefore, we conclude from Corollary 7.9 that the hybrid system (7.68) is asymptotically stable. The phase portrait in Fig. 7.2a confirms our results. To delineate the difference between hybrid systems and ordinary differential equations, we also present the phase portraits of (7.69) and (7.70) in Fig. 7.2b. The phase portraits reveal the interface of continuous and discrete components. They illustrate the hybrid characteristics in an illuminating way.

Example 7.17 In this example, we consider a nonlinear hybrid system:

$$\dot{X}(t) = f(X(t), \alpha(t)), \tag{7.71}$$

where $X(t)$ is a two-dimensional state trajectory and $\alpha(t)$ is a jump process taking value in $\mathcal{M} = \{1, 2, 3\}$ with generator

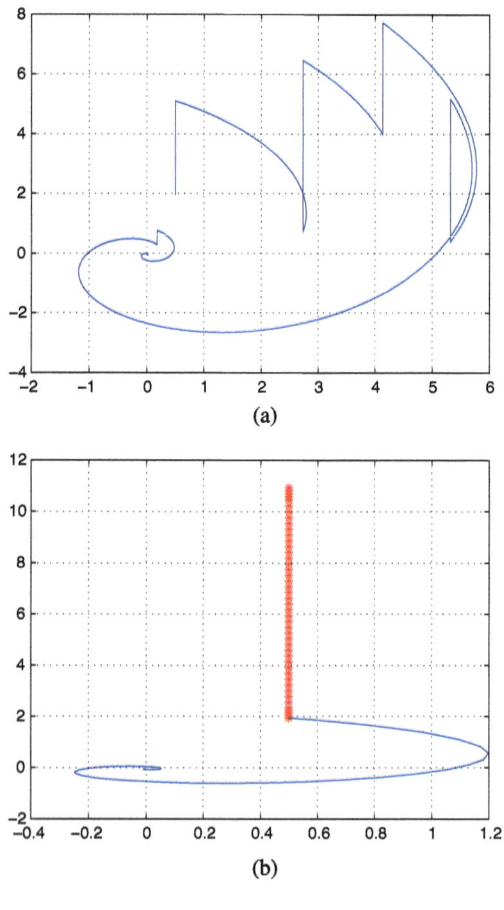

Fig. 7.2 Comparisons of switching system and the associated ordinary differential equations. (**a**) Hybrid linear system: Phase portrait of (7.68) with initial condition $(x, \alpha) = ([0.5, 2]', 1)$. (**b**) Phase portraits of (7.69) (starred line) and (7.70) (solid line) with the same initial condition $x = [0.5, 2]'$

$$Q(x_1, x_2) = \begin{bmatrix} \sin x_2 - 2 - \cos x_1 & 1 - \sin x_2 & 1 + \cos x_1 \\ 1 - \frac{\sin(x_1 x_2)}{2} & \frac{\sin(x_1 x_2)}{2} - 1 - \cos^2(x_1 x_2) & \cos^2(x_1 x_2) \\ 0 & 3 + \sin x_1 + \sin x_2 \cos x_2 & -3 - \sin x_1 - \sin x_2 \cos x_2 \end{bmatrix}.$$

The functions $f(x, i), i = 1, 2, 3$ are defined as

$$f(x, 1) = \begin{bmatrix} x_1 + \frac{x_1^2}{1+x_1^2+x_2^2} \\ -2x_2 + \sin(x_1 x_2) \cos x_2 \end{bmatrix},$$

$$f(x, 2) = \begin{bmatrix} -2x_1 + x_2 - 2x_1 \sin x_2 \\ -x_1 - x_2 + x_1 \cos x_1 \sin x_2 \end{bmatrix},$$

$$f(x, 3) = \begin{bmatrix} x_2 + 2x_1 \cos x_1 \sin x_2 \\ -x_1 + x_2 \sin x_1 \end{bmatrix}.$$

Note that the matrices $\widehat{Q}$ and $A_i, i \in \mathcal{M}$ in Assumption (H7.4) can be obtained as follows:

$$\widehat{Q} = \begin{pmatrix} -3 & 1 & 2 \\ 1 & -1 & 0 \\ 0 & 3 & -3 \end{pmatrix} \tag{7.72}$$

and

$$A_1 = \begin{pmatrix} 1 & 0 \\ 0 & -2 \end{pmatrix}, \quad A_2 = \begin{pmatrix} -2 & 1 \\ -1 & -1 \end{pmatrix}, \quad \text{and } A_3 = \begin{pmatrix} 0 & 1 \\ -1 & 0 \end{pmatrix}. \tag{7.73}$$

For a system of differential equations without switching, the well-known Hartman-Grobman theorem holds. It indicates that for hyperbolic equilibria, a linear system arising from approximation is topologically equivalent to the associated nonlinear system, whereas a system with a center is not. Here, we demonstrate that this phenomenon changes a little. We show that the nonlinear system and its approximation could be equivalent even for equilibria with a center in one or more of its components as long as the component with hyperbolic equilibrium dominates the rest of them.

We proceed to use Theorem 7.8 and Corollary 7.9 to verify that the hybrid system (7.71) is asymptotically stable. The stationary distribution of the Markov chain $\widehat{\alpha}(t)$ is $\pi = (3/14,\ 9/14,\ 1/7)$. Now, consider the symmetric and positive definite matrix

$$G = \begin{pmatrix} 3 & -1 \\ -1 & 3 \end{pmatrix}$$

and compute the maximum eigenvalues of the matrices $GA_iG^{-1} + G^{-1}A_i'G$ with $i = 1, 2, 3$,

$$\lambda_{\max}(GA_1G^{-1} + G^{-1}A_1'G) = \frac{11}{4},$$

$$\lambda_{\max}(GA_2G^{-1} + G^{-1}A_2'G) = -\frac{11}{4},$$

and

$$\lambda_{\max}(GA_3G^{-1} + G^{-1}A_3'G) = \frac{3}{2}.$$

Thus, we obtain

$$\sum_{i=1}^{3} \pi_i \lambda_{\max}(GA_iG^{-1} + G^{-1}A_i'G) = -\frac{27}{28} < 0. \tag{7.74}$$

Hence, Theorem 7.8 implies that the hybrid system (7.71) is asymptotically stable.

It is interesting to note that the linear approximation

$$\dot{X}(t) = A(\widehat{\alpha}(t))X(t), \tag{7.75}$$

is also asymptotically stable, where the matrices $A(i) = A_i, i \in \mathcal{M}$ are as in (7.73), and the Markov chain $\widehat{\alpha}(t)$ is generated by $\widehat{Q}$ in (7.72). To demonstrate, we present the phase portrait of (7.71) in Fig. 7.3a, whereas Fig. 7.3b presents the phase portrait of its first-order linear approximation (7.75).

7.7 Notes

Consideration of switching ordinary differential equations of the form (7.1) stems from a wide variety of applications including control, optimization, estimation, and tracking. For example, in [294], with motivation of using stochastic recursive algorithms for tracking Markovian parameters such as those in spreading code optimization in CDMA (Code Division Multiple Access) wireless communication, we used an adaptive algorithm with constant stepsize to construct a sequence of estimates of the time-varying distribution. It was shown that under simple conditions, a continuous-time interpolation of the iteration converges weakly not to an ODE as is widely known in the literature of stochastic approximation [177] but to a system of ODEs with regime switching. Subsequently, treating least-squares-type algorithms involving Markovian jump processes in [293], random-switching ODEs were also obtained. Thus, not only is the study of systems given by (7.1) and (7.2) of mathematical interest, but also it provides practical guidance for many applications.

Fig. 7.3 Comparisons of nonlinear system and its linear approximation. (**a**) Hybrid nonlinear system: Phase portrait of (7.71) with initial condition $(x, \alpha) = ([0.8, 1]', 1)$. (**b**) Hybrid linear system: Phase portrait of (7.75) with initial condition $(x, \alpha) = ([0.8, 1]', 1)$

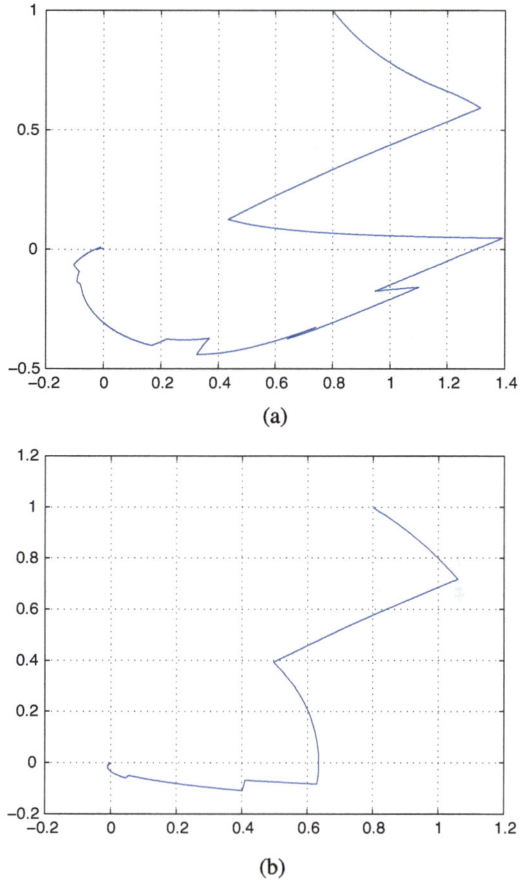

Taking into consideration that numerous real-world systems are in operation for a long period of time, longtime behavior of such systems is of foremost importance. Recently, much attention has been drawn to the study of stability of such systems; see [11, 143, 199, 311] among others. Much of the contemporary study of stochastic stability of dynamic systems can be traced back to the original work [146] by Kac and Krasovskii, in which a systematic approach was developed for stability of systems with Markovian switching using Lyapunov function methods. This important work stimulated much of the subsequent development.

This chapter focuses on stability and instability of random switching systems of differential equations. Sufficient conditions have been derived. These conditions are easily verifiable and are based on coefficients of the systems. For systems that are linear in the continuous component, certain necessary and sufficient conditions are derived using a transformation technique, which close the gap in using maximal and minimal eigenvalues of certain matrices. Somewhat remarkable, a particularly interesting discovery is: Different from a single system of differential equations, in which the Hartman-Grobman theorem is in force, for switching systems of differential equations modulated by a random process, even if some of the equilibria are not hyperbolic (e.g., center), the original system and that of the "linearized" system (with respect to the continuous variable) could still have the same asymptotic behavior. This is demonstrated by our analytical results as well as computations with the use of phase portraits. The stability study is based on the results from Zhu et al. [324]; the discussion on Lyapunov exponent is based on He and Yin [115].

Owing to the increasing needs of random environment models, stability of hybrid systems has received resurgent attention lately. Effort has been placed on deriving more easily verifiable conditions for stability and instability. This work also aims to contribute in this direction. One of the main features of [146] is to use quadratic Lyapunov functions to obtain verifiable conditions for stability of the system $\dot{X}(t) = A(\alpha(t))X(t)$, where $\alpha(t)$ is a continuous-time Markov chain with a constant generator Q. Their conditions amount to solving a system of linear nonhomogeneous equations, which is generally complicated. Here, we presented easily verifiable conditions for stability and instability with the aid of nonquadratic Lyapunov functions. Compared with the conditions in [146], our conditions in Theorem 7.8, Corollary 7.9, and Corollary 7.10 are simpler and easier to verify. Moreover, our results can be applied to more general models.

Chapter 8
Invariance Principles

8.1 Introduction

In the previous two chapters, we have studied stability of switching diffusions and random switching ordinary differential equations. Continuing our effort, this chapter is concerned with invariance principles of switching diffusion processes. This chapter together with the previous two chapters delineates longtime behavior and gives a complete picture of the switching-diffusion processes under consideration.

The rest of the chapter is arranged as follows. Section 8.2 begins with the formulation of the problem. Section 8.3 is devoted to the invariance principles using sample paths and kernels of the associated Lyapunov functions. Here, Lyapunov function-type criteria are obtained first. Then, linear (in x) systems are treated. Finally, a few more remarks are made in Section 8.4 to conclude the chapter.

8.2 Formulation

As in the previous chapters, we use z' to denote the transpose of $z \in \mathbb{R}^{\ell_1 \times \ell_2}$ with $\ell_1, \ell_2 \geq 1$, whereas $\mathbb{R}^{\ell \times 1}$ is simply written as $\mathbb{R}^{\ell}$; $\mathbb{1} = (1, 1, \ldots, 1)' \in \mathbb{R}^{m_0}$ is a column vector with all entries being 1; the Euclidean norm for a row or a column vector x is denoted by $|x|$. As usual, I denotes the identity matrix with suitable dimension. For a matrix A, its trace norm is denoted by $|A| = \sqrt{\text{tr}(A'A)}$. If a matrix A is real and symmetric, we use $\lambda_{\max}(A)$ and $\lambda_{\min}(A)$ to denote the maximal and minimal eigenvalues of A, respectively, and set $\rho(A) := \max\{|\lambda_{\max}(A)|, |\lambda_{\min}(A)|\}$. When B is a set, $I_B(\cdot)$ denotes the indicator function of B.

We work with $(\Omega, \mathcal{F}, \mathbb{P})$, a complete probability space, and consider a two-component Markov process $(X(t), \alpha(t))$, where $X(\cdot)$ is a continuous component taking values in $\mathbb{R}^r$ and $\alpha(\cdot)$ is a jump component taking values in a finite set

H.-D. Nguyen et al., *Hybrid Switching Diffusions*, Probability Theory and Stochastic Modelling 63, https://doi.org/10.1007/978-3-031-93303-5_8

$\mathcal{M} = \{1, 2, \ldots, m_0\}$. The process $(X(t), \alpha(t))$ has a generator $\mathcal{L}$ given as follows. For each $i \in \mathcal{M}$ and any twice continuously differentiable function $g(\cdot, i)$

$$\begin{aligned}\mathcal{L}g(x,i) &= \frac{1}{2}\sum_{j,k=1}^{r} a_{jk}(x,i)\frac{\partial^2 g(x,i)}{\partial x_j \partial x_k} + \sum_{j=1}^{r} b_j(x,i)\frac{\partial g(x,i)}{\partial x_j} + Q(x)g(x,\cdot)(i),\\ &= \frac{1}{2}\mathrm{tr}(a(x,i)\nabla^2 g(x,i)) + b'(x,i)\nabla g(x,i) + Q(x)g(x,\cdot)(i),\end{aligned} \tag{8.1}$$

where $x \in \mathbb{R}^r$ and $Q(x) = (q_{ij}(x))$ is an $m_0 \times m_0$ matrix depending on x satisfying $q_{ij}(x) \geq 0$ for $i \neq j$ and $\sum_{j \in \mathcal{M}} q_{ij}(x) = 0$ for each $i \in \mathcal{M}$

$$Q(x)g(x,\cdot)(i) = \sum_{j \in \mathcal{M}} q_{ij}(x)g(x,j) = \sum_{j \in \mathcal{M}, j \neq i} q_{ij}(x)(g(x,j) - g(x,i)),$$

and $\nabla g(\cdot, i)$ and $\nabla^2 g(\cdot, i)$ denote the gradient and Hessian of $g(\cdot, i)$, respectively.

The process $(X(t), \alpha(t))$ can be described by

$$\begin{aligned} dX(t) &= b(X(t),\alpha(t))dt + \sigma(X(t),\alpha(t))dw(t),\\ X(0) &= x,\ \alpha(0) = \alpha, \end{aligned} \tag{8.2}$$

and

$$\begin{aligned}\mathbb{P}\{\alpha(t+\Delta) = j | \alpha(t) = i, (X(s),\alpha(s)), s \leq t\} \qquad\qquad (8.3)\\ = \begin{cases} q_{ij}(X(t))\Delta + o(\Delta), & \text{if } i \neq j,\\ 1 + q_{ii}(X(t))\Delta + o(\Delta), & \text{if } j = i, \end{cases}\end{aligned}$$

where $w(t)$ is a d-dimensional standard Brownian motion, $b(\cdot,\cdot) : \mathbb{R}^r \times \mathcal{M} \mapsto \mathbb{R}^r$, and $\sigma(\cdot,\cdot) : \mathbb{R}^r \times \mathcal{M} \mapsto \mathbb{R}^{r \times d}$ satisfies $\sigma(x,i)\sigma'(x,i) = a(x,i)$.

Throughout the chapter, we assume that both $b(\cdot, i)$ and $\sigma(\cdot, i)$ satisfy the usual Lipschitz condition and linear growth condition for each $i \in \mathcal{M}$ and that $Q(\cdot)$ is bounded and continuous, and satisfies (2.21) for some concave function $\gamma : \mathbb{R}_+ \mapsto \mathbb{R}_+$ with $\gamma(0) = 0$. Under these conditions, the system (8.2)–(8.3) has a unique strong solution for any initial condition $(X(0), \alpha(0)) = (x, \alpha) \in \mathbb{R}^r \times \mathcal{M}$; see Chapter 2 and also [141] or [297] for details. From time to time, we often wish to emphasize the initial data $(X(0), \alpha(0)) = (x, \alpha)$ dependence of the solution of (8.2)–(8.3), which is denoted by $(X^{x,\alpha}(t), \alpha^{x,\alpha}(t))$.

8.3 Invariance: A Sample Path Approach

Recall that the evolution of the discrete component $\alpha(\cdot)$ can be represented as a stochastic integral with respect to a Poisson random measure; see Chapter 2 and [100, 262]. Indeed, for $x \in \mathbb{R}^r$ and $i, j \in \mathcal{M}$ with $j \neq i$, let $\Delta_{ij}(x)$ be consecutive (with respect to the lexicographic ordering on $\mathcal{M} \times \mathcal{M}$), left-closed, right-open intervals of the real line, each having length $q_{ij}(x)$. Define a function $h : \mathbb{R}^r \times \mathcal{M} \times \mathbb{R} \mapsto \mathbb{R}$ by

$$h(x, i, z) = \sum_{j=1}^{m_0} (j - i) I_{\{z \in \Delta_{ij}(x)\}}. \tag{8.4}$$

Then, (8.3) is equivalent to

$$d\alpha(t) = \int_{\mathbb{R}} h(X(t), \alpha(t-), z)\mathfrak{p}(dt, dz), \tag{8.5}$$

where $\mathfrak{p}(dt, dz)$ is a Poisson random measure with intensity $dt \times \mathfrak{m}(dz)$ and $\mathfrak{m}$ is the Lebesgue measure on $\mathbb{R}$. The Poisson random measure $\mathfrak{p}(\cdot, \cdot)$ is independent of the Brownian motion $w(\cdot)$. Denote the natural filtration by $\mathcal{F}_t := \sigma\{(X(s), \alpha(s)), s \le t\}$. Without loss of generality, upon augmentation, we can assume that the filtration $\{\mathcal{F}_t\}_{t \ge 0}$ satisfies the usual condition. That is, it is right continuous with $\mathcal{F}_0$ containing all $\mathbb{P}$-null sets.

For any $(x, i), (y, j) \in \mathbb{R}^r \times \mathcal{M}$, define

$$d((x, i), (y, j)) = |x - y| + \delta_{ij}, \tag{8.6}$$

where $\delta_{ij} = 1$ if $i = j$ and 0 if $i \neq j$. It is easy to verify that for any (x, i), (y, j), and (z, l):

(i) $d((x, i), (y, j)) \ge 0$ and $d((x, i), (y, j)) = 0$ if and only if $(x, i) = (y, j)$,
(ii) $d((x, i), (y, j)) = d((y, j), (x, i))$,
(iii) $d((x, i), (y, j)) \le d((x, i), (z, l)) + d((z, l), (y, j))$.

Thus, d is a distance function of $\mathbb{R}^r \times \mathcal{M}$. Also, if U is a subset of $\mathbb{R}^r \times \mathcal{M}$, we define

$$d((x, i), U) = \inf\{d((x, i), (y, j)) : (y, j) \in U\}. \tag{8.7}$$

Let $\mathcal{M}$ be endowed with the trivial topology. Then, for a fixed $U \in \mathbb{R}^r \times \mathcal{M}$, the function $d(\cdot, U)$ is continuous. As usual, we denote by $d(x, D)$ the distance between $x \in \mathbb{R}^r$ and $D \subset \mathbb{R}^r$; that is,

$$d(x, D) = \inf\{|x - y| : y \in D\}. \tag{8.8}$$

In addition, we use $\mathbb{P}_{x,\alpha}$ and $\mathbb{E}_{x,\alpha}$ to denote the probability and expectation with $(X(0), \alpha(0)) = (x, \alpha)$, respectively.

8.3.1 Invariant Sets

Inspired by the study in [94], we define the invariant set as follows.

Definition 8.1 A Borel measurable set $U \subset \mathbb{R}^r \times \mathcal{M}$ is said to be *invariant* with respect to the solutions of (8.2)–(8.5), or simply, U is *invariant* with respect to the process $(X(t), \alpha(t))$ if

$$\mathbb{P}_{x,i}\,\{(X(t), \alpha(t)) \in U, \quad \text{for all } t \geq 0\} = 1, \text{ for any } (x, i) \in U.$$

That is, a process starting from U will remain in U a.s.

As shown in Lemma 6.1, when the coefficients of (8.2)–(8.5) satisfy

$$b(0, \alpha) = \sigma(0, \alpha) = 0, \text{ for each } \alpha \in \mathcal{M},$$

then any solution with initial condition (x, i) satisfying $x \neq 0$ will never reach the origin almost surely; in other words, the set $(\mathbb{R}^r - \{0\}) \times \mathcal{M}$ is invariant with respect to the solutions of (8.2)–(8.5).

Using the terminologies in [94, 156], we recall the definitions of stability and asymptotic stability of a set. Then, general results in terms of the Lyapunov function are provided.

Definition 8.2 A closed and bounded set $K \subset \mathbb{R}^r \times \mathcal{M}$ is said to be

(i) *stable in probability* if for any $\varepsilon > 0$ and $\rho > 0$, there is a $\delta > 0$ such that

$$\mathbb{P}_{x,i}\left\{\sup_{t\geq 0} d((X(t), \alpha(t)), K) < \rho\right\} \geq 1 - \varepsilon, \text{ whenever } d((x, i), K) < \delta,$$

(ii) *asymptotically stable in probability* if it is stable in probability, and moreover

$$\mathbb{P}_{x,i}\left\{\lim_{t\to\infty} d((X(t), \alpha(t)), K) = 0\right\} \to 1, \quad \text{as } d((x, i), K) \to 0,$$

(iii) *stochastically asymptotically stable in the large* if it is stable in probability, and

$$\mathbb{P}_{x,i}\left\{\lim_{t\to\infty} d((X(t), \alpha(t)), K) = 0\right\} = 1, \quad \text{for any } (x, i) \in \mathbb{R}^r \times \mathcal{M},$$

(iv) *asymptotically stable with probability one* if

$$\lim_{t\to\infty} d((X(t), \alpha(t)), K) = 0 \text{ a.s.}$$

Theorem 8.3 *Assume that there exists a nonnegative function* $V(\cdot, \cdot) : \mathbb{R}^r \times \mathcal{M} \mapsto \mathbb{R}_+$ *such that*

$$\mathrm{Ker}(V) := \{(x, i) \in \mathbb{R}^r \times \mathcal{M} : V(x, i) = 0\} \tag{8.9}$$

is nonempty and bounded, $V(\cdot, \alpha) \in C^2(\mathbb{R}^r)$ *for each* $\alpha \in \mathcal{M}$*, and that*

$$\mathcal{L}V(x, i) \le 0, \quad \text{for all } (x, i) \in \mathbb{R}^r \times \mathcal{M}. \tag{8.10}$$

Then

(i) $\mathrm{Ker}(V)$ *is an invariant set for the process* $(X(t), \alpha(t))$*, and*
(ii) $\mathrm{Ker}(V)$ *is stable in probability.*

Proof Let $(x_0, i_0) \in \mathrm{Ker}(V)$. By virtue of generalized Itô's lemma (2.5) (see also [262, Chapter 2]), we have for any $t \ge 0$

$$V(X(t), \alpha(t)) = V(x_0, i_0) + \int_0^t \mathcal{L}V(X(s), \alpha(s))ds + M(t), \tag{8.11}$$

where $M(t) = M_1(t) + M_2(t)$ is a local martingale with

$$\begin{aligned} M_1(t) &= \int_0^t \big\langle \nabla V(X(s), \alpha(s)), \sigma(X(s), \alpha(s))dw(s)\big\rangle, \\ M_2(t) &= \int_0^t \int_{\mathbb{R}} \big[V(X(s-), \alpha(s-) + h(X(s-), \alpha(s-), z)) \\ &\qquad - V(X(s-), \alpha(s-))\big]\widetilde{\mathfrak{p}}(ds, dz), \end{aligned}$$

where $\langle \cdot, \cdot \rangle$ denotes the usual inner product and $\widetilde{\mathfrak{p}}(ds, dz) = \mathfrak{p}(ds, dz) - ds \times \mathfrak{m}(dz)$ is a martingale measure. Taking expectations on both sides of (8.11) (using a sequence of stopping times and Fatou's lemma, if necessary as in the argument of Theorem 3.14), it follows from (8.10) that

$$\mathbb{E}_{x_0, i_0}[V(X(t), \alpha(t))] \le V(x_0, i_0) = 0;$$

the last equality above holds because $(x_0, i_0) \in \mathrm{Ker}(V)$. But V is nonnegative, so we must have $V(X(t), \alpha(t)) = 0$ a.s. for any $t \ge 0$. Since the process $(X(t), \alpha(t))$ is cádlág (sample paths being right continuous and having left limits), we obtain

$$\mathbb{P}_{x_0, i_0}\left\{\sup_{t \ge 0} V(X(t), \alpha(t)) = 0\right\} = 1.$$

That is,

$$\mathbb{P}_{x_0, i_0}\{(X(t), \alpha(t)) \in \mathrm{Ker}(V), \quad \text{for all } t \ge 0\} = 1.$$

This proves the first assertion of the theorem.

We proceed to prove the second assertion. For any $\delta > 0$, let U_δ be a neighborhood of $\mathrm{Ker}(V)$ such that

$$U_\delta := \left\{(x,i) \in \mathbb{R}^r \times \mathcal{M} : d((x,i), \mathrm{Ker}(V)) < \delta\right\}. \tag{8.12}$$

Let the initial condition $(x,i) \in U_\delta - \mathrm{Ker}(V)$ and τ be the first exit time of the process from U_δ. That is, $\tau = \inf\{t : (X(t), \alpha(t)) \notin U_\delta\}$. Then, for any $t \geq 0$, by virtue of generalized Itô's lemma

$$V(X(t\wedge\tau), \alpha(t\wedge\tau)) = V(x,i) + \int_0^{t\wedge\tau} \mathcal{L}V(X(s), \alpha(s))ds + M(t\wedge\tau),$$

where M is a local martingale defined earlier in the proof. As argued in the previous paragraph, by virtue of (8.10), we can use a sequence of stopping times and Fatou's lemma, if necessary, to obtain

$$\mathbb{E}_{x,i}[V(X(t\wedge\tau), \alpha(t\wedge\tau))] \leq V(x,i) + \mathbb{E}_{x,i}\int_0^{t\wedge\tau} \mathcal{L}V(X(s), \alpha(s))ds \leq V(x,i).$$

Because V is nonnegative, we further have

$$\begin{aligned} V(x,i) &\geq \mathbb{E}_{x,i}[V(X(\tau), \alpha(\tau))I_{\{\tau<t\}}] + \mathbb{E}_{x,i}[V(X(t), \alpha(t))I_{\{t\leq\tau\}}] \\ &\geq \mathbb{E}_{x,i}[V(X(\tau), \alpha(\tau))I_{\{\tau<t\}}]. \end{aligned} \tag{8.13}$$

For notational simplicity, denote $(\xi, \ell) = (X(\tau), \alpha(\tau))$. We claim that

$$V(\xi, \ell) > \rho, \quad \text{for some constant } \rho > 0. \tag{8.14}$$

To prove this claim, write $\mathrm{Ker}(V) = \bigcup_{l=1}^{k}(N_{j_l} \times \{j_l\})$, where $k \leq m_0$, $N_{j_l} \subset \mathbb{R}^r$ and $j_l \in \mathcal{M}$, for $l = 1, \ldots, k$. We denote further $J = \{j_1, \ldots, j_k\} \subset \mathcal{M}$. Let us first consider the case when $\ell \notin J$. Note that $\xi \in D$, where D is a bounded neighborhood of $\bigcup_{l=1}^{k} N_l$ (such a neighborhood D exists because $\mathrm{Ker}(V)$ is bounded by the assumption of the theorem). Then, we have

$$\inf\{V(x,\ell) : x \in D\} \geq \rho_1 > 0. \tag{8.15}$$

If (8.15) were false, there would exist a sequence $\{x_n\} \subset D$ with $\lim_{n\to\infty} V(x_n, \ell) = 0$. Because $\{x_n\}$ is bounded, there exists a subsequence $\left\{x_{n_k}\right\}$ with $x_{n_k} \to \widetilde{x} \in \mathbb{R}^r$. Thus, by the continuity of $V(\cdot, \ell)$, we have

$$V(\widetilde{x}, \ell) = \lim_{k\to\infty} V(x_{n_k}, \ell) = 0.$$

That is, $(\widetilde{x}, \ell) \in \mathrm{Ker}(V)$. This is a contradiction to the assumption that $\ell \notin J$. Thus, (8.15) is true and hence $V(\xi, \ell) \geq \rho_1$.

Now, let us consider the case $\ell \in J$. It follows that $\delta \le d(\xi, N_\ell) \le \widetilde{\delta} < \infty$. A similar argument using contradiction as in the previous case shows that

$$\inf\left\{V(x,\ell) : \delta \le d(x, N_\ell) \le \widetilde{\delta}\right\} \ge \rho_2 > 0.$$

Thus, it follows that $V(\xi, \ell) \ge \rho_2$. A combination of the two cases gives us $V(\xi, \ell) \ge \rho$, where $\rho = \rho_1 \wedge \rho_2$. Hence, the claim (8.14) follows.

Finally, we have from (8.13) and (8.14) that $\mathbb{P}_{x,i}\{\tau < t\} \le \frac{1}{\rho} V(x, i)$. Letting $t \to \infty$ to obtain

$$\mathbb{P}_{x,i}\{\tau < \infty\} \le \frac{1}{\rho} V(x, i).$$

Note that

$$\{\tau < \infty\} = \left\{\sup_{0 \le t < \infty} d((X(t), \alpha(t)), \operatorname{Ker}(V)) \ge \delta\right\}.$$

Therefore, it follows that

$$\mathbb{P}_{x,i}\left\{\sup_{0 \le t < \infty} d((X(t), \alpha(t)), \operatorname{Ker}(V)) \ge \delta\right\} \le \frac{1}{\rho} V(x, i) \to 0,$$

as $d((x, i), \operatorname{Ker}(V)) \to 0$. This finishes the proof of the theorem. □

Next, we consider asymptotic stability. To this end, we need the following lemma.

Lemma 8.4 *Assume that there exists a nonnegative function* $V : \mathbb{R}^r \times \mathcal{M} \mapsto \mathbb{R}_+$ *with nonempty and bounded* $\operatorname{Ker}(V)$ *such that for each* $\alpha \in \mathcal{M}$, $V(\cdot, \alpha)$ *is twice continuously differentiable with respect to* x *and that for any* $\varepsilon > 0$

$$\mathcal{L}V(x, i) \le -\kappa_\varepsilon < 0, \quad \text{for any } (x, i) \in (\mathbb{R}^r \times \mathcal{M}) \setminus \overline{U}_\varepsilon, \tag{8.16}$$

where κ_ε *is a positive constant depending on* ε, U_ε *is a neighborhood of* $\operatorname{Ker}(V)$ *as defined in* (8.12), *and* $\overline{U}_\varepsilon$ *denotes the closure of* U_ε. *Then, for any* $0 < \varepsilon < r_0$, *we have*

$$\mathbb{P}_{x,i}\left\{\tau_{\varepsilon, r_0} < \infty\right\} = 1, \quad \text{for any } (x, i) \in U_{\varepsilon, r_0},$$

where

$$U_{\varepsilon, r_0} = \left\{(y, j) \in \mathbb{R}^r \times \mathcal{M} : \varepsilon < d((y, j), \operatorname{Ker}(V)) < r_0\right\},$$

and τ_{ε,r_0} *is the first exit time from* U_{ε,r_0}*; that is,*

$$\tau_{\varepsilon,r_0} := \inf\left\{t \geq 0 : (X(t), \alpha(t)) \notin U_{\varepsilon,r_0}\right\}.$$

Proof Fix any $(x, i) \in U_{\varepsilon,r_0}$. By virtue of generalized Itô's lemma, we have that for any $t \geq 0$

$$\begin{aligned} &V(X(t \wedge \tau_{\varepsilon,r_0}), \alpha(t \wedge \tau_{\varepsilon,r_0})) \\ &= V(x, i) + \int_0^{t\wedge\tau_{\varepsilon,r_0}} \mathcal{L}V(X(s), \alpha(s))ds + M(t \wedge \tau_{\varepsilon,r_0}), \end{aligned}$$

where $M(t \wedge \tau_{\varepsilon,r_0})$ is a martingale with mean zero. Thus, by taking expectations on both sides, and using (8.16), we obtain

$$\begin{aligned} \mathbb{E}_{x,i}\Big[V(X(t \wedge \tau_{\varepsilon,r_0}), \alpha(t \wedge \tau_{\varepsilon,r_0}))\Big] &\leq V(x, i) - \mathbb{E}_{x,i} \int_0^{t\wedge\tau_{\varepsilon,r_0}} \kappa_\varepsilon ds \\ &= V(x, i) - \kappa_\varepsilon \mathbb{E}_{x,i}[t \wedge \tau_{\varepsilon,r_0}]. \end{aligned}$$

Note that V is nonnegative; hence, we have $\mathbb{E}_{x,i}[t \wedge \tau_{\varepsilon,r_0}] \leq \frac{1}{\kappa_\varepsilon} V(x, i)$. But

$$\begin{aligned} \mathbb{E}_{x,i}[t \wedge \tau_{\varepsilon,r_0}] &= t\mathbb{P}_{x,i}\left\{\tau_{\varepsilon,r_0} > t\right\} + \mathbb{E}_{x,i}[\tau_{\varepsilon,r_0} I_{\{\tau_{\varepsilon,r_0} \leq t\}}] \\ &\geq t\mathbb{P}_{x,i}\left\{\tau_{\varepsilon,r_0} > t\right\}. \end{aligned}$$

Thus, it follows that

$$t\mathbb{P}_{x,i}\left\{\tau_{\varepsilon,r_0} > t\right\} \leq \frac{1}{\kappa_\varepsilon} V(x, i).$$

Now, letting $t \to \infty$, we have

$$\mathbb{P}_{x,i}\left\{\tau_{\varepsilon,r_0} = \infty\right\} = 0 \quad \text{or} \quad \mathbb{P}_{x,i}\left\{\tau_{\varepsilon,r_0} < \infty\right\} = 1.$$

The assertion thus follows. □

Theorem 8.5 *Assume that there exists a function* $V : \mathbb{R}^r \times \mathcal{M} \mapsto \mathbb{R}_+$ *satisfying the conditions of Lemma* 8.4*. Then,* $\operatorname{Ker}(V)$ *is an invariant set for the process* $(X(t), \alpha(t))$ *and* $\operatorname{Ker}(V)$ *is asymptotically stable in probability.*

Proof Motivated by Mao and Yuan [205, Theorem 5.36], we use similar ideas and proceed as follows. By virtue of Theorem 8.3, we know that $\operatorname{Ker}(V)$ is an invariant set for the process $(X(t), \alpha(t))$ and that $\operatorname{Ker}(V)$ is stable in probability. Hence, it remains to show that

$$\mathbb{P}_{x,i}\left\{\lim_{t\to\infty} d((X(t), \alpha(t)), \operatorname{Ker}(V)) = 0\right\} \to 1 \text{ as } d((x, i), \operatorname{Ker}(V)) \to 0.$$

Because $\text{Ker}(V)$ is stable in probability, for any $\varepsilon > 0$ and any $\theta > 0$, there exists some $\delta > 0$ (without loss of generality, we may assume that $\delta < \theta$) such that

$$\mathbb{P}_{x,i}\left\{\sup_{t\geq 0} d((X(t),\alpha(t)),\text{Ker}(V)) < \theta\right\} \geq 1 - \frac{\varepsilon}{2}, \tag{8.17}$$

for any $(x, i) \in U_\delta$, where U_δ is defined in (8.12). Now, fix any $(x, \alpha) \in U_\delta - \text{Ker}(V)$ and let $\rho > 0$ be arbitrary satisfying $0 < \rho < d((x, i), \text{Ker}(V))$ and choose some $\varrho \in (0, \rho)$. Define

$$\begin{aligned}\tau_\varrho &:= \inf\{t \geq 0 : d((X(t),\alpha(t)),\text{Ker}(V)) \leq \varrho\},\\ \tau_\theta &:= \inf\{t \geq 0 : d((X(t),\alpha(t)),\text{Ker}(V)) \geq \theta\}.\end{aligned}$$

Then, it follows from Lemma 8.4 that

$$\mathbb{P}_{x,\alpha}\left\{\tau_\varrho \wedge \tau_\theta < \infty\right\} = \mathbb{P}_{x,\alpha}\left\{\tau_{\varrho,\theta} < \infty\right\} = 1, \tag{8.18}$$

where $\tau_{\varrho,\theta}$ is the first exit time from $U_{\varrho,\theta}$, where $U_{\varrho,\theta}$ is defined as

$$U_{\varrho,\theta} := \left\{(y, j) \in \mathbb{R}^r \times \mathcal{M} : \varrho < d((X(t),\alpha(t)),\text{Ker}(V)) < \theta\right\}.$$

But (8.17) implies that $\mathbb{P}_{x,\alpha}\{\tau_\theta < \infty\} \leq \varepsilon/2$. Note also

$$\mathbb{P}_{x,\alpha}\left\{\tau_\varrho \wedge \tau_\theta < \infty\right\} \leq \mathbb{P}_{x,\alpha}\left\{\tau_\varrho < \infty\right\} + \mathbb{P}_{x,\alpha}\{\tau_\theta < \infty\}.$$

Thus, it follows that

$$\mathbb{P}_{x,\alpha}\left\{\tau_\varrho < \infty\right\} \geq \mathbb{P}_{x,\alpha}\left\{\tau_\varrho \wedge \tau_\theta < \infty\right\} - \mathbb{P}_{x,\alpha}\{\tau_\theta < \infty\} \geq 1 - \frac{\varepsilon}{2}. \tag{8.19}$$

Now, let

$$\tau_\rho := \inf\left\{t \geq \tau_\varrho : d((X(t),\alpha(t)),\text{Ker}(V)) \geq \rho\right\}.$$

We have used the convention that $\inf\{\emptyset\} = \infty$. For any $t \geq 0$, we apply generalized Itô's lemma and (8.16) to obtain

$$\begin{aligned}\mathbb{E}_{x,\alpha} V(X(\tau_\rho \wedge t), \alpha(\tau_\rho \wedge t)) &\leq \mathbb{E}_{x,\alpha} V(X(\tau_\varrho \wedge t), \alpha(\tau_\varrho \wedge t))\\ &\quad + \mathbb{E}_{x,\alpha}\int_{\tau_\varrho\wedge t}^{\tau_\rho\wedge t} \mathcal{L}V(X(s),\alpha(s))ds\\ &\leq \mathbb{E}_{x,\alpha} V(X(\tau_\varrho \wedge t), \alpha(\tau_\varrho \wedge t)).\end{aligned} \tag{8.20}$$

Note that $\tau_\varrho \geq t$ implies $\tau_\rho \geq t$ by the definition of τ_ρ. As a result, we have

$$\mathbb{E}_{x,\alpha}\big[I_{\{\tau_\varrho \geq t\}} V(X(\tau_\rho \wedge t), \alpha(\tau_\rho \wedge t))\big] \leq \mathbb{E}_{x,\alpha}\big[I_{\{\tau_\rho \geq t\}} V(X(\tau_\rho \wedge t), \alpha(\tau_\rho \wedge t))\big]. \tag{8.21}$$

Therefore, it follows from (8.20) and (8.21) that

$$\begin{aligned}
&\mathbb{E}_{x,\alpha}\big[I_{\{\tau_\rho < t\}} V(X(\tau_\rho \wedge t), \alpha(\tau_\rho \wedge t))\big] \\
&\quad = \mathbb{E}_{x,\alpha}\big[V(X(\tau_\rho \wedge t), \alpha(\tau_\rho \wedge t))\big] - \mathbb{E}_{x,\alpha}\big[I_{\{\tau_\rho \geq t\}} V(X(\tau_\rho \wedge t), \alpha(\tau_\rho \wedge t))\big] \\
&\quad \leq \mathbb{E}_{x,\alpha}[V(X(\tau_\varrho \wedge t), \alpha(\tau_\varrho \wedge t))] - \mathbb{E}_{x,\alpha}\big[I_{\{\tau_\varrho \geq t\}} V(X(\tau_\rho \wedge t), \alpha(\tau_\rho \wedge t))\big] \\
&\quad = \mathbb{E}_{x,\alpha}[V(X(\tau_\varrho \wedge t), \alpha(\tau_\varrho \wedge t))] - \mathbb{E}_{x,\alpha}\big[I_{\{\tau_\varrho \geq t\}} V(X(\tau_\varrho \wedge t), \alpha(\tau_\varrho \wedge t))\big] \\
&\quad = \mathbb{E}_{x,\alpha}\big[I_{\{\tau_\varrho < t\}} V(X(\tau_\varrho), \alpha(\tau_\varrho))\big] \\
&\quad \leq \widehat{V}_\varrho,
\end{aligned}$$

where $\widehat{V}_\varrho := \sup\{V(y, j) : d((y, j), \mathrm{Ker}(V)) = \varrho\}$. Then

$$\begin{aligned}
\widehat{V}_\varrho &\geq \mathbb{E}_{x,\alpha}\big[I_{\{\tau_\rho < t\}} V(X(\tau_\rho \wedge t), \alpha(\tau_\rho \wedge t))\big] \\
&= \mathbb{E}_{x,\alpha}\big[I_{\{\tau_\rho < t\}} V(X(\tau_\rho), \alpha(\tau_\rho))\big] \\
&\geq V_\rho \mathbb{P}_{x,\alpha}\left\{\tau_\rho < t\right\},
\end{aligned}$$

where $V_\rho := \inf\{V(y, j) : \rho \leq d((y, j), \mathrm{Ker}(V)) \leq \widetilde{\rho}\}$, with $\widetilde{\rho} > 0$ being some constant. Recall that we showed in the proof of Theorem 8.3 that $V_\rho > 0$. Because V is continuous, we may choose ϱ sufficiently small so that

$$\mathbb{P}_{x,\alpha}\left\{\tau_\rho < t\right\} \leq \frac{\widehat{V}_\varrho}{V_\rho} \leq \frac{\varepsilon}{2}.$$

Letting $t \to \infty$, we obtain

$$\mathbb{P}_{x,\alpha}\left\{\tau_\rho < \infty\right\} \leq \frac{\varepsilon}{2}. \tag{8.22}$$

Finally, it follows from (8.19) and (8.22) that

$$\begin{aligned}
\mathbb{P}_{x,\alpha}\left\{\tau_\varrho < \infty, \tau_\rho = \infty\right\} &\geq \mathbb{P}_{x,\alpha}\left\{\tau_\varrho < \infty\right\} - \mathbb{P}_{x,\alpha}\left\{\tau_\rho < \infty\right\} \\
&\geq 1 - \frac{\varepsilon}{2} - \frac{\varepsilon}{2} = 1 - \varepsilon.
\end{aligned}$$

This implies that

$$\mathbb{P}_{x,\alpha}\left\{\limsup_{t\to\infty} d((X(t), \alpha(t)), \mathrm{Ker}(V)) \leq \rho\right\} \geq 1 - \varepsilon.$$

But $\rho > 0$ can be chosen to be arbitrarily small. Therefore, we have

$$\mathbb{P}_{x,\alpha}\left\{\lim_{t\to\infty} d((X(t),\alpha(t)),\mathrm{Ker}(V)) = 0\right\} \geq 1-\varepsilon.$$

This finishes the proof of the theorem. □

Theorem 8.6 *Assume there exists a function $V : \mathbb{R}^r \times \mathcal{M} \mapsto \mathbb{R}_+$ satisfying the conditions of Lemma* 8.4. *If V also satisfies*

$$\lim_{|x|\to\infty} \inf_{\alpha\in\mathcal{M}} V(x,\alpha) = \infty, \tag{8.23}$$

then $\mathrm{Ker}(V)$ *is asymptotically stable in probability in the large; that is,* $\mathrm{Ker}(V)$ *is stable in probability and*

$$\mathbb{P}_{x,\alpha}\left\{\lim_{t\to\infty} d((X(t),\alpha(t)),\mathrm{Ker}(V)) = 0\right\} = 1, \tag{8.24}$$

for any $(x,\alpha) \in \mathbb{R}^r \times \mathcal{M}$.

Proof By virtue of Theorem 8.3, $\mathrm{Ker}(V)$ is stable in probability. Thus, it remains to verify (8.24). To this end, as in the proof of Theorem 8.3, we write $\mathrm{Ker}(V) = \bigcup_{l=1}^{k}(N_{j_l} \times \{j_l\})$, where $k \leq m_0$, $N_{j_l} \subset \mathbb{R}^r$, and $j_l \in \mathcal{M}$ for $l = 1,\ldots,k$. Since $\mathrm{Ker}(V)$ is bounded by assumption, in particular, $\bigcup_{l=1}^{k} N_{j_l}$ is bounded, there exists some $R > 0$ such that

$$\sup\left\{|y| : y \in \bigcup_{l=1}^{k} N_{j_l}\right\} \leq R. \tag{8.25}$$

Let $\varepsilon > 0$ and fix any $(x,\alpha) \in \mathbb{R}^r \times \mathcal{M}$. Then, (8.23) implies that there exists some positive constant $\beta > (R+2) \vee d((x,\alpha),\mathrm{Ker}(V))$ such that

$$\inf\{V(y,j) : |y| \geq \beta, j \in \mathcal{M}\} \geq \frac{2V(x,\alpha)}{\varepsilon}. \tag{8.26}$$

Define

$$\tau_\beta := \inf\{t \geq 0 : d((X(t),\alpha(t)),\mathrm{Ker}(V)) \geq 2\beta\}.$$

For any $t \geq 0$, we have by virtue of generalized Itô's lemma and (8.16) that

$$\mathbb{E}_{x,\alpha} V(X(t\wedge\tau_\beta),\alpha(t\wedge\tau_\beta)) \leq V(x,\alpha). \tag{8.27}$$

We claim that $\left|X(\tau_\beta)\right| \geq \beta$. If this were not true, it would follow from (8.25) that for any $(y,j) \in \mathrm{Ker}(V)$

$$
\begin{aligned}
d((X(\tau_\beta), \alpha(\tau_\beta)), (y, j)) &\le \left|X(\tau_\beta) - y\right| + 1 \\
&\le \left|X(\tau_\beta)\right| + |y| + 1 < \beta + R + 1 < 2\beta - 1,
\end{aligned}
$$

where in the last inequality above, we used the fact that $\beta > R + 2$. Then, we have $d((X(\tau_\beta), \alpha(\tau_\beta)), \text{Ker}(V)) \le 2\beta - 1 < 2\beta$. This contradicts the definition of τ_β. Thus, we must have $\left|X(\tau_\beta)\right| \ge \beta$. Then, it follows from (8.27) that

$$
\begin{aligned}
V(x, \alpha) &\ge \mathbb{E}_{x,\alpha}\Big[V(X(\tau_\beta), \alpha(\tau_\beta)) I_{\{\tau_\beta < t\}}\Big] \\
&\ge \inf\{V(y, j) : |y| \ge \beta, j \in \mathcal{M}\} \, \mathbb{P}_{x,\alpha}\left\{\tau_\beta < t\right\},
\end{aligned}
$$

and hence (8.26) implies that $\mathbb{P}_{x,\alpha}\left\{\tau_\beta < t\right\} \le \frac{\varepsilon}{2}$. This, in turn, implies that $\mathbb{P}_{x,\alpha}\left\{\tau_\beta < \infty\right\} \le \frac{\varepsilon}{2}$ by passing to the limit as $t \to \infty$. Then, we can finish the proof using the same argument as that in the proof of Theorem 8.5. □

The following theorem provides a criterion for asymptotic stability with probability 1.

Theorem 8.7 *Assume that there exists a nonnegative function $V : U \mapsto \mathbb{R}_+$ such that for each $\alpha \in \mathcal{M}$, $V(\cdot, \alpha)$ is twice continuously differentiable with respect to x and that there exists a continuous function $\widehat{W} : \mathbb{R}^r \times \mathcal{M} \mapsto \mathbb{R}_+$ satisfying*

$$
\mathcal{L}V(x, i) \le -\widehat{W}(x, i), \quad \textit{for any } (x, i) \in U, \tag{8.28}
$$

where $U \subset \mathbb{R}^r \times \mathcal{M}$ is an invariance set for the process $(X(t), \alpha(t))$. Assume also that either U is bounded or

$$
\lim_{|x| \to \infty, \ (x,\alpha) \in U} V(x, \alpha) = \infty. \tag{8.29}
$$

Then, for any initial condition $(x, \alpha) \in \mathbb{R}^r \times \mathcal{M}$, the following assertions hold:

(i) $\limsup_{t \to \infty} V(X^{x,\alpha}(t), \alpha^{x,\alpha}(t)) < \infty$ *a.s.*
(ii) $\text{Ker}(\widehat{W}) \ne \emptyset$.
(iii) $\lim_{t \to \infty} d((X^{x,\alpha}(t), \alpha^{x,\alpha}(t)), \text{Ker}(\widehat{W})) = 0$ *a.s.*
(iv) *If moreover,* $\text{Ker}(\widehat{W}) = \{0\} \times \mathcal{M}$, *then*

$$
\lim_{t \to \infty} X^{x,\alpha}(t) = 0 \ \ a.s.
$$

Proof This theorem can be proved using the arguments in [200, Theorem 2.1] although some modifications are needed. □

8.3.2 Linear Systems

We end this section with the following results on linear systems. Again, by linear systems, we mean systems that are linear in the x variable. Consider

$$dX(t) = b(\alpha(t))X(t)dt + \sum_{j=1}^{d} \sigma_j(\alpha(t))X(t)dw_j(t), \tag{8.30}$$

where $b(i), \sigma_j(i)$ are $r \times r$ constant matrices and $w_j(t)$ are independent one-dimensional standard Brownian motions for $i = 1, 2, \ldots, m_0$ and $j = 1, 2, \ldots, d$.

Note that 0 is an equilibrium point for the system given by (8.30) and (8.3). As we indicated earlier, it was shown in [164, 199] that the set $\mathbb{R}^r \setminus \{0\} \times \mathcal{M}$ is invariant with respect to the process $(X(t), \alpha(t))$.

Theorem 8.8 *Assume that the discrete component $\alpha(\cdot)$ is ergodic with constant generator $Q = (q_{ij})$ and invariant distribution $\pi = (\pi_1, \ldots, \pi_{m_0}) \in \mathbb{R}^{1\times m_0}$. Then, the equilibrium point $x = 0$ of the system given by* (8.30) *and* (8.3)

(i) *is asymptotically stable with probability one if*

$$\sum_{i=1}^{m} \pi_i \lambda_{\max}\left(b(i) + b'(i) + \sum_{j=1}^{d} \sigma_j(i)\sigma_j'(i) \right) < 0, \text{ and} \tag{8.31}$$

(ii) *is unstable in probability if*

$$\sum_{i=1}^{m} \pi_i \left[\lambda_{\min}\left(b(i) + b'(i) + \sum_{j=1}^{d} \sigma_j(i)\sigma_j'(i) \right) - \frac{1}{2}\sum_{j=1}^{d} \left(\rho(\sigma_j(i) + \sigma_j'(i)) \right)^2 \right] > 0. \tag{8.32}$$

Proof We only need to prove assertion (i), because assertion (ii) was established in Theorem 6.21 in Chapter 6. For notational simplicity, define the column vector $\mu = (\mu_1, \mu_2, \ldots, \mu_{m_0})' \in \mathbb{R}^{m_0}$ with

$$\mu_i = \frac{1}{2}\lambda_{\max}\left(b(i) + b'(i) + \sum_{j=1}^{d} \sigma_j(i)\sigma_j'(i) \right).$$

Also, let $\beta := -\pi\mu$. Note that $\beta > 0$ by (8.31). As in Chapter 6, it follows that the equation $Qc = \mu + \beta\mathbb{1}$ has a solution $c = (c_1, c_2, \ldots, c_{m_0})' \in \mathbb{R}^{m_0}$. Thus, we have

$$\mu_i - \sum_{j=1}^{m_0} q_{ij}c_j = -\beta, \quad i \in \mathcal{M}. \tag{8.33}$$

For each $i \in \mathcal{M}$, consider the Lyapunov function $V(x,i) = (1-\gamma c_i)|x|^\gamma$, where $0 < \gamma < 1$ is sufficiently small so that $1-\gamma c_i > 0$ for each $i \in \mathcal{M}$. It is readily seen that for each $i \in \mathcal{M}$, $V(\cdot,i)$ is continuous and nonnegative, vanishes only at $x=0$, and satisfies (8.29). Detailed calculations as in the proof of Theorem 6.21 in Chapter 6 reveal that for $x \neq 0$, $i \in \mathcal{M}$, and γ sufficiently small, we have

$$\begin{aligned}\mathcal{L}V(x,i) &\le \gamma(1-\gamma c_i)|x|^\gamma \Big\{\mu_i - \sum_{j=1}^m q_{ij}c_j + O(\gamma)\Big\}\\ &= \gamma(1-\gamma c_i)|x|^\gamma \Big(-\beta + O(\gamma)\Big)\\ &\le -\frac{\beta}{2}\gamma(1-\gamma c_i)\,|x|^\gamma := -\widehat{W}(x,i),\end{aligned}$$

Note that $\mathrm{Ker}(\widehat{W}) = \{0\} \times \mathcal{M}$. Thus, we conclude from Theorem 8.7 that the equilibrium point $x=0$ is asymptotically stable with probability 1. □

8.4 Notes

For systems running for a long time, it is crucial to learn their long-run behavior; see [164, 199, 311] for recent progress on stability of such systems. The rapid progress in natural science, life science, engineering, as well as in social science demands the consideration of stability of such systems. In fact, the advent of switching diffusions is largely because of the practical needs in modeling complex dynamic systems; see [12, 100, 138, 164, 199, 294, 308, 311, 321, 323] for some of the recent studies.

Most works to date have been concentrated on Markov-modulated diffusions, in which the Brownian motion and the switching process are independent, whereas less is known for systems with continuous-component dependent switching processes. As demonstrated in Chapter 2 (see also [283, 323]), when x-dependent switching diffusions are encountered, even such properties as continuity and smooth dependence on the initial data are nontrivial and fairly difficult to establish. Nevertheless, studying such systems is both practically useful and theoretically interesting. In our works, basic properties such as recurrence, positive recurrence, and ergodicity were studied in [321]; stability was treated in [164]; stability of randomly switching ordinary differential equations was treated in [324].

This chapter has taken up the issue of examination of the invariance principle akin to LaSalle's theorem for deterministic systems [110, 111, 182]. Previous study of invariance principles for stochastic systems can be found in [174, 200]. Inspired by the work of Mao [200], we use kernels of Lyapunov functions in this chapter to treat the invariance. There is another approach using the measure-theoretic viewpoint; see Kushner [174]. We remark that the techniques used here can also be adopted to treat random switching ordinary differential equations.

Finally, we add that in our more recent work [58] and [66], in which somewhat different definition of ω-limit set was given. Rather than using measures, solution paths are used. In studying Kolmogorov systems, it was found that the ω limit sets are the same as the set of recurrent points; the smallest invariant sets are the support of invariants measures. We refer the reader to the two papers mentioned above for further reading. Some related references can also be found in Chapter 12 of this book.

Chapter 9
Two-Time-Scale Switching Diffusions

9.1 Introduction

This chapter focuses on switching diffusions with two-time scales. It consists of two parts. The first part aims to model stochastic volatility using regime-switching diffusions for option pricing. Effort is devoted to developing asymptotic expansions of a system of coupled differential equations with applications to option pricing under regime-switching diffusions. By focusing on fast mean reversion, we aim at finding the "effective volatility." The main techniques used are singular perturbation methods. Under simple conditions, asymptotic expansions are developed with uniform asymptotic error bounds. The leading term in the asymptotic expansions satisfies a Black-Scholes equation in which the mean return rate and volatility are averaged out with respect to the stationary measure of the switching process. In addition, the full asymptotic series is developed, which helps us to gain insight on the behavior of the option price when time t approaches maturity. The asymptotic expansions obtained in this chapter are interesting in their own right and can be used for other problems in control optimization of systems involving fast-varying switching processes.

To study the positive recurrence and ergodicity, one of the conditions used in the previous chapters is that the states of the switching process belong to only one ergodic class. In the second part of this chapter, we further our study by treating the case that the states of the discrete event process belong to several "ergodic" classes that are weakly connected. This notion is made more precise in what follows. A key idea is the use of two-time-scale formulation; see [303, 304] and many references therein.

The rest of the chapter is arranged as follows. Sections 9.2–9.4 present the motivation, formulation, and methods of treating stochastic volatility using switching diffusions. Our main focus is on constructing asymptotic expansions to resolve the issue. More specifically, Section 9.3 constructs asymptotic expansions. Section 9.4 proceeds with validation of the expansions. Then, Section 9.5 switches gears to

H.-D. Nguyen et al., *Hybrid Switching Diffusions*, Probability Theory and Stochastic Modelling 63, https://doi.org/10.1007/978-3-031-93303-5_9

treat the second part of this chapter, namely, hybrid diffusions whose discrete component lives in weakly connected "ergodic" (irreducible) classes. Ergodicity of such systems is considered. Finally, the chapter is concluded in Section 9.6 with a few more remarks.

9.2 Stochastic Volatility Using Regime-Switching Diffusions

Nowadays, financial derivatives such as options are used widely. The Nobel prize winning Black-Scholes (also called Black-Scholes-Merton) formula provides an important tool for pricing options on a basic equity. It has encouraged and facilitated the union of mathematics, finance, computational sciences, and economics. On the other hand, it has been recognized, especially by practitioners in the financial market, that the assumption of constant volatility, which is essential in the Black-Scholes formula, is a less-than-perfect description of the real world. To capture the behavior of stock prices and other derivatives, there has been much effort in taking into account frequent volatility changes. It has been recognized that it is more suitable to use a stochastic process to model volatility variations. In [132], instead of the usual GBM (geometric Brownian motion) model, a second stochastic differential equation was introduced to describe the random environments. Such a formulation is known as a stochastic volatility model.

What happens if the stochastic volatility undergoes fast mean reversion? To answer this question, in [89] and the subsequent papers [90, 91], a class of volatility models was studied in detail. Under the setup of mean reversion, two-time-scale methods were used. The rationale is to identify the important groupings of market parameters. It also reveals that the Black-Scholes formula is a "first approximation" to such fast-varying volatility models by assuming the volatility being a function $f(\cdot)$ of a fast-varying diffusion that is mean reverting (or ergodic). The mean reversion implies although rapidly varying, the volatility does not blow up. By exploiting the time-scale separation, it was shown in [89, 91] that the "slow" component (the leading term or the zeroth-order outer expansion term in the approximation) of the option prices can be approximated by a Black-Scholes differential equation with constant volatility $\overline{f}$ where $\overline{f}^2$ is the average of $f^2(\cdot)$ with respect to the stationary measure of the "fast" component. Moreover, using a singular perturbation approach, the next term (the first-order outer expansion term) in the asymptotic expansion was also found. For convenience, the volatility was assumed to be driven by an Ornstein–Uhlenbeck (OU) process in [89, 90]. The fast mean reversion has been further examined in [162] with more general models. A full asymptotic series with uniform error bounds was obtained; see also related diffusion approximation in [230] and related asymptotic expansions for diffusion processes [160, 161].

Along another line, increasing attention has been drawn to modeling, analysis, and computing using regime-switching models [289], which are alternatives to the stochastic volatility models mentioned above. They present an effective way to model stochastic volatility with simpler structures. The use of the Markov chains is much simpler than the use of a second stochastic differential equation. This is motivated by the desire to use regime switching to describe uncertainty and stochastic volatility. Nowadays, it has been well recognized that due to stochastic volatility, a phenomenon known as the volatility smile arises. There has been an effort to provide better models to replicate the "smile." In [289], the authors reproduced the volatility smile successfully by using regime-switching models easily. Earlier efforts in modeling and analysis of regime-switching models can be found in [10, 62, 314] among others.

To some large extent, our work here is motivated by [89] for fast mean reversion processes in the driving force of stochastic volatility. Nevertheless, rather than using an additional SDE to represent the stochastic volatility, we use the same setup of a regime-switching model as that of [289]. Therefore, it accommodates the desired goal from a different angle. Here, the fast mean reversion is captured by a fast-varying continuous-time Markov chain. We demonstrate that the model under consideration leads to "effective volatility," which is an average with respect to the stationary distribution of the fast-changing jump process. In mathematical terms, the fast reversion corresponds to the Markov chain being ergodic with fast variations. Using a two-time-scale formulation, the problem itself centers around fast-varying switching diffusion processes. The approach that we are using is analytical. We focus on developing asymptotic expansions that are approximation of solutions of systems of parabolic equations and aim to obtain such expansions with uniform asymptotic error bounds on a compact set for the continuous component. It should be noted that if one is only interested in convergence in the pointwise sense (termed pointwise asymptotic expansions henceforth) rather than obtaining uniform error bounds, then one can proceed similarly to the diffusion counterpart in [89, 91] to obtain "effective volatility." It should also be noted that the asymptotic expansions presented in this chapter give some new insight on the construction of approximation of solutions of backward type systems of PDEs, which is different from [303] where probability distributions were considered. The methods presented are constructive; the asymptotic analysis techniques are interesting in their own right.

9.2.1 Time-Scale Separation

In the rest of the book, r is often used as the dimension of the continuous state variables (i.e., $\mathbb{R}^r$ is used as the space for the continuous state variable). In this chapter, however, to adopt the traditional convention, we use r as the interest rate throughout. We consider a basic equity, a stock whose price is given by $S(t)$. Different from the traditional geometric Brownian motion setup, we assume that the price follows a switching diffusion model. Suppose that $\alpha(t)$ is a continuous-time Markov chain with generator $Q(t)$ and a finite state space $\mathcal{M} = \{1, \dots, m_0\}$, where

m_0 is a finite positive integer. The price of the stock is a solution of the stochastic differential equation:

$$dS(t) = \mu(\alpha(t))S(t)dt + \sigma(\alpha(t))S(t)dw(t), \tag{9.1}$$

where $w(\cdot)$ is a standard Brownian motion independent of $\alpha(\cdot)$, and $\mu(\cdot)$ and $\sigma(\cdot)$ are the appreciation rate and the volatility rate, respectively. Such a model is frequently referred to as a regime-switching asset model. Note that in the above, both the appreciation rate and volatility rate are functions of the Markov chain.

Consider a European type of call option. The time of maturity is denoted by T, and the payoff at time T is given by $H(S)$, a nonnegative function. Denote $h(S) = H(e^S)$. Suppose that the associated risk-free interest rate is given by r. A standard approach in option pricing is risk-neutral valuation. The rationale is to derive a suitable probability space on which the expected rate of return of all securities is equal to the risk-free interest rate. The mathematical requirement is that the discounted asset price is a martingale. The associated probability space is referred to as the risk-neutral world. The price of the option on the asset is then the expected value, with respect to this martingale measure, of the discounted option payoff. In what follows, we use the so-called risk-neutral probability measure; see, e.g., [89] and [131] among others for diffusion processes and [289] for switching diffusion processes. To proceed, we first present a lemma, which is essentially a generalized Girsanov theorem for Markov-modulated processes. The results of this type are known; a proof can be found in [289].

Lemma 9.1 *The following assertions hold.*

(a) *Suppose that $\sigma(i) > 0$ for each $i \in \mathcal{M}$, and let*

$$\widetilde{w}(t) := w(t) - \int_0^t \frac{r - \mu(\alpha(u))}{\sigma(\alpha(u))} du.$$

Then, $\widetilde{w}(\cdot)$ is a $\widetilde{\mathbb{P}}$-Brownian motion, where $\widetilde{\mathbb{P}}$ is known as the risk-neutral probability measure; see [89, 289] *among others.*

(b) *$S(0)$, $\alpha(\cdot)$, and $\widetilde{w}(\cdot)$ are mutually independent under $\widetilde{\mathbb{P}}$.*

(c) *(Itô's lemma) For each $i \in \mathcal{M}$ and each $g(\cdot, \cdot, i) \in C^{2,1}$, we have*

$$\begin{aligned} g(S(s), s, \alpha(s)) = g(S(t), t, \alpha(t)) &+ \int_t^s \mathcal{L}g(S(u), u, \alpha(u))du \\ &+ M(s) - M(t), \end{aligned}$$

where $M(\cdot)$ is a $\widetilde{\mathbb{P}}$ local martingale and $\mathcal{L}$ is a generator given by

$$\begin{aligned} \mathcal{L}g(S, t, i) = \frac{\partial}{\partial t} g(S, t, i) &+ \frac{1}{2} S^2 \sigma^2(i) \frac{\partial^2}{\partial S^2} g(S, t, i) \\ &+ rS \frac{\partial}{\partial S} g(S, t, i) + Q(t) g(S, t, \cdot)(i), \end{aligned} \tag{9.2}$$

with

$$Q(t)g(S,t,\cdot)(i) = \sum_{j=1}^{m_0} q_{ij}(t)g(S,t,j). \tag{9.3}$$

Note that $(S(t), \alpha(t))$ is a Markov process with generator $\mathcal{L}$ (with a slight abuse of notation). Strictly speaking, we would need to write $\mathcal{L}(t)$ because $Q(t)$ is time dependent; we suppress the t dependence for simplicity. To proceed, as alluded to in the introduction, in reference to [89], we also consider a fast mean reverting driving process, but the driving process is a continuous-time Markov chain not a diffusion.

By introducing a small parameter $\varepsilon > 0$, we aim to show that the regime-switching model is also a good approximation of the Black-Scholes model. To this end, suppose that the generator of the Markov chain is given by $Q^\varepsilon(t) = (q^\varepsilon_{ij}(t)) \in \mathbb{R}^{m_0 \times m_0}$:

$$Q^\varepsilon(t) = \frac{Q(t)}{\varepsilon}, \tag{9.4}$$

where $Q(t)$ is the generator of a continuous-time Markov chain. To highlight the ε-dependence, we denote the Markov chain by $\alpha(t) = \alpha^\varepsilon(t)$.

We are mainly interested in the uniform asymptotic expansions for the option price. To obtain such uniform asymptotic expansions, we need to have the continuous component of the switching diffusion be in a compact set. For simplicity, we take the compact set to be $[0, 1]$; see [162] for a comment on the corresponding stochastic volatility model without switching. We use this as a standing assumption throughout this chapter. Next, we state a couple of additional conditions.

(H9.1) There is a $T_1 \geq T > 0$, and for all $t \in [0, T_1]$, the generator $Q(t)$ is weakly irreducible, where T is the time of maturity of the option price as indicated earlier. There is an $n \geq 1$ such that $Q(\cdot) \in C^{n+2}[0, T_1]$.

(H9.2) The function $h(\cdot)$ is sufficiently smooth and vanishes outside of a compact set.

Remark 9.2 Recall that by weak irreducibility of the switching process, we mean that the system of equations

$$\begin{cases} \nu(t)Q(t) = 0 \\ \nu(t)\mathbb{1} = 1 \end{cases} \tag{9.5}$$

has a unique solution $\nu(t) = (\nu_1(t), \dots, \nu_{m_0}(t)) \in \mathbb{R}^{1\times m_0}$ satisfying $\nu_i(t) \geq 0$ for each $i \in \mathcal{M}$. Such a nonnegative solution is termed a quasi-stationary distribution; see [303].

Condition (H9.1) indicates that the Markov chain has a unique quasi-stationary distribution $\nu(t) = (\nu_1(t), \dots, \nu_{m_0}(t)) \in \mathbb{R}^{1\times m_0}$. Our assumption on the operator $\mathcal{L}^\varepsilon$ then leads to the equity undergoing fast mean reverting switching.

Following [89, §5.4], we have assumed $h(S)$ to be a bounded and smooth function in (H9.2). For European options, if it is a call option, $h(S) = (S - K_0)^+$, and if it is a put option, $h(S) = (K_0 - S)^+$ where K_0 is the exercise price. Thus, the function $h(\cdot)$ is not smooth and is unbounded. In [91], using regularization or smoothing techniques, asymptotic expansions up to the order $O(\varepsilon)$ were obtained without assuming smoothness and boundedness of the payoff function $h(S)$. Note that the accuracy of the approximation was obtained only for a fixed state variable x and time variable t with $t < T$ in [91], however. In this chapter, as in [89], we confine our attention to the smooth and bounded payoff $h(S)$ and prove the *uniform* in S accuracy of approximation. It appears that the unbounded and nonsmooth payoff function $h(S)$ can be handled and the formal asymptotic expansions can be obtained, but such expansions are not uniform in S. Moreover, different methods and ideas must be used to justify the expansions as commented on in [162].

Let

$$\begin{aligned} V^\varepsilon(S,t,i) &= \mathbb{E}_{S,i}[\exp(-r(T-t))h(S^\varepsilon(T))] \\ &= \mathbb{E}[\exp(-r(T-t))h(S^\varepsilon(T))|S^\varepsilon(t) = S, \alpha^\varepsilon(t) = i]. \end{aligned} \tag{9.6}$$

The option price can be characterized by the following system of partial differential equations:

$$\begin{aligned} &\frac{\partial V^\varepsilon(S,t,i)}{\partial t} + \frac{1}{2}\sigma^2(i)S^2\frac{\partial^2 V^\varepsilon(S,t,i)}{\partial S^2} + rS\frac{\partial V^\varepsilon(S,t,i)}{\partial S} \\ &\qquad - rV^\varepsilon(S,t,i) + Q^\varepsilon(t)V^\varepsilon(S,t,\cdot)(i) = 0, \ i \in \mathcal{M}, \end{aligned} \tag{9.7}$$

which is a generalization of the usual Black-Scholes PDE. Associated with the above system of PDEs, we define an operator $\mathcal{L}^\varepsilon$. For $i \in \mathcal{M}$, and each $g(\cdot,\cdot,i) \in C^{2,1}$, let

$$\begin{aligned} \mathcal{L}^\varepsilon g(S,t,i) =& \frac{\partial g(S,t,i)}{\partial t} + \frac{1}{2}\sigma^2(i)S^2\frac{\partial^2 g(S,t,i)}{\partial S^2} \\ &+ rS\frac{\partial g(S,t,i)}{\partial S} - rg(S,t,i) + Q^\varepsilon(t)g(S,t,\cdot)(i). \end{aligned} \tag{9.8}$$

Now, the setup of the problem is complete. We proceed to obtain the approximation of the option price by means of asymptotic expansions.

9.3 Asymptotic Expansions

What can one say about the effect of $\alpha^\varepsilon(\cdot)$? Is there an "effective volatility?" Is the Black-Scholes formula still a reasonable approximation to such regime-switching models? To answer these questions, we develop an asymptotic series using analytic techniques. We seek asymptotic expansions of $V^\varepsilon(S,t,i)$ of the form

$$\Phi_n^\varepsilon(S,t,i)+\Psi_n^\varepsilon(S,\tau,i)=\sum_{k=0}^{n}\varepsilon^k\varphi_k(S,t,i)+\sum_{k=0}^{n}\varepsilon^k\psi_k(S,\tau,i),\ \ i\in\mathcal{M}, \tag{9.9}$$

where τ is a stretched-time variable defined by

$$\tau=\frac{T-t}{\varepsilon}.$$

The $\varphi_k(\cdot)$ are called regular terms or outer expansion terms, and the $\psi_k(\cdot)$ are the terminal layer corrections (also referred to as boundary layer correction terms by many people in singular perturbations). In this problem, the terminal layer correction terms are particularly useful for behavior of the option price near the time of maturity.

We aim to obtain asymptotic expansions of order n and derive the uniform error bounds. For the purposes of error estimates, we need to calculate a couple of more terms for analysis reasons.

First, let us look at $\Phi_n^\varepsilon(S,t,i)$, the regular part of the asymptotic expansions. Substituting it into (9.7), and comparing coefficients of like powers of ε^k for $0\le k\le n+1$, we obtain

$$\begin{aligned}
&Q(t)\varphi_0(S,t,\cdot)(i)=0,\\
&\frac{\partial\varphi_0(S,t,i)}{\partial t}+\frac{1}{2}\sigma^2(i)S^2\frac{\partial^2\varphi_0(S,t,i)}{\partial S^2}+rS\frac{\partial\varphi_0(S,t,i)}{\partial S}\\
&\quad-r\varphi_0(S,t,i)+Q(t)\varphi_1(S,t,\cdot)(i)=0,\\
&\quad\vdots\\
&\frac{\partial\varphi_k(S,t,i)}{\partial t}+\frac{1}{2}\sigma^2(i)S^2\frac{\partial^2\varphi_k(S,t,i)}{\partial S^2}+rS\frac{\partial\varphi_k(S,t,i)}{\partial S}\\
&\quad-r\varphi_k(S,t,i)+Q(t)\varphi_{k+1}(S,t,\cdot)(i)=0.
\end{aligned} \tag{9.10}$$

To ensure that the terminal conditions are met, we choose

$$\begin{aligned}
&\varphi_0(S,T,i)+\psi_0(S,0,i)=h(S),\quad i\in\mathcal{M},\\
&\varphi_k(S,T,i)+\psi_k(S,0,i)=0,\quad i\in\mathcal{M}\text{ and }k\ge 1.
\end{aligned} \tag{9.11}$$

Taking a Taylor expansion of $Q(\cdot)$ around T, we obtain

$$Q(t)=Q(T-\varepsilon\tau)=\sum_{k=0}^{n+1}\frac{(-1)^k(\varepsilon\tau)^k}{k!}\frac{d^kQ(T)}{dt^k}+R_{n+1}(T-\varepsilon\tau), \tag{9.12}$$

where by assumption (H9.1), it can be shown that $R_{n+1}(t)=O(t^{n+1})$ uniformly in $t\in[0,T]$.

Similar to the equations in (9.10) for $\Phi^{\varepsilon}_{n+1}(S,t,i)$, we can substitute $\Psi^{\varepsilon}_{n+1}(S,\tau,i)$ into (9.7). This results in

$$
\begin{aligned}
&-\frac{\partial\psi_0(S,\tau,i)}{\partial\tau}+Q(T)\psi_0(S,\tau,\cdot)(i)=0,\\
&-\frac{\partial\psi_1(S,\tau,i)}{\partial\tau}+Q(T)\psi_1(S,\tau,\cdot)(i)+\Big[-\varepsilon\tau Q^{(1)}(T)\psi_0(S,\tau,\cdot)(i)\\
&\quad+\frac{1}{2}\sigma^2(i)S^2\frac{\partial^2\psi_0(S,\tau,i)}{\partial S^2}+rS\frac{\partial\psi_0(S,\tau,i)}{\partial S}-r\psi_0(S,\tau,i)\Big]=0,\\
&\quad\vdots\\
&-\frac{\partial\psi_k(S,\tau,i)}{\partial\tau}+Q(T)\psi_k(S,\tau,\cdot)(i)+\widetilde{R}_k(S,\tau,i)=0,\ \text{for } k=2,\dots,n+1,
\end{aligned}
\tag{9.13}
$$

where the remainder term is given by

$$
\begin{aligned}
\widetilde{R}_k(S,\tau,i)=&\sum_{j=1}^{k}\frac{(-1)^j(\tau)^j}{j!}\frac{d^jQ(T)}{dt^j}\psi_{k-j}(S,\tau,\cdot)(i)-r\psi_{k-1}(S,\tau,i)\\
&+\frac{1}{2}\sigma^2(i)S^2\frac{\partial^2\psi_{k-1}(S,\tau,i)}{\partial S^2}+rS\frac{\partial\psi_{k-1}(S,\tau,i)}{\partial S}.
\end{aligned}
\tag{9.14}
$$

Our task to follow is to construct the sequences $\{\varphi_k(S,t)\}$ and $\{\psi_k(S,\tau)\}$.

9.3.1 Construction of $\varphi_0(S,t,i)$ and $\psi_0(S,\tau,i)$

We claim that $\varphi_0(S,t,i)$ must be independent of i. To see this, write

$$\varphi_0(S,t)=(\varphi_0(S,t,1),\dots,\varphi_0(S,t,m_0))'\in\mathbb{R}^{m_0\times 1}.$$

Then, the first equation in (9.10) may be written as $Q(t)\varphi_0(S,t)=0$. This equation in turn implies that $\varphi_0(S,t)$ is in the null space of $Q(t)$. Because $Q(t)$ is weakly irreducible, $\text{rank}(Q(t))=m_0-1$, so the null space is one dimensional. Consequently, the null space is spanned by $\mathbb{1}=(1,\dots,1)'\in\mathbb{R}^{m_0\times 1}$. This yields $\varphi_0(S,t)=\gamma_0(S,t)\mathbb{1}$, where $\gamma_0(S,t)$ is a real-valued function and hence $\varphi_0(S,t,i)$ is a function independent of i.

Using the above argument in the second equation of (9.10), and multiplying from the left through the equation by $\nu_i(t)$, where $(\nu_1(t),\dots,\nu_{m_0}(t))$ is the quasi-stationary distribution associated with $Q(t)$, and summing over $i\in\mathcal{M}$, we obtain

$$\frac{\partial\gamma_0(S,t)}{\partial t}+\frac{1}{2}\overline{\sigma}^2(t)S^2\frac{\partial^2\gamma_0(S,t)}{\partial S^2}+rS\frac{\partial\gamma_0(S,t)}{\partial S}-r\gamma_0(S,t)=0,\tag{9.15}$$

where

$$\overline{\sigma}^2(t) = \sum_{i=1}^{m_0} \nu_i(t)\sigma^2(i).$$

The terminal condition is given by

$$\gamma_0(S, T) = h(S). \tag{9.16}$$

Then, (9.15) together with the terminal condition (9.16) has a unique solution.

Remark 9.3 Note that $\gamma_0(S, t)$ satisfies the Black-Scholes partial differential equation, in which the coefficients are averaged out with respect to the stationary distributions of the Markov chain. The result reveals that the Black-Scholes formula, indeed, is a first approximation to the option model under regime switching. Thus, the regime-switching model can be thought of as another stochastic volatility model.

In view of (9.11) and (9.13), $\psi_0(S, \tau, i)$ is obtained from the first equation in (9.13) together with the condition $\psi_0(S, 0, i) = 0$. It follows that $\psi_0(S, \tau, i) = 0$ for each $i \in \mathcal{M}$.

9.3.2 Construction of $\varphi_1(S, t, i)$ and $\psi_1(S, \tau, i)$

Now, we proceed to obtain $\varphi_1(S, t, i)$ and $\psi_1(S, \tau, i)$. Similar to $\varphi_0(S, t)$, for $i \le n + 1$, define

$$\begin{aligned}
\varphi_i(S, t) &= (\varphi_i(S, t, 1), \ldots, \varphi_i(S, t, m_0))' \in \mathbb{R}^{m_0}, \\
\psi_i(S, \tau) &= (\psi_i(S, \tau, 1), \ldots, \psi_i(S, \tau, m_0))' \in \mathbb{R}^{m_0}.
\end{aligned}$$

Consider $\varphi_1(S, t)$ in (9.10). It is a solution of a nonhomogeneous algebraic equation as in (A.10) in the appendix of this book. By virtue of Lemma A.12, we can write $\varphi_1(S, t, i)$ as the sum of a general solution of the associated homogeneous equation and the unique solution of the inhomogeneous equation that is orthogonal to the stationary distribution $\nu(t)$. That is,

$$\varphi_1(S, t, i) = \gamma_1(S, t) + \varphi_1^0(S, t, i), \quad i \in \mathcal{M},$$

where $\{\varphi_1^0(S, t, i) : i \in \mathcal{M}\}$ is the unique solution of (9.10) satisfying

$$\sum_{i=1}^{m_0} \nu_i(t)\varphi_1^0(S, t, i) = 0.$$

Equivalently, denote

$$\varphi_1^0(S,t) = (\varphi_1^0(S,t,1), \ldots, \varphi_1^0(S,t,m_0))' \in \mathbb{R}^{m_0}.$$

Then

$$\varphi_1(S,t) = \gamma_1(S,t)\mathbb{1} + \varphi_1^0(S,t), \tag{9.17}$$

where $\varphi_1^0(S,t)$ is the unique solution of the system of equations

$$\begin{aligned} Q(t)\varphi_1^0(S,t) &= F_0(S,t) \\ \nu(t)\varphi_1^0(S,t) &= 0, \end{aligned} \tag{9.18}$$

where

$$F_0(S,t) = (F_0(S,t,1), \ldots, F_0(S,t,m_0))' \in \mathbb{R}^{m_0},$$

with

$$F_0(S,t,i) = -\Big[\frac{\partial \gamma_0(S,t)}{\partial t} + \frac{\sigma^2(i)S^2}{2}\frac{\partial \gamma_0(S,t)}{\partial S^2} + rS\frac{\partial \gamma_0(S,t)}{\partial S} - r\gamma_0(S,t)\Big], \tag{9.19}$$

by Lemma A.12. That is, $\varphi_1(S,t)$ is the sum of a general solution of the homogeneous equation plus a particular solution that satisfies the orthogonality condition.

Note that $\gamma_1(S,t)$ has not been determined yet. We proceed to obtain it from the next equation in (9.10). By substituting (9.17) into the next equation in (9.10) and premultiplying it by ν, we obtain

$$\frac{\partial \gamma_1(S,t)}{\partial t} + \frac{1}{2}\overline{\sigma}^2(t)S^2\frac{\partial^2 \gamma_1(S,t)}{\partial S^2} + rS\frac{\partial \gamma_1(S,t)}{\partial S} - r\gamma_1(S,t) = \overline{F}_1(S,t), \tag{9.20}$$

where $\overline{F}_1(S,t) = \nu(t)F_1(S,t)$, with

$$F_1(S,t) = (F_1(S,t,1), \ldots, F_1(S,t,m_0))' \in \mathbb{R}^{m_0},$$

and

$$F_1(S,t,i) = -\Big[\frac{\partial \varphi_1^0(S,t,i)}{\partial t}\frac{\sigma^2(i)S^2}{2}\frac{\partial \varphi_1^0(S,t,i)}{\partial S^2} + rS\frac{\partial \varphi_1^0(S,t,i)}{\partial S} - r\varphi_1^0(S,t,i)\Big]. \tag{9.21}$$

It is easily seen that (9.20) is a uniquely solvable Cauchy problem if the terminal condition $\gamma_1(S,T) \in \mathbb{R}$ is specified. We determine this by matching the terminal condition with the terminal layer term $\psi_1(S,0)$.

Since $\psi_0(S,\tau)=0$, from the second equation in (9.13), $\psi_1(S,\tau)$ satisfies

$$\frac{\partial\psi_1(S,\tau)}{\partial\tau}=Q(T)\psi_1(S,\tau). \tag{9.22}$$

Choose $\psi_1(S,0)=\psi_1^0(S)$ such that $\overline{\psi}_1^0(S)=\nu\psi_1^0(S)=0$, where $\nu=\nu(T)$, the stationary distribution associated with $Q(T)$. Here and hereafter, we always use ν to denote $\nu(T)$ for notational simplicity. Multiplying (9.22) from the left by ν, we obtain

$$\begin{aligned}&\frac{\partial\overline{\psi}_1(S,\tau)}{\partial\tau}=0\in\mathbb{R},\\&\overline{\psi}_1(S,0)=\overline{\psi}_1^0(S)=0,\end{aligned} \tag{9.23}$$

where $\overline{\psi}_1(S,\tau)=\nu\psi_1(S,\tau)$. The solution of (9.23) is thus given by

$$\overline{\psi}_1(S,\tau)=0.$$

Remark 9.4 The above may also be seen as

$$\psi_1(S,\tau)=\exp(Q(T)\tau)\psi_1^0(S)$$

for a chosen initial $\psi_1^0(S)$. Using

$$\exp(Q(T)\tau)=\sum_{j=0}^{\infty}(Q(T)\tau)^j/j!,$$

the orthogonality of $\nu Q(T)=0$ leads to

$$\nu\exp(Q(T)\tau)\psi_1^0(S)=\nu I\psi_1^0(S)=\nu\psi_1^0(S)=\overline{\psi}_1^0(S)$$

for each $\tau\geq 0$.

Note that once $\psi_1^0(S)$ is chosen, the $\varphi_1(S,t)$ and $\psi_1(S,\tau)$ will be determined. The choice of $\psi_1^0(S,0)$ enables us to obtain the exponential decay of $\psi_1(S,\tau)$ easily.

In view of the defining equation for $\varphi_1(S,t)$, Eqs. (9.17)–(9.20), and the terminal condition (9.11), Eq. (9.20) together with the terminal condition $\gamma_1(S,T)=-\overline{\psi}_1^0(S)=0$ has a unique solution. Up to now, $\varphi_1(S,t)$ has been completely determined. Next, we proceed to find $\psi_1(S,\tau)$. Condition (9.11) implies that

$$\psi_1(S,0)=-\varphi_1(S,T)=\overline{\psi}_1^0(S)\mathbb{1}-\varphi_1^0(S,T)=-\varphi_1^0(S,T). \tag{9.24}$$

It follows that the Cauchy problem given by (9.22) and (9.24) has a unique solution. We are in a position to derive the exponential decay property of $\psi_1(S,\tau)$.

Lemma 9.5 *For $\psi_1(S, \tau)$ obtained from the solution of* (9.22) *and* (9.24)*, we have that for some $K > 0$ and $\kappa_0 > 0$*

$$\sup_{S\in[0,1]} |\psi_1(S, \tau)| \le K \exp(-\kappa_0 \tau). \tag{9.25}$$

Proof The weak irreducibility of $Q(T)$ implies that $\exp(Q(T)\tau) \to \mathbb{1}\nu$ as $\tau \to \infty$, and

$$|\exp(Q(T)\tau) - \mathbb{1}\nu| \le \exp(-\kappa_0 \tau) \text{ for some } \kappa_0 > 0.$$

The solution of (9.22) and (9.24) yields that for each $S \in [0, 1]$ and for some $K > 0$

$$\begin{aligned} |\psi_1(S, \tau)| &= |\psi_1(S, \tau) - \mathbb{1}\nu\psi_1(S, 0)| \\ &= |[\exp(Q(T)\tau) - \mathbb{1}\nu]\psi_1(S, 0)| \\ &\le |\exp(Q(T)\tau) - \mathbb{1}\nu||\psi_1(S, 0)| \\ &\le K \exp(-\kappa_0 \tau). \end{aligned}$$

Furthermore, it is readily seen that the above estimate holds uniformly for $S \in [0, 1]$. The desired result thus follows. □

Remark 9.6 In what follows, for notational simplicity, K and κ_0 are generic positive constants. Their values may change for different appearances.

We are now in a position to obtain a priori bounds for the derivatives of $\psi_1(S, \tau)$. This is presented in the next lemma.

Lemma 9.7 *The function $\psi_1(S, \tau)$ satisfies*

$$\begin{aligned} \sup_{S\in[0,1]} \left| \frac{\partial \psi_1(S, \tau)}{\partial S} \right| &\le K \exp(-\kappa_0 \tau), \\ \sup_{S\in[0,1]} \left| \frac{\partial^2 \psi_1(S, \tau)}{\partial S^2} \right| &\le K \exp(-\kappa_0 \tau). \end{aligned}$$

Proof Consider

$$U(S, \tau) = \frac{\partial \psi_1(S, \tau)}{\partial S} \quad \text{and} \quad V(S, \tau) = \frac{\partial^2 \psi_1(S, \tau)}{\partial S^2}.$$

Then, U and V satisfy

$$\begin{aligned} \frac{\partial U(S, \tau)}{\partial \tau} &= Q(T)U(S, \tau), \quad U(S, 0) = \frac{\partial \psi_1(S, 0)}{\partial S}, \\ \frac{\partial V(S, \tau)}{\partial \tau} &= Q(T)V(S, \tau), \quad V(S, 0) = \frac{\partial^2 \psi_1(S, 0)}{\partial S^2}, \end{aligned} \tag{9.26}$$

$$\frac{\partial \overline{U}(S,\tau)}{\partial \tau} = 0, \quad \overline{U}(S,0) = 0,$$
$$\frac{\partial \overline{V}(S,\tau)}{\partial \tau} = 0, \quad \overline{V}(S,0) = 0,$$

respectively, where $\overline{U}(S,\tau) = \nu U(S,\tau)$ and $\overline{V}(S,\tau) = \nu V(S,\tau)$. Thus, we have

$$\overline{U}(S,\tau) = 0, \quad \overline{V}(S,\tau) = 0,$$

and

$$\begin{aligned} U(S,\tau) &= \exp(Q(T)\tau)U(S,0), \\ V(S,\tau) &= \exp(Q(T)\tau)V(S,0). \end{aligned}$$

It the follows that

$$|U(S,\tau)| = |\exp(Q(T)\tau)U(S,0) - \mathbb{1}\nu U(S,0)| \le K\exp(-\kappa_0\tau).$$

Likewise, $|V(S,\tau)| \le K\exp(-\kappa_0\tau)$ as desired. □

9.3.3 Construction of $\varphi_k(S,t)$ and $\psi_k(S,\tau)$

Denote $\varphi_k(S,t) \in \mathbb{R}^{m_0}$, $\varphi_k^0(S,t) \in \mathbb{R}^{m_0}$, $\gamma_k(S,t) \in \mathbb{R}$, and $\psi_k(S,\tau) \in \mathbb{R}^{m_0}$ as that of $\varphi_1(S,t)$, $\varphi_1^0(S,t)$, $\gamma_1(S,t)$, and $\psi_1(S,\tau)$, respectively. By induction, we proceed to obtain the terms $\varphi_k(S,t)$ and $\psi_k(S,\tau)$ for $1 < k \le n+1$. Similar to the last section, for $1 < k \le n+1$, denote

$$\begin{aligned} F_k(S,t) &= (F_k(S,t,1),\ldots,F_k(S,t,m_0))' \in \mathbb{R}^{m_0}, \\ F_k(S,t,i) &= -\Big[\frac{\partial \varphi_k^0(S,t,i)}{\partial t} + \frac{\sigma^2(i)S^2}{2}\frac{\partial \varphi_k^0(S,t,i)}{\partial S^2} \\ &\quad + rS\frac{\partial \varphi_k^0(S,t,i)}{\partial S} - r\varphi_k^0(S,t,i)\Big], \\ \overline{F}_k(S,t) &= \nu(t)F_k(S,t). \end{aligned} \tag{9.27}$$

Suppose that we have constructed $\varphi_{k-1}(S,t)$ and $\psi_{k-1}(S,\tau)$ for $1 < k \le n+1$. Then, similar to $\varphi_1(S,t)$ and $\psi_1(S,\tau)$, we can define $\varphi_k(S,t)$ and $\psi_k(S,\tau)$. Next, write $\varphi_k(S,t)$ as

$$\varphi_k(S,t) = \gamma_k(S,t)\mathbb{1} + \varphi_k^0(S,t), \tag{9.28}$$

where $\varphi_k^0(S,t)$ is the unique solution of the system of equations

$$\begin{aligned} &Q(t)\varphi_k^0(S,t) = F_{k-1}(S,t) \\ &\nu(t)\varphi_k^0(S,t) = 0, \end{aligned} \tag{9.29}$$

$\nu(t)$ is the quasi-stationary distribution associated with $Q(t)$, and $\gamma_k(S,t)$ satisfies

$$\frac{\partial \gamma_k(S,t)}{\partial t} + \frac{1}{2}\overline{\sigma}^2(t)S^2\frac{\partial^2 \gamma_k(S,t)}{\partial S^2} + rS\frac{\partial \gamma_k(S,t)}{\partial S} - r\gamma_k(S,t) = \overline{F}_k(S,t). \tag{9.30}$$

By virtue of Lemma A.12, there is a unique solution to (9.29). To determine the solution of (9.30) as a solution of the Cauchy problem, we select the terminal condition from the terminal layer $\psi_k(S,\tau)$ via the matching condition (9.11). Note that $\psi_k(S,\tau)$ satisfies

$$-\frac{\partial \psi_k(S,\tau)}{\partial \tau} + Q(T)\psi_k(S,\tau) + \widetilde{R}_k(S,\tau) = 0, \tag{9.31}$$

where $\widetilde{R}_k(S,\tau) = (\widetilde{R}_k(S,\tau,1),\ldots,\widetilde{R}_k(S,\tau,m_0))' \in \mathbb{R}^{m_0}$ with $\widetilde{R}_k(S,\tau,i)$ given by (9.14).

Since the construction of $\psi_2(S,\tau)$ is somewhat different from $\psi_1(S,\tau)$, we single it out first. In view of (9.13) and noting $\psi_0(S,\tau)=0$, $\psi_2(S,\tau)$ satisfies

$$\begin{aligned} \frac{\partial \psi_2(S,\tau)}{\partial \tau} &= Q(T)\psi_2(S,\tau) + \widetilde{R}_2(S,\tau) \\ &= Q(T)\psi_2(S,\tau) + \Big[-\tau Q^{(1)}(T)\psi_1(S,\tau) + \frac{1}{2}S^2\Sigma\frac{\partial^2\psi_1(S,\tau)}{\partial S^2} \\ &\qquad + rS\frac{\partial \psi_1(S,\tau)}{\partial S} - r\psi_1(S,\tau)\Big], \end{aligned} \tag{9.32}$$

where

$$\Sigma = \mathrm{diag}(\sigma(1),\ldots,\sigma(m_0)) \in \mathbb{R}^{m_0\times m_0}. \tag{9.33}$$

With given initial data $\psi_2(S,0)$, the solution of (9.32) is given by

$$\psi_2(S,\tau) = \exp(Q(T)\tau)\psi_2(S,0) + \int_0^\tau \exp(Q(T)(\tau-s))\widetilde{R}_2(S,s)ds. \tag{9.34}$$

The solution of (9.34) can be further expanded as

$$\begin{aligned} \psi_2(S,\tau) &= \mathbb{1}\nu\psi_2(S,0) + \mathbb{1}\int_0^\infty \nu\widetilde{R}_2(S,s)ds + [\exp(Q(T)\tau) - \mathbb{1}\nu]\psi_2(S,0) \\ &\quad - \mathbb{1}\nu\int_\tau^\infty \widetilde{R}_2(S,s)ds + \int_0^\tau [\exp(Q(T)(\tau-s)) - \mathbb{1}\nu]\widetilde{R}_2(S,s)ds. \end{aligned} \tag{9.35}$$

By virtue of Lemmas 9.5 and 9.7, $\widetilde{R}_2(S,\tau)$ decays exponentially fast to 0 so the integral $\int_0^\infty \widetilde{R}_2(S,s)ds$ is well defined. Furthermore

$$\Big|\int_\tau^\infty \widetilde{R}_2(S,s)ds\Big| \le K\exp(-\kappa_0\tau).$$

A similar argument to that of Lemma 9.5 shows that

$$|[\exp(Q(T)\tau) - \mathbb{1}\nu]\psi_2(S,0)| \le K\exp(-\kappa_0\tau).$$

In addition,

$$\begin{aligned}
&\Big|\int_0^\tau [\exp(Q(T)(\tau-s)) - \mathbb{1}\nu]\widetilde{R}_2(S,s)ds\Big| \\
&\le K\int_0^\tau |\exp(Q(T)(\tau-s)) - \mathbb{1}\nu||\widetilde{R}_2(S,s)|ds \\
&\le K\int_0^\tau \exp(-\kappa_0(\tau-s))\exp(-\kappa_0 s)ds \\
&\le K\tau\exp(-\kappa_0\tau).
\end{aligned}$$

Thus, $\psi_2(S,\tau)$ will decay to 0 exponentially fast if we choose

$$\mathbb{1}\overline{\psi}_2^0(S) = \mathbb{1}\nu\psi_2(S,0) = -\mathbb{1}\int_0^\infty \nu\widetilde{R}_2(S,s)ds. \tag{9.36}$$

Note that there is only one unknown in (9.36), namely, $\overline{\psi}_2(S,0) = \overline{\psi}_2^0(S)$. That is, (9.36) enables us to obtain $\overline{\psi}_2(S,0)$ uniquely. Using $\gamma_2(S,T) = -\overline{\psi}_2(S,0)$ together with (9.30) enables us to find the unique solution for the Cauchy problem (9.30) that satisfies $\overline{\psi}_2(S,0)$. Therefore, $\varphi_2(S,t)$ and $\psi_2(S,\tau)$ are completely determined. Moreover, the construction ensures that

$$\sup_{S\in[0,1]} |\psi_2(S,\tau)| \le K\exp(-\kappa_0\tau).$$

In addition, it can be shown that

$$\sup_{S\in[0,1]} \Big|\frac{\partial^i\psi_2(S,\tau)}{\partial S^i}\Big| \le K\exp(-\kappa_0\tau),\quad i=1,2.$$

Recall that for simplicity, we have used the convention that K and κ_0 are some positive real numbers. Their values may vary. Their precise values are not important but only the exponential decay property is crucial.

Proceeding in a similar way, we obtain $\psi_k(S,\tau)$ as follows. Suppose that $\psi_1(S,\tau),\ldots,\psi_{k-1}(S,\tau)$ have been constructed so that $\psi_j(S,\tau)$ for $j = 1,\ldots,k-1$ decay exponentially fast together with their first and second derivatives $(\partial/\partial S)\psi_j(S,\tau)$ and $(\partial^2/\partial S^2)\psi_j(S,\tau)$, respectively.

Consider

$$\frac{\partial \psi_k(S,\tau)}{\partial \tau} = Q(T)\psi_k(S,\tau) + \widetilde{R}_k(S,\tau). \tag{9.37}$$

The solution is then given by

$$\begin{aligned}
\psi_k(S,\tau) &= \exp(Q(T)\tau)\psi_k(S,0) + \int_0^\tau \exp(Q(T)(\tau - s))\widetilde{R}_k(S,s)ds \\
&= \mathbb{1}\nu\psi_k(S,0) + \mathbb{1}\int_0^\infty \nu\widetilde{R}_k(S,s)ds + [\exp(Q(T)\tau) - \mathbb{1}\nu]\psi_k(S,0) \\
&\quad - \mathbb{1}\nu\int_\tau^\infty \widetilde{R}_k(S,s)ds + \int_0^\tau [\exp(Q(T)(\tau - s)) - \mathbb{1}\nu]\widetilde{R}_k(S,s)ds.
\end{aligned} \tag{9.38}$$

Choose

$$\begin{aligned}
&\overline{\psi}_k^0(S) := \mathbb{1}\nu\psi_k(S,0) = -\mathbb{1}\int_0^\infty \nu\widetilde{R}_k(S,s)ds, \\
&\gamma_k(S,T) = -\overline{\psi}_k^0(S), \\
&\psi_k(S,0) = -\varphi_k(S,T) = \overline{\psi}_k^0(S)\mathbb{1} - \varphi_k(S,T).
\end{aligned} \tag{9.39}$$

Then, $\varphi_k(S,t)$ and $\psi_k(S,\tau)$ are completely specified. Moreover, we can establish the following lemmas.

Lemma 9.8 *The following assertions hold.*

- $\psi_k(S,\tau)$ *can be constructed by using the first line of* (9.38) *with terminal data* (9.39).
- ψ_k *decay exponentially fast in the sense*

$$\sup_{S\in[0,1]} |\psi_k(S,\tau)| \le K\exp(-\kappa_0\tau) \tag{9.40}$$

for some $K > 0$ *and* $\kappa_0 > 0$.

Lemma 9.9 *For* $k > 1$, *the functions* $\psi_k(S,\tau)$ *satisfy*

$$\begin{aligned}
&\sup_{S\in[0,1]} \left|\frac{\partial \psi_k(S,\tau)}{\partial S}\right| \le K\exp(-\kappa_0\tau), \quad \textit{and} \\
&\sup_{S\in[0,1]} \left|\frac{\partial^2 \psi_k(S,\tau)}{\partial S^2}\right| \le K\exp(-\kappa_0\tau).
\end{aligned}$$

Proof Redefine

$$U(S,\tau) = \frac{\partial \psi_k(S,\tau)}{\partial S} \quad \text{and} \quad V(S,\tau) = \frac{\partial^2 \psi_k(S,\tau)}{\partial S^2}.$$

Then, U and V satisfy

$$\begin{aligned}\frac{\partial U(S,\tau)}{\partial \tau} &= Q(T)U(S,\tau), \quad U(S,0)=\frac{\partial \psi_k(S,0)}{\partial S},\\ \frac{\partial V(S,\tau)}{\partial \tau} &= Q(T)V(S,\tau), \quad V(S,0)=\frac{\partial^2 \psi_k(S,0)}{\partial S^2},\end{aligned} \tag{9.41}$$

respectively. The rest of the proof is similar to that of Lemma 9.7. □

We summarize what we have obtained thus far and put it in the following theorem.

Theorem 9.10 *Under conditions* (H9.1) *and* (H9.2), *we can construct sequences* $\{\varphi_k(S,t,i) : i \in \mathcal{M}; k = 0,\dots,n\}$ *and* $\{\psi_k(S,\tau,i) : i \in \mathcal{M}; k = 0,\dots,n\}$ *such that*

- $\varphi_0(S,t) = \gamma_0(S,t)\mathbb{1}$ *with* $\gamma_0(S,t)$ *being the solution of* (9.15) *satisfying the terminal condition* (9.16); $\psi_0(S,\tau,i) = 0$ *for each* $i \in \mathcal{M}$,
- $\varphi_1(S,t,i)$ *is given by* (9.17) *with* $\varphi_1^0(S,t)$ *being the unique solution of* (9.18) *and* $\gamma_1(S,t)$ *given by* (9.20) *with* $\gamma_1(S,T) = 0$ *and the terminal layer term* $\psi_1(S,\tau,i)$ *specified in Lemma* 9.5,
- $\varphi_k(S,t)$ *is given by* (9.28) *with* $\varphi_k^0(S,t)$ *being the unique solution of* (9.29) *and* $\gamma_k(S,t)$ *being the solution of* (9.30) *with* $\gamma(S,T)$ *given in Lemma* 9.8 *and the terminal layer term* $\psi_k(S,\tau,i)$ *specified in Lemma* 9.8.

9.4 Asymptotic Error Bounds

We have constructed the formal asymptotic expansions of the option price. Here, we validate the expansions by deriving

$$\max_{i\in\mathcal{M}} \sup_{(S,t)\in[0,1]\times[0,T]} \left| \sum_{k=0}^{n} \varepsilon^k \varphi_k(S,t,i) + \sum_{k=0}^{n} \varepsilon^k \psi_k(S,\tau,i) - V^\varepsilon(S,t,i) \right| = O(\varepsilon^{n+1}).$$

We first deduce a couple of lemmas.

Lemma 9.11 *For each* $i \in \mathcal{M}$, *let* $u(\cdot,\cdot,i) \in C^{2,1}([0,1]\times[0,T];\mathbb{R})$ *such that* $v(S,t) = (v(S,t,1),\dots,v(S,t,m_0))'$ *and*

$$\begin{aligned}&\mathcal{L}^\varepsilon u(S,t,i) = v(S,t,i), \ t < s \le T,\\ &u(S,s,i) = 0, \ S \in [0,1],\end{aligned}$$

where $\mathcal{L}^\varepsilon$ *is defined in* (9.8).

Assume (H9.1) *and* (H9.2). *Then*

$$u(S,t,i) = -\mathbb{E}\int_t^S v(S^\varepsilon(\xi),\xi,\alpha^\varepsilon(\xi))d\xi, \tag{9.42}$$

where $(S^{\varepsilon,S}(t), \alpha^{\varepsilon,i}(t)) = (S,i)$.

Proof Note that $S \in [0,1]$ and $t \in [0,T]$ for some finite $T > 0$ so $v(\cdot)$ is bounded together with its derivatives with respect to S up to the second order and its derivative with respect to t. In view of (9.2) and (9.3), $u(\cdot)$ is integrable. Moreover, the local martingale in Lemma 9.1 is in fact a martingale now. The desired result then follows from Dynkin's formula. □

Lemma 9.12 *Suppose that* (H9.1) *and* (H9.2) *hold and for each* $i \in \mathcal{M}$, $e^\varepsilon(\cdot,\cdot,i)$ *is a suitable function such that*

$$\sup_{(S,t)\in[0,1]\times[0,T]} |e^\varepsilon(S,t,i)| = O(\varepsilon^\ell) \text{ for } \ell \le n+1. \tag{9.43}$$

Then, for each $i \in \mathcal{M}$, *the solution of*

$$\mathcal{L}^\varepsilon u^\varepsilon(S,t,i) = e^\varepsilon(S,t,i),\ \ u^\varepsilon(S,T,i) = 0 \tag{9.44}$$

satisfies

$$\sup_{(S,t)\in[0,1]\times[0,T]} |u^\varepsilon(S,t,i)| = O(\varepsilon^\ell).$$

Proof In view of Lemma 9.11, the solution of (9.44) can be written as

$$u^\varepsilon(S,t,i) = -\mathbb{E}\int_0^t e^\varepsilon(S^\varepsilon(\xi),\xi,\alpha^\varepsilon(\xi))d\xi.$$

Thus, (9.43) leads to

$$\sup_{(S,t)\in[0,1]\times[0,T]} |u^\varepsilon(S,t,i)| \le K\int_0^T O(\varepsilon^\ell) \le O(\varepsilon^\ell).$$

The desired result thus follows. □

With the preparation above, we proceed to obtain the desired upper bounds on the approximation errors. For $k = 0,\ldots,n+1$, define a sequence of approximation errors:

$$e_k^\varepsilon(S,t,i) = \Phi_k^\varepsilon(S,t,i) + \Psi_k^\varepsilon(S,\tau,i) - V^\varepsilon(S,t,i),\ i \in \mathcal{M},$$

where for each $i \in \mathcal{M}$, $V^\varepsilon(S,t,i)$ is the option price given by (9.6) and $\Phi_k^\varepsilon(S,t,i)+\Psi_k^\varepsilon(S,\tau,i)$ is the kth-order approximation to the option price. We proceed to obtain the order of magnitude estimates of $e_n^\varepsilon(S,t,i)$.

Theorem 9.13 *Assume* (H9.1) *and* (H9.2). *Then, for the asymptotic expansions constructed in Theorem* 9.10,

$$\max_{i\in\mathcal{M}}\sup_{(S,t)\in[0,1]\times[0,T]}|e_n^\varepsilon(S,t,i)| = O(\varepsilon^{n+1}). \tag{9.45}$$

Proof The proof is divided in two steps. In the first step, we obtain an estimate on $\mathcal{L}^\varepsilon e_{n+1}^\varepsilon(S,t,i)$, and in the second step, we derive the desired order estimate.

Step 1. We show that $\mathcal{L}^\varepsilon e_{n+1}^\varepsilon(S,t,i) = O(\varepsilon^{n+1})$. To obtain this, first note that $\mathcal{L}^\varepsilon V^\varepsilon(S,t,i) = 0$. Thus

$$\begin{aligned}\mathcal{L}^\varepsilon e_{n+1}^\varepsilon(S,t,i) &= \mathcal{L}^\varepsilon \Phi_{n+1}^\varepsilon(S,t,i) + \mathcal{L}^\varepsilon \Psi_{n+1}^\varepsilon(S,t,i)\\ &= \sum_{k=0}^{n+1}\varepsilon^k\mathcal{L}^\varepsilon\varphi_k(S,t,i) + \sum_{j=1}^{n+1}\varepsilon^k\mathcal{L}^\varepsilon\psi_k(S,\tau,i),\end{aligned}$$

where in the last line above, we have used $\psi_0(S,\tau,i) = 0$ for each $i \in \mathcal{M}$. For the outer expansions, we have

$$\begin{aligned}&\mathcal{L}^\varepsilon\sum_{k=0}^{n+1}\varepsilon^k\varphi_k(S,t,i)\\
&= \sum_{k=0}^{n+1}\varepsilon^k\Big[\frac{\partial\varphi_k(S,t,i)}{\partial t} + \frac{1}{2}\sigma^2(i)S^2\frac{\partial^2\varphi_k(S,t,i)}{\partial S^2}\\
&\quad + rS\frac{\partial\varphi_k(S,t,i)}{\partial S} - r\varphi_k(S,t,i) + \frac{Q(t)}{\varepsilon}\varphi_k(S,t,\cdot)(i)\Big]\\
&= \sum_{k=0}^{n}\varepsilon^k\Big[\frac{\partial\varphi_k(S,t,i)}{\partial t} + \frac{1}{2}\sigma^2(i)S^2\frac{\partial^2\varphi_k(S,t,i)}{\partial S^2}\\
&\quad + rS\frac{\partial\varphi_k(S,t,i)}{\partial S} - r\varphi_k(S,t,i) + Q(t)\varphi_{k+1}(S,t,\cdot)(i)\Big]\\
&\quad + \frac{Q(t)}{\varepsilon}\varphi_0(S,t,\cdot)(i) + \varepsilon^{n+1}\Big[\frac{\partial\varphi_{n+1}(S,t,i)}{\partial t}\\
&\quad + \frac{1}{2}\sigma^2(i)S^2\frac{\partial^2\varphi_{n+1}(S,t,i)}{\partial S^2} + rS\frac{\partial\varphi_{n+1}(S,t,i)}{\partial S} - r\varphi_{n+1}(S,t,i)\Big]\\
&= \varepsilon^{n+1}\Big[\frac{\partial\varphi_{n+1}(S,t,i)}{\partial t} + \frac{1}{2}\sigma^2(i)S^2\frac{\partial^2\varphi_{n+1}(S,t,i)}{\partial S^2}\\
&\quad + rS\frac{\partial\varphi_{n+1}(S,t,i)}{\partial S} - r\varphi_{n+1}(S,t,i)\Big].\end{aligned} \tag{9.46}$$

The boundedness of $\varphi_{n+1}(S,t,i)$ and their derivatives up to the order 2 together with (9.46) then lead to

$$\max_{i\in\mathcal{M}}\sup_{(S,t)\in[0,1]\times[0,T]}|\mathcal{L}^{\varepsilon}\Phi^{\varepsilon}_{n+1}(S,t,i)| = O(\varepsilon^{n+1}). \tag{9.47}$$

As for the terminal layer terms, we have

$$\begin{aligned}
&\mathcal{L}^{\varepsilon}\sum_{k=1}^{n+1}\varepsilon^k\psi_k(S,\tau,i)\\
&=\sum_{k=1}^{n+1}\varepsilon^{k-1}\Big[\frac{\partial\psi_k(S,\tau,i)}{\partial\tau}+\frac{1}{2}\varepsilon\sigma^2(i)S^2\frac{\partial^2\psi_k(S,\tau,i)}{\partial S^2}+\varepsilon rS\frac{\partial\psi_k(S,\tau,i)}{\partial S}\\
&\quad-\varepsilon r\psi_k(S,\tau,i)+Q(t)\psi_k(S,\tau,\cdot)(i)\Big]\\
&=\sum_{k=1}^{n+1}\varepsilon^{k-1}\Big\{\Big[\frac{\partial\psi_k(S,\tau,i)}{\partial\tau}+\frac{1}{2}\varepsilon\sigma^2(i)S^2\frac{\partial^2\psi_k(S,\tau,i)}{\partial S^2}+\varepsilon rS\frac{\partial\psi_k(S,\tau,i)}{\partial S}\\
&\quad-\varepsilon r\psi_k(S,\tau,i)+Q(T)\psi_k(S,\tau,\cdot)(i)+\sum_{j=1}^{k}\frac{(-1)^j(\varepsilon\tau)^j}{j!}\frac{d^jQ(T)}{dt^j}\psi_k(S,\tau,i)\\
&\quad+\Big[Q(t)-\sum_{j=1}^{k}\frac{(-1)^j(\varepsilon\tau)^j}{j!}\frac{d^jQ(T)}{dt^j}\Big]\psi_k(S,\tau,\cdot)(i)\Big\}.
\end{aligned} \tag{9.48}$$

Note that for $k=1,\ldots,n+1$,

$$\Big|Q(t)-\sum_{j=1}^{k}\frac{(-1)^j(\varepsilon\tau)^j}{j!}\frac{d^jQ(T)}{dt^j}\Big| = |R_k(T-\varepsilon\tau)|\le Kt^{k+1},$$

and that

$$\Big|\sum_{j=1}^{n+1}\varepsilon^{j-1}\psi_j(S,\tau,i)O(t^{n+1-j})\Big|\le K\sum_{j=1}^{n+1}\varepsilon^{n+1-j}t^j\exp(-\kappa_0\tau)\le K\varepsilon^{n+1}.$$

The above observation together with (9.48) and

$$-\frac{\partial\psi_1(S,\tau,i)}{\partial\tau}+Q(T)\psi_1(S,\tau,i)=0$$

gives us

$$
\begin{aligned}
&\mathcal{L}^\varepsilon \sum_{k=1}^{n+1} \varepsilon^k \psi_k(S,\tau,i) \\
&= \sum_{k=2}^{n+1} \varepsilon^{k-1}\Big[-\frac{\partial \psi_k(S,\tau,i)}{\partial \tau} + Q(T)\psi_k(S,\tau,i) + \widetilde{R}_k \Big] + O(\varepsilon^{n+1}) \qquad (9.49)\\
&= O(\varepsilon^{n+1}),
\end{aligned}
$$

which holds uniformly in $(S,t) \in [0,1] \times [0,T]$. Thus, we obtain

$$
\max_{i\in\mathcal{M}} \sup_{(S,t)\in[0,1]\times[0,T]} |\mathcal{L}^\varepsilon[\Phi_{n+1}(S,t,i) + \Psi_{n+1}(S,\tau,i)]| = O(\varepsilon^{n+1}). \qquad (9.50)
$$

Step 2. Obtain estimate (9.45). Note that the definition of $e^\varepsilon_{n+1}(S,t,i)$ and (9.11) yield that $e^\varepsilon_{n+1}(S,T,i) = 0$. Because (9.50) holds, it follows from Lemma 9.12 that

$$
\max_{i\in\mathcal{M}} \sup_{(S,t)\in[0,1]\times[0,T]} |e^\varepsilon_{n+1}(S,t,i)| = O(\varepsilon^{n+1}).
$$

Note that

$$
e^\varepsilon_{n+1}(S,t,i) = e^\varepsilon_n(S,t,i) + \varepsilon^{n+1}\varphi_{n+1}(S,t,i) + \varepsilon^{n+1}\psi_{n+1}(S,\tau,i). \qquad (9.51)
$$

The smoothness of $\varphi_{n+1}(S,t,i)$ and the exponential decay of $\psi_{n+1}(S,\tau,i)$ imply that

$$
\max_{i\in\mathcal{M}} \sup_{(S,t)\in[0,1]\times[0,T]} |\varepsilon^{n+1}\varphi_{n+1}(S,t,i) + \varepsilon^{n+1}\psi_{n+1}(S,\tau,i)| = O(\varepsilon^{n+1}).
$$

Substituting the above into (9.51), we obtain (9.45) as desired.

□

9.5 Weakly Connected, Multiergodic-Class Switching Processes

Let $x \in \mathbb{R}^r$, $\mathcal{M} = \{1, \ldots, m_0\}$ with m_0 being a finite positive integer, and $Q = (q_{ij})$ an $m_0 \times m_0$ matrix independent of x satisfying $i \in \mathcal{M}$, $q_{ij} \geq 0$ for $i \neq j$ and $\sum_{j=1}^{m_0} q_{ij} = 0$. Consider a switching diffusion process $Y(t) = (X(t), \alpha(t))$, which has two components, an r-dimensional diffusion component $X(t)$ and a switching component $\alpha(t)$ taking value in $\mathcal{M}$ representing the discrete events. Because the Q matrix is independent of x, $\alpha(t)$ is simply a continuous-time Markov chain with state space $\mathcal{M}$ and generator Q. The switching diffusion process $Y(t) = (X(t), \alpha(t))$ satisfies

$$\begin{aligned}
&dX(t) = b(X(t), \alpha(t))dt + \sigma(X(t), \alpha(t))dw(t), \\
&X(0) = x, \ \alpha(0) = \alpha, \ \text{ and} \\
&\mathbb{P}(\alpha(t+\Delta t) = j | \alpha(t) = i, \alpha(s), s \le t) = q_{ij}\Delta t + o(\Delta t), \ i \ne j,
\end{aligned} \tag{9.52}$$

where $w(t)$ is a d-dimensional standard Brownian motion independent of $\alpha(t)$, $b(\cdot,\cdot) : \mathbb{R}^r \times \mathcal{M} \mapsto \mathbb{R}^r$, and $\sigma(\cdot,\cdot) : \mathbb{R}^r \times \mathcal{M} \mapsto \mathbb{R}^{r\times d}$ satisfying $b(x,i) = (b_j(x,i)) \in \mathbb{R}^r$, $\sigma(x,i)\sigma'(x,i) = a(x,i) = (a_{jk}(x,i)) \in \mathbb{R}^{r\times r}$, with z' denoting the transpose of z for $z \in \mathbb{R}^{\iota_1\times\iota_2}$ and $\iota_1, \iota_2 \ge 1$. Associated with the process given by (9.52), there is a generator $\mathcal{L}_0$ defined as follows. For any $g : \mathbb{R}^r \times \mathcal{M} \mapsto \mathbb{R}$ with $g(\cdot,i) \in \mathscr{C}^2(\mathbb{R}^r)$ for each $i \in \mathcal{M}$, we define

$$\begin{aligned}
&\mathcal{L}_0 g(x,i) = \frac{1}{2}\mathrm{tr}(a(x,i)\nabla^2 g(x,i)) + b'(x,i)\nabla g(x,i) + Qg(x,\cdot)(i), \\
&Qg(x,\cdot)(i) = \sum_{j=1}^{m_0} q_{ij} g(x,j) = \sum_{j\ne i, j\in\mathcal{M}} q_{ij}(g(x,j) - g(x,i)), \quad i \in \mathcal{M}.
\end{aligned} \tag{9.53}$$

where $\nabla g(\cdot,i)$ and $\nabla^2 g(\cdot,i)$ denote the gradient and the Hessian of $g(\cdot,i)$, respectively. For further references on stochastic differential equations involving Poisson measures describing the evolution of the jump processes, we refer the reader to Chapter 2 of this book; see also Skorohod [262].

In this chapter, we assume that for each $i \in \mathcal{M}$, both $b(\cdot,i)$ and $\sigma(\cdot,i)$ satisfy the usual local Lipschitz and linear growth conditions. Then, the system (9.52) has a unique solution; see Chapter 2 of this book and also [262] for details.

As previously mentioned, one of the conditions used in the ergodicity study in the earlier chapters is that the states of the jump component belong to a single ergodic class. This chapter addresses the situation where multiple weakly connected ergodic classes are present. We assume that the switching component is generated by

$$Q^\varepsilon = \frac{1}{\varepsilon}\widetilde{Q} + \widehat{Q}, \tag{9.54}$$

where $0 < \varepsilon \ll 1$ and $\widetilde{Q}$ and $\widehat{Q}$ are themselves generators of certain Markov chains. Corresponding to (9.54), the states of the switching process live in a number of ergodic classes that are connected through the generator $\widehat{Q}$. We say that switching processes with such a structure having multiergodic classes are connected. This section is divided into two parts. We first concern ourselves with the case that the jump components are divided into l recurrent classes. Later, we consider the case that transient states are also included.

9.5.1 Preliminary

Before proceeding further, let us give the motivation for using such models. First, we note that Q^ε is a constant matrix independent of x. The rationale is similar to those considered in the previous chapters, which can be considered as linearizing an x-dependent matrix $Q^\varepsilon(x)$ at "point of ∞." We may begin with an x-dependent matrix $Q^\varepsilon(x)$. Then, for x large enough, it can be replaced by a constant Q^ε. That is, we may assume

$$Q^\varepsilon(x) = Q^\varepsilon + o(1), \quad \text{as } |x| \to \infty, \tag{9.55}$$

where Q^ε has the form (9.54). Note that in the previous chapters, a condition similar to (9.55) was used without the ε-dependence, but the corresponding constant matrix (the limit at $|x| \to \infty$) is a generator of an ergodic Markov chain. Here, we are mostly concerned with the case that the generator could possibly be reducible with several ergodic classes. Nevertheless, the states belonging to different ergodic classes are not completely separable. They are linked together through weak interaction due to the presence of the slow part of the generator $\widehat{Q}$.

The formulation of Q^ε being a generator of an ε-dependent Markov chain for a small parameter ε stems from an effort of using two-time-scale models to reduce the complexity of the underlying systems. It has been observed in [260] that there are natural hierarchical structures in many large-scale and complex systems. The formulation in (9.54) is an effort to highlight the different parts of subsystems varying at different rates. It is often possible to partition the system states into a number of groups so that within each group, the state transitions take place rapidly, whereas among different groups, the changes are relatively infrequent. Such scenarios, in fact, appear in many applications. Thus, an effective way is to treat the systems through decompositions and aggregations. Loosely, one can use the natural scales shown in the system to aggregate the states in each ergodic class into one state. In this way, the total number of states is much reduced for the aggregated system. This point of view was extensively discussed in Yin and Zhang [303]. To begin, one may not have an ε in the system, but it is brought into the formulation to highlight the different rates of change so as to separate the fast and slow motions.

9.5.2 Weakly Connected, Multiple Ergodic Classes

Suppose

$$\widetilde{Q} = \text{diag}(\widetilde{Q}^1, \ldots, \widetilde{Q}^l). \tag{9.56}$$

In view of (9.56), the state space $\mathcal{M}$ of the underlying Markov chain is decomposable into l subspaces. That is, we can relabel the states so that

$$\mathcal{M} = \mathcal{M}_1 \cup \mathcal{M}_2 \cup \cdots \cup \mathcal{M}_l, \tag{9.57}$$

with $\mathcal{M}_i = \{s_{i1}, \dots, s_{im_i}\}$ and $m_0 = m_1 + m_2 + \cdots + m_l$ such that $\widetilde{Q}^i$, the generator associated with the subspace $\mathcal{M}_i$ for each $i = 1, \dots, l$ is irreducible. Thus, the corresponding $\mathcal{M}_i$ for $i = 1, \dots, l$ consist of recurrent states belonging to l ergodic classes. To signify the ε-dependence, we index the process by ε and write it as $Y^\varepsilon(t) = (X^\varepsilon(t), \alpha^\varepsilon(t))$. Then, (9.52) becomes

$$\begin{aligned} &dX^\varepsilon(t) = b(X^\varepsilon(t), \alpha^\varepsilon(t))dt + \sigma(X^\varepsilon(t), \alpha^\varepsilon(t))dw(t), \\ &X^\varepsilon(0) = x,\ \alpha^\varepsilon(0) = \alpha, \\ &\mathbb{P}(\alpha^\varepsilon(t+\Delta) = j | \alpha^\varepsilon(t) = i, \alpha^\varepsilon(s), s \le t) = q^\varepsilon_{ij}\Delta + o(\Delta), \quad i \ne j. \end{aligned} \tag{9.58}$$

The associated operator for the switching diffusion is given by

$$\begin{aligned} \mathcal{L}^\varepsilon g(x,\iota) = {} & \frac{1}{2}\mathrm{tr}(a(x,\iota)\nabla^2 g(x,\iota)) + b'(x,\iota)\nabla g(x,\iota) \\ & + Q^\varepsilon g(x,\cdot)(\iota),\ (x,\iota) \in \mathbb{R}^r \times \mathcal{M}. \end{aligned} \tag{9.59}$$

To proceed, lump the states of the switching component in each $\mathcal{M}_i$ into a single state and define

$$\overline{\alpha}^\varepsilon(t) = i \ \text{ if } \alpha^\varepsilon(t) \in \mathcal{M}_i. \tag{9.60}$$

Denote the state space of $\overline{\alpha}^\varepsilon(\cdot)$ by $\overline{\mathcal{M}} = \{1, \dots, l\}$ and $\tilde{\nu} = \mathrm{diag}(\nu^1, \dots, \nu^l)$, where ν^i is the stationary distribution corresponding to $\widetilde{Q}^i$. Define

$$\overline{Q} = \tilde{\nu}\widehat{Q}\mathbb{1} \tag{9.61}$$

with $\mathbb{1} = \mathrm{diag}(\mathbb{1}_{m_1}, \dots, \mathbb{1}_{m_l})$ and $\mathbb{1}_\ell = (1, \dots, 1)' \in \mathbb{R}^\ell$. The essence of the aggregated process is to treat all the states in $\mathcal{M}_i$ as one state, so the total number of states in the "effective" state space is much reduced. We need the following assumption about the generator of $\alpha^\varepsilon(\cdot)$.

(H9.3) For each $i = 1, \dots, l$, $\widetilde{Q}^i$ is irreducible.

Lemma 9.14 *Under* (H9.3), *the following assertions hold.*

(a) *The probability vector* $p^\varepsilon(t) \in \mathbb{R}^{1\times m_0}$ *with*

$$p^\varepsilon(t) = (\mathbb{P}(\alpha^\varepsilon(t) = s_{ij}), i = 1, \dots, l, j = 1, \dots, m_i),$$

satisfies

$$p^\varepsilon(t) = \theta(t)\tilde{\nu} + O(\varepsilon(t+1) + e^{-\kappa_0 t/\varepsilon})$$

for some $\kappa_0 > 0$, *and* $\theta(t) = (\theta_1(t), \ldots, \theta_l(t)) \in \mathbb{R}^{1\times l}$ *satisfies*

$$\frac{d\theta(t)}{dt} = \theta(t)\overline{Q},\ \theta(0) = p(0)\mathbb{1}.$$

(b) *The transition matrix satisfies*

$$P^{\varepsilon}(t) = P^{(0)}(t) + O(\varepsilon(t+1) + e^{-\kappa_0 t/\varepsilon}),$$

where $P^{(0)}(t) = \mathbb{1}\Theta(t)\tilde{\nu}$ *and*

$$\frac{d\Theta(t)}{dt} = \Theta(t)\overline{Q},\ \Theta(0) = I.$$

(c) *The aggregated process* $\overline{\alpha}^{\varepsilon}(\cdot)$ *converges weakly to* $\overline{\alpha}(\cdot)$ *as* $\varepsilon \to 0$, *where* $\overline{\alpha}(\cdot)$ *is a Markov chain generated by* $\overline{Q}$.

(d) *For* $i = 1, \ldots, l,\ j = 1, \ldots, m_i$

$$\mathbb{E}\left[\int_0^{\infty} e^{-t}(I_{\{\alpha^{\varepsilon}(t)=s_{ij}\}} - \nu_j^i I_{\{\overline{\alpha}^{\varepsilon}(t)=i\}})dt\right]^2 = O(\varepsilon),$$

where ν_j^i *denotes the* j*th component of* ν^i *for* $i = 1, \ldots, l$ *and* $j = 1, \ldots, m_i$.

Proof The proofs of (a) and (b) in Lemma 9.14 can be found in [303, Corollary 6.12, p. 130]. The proof of (c) is based on the martingale averaging method [303, Theorem 7.4, p. 172]; an outline of the idea is in [9], and a discrete version of the approximation may be found in [304].

As for (d), using (a) and (b), direct calculation reveals that

$$\begin{aligned}
&\mathbb{E}\left[\int_0^{\infty} e^{-t}(I_{\{\alpha^{\varepsilon}(t)=s_{ij}\}} - \nu_j^i I_{\{\overline{\alpha}^{\varepsilon}(t)=i\}})dt\right]^2 \\
&= \int_0^{\infty}\int_0^{t} e^{-t-s} O\left(\varepsilon(t+1) + e^{-\kappa_0(t-s)/\varepsilon}\right) dsdt \\
&\quad + \int_0^{\infty}\int_0^{s} e^{-t-s} O\left(\varepsilon(t+1) + e^{-\kappa_0(s-t)/\varepsilon}\right) dtds.
\end{aligned}$$

Detailed calculations then yield the desired result. □

Due to the aggregation, certain averages take place. When ε goes to 0, we obtain a limit system. The following lemma records this fact, whose proof can be found in, for example, Yin and Zhang [303] and Yin [290]. The basic idea is to use martingale averaging; we omit the details here.

Lemma 9.15 *Assume* (H9.3). *Then,* $(X^\varepsilon(\cdot), \overline{\alpha}^\varepsilon(\cdot))$ *converges weakly to a process* $(X(\cdot), \overline{\alpha}(\cdot))$, *whose operator is given by*

$$\mathcal{L}g(x,i) = \frac{1}{2}\mathrm{tr}(\overline{a}(x,i)\nabla^2 g(x,i)) + \overline{b}'(x,i)\nabla g(x,i) + \overline{Q}g(x,\cdot)(i), \text{ where}$$

$$\overline{b}(x,i) = \sum_{j=1}^{m_i} \nu_j^i b(x,s_{ij}),\ \overline{a}(x,i) = \sum_{j=1}^{m_i} \nu_j^i a(x,s_{ij}),\ i = 1,\ldots,l. \tag{9.62}$$

Remark 9.16 In view of the weak convergence result, the limit stochastic differential equation for (9.58) is

$$dX(t) = \overline{b}(X(t), \overline{\alpha}(t))dt + \overline{\sigma}(X(t), \overline{\alpha}(t))dw, \tag{9.63}$$

where $\overline{\sigma}(x,i)$ is defined as $\overline{\sigma}(x,i)\overline{\sigma}'(x,i) = \overline{a}(x,i)$.

(H9.4) $\overline{Q}$ is irreducible. For each $i \in \overline{\mathcal{M}}$, $\overline{a}(x,i)$ satisfies $\kappa_1|\xi|^2 \le \xi'\overline{a}(x,i)\xi$ for all $\xi \in \mathbb{R}^r$, with some constant $\kappa_1 \in (0,1]$ for all $x \in \mathbb{R}^r$.

Under the conditions outlined above, and using the techniques from Chapter 3—specifically, Theorem 3.26—we establish the following assertion.

Proposition 9.17 *Under* (H9.3) *and* (H9.4), *the switching diffusion with generator given by* (9.62) *is positive recurrent.*

We now present the result on positive recurrence of the underlying process. The main idea is that although the discrete events, described by a continuous-time Markov chain may have several weakly connected ergodic classes, when ε is sufficiently small, we still have positive recurrence for the process $(X^\varepsilon(\cdot), \alpha^\varepsilon(\cdot))$. The proof relies on the use of perturbed Lyapunov function methods, initially employed to address diffusion approximations by Papanicolaou, Stroock, and Varadhan in [230]. These methods have since been successfully applied in stochastic systems theory (see Kushner [175]) and stochastic approximation (see Kushner and Yin [177]), among others. The basic idea is to introduce perturbations of Lyapunov functions that are small in magnitude and that result in the desired cancellation of unwanted terms.

Theorem 9.18 *Suppose that Assumptions* (H9.3) *and* (H9.4) *hold. Then, for sufficiently small* $\varepsilon > 0$, *the process* $(X^\varepsilon(\cdot), \alpha^\varepsilon(\cdot))$ *is positive recurrent.*

Remark 9.19 First note that by Proposition 9.17, the process $(X(\cdot), \overline{\alpha}(\cdot))$ of (9.63) is positive recurrent. Then, it follows that there exists a Lyapunov function $V(x,i)$ for the limit system (9.63) such that

$$\mathcal{L}V(x,i) \le -\gamma \quad \text{for some } \gamma > 0. \tag{9.64}$$

In Theorem 9.18, by the property holds for sufficiently small ε, we meant that there exists an $\varepsilon_0 > 0$ such that for all $0 < \varepsilon \le \varepsilon_0$, the property holds.

Proof We begin with the Lyapunov function in (9.64) for the limit system. Choose n_0 to be an integer large enough so that D is contained in the ball $\{|x| < n_0\}$. For any $(x, i) \in D^c \times \mathcal{M}$, any $t > 0$, any positive integer $n > n_0$, define

$$\tau_D = \inf\{t : X^\varepsilon(t) \in D\}, \quad \text{and} \quad \tau_{D,n}(t) = t \wedge \tau_D \wedge \beta_n, \tag{9.65}$$

where β_n comes from the regularity consideration and is the first exit time of the process $(X(t), \overline{\alpha}(t))$ from the set $\{\widetilde{x} : |\widetilde{x}| < n\} \times \mathcal{M}$; this is, $\beta_n = \inf\{t : |X(t)| = n\}$. For $i \in \overline{\mathcal{M}}$, use the Lyapunov function $V(x, i)$ in (9.64) to define

$$\overline{V}(x, \alpha) = \sum_{i=1}^{l} V(x, i) I_{\{\alpha \in \mathcal{M}_i\}} = V(x, i) \text{ if } \alpha \in \mathcal{M}_i. \tag{9.66}$$

Note that

$$\overline{V}(X^\varepsilon(t), \alpha^\varepsilon(t)) = V(X^\varepsilon(t), \overline{\alpha}^\varepsilon(t)), \tag{9.67}$$

so these two expressions are to be used interchangeably in what follows. Observe that $\overline{V}(x, \alpha)$ is orthogonal to $\widetilde{Q}$; that is, $\widetilde{Q}\overline{V}(x, \cdot)(\alpha) = 0$. Then

$$\begin{aligned} \mathcal{L}^\varepsilon \overline{V}(X^\varepsilon(t), \alpha^\varepsilon(t)) = {} & \overline{V}'_x(X^\varepsilon(t), \alpha^\varepsilon(t)) b(X^\varepsilon(t), \alpha^\varepsilon(t)) \\ & + \frac{1}{2} \operatorname{tr}[\overline{V}_{xx}(X^\varepsilon(t), \alpha^\varepsilon(t)) a(X^\varepsilon(t), \alpha^\varepsilon(t))] \\ & + \overline{Q}\overline{V}(X^\varepsilon(t), \cdot)(\alpha^\varepsilon(t)). \end{aligned} \tag{9.68}$$

By virtue of Dynkin's formula

$$\begin{aligned} & \mathbb{E}_{x,i} \overline{V}(X^\varepsilon(\tau_{D,n}(t), \alpha^\varepsilon(\tau_{D,n}(t)) - \overline{V}(x, i) \\ & = \mathbb{E}_{x,i} \int_0^{\tau_{D,n}(t)} \mathcal{L}^\varepsilon \overline{V}(X^\varepsilon(s), \alpha^\varepsilon(s)) ds. \end{aligned} \tag{9.69}$$

Note that there are unwanted terms on the right-hand side in view of (9.68). To get rid of these terms, we use methods of perturbed Lyapunov functions [177, 303] to average out the "bad" terms. Note that the process $(X^\varepsilon(t), \alpha^\varepsilon(t))$ is Markov. Thus, for a suitable function $\xi(\cdot)$, $\mathcal{L}^\varepsilon \xi(t)$ can be calculated by

$$\mathcal{L}^\varepsilon \xi(t) = \lim_{\delta \to 0} \frac{\mathbb{E}_t^\varepsilon \xi(X^\varepsilon(t+\delta), \alpha^\varepsilon(t+\delta)) - \xi(X^\varepsilon(t), \alpha^\varepsilon(t))}{\delta}, \tag{9.70}$$

where $\mathbb{E}_t^\varepsilon$ denotes the conditional expectation with respect to the σ-algebra $\mathcal{F}_t^\varepsilon = \sigma\{(X^\varepsilon(s), \alpha^\varepsilon(s)) : s \le t\}$. To obtain the desired results, we introduce three perturbations. The integrand of each of these perturbations is formed by the difference of two terms, an original term and its "average." The goal is to use the averages in the final form for the evaluation of the Lyapunov function. To ensure the integrability in an infinite horizon, exponential discounting is used in the integrals.

Define

$$
\begin{aligned}
V_1^\varepsilon(x,\alpha,t) &= \mathbb{E}_t^\varepsilon \int_t^\infty e^{t-u}\overline{V}_x'(x,\alpha)\left[b(x,\alpha^\varepsilon(u)) - \overline{b}(x,\overline{\alpha}^\varepsilon(u))\right] du,\\
V_2^\varepsilon(x,\alpha,t) &= \mathbb{E}_t^\varepsilon \int_t^\infty e^{t-u}\frac{1}{2}\mathrm{tr}[\overline{V}_{xx}(x,\alpha)(a(x,\alpha^\varepsilon(u)) - \overline{a}(x,\overline{\alpha}^\varepsilon(u)))]du,\\
V_3^\varepsilon(x,t) &= \mathbb{E}_t^\varepsilon \int_t^\infty e^{t-u}[\widehat{Q}\overline{V}(x,\cdot)(\alpha^\varepsilon(u)) - \overline{Q}V(x,\cdot)(\overline{\alpha}^\varepsilon(u))]du.
\end{aligned} \tag{9.71}
$$

To proceed, we first state a lemma. □

Lemma 9.20 *Assume the conditions of Theorem* 9.18*. Then, the following assertions hold.*

(a) *For* V_i^ε *with* $i = 1, 2, 3$, *we have the following estimates:*

$$
\begin{aligned}
&V_1^\varepsilon(X^\varepsilon(t),\alpha^\varepsilon(t),t) = O(\varepsilon)(\overline{V}(X^\varepsilon(t),\alpha^\varepsilon(t)) + 1),\\
&V_2^\varepsilon(X^\varepsilon(t),\alpha^\varepsilon(t),t) = O(\varepsilon)(\overline{V}(X^\varepsilon(t),\alpha^\varepsilon(t)) + 1),\\
&V_3^\varepsilon(X^\varepsilon(t),t) = O(\varepsilon)(\overline{V}(X^\varepsilon(t),\alpha^\varepsilon(t)) + 1).
\end{aligned} \tag{9.72}
$$

(b) *Moreover,*

$$
\begin{aligned}
&\mathcal{L}^\varepsilon V_1^\varepsilon(X^\varepsilon(t),\alpha^\varepsilon(t),t)\\
&= -\overline{V}_x'(X^\varepsilon(t),\alpha^\varepsilon(t))[b(X^\varepsilon(t),\alpha^\varepsilon(t)) - \overline{b}(X^\varepsilon(t),\overline{\alpha}^\varepsilon(t))]\\
&\quad + O(\varepsilon)(\overline{V}(X^\varepsilon(t),\alpha^\varepsilon(t)) + 1),\\
&\mathcal{L}^\varepsilon V_2^\varepsilon(X^\varepsilon(t),\alpha^\varepsilon(t),t)\\
&= -\frac{1}{2}\mathrm{tr}[\overline{V}_{xx}(X^\varepsilon(t),\alpha^\varepsilon(t))(a(X^\varepsilon(t),\alpha^\varepsilon(t)) - \overline{a}(X^\varepsilon(t),\overline{\alpha}^\varepsilon(t))]\\
&\quad + O(\varepsilon)(\overline{V}(X^\varepsilon(t),\alpha^\varepsilon(t)) + 1),\\
&\mathcal{L}^\varepsilon V_3^\varepsilon(X^\varepsilon(t),t)\\
&= -\widehat{Q}\overline{V}(X^\varepsilon(t),\cdot)(\alpha^\varepsilon(t)) + \overline{Q}V(X^\varepsilon(t),\cdot)(\overline{\alpha}^\varepsilon(t))\\
&\quad + O(\varepsilon)(\overline{V}(X^\varepsilon(t),\alpha^\varepsilon(t)) + 1).
\end{aligned} \tag{9.73}
$$

Proof The proof is inspired by the technique of perturbed Lyapunov function methods of Kushner and Yin in [177, Chapter 6], and the construction of the Lyapunov functions is along the line of Badowski and Yin [9]. By the definition of $\overline{b}(\cdot)$

$$
\begin{aligned}
&b(X^\varepsilon(t),\alpha^\varepsilon(u)) - \overline{b}(X^\varepsilon(t),\overline{\alpha}^\varepsilon(u))\\
&= \sum_{i=1}^{l}\sum_{j=1}^{m_i} b(X^\varepsilon(t),s_{ij})[I_{\{\alpha^\varepsilon(u)=s_{ij}\}} - \nu_j^i I_{\{\alpha^\varepsilon(u)\in\mathcal{M}_i\}}].
\end{aligned} \tag{9.74}
$$

It can be argued by using the Markov property and the two-time-scale structure (see [303, p. 187]) that for $u \ge t$

$$\mathbb{E}_t^\varepsilon[I_{\{\alpha^\varepsilon(u)=s_{ij}\}} - \nu_j^i I_{\{\overline{\alpha}^\varepsilon(u)=i\}}] = O\left(\varepsilon + e^{-\kappa_0(u-t)/\varepsilon}\right). \tag{9.75}$$

Thus, using (9.67), (9.74), and (9.75) in (9.71), we obtain

$$\begin{aligned}|V_1^\varepsilon(X^\varepsilon(t), \alpha^\varepsilon(t), t)| &\le \sum_{i=1}^{l}\sum_{j=1}^{m_i} \left|\overline{V}_x'(X^\varepsilon(t), \alpha^\varepsilon(t)) b(X^\varepsilon(t), s_{ij})\right| \\ &\quad \times \left|\int_t^\infty e^{t-u} O\left(\varepsilon + e^{-\kappa_0(u-t)/\varepsilon}\right) du\right| \\ &= O(\varepsilon)(\overline{V}(X^\varepsilon(t), \alpha^\varepsilon(t)) + 1).\end{aligned} \tag{9.76}$$

Likewise, we obtain the estimates for $V_2^\varepsilon(X^\varepsilon(t), \alpha^\varepsilon(t), t)$, and $V_3^\varepsilon(X^\varepsilon(t), t)$ in (9.72). This establishes statement (a).

Next, we prove (b). For convenience, introduce a notation:

$$\Gamma(x, \alpha, \alpha_1) = \overline{V}_x'(x, \alpha)[b(x, \alpha_1) - \overline{b}(x, \overline{\alpha}_1)]. \tag{9.77}$$

By virtue of the definition of $\mathcal{L}^\varepsilon$, we have

$$\begin{aligned}&\mathcal{L}^\varepsilon\ V_1^\varepsilon(X^\varepsilon(t), \alpha^\varepsilon(t), t) \\ &= -\lim_{\delta\to 0}\frac{1}{\delta}\int_t^{t+\delta} \mathbb{E}_t^\varepsilon e^{t-u}\Gamma(X^\varepsilon(t), \alpha^\varepsilon(t), \alpha^\varepsilon(u))du \\ &\quad + \lim_{\delta\to 0}\frac{1}{\delta}\int_{t+\delta}^\infty \mathbb{E}_t^\varepsilon[e^{t+\delta-u} - e^{t-u}]\Gamma(X^\varepsilon(t), \alpha^\varepsilon(t), \alpha^\varepsilon(u))du \\ &\quad + \lim_{\delta\to 0}\int_{t+\delta}^\infty e^{t+\delta-u}\mathbb{E}_t^\varepsilon[\Gamma(X^\varepsilon(t+\delta), \alpha^\varepsilon(t), \alpha^\varepsilon(u)) \\ &\quad - \Gamma(X^\varepsilon(t), \alpha^\varepsilon(t), \alpha^\varepsilon(u))]du \\ &\quad + O(\varepsilon)(\overline{V}(X^\varepsilon(t), \alpha^\varepsilon(t)) + 1).\end{aligned}$$

Moreover

$$\begin{aligned}&-\lim_{\delta\to 0}\frac{1}{\delta}\int_t^{t+\delta} \mathbb{E}_t^\varepsilon e^{t-u}\Gamma(X^\varepsilon(t), \alpha^\varepsilon(t), \alpha^\varepsilon(u))du \\ &= -\overline{V}_x'(X^\varepsilon(t), \alpha^\varepsilon(t))\Big[b(X^\varepsilon(t), \alpha^\varepsilon(t)) - \overline{b}(X^\varepsilon(t), \overline{\alpha}^\varepsilon(t))\Big],\end{aligned} \tag{9.78}$$

and

$$\lim_{\delta\to 0}\frac{1}{\delta}\int_{t+\delta}^{\infty}\mathbb{E}_t^{\varepsilon}[e^{t+\delta-u}-e^{t-u}]\Gamma(X^{\varepsilon}(t),\alpha^{\varepsilon}(t),\alpha^{\varepsilon}(u))du$$

$$=\int_t^{\infty}e^{t-u}\mathbb{E}_t^{\varepsilon}\Gamma(X^{\varepsilon}(t),\alpha^{\varepsilon}(t),\alpha^{\varepsilon}(u))du=V_1^{\varepsilon}(X^{\varepsilon}(t),\alpha^{\varepsilon}(t)). \tag{9.79}$$

The independence of $\alpha^{\varepsilon}(\cdot)$ and $w(\cdot)$, (9.74), and (9.77) leads to that for $u\geq t$

$$\begin{aligned}&\mathbb{E}_t^{\varepsilon}[\Gamma(X^{\varepsilon}(t+\delta),\alpha^{\varepsilon}(t),\alpha^{\varepsilon}(u))-\Gamma(X^{\varepsilon}(t),\alpha^{\varepsilon}(t),\alpha^{\varepsilon}(u))]\\&=\sum_{i=1}^{l}\sum_{j=1}^{m_i}\mathbb{E}_t^{\varepsilon}[\overline{V}_x'(X^{\varepsilon}(t+\delta),\alpha^{\varepsilon}(t))b(X^{\varepsilon}(t+\delta),s_{ij})\\&\quad-\overline{V}_x'(X^{\varepsilon}(t),\alpha^{\varepsilon}(t))b(X^{\varepsilon}(t),s_{ij})]\mathbb{E}_{t+\delta}^{\varepsilon}[I_{\{\alpha^{\varepsilon}(u)=s_{ij}\}}-\nu_j^iI_{\{\overline{\alpha}^{\varepsilon}(u)\in\mathcal{M}_i\}}].\end{aligned} \tag{9.80}$$

Furthermore, we have

$$\begin{aligned}&\lim_{\delta\to 0}\frac{1}{\delta}\int_{t+\delta}^{\infty}e^{t+\delta-u}\mathbb{E}_t^{\varepsilon}[\Gamma(X^{\varepsilon}(t+\delta),\alpha^{\varepsilon}(t),\alpha^{\varepsilon}(u))-\Gamma(X^{\varepsilon}(t),\alpha^{\varepsilon}(t),\alpha^{\varepsilon}(u))]du\\&=\lim_{\delta\to 0}\frac{1}{\delta}\int_{t+\delta}^{\infty}e^{t-u}\mathbb{E}_t^{\varepsilon}[\Gamma(X^{\varepsilon}(t+\delta),\alpha^{\varepsilon}(t),\alpha^{\varepsilon}(u))\\&\quad-\Gamma(X^{\varepsilon}(t),\alpha^{\varepsilon}(t),\alpha^{\varepsilon}(u))]du+o(1),\end{aligned}$$

where $o(1)\to 0$ in probability uniformly in t. Using (9.75), (9.77), and (9.80),

$$\begin{aligned}&\sum_{i=1}^{l}\sum_{j=1}^{m_i}\Big(\overline{V}_x'(X^{\varepsilon}(t),\alpha^{\varepsilon}(t))b(X^{\varepsilon}(t),s_{ij})\Big)_x b(X^{\varepsilon}(t),\alpha^{\varepsilon}(t))\\&\quad\times\mathbb{E}_t^{\varepsilon}\int_t^{\infty}e^{t-u}\Big[I_{\{\alpha^{\varepsilon}(u)=s_{ij}\}}-\nu_j^iI_{\{\overline{\alpha}^{\varepsilon}(u)=i\}}\Big]du\\&=O(\varepsilon)(\overline{V}(X^{\varepsilon}(t),\alpha^{\varepsilon}(t))+1),\end{aligned} \tag{9.81}$$

and

$$\begin{aligned}&\sum_{i=1}^{l}\sum_{j=1}^{m_i}\frac{1}{2}\mathrm{tr}[\big(V_x'(X^{\varepsilon}(t),\alpha^{\varepsilon}(t))b(X^{\varepsilon}(t),s_{ij})\big)_{xx}a(X^{\varepsilon}(t),\alpha^{\varepsilon}(t))]\\&\quad\times\mathbb{E}_t^{\varepsilon}\int_t^{\infty}e^{t-u}\Big[I_{\{\alpha^{\varepsilon}(u)=s_{ij}\}}-\nu_j^iI_{\{\overline{\alpha}^{\varepsilon}(u)=i\}}\Big]du\\&=O(\varepsilon)(\overline{V}(X^{\varepsilon}(t),\alpha^{\varepsilon}(t))+1).\end{aligned} \tag{9.82}$$

Thus, (9.78), (9.79), (9.81), and (9.82) lead to

$$\begin{aligned}
&\mathcal{L}^{\varepsilon} V_1^{\varepsilon}(X^{\varepsilon}(t), \alpha^{\varepsilon}(t), t) \\
&\quad = -\overline{V}_x'(X^{\varepsilon}(t), \alpha^{\varepsilon}(t))[b(X^{\varepsilon}(t), \alpha^{\varepsilon}(t)) - \overline{b}(X^{\varepsilon}(t), \overline{\alpha}^{\varepsilon}(t))] \\
&\qquad + O(\varepsilon)(\overline{V}(X^{\varepsilon}(t), \alpha^{\varepsilon}(t)) + 1).
\end{aligned}$$

Similar calculations enable us to obtain the rest of the estimates in part (b) of the lemma. The proof of the lemma is concluded.

Using Lemma 9.20, we proceed to eliminate the unwanted terms and to obtain the detailed estimates. Define

$$\begin{aligned}
V^{\varepsilon}(t) =& \overline{V}(X^{\varepsilon}(t), \alpha^{\varepsilon}(t)) + V_1^{\varepsilon}(X^{\varepsilon}(t), \alpha^{\varepsilon}(t), t) \\
&+ V_2^{\varepsilon}(X^{\varepsilon}(t), \alpha^{\varepsilon}(t), t) + V_3^{\varepsilon}(X^{\varepsilon}(t), t).
\end{aligned} \tag{9.83}$$

It follows that

$$V^{\varepsilon}(t) = V(X^{\varepsilon}(t), \overline{\alpha}^{\varepsilon}(t)) + O(\varepsilon)(\overline{V}(X^{\varepsilon}(t), \alpha^{\varepsilon}(t)) + 1), \tag{9.84}$$

and

$$\begin{aligned}
\mathcal{L}^{\varepsilon} V^{\varepsilon}(t) =& V_x'(X^{\varepsilon}(t), \overline{\alpha}^{\varepsilon}(t))\overline{b}(X^{\varepsilon}(t), \overline{\alpha}^{\varepsilon}(t)) \\
&+ \frac{1}{2}\mathrm{tr}\left[V_{xx}(X^{\varepsilon}(t), \overline{\alpha}^{\varepsilon}(t))\overline{a}(X^{\varepsilon}(t), \overline{\alpha}^{\varepsilon}(t))\right] \\
&+ \overline{Q}V(X^{\varepsilon}(t), \cdot)(\overline{\alpha}^{\varepsilon}(t)) + O(\varepsilon)(\overline{V}(X^{\varepsilon}(t), \alpha^{\varepsilon}(t)) + 1).
\end{aligned} \tag{9.85}$$

Therefore,

$$\mathcal{L}^{\varepsilon} V^{\varepsilon}(t) = \mathcal{L}V(X^{\varepsilon}(t), \overline{\alpha}^{\varepsilon}(t)) + O(\varepsilon)(\overline{V}(X^{\varepsilon}(t), \alpha^{\varepsilon}(t)) + 1). \tag{9.86}$$

Note that through the use of perturbations, the first term on the right-hand side above involves the limit operator and the Lyapunov function of the limit system, which is crucially important.

For fixed but arbitrary $T > 0$, we then obtain

$$\begin{aligned}
&\mathbb{E}_{x,i} V^{\varepsilon}(\tau_{D,n}(t) \wedge T) \\
&= \overline{V}(x, i) + \mathbb{E}_{x,i} V_1^{\varepsilon}(x, i, 0) + \mathbb{E}_{x,i} \int_0^{\tau_{D,n}(t)\wedge T} \mathcal{L}^{\varepsilon} V^{\varepsilon}(s) ds \\
&= \overline{V}(x, i) + \mathbb{E}_{x,i} V_1^{\varepsilon}(x, i, 0) + \mathbb{E}_{x,i} \int_0^{\tau_{D,n}(t)\wedge T} [\mathcal{L}V(X^{\varepsilon}(s), \overline{\alpha}^{\varepsilon}(s)) \\
&\quad + O(\varepsilon)(\overline{V}(X^{\varepsilon}(s), \alpha^{\varepsilon}(s)) + 1)] ds \\
&= \overline{V}(x, i) + \mathbb{E}_{x,i} V_1^{\varepsilon}(x, i, 0) \\
&\quad + \mathbb{E}_{x,i} \int_0^{\tau_{D,n}(t)\wedge T} [\mathcal{L}V(X^{\varepsilon}(s), \overline{\alpha}^{\varepsilon}(s)) + O(\varepsilon)(V^{\varepsilon}(s) + 1)] ds \\
&\le \overline{V}(x, i) + \mathbb{E}_{x,i} V_1^{\varepsilon}(x, i, 0) \\
&\quad + [O(\varepsilon) - \gamma]\mathbb{E}_{x,i}[\tau_{D,n}(t) \wedge T] + O(\varepsilon)\mathbb{E}_{x,i} \int_0^{\tau_{D,n}(t)\wedge T} V^{\varepsilon}(s) ds.
\end{aligned} \tag{9.87}$$

The expression after the third equality sign follows from the estimates in (9.72), and the expression after the last inequality sign is a consequence of (9.64).

Because $\varepsilon > 0$ is sufficiently small, there is a γ_1 such that $-\gamma_1 \geq -\gamma + O(\varepsilon)$. It is clear that for the fixed T, $\tau_{D,n}(t) \wedge T \leq T$. Thus, the foregoing together with [198, Theorem 2.6.1, p. 68] and [110, p. 36] yields that

$$\begin{aligned} &\mathbb{E}_{x,i} V^{\varepsilon}(\tau_{D,n}(t) \wedge T) \\ &\leq \left[\overline{V}(x,i) + \mathbb{E}_{x,i} \sum_{\iota=1}^{3} V_{\iota}^{\varepsilon}(x,i,0) - \gamma_1 \mathbb{E}_{x,i}(\tau_{D,n}(t) \wedge T) \right] \exp(\varepsilon T). \end{aligned} \tag{9.88}$$

Consequently

$$\begin{aligned} &\exp(-\varepsilon T)\mathbb{E}_{x,i} V^{\varepsilon}(\tau_{D,n}(t) \wedge T) + \gamma_1 \mathbb{E}_{x,i}[\tau_{D,n}(t) \wedge T] \\ &\quad \leq \overline{V}(x,i) + \mathbb{E}_{x,i} \sum_{\iota=1}^{3} V_{\iota}^{\varepsilon}(x,i,0). \end{aligned} \tag{9.89}$$

Taking limit $T \to \infty$ in (9.89) leads to

$$\gamma_1 \mathbb{E}_{x,i} \tau_{D,n}(t) \leq \overline{V}(x,i) + \mathbb{E}_{x,i} \sum_{\iota=1}^{3} V_{\iota}^{\varepsilon}(x,i,0). \tag{9.90}$$

By means of regularity of the underlying process, passing to the limit as $n \to \infty$, (9.90) together with Fatou's lemma yields that

$$\gamma_1 \mathbb{E}_{x,i}[t \wedge \tau_D] \leq \overline{V}(x,i) + \mathbb{E}_{x,i} \sum_{\iota=1}^{3} V_{\iota}^{\varepsilon}(x,i,0).$$

Finally, it is easily seen that $\mathbb{E}_{x,i} \sum_{\iota=1}^{3} V_{\iota}^{\varepsilon}(x,i,0) < \infty$. In addition

$$\mathbb{E}_{x,i}[t \wedge \tau_D] = \mathbb{E}_{x,i} \tau_D I_{\{\tau_D \leq t\}} + \mathbb{E}_{x,i} t I_{\{\tau_D > t\}}. \tag{9.91}$$

Because

$$\mathbb{P}_{x,i}(\tau_D > t) \leq \frac{1}{\gamma_1 t} \left[\overline{V}(x,i) + \mathbb{E}_{x,i} \sum_{\iota=1}^{3} V_{\iota}^{\varepsilon}(x,i,0) \right] \to 0 \text{ as } t \to \infty,$$

we obtain $\mathbb{E}_{x,i} \tau_D < \infty$ as desired. This establishes the positive recurrent. □

9.5.3 *Inclusion of Transient Discrete Events*

Here, we extend the result to the case that $\alpha^\varepsilon(t)$ includes transient states in addition to the states of ergodic classes. Let $\alpha^\varepsilon(\cdot)$ be a Markov chain with generator given by (9.54) with

$$\widetilde{Q} = \begin{pmatrix} \widetilde{Q}^1 & & & & \\ & \widetilde{Q}^2 & & & \\ & & \ddots & & \\ & & & \widetilde{Q}^l & \\ \widetilde{Q}^1_* & \widetilde{Q}^2_* & \cdots & \widetilde{Q}^l_* & \widetilde{Q}_* \end{pmatrix}. \tag{9.92}$$

Now, the switching process is again nearly completely decomposable. However, in addition to the l recurrent classes, there are a number of transient states. That is, $\mathcal{M} = \mathcal{M}_1 \cup \cdots \cup \mathcal{M}_l \cup \mathcal{M}_*$, where $\mathcal{M}_* = \{s_{*1}, \dots, s_{*m_*}\}$ contains a collection of transient states. We replace (H9.3) by (H9.5) given below.

(H9.5) For $i = 1, \dots, l$, $\widetilde{Q}^i$ are irreducible, and $\widetilde{Q}_*$ is Hurwitz (i.e., all of its eigenvalues have negative real parts).

The Hurwitz condition implies that the states in $\mathcal{M}_*$ are transient. Within a short period of time, they will enter one of the recurrent classes. To proceed, we define the aggregate process. Note, however, we only lump the states in each recurrent class, not in the transient class. Define

$$\begin{aligned} \nu_* &= \operatorname{diag}(\tilde{\nu}, 0_{m_* \times m_*}) \in \mathbb{R}^{(l+m_*)\times m_0}, \\ \widetilde{a}_i &= -\widetilde{Q}_*^{-1}\widetilde{Q}^i_* \mathbb{1}_{m_i} \in \mathbb{R}^{m_* \times 1}, \quad \text{for } i = 1, \dots, l, \\ \widetilde{A} &= (\widetilde{a}_1, \dots, \widetilde{a}_l) \in \mathbb{R}^{m_* \times l}, \\ \mathbb{1}_* &= \begin{pmatrix} \mathbb{1} & 0_{(m_0-m_*)\times m_*} \\ \widetilde{A} & 0_{m_* \times m_*} \end{pmatrix} \in \mathbb{R}^{m_0 \times (l+m_*)}. \end{aligned} \tag{9.93}$$

Write $\widehat{Q} = \begin{pmatrix} \widehat{Q}^{11} & \widehat{Q}^{12} \\ \widehat{Q}^{21} & \widehat{Q}^{22} \end{pmatrix}$ so that $\widehat{Q}^{11} \in \mathbb{R}^{(m_0-m_*)\times(m_0-m_*)}$, $\widehat{Q}^{22} \in \mathbb{R}^{m_* \times m_*}$, and $\widehat{Q}^{12}$ and $\widehat{Q}^{21}$ have appropriate dimensions. Denote

$$\overline{Q} = \tilde{\nu}(\widehat{Q}^{11}\mathbb{1} + \widehat{Q}^{12}\widetilde{A}), \quad \overline{Q}_* = \operatorname{diag}(\overline{Q}, 0_{m_* \times m_*}). \tag{9.94}$$

Define the aggregated process $\overline{\alpha}^\varepsilon(\cdot)$ by

$$\overline{\alpha}^\varepsilon(t) = \begin{cases} i, & \text{if } \alpha^\varepsilon(t) \in \mathcal{M}_i, \\ U_j, & \text{if } \alpha^\varepsilon(t) = s_{*j}, \end{cases} \tag{9.95}$$

where

$$U_j = \sum_{i=1}^{l} i I_{\{\sum_{j_0=1}^{i-1} \widetilde{a}_{j_0 j} < U \le \sum_{j_0=1}^{i} \widetilde{a}_{j_0 j}\}},$$

and U is a random variable uniformly distributed on [0, 1], independent of $\alpha^\varepsilon(\cdot)$. We only lump the states in each irreducible class; thus, the state space of the aggregated process $\overline{\alpha}^\varepsilon(\cdot)$ is again $\overline{\mathcal{M}} = \{1, \dots, l\}$. The detailed proofs are omitted; see [303] and the references therein.

Lemma 9.21 *Assume condition* (H9.5). *Then, the following assertions hold.*

(a) *The probability vector* $p^\varepsilon(t) = (p^{\varepsilon,1}(t), p^{\varepsilon,2}(t), \dots, p^{\varepsilon,l}(t), p^{\varepsilon,*}(t))$, *with* $p^{\varepsilon,i}(t) \in \mathbb{R}^{1\times m_i}$ *and*

$$p^{\varepsilon,*}(t) = (\mathbb{P}(\alpha^\varepsilon(t) = s_{*1}), \dots, \mathbb{P}(\alpha^\varepsilon(t) = s_{*m_*})) \in \mathbb{R}^{1\times m_*},$$

satisfies

$$p^\varepsilon(t) = (\theta(t)\tilde{\nu}, 0_{m_*}) + O\left(\varepsilon(t+1) + \exp(-\kappa_0 t/\varepsilon)\right)$$

for some $\kappa_0 > 0$, *where*

$$\begin{aligned}
&\frac{d\theta(t)}{dt} = \theta(t)\overline{Q},\\
&\theta(0) = (p^1(0)\mathbb{1}_{m_i} + p^*(0)\widetilde{a}_1, \dots, p^l(0)\mathbb{1}_{m_l} + p^*(0)\widetilde{a}_{m_l}).
\end{aligned}$$

(b) *The transition probability matrix* $P^\varepsilon(t)$ *satisfies*

$$P^\varepsilon(t) = P^{(0)}(t) + O(\varepsilon(t+1) + e^{-\kappa_0 t/\varepsilon}),$$

where $P^{(0)}(t) = \mathbb{1}_*\Theta_*(t)\nu_*$ *with* $\Theta_*(t) = \mathrm{diag}(\Theta(t), I_{m_*\times m_*})$, *where* $\Theta(t)$ *satisfies*

$$\frac{d\Theta(t)}{dt} = \Theta(t)\overline{Q}, \quad \Theta(0) = I.$$

(c) $\overline{\alpha}^\varepsilon(\cdot)$ *converges weakly to* $\overline{\alpha}(\cdot)$, *a Markov chain generated by* $\overline{Q}$.

(d) *For* $i = 1, \dots, l$, $j = 1, \dots, m_i$

$$\mathbb{E}\left[\int_0^\infty e^{-t}(I_{\{\alpha^\varepsilon(t)=s_{ij}\}} - \nu^i_j I_{\{\overline{\alpha}^\varepsilon(t)=i\}})dt\right]^2 = O(\varepsilon),$$

and for $i = * \ j = 1, \ldots, m_*$,

$$\mathbb{E}\left[\int_0^{\infty} e^{-t} I_{\{\alpha^{\varepsilon}(t)=s_{*j}\}} dt\right]^2 = O(\varepsilon).$$

(e) $(X^{\varepsilon}(\cdot), \overline{\alpha}^{\varepsilon}(\cdot))$ *converges weakly to* $(X(\cdot), \overline{\alpha}(\cdot))$ *such that the limit operator is as given in Lemma* 9.15.

Remark 9.22 Although a collection of transient states of the discrete events are included, the limit system is still an average with respect to the stationary measures of those ergodic classes only. Asymptotically, the transient states can be discarded because the probabilities go to 0 rapidly.

Theorem 9.23 *Assume* (H9.4) *and* (H9.5), *and for each* $i \in \overline{\mathcal{M}}$, *there is a Lyapunov function* $V(x, i)$ *such that* $\mathcal{L}V(x, i) \le -\gamma$ *for some* $\gamma > 0$. *Then, for sufficiently small* $\varepsilon > 0$, *the process* $(X^{\varepsilon}(t), \alpha^{\varepsilon}(t))$ *with inclusion of transient discrete events is still positive recurrent.*

Idea of Proof Since the proof is along the line of that of Theorem 9.18, we only note the differences compared with that theorem. For $i = 1, \ldots, l$, let $V(x, i)$ be the Lyapunov function associated with the limit process given by Lemma 9.21(v). The perturbed Lyapunov function method is used again. This time, redefine

$$\overline{V}(x, \alpha) = \sum_{i=1}^{l} V(x, i) I_{\{\alpha \in \mathcal{M}_i\}} + \sum_{i=1}^{m_*} V(x, i) \widetilde{a}_{i,j} I_{\{\alpha = s_{*j}\}}. \tag{9.96}$$

It is easy to check that $\widetilde{Q}\overline{V}(x, \cdot)(\alpha) = 0$, where $\widetilde{Q}$ is defined in (9.92). Moreover

$$\begin{aligned} &b(x, \alpha^{\varepsilon}(t)) - \overline{b}(x, \overline{\alpha}^{\varepsilon}(t)) \\ &= \sum_{i=1}^{i} \sum_{j=1}^{m_i} b(x, s_{ij}) [I_{\{\alpha^{\varepsilon}(t)=s_{ij}\}} - \nu_j^i I_{\{\overline{\alpha}^{\varepsilon}(t)=i\}}] + \sum_{j=1}^{m_*} b(x, s_{*j}) I_{\{\alpha^{\varepsilon}(t)=s_{*j}\}}. \end{aligned}$$

Therefore, we can carry out the proof in a similar manner to that of Theorem 9.18. The details are omitted.

9.6 Notes

In the first part of this chapter, we developed asymptotic expansions for a European-type option price. The essence is the use of two-time-scale formulation to deal with solutions of systems of parabolic PDEs. The result is based on the work of Yin [291]. The approach we are using is constructive. Thus, it sheds more light on how these approximations can be carried out. Full asymptotic expansions have been obtained with uniform asymptotic error bounds for the continuous component belonging to a compact set. If one is only interested in getting asymptotic expansions with certain fixed state variables (as in [89, 91]), then one can work with the entire space $\mathbb{R}$ rather than a compact set. In lieu of $Q^{\varepsilon}(t)$ considered thus far, we may treat a slightly more complex model with

$$Q^{\varepsilon}(t) = \frac{Q_0(t)}{\varepsilon} + Q_1(t),$$

where both $Q_0(\cdot)$ and $Q_1(\cdot)$ are generators of continuous-time Markov chains such that $Q_0(t)$ is weakly irreducible. Then, we can still obtain asymptotic expansions. The notation, however, will be a bit more complex due to the addition of $Q_1(t)$. In view of the work by Il'in, Khasminskii, and Yin [137, 138], the results of this chapter can be extended to switching diffusions in which the switching process has generator $Q^{\varepsilon}(x, t)$ that depends on x as well.

For risk-neutral valuation, it is natural to let r be a constant. Nevertheless, the techniques presented here carry over for the more general α-dependent process; that is, $r(t) = r(\alpha(t))$. Although the main motivation is from mathematical finance, the techniques developed here can be used for other systems of coupled differential equations where a fast-varying switching process is a driving force.

The second part of this chapter continues our study of positive recurrence for switching diffusions. In contrast to the previous chapters, this chapter takes up the issue that the switching component may have multiple ergodic classes that are weakly connected. The main idea is the use of a two-time-scale approach. Roughly, if the discrete events or certain components of the discrete events change sufficiently quickly, the positive recurrence can still be guaranteed. The ingredient is the use of perturbed Lyapunov function methods.

Chapter 10
Switching Jump Diffusions: Time-Scale Separations

10.1 Introduction

This chapter is concerned with jump diffusions modulated by a Markovian switching process. In the models, there are a finite set of regimes or configurations and a switching process that dictates which regime to take at a given instance. At each time t, once the configuration is determined by the switching process, the dynamics of the system follow a jump diffusion process. It evolves until the next jump takes place. Then, the post-jump location is determined and the process sojourns in the new location following the evolution of another jump-diffusion process and so on. The entire system consists of random switches and jump-diffusive motions.

One of our motivations stems from insurance risk theory. To capture the features of insurance policies that are subject to economic or political environment changes, generalized hybrid risk models may be considered. To reduce the complexity of the systems, time-scale separation may be used. Under the classical insurance risk model, the surplus $U(t)$ of an insurance company at $t \geq 0$ is given by

$$U(t) = u + ct - S(t), \tag{10.1}$$

where u is the initial surplus, $c > 0$ is the rate at which the premiums are received, and $S(t)$, a compound Poisson process, is the total claim in the duration $[0, t]$. In [72], Dufresne and Gerber extended the classical risk model by adding an independent diffusion process so that the surplus is given by

$$U(t) = u + ct - S(t) + \sigma w(t), \tag{10.2}$$

where $w(t)$ is a standard real-valued Brownian motion that represents uncertainty (often referred to as oscillations) of premium incomes and claims. Subsequently, much work has been devoted to such jump-diffusion models; see also variants of the models in [215, 248] and the references therein.

H.-D. Nguyen et al., *Hybrid Switching Diffusions*, Probability Theory and Stochastic Modelling 63, https://doi.org/10.1007/978-3-031-93303-5_10

Recently, growing attention has been drawn to the use of switching models in finance and insurance industry. For instance, taking the opportunity provided by using a switching process to represent the underlying economy switching among a finite number of discrete states, the European options under the Black-Scholes formulation of the stock market were considered in [62]; the American options were dealt with in [32]; algorithms for liquidation of a stock were constructed in [295]. Using a random pure jump process to represent a random environment, we proposed a Markovian regime-switching formulation in [287] to model the insurance surplus process. Suppose that there is a finite set $\mathcal{M} = \{1, \ldots, m_0\}$, representing the possible regimes (configurations) of the environment. At each $i \in \mathcal{M}$, assume that the premium is payable at the rate $c(i)$ continuously. Let $U(t, i)$ be the surplus process given the initial surplus $u > 0$ and initial state $\alpha(0) = i$:

$$U(t, i) = u + \int_0^t c(\alpha(s))ds - S(t), \tag{10.3}$$

where $S(t)$, as in the classical risk model, is a compound Poisson process and $\alpha(t)$ is a continuous-time Markov chain with state space $\mathcal{M}$ representing the random environment. Under suitable conditions, we obtained Lunderberg-type upper bounds and nonexponential upper bounds for the ruin probability and treated a renewal-type system of equations for ruin probability when the claim sizes are exponentially distributed. One of the main features of [287] is that there is an additional Markov chain, which enables the underlying surplus to vary in accordance with different regimes. Likewise, we can consider models with an additional Brownian motion term as in (10.2).

Here, we are concerned with a class of jump diffusions with regime switching to prepare us for treating applications involving more general risk models. In this chapter, we consider jump diffusions modulated by a continuous-time Markov chain. Because the dynamic systems are complex, it is of foremost importance to reduce the complexity. Taking into consideration the inherent hierarchy in a complex system [260] and different rates of variations of subsystems and components, we use the two-time-scale method leading to systems in which the fast and slow rates of change are in sharp contrast. Then, we proceed to reduce the system complexity by aggregation/decomposition and averaging methods. We demonstrate that under broad conditions, associated with the original systems, there are limit or reduced systems, which are averages with respect to certain invariant measures. Using weak convergence methods [175, 176], we obtain the limit system (jump diffusion with regime switching) using martingale problem formulation.

10.1.1 Hybrid Jump Diffusion

For notational simplicity, we will consider only real-valued process throughout this chapter. Let us recall the notion of a switching jump diffusion or hybrid jump diffusion. A hybrid jump diffusion is a collection of jump diffusions modulated by an additional continuous-time switching process. In what follows, we confine ourselves to the case where the switching process is a continuous-time Markov chain. In this case, in lieu of one jump diffusion, we have a number of jump diffusions, which switch back and forth modulated by the continuous-time Markov chain. The description of the jump-diffusion system below is a modification of that of [176] due to the appearance of the switching process. Suppose that $\alpha(\cdot)$ is a continuous-time Markov chain with state space $\mathcal{M} = \{1, \ldots, m\}$ and generator $Q(t) = (q_{ij}(t))$ [303, Sections 2.3–2.5]. Let $\{\rho_n\}$ be an increasing sequence of stopping times independent of $\alpha(\cdot)$. Let $\{\psi_n\}$ be a sequence of random variables representing the "impulses" and $\psi(\cdot)$ be a random process defined by

$$\psi(t) = \begin{cases} \psi_n & \text{if } t = \rho_n, \\ 0, & \text{otherwise.} \end{cases}$$

The process is termed a point process if $\{\psi_n, \rho_n\}$ is a random sequence and $\{\rho_n\}$ has no finite accumulation point. In the above, the ψ_n are referred to as impulses.

Let $\Gamma \subset \mathbb{R} \setminus \{0\}$ be a compact set of the range space of $\psi(\cdot)$ (the impulsive jumps). Note that Γ does not include the origin of the Euclidean space. We will comment on some more general setups in the final section of this chapter. Denote the σ-algebra of Borel sets of Γ by $\mathcal{B}(\Gamma)$. Throughout this chapter, we assume that ψ is an $\mathcal{F}_t$-adapted stationary Poisson point process; we refer to Section I.9 of [136] for more description on Poisson point process. Consequently, ψ induces a Poisson random measure on $(0, \infty) \times \mathcal{B}(\Gamma)$ given by

$$\begin{aligned} N([0,t], H) &:= \sum_n I_{\{\rho_n \le t, \psi_n \in H\}} \\ &= \{\#\text{ of impulses of } \psi \text{ on } [0,t] \text{ with values in } H\}, \quad t > 0, H \in \mathcal{B}(\Gamma). \end{aligned}$$

Note that for each $H \in \mathcal{B}(\Gamma)$, $\{N(t, H) := N([0,t], H), t \ge 0\}$ is a counting process that counts the number of impulses on $[0, t]$ with values in the set H. In addition, for each $t > 0$ and $H \in \mathcal{B}(\Gamma)$, the process $N(t + \cdot, H) - N(t, H)$ is independent of $\mathcal{F}_t$. Suppose throughout this chapter that $\mathbb{E}N(t, \Gamma) < \infty$ for each $t > 0$. Consequently, there exists a $\lambda > 0$ and probability measure $\pi(\cdot)$ on $\mathcal{B}(\Gamma)$ such that

$$\mathbb{E}[N(t+s, H) - N(t, H) \big| \mathcal{F}_t] = s\pi(H)\lambda, \tag{10.4}$$

where λ is known as the impulse rate of $\psi(\cdot)$ and/or the jump rate of $N(\cdot, \Gamma)$ and $\pi(H)$ is the jump distribution in the sense that

$$\mathbb{P}(\psi(t) \in H | \psi(t) \ne 0, \ \psi(u), \ u < t) = \pi(H).$$

Moreover, for any bounded and $\mathcal{F}_t$-predictable process $f : [0, \infty) \times \Gamma \times \Omega \mapsto \mathbb{R}$, we have

$$\int_0^t f(s, \gamma) N(ds, d\gamma) = \sum_{s \le t, \psi(s) \ne 0} f(s, \psi(s)).$$

Consider the following jump-diffusion process with regime switching

$$\begin{aligned} X(t) = x &+ \int_0^t f(s, X(s), \alpha(s))ds + \int_0^t \sigma(s, X(s), \alpha(s))dw(s) \\ &+ \int_0^t \int_\Gamma g(\gamma, X(s-), \alpha(s-))N(ds, d\gamma). \end{aligned} \tag{10.5}$$

where $w(\cdot)$ is a real-valued standard Brownian motion, $\alpha(\cdot)$ is a continuous-time Markov chain with state space $\mathcal{M} = \{1, \dots, m\}$ and generator $Q(t) = (q_{ij}(t))$, and $N(\cdot, \cdot)$ is the Poisson random measure introduced in the previous paragraph. In (10.5), $f : [0, T]\times\mathbb{R}\times\mathcal{M} \mapsto \mathbb{R}$, $\sigma : [0, T]\times\mathbb{R}\times\mathcal{M} \mapsto \mathbb{R}$, and $g : \Gamma\times\mathbb{R}\times\mathcal{M} \mapsto \mathbb{R}$ are Borel measurable functions. Throughout the chapter, we assume that $w(\cdot)$, $N(\cdot)$, and $\alpha(\cdot)$ are mutually independent. Compared with the traditional jump-diffusion processes, the coefficients involved in (10.5) all depend on an additional switching process, namely, the Markov chain $\alpha(\cdot)$.

In the context of risk theory, $X(t)$ can be considered as the surplus of the insurance company at time t, x is the initial surplus, $f(t, X(t), \alpha(t))$ represents the premium rate (assumed to be ≥ 0), $g(\gamma, X(t), \alpha(t))$ is the amount of the claim whenever there is one (assumed to be ≤ 0), and the diffusion is used to model additional uncertainty of the claims and/or premium incomes. Similar to the volatility in stock market models, $\sigma(\cdot, \cdot, i)$ represents the amount of oscillations or volatility in an appropriate sense. The model is sufficiently general to cover the traditional compound Poisson models as well as the diffusion perturbed ruin models. It may also be used to represent security price in finance (see [209, Chapter 3]). The process $\alpha(t)$ may be viewed as an environment variable dictating the regime. The use of the Markov chain results from consideration of the general trend of the market environment as well as other economic factors. The economic and/or political environment changes lead to the changes of regime of the surplus, resulting in markedly different behavior of the system across regimes.

Define a centered (or compensated) Poisson measure as

$$\widetilde{N}(t, H) = N(t, H) - \lambda t \pi(H) \text{ for } H \subset \Gamma. \tag{10.6}$$

In this chapter, we assume $0 < \lambda < \infty$. The generator of the underlying process can be written as

$$
\begin{aligned}
\mathcal{G}F(t,x,\iota) = & \left(\frac{\partial}{\partial t} + \mathcal{L}\right) F(t,x,\iota) \\
& + \int_{\Gamma} \lambda[F(t,x+g(\gamma,x,\iota),\iota) - F(t,x,\iota)]\pi(d\gamma) \\
& + Q(t)F(t,x,\cdot)(\iota), \quad \text{for each } \iota \in \mathcal{M},
\end{aligned} \tag{10.7}
$$

where

$$
\begin{aligned}
\mathcal{L}F(t,x,\iota) &= \frac{1}{2}\sigma^2(t,x,\iota)\frac{\partial^2}{\partial x^2}F(t,x,\iota) + f(t,x,\iota)\frac{\partial}{\partial x}F(t,x,\iota), \\
Q(t)F(t,x\cdot)(\iota) &= \sum_{\ell=1}^{m_0} a_{\iota\ell}(t)F(t,x,\ell) = \sum_{\ell\neq\iota} a_{\iota\ell}(t)[F(t,x,\ell) - F(t,x,\iota)].
\end{aligned} \tag{10.8}
$$

By concentrating on time-scale separations, in this chapter, we treat two cases. In the first one, the regime switching is significantly faster than the dynamics of the jump diffusions, whereas in the second case, the jump diffusion varies an order of magnitude faster than the switching processes.

10.1.2 Chapter Outline

The rest of the chapter is arranged as follows. Section 10.2 is devoted to the case of fast switching. It begins with the precise formulation of the problem. Then, we derive weak convergence results and demonstrate that a complex problem can be "replaced" by a limit problem in which the system coefficients are averaged out with respect to the stationary measures of the switching process. In Section 10.3, we continue our study for the case of fast-varying diffusions. Again, by means of weak convergence methods, we obtain a limit system. Section 10.4 provides remarks on specialization and generalization of the asymptotic results. Section 10.5 remarks on numerical approximation for switching-jump-diffusion processes. Section 10.6 concludes the chapter.

10.2 Fast-Varying Switching

10.2.1 Fast-Varying Markov Chain Model

Consider a continuous-time inhomogeneous Markov chain $\alpha(t)$ with generator $Q(t)$. Recall that the Markov chain $\alpha(t)$ or the generator $Q(t)$ is weakly irreducible if the system of equations

$$\begin{cases} \nu(t)Q(t) = 0, \\ \displaystyle\sum_{i=1}^{m_0} \nu_i(t) = 1 \end{cases}$$

has a unique nonnegative solution. The nonnegative solution (row-vector-valued function) $\nu(t) = (\nu_1(t), \ldots, \nu_{m_0}(t))$ is termed a quasi-stationary distribution.

For the fast-varying Markov chain model, by introducing a small parameter $\varepsilon > 0$ into the problem, suppose that $\alpha(t) = \alpha^\varepsilon(t)$ with the generator of the Markov chain given by

$$Q^\varepsilon(t) = \frac{1}{\varepsilon}\widetilde{Q}(t) + \widehat{Q}(t). \tag{10.9}$$

Both $\widetilde{Q}(t)$ and $\widehat{Q}(t)$ are generators, where $\widetilde{Q}(t)/\varepsilon$ represents the rapidly changing part and $\widehat{Q}(t)$ describes the slowly varying motion. The slow and fast processes are coupled through weak and strong interactions in the sense that the underlying Markov chain fluctuates rapidly within a single group $\mathcal{M}_k$ of states and jumps less frequently from group $\mathcal{M}_k$ to $\mathcal{M}_j$ for $k \neq j$. Suppose that the generator $\widetilde{Q}(t)$ has the form

$$\widetilde{Q}(t) = \operatorname{diag}\left(\widetilde{Q}^1(t), \ldots, \widetilde{Q}^l(t)\right), \tag{10.10}$$

where for each $k = 1, \ldots, l$. $\widetilde{Q}^k(t)$ is a generator corresponding to the states in $\mathcal{M}_k = \{s_{k1}, \ldots, s_{km_k}\}$. Naturally, the state space can be decomposed to

$$\mathcal{M} = \mathcal{M}_1 \cup \cdots \cup \mathcal{M}_l = \{s_{11}, \ldots, s_{1m_1}\} \cup \cdots \cup \{s_{l1}, \ldots, s_{lm_l}\}. \tag{10.11}$$

The corresponding system can be written as

$$\begin{aligned} X^\varepsilon(t) = x &+ \int_0^t f(s, X^\varepsilon(s), \alpha^\varepsilon(s))ds + \int_0^t \sigma(s, X^\varepsilon(s), \alpha^\varepsilon(s))dw(s) \\ &+ \int_0^t \int_\Gamma g(\gamma, X^\varepsilon(s-), \alpha^\varepsilon(s-))N(ds, d\gamma), \end{aligned} \tag{10.12}$$

Using the centered Poisson measure $\widetilde{N}(t, H)$ defined in (10.6), we can rewrite (10.12) as

$$\begin{aligned} X^\varepsilon(t) = x &+ \int_0^t f(s, X^\varepsilon(s), \alpha^\varepsilon(s))ds \\ &+ \lambda \int_0^t \int_\Gamma g(\gamma, X^\varepsilon(s-), \alpha^\varepsilon(s-))\pi(d\gamma)ds \\ &+ \int_0^t \sigma(s, X^\varepsilon(s), \alpha^\varepsilon(s))dw(s) \\ &+ \int_0^t \int_\Gamma g(\gamma, X^\varepsilon(s-), \alpha^\varepsilon(s-))\widetilde{N}(ds, d\gamma). \end{aligned} \tag{10.13}$$

Note that the last two terms are martingales. The operator of the regime-switching jump-diffusion process is given by

$$\mathcal{G}^{\varepsilon} F(t, x, \iota) = \left(\frac{\partial}{\partial t} + \mathcal{L}\right) F(t, x, \iota) + J(t, x, \iota) + Q^{\varepsilon}(t) F(t, x, \cdot)(\iota), \tag{10.14}$$

where

$$J(t, x, \iota) = \int_{\Gamma} \lambda[F(t, x + g(\gamma, x, \iota), \iota) - F(t, x, \iota)]\pi(d\gamma).$$

Remark 10.1 The use of the two-time-scale formulation with a small parameter $\varepsilon > 0$ stems from an effort to reduce complexity. Taking into consideration of various factors, the number of states of the Markov chain is usually large; the system becomes complex; dealing with it is difficult. Nevertheless, we demonstrate by letting $\varepsilon \to 0$, a limit system can be obtained, which is an average of the original system with respect to the quasi-stationary distribution of the Markov chain.

To carry out the desired analysis, we postulate the following conditions.

(H10.1) $\mathbb{E}|x|^2 < \infty$, and the following conditions hold.

(a) The functions $f, \sigma : [0, T] \times \mathbb{R} \times \mathcal{M} \mapsto \mathbb{R}$ are Borel measurable and satisfy

$$|f(t, x, \alpha)| + |\sigma(t, x, \alpha)| \leq K(1 + |x|),$$
$$|f(t, z, \alpha) - f(t, x, \alpha)| + |\sigma(t, z, \alpha) - \sigma(t, x, \alpha)| \leq K|z - x|,$$

for all $t \in [0, T]$, $x, z \in \mathbb{R}$, and $\alpha \in \mathcal{M}$, where K is a positive constant.

(b) For each $\alpha \in \mathcal{M}$, the function $g(\cdot, \cdot, \alpha)$ is bounded and continuous with $g(0, x, \alpha) = 0$.

(H10.2) Both $\widetilde{Q}(t)$ and $\widehat{Q}(t)$ are generators that are bounded and Borel measurable such that for each $k = 1, \ldots, l$ and $t \in [0, T]$, $\widetilde{Q}^k(t)$ is weakly irreducible with the associated quasi-stationary distribution $\nu^k(t) = (\nu_1^k(t), \ldots, \nu_{m_k}^k(t)) \in \mathbb{R}^{1 \times m_k}$.

Remark 10.2 In the above and throughout the chapter, K is used as a generic positive constant, whose values may change for different usages, so the conventions $K + K = K$ and $KK = K$ are understood.

For each fixed $\alpha \in \mathcal{M}$, the conditions on $f(\cdot, \cdot, \alpha)$ and $\sigma(\cdot, \cdot, \alpha)$ are the "Itô conditions" that ensure the existence and uniqueness of the solution of the stochastic differential equation. The assumption on $\widetilde{Q}(t)$ leads to the partition of $\mathcal{M}$, the state space of $\alpha^{\varepsilon}(\cdot)$ as in (10.11), which in turn leads to the natural definition of an aggregated process. That is, by aggregating all the states s_{kj} in $\mathcal{M}_k$ as one state, we obtain an aggregated process $\overline{\alpha}^{\varepsilon}(\cdot)$ defined by

$$\overline{\alpha}^{\varepsilon}(t) = k \quad \text{if} \quad \alpha^{\varepsilon}(t) \in \mathcal{M}_k. \tag{10.15}$$

Thus, lump all the states in each weakly irreducible class into one state resulting in a process with considerably smaller state space. Note that the process $\overline{\alpha}^\varepsilon(\cdot)$ is not necessarily Markovian.

Condition (H10.2) is concerned with the Markov chain $\alpha^\varepsilon(\cdot)$, which is modeled after that of [303, Chapter 7.5, p. 210] and allows the generators to be time dependent. Note that no continuity of $\widetilde{Q}(\cdot)$ and $\widehat{Q}(\cdot)$ is required, but merely boundedness and measurability are assumed; see also [305] for the inclusion of transient states. For our applications, the only requirement is the partitioned form $\widetilde{Q}(\cdot)$ and the weak irreducibility of each $\widetilde{Q}^k(\cdot)$.

The following idea uses an averaging approach by aggregating the states in each weakly irreducible class into a single state, and replacing the original complex system by its limit, an average with respect to the quasi-stationary distributions. Using certain probabilistic arguments, we have shown in [303, Section 7.5] (see also [305]) that

(i) $\overline{\alpha}^\varepsilon(\cdot)$ converges weakly to $\overline{\alpha}(\cdot)$ whose generator is given by

$$\overline{Q}(t) = \mathrm{diag}(\nu^1(t), \ldots, \nu^l(t))\widehat{Q}(t)\mathrm{diag}(\mathbb{1}_{m_1}, \ldots, \mathbb{1}_{m_l}), \tag{10.16}$$

where $\nu^k(t)$ is the quasi-stationary distribution of $\widetilde{Q}^k(t)$, for $k = 1, \ldots, l$, and $\mathbb{1}_\iota = (1, \ldots, 1)' \in \mathbb{R}^\iota$ is an ι-dimensional column vector with all components being equal to 1.

(ii) for $k = 1, \ldots, l$ and $j = 1, \ldots, m_k$, as $\varepsilon \to 0$

$$\sup_{0 \le t \le T} \mathbb{E}\left(\int_0^t \left(I_{\{\alpha^\varepsilon(s)=s_{kj}\}} - \nu_j^k I_{\{\overline{\alpha}^\varepsilon(s)=k\}}\right) ds\right)^2 \to 0, \tag{10.17}$$

where I_A is the indicator function of the set A.

10.2.2 Limit System

Working with the pair $Y^\varepsilon(\cdot) = (X^\varepsilon(\cdot), \overline{\alpha}^\varepsilon(\cdot))$, we obtain the following theorem. It indicates that there is a limit system associated with the original process leading to a reduction of complexity.

Theorem 10.3 *Assuming* (H10.1) *and* (H10.2), $Y^\varepsilon(\cdot) = (X^\varepsilon(\cdot), \overline{\alpha}^\varepsilon(\cdot))$ *converges weakly to* $Y(\cdot) = (X(\cdot), \overline{\alpha}(\cdot))$*, which is a solution of the martingale problem with operator*

$$\overline{\mathcal{G}}F(t, x, i) = \left(\frac{\partial}{\partial t} + \overline{\mathcal{L}}\right) F(t, x, i) + \overline{\mathcal{J}}(t, x, i) + \overline{Q}(t)F(t, x, \cdot)(i), \tag{10.18}$$

for $i \in \overline{\mathcal{M}} = \{1, \ldots, l\}$, *where*

$$\overline{Q}(t)F(t,x,\cdot)(i) = \sum_{j=1}^{l} \overline{q}_{ij}(t)F(t,j,x) = \sum_{\substack{j\in\overline{\mathcal{M}}\\ j\neq i}} \overline{q}_{ij}(t)(F(t,x,j) - F(t,x,i)), \tag{10.19}$$

$\overline{\mathcal{L}}$ *is a second-order differential operator given by* (10.8) *with* $\sigma^2(t,x,\iota)$, $J(t,x,\iota)$, *and* $f(t,x,\iota)$ *replaced by* $\overline{\sigma}^2(t,x,i)$, $\overline{J}(t,x,i)$, *and* $\overline{f}(t,x,i)$, *respectively, and*

$$\begin{aligned} \overline{f}(t,x,i) &= \sum_{j=1}^{m_i} \nu_j^i(t) f(t,x,s_{ij}), \\ \overline{J}(t,x.i) &= \sum_{j=1}^{m_i} \nu_j^i(t) J(t,x,s_{ij}) \\ \overline{\sigma}^2(t,x,i) &= \sum_{j=1}^{m_i} \nu_j^i(t)\sigma^2(t,x,s_{ij}). \end{aligned} \tag{10.20}$$

Remark 10.4 Theorem 10.3 characterizes the limit as a solution of the associated martingale problem with operator $\overline{\mathcal{G}}$. It can also be described by the limit stochastic differential equation given by

$$\begin{aligned} X(t) = x &+ \int_0^t \overline{f}(s, X(s), \overline{\alpha}(s))ds + \int_0^t \overline{\sigma}(s, X(s), \overline{\alpha}(s))dw(s) \\ &+ \int_0^t \int_\Gamma \overline{g}(\gamma, X(s-), \overline{\alpha}(s-))N(ds, d\gamma), \end{aligned} \tag{10.21}$$

where $\overline{\alpha}(\cdot)$ is a Markov chain generated by $\overline{Q}(\cdot)$. In particular, if the Markov chain corresponding to the generator $\widetilde{Q}(t)$ consists of only one weakly irreducible block, then the limit or the averaged system becomes a jump-diffusion process. We state the result as follows.

Corollary 10.5 *Under the conditions of Theorem* 10.3 *with the modification that* $Q^\varepsilon(t) = Q(t)/\varepsilon + \widehat{Q}(t)$ *such that* $\mathcal{M} = \{1, \ldots, m\}$ *and that* $Q(t)$ *is weakly irreducible for each* $t \in [0, T]$ *with the associated quasi-stationary distribution* $\nu(t) = (\nu_1(t), \ldots, \nu_{m_0}(t))$, *then* $Y^\varepsilon(\cdot)$ *converges weakly to* $Y(\cdot)$ *such that*

$$\begin{aligned} X(t) = x &+ \int_0^t \sum_{\iota=1}^{m_0} f(s, X(s), \iota)\nu_\iota(s)ds \\ &+ \int_0^t \sqrt{\sum_{\iota=1}^{m_0} \sigma^2(s, X(s), \iota)\nu_\iota(s)}dw(s) \\ &+ \int_0^t \sum_{\iota=1}^{m_0} g(\gamma, X(s-), \iota)\nu_\iota(s-)N(ds, d\gamma). \end{aligned} \tag{10.22}$$

***Proof of Theorem* 10.3** To proceed, the detailed proof is given by establishing a series of lemmas. We first show that an a priori bound holds.

Lemma 10.6 *Under the conditions of Theorem* 10.3

$$\sup_{0\le t\le T} \mathbb{E}|X^\varepsilon(t)|^2 = O(1).$$

Proof It follows from (10.12) that

$$\begin{aligned}\mathbb{E}|X^\varepsilon(t)|^2 \le K\bigg(\mathbb{E}|x|^2 &+ \mathbb{E}\bigg|\int_0^t f(u, X^\varepsilon(u), \alpha^\varepsilon(u))du\bigg|^2 \\ &+ \mathbb{E}\bigg|\int_0^t \sigma(u, X^\varepsilon(u), \alpha^\varepsilon(u))dw(u)\bigg|^2 \\ &+ \mathbb{E}\bigg|\int_0^t\int_\Gamma g(\gamma, X^\varepsilon(u-), \alpha^\varepsilon(u-))N(du, d\gamma)\bigg|^2\bigg).\end{aligned}$$

By virtue of the boundedness assumption on g and the Poisson random measure N, we have

$$\mathbb{E}\sup_{0\le t\le T}\bigg|\int_0^t\int_\Gamma g(\gamma, X^\varepsilon(u-), \alpha^\varepsilon(u-))N(du, d\gamma)\bigg|^2 = O(1).$$

Then, using the linear growth of $f(t, x, \alpha)$ and $\sigma(t, x, \alpha)$ given in (H10.1) together with properties of stochastic integrals, we obtain

$$\mathbb{E}|X^\varepsilon(t)|^2 \le K + K\int_0^t \mathbb{E}|X^\varepsilon(u)|^2 du.$$

The well-known Gronwall inequality yields

$$\sup_{t\in[0,T]} \mathbb{E}|X^\varepsilon(t)|^2 \le K\exp(KT) < \infty$$

as desired. □

Next, we derive the tightness of $Y^\varepsilon(\cdot)$.

Lemma 10.7 *Assume that the conditions of Theorem* 10.3 *are satisfied. Then,* $Y^\varepsilon(\cdot)$ *is tight in* $D([0, T] : \mathbb{R}\times\mathcal{M})$*, the space of functions that are right continuous and have left limits endowed with the Skorokhod topology.*

Proof Because $\overline{\alpha}^\varepsilon(\cdot)$ converges weakly to $\overline{\alpha}(\cdot)$ [303, p. 172], $\{\overline{\alpha}^\varepsilon(\cdot)\}$ is tight. Therefore, to obtain the tightness of $\{Y^\varepsilon(\cdot)\}$, it suffices to derive the tightness of $\{X^\varepsilon(\cdot)\}$.

Note that for any $t > 0$, $s > 0$, and any $\delta > 0$ with $0 < s \le \delta$, we have

$$
\begin{aligned}
&\mathbb{E}_t^\varepsilon |X^\varepsilon(t+s)-X^\varepsilon(t)|^2 \\
&\le K\mathbb{E}_t^\varepsilon\bigg[\bigg|\int_t^{t+s}|f(u,X^\varepsilon(u),\alpha^\varepsilon(u))|du\bigg|^2+\bigg|\int_t^{t+s}\sigma(u,X^\varepsilon(u),\alpha^\varepsilon(u))dw(u)\bigg|^2\bigg] \\
&\quad + K\mathbb{E}_t^\varepsilon\bigg|\int_t^{t+s}\int_\Gamma g(\gamma,X^\varepsilon(u-),\alpha^\varepsilon(u-))N(du,d\gamma)\bigg|^2,
\end{aligned}
\tag{10.23}
$$

where $\mathbb{E}_t^\varepsilon$ denotes the conditional expectation with respect to the σ-algebra generated by $\{\alpha^\varepsilon(u), X^\varepsilon(u) : u \le t\}$. Using the argument as in [176, p. 39], since $s \le \delta$

$$
\mathbb{E}_t^\varepsilon\bigg|\int_t^{t+s}\int_\Gamma g(\gamma,X^\varepsilon(u-),\alpha^\varepsilon(u-))N(du,d\gamma)\bigg|^2 \le \beta^\varepsilon(\delta)
$$

where $\beta^\varepsilon(\delta)$ is $\mathcal{F}_t^\varepsilon$-measurable and $\beta^\varepsilon(\delta)$ is bounded by $O(\delta)$. In view of (H10.1) and Lemma 10.6, the usual arguments lead to

$$
\begin{aligned}
\mathbb{E}_t^\varepsilon\bigg|\int_t^{t+s}|f(u,X^\varepsilon(u),\alpha^\varepsilon(u))|du\bigg|^2+\mathbb{E}_t^\varepsilon\bigg|\int_t^{t+s}\sigma(u,X^\varepsilon(u),\alpha^\varepsilon(u))dw(u)\bigg|^2& \\
\le \beta^\varepsilon(\delta) \le O(\delta).&
\end{aligned}
$$

As a result, we have

$$
\lim_{\delta\to 0}\limsup_{\varepsilon\to 0}\mathbb{E}_t^\varepsilon|X^\varepsilon(t+s)-X^\varepsilon(t)|^2 \le 2\lim_{\delta\to 0}\limsup_{\varepsilon\to 0}\mathbb{E}\beta^\varepsilon(\delta)=0.
$$

By virtue of the tightness criterion (see Lemma A.29 of this book, and also [81, Section 3.8, p. 132] or [175, p. 47] or [28]), the tightness of $\{X^\varepsilon(\cdot)\}$ follows. □

Lemma 10.8 *Assume the conditions of Theorem* 10.3 *are fulfilled. Suppose that* $\zeta(t,z)$ *defined on* $[0,T]\times\mathbb{R}$ *is a real-valued function that is Lipschitz continuous in both variables, and that for each* $x\in\mathbb{R}^r$, $|\zeta(t,x)|\le K(1+|x|)$. *Denote*

$$
v_{ij}^\varepsilon(t)=v_{ij}(t,\alpha^\varepsilon(t)) \ \textit{with} \ v_{ij}(t,\alpha)=I_{\{\alpha=s_{ij}\}}-\nu_j^i(t)I_{\{\alpha\in\mathcal{M}_i\}}.
$$

Then, for any $i=1,\ldots,l$ $j=1,\ldots,m_i$,

$$
\sup_{0<t\le T}\mathbb{E}\bigg|\int_0^t\zeta(u,X^\varepsilon(u))v_{ij}^\varepsilon(u,\alpha^\varepsilon(u))du\bigg|^2 \to 0 \ \textit{as} \ \varepsilon\to 0. \tag{10.24}
$$

Proof The proof of this lemma is similar to that of Lemma 7.14 in [303]. Pick out $0<\Delta<1$. For any $t\in[0,T]$, partition $[0,t]$ into subintervals of equal length $\varepsilon^{1-\Delta}$ (without loss of generality, assume that $\ell_0=t/\varepsilon^{1-\Delta}$ is an integer, otherwise, we can always take its integer part). Denote the partition boundaries by $t_k=k\varepsilon^{1-\Delta}$ for $0\le k\le\ell_0$. Define

$$
\widetilde{\zeta}^\varepsilon(u)=\zeta(t_k,X^\varepsilon(t_k)),\ \ u\in[t_k,t_{k+1}),\ 0\le k\le\ell_0-1. \tag{10.25}
$$

In view of the process $N(\cdot)$, $\alpha^\varepsilon(\cdot)$, and $w(\cdot)$, similar argument as in the proof of the tightness yields

$$\mathbb{E}|X^\varepsilon(t) - X^\varepsilon(t_k)|^2 = O(t - t_k) = O(\varepsilon^{1-\Delta}) \to 0 \text{ as } \varepsilon \to 0 \tag{10.26}$$

for $t \in [t_k, t_{k+1}]$, $0 \le k \le \ell_0 - 1$. It follows that

$$\begin{aligned} &\mathbb{E}\left|\int_0^t \zeta(u, X^\varepsilon(u)))v_{ij}^\varepsilon(u, \alpha^\varepsilon(u))du\right|^2 \\ &\le 2\mathbb{E}\left|\int_0^t [\zeta(u, X^\varepsilon(u)) - \widetilde{\zeta}^\varepsilon(u)]v_{ij}^\varepsilon(u, \alpha^\varepsilon(u))du\right|^2 \\ &\quad + 2\mathbb{E}\left|\int_0^t \widetilde{\zeta}^\varepsilon(u)v_{ij}^\varepsilon(u, \alpha^\varepsilon(u))du\right|^2. \end{aligned} \tag{10.27}$$

We claim that the term on the second line of (10.27) goes to 0. To see this (recall that K is a generic positive constant), by using the Cauchy-Schwarz inequality and the Lipschitz continuity of $\zeta(\cdot)$ and (10.26)

$$\begin{aligned} &\mathbb{E}\left|\int_0^t [\zeta(u, X^\varepsilon(u)) - \widetilde{\zeta}^\varepsilon(u)]v_{ij}^\varepsilon(u, \alpha^\varepsilon(u))du\right|^2 \\ &\le T\int_0^t \mathbb{E}|\zeta(u, X^\varepsilon(u)) - \widetilde{\zeta}^\varepsilon(u)|^2 du \\ &\le K\sum_{k=0}^{\ell_0-1}\int_{t_k}^{t_{k+1}} \mathbb{E}[(u - t_k)^2 + |X^\varepsilon(u) - X^\varepsilon(t_k)|^2]du \\ &\le K\sum_{k=0}^{\ell_0-1}\int_{t_k}^{t_{k+1}} O(\varepsilon^{1-\Delta})du \\ &\to 0 \text{ as } \varepsilon \to 0. \end{aligned}$$

To estimate the term on the last line of (10.27), for each $i = 1, \ldots, l$, and $j = 1, \ldots, m_i$, define

$$\eta_{ij}^\varepsilon(t) = \mathbb{E}\left|\int_0^t \widetilde{\zeta}^\varepsilon(u)v_{ij}^\varepsilon(u, \alpha^\varepsilon(u))du\right|^2.$$

Then, similar to the derivation of [303, pp. 191–192]

$$\frac{d}{dt}\eta_{ij}^\varepsilon(t) = O(\varepsilon^{1-\Delta}), \quad \eta^\varepsilon(0) = 0,$$

as $\varepsilon \to 0$. Thus, solving the above initial value problem leads to

$$\sup_{0\le t\le T} \eta^{\varepsilon}_{ij}(t) = \sup_{0\le t\le T} \int_0^t O(\varepsilon^{1-\Delta})ds = O(\varepsilon^{1-\Delta}) \to 0 \text{ as } \varepsilon \to 0$$

as desired. □

To proceed, for each $i \in \overline{\mathcal{M}} = \{1, \dots, l\}$ and each $F(\cdot, \cdot, i) \in C_0^{1,2}$ ($C_0^{1,2}$ represents the class of functions that have compact support and that are continuously differentiable with respect to t and twice continuously differentiable with respect to x), consider the operator defined in (10.18). The next lemma gives the characterization of the limit process as a solution of a martingale problem.

Lemma 10.9 *Under the conditions of Theorem* 10.3, *the limit process* $\{Y(\cdot)\}$ *is the solution of the martingale problem with operator* $\overline{\mathcal{G}}$ *given by* (10.18).

Proof Theorem A.25 establishes the equivalence between solutions of martingale problems and the corresponding stochastic differential equations. Combined with the uniqueness of solutions to stochastic differential equations stated in [136, Theorem 9.1, p. 245], this implies that the martingale problem with operator $\overline{\mathcal{G}}$ has a unique solution for each initial condition.

To obtain the desired result, it suffices to show that for each $i \in \overline{\mathcal{M}}$ and $F(\cdot, \cdot, i) \in C_0^{1,2}$

$$F(t, X(t), \overline{\alpha}(t)) - \int_0^t \overline{\mathcal{G}}F(u, X(u), \overline{\alpha}(u))du$$

is a martingale. To this end, we show that for any positive integer n_0, any bounded and continuous functions $h_\ell(\cdot)$, $\ell \le n_0$, and any $t, s, t_\ell \ge 0$ with $t_\ell \le t < t+s \le T$

$$\begin{aligned}\mathbb{E}\prod_{\ell=1}^{n_0} h_\ell(X(t_\ell), \overline{\alpha}(t_\ell))\Big(& F(t+s, X(t+s), \overline{\alpha}(t+s)) - F(t, X(t), \overline{\alpha}(t)) \\ & - \int_t^{t+s} \overline{\mathcal{G}}F(u, X(u), \overline{\alpha}(u))du\Big) = 0.\end{aligned} \tag{10.28}$$

Let us begin with the process $Y^\varepsilon(\cdot)$. Define

$$\widehat{F}(t, x, \alpha) = \sum_{i=1}^{l} F(t, x, i) I_{\{\alpha\in\mathcal{M}_i\}} \text{ for each } \alpha \in \mathcal{M}.$$

Clearly, $\widehat{F}(t, X^\varepsilon(t), \alpha^\varepsilon(t)) = F(t, X^\varepsilon(t), \overline{\alpha}^\varepsilon(t))$. Moreover, for each $\iota \in \mathcal{M}$, $\widehat{F}(\cdot, \cdot, \iota) \in C_0^{1,2}$. The function $\widehat{F}(\cdot)$ allows us to conveniently use the available $\alpha^\varepsilon(\cdot)$ process in lieu of the aggregated process $\overline{\alpha}^\varepsilon(\cdot)$. Consider the operator $\mathcal{G}^\varepsilon$ defined in (10.14). Because $Y^\varepsilon(\cdot)$ is a Markov process

$$\widehat{F}(t, X^\varepsilon(t), \alpha^\varepsilon(t)) - \widehat{F}(0, x, \alpha^\varepsilon(0)) - \int_0^t \mathcal{G}^\varepsilon \widehat{F}(u, X^\varepsilon(u), \alpha^\varepsilon(u))du$$

is a martingale. Consequently

$$\mathbb{E}\prod_{\ell=1}^{n_0} h_\ell(X^\varepsilon(t_\ell),\overline{\alpha}^\varepsilon(t_\ell))\Big[\widehat{F}(t+s,X^\varepsilon(t+s),\alpha^\varepsilon(t+s))-\widehat{F}(t,X^\varepsilon(t),\alpha^\varepsilon(t))\\ -\int_t^{t+s}\mathcal{G}^\varepsilon\widehat{F}(u,X^\varepsilon(u),\alpha^\varepsilon(u))du\Big]=0. \tag{10.29}$$

We proceed to obtain the limit in (10.29) as $\varepsilon\to 0$.

First, by the weak convergence of $Y^\varepsilon(\cdot)$ to $Y(\cdot)$, the definition of $\widehat{F}(\cdot)$, and the Skorokhod representation, as $\varepsilon\to 0$

$$\begin{aligned}&\mathbb{E}\prod_{\ell=1}^{n_0} h_\ell(X^\varepsilon(t_\ell),\overline{\alpha}^\varepsilon(t_\ell))[\widehat{F}(t+s,X^\varepsilon(t+s),\alpha^\varepsilon(t+s))-\widehat{F}(t,X^\varepsilon(u),\alpha^\varepsilon(t))]\\ &\to \mathbb{E}\prod_{\ell=1}^{n_0} h_\ell(X(t_\ell),\overline{\alpha}(t_\ell))[F(t+s,X(t+s),\overline{\alpha}(t+s))-F(t,X(t),\overline{\alpha}(t))].\end{aligned} \tag{10.30}$$

The definition of $\mathcal{G}^\varepsilon$ leads to

$$\begin{aligned}&\mathbb{E}\prod_{\ell=1}^{n_0} h_\ell(X^\varepsilon(t_\ell),\overline{\alpha}^\varepsilon(t_\ell))\left[\int_t^{t+s}\mathcal{G}^\varepsilon\widehat{F}(u,X^\varepsilon(u),\alpha^\varepsilon(u))du\right]\\ &=\mathbb{E}\prod_{\ell=1}^{n_0} h_\ell(X^\varepsilon(t_\ell),\overline{\alpha}^\varepsilon(t_\ell))\Big[\int_t^{t+s}\frac{\partial}{\partial u}\widehat{F}(u,X^\varepsilon(u),\alpha^\varepsilon(u))du\\ &\quad+\int_t^{t+s}\widehat{F}_x(u,X^\varepsilon(u),\alpha^\varepsilon(u))f(u,X^\varepsilon(u),\alpha^\varepsilon(u))du\\ &\quad+\int_t^{t+s}\frac{1}{2}[\widehat{F}_{xx}(u,X^\varepsilon(u),\alpha^\varepsilon(u))\sigma^2(u,X^\varepsilon(u),\alpha^\varepsilon(u))]du\\ &\quad+\int_t^{t+s}Q^\varepsilon(u)\widehat{F}(u,X^\varepsilon(u),\cdot)(\alpha^\varepsilon(u))du+\int_t^{t+s}J(u,X^\varepsilon(u),\alpha^\varepsilon(u))du\Big].\end{aligned}$$

Note that

$$\begin{aligned}&\mathbb{E}\prod_{\ell=1}^{n_0} h_\ell(X^\varepsilon(t_\ell),\overline{\alpha}^\varepsilon(t_\ell))\left[\int_t^{t+s}\widehat{F}_x(u,X^\varepsilon(u),\alpha^\varepsilon(u))f(u,X^\varepsilon(u),\alpha^\varepsilon(u))du\right]\\ &=\sum_{i=1}^{l}\sum_{j=1}^{m_i}\mathbb{E}\prod_{\ell=1}^{n_0} h_\ell(X^\varepsilon(t_\ell),\overline{\alpha}^\varepsilon(t_\ell))\\ &\quad\times\left[\int_t^{t+s}\widehat{F}_x(u,X^\varepsilon(u),s_{ij})f(u,X^\varepsilon(u),s_{ij})I_{\{\alpha^\varepsilon(u)=s_{ij}\}}du\right]\end{aligned}$$

$$
\begin{aligned}
&= \sum_{i=1}^{l} \sum_{j=1}^{m_i} \mathbb{E} \prod_{\ell=1}^{n_0} h_\ell(X^\varepsilon(t_\ell), \overline{\alpha}^\varepsilon(t_\ell)) \\
&\quad \times \left[\int_t^{t+s} \widehat{F}_x(u, X^\varepsilon(u), s_{ij}) f(u, X^\varepsilon(u), s_{ij}) \nu_j^i(u) I_{\{\overline{\alpha}^\varepsilon(u)=i\}} du \right] \\
&\quad + \sum_{i=1}^{l} \sum_{j=1}^{m_i} \mathbb{E} \prod_{\ell=1}^{n_0} h_\ell(X^\varepsilon(t_\ell), \overline{\alpha}^\varepsilon(t_\ell)) \\
&\quad \times \left[\int_t^{t+s} \widehat{F}_x(u, X^\varepsilon(u), s_{ij}) f(u, X^\varepsilon(u), s_{ij}) \big(I_{\{\alpha^\varepsilon(u)=s_{ij}\}} - \nu_j^i(u) I_{\{\overline{\alpha}^\varepsilon(u)=i\}} \big) du \right].
\end{aligned}
$$

By virtue of Lemma 10.8, the use of the Cauchy-Schwarz inequality, and noting the boundedness of $h_\ell(\cdot)$, for each $i = 1, \ldots, l$ $j = 1, \ldots, m_i$,

$$
\begin{aligned}
&\mathbb{E} \Big| \prod_{\ell=1}^{n_0} h_\ell(X^\varepsilon(t_\ell), \overline{\alpha}^\varepsilon(t_\ell)) \int_t^{t+s} \widehat{F}_x(u, X^\varepsilon(u), s_{ij}) f(u, X^\varepsilon(u), s_{ij}) \\
&\qquad \times \left(I_{\{\alpha^\varepsilon(u)=s_{ij}\}} - \nu_j^i(u) I_{\{\overline{\alpha}^\varepsilon(u)=i\}} \right) du \Big|^2 \\
&\leq K \mathbb{E} \Big| \int_t^{t+s} \widehat{F}_x(u, X^\varepsilon(u), s_{ij}) f(u, X^\varepsilon(u), s_{ij}) \\
&\qquad \times \left(I_{\{\alpha^\varepsilon(u)=s_{ij}\}} - \nu_j^i(u) I_{\{\overline{\alpha}^\varepsilon(u)=i\}} \right) du \Big|^2 \\
&\to 0 \ \text{as} \ \varepsilon \to 0.
\end{aligned}
$$

In view of [303, Lemma 2.4] and similar to [303, Theorem 7.30], it can be shown that

$$
(I_{\{\overline{\alpha}^e(\cdot)=1\}}, \ldots, I_{\{\overline{\alpha}^e(\cdot)=l\}}) \ \text{converges weakly to} \ (I_{\{\overline{\alpha}(\cdot)=1\}}, \ldots, I_{\{\overline{\alpha}(\cdot)=l\}}).
$$

By means of Cramér-Wold's device [147, Corollary 6.5], for each $i \in \mathcal{M}$, $I_{\{\overline{\alpha}^\varepsilon(\cdot)=i\}}$ converges to $I_{\{\overline{\alpha}(\cdot)=i\}}$ weakly. By the Skorokhod representation (without changing notation and with a slight abuse of notation), we may assume $I_{\{\overline{\alpha}^\varepsilon(\cdot)=i\}} \to I_{\{\overline{\alpha}(\cdot)=i\}}$ w.p.1. Consequently, the weak convergence of $Y^\varepsilon(\cdot)$ to $Y(\cdot)$, the Skorokhod representation, and the convergence of $I_{\{\overline{\alpha}^\varepsilon(\cdot)=i\}}$ to $I_{\{\overline{\alpha}(\cdot)=i\}}$ imply that

$$
\begin{aligned}
&\mathbb{E} \prod_{\ell=1}^{n_0} h_\ell(X^\varepsilon(t_\ell), \overline{\alpha}^\varepsilon(t_\ell)) \left[\int_t^{t+s} \widehat{F}_x(u, X^\varepsilon(u), \alpha^\varepsilon(u)) f(u, X^\varepsilon(u), \alpha^\varepsilon(u)) du \right] \\
&\overset{\varepsilon \to 0}{\longrightarrow} \sum_{i=1}^{l} \sum_{j=1}^{m_i} \mathbb{E} \prod_{\ell=1}^{n_0} h_\ell(X(t_\ell), \overline{\alpha}(t_\ell))
\end{aligned}
$$

$$\times \left[\int_t^{t+s} \widehat{F}_x(u, X(u), s_{ij}) f(u, X(u), s_{ij}) \nu_j^i(u) I_{\{\overline{\alpha}(u)=i\}} du \right]$$
$$= \sum_{i=1}^{l} \mathbb{E} \prod_{\ell=1}^{n_0} h_\ell(X(t_\ell), \overline{\alpha}(t_\ell)) \left[\int_t^{t+s} F_x(u, X(u), i) \overline{f}(u, X(u), i) I_{\{\overline{\alpha}(u)=i\}} du \right]$$
$$= \mathbb{E} \prod_{\ell=1}^{n_0} h_\ell(X(t_\ell), \overline{\alpha}(t_\ell)) \left[\int_t^{t+s} F_x(u, X(u), \overline{\alpha}(u)) \overline{f}(u, X(u), \overline{\alpha}(u)) du \right]. \tag{10.31}$$

Exactly the same argument as in the derivation of (10.31) yields

$$\mathbb{E} \prod_{\ell=1}^{n_0} h_\ell(X^\varepsilon(t_\ell), \overline{\alpha}^\varepsilon(t_\ell)) \left[\int_t^{t+s} [\widehat{F}_{xx}(u, X^\varepsilon(u), \alpha^\varepsilon(u)) \sigma^2(u, X^\varepsilon(u), \alpha^\varepsilon(u))] du \right]$$
$$\stackrel{\varepsilon \to 0}{\longrightarrow} \mathbb{E} \prod_{\ell=1}^{n_0} h_\ell(X(t_\ell), \overline{\alpha}(t_\ell)) \left[\int_t^{t+s} [F_{xx}(u, X(u), \overline{\alpha}(u)) \overline{\sigma}^2(u, X(u), \overline{\alpha}(u))] du \right], \tag{10.32}$$

and

$$\begin{aligned} &\mathbb{E} \prod_{\ell=1}^{n_0} h_\ell(X^\varepsilon(t_\ell), \overline{\alpha}^\varepsilon(t_\ell)) \left[\int_t^{t+s} \frac{\partial}{\partial u} \widehat{F}(u, X^\varepsilon(u), \alpha^\varepsilon(u)) du \right] \\ &\quad \stackrel{\varepsilon \to 0}{\longrightarrow} \prod_{\ell=1}^{n_0} h_\ell(\overline{\alpha}(t_\ell), X(t_\ell)) \left[\int_t^{t+s} \frac{\partial}{\partial u} F(u, \overline{\alpha}(u), X(u), \overline{\alpha}(u)) du \right]. \end{aligned} \tag{10.33}$$

Next, since

$$\widetilde{Q}^i(u) \mathbb{1}_{m_i} = 0 \ \text{ for each } \ i = 1, \ldots, l,$$

the definition of $\widehat{F}(\cdot)$ yields

$$\widetilde{Q}(u) \widehat{F}(u, X^\varepsilon(u), \cdot)(\alpha^\varepsilon(u)) = 0.$$

Therefore, we have

$$\mathbb{E} \prod_{\ell=1}^{n_0} h_\ell(X^\varepsilon(t_\ell), \overline{\alpha}^\varepsilon(t_\ell)) \left[\int_t^{t+s} Q^\varepsilon(u) \widehat{F}(u, X^\varepsilon(u), \cdot)(\alpha^\varepsilon(u)) du \right]$$
$$= \mathbb{E} \prod_{\ell=1}^{n_0} h_\ell(X^\varepsilon(t_\ell), \overline{\alpha}^\varepsilon(t_\ell)) \left[\int_t^{t+s} \widehat{Q}(u) \widehat{F}(u, X^\varepsilon(u), \cdot)(\alpha^\varepsilon(u)) du \right]$$

$$
\begin{aligned}
&= \sum_{i=1}^{l}\sum_{j=1}^{m_i} \mathbb{E}\prod_{\ell=1}^{n_0} h_\ell(X^\varepsilon(t_\ell), \overline{\alpha}^\varepsilon(t_\ell)) \\
&\quad \times \left[\int_t^{t+s} \widehat{Q}(u)\widehat{F}(u, X^\varepsilon(u), \cdot)(s_{ij})\nu_j^i(u) I_{\{\overline{\alpha}^\varepsilon(u)=i\}} du\right] \\
&\quad + \sum_{i=1}^{l}\sum_{j=1}^{m_i} \mathbb{E}\prod_{\ell=1}^{n_0} h_\ell(X^\varepsilon(t_\ell), \overline{\alpha}^\varepsilon(t_\ell)) \\
&\quad \times \left[\int_t^{t+s} \widehat{Q}(u)\widehat{F}(u, X^\varepsilon(u), \cdot)(s_{ij})[I_{\{\alpha^\varepsilon(u)=s_{ij}\}} - \nu_j^i(u) I_{\{\overline{\alpha}^\varepsilon(u)=i\}}] du\right].
\end{aligned}
$$

By virtue of Lemma 10.8 again, the last term above goes to 0 as $\varepsilon \to 0$. For the next to the last term, we have

$$
\begin{aligned}
&\sum_{i=1}^{l}\sum_{j=1}^{m_i} \mathbb{E}\prod_{\ell=1}^{n_0} h_\ell(X^\varepsilon(t_\ell), \overline{\alpha}^\varepsilon(t_\ell)) \\
&\quad \times \left[\int_t^{t+s} \widehat{Q}(u)\widehat{F}(u, X^\varepsilon(u), \cdot)(s_{ij})\nu_j^i(u) I_{\{\overline{\alpha}^\varepsilon(u)=i\}} du\right] \\
&= \sum_{i=1}^{l}\sum_{j=1}^{m_i} \mathbb{E}\prod_{\ell=1}^{n_0} h_\ell(X^\varepsilon(t_\ell), \overline{\alpha}^\varepsilon(t_\ell)) \\
&\quad \times \left[\int_t^{t+s} \widehat{Q}(u) F(u, X^\varepsilon(u), \cdot)(i) I_{\{\overline{\alpha}^\varepsilon(u)=i\}} du\right] \\
&\to \sum_{i=1}^{l} \mathbb{E}\prod_{\ell=1}^{n_0} h_\ell(X(t_\ell), \overline{\alpha}(t_\ell)) \left[\int_t^{t+s} \overline{Q}(u) F(u, X(u), \cdot)(i) I_{\{\overline{\alpha}(u)=i\}} du\right] \\
&= \mathbb{E}\prod_{\ell=1}^{n_0} h_\ell(X(t_\ell), \overline{\alpha}(t_\ell)) \left[\int_t^{t+s} \overline{Q}(u) F(u, X(u), \cdot)(\overline{\alpha}(u)) du\right],
\end{aligned}
$$

as $\varepsilon \to 0$.

Arguing along the same line as in the above estimates, we obtain

$$
\begin{aligned}
&\int_t^{t+s}\int_\Gamma \lambda[\widehat{F}(u, X^\varepsilon(u-) + g(\gamma, X^\varepsilon(u-), \alpha^\varepsilon(u-)), \alpha^\varepsilon(u-)) \\
&\quad - \widehat{F}(u, X^\varepsilon(u-), \alpha^\varepsilon(u-))]\pi(d\gamma)du \\
&= \sum_{i=1}^{l}\sum_{j=1}^{m_i} \int_t^{t+s}\int_\Gamma \lambda[\widehat{F}(u, X^\varepsilon(u-) + g(\gamma, X^\varepsilon(u-), s_{ij}), s_{ij}) \\
&\quad - \widehat{F}(u, X^\varepsilon(u-), s_{ij})]\pi(d\gamma) I_{\{\alpha^\varepsilon(u)=s_{ij}\}}
\end{aligned}
$$

$$
\begin{aligned}
&= \sum_{i=1}^{l}\sum_{j=1}^{m_i}\int_t^{t+s}\int_\Gamma \lambda[\widehat{F}(u, X^\varepsilon(u-) + g(\gamma, X^\varepsilon(u-), s_{ij}), s_{ij}) \\
&\quad - \widehat{F}(u, X^\varepsilon(u-), s_{ij})]\pi(d\gamma)\nu_j^i(u-)I_{\{\alpha^\varepsilon(u-)\in\mathcal{M}_i\}}du \\
&\quad + \sum_{i=1}^{l}\sum_{j=1}^{m_i}\int_t^{t+s}\int_\Gamma \lambda[\widehat{F}(u, X^\varepsilon(u-) + g(\gamma, X^\varepsilon(u-), s_{ij}), s_{ij}) \\
&\quad - \widehat{F}(u, X^\varepsilon(u-), s_{ij})]\pi(d\gamma)[I_{\{\alpha^\varepsilon(u-)=s_{ij}\}} - \nu_j^i(u-)I_{\{\alpha^\varepsilon(u-)\in\mathcal{M}_i\}}]du \\
&= \sum_{i=1}^{l}\sum_{j=1}^{m_i}\int_t^{t+s}\int_\Gamma \lambda[\widehat{F}(u, X^\varepsilon(u-) + g(\gamma, X^\varepsilon(u-), s_{ij}), s_{ij}) \\
&\quad - \widehat{F}(u, X^\varepsilon(u-), s_{ij})]\pi(d\gamma)\nu_j^i(u-)I_{\{\alpha^\varepsilon(u-)\in\mathcal{M}_i\}}du + o(1),
\end{aligned}
$$

where $o(1) \to 0$ in probability uniformly in t on any bounded set. By virtue of the weak convergence of $Y^\varepsilon(\cdot)$ to $Y(\cdot)$, the Skorokhod representation, and the dominated convergence theorem, we have

$$
\begin{aligned}
&\mathbb{E}\prod_{\ell=1}^{n_0} h_\ell(X^\varepsilon(t_\ell), \overline{\alpha}^\varepsilon(t_\ell)) \\
&\quad\times\Bigg[\int_t^{t+s}\int_\Gamma \lambda[\widehat{F}(u, X^\varepsilon(u-) + g(\gamma, X^\varepsilon(u-), \alpha^\varepsilon(u-)), \alpha^\varepsilon(u-)) \\
&\quad - \widehat{F}(u, X^\varepsilon(u-), \alpha^\varepsilon(u-))]\pi(d\gamma)du\Bigg] \\
&\to \mathbb{E}\prod_{\ell=1}^{n_0} h_\ell(X(t_\ell), \overline{\alpha}(t_\ell)) \\
&\quad\times\Bigg[\sum_{i=1}^{l}\sum_{j=1}^{m_i}\int_t^{t+s}\int_\Gamma \lambda[\widehat{F}(u, X(u-) + g(\gamma, X(u-), s_{ij}), s_{ij}) \\
&\quad - \widehat{F}(u, X(u-), s_{ij})]\pi(d\gamma)\nu_j^i(u-)I_{\{\overline{\alpha}(u-)=i\}}du\Bigg],
\end{aligned}
\tag{10.34}
$$

as $\varepsilon \to 0$. Combining (10.30)–(10.34), we obtain the desired result. □

10.3 Fast-Varying Diffusion

This section presents another hybrid jump-diffusion model. Compared with the Markov modulated jump-diffusion model with fast switching discussed in the last section, there is an additional periodic fast-varying diffusion. From the motivation of an insurance point of view, the added periodic diffusion may be seen as a way of handling seasonal effects of uncertainty due to claims and premium incomes. This periodic diffusion varies at a faster pace compared with the other random effects.

Consider the system given by

$$\begin{aligned} X^{\varepsilon}(t) &= x + \int_0^t f(X^{\varepsilon}(s), \alpha(s), z^{\varepsilon}(s))ds + \int_0^t \sigma(X^{\varepsilon}(s), \alpha(s), z^{\varepsilon}(s))dw(s) \\ &\quad + \int_0^t \int_{\Gamma} g(\gamma, X^{\varepsilon}(s-), \alpha(s-), z^{\varepsilon}(s-))N(ds, d\gamma), \qquad (10.35) \\ z^{\varepsilon}(t) &= z_0 + \frac{1}{\varepsilon}\int_0^t f_1(X^{\varepsilon}(s), z^{\varepsilon}(s))ds + \frac{1}{\sqrt{\varepsilon}}\int_0^t \sigma_1(X^{\varepsilon}(s), z^{\varepsilon}(s))dv(s), \end{aligned}$$

where $w(\cdot)$ and $v(\cdot)$ are independent standard Brownian motions. In the above, $z^{\varepsilon}(\cdot)$ represents a fast-varying diffusion. Relative to $z^{\varepsilon}(\cdot)$, $X^{\varepsilon}(\cdot)$ is a slowly varying jump-diffusion process, which is modulated by a continuous-time Markov chain. Due to its scaling, the process $z^{\varepsilon}(\cdot)$ does not blow up. Under suitable conditions, we show that $X^{\varepsilon}(\cdot)$ converges weakly to a jump-diffusion process modulated by the Markov chain $\alpha(\cdot)$, in which the system dynamics are averaged out with respect to the stationary measure of the fast process $z^{\varepsilon}(\cdot)$. Because we have treated time-dependent generator in the last section and for notational simplicity, we shall assume that the generator of the Markov chain is independent of time (i.e., $Q = (q_{ij})$ a constant matrix) and that there is no explicit time dependence in the coefficients $f(\cdot)$, $f_1(\cdot)$, $\sigma(\cdot)$, and $\sigma_1(\cdot)$ in this section. That is, both the diffusion process $z^{\varepsilon}(\cdot)$ and the jump diffusion $X^{\varepsilon}(\cdot)$ are time homogeneous. Treating z^{ε} as a parameter, the generator for $X^{\varepsilon}(\cdot)$ can be written as

$$\begin{aligned} \mathcal{G}^{\varepsilon} F(x, \iota) = \mathcal{L}^{\varepsilon} F(x, \iota) &+ \int_{\Gamma} \lambda[F(x + g(\gamma, x, z^{\varepsilon}, \iota), \iota) - F(x, \iota)]\pi(d\gamma) \\ &+ QF(x, \cdot)(\iota), \quad \text{for each } \iota \in \mathcal{M}, \end{aligned} \qquad (10.36)$$

where

$$\begin{aligned} \mathcal{L}^{\varepsilon} F(x, \iota) &= \frac{1}{2}\sigma^2(x, \iota, z^{\varepsilon})\frac{\partial^2}{\partial x^2}F(x, \iota) + f(x, \iota, z^{\varepsilon})\frac{\partial}{\partial x}F(x, \iota), \\ QF(x, \cdot)(\iota) &= \sum_{\ell=1}^{m_0} q_{\iota\ell}F(x, \ell) = \sum_{\ell \neq \iota} q_{\iota\ell}[F(x, \ell) - F(x, \iota)]. \end{aligned} \qquad (10.37)$$

We need the following conditions.

(H10.3) $\mathbb{E}|x|^2 < \infty$ and $\mathbb{E}|z_0|^2 < \infty$. Moreover:

(a) For each $\alpha \in \mathcal{M}$, $f(\cdot, \alpha, \cdot)$ and $\sigma(\cdot, \alpha, \cdot)$ are continuous functions satisfying

$$\begin{aligned} |f(x, \alpha, z)| + |\sigma(x, \alpha, z)| &\le K(1 + |x|), \\ |f(y, \alpha, z) - f(x, \alpha, z)| + |\sigma(y, \alpha, z) - \sigma(x, \alpha, z)| &\le K|y - x|, \end{aligned}$$

for all $x, y \in \mathbb{R}$ and $\alpha \in \mathcal{M}$, uniformly in $z \in \mathbb{R}$. For each $\alpha \in \mathcal{M}$, $\lambda < \infty$, $g(\cdot, \cdot, \alpha, \cdot)$ is bounded and continuous, $g(0, x, \alpha, z) = 0$ for all x, z, and α.

(b) Both $f_1(x, \cdot)$ and $\sigma_1(x, \cdot)$ are periodic functions with period 1 such that $\sigma_1(x, z) > 0$ for each x, z and

$$|f_1(x, z)| + |\sigma_1(x, z)| \le K(1 + |z|),$$
$$|f_1(x, z) - f_1(x, y)| + |\sigma_1(x, z) - \sigma_1(x, y)| \le K|z - y|,$$

for all $y, z \in \mathbb{R}$ uniformly in $x \in \mathbb{R}$.

Remark 10.10 Part (b) yields that the fast-varying process $z^\varepsilon(\cdot)$ is a so-called periodic diffusion; see [21, 158] among others. This condition guarantees that there is an invariant density $\mu(x, z)$ (for a fixed x). In [158], under suitable conditions, it was proved that not only an invariant measure exists but also asymptotic expansions of the transition density can be constructed. The choice of periodicity 1 is more or less for convenience; we could in fact use other positive constants as the periodicity. In fact, in lieu of periodic diffusion, what we really need is the assumption that the fast-varying $z^\varepsilon(\cdot)$ has an invariant density. It seems to be more instructive to use simpler conditions as in the current setup.

Lemma 10.11 *Under* (H10.3) $\{X^\varepsilon(\cdot)\}$ *is tight in* $D([0, T]; \mathbb{R})$.

Idea of Proof The proof is similar to the fast switching case. Therefore, we will not spell out all the details. Similar to Lemma 10.6, it can be shown that $\sup_{t\in[0,T]} \mathbb{E}|X^\varepsilon(t)|^2 \le K$. For any $\delta > 0$, and any $t > 0$, $s > 0$ with $s \le \delta$, we still have (10.23). Deriving a similar estimate as in that of Lemma 10.7 for the last term (10.23), using the estimate, the linear growth in x, for the first two terms on the right side of (10.23), and by virtue of Gronwall's inequality, we can show that there exists a $\widehat{\beta}^\varepsilon(\delta)$ being $\mathcal{F}_t^\varepsilon$-measurable such that $\mathbb{E}_t^\varepsilon |X^\varepsilon(t+s) - X^\varepsilon(t)|^2 \le \widehat{\beta}^\varepsilon(\delta)$ and that

$$\lim_{\delta\to 0} \limsup_{\varepsilon\to 0} \mathbb{E}\widehat{\beta}^\varepsilon(\delta) = 0.$$

Hence, the tightness is obtained.

Theorem 10.12 *Assume* (H10.3) *holds. Then,* $(X^\varepsilon(\cdot), \alpha(\cdot))$ *converges weakly to* $(X(\cdot), \alpha(\cdot))$*, a jump-diffusion process modulated by the Markov chain* $\alpha(\cdot)$ *(with generator* Q*), which is a solution of the equation*

$$\begin{aligned} X(t) = x &+ \int_0^t \widehat{f}(X(u), \alpha(u))du + \int_0^t \widehat{\sigma}(X(u), \alpha(u))dw(s) \\ &+ \int_0^t \int_\Gamma \widehat{g}(\gamma, X(u), \alpha(u))N(du, d\gamma), \end{aligned} \tag{10.38}$$

where

$$\widehat{f}(x,\alpha) = \int_0^1 f(x,\alpha,z)\mu(x,z)dz,$$

$$\widehat{g}(\gamma,x,\alpha) = \int_0^1 g(\gamma,x,\alpha,z)\mu(x,z)dz, \tag{10.39}$$

$$\widehat{\sigma}^2(x,\alpha) = \int_0^1 \sigma^2(x,\alpha,z)\mu(x,z)dz, \quad \textit{and} \quad \widehat{\sigma}(x,\alpha) = \sqrt{\widehat{\sigma}^2(x,\alpha)}.$$

Remark 10.13 To proceed, a pertinent way of carrying out the averaging is to work with truncated processes (see [177, p. 248], for instance). The basic idea is as follows. Let $M > 0$ be given and S_M be an interval with length $2M$ centered at the origin (for a multidimensional case, replace the interval by a sphere with radius M centered at the origin). Define $X^{\varepsilon,M}(t) = X^{\varepsilon}(t)$ up until the first exit from the M-sphere. Use a smooth truncation function $q_M(x)$ that is equal to 1 when it is inside the M-interval and is 0 if it is outside the $(M+1)$-interval. Then, rewrite the dynamics with the use of the M-truncated process and the truncation function. In this process, in lieu of $f(x,\alpha,z)$, $g(\gamma,x,\alpha,z)$, and $\sigma(x,\alpha,z)$, we use

$$\begin{aligned} f^M(x,\alpha,z) &\stackrel{\text{def}}{=} f(x,\alpha,z)q_M(x), \\ g^M(\gamma,x,\alpha,z) &\stackrel{\text{def}}{=} g(\gamma,x,\alpha,z)q_M(x), \\ \sigma^M(x,\alpha,z) &\stackrel{\text{def}}{=} \sigma(x,\alpha,z)q_M(x), \end{aligned}$$

respectively. Then, we proceed to derive the weak convergence of $X^{\varepsilon,M}(\cdot)$ to $X^M(\cdot)$. Finally, using the uniqueness of the martingale problem letting $M \to \infty$, we conclude that $X^{\varepsilon}(\cdot)$ also converges to $X(\cdot)$. However, for notational simplicity, in what follows, we will not use the truncation notation, but simply assume that the process itself is bounded.

Proof Starting from the fast-varying process, define

$$\widetilde{z}^{\varepsilon}(t) = z^{\varepsilon}(\varepsilon t), \ \widetilde{v}(t) = v(\varepsilon t)/\sqrt{\varepsilon}, \ \tau = t/\varepsilon,$$

where $v(\cdot)$ is the Brownian motion given in (10.35). Then, the equation for $z^{\varepsilon}(t)$ may be written as

$$\widetilde{z}^{\varepsilon}(\tau) = z_0 + \int_0^{\tau} f_1(X^{\varepsilon}(\varepsilon u), \widetilde{z}^{\varepsilon}(u))du + \int_0^{\tau} \sigma_1(X^{\varepsilon}(\varepsilon u), \widetilde{z}^{\varepsilon}(u))d\widetilde{v}(u). \tag{10.40}$$

Note that as $\varepsilon \to 0$, $\tau \to \infty$, where τ is known as the stretched time. Note also that $X^{\varepsilon}(\varepsilon\tau)$ is slowly varying, whereas $\widetilde{z}^{\varepsilon}(\tau)$ is fast changing. Intuitively, it tells us that we can treat $X^{\varepsilon}(\varepsilon\tau)$ as if it were a "constant" in a small interval. Taking x as a fixed parameter, in what follows, we consider the following fixed-x process (see [176] for an explanation of the fixed-x process):

$$\widetilde{z}^{\varepsilon,x}(\tau) = z_0 + \int_0^{\tau} f_1(x, \widetilde{z}^{\varepsilon,x}(u))du + \int_0^{\tau} \sigma_1(x, \widetilde{z}^{\varepsilon,x}(u))d\widetilde{v}(u),$$

and the associated generator is

$$\widetilde{L} = \frac{1}{2}\sigma_1^2(x,z)\frac{\partial^2}{\partial z^2} + f_1(x,z)\frac{\partial}{\partial z}. \tag{10.41}$$

Let $p(t, x, z_1, z)$ denote the transition density associated with the diffusion generated by $\widetilde{L}$. Then, $p(t, x, z_1, z)$ satisfies the Kolmogorov forward equation:

$$\begin{aligned} &\frac{\partial p}{\partial t} = \widetilde{L}^* p, \\ &\lim_{t \to 0^+} p(t, x, z_1, z) = \delta(z - z_1), \end{aligned}$$

where $\widetilde{L}^*$ is the adjoint of $\widetilde{L}$. Because this diffusion is on a compact set thanks to the periodic diffusion, the classical theory of diffusion processes and related results on partial differential equations (see, e.g., Ikeda and Watanabe [136], and Agranovich [1]) yields that there is a unique invariant transition density $\mu(x,z)$ such that $p(t, x, z_1, z) \to \mu(x,z)$ as $t \to \infty$, where $\mu(x,z)$ is the unique solution of

$$\begin{cases} \widetilde{L}^*\mu(x,z) = 0, \\ \mu(x,z) \geq 0 \text{ and } \displaystyle\int_0^1 \mu(x,z)dz = 1. \end{cases} \tag{10.42}$$

We note that $(X^\varepsilon(\cdot), \alpha(\cdot), z^\varepsilon(\cdot))$ is a Markov process and that $z^\varepsilon(\cdot)$ can be treated as a fast-varying noise process that will be averaged out in the limit process. The average is taken with respect to the stationary measure of $\widetilde{z}^\varepsilon(\cdot)$. Because $(X^\varepsilon(\cdot), \alpha(\cdot))$ is tight in $D([0,T]; \mathbb{R} \times \mathcal{M})$, by Prokhorov's theorem, we can extract a weakly convergent subsequence. Select such a subsequence and denote the limit by $(X(\cdot), \alpha(\cdot))$. For notational simplicity, still denote the subsequence by $(X^\varepsilon(\cdot), \alpha(\cdot))$. By virtue of the Skorokhod representation, $(X^\varepsilon(\cdot), \alpha(\cdot))$ converges to $(X(\cdot), \alpha(\cdot))$ w.p.1 and the convergence is uniform on each bounded set. We proceed to show that the limit is a solution of the martingale problem with generator $\widehat{\mathcal{G}}$ given by

$$\begin{aligned} \widehat{\mathcal{G}}F(x,\alpha) = \frac{1}{2}\widehat{\sigma}^2(x,\alpha)\frac{\partial^2 F(x,\alpha)}{\partial x^2} &+ \widehat{f}(x,\alpha)\frac{\partial F(x,\alpha)}{\partial x} + \widehat{J}(x,\alpha) \\ &+ QF(x,\cdot)(\alpha), \ \alpha \in \mathcal{M}, \end{aligned} \tag{10.43}$$

where

$$\widehat{J}(x,\alpha) = \int_\Gamma \lambda[F(x + \widehat{g}(\gamma, x, \alpha), \alpha) - F(x,\alpha)]\pi(d\gamma),$$

and $\widehat{\varphi}(\cdot)$ is an average of $\varphi(\cdot)$ as defined in (10.39) for $\varphi(\cdot)$ being $f(\cdot)$, $\sigma^2(\cdot)$, and $g(\cdot)$, respectively.

Again, Theorem A.25 and [136, Theorem 9.1, p. 245] show that the martingale problem with operator $\widehat{\mathcal{G}}$ has a unique solution. To show that $(X(\cdot), \alpha(\cdot))$ is indeed the solution of the martingale problem with operator $\widehat{\mathcal{G}}$, it suffices to verify that for each $\alpha \in \mathcal{M}$ and for any $F(\cdot, \alpha) \in C_0^2$,

$$F(X(t), \alpha(t)) - F(x, \alpha) - \int_0^t \widehat{\mathcal{G}} F(X(u), \alpha(u)) du \tag{10.44}$$

is a martingale. As in the previous section, we begin with the $X^\varepsilon(\cdot)$ process. For any positive integer n_0, and $0 \le t_\ell \le t$ and $\ell \le n_0$, and any bounded and continuous function $h_\ell(\cdot)$, the weak convergence and the Skorokhod representation imply that

$$\begin{aligned}
&\mathbb{E} \prod_{\ell=1}^{n_0} h_\ell(X^\varepsilon(t_\ell))[F(X^\varepsilon(t+s), \alpha(t+s)) - F(X^\varepsilon(t), \alpha(t))] \\
&\quad \to \mathbb{E} \prod_{\ell=1}^{n_0} h_\ell(X(t_\ell))[F(X(t+s), \alpha(t+s)) - F(X(t), \alpha(t))] \ \text{ as } \varepsilon \to 0.
\end{aligned} \tag{10.45}$$

In addition

$$\begin{aligned}
\mathbb{E} \prod_{\ell=1}^{n_0} h_\ell(X^\varepsilon(t_\ell)) \Big[& F(X^\varepsilon(t+s), \alpha(t+s)) - F(X^\varepsilon(t), \alpha(t)) \\
& - \int_t^{t+s} \mathcal{G}^\varepsilon F(X^\varepsilon(u), \alpha(u)) du \Big] = 0.
\end{aligned}$$

Thus, in view of (10.45), it suffices to consider the limit of

$$\begin{aligned}
&\mathbb{E} \prod_{\ell=1}^{n_0} h_\ell(X^\varepsilon(t_\ell)) \left[\int_t^{t+s} \mathcal{G}^\varepsilon F(X^\varepsilon(u), \alpha(u)) du \right] \\
&= \mathbb{E} \prod_{\ell=1}^{n_0} h_\ell(X^\varepsilon(t_\ell)) \Bigg[\int_t^{t+s} \Bigg[f(X^\varepsilon(u), \alpha(u), z^\varepsilon(u)) \frac{\partial F(X^\varepsilon(u), \alpha(u))}{\partial x} \\
&\quad + \frac{1}{2} \sigma^2(X^\varepsilon(u), \alpha(u), z^\varepsilon(u)) \frac{\partial^2 F(X^\varepsilon(u), \alpha(u), z^\varepsilon(u))}{\partial x^2} \\
&\quad + \int_\Gamma \lambda [F(X^\varepsilon(u) + g(\gamma, X^\varepsilon(u), \alpha(u), z^\varepsilon(u)), \alpha(u)) - F(X^\varepsilon(u), \alpha(u))] \pi(d\gamma) \\
&\quad + QF(X^\varepsilon(u), \alpha(u)) \Bigg] du \Bigg].
\end{aligned} \tag{10.46}$$

Consider the last term in (10.46). Using the weak convergence, the Skorokhod representation, and the continuity of the function $F(\cdot, \alpha)$ for each $\alpha \in \mathcal{M}$, it can be shown that as $\varepsilon \to 0$,

$$\begin{aligned}&\mathbb{E}\prod_{\ell=1}^{n_0} h_\ell(X^\varepsilon(t_\ell))\left[\int_t^{t+s} QF(X^\varepsilon(u),\alpha(u))du\right]\\&\to \mathbb{E}\prod_{\ell=1}^{n_0} h_\ell(X(t_\ell))\left[\int_t^{t+s} QF(X(u),\alpha(u))du\right].\end{aligned} \tag{10.47}$$

Choose $0 \le \delta_\varepsilon$ such that $\delta_\varepsilon \to 0$ as $\varepsilon \to 0$ but $\delta_\varepsilon/\varepsilon \to \infty$. For any $t, s \ge 0$ with $0 \le t+s \le T$, by partitioning the interval $[t, t+s]$ into subintervals of length δ_ε, we can rewrite the terms (except the last one) in (10.46) as

$$\int_t^{t+s} \widehat{H}^\varepsilon(u)du = \sum_{l=t/\delta_\varepsilon}^{(t+s)/\delta_\varepsilon-1} \delta_\varepsilon \frac{1}{\delta_\varepsilon}\int_{l\delta_\varepsilon}^{l\delta_\varepsilon+\delta_\varepsilon} \widehat{H}^\varepsilon(u)du, \tag{10.48}$$

where $\widehat{H}^\varepsilon(u)$ is a representation of any of the functions that appeared in the integrand of (10.46). We then work with each of the terms and find the corresponding limits.

Noting $t_\ell \le t$, $h_\ell(X^\varepsilon(t_\ell))$ is $\mathcal{F}^\varepsilon_{l\delta_\varepsilon}$-measurable, where

$$\mathcal{F}^\varepsilon_{l\delta_\varepsilon} = \{X^\varepsilon(u), \alpha(u), z^\varepsilon(u) : 0 \le u \le l\delta_\varepsilon\}.$$

Making a change of variable from u to εu leads to

$$\begin{aligned}&\mathbb{E}\prod_{\ell=1}^{n_0} h_\ell(X^\varepsilon(t_\ell))\left[\int_t^{t+s} f(X^\varepsilon(u),\alpha(u),z^\varepsilon(u))\frac{\partial F(X^\varepsilon(u),\alpha(u))}{\partial x}du\right]\\&=\mathbb{E}\prod_{\ell=1}^{n_0} h_\ell(X^\varepsilon(t_\ell))\Bigg[\sum_{l=t/\delta_\varepsilon}^{(t+s)/\delta_\varepsilon-1}\delta_\varepsilon\frac{\varepsilon}{\delta_\varepsilon}\int_{\frac{l\delta_\varepsilon}{\varepsilon}}^{\frac{l\delta_\varepsilon+\delta_\varepsilon}{\varepsilon}} \mathbb{E}^\varepsilon_{l\delta_\varepsilon} f(X^\varepsilon(\varepsilon u),\alpha(\varepsilon u),\widetilde{z}^\varepsilon(u))]\\&\quad\times\frac{\partial F(X^\varepsilon(\varepsilon u),\alpha(\varepsilon u))}{\partial x}du\Bigg],\end{aligned}$$

where $\mathbb{E}^\varepsilon_{l\delta_\varepsilon}$ denotes the conditioning on the σ-algebra $\mathcal{F}^\varepsilon_{l\delta_\varepsilon}$. Denote $t^l_\varepsilon = l\delta_\varepsilon/\varepsilon$ and $T_\varepsilon = \delta_\varepsilon/\varepsilon$. Then, as $\varepsilon \to 0$, $T_\varepsilon \to \infty$. By the Lipschitz continuity of $f(\cdot,\alpha,z)$ and the boundedness of $(\partial/\partial x)F(\cdot,\alpha)$,

$$\begin{aligned}&\mathbb{E}\Bigg|\frac{\varepsilon}{\delta_\varepsilon}\int_{(l\delta_\varepsilon)/\varepsilon}^{(l\delta_\varepsilon+\delta_\varepsilon)/\varepsilon} \mathbb{E}^\varepsilon_{l\delta_\varepsilon}[f(X^\varepsilon(\varepsilon u),\alpha(\varepsilon u),\widetilde{z}^\varepsilon(u)) - f(X^\varepsilon(l\delta_\varepsilon),\alpha(\varepsilon u),\widetilde{z}^\varepsilon(u))]\\&\quad\times\frac{\partial F(X^\varepsilon(\varepsilon u),\alpha(\varepsilon u))}{\partial x}du\Bigg|\\&\le K\frac{1}{T_\varepsilon}\int_{t^l_\varepsilon}^{t^l_\varepsilon+T_\varepsilon}\mathbb{E}|X^\varepsilon(\varepsilon u)-X^\varepsilon(l\delta_\varepsilon)|du\\&\le K\sup_{l\delta_\varepsilon\le\varepsilon u\le l\delta_\varepsilon+\delta_\varepsilon}\mathbb{E}|X^\varepsilon(\varepsilon u)-X^\varepsilon(l\delta_\varepsilon)| \to 0 \text{ as } \varepsilon\to 0.\end{aligned}$$

Similarly

$$\begin{aligned}
&\mathbb{E}\bigg|\frac{1}{T_\varepsilon}\int_{t_\varepsilon^l}^{t_\varepsilon^l+T_\varepsilon}\mathbb{E}_{l\delta_\varepsilon}^\varepsilon f(X^\varepsilon(l\delta_\varepsilon),\alpha(\varepsilon u),\widetilde{z}^\varepsilon(u))] \\
&\quad\times\left[\frac{\partial F(X^\varepsilon(\varepsilon u),\alpha(\varepsilon u))}{\partial x}-\frac{\partial F(X^\varepsilon(l\delta_\varepsilon),\alpha(\varepsilon u))}{\partial x}\right]du\bigg| \\
&\le K\frac{1}{T_\varepsilon}\int_{t_\varepsilon^l}^{t_\varepsilon^l+T_\varepsilon}\mathbb{E}|X^\varepsilon(\varepsilon u)-X^\varepsilon(l\delta_\varepsilon)|du \\
&\le K\sup_{l\delta_\varepsilon\le\varepsilon u\le l\delta_\varepsilon+\delta_\varepsilon}\mathbb{E}|X^\varepsilon(\varepsilon u)-X^\varepsilon(l\delta_\varepsilon)|\to 0\ \text{ as }\ \varepsilon\to 0.
\end{aligned}$$

Thus

$$\begin{aligned}
&\frac{1}{\delta_\varepsilon}\int_{l\delta_\varepsilon}^{l\delta_\varepsilon+\delta_\varepsilon}\mathbb{E}_{l\delta_\varepsilon}^\varepsilon f(X^\varepsilon(u),\alpha(u),z^\varepsilon(u))\frac{\partial F(X^\varepsilon(u),\alpha(u))}{\partial x}du \\
&=\frac{1}{T_\varepsilon}\int_{t_\varepsilon^l}^{t_\varepsilon^l+T_\varepsilon}\mathbb{E}_{l\delta_\varepsilon}^\varepsilon f(X^\varepsilon(l\delta_\varepsilon),\alpha(\varepsilon u),\widetilde{z}^\varepsilon(u))\frac{\partial F(X^\varepsilon(l\delta_\varepsilon),\alpha(\varepsilon u))}{\partial x}du+o(1),
\end{aligned}$$

where $o(1)\to 0$ in probability uniformly on any bounded t-set.

By virtue of the measurability of $(\partial/\partial x)F(X^\varepsilon(l\delta_\varepsilon),j)$ with respect to $\mathcal{F}_{l\delta_\varepsilon}^\varepsilon$, and the independence of $\alpha(\cdot)$ and $z^\varepsilon(\cdot)$

$$\begin{aligned}
&\frac{1}{T_\varepsilon}\int_{t_\varepsilon^l}^{t_\varepsilon^l+T_\varepsilon}\mathbb{E}_{l\delta_\varepsilon}^\varepsilon f(X^\varepsilon(l\delta_\varepsilon),\alpha(\varepsilon u),\widetilde{z}^\varepsilon(u))\frac{\partial F(X^\varepsilon(l\delta_\varepsilon),\alpha(\varepsilon u))}{\partial x}du \\
&=\sum_{j\in\mathcal{M}}\frac{1}{T_\varepsilon}\int_{t_\varepsilon^l}^{t_\varepsilon^l+T_\varepsilon}\mathbb{E}_{l\delta_\varepsilon}^\varepsilon f(X^\varepsilon(l\delta_\varepsilon),j,\widetilde{z}^\varepsilon(u))\frac{\partial F(X^\varepsilon(l\delta_\varepsilon),j)}{\partial x}I_{\{\alpha(\varepsilon u)=j\}}du \\
&=\sum_{i,j\in\mathcal{M}}\frac{1}{T_\varepsilon}\int_{t_\varepsilon^l}^{t_\varepsilon^l+T_\varepsilon}\mathbb{E}_{l\delta_\varepsilon}^\varepsilon f(X^\varepsilon(l\delta_\varepsilon),j,\widetilde{z}^\varepsilon(u))\frac{\partial F(X^\varepsilon(l\delta_\varepsilon),j)}{\partial x} \\
&\quad\times\mathbb{P}(\alpha(\varepsilon u)=j|\alpha(l\delta_\varepsilon)=i)I_{\{\alpha(l\delta_\varepsilon)=i\}}du \\
&=\sum_{i\in\mathcal{M}}\frac{1}{T_\varepsilon}\int_{t_\varepsilon^l}^{t_\varepsilon^l+T_\varepsilon}\mathbb{E}_{l\delta_\varepsilon}^\varepsilon f(X^\varepsilon(l\delta_\varepsilon),i,\widetilde{z}^\varepsilon(u))\frac{\partial F(X^\varepsilon(l\delta_\varepsilon),i)}{\partial x} \\
&\quad\times I_{\{\alpha(l\delta_\varepsilon)=i\}}du+o(1),
\end{aligned}$$

where $o(1)\to 0$ in probability as $\varepsilon\to 0$ uniformly on any bounded t-set. The last step above follows from the well-known fact of continuous-time Markov chain: Because $\varepsilon u-l\delta_\varepsilon\to 0$ as $\varepsilon\to 0$

$$\mathbb{P}(\alpha(\varepsilon u)=j|\alpha(l\delta_\varepsilon)=i)\to\delta_{ij}=\begin{cases}1, & \text{if } i=j,\\ 0, & \text{otherwise.}\end{cases}$$

The above estimates indicate that we need only consider the term

$$\rho^\varepsilon = \frac{1}{T_\varepsilon}\int_{t_\varepsilon^l}^{t_\varepsilon^l+T_\varepsilon} \mathbb{E}^\varepsilon_{l\delta_\varepsilon} f(X^\varepsilon(l\delta_\varepsilon), i, \widetilde{z}^\varepsilon(u)) \frac{\partial F(X^\varepsilon(l\delta_\varepsilon), i)}{\partial x} I_{\{\alpha(l\delta_\varepsilon)=i\}} du$$

in the averaging.

We approximate the $X^\varepsilon(l\delta_\varepsilon)$ by a process taking finitely many values. To be more specific, as in [177, Section 6.1, p. 143 and Section 8.2, p. 227], for any $\Delta > 0$, let $\{B_n^\Delta : n \le n_\Delta\}$ be a finite collection of disjoint sets that satisfies $\mathbb{P}(X^\varepsilon(l\delta_\varepsilon) \in \partial B_n^\Delta) = 0$ and that covers the range of $X^\varepsilon(l\delta_\varepsilon)$. Recall that we have assumed the boundedness of $X^\varepsilon(\cdot)$ for notational simplicity (see Remark 10.13). Select a point $X_n^\Delta \in B_n^\Delta$ and rewrite ρ^ε as

$$\begin{aligned}
\rho^\varepsilon &= \sum_{n=1}^{n_\Delta} I_{\{X^\varepsilon(l\delta_\varepsilon)\in B_n^\Delta\}} I_{\{\alpha(l\delta_\varepsilon)=i\}} \\
&\quad \times \frac{1}{T_\varepsilon}\int_{t_\varepsilon^l}^{t_\varepsilon^l+T_\varepsilon} \mathbb{E}^\varepsilon_{l\delta_\varepsilon} f(X_n^\Delta, i, \widetilde{z}^\varepsilon(u)) \frac{\partial F(X_n^\Delta, i)}{\partial x} du \\
&\quad + \sum_{n=1}^{n_\Delta} I_{\{X^\varepsilon(l\delta_\varepsilon)\in B_n^\Delta\}} I_{\{\alpha(l\delta_\varepsilon)=i\}} \\
&\quad \times \frac{1}{T_\varepsilon}\int_{t_\varepsilon^l}^{t_\varepsilon^l+T_\varepsilon} \mathbb{E}^\varepsilon_{l\delta_\varepsilon} \bigg[f(X^\varepsilon(l\delta_\varepsilon), i, \widetilde{z}^\varepsilon(u)) \frac{\partial F(X^\varepsilon(l\delta_\varepsilon), i)}{\partial x} \\
&\quad - f(X_n^\Delta, i, \widetilde{z}^\varepsilon(u)) \frac{\partial F(X_n^\Delta, i)}{\partial x} \bigg] du.
\end{aligned} \tag{10.49}$$

The term in the last three lines of (10.49) goes to 0 in probability as $\varepsilon \to 0$ and then $\Delta \to 0$ by the weak convergence, the Skorokhod representation, the Lipschitz continuity of $f(\cdot, i, z)$, and the smoothness of $F(\cdot, i)$. We proceed to figure out the limit of the term in the first two lines of the right-hand side of (10.49).

Note that as $\varepsilon \to 0$

$$I_{\{X^\varepsilon(l\delta_\varepsilon)\in B_n^\Delta\}} \to \begin{cases} 1, & \text{if } X^\varepsilon(l\delta_\varepsilon) - X_n^\Delta \to 0, \\ 0, & \text{otherwise.} \end{cases}$$

In view of (10.40) and the existence of the unique invariant density $\mu(x, \cdot)$ for each x, we have that

$$\begin{aligned}
&\sum_{i\in\mathcal{M}}\sum_{n=1}^{n_\Delta} I_{\{X^\varepsilon(l\delta_\varepsilon)\in B_n^\Delta\}} I_{\{\alpha(l\delta_\varepsilon)=i\}} \frac{1}{T_\varepsilon}\int_{t_\varepsilon^l}^{t_\varepsilon^l+T_\varepsilon} \mathbb{E}^\varepsilon_{l\delta_\varepsilon} f(X_n^\Delta, i, \widetilde{z}^\varepsilon(u)) \frac{\partial F(X_n^\Delta, i)}{\partial x} du\\
&= \sum_{i\in\mathcal{M}}\sum_{n=1}^{n_\Delta} I_{\{X^\varepsilon(l\delta_\varepsilon)\in B_n^\Delta\}} I_{\{\alpha(l\delta_\varepsilon)=i\}}\\
&\quad\times \frac{1}{T_\varepsilon}\int_{t_\varepsilon^l}^{t_\varepsilon^l+T_\varepsilon}\int_0^1 f(X_n^\Delta, i, z)\mathbb{E}^\varepsilon_{l\delta_\varepsilon} I_{\{\widetilde{z}^{\varepsilon,X_n^\Delta}(u)\in dz\}} \frac{\partial F(X_n^\Delta, i)}{\partial x} du\\
&= \sum_{i\in\mathcal{M}}\sum_{n=1}^{n_\Delta} I_{\{X^\varepsilon(l\delta_\varepsilon)\in B_n^\Delta\}} I_{\{\alpha(l\delta_\varepsilon)=i\}}\\
&\quad\times \frac{1}{T_\varepsilon}\int_{t_\varepsilon^l}^{t_\varepsilon^l+T_\varepsilon}\int_0^1 f(X_n^\Delta, i, z)\mathbb{P}(\widetilde{z}^{\varepsilon,X_n^\Delta}(u)\in dz|\widetilde{z}^{\varepsilon,X_n^\Delta}(t_\varepsilon^l)) \frac{\partial F(X_n^\Delta, i)}{\partial x} du\\
&\to \sum_{i\in\mathcal{M}}\int_0^1 I_{\{\alpha(u)=i\}} f(X(u), i, z)\mu(X(u), z)dz \frac{\partial F(X(u), i)}{\partial x}\\
&= \widehat{f}(X(u), \alpha(u))\frac{\partial F(X(u), \alpha(u))}{\partial x}.
\end{aligned}$$

Thus, as $\varepsilon \to 0$

$$\begin{aligned}
&\mathbb{E}\prod_{\ell=1}^{n_0} h_\ell(X^\varepsilon(t_\ell))\left[\int_t^{t+s} f(X^\varepsilon(u), \alpha(u), z^\varepsilon(u))\frac{\partial F(X^\varepsilon(u), \alpha(u))}{\partial x}\right] du\\
&\quad\to \mathbb{E}\prod_{\ell=1}^{n_0} h_\ell(X(t_\ell))\left[\int_t^{t+s} \widehat{f}(X(u), \alpha(u))\frac{\partial F(X(u), \alpha(u))}{\partial x}\right] du.
\end{aligned}$$

Likewise, we obtain that as $\varepsilon \to 0$

$$\begin{aligned}
&\mathbb{E}\prod_{\ell=1}^{n_0} h_\ell(X^\varepsilon(t_\ell))\\
&\qquad\times\left[\frac{1}{2}\int_t^{t+s} \sigma^2(\alpha(u), X^\varepsilon(u), z^\varepsilon(u))\frac{\partial^2 F(X^\varepsilon(u), \alpha(u), z^\varepsilon(u))}{\partial x^2}\right] du\\
&\quad\to \mathbb{E}\prod_{\ell=1}^{n_0} h_\ell(X(t_\ell))\left[\int_t^{t+s} \widehat{\sigma}^2(X(u), \alpha(u))\frac{\partial^2 F(X(u), \alpha(u))}{\partial x^2}\right] du,
\end{aligned}$$

and

$$\begin{aligned}
&\mathbb{E}\prod_{\ell=1}^{n_0} h_\ell(X^\varepsilon(t_\ell))\Big[\int_t^{t+s}\int_\Gamma \lambda[F(X^\varepsilon(u)+g(\gamma, X^\varepsilon(u), \alpha(u), z^\varepsilon(u)), \alpha(u))\\
&\quad - F(X^\varepsilon(u), \alpha(u))]\pi(d\gamma)\Big]
\end{aligned}$$

$$\to \mathbb{E}\prod_{\ell=1}^{n_0} h_\ell(X(t_\ell))\Big[\int_t^{t+s}\int_\Gamma \lambda[F(X(u)+\widehat{g}(\gamma, X(u), \alpha(u)), \alpha(u))$$
$$- F(X(u), \alpha(u))]\pi(d\gamma)\Big].$$

Combining these results, we obtain

$$\mathbb{E}\prod_{\ell=1}^{n_0} h_\ell(X(t_\ell)\Big[F(X(t+s), \alpha(t+s)) - F(X(t), \alpha(t))$$
$$- \int_t^{t+s} \mathcal{L}F(X(u), \alpha(u))du\Big] = 0.$$

Hence, (10.44) holds. Consequently, the desired result follows. □

10.4 Discussion and Remarks

We have studied two-time-scale hybrid jump diffusions. The motivational insurance risk models are more general than the classical compound Poisson models and compound Poisson models under diffusion perturbations. In our models, the rate at which the premiums are received and the rate of oscillation due to diffusion depend on the surplus level. Moreover, the models also take into consideration Markovian regime switching. Under suitable conditions, we have derived limit systems for fast switching and fast diffusion, respectively. The results obtained can be specialized to a number of cases.

Example 10.14 Consider (10.12) with $\sigma(t, x, \alpha) \equiv 0$. Under such a condition, Theorem 10.3 still holds with $\overline{\sigma}^2(t, x, \alpha) \equiv 0$. Then, the limit system becomes

$$X(t) = x + \int_0^t \overline{f}(s, X(s), \overline{\alpha}(s))ds + \int_0^t \int_\Gamma \overline{g}(\gamma, X(s-), \overline{\alpha}(s-))N(ds, d\gamma).$$

Under further specialized coefficients, it reduces to the surplus model involving regime switching [287].

Example 10.15 We consider the model in (10.12) again with $\widetilde{Q}(t) = Q(t)$ that is weakly irreducible. Then, the limit system is given by Corollary 10.5. The limit does not involve the Markov switching process. Thus, the complexity reduction is more pronounced. Further specification leads to the classical risk model.

Example 10.16 Consider the system given by (10.35) with $\sigma(t, x, \alpha) \equiv 0$. Then, similar to Example 10.14, Theorem 10.3 still holds with the limit system being a Markovian modulated jump process.

Remark 10.17 In the fast diffusion case, for simplicity, we concentrated on the case that $Q(t) = Q$ a constant matrix. Time-varying generator can be considered. Essentially, the same ideas work. Only the notation needs to be changed.

10.5 Remarks on Numerical Solutions for Switching Jump Diffusions

This chapter focuses on switching jump diffusions. The central theme is the treatment of two-time-scale systems. Most of the jump-diffusion processes with regime switching are nonlinear. Even without two-time-scales, closed-form solutions are virtually impossible to obtain. As a viable alternative, one has to find feasible numerical schemes. In what follows, we suggest numerical algorithms for (10.5). We present the algorithm and the corresponding convergence properties. The approach we are taking is again the martingale problem formulation, which brings out the profile and dynamic behavior of the process rather than dealing with the iterations directly. Because the detailed proof of convergence of numerical approximation algorithms has a certain similarity to that of Chapter 5, we omit the verbatim details.

We use the notation as in Section 10.1.1 of this book. Please consult that section for various definitions such as τ_n, ψ_n, $\psi(t)$ and the like. Consider the case where the coefficients have no explicit dependence on the time variable. For simplicity, we take $Q(t) = Q$, a constant matrix. We describe the algorithm as follows:

1. Choose $\Delta > 0$, a small parameter, as the step size.
2. Using the Δ-skeleton process of the continuous-time Markov chain $\alpha(t)$, given in (10.5), we obtain the discrete-time Markov chain $\alpha_n = \alpha(\Delta n)$ for each n.
3. To approximate the Brownian motion $w(\cdot)$, a usual practice in numerical solutions for stochastic differential equations is to use the Brownian increments:

$$\Delta w_n = \frac{w(\Delta(n+1)) - w(\Delta n)}{\sqrt{\Delta}} := \xi_n$$

   to approximate dw. Because $w(\cdot)$ has independent increments, $\{\Delta w_k\}$ is a sequence of independent and identically distributed random variables with mean 0 and variance 1.
4. Let $\{\rho_n\}$ and $\{\psi_n\}$ be sequences of independent and identically distributed random variables such that ρ_n has an exponential distribution with parameter $\lambda > 0$, and ψ_n is the impulse having distribution $\pi(\cdot)$. Define

$$\widetilde{\tau}_{n+1} = \widetilde{\tau}_n + \rho_n, \quad \text{with } \widetilde{\tau}_0 = 0.$$

   It follows from the independence assumption, $\{\Delta w_k\}$, $\{\alpha_k\}$, $\{\widetilde{\tau}_n\}$, and $\{\psi_n\}$ are also independent.
5. Construct the approximation algorithm:

$$\begin{aligned} X_{n+1} &= X_0 + \Delta \sum_{k=0}^{n} f(X_k, \alpha_k) + \sqrt{\Delta} \sum_{k=0}^{n} \sigma(X_k, \alpha_k)\xi_k \\ &\quad + \sum_{\widetilde{\tau}_j \le \Delta n} g(\psi_j, X_{\lfloor \widetilde{\tau}_j/\Delta \rfloor}, \alpha_{\lfloor \widetilde{\tau}_j/\Delta \rfloor}), \end{aligned} \tag{10.50}$$

where $\lfloor y \rfloor$ denotes the integer part of a real number y.

Remark 10.18 For construction of a discrete-time Markov chain, see [304, pp. 315–316]. Thanks to the use of Δ-skeleton process, the approximation to the Markov chain is further simplified. For further discussion on a skeleton process, we refer the reader to [52].

To approximate the Brownian motion, in lieu of the usual approach as given above, we could generate a sequence of independent and identically distributed random variables $\{\xi_n\}$ such that $E\xi_n = 0$ and $E\xi_n = I$ to further simplify the computation. The functional central limit theorem ensures the approximation to the Brownian motion.

To proceed, define

$$J_n = \sum_{\widetilde{\tau}_j \le \Delta n} g(\psi_j, X_{\lfloor \widetilde{\tau}_j/\Delta \rfloor}, \alpha_{\lfloor \widetilde{\tau}_j/\Delta \rfloor}). \tag{10.51}$$

It is often desirable to write (10.50) recursively. This can be done as follows:

$$X_{n+1} = X_n + \Delta f(X_n, \alpha_n) + \sqrt{\Delta}\sigma(X_n, \alpha_n)\xi_n + \Delta J_n, \tag{10.52}$$

where $\Delta J_n = J_n - J_{n-1}$.

Note that the sequences $\{\rho_n\} = \{\widetilde{\tau}_{n+1} - \widetilde{\tau}_n\}$ and $\{\psi_n\}$ are as in those discussed in the last section. The process $\widetilde{\tau}_n$, in fact, represents the jump times of the underlying process.

To proceed, let us define the interpolated processes via piecewise constant interpolations as

$$\left.\begin{aligned} X^\Delta(t) &= X_n, \\ \alpha^\Delta(t) &= \alpha_n, \\ w^\Delta(t) &= \sqrt{\Delta} \sum_{k=0}^{t/\Delta - 1} \xi_k, \\ J^\Delta(t) &= J_n, \end{aligned}\right\} \quad \text{for } t \in [\Delta n, \Delta n + \Delta). \tag{10.53}$$

For convenience, with a slight abuse of notation, we omitted the floor function notation above. Henceforth, for instance, we use t/Δ to denote the integer part of t/Δ here. We state a result, whose proof is along the line of martingale averaging. Further details can be found in [298].

Theorem 10.19 *Under the conditions in* (H10.1), $(X^\Delta(\cdot), \alpha^\Delta(\cdot))$ *converges weakly to* $(X(\cdot), \alpha(\cdot))$ *as* $\Delta \to 0$ *such that* $(X(\cdot), \alpha(\cdot))$ *is the solution of the martingale problem with operator* $\mathcal{L}$.

As in Chapter 5, Theorem 10.19 implies that the algorithm we constructed is convergent. The limit is nothing but the solution of (10.5). One of the variations of the algorithm is to use a sequence of decreasing step sizes. The modifications are as follows. Let $\{\Delta_n\}$ be a sequence of nonnegative real numbers such that $\Delta_n \to 0$ and $\sum_{n=0}^{\infty} \Delta_n = \infty$. For example, we may take $\Delta_n = 1/n$, or $\Delta_n = 1/n^\gamma$ for some $0 < \gamma < 1$. Define

$$t_n = \sum_{l=0}^{n-1} \Delta_l, \quad \text{and } \alpha_n = \alpha(t_n), \quad n \geq 0.$$

It then can be verified that $\{\alpha_n\}$ so defined is a discrete-time Markov chain with transition probability matrix:

$$P^{n,n+1} = (p_{ij}^{n,n+1})_{m_0 \times m_0} = \exp((t_{n+1} - t_n)Q) = \exp(\Delta_n Q). \tag{10.54}$$

Using ideas in stochastic approximation [177], define

$$m(t) = \max\{n : t_n \leq t\}. \tag{10.55}$$

It is clear that now the Markov chain α_n is not time homogeneous. The approximate solution for the stochastic differential equation with jumps and regime switching (10.5) is given by

$$\begin{cases} X_{n+1} = X_0 + \sum\limits_{k=0}^{n} f(X_k, \alpha_k)\Delta_k + \sum\limits_{k=0}^{n} \sqrt{\Delta_k}\sigma(X_k, \alpha_k)\xi_k \\ \qquad + \sum\limits_{\widetilde{\tau}_j \leq n} g(\psi_j, X_{j-1}, \alpha_j), \\ X_0 = x, \ \alpha_0 = \alpha. \end{cases} \tag{10.56}$$

With the algorithm proposed, we can then proceed to study its performance. The proof is along the same lines as that of the constant stepsize algorithm. The interested reader is referred to [298]; see also Chapter 5 of this book for further reading.

10.6 Notes

As a continuation of our study, based on the work Yin and Yang [301], this chapter has been devoted to two-time-scale jump diffusions with regime switching. Roughly, the fast-changing driving processes can be treated as a noise, whose stationary measure does exist. Under suitable conditions, the slow process is averaged out with respect to the stationary measure of the fast-varying process. Such an idea has been used in the literature. We refer the reader to Khasminskii [114], Papanicolaou et al. [230], Khasminskii and Yin [159–161], Pardoux and Veretennikov [231], and references therein for related work in diffusions, and Yin [290] for that of switching diffusions.

The limit results obtained in this chapter can be useful for applications. For example, one may use such results in a subsequent study for obtaining bounds on ruin probability in risk management. It is conceivable that this approach will lead to approximations with tighter error bounds. Certain generalizations are possible. For example, the finite-state Markov chain may also include transient states. Treating (10.35), time-inhomogeneous systems may be considered. Further investigation may also include the replacement of the diffusion by a wideband noise [175] yielding more realistic systems. Finally, for notational simplicity, only scalar $X^{\varepsilon}(t)$ is treated in this chapter. The approach presented can be carried over to multidimensional cases. In fact, the main arguments will all follow through with no essential difficulties.

Toward the end of the chapter, we also outlined numerical schemes for solutions of switching jump diffusions. For detailed development, the reader is referred to Yin et al. [298]. For simplicity, the switching process is assumed to be a continuous-time Markov chain independent of the Brownian motion. By combining the treatment of jump diffusions with that of the switching diffusions with x-dependent switching, x-dependent switching jump diffusions can also be treated.

Our discussion thus far requires the range space of the jump, namely, Γ being compact. The jump measure is assumed to be finite. Such models are suitable for many control and optimization problems; see, for example, [176] without the switching. Switching jump diffusions with more general setup have been provided in Chen et al. [44, 45], Kunwai and Zhu [169], and Xi et al. [283]. In these references, in addition to the aforementioned generalization. the switching process is allowed to depend on the jump diffusions. Some basic properties such as maximum principles and Harnack inequalities were treated in [44], recurrence and stability were considered in [45], whereas Feller and strong Feller properties were explored in [169, 283].

Chapter 11
Past-Dependent Switching Diffusion

11.1 Introduction

Owing to the demand of modeling, analysis, and computation of complex networked systems, much attention has been devoted to building more realistic dynamic system models. Thus far, this book has treated hybrid diffusions in which the switching rates of the discrete component $\alpha(t)$ depend only on the current state of the continuous component $X(t)$ and $\alpha(t)$ takes values in a finite set. Questions naturally arise. What happens if the switching process depends on the past history of the continuous states? What if the switching also takes places in a countable state space?

There are plenty of real-world applications involving such switching diffusions. Perhaps, one of the most widely used control models in the literature is the so-called LQG (linear quadratic Gaussian regulator) problem; see [86, pp.165–166] for a traditional model. However, for many new applications in networked systems, it has been found that in addition to the random noise represented by Brownian type of disturbances, there is a source of randomness owing to the presence of random environment that can be modeled by a continuous-time Markov chain. Let $\alpha(t)$ be a continuous-time Markov chain with state space $\mathcal{M} = \{1, 2, \cdots, m_0\}$ where m_0 can be infinity and generator Q. Consider the controlled dynamic system

$$\begin{aligned} dX(t) &= [A(\alpha(t))X(t) + B(\alpha(t))u(t)]dt + \sigma(\alpha(t))dW(t), \\ X(s) &= x, \ \text{ for } s \le t \le T, \end{aligned} \tag{11.1}$$

where $X(t) \in \mathbb{R}^{n_1}$ is the continuous state variable, $u(t) \in \mathbb{R}^{n_2}$ is the control, $A(i) \in \mathbb{R}^{n_1 \times n_1}$, and $B(i) \in \mathbb{R}^{n_1 \times n_2}$ are well defined and have finite values for each $i \in \mathcal{M}$. One may wish to find the optimal control $u(\cdot)$ so that the expected quadratic cost function

H.-D. Nguyen et al., *Hybrid Switching Diffusions*, Probability Theory and Stochastic Modelling 63, https://doi.org/10.1007/978-3-031-93303-5_11

$$J(s,i,x,u(\cdot))=\mathbb{E}\Big[\int_s^T [X'(t)M(\alpha(t))X(t)+u'(t)N(\alpha(t))u(t)]dt+X'(T)DX(T)\Big] \tag{11.2}$$

is minimized. The use of $\alpha(t)$ stems from the formulation of discrete events, and the use of $\mathcal{M}$ enlarges the applicability of previous consideration of finite state space cases. Switched dynamic systems can also be found in, for example, modeling impatient customers and customer abandonment of Markov-modulated service speeds in the heavy-traffic regime and the many-server systems in the Halfin-Whitt regime and the non-degenerate slowdown regime; see [130]. We also refer the reader to Whitt [274] for further reading on limit results in queueing theory and many references therein. In fact, in most of the queueing models, the discrete set is countable rather than finite.

In another example, we consider an extension of the Markov-modulated-rate fluid models treated in [302]. Stemming from queueing systems, this example is simple in that it is even without the Brownian motion part, but it explains the modeling view point of the past-dependent switching with a countable state space. Consider the fluid buffer model with an infinite capacity. Let $X(t)$ be the amount of fluids in the buffer at time t, known as buffer content or buffer level. The fluids enter and leave the buffer at random rates. The input and output of fluids are modulated by a switching process $\alpha(t)$ with state space $\mathbb{N} = \{1, 2, \ldots\}$, known as a stochastic external environment. Using $\alpha(t)$ to determine the input and output rates, we introduce a drift function $f(\cdot)$ (see Kulkarni [168]): $f(\cdot) : \mathcal{M} \mapsto (-\infty, \infty)$. Different from [302], we do not assume $\alpha(t)$ to be a Markov chain, but rather, we assume that the transition rates satisfy (11.6). That is, the transition rates depend on the history of the process $X(t)$.

Note that formally, the net rate that is the difference of the input and the output rates at time t is given by $f(\alpha(t))$. The dynamics of the buffer content $\{X(t) : t \geq 0\}$ can be described by the following differential equation:

$$\begin{aligned}\frac{d}{dt}X(t) &= \begin{cases} f(\alpha(t)), & \text{if } X(t) > 0 \\ (f(\alpha(t)))^+, & \text{if } X(t) = 0, \end{cases} \\ &= f(\alpha(t))I_{\{X(t)>0\}} + (f(\alpha(t)))^+ I_{\{X(t)=0\}},\end{aligned} \tag{11.3}$$

where $x^+ = \max\{x, 0\}$. Note that $X(t)$ can be rewritten as

$$X(t) = X(0) + Y(t) - \Big(\inf_{0\leq s\leq t} \{Y(s) + X(0)\}\Big) \wedge 0 \tag{11.4}$$

with $Y(s) = \int_0^s f(\alpha(v))dv$, where $a \wedge b = \min(a, b)$ for two real numbers a and b, and $-(\inf_{0\leq s\leq t}\{Y(s) + X(0)\}) \wedge 0$ measures the amount of potential output lost up to time t due to the emptiness of the buffer. Many of the current interests are concerned with the fluid model above such as long-run average control problems or stability of the systems.

Two more dynamic systems are to be presented in the next section, in which the main interests are to find long-term behavior and control design in an ecological system and to find optimal strategies under long-run average criteria for a pollution management problem. In order to study the aforesaid problems, we first need to ensure that the systems under consideration have unique solutions and that the solutions possess good properties. Our study will build a bridge for future study on related control systems.

The rest of this chapter is organized as follows. The formulation and some more examples of switching diffusions with past-dependent switching and countably many possible switching locations is given in Section 11.2. The existence and uniqueness of solutions to the stochastic equations are then proved under suitable conditions in Section 11.3. Section 11.4 studies the Markov and Feller properties of a functional stochastic process associated with our equation. The proof for the Feller property is rather complex because the state space of $\alpha(t)$ is infinite, the space of continuous functions is not locally compact, and we do not assume uniform continuity of the switching intensities. In Section 11.5, we consider the recurrence and ergodicity of the process. Section 11.6 provides conditions for stability and instability of the system. Section 11.7 is devoted to the proofs of some technical results needed in the preceding sections.

11.2 Formulation and Examples

11.2.1 Formulation

Let r be a fixed positive number. Denote by $\mathscr{C}([a,b],\mathbb{R}^n)$ the set of $\mathbb{R}^n$-valued continuous functions defined on $[a,b]$. In what follows, we mainly work with $\mathscr{C}([-r,0],\mathbb{R}^n)$, and simply denote it by $\mathscr{C} := \mathscr{C}([-r,0],\mathbb{R}^n)$. Denote by $|x|$ the Euclidean norm of $x \in \mathbb{R}^n$. For $\phi \in \mathscr{C}$, we use the norm $\|\phi\| = \sup\{|\phi(t)| : t \in [-r,0]\}$. For $y(\cdot) \in \mathscr{C}([-r,\infty),\mathbb{R}^n)$ and $t \ge 0$, denote by y_t the so-called segment function (or memory segment function) $y_t(\cdot) := y(t+\cdot) \in \mathscr{C}$. Let $(\Omega, \mathcal{F}, \{\mathcal{F}_t\}_{t\ge 0}, \mathbb{P})$ be a complete filtered probability space with the filtration $\{\mathcal{F}_t\}_{t\ge 0}$ satisfying the usual condition, i.e., it is increasing and right continuous while $\mathcal{F}_0$ contains all $\mathbb{P}$-null sets. Let $W(t)$ be an $\mathcal{F}_t$-adapted and $\mathbb{R}^d$-valued Brownian motion. Suppose $b(\cdot,\cdot) : \mathbb{R}^n \times \mathcal{M} \to \mathbb{R}^n$ and $\sigma(\cdot,\cdot) : \mathbb{R}^n \times \mathcal{M} \to \mathbb{R}^{n\times d}$, where $\mathcal{M} = \{1, 2, \dots\}$, the set of positive integers. Consider the two-component process $(X(t), \alpha(t))$ where $\alpha(t)$ is a pure jump process taking value in $\mathcal{M}$, and $X(t)$ satisfies

$$dX(t) = b(X(t),\alpha(t))dt + \sigma(X(t),\alpha(t))dW(t). \tag{11.5}$$

We assume that if $\alpha(t-) := \lim_{s\to t^-}\alpha(s) = i$, then it can switch to $j \neq i$ at t with intensity $q_{ij}(X_t)$, where $q_{ij} : \mathscr{C} \to \mathbb{R}$ is a measurable function. When $q_i(\phi) := \sum_{j=1,j\neq i}^{\infty} q_{ij}(\phi)$ is uniformly bounded in $(\phi, i) \in \mathscr{C} \times \mathcal{M}$, and $q_i(\cdot)$ and $q_{ij}(\cdot)$ are continuous, one may view the switching intensities as

$$\mathbb{P}\{\alpha(t+\Delta) = j|\alpha(t) = i, X_s, \alpha(s), s \leq t\} = \begin{cases} q_{ij}(X_t)\Delta + o(\Delta), & \text{if } i \neq j, \\ 1 - q_i(X_t)\Delta + o(\Delta), & \text{if } j = i. \end{cases} \tag{11.6}$$

However, when $q_i(\phi)$ and $q_{ij}(\phi)$ are either discontinuous or unbounded, it does not seem appropriate to use (11.6) to model the switching intensity. To formulate the problem in a general setting without the boundedness and continuity assumptions mentioned above, we construct $\alpha(t)$ as the solution to a stochastic differential equation with respect to a Poisson random measure as in Chapter 2. Let $\mathfrak{p}(dt, dz)$ be a Poisson random measure with intensity $dt \times \mathfrak{m}(dz)$, where $\mathfrak{m}$ is the Lebesgue measure on $\mathbb{R}$. For each function $\phi : [-r, 0] \to \mathbb{R}^n$, and $i \in \mathcal{M}$, let $\Delta_{ij}(\phi)$, $j \neq i$ be disjoint intervals on $\mathbb{R}$ with length $q_{ij}(\phi)$ such that $\{\Delta_{ij}(\phi), j \neq i\}$'s are mutually disjoint. Define $h : \mathscr{C} \times \mathcal{M} \times \mathbb{R} \mapsto \mathbb{R}$ by $h(\phi, i, z) = \sum_{j=1,j\neq i}^{\infty}(j - i)I_{\{z\in\Delta_{ij}(\phi)\}}$, where $I_{\{z\in\Delta_{ij}(\phi)\}} = 1$ if $z \in \Delta_{ij}$; otherwise $I_{\{z\in\Delta_{ij}(\phi)\}} = 0$ is the indicator function. The process $\alpha(t)$ can be defined as a solution to

$$d\alpha(t) = \int_{\mathbb{R}} h(X_t, \alpha(t-), z)\mathfrak{p}(dt, dz).$$

The pair $(X(t), \alpha(t))$ is therefore a solution to the system of equations

$$\begin{cases} dX(t) = b(X(t), \alpha(t))dt + \sigma(X(t), \alpha(t))dW(t) \\ d\alpha(t) = \displaystyle\int_{\mathbb{R}} h(X_t, \alpha(t-), z)\mathfrak{p}(dt, dz). \end{cases} \tag{11.7}$$

A strong solution to (11.7) on $[0, T]$ with initial data (ϕ, i_0) being $\mathscr{C}\times\mathcal{M}$-valued and $\mathcal{F}_0$-measurable random variable is an $\mathcal{F}_t$-adapted process $(X(t), \alpha(t))$ such that:

- $X(t)$ is continuous and $\alpha(t)$ is càdlàg (right continuous with left limits) almost surely (a.s.).
- $X(t) = \phi(t)$ for $t \in [-r, 0]$ and $\alpha(0) = i_0$,
- $(X(t), \alpha(t))$ satisfies (11.7) for all $t \in [0, T]$ a.s.

We will show in Section 11.7.3 that the solution $(X(t), \alpha(t))$ to (11.7), satisfies (11.6) under suitable conditions. Let $f(\cdot, \cdot) : \mathbb{R}^n \times \mathcal{M} \mapsto \mathbb{R}$ be a function that is twice continuously differentiable in x and bounded in $(x, i) \in \mathbb{R}^n \times \mathcal{M}$. We define the "operator" $\mathbb{L}f(\cdot, \cdot) : \mathscr{C} \times \mathcal{M} \mapsto \mathbb{R}$ as

$$
\begin{aligned}
\mathbb{L}f(\phi,i) =& \nabla f(\phi(0),i)b(\phi(0),i) + \frac{1}{2}\text{trace}\Big(\nabla^2 f(\phi(0),i)A(\phi(0),i)\Big) \\
&+ \sum_{j=1,j\neq i}^{\infty} q_{ij}(\phi)\big[f(\phi(0),j) - f(\phi(0),i)\big] \\
=& \sum_{k=1}^{n} b_k(\phi(0),i)f_k(\phi(0),i) + \frac{1}{2}\sum_{k,l=1}^{n} a_{kl}(\phi(0),i)f_{kl}(\phi(0),i) \\
&+ \sum_{j=1,j\neq i}^{\infty} q_{ij}(\phi)\big[f(\phi(0),j) - f(\phi(0),i)\big],
\end{aligned}
\tag{11.8}
$$

where $b(x,i) = (b_1(x,i),\dots,b_n(x,i))'$, $\nabla f(x,i) = (f_1(x,i),\dots,f_n(x,i)) \in \mathbb{R}^{1\times n}$ and $\nabla^2 f(x,i) = (f_{ij}(x,i))_{n\times n}$ are the gradient and Hessian of $f(x,i)$ with respect to x, respectively, with

$$
\begin{aligned}
&f_k(x,i) = (\partial/\partial x_k)f(x,i), \quad f_{kl}(x,i) = (\partial^2/\partial x_k\partial x_l)f(x,i), \quad \text{and} \\
&A(x,i) = (a_{kl}(x,i))_{n\times n} = \sigma(x,i)\sigma'(x,i),
\end{aligned}
$$

with z' denoting the transpose of z. Similar to (2.5), we have Itô's formula:

$$
f(X(t),\alpha(t)) - f(X(0),\alpha(0)) = \int_0^t \mathbb{L}f(X_s,\alpha(s-))ds + M_1(t) + M_2(t) \text{ a.s.,}
\tag{11.9}
$$

where $M_1(\cdot)$ and $M_2(\cdot)$ are local martingales, defined by

$$
\begin{aligned}
M_1(t) &= \int_0^t \big\langle \nabla f(X(s),\alpha(s-)), \sigma(X(s),\alpha(s-))dW(s)\big\rangle, \\
M_2(t) &= \int_0^t \int_{\mathbb{R}} \big[f\big(X(s),\alpha(s-)+h(X_s,\alpha(s-),z)\big) - f(X(s),\alpha(s-))\big]\widetilde{\mathfrak{p}}(ds,dz).
\end{aligned}
\tag{11.10}
$$

It should be noted that $\mathbb{L}$ is not the generator of the Markov process $(X_t,\alpha(t))$. However this operator is useful for analyzing the process $(X(t),\alpha(t))$. In view of (11.10), if $\tau_1 \le \tau_2$ are stopping times that are bounded above by T a.s., and $f(\cdot,\cdot)$ and $\mathbb{L}f(\cdot,\cdot)$ are bounded and $q_{\alpha(t)}(X_t) < \infty$ on $[\tau_1,\tau_2]$, then

$$
\mathbb{E}f(X(\tau_2),\alpha(\tau_2)) = \mathbb{E}f(X(\tau_1),\alpha(\tau_1)) + \mathbb{E}\int_{\tau_1}^{\tau_2} \mathbb{L}f(X_t,\alpha(t-))dt.
$$

Remark 11.1 If $\alpha(t)$ depends on the continuous state, but there is no past dependence (i.e., the segment process X_t is replaced by $X(t)$ in (11.6), and ϕ and $\phi(0)$ are replaced by the current state $X(t) = x$ in (11.8), respectively), then $\mathbb{L}$ is indeed the generator of the process $(X(t), \alpha(t))$; c.f. (2.6). Even in this case, the investigation of recurrence, ergodicity, and stability for past-independent switching processes with a countable switching state space generalizes and extends the results discussed in the previous chapters.

11.2.2 Examples

Example 11.2 This example stems from applications in ecological systems and biological control. Consider the evolution of two interacting species. One is micro described by a logistic differential equation perturbed by a white noise. The other is macro assumed to be a number of individuals following a birth-death process. Let $X(t)$ be the density of the micro species and $\alpha(t)$ the population of the macro species. The life cycle of a micro species is usually very short. It is reasonable to assume that the evolution of $X(t)$ can be described by the following past-independent equation:

$$dX(t) = X(t)\big[a(\alpha(t)) - b(\alpha(t))X(t)\big]dt + \sigma(\alpha(t))X(t)dW(t), \tag{11.11}$$

where $a(i), b(i), \sigma(i)$ are positive constants for each $i \in \mathcal{M}$.

On the other hand, the reproduction process of $\alpha(t)$ is assumed to be non-instantaneous. More precisely, suppose the reproduction depends on the period of time from egg formation to hatching, say r. Then we have

$$d\alpha(t) = \int_{\mathbb{R}} h(X_t, \alpha(t-), z)\mathfrak{p}(dt, dz), \tag{11.12}$$

where $h(\phi, i, z) = \sum_{j=1, j\neq i}^{\infty}(j - i)I_{\{z\in\Delta_{ij}(\phi)\}}$, $\Delta_{i,i+1}(\phi) = [0, \beta_i(\phi))$, $\Delta_{i,i-1}(\phi) = [0, \delta_i(\phi))$, $\Delta_{i,j}(\phi) = \emptyset$ if $j \notin \{i-1, i, i+1\}$ or $i = 0$. Usually, $\beta_i(\phi), \delta_i(\phi)$ can be given in the integral from $\beta_i(\phi) = \int_{-r}^{0}\tilde{\beta}_i(t)\phi(t)dt$, $\delta_i(\phi) = \int_{-r}^{0}\tilde{\delta}_i(t)\phi(t)dt$, for some appropriate weighting functions $\tilde{\beta}_i, \tilde{\delta}_i$. As can be seen from the above, the switching process at t, in fact, depends on past history of the state $X(\cdot)$. Investigating the interactions between the two species are very important for biological control. A basic biological control problem aims to choose a suitable living organism to control a particular pest (see, e.g., [129, 196]). The chosen organism might be a predator, parasite, or disease, which will attack the harmful insect. To design and evaluate effectiveness of a biological control, some questions should be answered first. For example, under which conditions the species will be permanent forever or they will extinct at some instance? Whether or not there is an invariant measure associated with the system under consideration. Mathematically, these questions are related to the stability and ergodicity of the corresponding stochastic systems.

Example 11.3 Pollution management is vitally important and has a significant impact on environment. A major issue is concerned with the trade-off of pollution accumulation and consumption, which affects environmental policy making. Following the seminal paper of Keeler et al. [153], much work has been devoted to the study of optimal control of dynamic economic systems. In [152], Kawaguchi and Morimoto treated a pollution accumulation problem of maximizing the long-run average welfare using a controlled diffusion model. Assume that an economy consumes some good and meanwhile generates pollution. The pollution stock is gradually degraded, and its instantaneous growth rate incorporates a random disturbance with mean zero and constant variance. The social welfare is defined by the utility of the consumption net of the disutility of pollution. The problem is to find optimal consumption strategies for the society in the long-run average sense. Departing from their formulation, we consider an extension of their model. Suppose that there is a switching process $\alpha(t)$ taking values in $\mathcal{M}$ such that $\alpha(t)$ represents the level of pollution at time t. Assume that the stock of pollution at time t is given by $X(t)$, a real-valued process, and there is a positive real-valued function $\rho(\cdot)$ so that for each $i \in \mathcal{M}$, the rate of pollution decay is $\rho(i)$. The consumption rate (or flow of pollution) is a control process, which is denoted by $c(t)$ at time t; the social utility function of the consumption c is denoted by $U(c)$, whereas the social disutility of the pollution stock x is $D(x)$. We say that the consumption rate is admissible if it is $\mathcal{F}_t$-measurable, where $\mathcal{F}_t = \{(X(s), \alpha(s)) : s \leq t\}$ such that $0 \leq c(t) \leq K_0$ for some $K_0 > 0$. The ultimate objective is to maximize the long-run average welfare

$$J(c(\cdot)) = \liminf_{T \to \infty} \frac{1}{T} \mathbb{E} \int_0^T [U(c(t)) - D(X(t))]dt, \tag{11.13}$$

subject to

$$dX(t) = [c(t) - \rho(\alpha(t))X(t)]dt + \sigma(X(t), \alpha(t))dW(t). \tag{11.14}$$

Assume that the pollution level $\alpha(\cdot)$ satisfies the conditions (11.6). First, it is reasonable that the level of pollution can be modeled by a continuous-time process taking values in $\mathcal{M}$. Second, to be more realistic, the pollution level depends on the pollution stock $X(t)$ as well as some past history as given in (11.6). As another generalization of [152], we assume that σ in fact depends on $(X(t), \alpha(t))$, and the switching rate depends on some past history of the pollution stock $X(\cdot)$ as in (11.6), and $\sigma^2(x, i) > 0$ for each $i \in \mathcal{M}$. Treating the optimal pollution management problem, it is natural to consider the replacement of the average in (11.13) by the average with respect to an invariant measure (if it exists) of the controlled systems. To do so, we need to make sure that (11.14) indeed has an invariant measure. Before this matter can be settled, we need to show that the system has a unique solution for each initial data, and the solution possesses certain desired properties such as Markov and Feller properties.

11.3 Existence and Uniqueness of Solutions

We are now in a position to prove the existence and uniqueness in the strong sense of a solution with given initial data under suitable conditions. We give several sets of conditions. The main reason is due to the past dependence and the use of $\mathcal{M}$. First in contrast to the case of switching process staying in a finite set, care needs to be exercised regarding uniformity with respect to the switching set. Second, the past dependence requires careful handling of the use of Lipschitz continuity, etc., and the uniformity with respect to the element in the corresponding function spaces. Depending on the preference, Assumption (H11.1) allows certain bounds to be dependent of the switching state i, but uniform in the variable in the function space, whereas (H11.2) requires uniformity in the bounds w.r.t. i, but requires the past-dependent part to be localized. Assumptions (H11.3) and (H11.4) relax the Lipschitz condition to local Lipschitz together with certain growth conditions presented by using bounds with the help of Lyapunov functions.

(H11.1) Assume the following conditions hold:

(i) For each $i \in \mathcal{M}$, there is a positive constant L_i such that

$$|b(x,i) - b(y,i)| + |\sigma(x,i) - \sigma(y,i)| \le L_i|x-y| \, \forall x, y \in \mathbb{R}^n.$$

(ii) $q_{ij}(\phi)$ is measurable in $\phi \in \mathscr{C}$ for all i and $j \in \mathcal{M}$. Moreover, $M := \sup_{\phi \in \mathscr{C}, i \in \mathcal{M}} \{q_i(\phi)\} < \infty$.

(H11.2) Assume the following conditions hold.

(i) There is a positive constant L such that

$$|b(x,i) - b(y,i)| + |\sigma(x,i) - \sigma(y,i)| \le L|x-y|, \ \forall x, y \in \mathbb{R}^n, i \in \mathcal{M}.$$

(ii) $q_{ij}(\phi)$ is measurable in $\phi \in \mathscr{C}$ for each $(i,j) \in \mathcal{M}^2$. Moreover, for any $H > 0$, $M_H := \sup_{\phi \in \mathscr{C}, \|\phi\| \le H, i \in \mathcal{M}} \{q_i(\phi)\} < \infty$.

Remark 11.4 We can use either (H11.1) or (H11.2) to obtain the existence and uniqueness of solutions to (11.7). Recall that now $\mathcal{M}$ is a countable set, so care must be taken to distinct it with a finite state case. In (H11.1), the Lipschitz constants of $b(\cdot, i), \sigma(\cdot, i)$ depend on i, and $q_i(\phi)$ is assumed to be bounded uniformly in $(\phi, i) \in \mathscr{C} \times \mathcal{M}$. In contrast, the uniform boundedness of $q_i(\phi)$ is relaxed, but the Lipschitz constant of $b(\cdot, i), \sigma(\cdot, i)$ is assumed to depend on $i \in \mathcal{M}$.

Theorem 11.5 *Under* (H11.1), *for each initial data* (ξ, i_0), *there exists a unique solution* $(X(t), \alpha(t))$ *to* (11.7).

Proof It is well-known that part (i) of (H11.1) guarantees the existence and uniqueness of strong solutions to the following diffusion:

$$dY(t) = b(Y(t), i)dt + \sigma(Y(t), i)dW(t) \ \text{ for each } i \in \mathcal{M}. \tag{11.15}$$

Then, given a stopping time τ and an $\mathcal{F}_\tau$-measurable $\mathbb{R}^n$-valued random variable $y = y(\tau)$ depending on τ, there exists a unique strong solution to (11.15) in $[\tau, \infty)$ satisfying $Y(\tau) = y(\tau)$; see Chapter 2 or [205, Remark 3.10]. We can now construct the solution to (11.7) with initial data (ξ, i_0) by the interlacing procedure similar to [5, Chapter 5]. Let $\tilde{Y}^{(0)}(t), t \geq 0$ be the solution with initial data $\xi(0)$ to

$$d\tilde{Y}^{(0)}(t) = b(\tilde{Y}^{(0)}(t), i_0)dt + \sigma(\tilde{Y}^{(0)}(t), i_0)dW(t),$$

and $\tilde{Y}^{(0)}(t) = \xi(t)$ for $t \in [-\tau, 0]$. Let

$$\begin{aligned} \tau_1 &= \inf\{t > 0 : \int_0^t \int_{\mathbb{R}} h(\tilde{Y}_s^{(0)}, i_0, z)\mathfrak{p}(ds, dz) \neq 0\} \text{ and} \\ i_1 &= i_0 + \int_0^{\tau_1} \int_{\mathbb{R}} h(\tilde{Y}_s^{(0)}, i_0, z)\mathfrak{p}(ds, dz), \end{aligned}$$

and $\tilde{Y}^{(1)}(t), t \geq \tau_1$ be the solution with $\tilde{Y}_{\tau_1}^{(1)} = \tilde{Y}_{\tau_1}^{(0)}$ to

$$d\tilde{Y}^{(1)}(t) = b(\tilde{Y}^{(1)}(t), i_1)dt + \sigma(\tilde{Y}^{(1)}(t), i_1)dW(t). \tag{11.16}$$

Define

$$\begin{aligned} \tau_2 &= \inf\{t > \tau_1 : \int_{\tau_1}^t \int_{\mathbb{R}} h(\tilde{Y}_s^{(1)}, i_1, z)\mathfrak{p}(ds, dz) \neq 0\} \text{ and} \\ i_2 &= i_1 + \int_{\tau_1}^{\tau_2} \int_{\mathbb{R}} h(\tilde{Y}_s^{(1)}, i_1, z)\mathfrak{p}(ds, dz). \end{aligned}$$

Note that, in the notation above, $\tilde{Y}_t^{(k)}$ is the function $s \in [-r, 0] \mapsto \tilde{Y}^{(k)}(t+s)$. Continuing this procedure, let $\tau_\infty = \lim_{k\to\infty} \tau_k$ and set

$$X(t) = \tilde{Y}^{(k)}(t), \ \alpha(t) = i_k \text{ if } \tau_k \leq t < \tau_{k+1}. \tag{11.17}$$

Clearly, $X(t)$ satisfies that for every $t \geq 0$,

$$\begin{cases} X(t \wedge \tau_k) = X(0) + \displaystyle\int_0^{t\wedge\tau_k} \big[b(X(s), \alpha(s))ds + \sigma(X(s), \alpha(s))dW(t)\big] \\ \alpha(t \wedge \tau_k) = i_0 + \displaystyle\int_0^{t\wedge\tau_k} \int_{\mathbb{R}} h(X_s, \alpha(s-), z)\mathfrak{p}(ds, dz). \end{cases} \tag{11.18}$$

To show that $X(t)$ is a global solution, we need only prove that $\tau_\infty = \infty$ a.s. For any $T > 0$,

$$
\begin{aligned}
\mathbb{P}\{\tau_k \le T\} =&\mathbb{P}\bigg\{\int_0^{T\wedge\tau_k}\int_{\mathbb{R}} I_{\{z\in[0,q_{\alpha(s-)}(X_s))\}}\mathfrak{p}(ds,dz)=k\bigg\}\\
\le&\mathbb{P}\bigg\{\int_0^{T\wedge\tau_k}\int_{\mathbb{R}} I_{\{z\in[0,M)\}}\mathfrak{p}(ds,dz)\ge k\bigg\}\\
\le&\mathbb{P}\bigg\{\int_0^{T}\int_{\mathbb{R}} I_{\{z\in[0,M)\}}\mathfrak{p}(ds,dz)\ge k\bigg\}\\
=&\sum_{l=k}^{\infty} e^{-MT}\frac{(MT)^l}{l!}.
\end{aligned}
\tag{11.19}
$$

It follows that $\mathbb{P}\{\tau_k \le T\} \to 0$ as $k \to \infty$. As a result, $\tau_\infty = \infty$ a.s. By the construction, it can be seen that $X(t)$ is continuous and $\alpha(t)$ is càdlàg a.s. The uniqueness of $(X(t), \alpha(t))$ follows from the uniqueness of $\tilde{Y}^{(k)}(t)$ on $[\tau_k, \tau_{k+1}]$ and the uniqueness of i_k defined by

$$
i_k = i_{k-1} + \int_{\tau_{k-1}}^{\tau_k}\int_{\mathbb{R}} h(\tilde{Y}_t^{(k-1)}, i_{k-1}, z)\mathfrak{p}(dt,dz).
$$

This concludes the proof. □

Theorem 11.6 *Under* (H11.2), *for each initial data* (ξ, i_0), *there exists a unique solution* $(X(t), \alpha(t))$ *to* (11.7).

Proof Without loss of generality, we may assume that (ξ, i_0) is bounded, since we can use the truncation method in [101, Theorem 3 in §6] to obtain the result for general (ξ, i_0) once we have proved for the case (ξ, i_0) being bounded. Construct the process $(X(t), \alpha(t))$ as in the proof of Theorem 11.5. We need to show that $\tau_\infty = \infty$ a.s. Following Proposition 2.9, there is a $K = K(T)$ such that

$$
\mathbb{E}\Big(\sup_{0\le t\le T\wedge\tau_k} |X(t)|^2\Big) \le K \quad \forall k \in \mathcal{M}.
$$

As a result, for any $\varepsilon > 0$, there is an H_ε such that

$$
\mathbb{P}\{\|X_t\| \le H_\varepsilon \,\forall t \in [0, T\wedge\tau_k]\} > 1 - \frac{\varepsilon}{2}. \tag{11.20}
$$

Let $\eta_{H_\varepsilon} = \inf\{t \ge 0 : \|X_t\| \ge H_\varepsilon\}$ and $M_{H_\varepsilon} = \sup_{\phi\in\mathscr{C},\|\phi\|\le H_\varepsilon, i\in\mathcal{M}}\{q_i(\phi)\} < \infty$. Then

$$
\begin{aligned}
\mathbb{P}\{\tau_k \le T\wedge\eta_{H_\varepsilon}\} =&\mathbb{P}\{\int_0^{T\wedge\tau_k\wedge\eta_{H_\varepsilon}}\int_{\mathbb{R}} I_{\{z\in[0,q_{\alpha(s-)}(X_s))\}}\mathfrak{p}(ds,dz)=k\}\\
\le&\mathbb{P}\{\int_0^{T\wedge\tau_k\wedge\eta_{H_\varepsilon}}\int_{\mathbb{R}} I_{\{z\in[0,M_{H_\varepsilon})\}}\mathfrak{p}(ds,dz)\ge k\}
\end{aligned}
$$

$$
\begin{aligned}
&\leq \mathbb{P}\Big\{\int_0^T \int_{\mathbb{R}} I_{\{z\in[0,M_{H_\varepsilon})\}}\mathfrak{p}(ds,dz) \geq k\Big\} \\
&= e^{-M_{H_\varepsilon}T} \sum_{l=k}^{\infty} \frac{(M_{H_\varepsilon}T)^l}{l!}.
\end{aligned}
$$

For sufficiently large k, we have

$$
\mathbb{P}\{\tau_k \leq T \wedge \eta_{H_\varepsilon}\} \leq e^{-M_{H_\varepsilon}T} \sum_{l=k}^{\infty} \frac{(M_{H_\varepsilon}T)^l}{l!} \leq \frac{\varepsilon}{2}. \tag{11.21}
$$

From (11.20) and (11.21), $\mathbb{P}\{\tau_k \geq T\} \geq \mathbb{P}(\{\tau_k \wedge T < \eta_{H_\varepsilon}\} \cap \{\tau_k > T \wedge \eta_{H_\varepsilon}\}) \geq 1-\varepsilon$ for sufficiently large k. Thus, we obtain that $\mathbb{P}\{\tau_\infty \geq T\} \geq 1-\varepsilon$. It holds for every $T > 0$ and $\varepsilon > 0$, so we obtain the desired result. □

Remark 11.7 To obtain the existence and uniqueness of solutions, (H11.1) and (H11.2) can be relaxed by replacing the global Lipschitz conditions with local Lipschitz conditions together with Lyapunov-type functions. To be specific, let $V(\cdot) : \mathbb{R}^n \mapsto \mathbb{R}$ be twice continuously differentiable in x. For each $i \in \mathcal{M}$, let $\mathbb{L}_i V(x) = \nabla V(x) b(x,i) + \frac{1}{2}\text{trace}\Big(\nabla^2 V(x) A(x,i)\Big)$. For instance (1) of (H11.1) and (1) of (H11.2) can be replaced by (H11.3) and (H11.4), respectively.

(H11.3) Assume the following conditions hold.

(i) For each $H > 0$, $i \in \mathcal{M}$, there is a positive constant $L_{H,i}$ such that

$$
|b(x,i) - b(y,i)| + |\sigma(x,i) - \sigma(y,i)| \leq L_{H,i}|x-y|, \ \forall |x|, |y| \leq H, i \in \mathcal{M}.
$$

(ii) For each $i \in \mathcal{M}$, there exist a twice continuously differentiable function $V_i(x)$ and a constant $C_i > 0$ such that

$$
\lim_{R\to\infty}\Big(\inf\{V_i(x) : |x| \geq R\}\Big) = \infty \ \text{ and } \ \mathbb{L}_i V_i(x) \leq C_i(1 + V_i(x)) \,\forall x \in \mathbb{R}^n.
$$

(H11.4) Assume the following conditions hold.

(i) For each $H > 0$, $i \in \mathcal{M}$, there is a positive constant $L_{H,i}$ such that

$$
|b(x,i) - b(y,i)| + |\sigma(x,i) - \sigma(y,i)| \leq L_{H,i}|x-y| \ \forall |x|, |y| \leq H, i \in \mathcal{M}.
$$

(ii) There exist a twice continuously differentiable function $V(x)$ and a constant $C > 0$ independent of $i \in \mathcal{M}$ such that

$$
\lim_{R\to\infty}\Big(\inf_{|x|\geq R}\{V(x)\}\Big) = \infty \ \text{ and } \ \mathbb{L}_i V(x) \leq C(1 + V(x)), \ x \in \mathbb{R}^n, i \in \mathcal{M}.
$$

Theorem 11.8 *For given initial data* (ξ, i_0)*, there exists a unique solution* $(X(t), \alpha(t))$ *to* (11.7) *if either of the following conditions is satisfied:*

- (H11.3) *and* (ii) *of* (H11.1)*, or* (H11.4) *and* (ii) *of* (H11.2).

Proof It is well-known that (H11.3) guarantees the existence and uniqueness of solutions to (11.15). Hence, if (ii) in (H11.1) is satisfied, we can prove the desired result by using the proof of Theorem 11.5. Now, suppose (H11.4) and (ii) of (H11.2) hold. Similar to the proof of Theorem 11.6, we can assume that (ξ, i_0) is bounded. Consider $X(t)$ and define τ_k as in the proof of Theorem 11.5. Then $X(t)$ is the solution with initial data (ξ, i_0) to (11.7) on $[0, T \wedge \tau_k)$ for any $T > 0, k \in \mathcal{M}$. We have from the generalized Itô formula that

$$\begin{aligned}\mathbb{E}V(X(T \wedge \tau_k \wedge \eta_H)) &= \mathbb{E}V(\xi(0), i_0) + \mathbb{E}\int_0^{T\wedge\tau_k\wedge\eta_H} \mathbb{L}_i V(X(t), \alpha(t-))dt \\ &\le \mathbb{E}V(\xi(0), i_0) + C\mathbb{E}\int_0^{T\wedge\tau_k\wedge\eta_H} (1 + V(X(t))dt,\end{aligned}$$

where $\eta_H = \inf\{t \ge 0 : |X(t)| > H\}$. Using the estimate above and the proof of Theorem 2.12, we can show that

$$\mathbb{E}V(T \wedge \tau_k \wedge \eta_H) \le K = K(\xi, T)\, \forall H > 0, k \in \mathcal{M}.$$

In view of the property $\lim_{R\to\infty}\Big(\inf\{V(x) : |x| \ge R\}\Big) = \infty$, for any $\varepsilon > 0$, there is $H_\varepsilon > 0$ such that

$$\mathbb{P}\{\eta_{H_\varepsilon} > T \wedge \tau_k\} > 1 - \frac{\varepsilon}{2}\ \forall k \in \mathcal{M}.$$

Then, proceeding similarly as in the proof of Theorem 11.6 yields the existence and uniqueness of solutions with initial data (ξ, i_0) to (11.7). □

Example 11.9 (Example 11.2 cont.) We come back to Example 11.2. We want to show that $X(t) > 0$ for all $t \ge 0$ under certain conditions. To proceed, we can set $Y(t) = \ln X(t)$ to obtain

$$dY(t) = [a(\alpha(t) - \frac{\sigma^2(\alpha(t))}{2} - b(\alpha(t))\exp(Y(t))]dt + \sigma(\alpha(t))dW(t). \qquad (11.22)$$

To demonstrate that (11.11) and (11.12) has a unique solution with $X(t) > 0$ for all $t \ge 0$, it is equivalent to show that (11.22) and (11.12) has a strong solution on $[0, \infty)$. Let $V(y) = e^y + e^{-y}$. By direct calculation,

$$\begin{aligned}\mathbb{L}_i V(y) =& b(i) + (\sigma^2(i) - a(i))e^{-y} + a(i)e^y - b(i)e^{2y} \\ \le& c(i) + (\sigma^2(i) - a(i))V(y),\end{aligned}$$

where $c(i) = \max_{y\in\mathbb{R}}\{b(i) + (2a(i) - \sigma^2(i))e^y - b(i)e^{2y}\}$. Applying Theorem 3.3, the equation has a unique solution if one of the following is satisfied

- $\beta_i(\phi) + \delta_i(\phi)$ is bounded uniformly in $\phi \in \mathscr{C}_+ := \{\psi \in \mathscr{C} : \psi(t) > 0\, \forall t \in [-r, 0]\}$ and $i \in \mathcal{M}$.
- $c(i)$ and $\sigma^2(i) - a(i)$ are bounded above uniformly and for each $i \in \mathcal{M}$, $\beta_i(\phi) + \delta_i(\phi)$ is bounded in each compact subset of $\phi \in \mathscr{C}_+$.

It can be shown by applying the result of the next section that the process $(Y_t, \alpha(t))$ has the Markov-Feller property if $\beta_i(\cdot)$ and $\delta_i(\cdot)$ are continuous in addition to one of the above conditions.

Example 11.10 (Example 11.3 cont.) To study the long-run average control problem in Example 11.3, it is important to make sure that the system under consideration processes ergodicity. Before the ergodicity can be verified, we need to establish that (11.14) has a unique solution for each initial condition. Denote the control set by $\widetilde{K}$ and assume it is a compact and convex set. Using a relaxed control representation $m_t(\cdot)$ (see [176]) to represent the consumption rate $c(\cdot)$, we can rewrite (11.14) as

$$dX(t) = \Big[\int_{\widetilde{K}} c(u)m_t(du) - \rho(\alpha(t))X(t)\Big]dt + \sigma(X(t), \alpha(t))dW(t). \qquad (11.23)$$

Assume that for each $i \in \mathcal{M}$, $\sigma(x, i)$ satisfies the conditions in (H11.1) (i), and $Q(\phi)$ satisfies (H11.1) (ii). Then the conditions of Theorem 11.5 are all verified. As a result, (11.23) has a unique solution for each initial condition.

11.4 Markov and Feller Properties

This section establishes the Markov and Feller properties of the process $(X_t, \alpha(t))$. Although the Markov property can be derived by the well-known arguments, it requires much more efforts to obtain the Feller property. As already seen in the previous section, the past dependence and the use of $\mathcal{M}$ make the analysis more complex than that of the switching diffusions with diffusion-dependent switching living in a finite set. To overcome the difficulties, in this section, we carry out the analysis by introducing some auxiliary or intermediate processes. First, it would be better if we could untangle the past dependence of the switching process and the infinity of the cardinality of its state space. For this purpose, we introduce a continuous-time Markov chain independent of the past and continuous state; we call this process $\gamma(t)$. Then naturally, associated with $\gamma(t)$, we examine a pair of process $(Z(t), \gamma(t))$. Even after this introduction, in the analysis, we still need to look into the details of the switching process $\alpha(t)$ such as when it jumps and the post jump location. To do so, we introduce another auxiliary process $Y(t)$, which is a "fixed"-i process. We then have another pair of processes $(Y(t), \beta(t))$ to deal with.

These auxiliary processes help us to establish the desired results. Their connections and interactions will be further specified in what follows.

First, note that the Brownian motion and the Poisson point process associated with $\mathfrak{p}(dt, dz)$ possess stationary strong Markov property, that is, for any finite stopping time η, $\{W^*(t)\}_{t\geq 0} = \{W(t+\eta) - W(\eta)\}_{t\geq 0}$ is an $\mathcal{F}^*_t$-Brownian motion and $\mathfrak{p}^*([t, t+s) \times U) = \mathfrak{p}([t+\eta, t+s+\eta) \times U)$ is a Poisson random measure with density $dt \times \mathfrak{m}(dz)$ (see [261, Theorem 101]). Hence, by standard arguments, we can obtain the following theorem whose proof is omitted. In fact, the theorem can be proved essentially by imitating the proof in [240, Chapter 5], [5, Chapter 6], or [227, Chapter 7].

Theorem 11.11 *Assume that the hypotheses of Theorem* 11.5, *or Theorem* 11.6, *or Theorem* 11.8 *are satisfied. Let* $(X(t), \alpha(t))$ *be a solution to* (11.7). *Then* $(X_t, \alpha(t))$ *is a homogeneous strong Markov process taking value in* $\mathscr{C} \times \mathcal{M}$ *with transition probabilities* $P(\phi, i, t, A \times \{j\}) = \mathbb{P}\{X^{\phi,i}_t \in A, \alpha(t) = j\}$, *where* $X^{\phi,i}(t)$ *is the solution to* (11.7) *with initial data* $(\phi, i) \in \mathscr{C} \times \mathcal{M}$.

We proceed with obtaining the Feller property of $(X_t, \alpha(t))$. Assuming that the hypotheses of Theorem 11.5, or Theorem 11.6, or Theorem 11.8 are satisfied leads to the existence and uniqueness of strong solutions. Next, we introduce an auxiliary hybrid diffusion with Markov switching. Let $\gamma^i(t)$ be a Markov chain starting at i with generator $\tilde{Q} = (\rho_{ij})$ for $(i, j) \in \mathcal{M} \times \mathcal{M}$, where $\rho_{ii} = -1$ and $\rho_{ij} = 2^{-j}$ if $j < i$ and $\rho_{ij} = 2^{-j+1}$ if $j > i$, that is, $\tilde{Q} = \begin{pmatrix} -1 & 1/2 & 1/4 & \cdots \\ 1/2 & -1 & 1/4 & \cdots \\ 1/2 & 1/4 & -1 & \cdots \\ \vdots & \vdots & \vdots & \ddots \end{pmatrix}$. We define recursively a sequence of stopping times $\{\theta^i_k\}$ with θ^i_k being the first jump time of $\gamma^i(t)$ after θ^i_{k-1} as follows

$$\theta^i_0 = 0,\ \theta^i_k = \inf\{t > \theta^i_{k-1} : \gamma^i(t) \neq \gamma^i(\theta^i_{k-1})\}, k \in \mathcal{M}.$$

For $(\phi, i) \in \mathscr{C} \times \mathcal{M}$, let $Z^{\phi,i}(t)$ be the solution to

$$dZ(t) = b(Z(t), \gamma(t))dt + \sigma(Z(t), \gamma(t))dW(t)\,, t \geq 0$$

satisfying $Z^{\phi,i}(t) = \phi(t)$ in $[-r, 0]$ and $\gamma(0) = i$.

Similar to Girsanov's theorem, which tells us how to convert an Itô process to a Brownian motion under a change of measure, we aim to establish a change of measure allowing us to "convert" a hybrid diffusion with past-dependent switching to a hybrid diffusion with Markov switching. To establish such a change of measure, we need to find the distribution of jump times of $\alpha(t)$. Because of the interactions between $\alpha(t)$ and $X(t)$, we need to introduce another auxiliary (or intermediate) process, which helps to examine the distribution of the jump times of $\alpha(t)$. Let $(Y^{\phi,i}(t), \beta^{\phi,i}(t))$ be the solution to

$$\begin{cases} dY(t) = b(Y(t), i)dt + \sigma(Y(t), i)dW(t), \ t \geq 0 \\ d\beta(t) = \displaystyle\int_{\mathbb{R}} h(Y_t, \beta(t-), z)\mathfrak{p}(dt, dz), t \geq 0 \end{cases} \tag{11.24}$$

satisfying $Y^{\phi,i}(t) = \phi(t)$ in $[-r, 0]$ and $\beta^{\phi,i}(0) = i$. By the definition, $\alpha^{\phi,i}(t) = \beta^{\phi,i}(t)$, $X^{\phi,i}(t) = Y^{\phi,i}(t)$ up to the first jump time of the two process $\alpha(t)$ and $\beta(t)$. There is an advantage working with $(Y^{\phi,i}(t), \beta^{\phi,i}(t))$. Unlike the pair $(X(t), \alpha(t))$ in which $\alpha(t)$ depends on the continuous state, the process $Y^{\phi,i}(t)$ evolving for a fixed discrete state i that does not depend on $\beta^{\phi,i}(t)$. Thus, it is easier to examine, for example, the first jump time of $\beta^{\phi,i}(t)$ (or $\alpha^{\phi,i}(t)$).

Next we recursively define sequences of stopping times associated with $\beta(t)$ and $\alpha(t)$ so that $\lambda_k^{\phi,i}$ and $\tau_k^{\phi,i}$ are the first jump times of the processes $\beta^{\phi,i}(t)$ and $\alpha^{\phi,i}(t)$ after $\lambda_{k-1}^{\phi,i}$ and $\tau_{k-1}^{\phi,i}$, respectively. More specifically, for $k \in \mathcal{M}$, let

$$\begin{aligned} \lambda_0^{\phi,i} &= 0, \lambda_k^{\phi,i} = \inf\{t > \lambda_{k-1}^{\phi,i} : \beta^{\phi,i}(t) \neq \beta^{\phi,i}(\lambda_{k-1}^{\phi,i})\}, \ i \in \mathcal{M} \text{ and} \\ \tau_0^{\phi,i} &= 0, \ \tau_k^{\phi,i} = \inf\{t > \tau_{k-1}^{\phi,i} : \alpha^{\phi,i}(t) \neq \alpha^{\phi,i}(\tau_{k-1}^{\phi,i})\}, \ i \in \mathcal{M}. \end{aligned}$$

To simplify the notation, we put

$$\begin{aligned} &\alpha_k^{\phi,i} := \alpha^{\phi,i}(\tau_k^{\phi,i}), \ \beta_k^{\phi,i} := \beta^{\phi,i}(\lambda_k^{\phi,i}), \ \gamma_k^i := \gamma^i(\theta_k^i), \text{ and} \\ &X_{(k)}^{\phi,i} := X_{\tau_k^{\phi,i}}^{\phi,i}, \ Y_{(k)}^{\phi,i} := Y_{\lambda_k^{\phi,i}}^{\phi,i}, \ Z_{(k)}^{\phi,i} := Z_{\theta_k^i}^{\phi,i}, \end{aligned}$$

where we use the subscript k with parentheses to avoid confusion with the function-valued processes $X_t^{\phi,i}$, $Y_t^{\phi,i}$, $Z_t^{\phi,i}$ at $t = k$.

Lemma 11.12 *Let $g : \mathscr{C} \times \mathbb{R}_+ \times \mathcal{M} \mapsto \mathbb{R}$ be a bounded and measurable function, and $\mathcal{F}_T^W$ be the σ-algebra generated by $\{W(t), t \in [0, T]\}$. The following assertions hold:*

(i) $\mathbb{P}(\{\lambda_1^{\phi,i} > t\} | \mathcal{F}_T^W) = \mathbb{E}\Big[I_{\{\lambda_1^{\phi,i} > t\}} \Big| \mathcal{F}_T^W\Big] = \exp\big(-\int_0^t q_i(Y_s^{\phi,i})ds\big) \ \forall t \in [0, T]$.

(ii) $\mathbb{E}\Big[g(Y_{(1)}^{\phi,i}, \lambda_1^{\phi,i}, \beta_1^{\phi,i}) I_{\{\lambda_1^{\phi,i} \leq T\}} \Big| \mathcal{F}_T^W\Big] = \displaystyle\sum_{j=1, j\neq i}^{\infty} \int_0^T g(Y_t, t, j) q_{ij}(Y_t) e^{-\int_0^t q_i(Y_s)ds} dt$.

As indicated previously, it is difficult to estimate the difference of $X_t^{\phi_1,i}$ and $X_t^{\phi_2,i}$ because the states of $\alpha^{\phi_1,i}(t)$ and $\alpha^{\phi_2,i}(t)$ may differ significantly due to the continuous state dependence. In contrast, it is considerably easier to compare $Z_t^{\phi_1,i}$ and $Z_t^{\phi_2,i}$ because of the continuous-state-dependent switching is replaced by the continuous-state-independent Markov chain. With help of the intermediate process $(Y(t), \beta(t))$ and Lemma 11.12, we obtain the following change of measure formula, which is a bridge to connect the continuous-state-dependent and continuous-state-independent processes.

Proposition 11.13 *For any $T > 0$, let $f(\cdot,\cdot) : \mathscr{C} \times \mathcal{M} \mapsto \mathbb{R}$ be a bounded continuous function. For any $l = 0, 1, \ldots,$ any $i_k \in \mathcal{M}$ with $i_k \neq i_{k+1}$ and $k = 1, \ldots, l+1$, and any $(\phi, i) \in \mathscr{C} \times \mathcal{M}$,*

$$\mathbb{E}\Big[f(X_T^{\phi,i}, \alpha^{\phi,i}(T)) I_{\{\tau_l^{\phi,i} \leq T < \tau_{l+1}^{\phi,i}\}} \prod_{k=1}^{l} I_{\{\alpha_k^{\phi,i} = i_k\}}\Big]$$
$$= \exp(T)\mathbb{E}\Big[f(Z_T^{\phi,i}, i_l) I_{\{\theta_l^i \leq T < \theta_{l+1}^i\}} \prod_{k=1}^{l} \Big(I_{\{\gamma_k^i = i_k\}} \frac{q_{i_k i_{k+1}}(Z_{(k)}^{\phi,i})}{\rho_{i_k i_{k+1}}}\Big) \tag{11.25}$$
$$\times \exp\Big\{-\int_0^T q_{\gamma^i(s)}(Z_s^{\phi,i}) ds\Big\}\Big].$$

Remark 11.14 The proofs of Lemma 11.12 and Proposition 11.13 will be given in Section 11.7. We are now in a position to prove the Feller property for the solution to (11.7). In addition to the sufficient conditions for the existence and uniqueness of solution, we prove the Feller property of the solution only with an additional condition that $q_{ij}(\phi)$ is continuous in ϕ for any $i, j \in \mathcal{M}$. There are some difficulties because the process $\{X_t\}$ takes value in an infinite dimensional Banach space and the switching $\{\alpha(t)\}$ has an infinite state space. Moreover, although we suppose that $q_{ij}(\phi)$ is continuous, neither the uniform continuity in $\phi \in \mathscr{C}$ nor equi-continuity in $i, j \in \mathcal{M}$ is assumed. Because of these difficulties, we divide the proof into several steps. First, we make the following assumptions, which will be relaxed later.

(H11.5) Assume the following conditions hold.

(i) For each $i \in \mathcal{M}$, $b(x, i)$ and $\sigma(x, i)$ are Lipschitz continuous functions that are vanishing outside $\{x : |x| \leq R\}$ for some $R > 0$.
(ii) $M := \sup\{q_i(\phi) : i \in \mathcal{M}, \phi \in \mathscr{C}\} < \infty$.
(iii) For each $i, j \in \mathcal{M}$, $j \neq i$, $q_i(\cdot)$ and $q_{ij}(\cdot)$ are continuous on $\mathscr{C}$.

Before applying (11.25) to prove the continuous dependence of $u_f(\phi, i) = \mathbb{E}_{\phi,i} f(X_T, \alpha(T))$ on (ϕ, i), we first need the following lemma.

Lemma 11.15 *Assume that* (H11.5) *is satisfied. Let $(\phi_0, i_0) \in \mathscr{C} \times \mathcal{M}$ with $\|\phi_0\| \leq R$ and $T > 0$. For each $\Delta > 0$, there exist $m = m(\Delta) \in \mathcal{M}$, $n_m = n_m(\Delta) \in \mathcal{M}$, and $d_m = d_m(\Delta) > 0$ such that*

$$\mathbb{P}\Big(\{\tau_{m+1}^{\phi,i_0} > T\} \cap \{\alpha^{\phi,i_0}(t) \in N_{n_m}, \forall t \in [0, T]\}\Big) \geq 1 - \Delta, \quad \forall \|\phi - \phi_0\| < d_m,$$

where $N_k = \{1, \ldots, k\}$.

This lemma allows us to confine our attention to a finite subset of $\mathcal{M}$ (the state space of $\alpha^{\phi,i_0}(\cdot)$) and a finite number of jumps when ϕ is close to ϕ_0. It is a crucial step in providing some uniform estimates because we do not assume the equi-continuity of $q_{ij}(\cdot)$ in either i or j. Since the switching intensity of $\alpha^{\phi,i_0}(t)$

depends on X_t^{ϕ,i_0}, in order to obtain Lemma 11.15, we need to show that with an arbitrarily large probability, X_t^{ϕ,i_0}, $t \in [0,T]$ belongs to a compact set in $\mathscr{C}$ for any ϕ sufficiently close to ϕ_0. Note that under some suitable conditions, sample paths of a diffusion process in a finite interval $[0,T]$ are Hölder continuous. Thus, it is easy to find a compact set in which sample paths of a diffusion process lie with a large probability. Our arguments rely on this fact. However, the initial data ϕ of our process $X(t)$ does not always satisfy the Hölder continuity. Moreover, $X(t)$ depends on the state of $\alpha(t)$. We therefore need to introduce the following operator, which is motivated by merging trajectories of $X(t)$ at jump times. For $\mathcal{A}, \mathcal{B} \subset \mathscr{C}$, we define the set of continuous functions that are formed by merging functions in $\mathcal{A}$ and $\mathcal{B}$ as follows:

$$\begin{aligned}\mathcal{A} \uplus \mathcal{B} := \mathcal{A} \cup \mathcal{B} \cup \{\psi \in \mathscr{C} : \exists \psi_1 \in A, \psi_2 \in B, s \in [0,r] \text{ such that}\\ \psi(t) = \psi_1(s+t)\forall t \in [-r,-s];\ \psi(t) = \psi_2(t+s-r)\forall t \in [-s,0]\}.\end{aligned}$$

By virtue of the Arzelá-Ascoli theorem, if $\mathcal{A}$ and $\mathcal{B}$ are compact, so is $\mathcal{A} \uplus \mathcal{B}$. Using this fact and the Hölder continuity of sample paths of a diffusion process, we can find a suitable compact set to which X_t^{ϕ,i_0}, $t \in [0,T]$, belongs with a large probability for any ϕ which is sufficiently close to ϕ_0. Then, Lemma 11.15 can be proved. The details of the proof are postponed to Section 11.7. Now, we point out some nice properties of the diffusion process with Markov switching $(Z(t), \gamma(t))$, which are useful to compare the sample paths of $Z(t)$ with different initial values.

Lemma 11.16 *Fix $i_0 \in \mathcal{M}$. For each $k \in \mathcal{M}$ and $\varepsilon > 0$, there is an $\hbar_k = \hbar_k(\varepsilon) > 0$ such that*

$$\mathbb{P}\left\{\sup_{t,s\in[0,T\wedge\iota_k],0<t-s<\hbar_k} \frac{|Z^{\phi,i_0}(t) - Z^{\phi,i_0}(s)|}{(s-t)^{0.25}} \le 4\right\} > 1-\varepsilon\ \forall\, |\phi(0)| \le R,$$

and

$$\mathbb{E}\Big[\sup_{t\in[0,T\wedge\iota_k]} |Z^{\phi,i_0} - Z^{\psi,i_0}|^2\Big] \le \bar{C}|\phi-\psi|^2,$$

where $\iota_k = \inf\{t > 0 : \gamma^{i_0}(t) > k\}$ and $\bar{C}$ is some positive constant.

Proof Since $b(x,i)$ and $\sigma(x,i)$ are Lipschitzian in x uniformly in N_k, by standard arguments (e.g., [205, Theorem 3.23]), we can show that

$$\mathbb{E}|x(t\wedge\iota_k) - x(s\wedge\iota_k)|^6 < \tilde{C}_k(t-s)^3, \forall 0 \le s \le t \le T.$$

Using the Kolmogorov continuity theorem, we obtain the first inequality. The details are similar to the proof of Lemma 11.15 in Section 11.7. The second claim is proved in the same manner as that of Lemma 2.24. □

Having Lemmas 11.15 and 11.16, we are ready to use the change of measure (11.25) to prove the Feller property of $(X_t, \alpha(t))$ under (H11.5).

Proposition 11.17 *Suppose that* (H11.5) *is satisfied. Let* $f(\cdot,\cdot) : \mathscr{C} \times \mathcal{M} \mapsto \mathbb{R}$ *be continuous and bounded. Then for any* $T > 0$, $u_f(\phi, i) = \mathbb{E} f(X_T^{\phi,i}, \alpha^{\phi,i}(T))$ *is a continuous function in* $\phi \in \mathscr{C}$.

Proof We suppose without loss of generality that $|f(\phi, i)| \le 1 \, \forall \, (\phi, i) \in \mathscr{C} \times \mathcal{M}$. Fix $(\phi_0, i_0) \in \mathscr{C} \times \mathcal{M}$. We show that for any $\Delta > 0$, there exists $d^* = d^*(\Delta, \phi_0, i_0) > 0$ such that

$$\big|\mathbb{E} f(X_T^{\phi,i_0}, \alpha^{\phi,i_0}(T)) - \mathbb{E} f(X_T^{\phi_0,i_0}, \alpha^{\phi_0,i_0}(T))\big| \le 3\Delta \,\forall\, \|\phi-\phi_0\| < d^*. \tag{11.26}$$

In view of Lemma 11.15, there are $m, n_m \in \mathcal{M}$, and $d_m > 0$ such that

$$\mathbb{P}\Big(\{\tau_{m+1}^{\phi,i_0} > T\} \cap \{\alpha^{\phi,i_0}(t) \in N_{n_m}, \forall t \in [0,T]\}\Big) \ge 1 - \Delta \,\forall \|\phi - \phi_0\| < \frac{d_m}{2}. \tag{11.27}$$

Let $\varepsilon = \varepsilon(\Delta) > 0$ (to be specified later). Let $\hbar_k$ be as in Lemma 11.16. Denote

$$\tilde{h} = \Big\{\psi(\cdot) \in \mathscr{C} : \|\psi\| \le R + 1 \text{ and } \sup_{t,s\in[-r,0], 0<t-s<\hbar_{n_m}} \frac{|\psi(s) - \psi(t)|}{(s-t)^{0.25}} \le 4\Big\}$$

and $\tilde{\mathcal{K}} = \{\phi_0\} \uplus \tilde{h}$. By the compactness of $\tilde{K}$, there is a $\tilde{d}_m > 0$ such that

$$\|q_{ij}(\psi) - q_{ij}(\phi)\| < \varepsilon, \; |f(\psi, i) - f(\phi, i)| < \varepsilon \tag{11.28}$$

if $\phi \in \tilde{\mathcal{K}}, i, j \in N_{n_m}$ and $\|\psi - \phi\| < \tilde{d}_m$. In view of Lemma 11.16, we can choose $\hat{d}_m > 0$ such that

$$\mathbb{P}\Big\{\sup_{t\in[0,T\wedge\iota_k]} \|Z_t^{\phi,i_0} - Z_t^{\phi_0,i_0}\| \le \tilde{d}_m\Big\} < \varepsilon \text{ if } \|\phi - \phi_0\| \le \hat{d}_m. \tag{11.29}$$

Let A^ϕ be the event $\{\tau_m^{\phi,i_0} \le T < \tau_{m+1}^{\phi,i_0}, \iota_{n_m} > T\}$ and $l(T)$ be the number of jumps up to time T of $\gamma^{i_0}(t)$. It follows from Proposition 11.13 that

$$\begin{aligned}
&\mathbb{E}\big[f(X_T^{\phi,i_0}, \alpha^{\phi,i_0}(T)) I_{A^\phi}\big] - \mathbb{E}\big[f(X_T^{\phi_0,i_0}, \alpha^{\phi_0,i_0}(T)) I_{A^{\phi_0}}\big] \\
&= \exp(T)\mathbb{E}\Big[I_{\{l(T)<m+1, \iota_{n_m}>T\}}\Big[g(Z^{\phi,i_0}(\cdot), \gamma^{i_0}(\cdot)) - g(Z^{\phi_0,i_0}(\cdot), \gamma^{i_0}(\cdot))\Big]\Big],
\end{aligned} \tag{11.30}$$

where

$$g(Z^{\phi,i_0}(\cdot),\gamma^{i_0}(\cdot)) = f(Z_T^{\phi,i_0},\gamma^{i_0}(T))\prod_{k=1}^{l(T)}\frac{q_{\gamma_k^{i_0}\gamma_{k+1}^{i_0}}(Z_{(k+1)}^{\phi,i_0})}{\rho_{\gamma_k^{i_0}\gamma_{k+1}^{i_0}}}e^{-\int_0^T q_{\gamma^{i_0}(s)}(Z_s^{\phi,i_0})ds}.$$

Let D_m^ϕ be the event

$$\begin{aligned} D_m^\phi :=& \Big\{\sup_{t\in[0,T\wedge\iota_k]}\|Z_t^{\phi,i_0}-Z_t^{\phi_0,i_0}\|\le\tilde d_m\Big\}\\ &\cap\Big\{\sup_{t,s\in[0,T\wedge\iota_k],0<t-s<\hbar_{nm}}\frac{|Z^{\phi_0,i_0}(s)-Z^{\phi_0,i_0}(t)|}{(s-t)^{0.25}}\le 4\Big\}.\end{aligned}$$

Using (11.30) and the elementary estimate

$$\left|\prod_{k=1}^n a_k-\prod_{k=1}^n b_k\right|\le n(\max_{\{k=1,\cdots n\}}\{|c_k|,|d_k|\})\}^{n-1}(\max_{\{k=1,\cdots n\}}\{|c_k-d_k|\},$$

we obtain for $l\ge 1$,

$$\begin{aligned}&\Big|\mathbb{E}\big[f(X_T^{\phi,i_0},\alpha^{\phi,i_0}(T))I_{A_l^\phi}\big]-\mathbb{E}\big[f(X_T^{\phi_0,i_0},\alpha^{\phi_0,i_0}(T))I_{A_l^{\phi_0}}\big]\Big|\\
\le& K\mathbb{E}\Big[I_{\{\theta_l^{i_0}\le T<\theta_{l+1}^{i_0},\iota_{nm}>T\}}\times\sup_{i\in N_{nm}}|f(Z_T^{\phi,i_0},i)-f(Z_T^{\phi_0,i_0},i)|\Big]\\
&+K\mathbb{E}\Big[I_{\{\theta_l^{i_0}\le T<\theta_{l+1}^{i_0},\iota_{nm}>T\}}\times\sup_{t\in[0,T],i,j\in N_{nm}}|q_{ij}(Z_t^{\phi,i_0})-q_{ij}(Z_t^{\phi_0,i_0})|\Big]\\
\le& K\mathbb{E}\Big[I_{\{\theta_l^{i_0}\le T<\theta_{l+1}^{i_0},\iota_{nm}>T\}}I_{D_m^\phi}\times\sup_{i\in N_{nm}}|f(Z_T^{\phi,i_0},i)-f(Z_T^{\phi_0,i_0},i)|\Big]\\
&+K\mathbb{E}\Big[I_{\{\theta_l^{i_0}\le T<\theta_{l+1}^{i_0},\iota_{nm}>T\}}I_{D_m^\phi}\times\sup_{t\in[0,T],i,j\in N_{nm}}|q_{ij}(Z_t^{\phi,i_0})-q_{ij}(Z_t^{\phi_0,i_0})|\Big]\\
&+2K(M+1)\mathbb{P}(\Omega\setminus D_m^\phi),\end{aligned}$$

where K is a constant depending only on T, m, n_m.

Note that if $\omega\in\{\theta_l^{i_0}\le T<\theta_{l+1}^{i_0},\iota_{n_m}>T\}\cap D_m^\phi$, then $Z_t^{\phi_0,i_0}\in\tilde{\mathcal{K}}$ and $\|Z_t^{\phi,i_0}-Z_t^{\phi_0,i_0}\|\le\tilde d_m\ \forall t\in[0,T]$, which implies in view of (11.28) that

$$\sup_{i\in N_{nm}}|f(Z_T^{\phi,i_0},i)-f(Z_T^{\phi_0,i_0},i)|+\sup_{t\in[0,T],i,j\in N_{nm}}|q_{ij}(Z_t^{\phi,i_0})-q_{ij}(Z_t^{\phi_0,i_0})|<2\varepsilon.$$

On the other hand, Lemma 11.16 and (11.29) imply that $\mathbb{P}(\Omega\setminus D_m^\phi)\le 3\varepsilon$ if $\|\phi-\phi_0\|\le\hat d_m$. Hence for $\|\phi-\phi_0\|\le\hat d_m$, we have that

$$\left|\mathbb{E}\big[f(X_T^{\phi,i_0},\alpha^{\phi,i_0}(T))I_{A^\phi}\big]-\mathbb{E}\big[f(X_T^{\phi_0,i_0},\alpha^{\phi_0,i_0}(T))I_{A^{\phi_0}}\big]\right|\le 2K(4+3M)\varepsilon, \tag{11.31}$$

Note that

$$\mathbb{P}\big(\Omega\setminus A^\phi\big)=\mathbb{P}\big(\{\tau_{m+1}^{\phi,i_0}<T\}\cup\{\alpha^{\phi,i_0}(t)\notin N_{n_m}\text{ for some }t\in[0,T]\}\big)<\Delta,$$

which implies

$$\left|\mathbb{E}\big[f(X_T^{\phi,i_0},\alpha^{\phi,i_0}(T))I_{\Omega\setminus A^\phi}\big]-\mathbb{E}\big[f(X_T^{\phi_0,i_0},\alpha^{\phi_0,i_0}(T))I_{\Omega\setminus A^\phi}\big]\right|\le 2\Delta. \tag{11.32}$$

Choosing $\varepsilon=\dfrac{\Delta}{2K(4+3M)}$, we have from (11.31) and (11.32) that

$$\left|\mathbb{E}\big[f(X_T^{\phi,i_0},\alpha^{\phi,i_0}(T))\big]-\mathbb{E}\big[f(X_T^{\phi_0,i_0},\alpha^{\phi_0,i_0}(T))\big]\right|\le 3\Delta$$

if $\|\phi-\phi_0\|<d^*:=\dfrac{d_m}{2}\wedge\hat{d}_m$. □

With the above technical preparations, we are now in a position to prove the main theorem of this section. By using truncation arguments, we can obtain the Feller property of $(X_t,\alpha(t))$ even if $b(\cdot,i),\sigma(\cdot,i)$ do not vanish outside a bounded region and $q_i(\phi)$ is not bounded. The precise condition is given below.

Theorem 11.18 *Assume that the hypotheses of Theorem* 11.5*, or Theorem* 11.6*, or Theorem* 11.8 *hold. Assume further that $q_{ij}(\cdot)$ is a continuous function for any $i,j\in\mathcal{M}$. Then the solution to* (11.7) *has the Feller property.*

Proof Let $f(\cdot,\cdot):\mathscr{C}\times\mathcal{M}\mapsto\mathbb{R}$ be a continuous function with $|f(\phi,i)|\le 1\,\forall(\phi,i)\in\mathscr{C}\times\mathcal{M}$. Fix $R>0$ and $T>0$. Suppose that $\|\phi_0\|<R$. Under the hypotheses of Theorem 11.5, or Theorem 11.6, or Theorem 11.8, it is shown in the proofs of those theorems that for any $\varepsilon>0$, there exists an $\tilde{R}>0$ such that

$$\mathbb{P}\{\|X_t^{\phi,i_0}\|\le\tilde{R}\}>1-\frac{\varepsilon}{8}\,\forall\|\phi\|\le R+1. \tag{11.33}$$

Let $\Phi(x):\mathbb{R}^n\mapsto\mathbb{R}$ be a twice continuously differentiable satisfying $\Phi(x)=1$ if $|x|\le\tilde{R}$ and $\Phi(x)=0$ if $|x|\ge\tilde{R}+1$. Let $(\tilde{X}_t^{\phi,i_0},\tilde{\alpha}^{\phi,i_0}(t))$ be the solution with initial data (ϕ,i_0) to

$$\begin{cases} d\tilde{X}(t)=\Phi(\tilde{X}(t))b(\tilde{X}(t),\tilde{\alpha}(t))dt+\Phi(\tilde{X}(t))\sigma(\tilde{X}(t),\tilde{\alpha}(t))dW(t)\\ d\tilde{\alpha}(t)=\displaystyle\int_{\mathbb{R}}h(\tilde{X}_t,\tilde{\alpha}(t-),z)\mathfrak{p}(dt,dz). \end{cases} \tag{11.34}$$

Then $(\tilde{X}^{\phi,i_0}(t),\tilde{\alpha}^{\phi,i_0}(t))=(X^{\phi,i_0}(t),\alpha^{\phi,i_0}(t))$ up to the time that $\|X_t^{\phi,i_0}\|>\tilde{R}$, which combined with (11.33) implies

$$\mathbb{P}\{\tilde{\Omega}_{\phi,i_0}\} > 1 - \frac{\varepsilon}{8}, \ \forall \, \|\phi\| < R$$

where $\tilde{\Omega}_{\phi,i_0} := \{\tilde{X}_T^{\phi,i_0} = X_T^{\phi,i_0}, \tilde{\alpha}^{\phi,i_0}(T) = \alpha^{\phi,i_0}(T)\}$. As a result, if $\|\phi\| < R$, we have

$$\begin{aligned}
&\left|\mathbb{E} f(X_T^{\phi,i_0}, \alpha^{\phi,i_0}(T)) - \mathbb{E} f(\tilde{X}_T^{\phi,i_0}, \tilde{\alpha}^{\phi,i_0}(T))\right| \\
\le & \mathbb{E}\left[I_{\tilde{\Omega}^c_{\phi,i_0}} \left| f(X_T^{\phi,i_0}, \alpha^{\phi,i_0}(T)) - f(\tilde{X}_T^{\phi,i_0}, \tilde{\alpha}^{\phi,i_0}(T))\right| \right] \ \text{(with } \tilde{\Omega}^c_{\phi,i_0} = \Omega \setminus \tilde{\Omega}_{\phi,i_0}) \\
\le & 2\mathbb{P}\left(I_{\tilde{\Omega}^c_{\phi,i_0}} \right) \le 2\frac{\varepsilon}{8} = \frac{\varepsilon}{4}.
\end{aligned} \tag{11.35}$$

It follows from Proposition 11.17 that there exists a $\delta \in (0, 1)$ such that if $\|\phi - \phi_0\| < \delta$, we have

$$\left|\mathbb{E} f(\tilde{X}_T^{\phi,i_0}, \tilde{\alpha}^{\phi,i_0}(T)) - \mathbb{E} f(\tilde{X}_T^{\phi_0,i_0}, \tilde{\alpha}^{\phi_0,i_0}(T))\right| < \frac{\varepsilon}{2}. \tag{11.36}$$

Since $\|f\| \le 1$, we can easily obtain from (11.35) and (11.36) that

$$\begin{aligned}
&\left|\mathbb{E} f(X_T^{\phi,i_0}, \alpha^{\phi,i_0}(T)) - \mathbb{E} f(X_T^{\phi_0,i_0}, \alpha^{\phi_0,i_0}(T))\right| \\
\le & \left|\mathbb{E} f(\tilde{X}_T^{\phi,i_0}, \tilde{\alpha}^{\phi,i_0}(T)) - \mathbb{E} f(\tilde{X}_T^{\phi_0,i_0}, \tilde{\alpha}^{\phi_0,i_0}(T))\right| \\
& + \left|\mathbb{E} f(X_T^{\phi,i_0}, \alpha^{\phi,i_0}(T)) - \mathbb{E} f(\tilde{X}_T^{\phi,i_0}, \tilde{\alpha}^{\phi,i_0}(T))\right| \\
& + \left|\mathbb{E} f(X_T^{\phi_0,i_0}, \alpha^{\phi_0,i_0}(T)) - \mathbb{E} f(\tilde{X}_T^{\phi_0,i_0}, \tilde{\alpha}^{\phi_0,i_0}(T))\right| \\
< & \frac{\varepsilon}{2} + \frac{\varepsilon}{4} + \frac{\varepsilon}{4} = \varepsilon, \ \text{if } \|\phi - \phi_0\| < \delta.
\end{aligned}$$

The proof of the theorem is complete. □

11.4.1 Strong Feller Property of Past-Independent Hybrid Diffusion

Now, we suppose that the q_{ij}, $i, j \in \mathcal{M}$ associated with $\alpha(t)$ depend only the current state of $X(t)$. To be more precise $q_{ij}(\cdot)$ is a function from $\mathbb{R}^n$ to $\mathbb{R}$ for each $(i, j) \in \mathcal{M} \times \mathcal{M}$. As a special case of the hybrid diffusion with past-dependent switching, we obtain the following theorem.

Theorem 11.19 *Assume that $q_{ij}(\cdot)$ is a continuous function for any $i, j \in \mathcal{M}$. Assume further that one of the following conditions is satisfied:*

(a) (H11.3) *and* $q_i(y) = \sum_{j\neq i} q_{ij}(y)$ *is bounded uniformly in* $(i, y) \in \mathcal{M} \times \mathbb{R}^n$.
(b) (H11.4) *and* $q_i(y)$ *is bounded uniformly in* $(i, y) \in \mathcal{M} \times K$ *for each compact subset* K *of* $\mathbb{R}^n$.

Then the unique solution to (11.7) *is a Markov process having the Feller property.*

Remark 11.20 If for each discrete state $i \in \mathcal{M}$, the diffusion $Y^{(i)}(t)$, which is the solution process to

$$dY^{(i)}(t) = b(Y^{(i)}(t), i)dt + \sigma(Y^{(i)}(t), i)dW(t) \tag{11.37}$$

has the strong Feller property, we do not need the continuity of $q_{ij}(\cdot)$ to get the Feller property of $(X(t), \alpha(t))$. In fact, we will obtain a stronger result, namely, the strong Feller property. The condition for the strong Feller property of $Y^{(i)}(t)$ is essentially the ellipticity of $A(x, i)$ or the Hörmander condition for hyperellipticity; see, e.g., [225, 266].

Theorem 11.21 *Assume that* $q_{ij}(\cdot)$ *is measurable for any* $i, j \in \mathcal{M}$ *and either* (a) *or* (b) *in Theorem* 11.19 *holds. If for each* $i \in \mathcal{M}$, *the solution process* $Y^{(i)}(t)$ *to* (11.37) *has the strong Feller property, then the unique solution to* (11.7) *has the strong Feller property, that is, for each bounded measurable function* $g(y, i) : \mathbb{R}^n \times \mathcal{M} \to \mathbb{R}$, *the function* $(x, i) \to \mathbb{E}g(X^{x,i}(T), \alpha^{x,i}(T))$ *is continuous for each* $T > 0$.

Proof We assume without loss of generality that $|g(z, i)| \leq 1\, \forall z \in \mathbb{R}^n, i \in \mathcal{M}$. Let $Y^{y,i}(t)$ be the solution with initial data y to (11.37). Fix $(x, i) \in \mathbb{R}^n \times \mathcal{M}$ and $\varepsilon > 0$. Under the assumption (a) or (b) of Theorem 11.19, we can show that for each $x \in \mathbb{R}^n$, there is a $K > 0$ satisfying

$$\mathbb{P}((\Omega_\varepsilon^{y,i})^c) < \frac{\varepsilon}{8}\, \forall y \in B(x, 1) := \{z : |x - z| < 1\}, \tag{11.38}$$

where $\Omega_\varepsilon^{y,i} = \{|Y^{y,i}(t)| \leq K, t \in [0, 1]\}$, $(\Omega_\varepsilon^{y,i})^c = \Omega \setminus \Omega_\varepsilon^{y,i}$.

Let $M = \sup_{i\in\mathcal{M},|z|\leq K} q_i(z)$ and $t_0 \in (0, 1)$ satisfying $1 - \exp\{-Mt_0\} < \dfrac{\varepsilon}{16}$. It follows from (11.38) and (i) of Lemma 11.12 that

$$\mathbb{P}\{\tau^{y,i} > t_0\} > 1 - \frac{3\varepsilon}{16}\, \forall y \in B(x, 1), \tag{11.39}$$

where

$$\tau^{y,i} := \inf\{t > 0 : \alpha^{y,i}(t) \neq i\} = \inf\Big\{t > 0 : \int_0^t \int_{\mathbb{R}} h(Y^{y,i}(s), i, u)\mathfrak{p}(ds, du) \neq 0\Big\}.$$

Denote $\Phi(y, i) := \mathbb{E}g\big(X^{y,i}(T - t_0), \alpha^{y,i}(T - t_0)\big)$. The condition $|g(y, i)| \leq 1$ implies $|\Phi(y, i)| \leq 1$ for all $y \in \mathbb{R}^n, i \in \mathcal{M}$. By the strong Feller property of $Y^{(i)}(t)$, there is a $\delta > 0$ such that

$$|\mathbb{E}\Phi(Y^{y,i}(t_0), i) - \mathbb{E}\Phi(Y^{x,i}(t_0), i)| \leq \frac{\varepsilon}{4} \forall y \in B(x, \delta). \tag{11.40}$$

By the strong Markov property of $X(t)$, we have

$$\begin{aligned}
\mathbb{E}g\big(X^{y,i}(T), \alpha^{y,i}(T)\big) &= \mathbb{E}\Phi(X^{y,i}(t_0), \alpha^{y,i}(t_0)) \\
&= \mathbb{E}\big[I_{\{\tau^{y,i} > t_0\}}\Phi(X^{y,i}(t_0), \alpha^{y,i}(t_0))\big] + \mathbb{E}\big[I_{\{\tau^{y,i} \leq t_0\}}\Phi(X^{y,i}(t_0), \alpha^{y,i}(t_0))\big].
\end{aligned} \tag{11.41}$$

Applying (i) of Lemma 11.12, we obtain

$$\begin{aligned}
&\mathbb{E}\big[I_{\{\tau^{y,i} > t_0\}}\Phi(X^{y,i}(t_0), \alpha^{y,i}(t_0))\big] \\
=&\mathbb{E}\big[\Phi(Y^{y,i}(t_0), \alpha^{y,i}(t_0))e^{-\int_0^{t_0} q_i(Y^{y,i}(s))ds}\big] \\
=&\mathbb{E}\big[\Phi(Y^{y,i}(t_0), \alpha^{y,i}(t_0))\big] \\
&+ \mathbb{E}\Big[I_{(\Omega_\varepsilon^{y,i})^c}\Phi(Y^{y,i}(t_0), \alpha^{y,i}(t_0))\Big(e^{-\int_0^{t_0} q_i(Y^{y,i}(s))ds} - 1\Big)\Big] \\
&+ \mathbb{E}\Big[I_{\Omega_\varepsilon^{y,i}}\Phi(Y^{y,i}(t_0), \alpha^{y,i}(t_0))\Big(e^{-\int_0^{t_0} q_i(Y^{y,i}(s))ds} - 1\Big)\Big].
\end{aligned} \tag{11.42}$$

Note that if $|g(z, i)| \leq 1 \forall z \in \mathbb{R}^n, i \in \mathcal{M}$ then $|\Phi(z, i)| \leq 1 \forall z \in \mathbb{R}^n, i \in \mathcal{M}$. We have the following estimates for $y \in B(x, \delta)$ using (11.38), (11.39), (11.40), and the fact that $|g(z, i)|, |\Phi(z, i)| \leq 1 \forall z \in \mathbb{R}^n, i \in \mathcal{M}$.

$$\Big|\mathbb{E}\big[I_{(\Omega_\varepsilon^{y,i})^c}\Phi(Y^{y,i}(t_0), \alpha^{y,i}(t_0))\big(e^{-\int_0^{t_0} q_i(Y^{y,i}(s))ds} - 1\big)\big]\Big| \leq \mathbb{P}((\Omega_\varepsilon^{y,i})^c) \leq \frac{\varepsilon}{8}, \tag{11.43}$$

$$\Big|\mathbb{E}\Big[I_{\Omega_\varepsilon^{y,i}}\Phi(Y^{y,i}(t_0), \alpha^{y,i}(t_0))\Big(e^{-\int_0^{t_0} q_i(Y^{y,i}(s))ds} - 1\Big)\Big]\Big| \leq 1 - \exp(-Mt_0) \leq \frac{\varepsilon}{16}, \tag{11.44}$$

$$\mathbb{E}\big[I_{\{\tau^{y,i} \leq t_0\}}\Phi(X^{y,i}(t_0), \alpha^{y,i}(t_0))\big]\Big| \leq \mathbb{P}\{\tau^{y,i} \leq t_0\} \leq \frac{3\varepsilon}{16}. \tag{11.45}$$

Applying estimates (11.40), (11.43), (11.44), and (11.45) to (11.41) and (11.42), we have

$$\Big|\mathbb{E}\Phi(X^{y,i}(T), \alpha^{y,i}(T)) - \mathbb{E}\Phi(X^{x,i}(T), \alpha^{x,i}(T))\Big| \leq \varepsilon, \ \forall y \in B(x, \delta).$$

The proof is complete. □

Remark 11.22 Sufficient conditions for the strong Feller property of $(X(t), \alpha(t))$, in which the rates of switching q_{ij} for $i, j \in \mathcal{M}$ depend only on the current continuous state $X(t)$, was obtained in Shao [256]. However, the conditions there are restrictive. To obtain the strong Feller property, it is assumed in [256] that $q_{ij}(x)$, $b(x, i)$ and $\sigma(x, i)$ are Lipschitz in x uniformly in $i \in \mathcal{M}$. The ellipticity of $A(x, i)$ is also assumed to be uniform in $(x, i) \in \mathbb{R}^n \times \mathcal{M}$. Moreover, it is assumed that $q_{ij}(x) = 0$ if $|i - j| < k$ for some constant k. It can be seen that our conditions in this paper are much more relaxed compared with the aforementioned reference.

11.5 Recurrence and Ergodicity

Throughout this section, we suppose that one of the assumptions for existence and uniqueness of solutions in Section 11.3 is satisfied, that is, any of the following holds:

- (H11.1), or (H11.2), or (H11.3) and (ii) of (H11.1), or (H11.4) and (ii) of (H11.2).

We also assume that $q_{i,j}(\phi)$ is continuous in $\mathscr{C}$ for each $(i, j) \in \mathcal{M}^2$. To proceed, we need Dupire's functional Itô formula.

11.5.1 Dupire's Functional Itô Formula

Now we state the functional Itô formula for our process; see [55] for more details. Let $\mathbf{D}$ be the space of càdlàg functions $f : [-r, 0] \mapsto \mathbb{R}^n$. For $\phi \in \mathbf{D}$, we define horizontal and vertical perturbations for $h \geq 0$ and $y \in \mathbb{R}^n$ as

$$\phi_h(s) = \begin{cases} \phi(s+h) \text{ if } s \in [-r, -h], \\ \phi(0) \text{ if } s \in [-h, -0], \end{cases}$$

and

$$\phi^y(s) = \begin{cases} \phi(s) \text{ if } s \in [-r, 0), \\ \phi(0) + y, \end{cases}$$

respectively. Let $V : \mathbf{D} \times \mathcal{M} \mapsto \mathbb{R}$. The horizontal derivative at (ϕ, i) and vertical partial derivative of V are defined as

$$V_t(\phi, i) = \lim_{h \to 0} \frac{V(\phi_h, i) - V(\phi)}{h} \tag{11.46}$$

and

$$\partial_i V(\phi, i) = \lim_{h \to 0} \frac{V(\phi^{he_i}, i) - V(\phi)}{h} \tag{11.47}$$

if these limits exist. In (11.47), e_i is the standard unit vector in $\mathbb{R}^n$ whose i-th component is 1 and other components are 0. Let $\mathbf{F}$ be the family of function $V(\cdot,\cdot):\mathbf{D}\times\mathcal{M}\mapsto\mathbb{R}$ satisfying that:

- V is continuous, that is, for any $\varepsilon>0$, $(\phi,i)\in\mathbf{D}\times\mathcal{M}$, there is a $\delta>0$ such that $|V(\phi,i)-V(\phi',i)|<\varepsilon$ as long as $\|\phi-\phi'\|<\delta$.
- The derivatives V_t, $V_x=(\partial_k V)$, and $V_{xx}=(\partial_{kl}V)$ exist and are continuous.
- V, V_t, $V_x=(\partial_k V)$ and $V_{xx}=(\partial_{kl}V)$ are bounded in each $B_R:=\{(\phi,i):\|\phi\|\le R, i\le R\}$, $R>0$.

Remark 11.23 Recently, a functional Itô formula was developed in [74], which encouraged subsequent development (e.g., [55, 229]). We briefly recall the main idea in what follows. Consider functions of the form

$$V(\phi,i)=f_1(\phi(0),i)+\int_{-r}^{0}g(t,i)f_2(\phi(t),i)dt,$$

where $f_2(\cdot,\cdot):\mathbb{R}^n\times\mathcal{M}\mapsto\mathbb{R}$ is a continuous function and $f_1(\cdot,\cdot):\mathbb{R}^n\times\mathcal{M}\mapsto\mathbb{R}$ is a function that is twice continuously differentiable in the first variable and $g(\cdot,\cdot):\mathbb{R}_+\times\mathcal{M}\mapsto\mathbb{R}$ be a continuously differentiable function in the first variable. Then at $(\phi,i)\in\mathscr{C}\times\mathcal{M}$ we have (see [229] for the detailed computations)

$$V_t(\phi,i)=g(0,i)f_2(\phi(0),i)-g(-r,i)f_2(\phi(-r),i)-\int_{-r}^{0}f_2(\phi(t),i)dg(t,i),$$

$$\partial_k V(\phi,i)=\frac{\partial f_1}{\partial x_k}(\phi(0),i),\qquad \partial_{kl}V(\phi,i)=\frac{\partial^2 f_1}{\partial x_k\partial x_l}(\phi(0),i).$$

Let $V(\cdot,\cdot)\in\mathbf{F}$, we define the operator

$$\begin{aligned}\mathbb{L}V(\phi,i)=&V_t(\phi,i)+V_x(\phi,i)b(\phi(0),i)+\frac{1}{2}\text{trace}\Big(V_{xx}(\phi,i)A(\phi(0),i)\Big)\\&+\sum_{j=1,j\neq i}^{\infty}q_{ij}(\phi)\big[V(\phi,j)-V(\phi,i\big]\\=&V_t(\phi,i)+\sum_{k=1}^{n}b_k(\phi(0),i)V_k(\phi,i)+\frac{1}{2}\sum_{k,l=1}^{n}a_{kl}(\phi(0),i)V_{kl}(\phi,i)\\&+\sum_{j=1,j\neq i}^{\infty}q_{ij}(\phi)\big[V(\phi,j)-V(\phi,i)\big],\end{aligned}\tag{11.48}$$

for any bounded stopping time $\tau_1\le\tau_2$, we have the functional Itô formula:

$$\mathbb{E}V(X_{\tau_2}, \alpha(\tau_2)) = \mathbb{E}V(X_{\tau_1}, \alpha(\tau_1)) + \mathbb{E}\int_{\tau_1}^{\tau_2} \mathbb{L}V(X_s, \alpha(s))ds \tag{11.49}$$

if the expectations involved exist. Equation (11.49) is obtained by applying the functional Itô formula for general semimartingales given in [54, 55] specialized to our processes.

11.5.2 Recurrence and Ergodicity of Past-Dependent Switching Diffusions

First, we need some conditions for irreducibility of the process $\{(X_t, \alpha(t)) : t \ge 0\}$.

(H11.6) Assume the following conditions:

a. For any $i \in \mathcal{M}$, $A(x, i)$ is elliptic uniformly on each compact set, that is, for any $R > 0$, there is a $\theta_{R,i} > 0$ such that

$$y'A(x, i)y \ge \theta_{R,i}|y|^2 \;\; \forall |x| \le R, \;\; y \in \mathbb{R}^d. \tag{11.50}$$

b. There is an i^* satisfying that for any $i \in \mathcal{M}$, there exist $i_1, \ldots, i_k \in \mathcal{M}$ and $\phi_1, \ldots, \phi_{k+1} \in \mathscr{C}$ such that $q_{ii_1}(\phi_1) > 0$, $q_{i_l,i_{l+1}}(\phi_{l+1}) > 0, l = 1, \ldots, k-1$, and $q_{i_k,i^*}(\phi_{k+1}) > 0$.

(H11.7) There exists an $i^* \in \mathcal{M}$ such that

a. $A(x, i^*)$ is elliptic uniformly on each compact set;
b. for any $(\phi, i) \in \mathscr{C} \times \mathcal{M}$, there exist positive integers $i = i_1, \ldots, i_k = i^*$ satisfying $q_{i_l,i_{l+1}}(\phi) > 0, l = 1, \ldots, k-1$.

Let $(X^{\phi,i}, \alpha^{\phi,i}(t))$ be the solution to (11.7) with initial data $(\phi, i) \in \mathscr{C} \times \mathcal{M}$. To simplify the notation, we denote by $\mathbb{P}_{\phi,i}$ the probability measure conditioned on the initial data (ϕ, i), that is, for any $t > 0$,

$$\mathbb{P}_{\phi,i}\{(X_t, \alpha(t)) \in \cdot\} = \mathbb{P}\{(X_t^{\phi,i}, \alpha^{\phi,i}(t)) \in \cdot\},$$

and $\mathbb{E}_{\phi,i}$ the expectation associated with $\mathbb{P}_{\phi,i}$. To proceed, we state some auxiliary lemmas.

Lemma 11.24 *Let $\phi \in \mathscr{C}$ and $q_{ij}(\phi) > 0$. For any $\varepsilon > 0$, there is a $\delta > 0$ such that*

$$\inf_{\psi \in \mathscr{C}: \|\psi - \phi\| < \delta} \mathbb{P}_{\psi,i}\{\|X_\delta - \phi\| < \varepsilon, \alpha(\delta) = j\} > 0.$$

Lemma 11.25 *For any $i > 0$, $R > 0$ and $\varepsilon > 0$, there is a compact set $\mathcal{A} \in \mathscr{C}$ such that*

$$\inf_{\|\phi\| \le R} \mathbb{P}_{\phi,i}\{X_r \in \mathcal{A}, \alpha(r) = i\} > 0. \tag{11.51}$$

Moreover, if (11.50) *holds for i, then for any $k > 0$, there is a $T = T(k, i, R) > 0$ such that*

$$\inf_{\|\phi\| \le R} \mathbb{P}_{\phi,i}\{\|X_t\| > k \text{ for some } t \in [0, T]\} > 0. \tag{11.52}$$

Lemma 11.26 *Assume that either* (H11.6) *or* (H11.7) *is satisfied. There is a nontrivial measure $\boldsymbol{\nu}(\cdot)$ on $\mathfrak{B}(\mathscr{C})$ such that $\boldsymbol{\nu}(\mathcal{D}) > 0$ if $\mathcal{D}$ is a nonempty open subset of $\mathscr{C}$ and that for any $R > 0$, $T > r$, there is a $d_{R,T} > 0$ satisfying*

$$\mathbb{P}_{\phi,i^*}\{X_T \in \mathcal{B} \text{ and } \alpha(T) = i^*\} \ge d_{R,T}\boldsymbol{\nu}(\mathcal{B}),\ \mathcal{B} \in \mathfrak{B}(\mathscr{C}) \text{ given that } \|\phi\| \le R. \tag{11.53}$$

The three lemmas above will be proved in Section 11.7.2

Lemma 11.27 *Assume that either* (H11.6) *or* (H11.7) *holds. For any $i \in \mathcal{M}$, there is a $T_i > 0$ such that for any $T > T_i$ and any open set $\mathcal{B} \subset \mathscr{C}$, we have $\mathbb{P}_{\phi,i}\{X_T \in \mathcal{B}, \alpha(T) = i^*\} > 0$, $\phi \in \mathscr{C}$, where i^* is as in* (H11.6) *or* (H11.7) *accordingly.*

Proof Suppose that (H11.6) holds with $i = i_1, \ldots, i_k = i^* \in \mathcal{M}$ and $\phi_1, \ldots, \phi_{k+1} \in \mathscr{C}$ such that $q_{i_l,i_{l+1}}(\phi_{l+1}) > 0, l = 1, \ldots, k-1$. Since $q_{i_l,i_{1+1}}(\phi_{l+1}) > 0$, it follows from Lemma 11.24 that

$$\mathbb{P}_{\psi,i_l}\{\|X_{\varepsilon_l} - \phi_{l+1}\| < 1, \alpha(\varepsilon_l) = i_{l+1}\} > 0 \text{ if } \|\psi - \phi_{l+1}\| < \varepsilon_l \tag{11.54}$$

for some $\varepsilon_l \in (0, 1)$. In view of Lemma 11.26,

$$\mathbb{P}_{\psi,i_l}\{\|X_{1+r} - \phi_{l+1}\| < \varepsilon_l, \alpha(1+r) = i_l\} > 0 \text{ for any } \psi \in \mathscr{C}, \tag{11.55}$$

and

$$\mathbb{P}_{\psi,i^*}\{X_{1+r+T'} \in \mathcal{B}, \alpha(1+r+T') = i^*\} > 0 \text{ for any } \psi \in \mathscr{C}, T' \ge 0. \tag{11.56}$$

By (11.54), (11.55), and the Markov property of $(X_t, \alpha(t))$, we have

$$\mathbb{P}_{\psi,i_l}\{\|X_{1+r+\varepsilon_l} - \phi_l\| < 1, \alpha(1+r+\varepsilon_l) = i_{l+1}\} > 0 \text{ for any } \psi \in \mathscr{C}. \tag{11.57}$$

Using (11.56), (11.57), and applying the Kolmogorov-Chapman equation again, we obtain

$$\mathbb{P}_{\psi,i}\left\{X_{k(1+r)+\sum\varepsilon_l+T'}\in\mathcal{B},\alpha\left(k(1+r)+\sum\varepsilon_l+T'\right)=i^*\right\}>0 \text{ for any } \psi\in\mathscr{C}. \tag{11.58}$$

The lemma is proved with $T_i = k(2+r) \geq k(1+r)+\sum\varepsilon_l$.

Now, suppose that (H11.7) holds, it follows from Lemma 11.24 and the Kolmogorov-Chapman equationthat

$$\mathbb{P}_{\phi,i}\{\|X_\varepsilon - \phi\| < 1, \alpha(\varepsilon) = i^*\} > 0 \tag{11.59}$$

for sufficiently small ε. The desired result follows from (11.56) and (11.59) with $T_i = 2+r$. □

Lemma 11.28 *Assume that either* (H11.6) *or* (H11.7) *holds. Let*

$$\eta_k = \inf\{t > 0 : \|X_t\| \vee \alpha(t) > k\}. \tag{11.60}$$

Then for any $(\phi, i) \in \mathscr{C}\times\mathcal{M}$, *we have* $\mathbb{P}_{\phi,i}\{\eta_k < \infty\} = 1,\ \forall k\in\mathcal{M}$.

Proof Suppose that $p_0 = \mathbb{P}\{\eta_k < \infty\} < 1$. Since $A(x, i^*)$ is elliptic, in view of (11.52), there is a $T > 0$ such that

$$\mathbb{P}_{\varphi',i^*}\{\eta_k < T\} > 0,\ \forall k > 1, \|\varphi'\| \leq 1. \tag{11.61}$$

In view of Lemma 11.27 and (11.61), $\forall(\psi, j) \in \mathscr{C}\times\mathcal{M}$ there are $T_{\psi,j}, p_{\psi,j} > 0$ such that

$$\mathbb{P}_{\psi,j}\{\eta_k < T_{\psi,j}\} > 2p_{\psi,j}. \tag{11.62}$$

Due to the Feller property of $(X_t, \alpha(t))$, there exists a $\delta_{\psi,j} > 0$ such that

$$\mathbb{P}_{\psi',i}\{\eta_k \leq T_{\phi,i^*}\} > p_{\psi,j},\ \forall\ \psi'\in\mathscr{C}, \|\psi-\psi'\| < \delta_{\psi,j}. \tag{11.63}$$

Since $\sigma(\cdot, i)$ and $b(\cdot, i)$ are locally compact for each $i \in \mathcal{M}$, similar to Lemma 11.16, we can show that there is an $h_k > 0$ such that for any $t > 0$,

$$\mathbb{P}_{\phi,i}\left\{\frac{|X(s)-X(s')|}{(s-s')^{0.25}} \leq h_k,\ \forall\, 0\vee(\eta_k\wedge t - r) \leq s' < s < \eta_k\wedge t\right\} > \frac{1+p_0}{2}. \tag{11.64}$$

Since the set

$$\mathcal{A}_k = \left\{\psi\in\mathscr{C} : |\psi| \leq k, \frac{|\psi(s)-\psi(s')|}{(s-s')^{0.25}} \leq h_k, \forall -r \leq s' < s \leq 0\right\}$$

is compact in $\mathscr{C}$, we have from (11.63) that there exist T_k and $\tilde{p}_k$ such that

$$\mathbb{P}_{\psi,j}\{\eta_n < T_k\} > \tilde{p}_k > 0,\ \forall\ \psi\in\mathcal{A}_k, j\leq k. \tag{11.65}$$

Since $\lim_{t\to\infty}\mathbb{P}_{\phi,i}\{\eta_k < t\} = p_0 < 1$, there is a $T' > 0$ such that

$$p_0 \geq \mathbb{P}_{\phi,i}\{\eta_k \leq T'\} \geq p_0 - \frac{1-p_0}{2}\tilde{p}_k. \tag{11.66}$$

In view of (11.64) and (11.66), we have $\mathbb{P}_{\phi,i}\{T' < \eta_k, X_{T'} \in \mathcal{A}_k\} > \frac{1-p_0}{2}$. By the Markov property and (11.65),

$$\begin{aligned}\mathbb{P}_{\phi,i}\{T' < \eta_k < \infty\} &\geq \mathbb{P}_{\phi,i}\{X_{T'} \in \mathcal{A}_k, T' < \eta_k\}\\ &\geq \mathbb{E}_{\phi,i}\left[I_{\{T'<\eta_k, X_{T'}\in\mathcal{A}_k\}}\mathbb{P}_{X_{T'},\alpha(T')}\{\eta_k < \infty\}\right]\\ &> \frac{1-p_0}{2}\tilde{p}_k.\end{aligned} \tag{11.67}$$

We have from (11.66) and (11.67) that

$$\begin{aligned}p_0 = \mathbb{P}_{\phi,i}\{\eta_k < \infty\} =&\mathbb{P}_{\phi,i}\{\eta_k \leq T'\} + \mathbb{P}_{\phi,i}\{T' < \eta_k < \infty\}\\ >& p_0 - \frac{1-p_0}{2}\tilde{p}_k + \frac{1-p_0}{2}\tilde{p}_k = p_0,\end{aligned}$$

which is a contradiction. Thus $p_0 = 1$. □

Definition 11.29 The process $\{(X_t, \alpha(t)) : t \geq 0\}$ is said to be recurrent (resp., positive recurrent) relative to a measurable set $\widetilde{\mathcal{E}} \in \mathcal{C} \times \mathcal{M}$ if $\mathbb{P}_{\phi,i}\{(X_t, \alpha(t)) \in \widetilde{\mathcal{E}}$ for some $t \geq 0\} = 1$ $\big($resp. $\mathbb{E}_{\phi,i}\left[\inf\{t > 0 : (X_t, \alpha(t)) \in \widetilde{\mathcal{E}}\}\right] < \infty\big)$ for any $(\phi, i) \in \mathcal{C} \times \mathcal{M}$.

Theorem 11.30 *Suppose that either* (H11.6) *or* (H11.7) *holds. Let $\mathcal{D}$ be a bounded open subset of $\mathcal{C}$ and N be a finite subset of $\mathcal{M}$. If $(X_t, \alpha(t))$ is recurrent relative to $\mathcal{D} \times N$ then $(X_t, \alpha(t))$ is recurrent relative to $\mathcal{D}' \times N'$ for any open set $\mathcal{D}' \subset \mathcal{C}$ and a finite set $N' \subset \mathcal{M}$ containing i^* with i^* given in either* (H11.6) *or* (H11.7) *according to which hypothesis is satisfied.*

Proof Let $(\phi_0, i_0) \in \mathcal{C} \times \mathcal{M}$. In view of Lemma 11.25, there exists a compact set $\mathcal{A}_{\mathcal{D}} \subset \mathcal{C}$ such that

$$\inf_{\{(\psi,j)\in\mathcal{D}\times N\}} \mathbb{P}_{\psi,j}\{X_r \in \mathcal{A}_{\mathcal{D}}, \alpha(r) \in N\} := p_{\mathcal{D},N} > 0. \tag{11.68}$$

Since $i^* \in N'$, by Lemma 11.27, the Feller property of $(X_t, \alpha(t))$ and the compactness of $\mathcal{A}_{\mathcal{D}}$, there is a $T > 0$ such that

$$\inf_{\{(\psi,j)\in\mathcal{A}_{\mathcal{D}}\times N\}} \mathbb{P}_{\psi,j}\{X_T \in \mathcal{D}' \times N'\} \geq \varepsilon_0. \tag{11.69}$$

Define the stopping times $\vartheta_0 = 0$, $\vartheta_{k+1} = \inf\{t > \vartheta_k + T : X_{\eta_{k+1}} \in \mathcal{D} \times N\}$. By the hypothesis of the theorem, $\mathbb{P}_{\phi_0,i_0}\{\vartheta_k < \infty\} = 1,\ \forall\, k \in \mathcal{M}$. On the other hand, it follows from (11.68) and (11.69) that

$$\mathbb{P}_{\psi,j}\{(X_T, \alpha(T) \in \mathcal{D}' \times N'\} \ge p_{\mathcal{D},N}\varepsilon_0, \text{ for all } (\psi, j) \in \mathcal{D} \times N. \tag{11.70}$$

Consider the events $A^k = \{(X_{\vartheta_k+T} \notin \mathcal{D}' \times N'\}$ for $k \in \mathcal{M}$. By the strong Markov property of $(X_t, \alpha(t))$, we have

$$\mathbb{P}_{\phi_0,i_0}\left(\bigcap_{k'=k}^{\infty} A^{k'}\right) = \lim_{l\to\infty} \mathbb{P}_{\phi_0,i_0}\left(\bigcap_{k'=k}^{l} A^{k'}\right) \le \lim_{l\to\infty} (1 - p_{\mathcal{D},N}\varepsilon_0)^{l-k} = 0.$$

Thus $\mathbb{P}_{\phi_0,i_0}\left(\cap_{k'=k}^{\infty} A^{k'}\right) = 0$. It indicates that the event $\{(X_{\vartheta_k+T}, \alpha(\vartheta_k + T)) \in \mathcal{D}' \times N\}$ must occur with probability 1. □

Theorem 11.31 *Suppose that either* (H11.6) *or* (H11.7) *holds. Let* $V(\cdot, \cdot) \in \mathbf{F}$ *such that* $\lim_{n\to\infty} \inf\{V(\phi, i) : |\phi(0)| \vee i \ge n\} = \infty$. *Suppose further that there are positive constants* C *and* H *such that*

$$\mathbb{L}V(\phi, i) \le CI_{\{V(\phi,i)\le H\}}. \tag{11.71}$$

Then the process $(X_t, \alpha(t))$ *is recurrent relative to* $\mathcal{D} \times N$*, where* $\mathcal{D}$ *is any open bounded subset of* $\mathscr{C}$ *and* $N \subset \mathcal{M}$ *contains* i^*.

Proof Let $\upsilon_H = \inf\{t > 0 : V(X_t, \alpha(t)) \le H\}$ and η_k be defined as in Lemma 11.28. In view of Lemma 11.28, $\mathbb{P}_{\psi,j}\{\eta_k < \infty\} = 1, \forall k \in \mathcal{M}$. Let $t > 0$. By Itô's formula $\mathbb{E}_{\psi,j} V\Big(X_{t\wedge\upsilon_H\wedge\eta_k}, \alpha(t \wedge \upsilon_H \wedge \eta_k)\Big) \le V(\psi, j)$. Letting $t \to \infty$, we obtain $\mathbb{E}_{\psi,j} V\big(X_{\upsilon_H\wedge\eta_k}, \alpha(\upsilon_H \wedge \eta_k)\big) \le V(\psi, j)$, which implies

$$\mathbb{P}_{\psi,j}\{\upsilon_H > \eta_k\} \le \frac{V(\psi, j)}{\inf\{V(\phi, i) : |\phi(0)| \vee i \ge k\}}.$$

Letting $k \to \infty$ yields $\mathbb{P}_{\psi,j}\{\upsilon_H > \eta_k\} \to 0$. Thus,

$$\mathbb{P}_{\psi,j}\{\upsilon_H < \infty\} = 1, \ \forall (\psi, j) \in \mathscr{C} \times \mathcal{M}. \tag{11.72}$$

Now, let $k_0 > 0$ such that $\inf\{V(\phi, i) : |\phi(0)| \vee i \ge k_0\} \ge 2(H + C + r)$. For any $(\psi, j) \in \mathscr{C} \times \mathcal{M}$ satisfying $V(\psi, j) \le H$. We have from (11.71) and Itô's formula that $\mathbb{E}_{\psi,j} V\big(X_{r\wedge\eta_{n_0}}, \alpha(r \wedge \eta_{n_0})\big) \le H + C + r$, which implies

$$\mathbb{P}_{\psi,j}\{\eta_{n_0} < r\} \le \frac{H + C + r}{2(H + C + r)} \le \frac{1}{2}. \tag{11.73}$$

Thus,

$$\mathbb{P}_{\psi,j}\{\|X_r\| < n_0, \alpha(r) < n_0\} \ge \mathbb{P}_{\psi,j}\{\eta_{n_0} > r\} > \frac{1}{2} \text{ provided } V(\psi, j) \le H. \tag{11.74}$$

Now, fix $(\phi_0, i_0) \in \mathscr{C} \times \mathcal{M}$. By (11.72) and Lemma 11.28, we can define almost surely finite stopping times

$$\begin{aligned}\zeta_1 &= \inf\{t \geq 0 : \ V(X_t, \alpha(t)) \leq H\},\\ \zeta_{2k} &= \inf\{t \geq \zeta_{2k-1} + r : \ |X_t| \vee \alpha(t) \geq n_0\},\\ \zeta_{2k+1} &= \inf\{t \geq \zeta_{2k} : \ V(X_t, \alpha(t)) \leq H\}.\end{aligned} \tag{11.75}$$

Define the events $B_k = \{\|X_{\zeta_{2k+1}+r}\| \vee \alpha(\zeta_{2k+1} + r) \leq n_0\}$. In view of (11.74) and the strong Markov property of $(X_t, \alpha(t))$, we can use standard arguments in Theorem 11.30 to show that $\mathbb{P}_{\phi_0, i_0}\{B_k \text{ occurs for some } k\} = 1$. Thus, $(X_t, \alpha(t))$ is recurrent relative to $\{(\phi, i) : \|\phi\| \vee i \leq n_0\}$. Combining this with Theorem 11.30 yields the desired result. □

Example 11.32 Let

$$\begin{aligned}&q_{12}(\phi) = 1, q_{1j}(\phi) = 0 \ \text{ for } j \geq 3;\\ &q_{i,i-1}(\phi) = C_i + (1 + \|\phi\|)^{-1}, q_{i,i+1}(\phi) = C_i + (1 + \|\phi\|)^{-1} \ \text{ for } i \geq 2, \ C_i \geq 0;\\ &q_{ij}(\phi) = 0 \ \text{ for } i \geq 2, \ j \notin \{i-1, i, i+1\}.\end{aligned}$$

Suppose the switching diffusion is given by

$$dX(t) = \sigma(X(t), \alpha(t))dW(t) - X(t)b(X(t), \alpha(t))dt$$

where $b(x, i) > 0$, $\sigma(x, i)$ are locally Lipchitz in x and uniformly bounded in $K \times \mathcal{M}$ for each compact set $K \in \mathbb{R}$. Let $f(x)$ be twice continuously differentiable such that $f(x) > 0$ and $f(x) = |x|$ if $|x| \geq 1$. Let

$$\kappa := \sup_{|x| \leq 1, i \in \mathcal{M}} \left| -\left[\frac{df}{dx}(x)\right] x b(x, i) + \frac{1}{2}\left[\frac{d^2 f}{dx^2} f(x)\right] \sigma^2(x, i)\right| < \infty. \tag{11.76}$$

Let $V(\phi, i) = f(\phi(0)) + 2\kappa i$. Direct computation leads to

$$\mathbb{L}V(\phi, i) = \begin{cases} -\left[\dfrac{df}{dx}(\phi(0))\right]\phi(0)b(\phi(0), i) + \frac{\sigma^2(i)}{2}\left[\frac{d^2 f}{dx^2}(\phi(0))\right] - 2\kappa(1 + \|\phi\|)^{-1}, \\ \qquad\qquad\qquad\qquad\qquad\qquad \text{if } i > 1, \\ -\left[\dfrac{df}{dx}(\phi(0))\right]\phi(0)b(\phi(0), i) + \left[\frac{d^2 f}{dx^2}(\phi(0))\right] + 2\kappa, \quad \text{if } i = 1. \end{cases} \tag{11.77}$$

In view of (11.76) and the fact that $\frac{d}{dx} f(x) = \text{sign}(x)$, $\frac{d^2}{dx^2} f(x) = 0$ for $|x| \geq 1$, $i > 1$ we have

$$\mathbb{L}V(\phi, i) \leq 0, \qquad \forall \phi \in \mathscr{C}, i > 1. \tag{11.78}$$

By (11.77), if we assume further $\lim_{x\to\infty} |x|b(x,1) = \infty$, then we can verify that

$$\mathbb{L}V(\phi, 1) \leq \tilde{C}_1 I_{\{|\phi(0)|<\tilde{H}\}} - \tilde{C}_2, \forall \phi \in \mathscr{C}, \tag{11.79}$$

where $\tilde{C}_1, \tilde{C}_2, \tilde{H}$ are some positive constants. In view of (11.78) and (11.79), we can easily check that (11.71) holds in this example, for $V(\phi, i)$ defined above and suitable C, H. Thus, if there exists $i^* \in \mathcal{M}$ such that $\sigma(x, i^*) \neq 0$ for any $x \in \mathbb{R}$, then the conclusion of Theorem 11.31 holds for this example.

To proceed, we need the following lemma on irreducibility and aperiodicity of the Markov chain $\{(X_k, \alpha(k)) : k \in \mathcal{M}\}$; see Appendix A.2 for the definitions of the concepts.

Lemma 11.33 *Assume either* (H11.6) *or* (H11.7) *holds. The Markov chain* $\{(X_k, \alpha(k)) : k \in \mathcal{M}\}$ *is irreducible and aperiodic. Moreover, for every bounded set* $\mathcal{D} \in \mathscr{C}$ *and a finite set* $N \in \mathcal{M}$, *the set* $\mathcal{D} \times N$ *is petite for* $\{(X_k, \alpha(k)) : k \in \mathcal{M}\}$.

Proof Similar to (11.70), there are $k_0 \in \mathcal{M}$, $k_0 > r$, $\tilde{d}_{\mathcal{D},N} > 0$ such that

$$\mathbb{P}_{\phi,i}\{X_t \in \mathcal{D}, \alpha_{k_0} = i^*\} \geq \tilde{d}_{\mathcal{D},N} \text{ for all } (\phi, i) \in \mathcal{D} \times N. \tag{11.80}$$

By the Markov property, we deduce from (11.53) and (11.80) that for any $k > r$, there exists a $\hat{d}_{\mathcal{D},N,k} > 0$ such that

$$\mathbb{P}_{\phi,i}\{X_{k+k_0} \in \mathcal{B} \text{ and } \alpha(k+k_0)i^*\} \geq \hat{d}_{\mathcal{D},N,k}\boldsymbol{\nu}(\mathcal{B}),\ \mathcal{B} \in \mathfrak{B}(\mathscr{C}),\ (\phi, i) \in \mathcal{D} \times N. \tag{11.81}$$

Let $\hat{\boldsymbol{\nu}}(\cdot)$ be the measure on $\mathfrak{B}(\mathscr{C} \times \mathcal{M})$ given by $\hat{\boldsymbol{\nu}}(\widetilde{\mathcal{E}}) = \boldsymbol{\nu}(\{\phi \in \mathscr{C} : (\phi, i^*) \in \widetilde{\mathcal{E}}\})$ for $\widetilde{\mathcal{E}} \in \mathfrak{B}(\mathscr{C} \times \mathcal{M})$. Then (11.81) can be rewritten as

$$\mathbb{P}_{\phi,i}\{(X_{k+k_0}, \alpha(k+k_0)) \in \widetilde{\mathcal{E}}\} \geq \hat{d}_{D,N,k}\hat{\boldsymbol{\nu}}(\widetilde{\mathcal{E}}),\ \widetilde{\mathcal{E}} \in \mathfrak{B}(\mathscr{C} \times \mathcal{M}),\ (\phi, i) \in \mathcal{D} \times N. \tag{11.82}$$

It can be checked that (11.82) implies that the Markov chain $\{(X_k, \alpha(k)) : k \in \mathcal{M}\}$ is $\hat{\boldsymbol{\nu}}$-irreducible and every nonempty bounded set in $\mathscr{C} \times \mathcal{M}$ is petite. Moreover, suppose that $(X_k, \alpha(k))$ is not aperiodic. Then, there are disjoint set $\widetilde{\mathcal{E}}_0, \ldots, \widetilde{\mathcal{E}}_{d-1}, d > 1$ such that

$$\hat{\boldsymbol{\nu}}\left((\mathscr{C} \times \mathcal{M}) \setminus \bigcup_{j=0}^{d-1} \widetilde{\mathcal{E}}_j\right) = 0 \tag{11.83}$$

and $\mathbb{P}_{\phi,i}\{(X_1, \alpha(1)) \in \widetilde{\mathcal{E}}_j\} = 1$ if $j = j' + 1 \pmod d$ if $(\phi, i) \in \widetilde{\mathcal{E}}_{j'}$, which results in

$$\mathbb{P}_{\phi,i}((X_m, \alpha(m)) \in \widetilde{\mathcal{E}}_j) = \begin{cases} 1 & \text{where } m = j+1 \pmod d \\ 0 & \text{otherwise} \end{cases} \quad \text{if } (\phi, i) \in \widetilde{\mathcal{E}}_j. \tag{11.84}$$

In view of (11.81), for any $m > k_0+r$ and $(\phi, i) \in \mathscr{C} \times \mathcal{M}$, there is a $\tilde{p}_{\phi,i,k} > 0$ such that

$$\mathbb{P}_{\phi,i}\{(X_m, \alpha(m)) \in \widetilde{\mathcal{E}}\} \ge \tilde{p}_{\phi,i,k}\hat{\boldsymbol{\nu}}(\widetilde{\mathcal{E}}) \tag{11.85}$$

for any measurable set $\widetilde{\mathcal{E}} \in \mathfrak{B}(\mathscr{C} \times \mathcal{M})$. By virtue of (11.84) and (11.85), we have that $\hat{\nu}(\widetilde{\mathcal{E}}_j) = 0$ for any $j = 0, \ldots, d-1$. Thus,

$$\hat{\nu}\left((\mathscr{C} \times \mathcal{M}) \setminus \bigcup_{j=0}^{d-1} \widetilde{\mathcal{E}}_j\right) = \hat{\nu}(\mathscr{C} \times \mathcal{M}) > 0, \tag{11.86}$$

which contradicts (11.83). This contradiction implies that $(X_k, \alpha(k))$ is aperiodic. □

Theorem 11.34 *Suppose that either* (H11.6) *or* (H11.7) *holds. Let* $V(\cdot,\cdot) \in \mathbf{F}$ *such that*

$$\lim_{n\to\infty} \inf\{V(\phi, i) : |\phi(0)| \vee i \ge n\} = \infty. \tag{11.87}$$

Suppose further that there are positive constants C_1, C_2, *and* H *such that*

$$\mathbb{L}V(\phi, i) \le -C_1 + C_2 I_{\{V(\phi,i)\ge H\}}. \tag{11.88}$$

Then, $(X_t, \alpha(t))$ *is positive recurrent relative to any set of the form* $\mathcal{D} \times N$ *where* $\mathcal{D}$ *is a nonempty open set of* $\mathscr{C}$ *and* $N \ni i^*$ *with* i^* *given in either* (H11.6) *or* (H11.7). *Moreover, there is a unique invariant probability measure* μ^*, *and for any* $(\phi, i) \in \mathscr{C} \times \mathcal{M}$ $\lim_{t\to\infty} \|P(t, (\phi, i), \cdot) - \mu^*\|_{TV} = 0$.

Proof Let $\upsilon_H = \inf\{t \ge 0 : V(X_t, \alpha(t)) \le H\}$. In view of the functional Itô formula,

$$\begin{aligned}\mathbb{E}_{\phi,i} V(X_{1\wedge\upsilon_H}, \alpha(1 \wedge \upsilon_H)) =& V(\phi, i) + \mathbb{E}_{\phi,i} \int_0^{1\wedge\upsilon_H} \mathbb{L}V(X_s, \alpha(s))ds \\ \le& V(\phi, i) - C_1\mathbb{E}_{\phi,i} 1 \wedge \upsilon_H \\ \le& V(\phi, i) - C_1\mathbb{P}_{\phi,i}\{\upsilon_H \ge 1\}.\end{aligned} \tag{11.89}$$

For any $t \le 1$ and $V(\phi, i) \le H$, we have

$$\begin{aligned}\mathbb{E}_{\phi,i} V(X_t, \alpha(t)) =& V(\phi, i) + \mathbb{E}_{\phi,i} \int_0^t \mathbb{L}V(X_s, \alpha(s))ds \\ \le& V(\phi, i) + C_2 t \\ \le& H + C_2.\end{aligned} \tag{11.90}$$

It follows from (11.90) and the strong Markov property of $(X_t, \alpha(t))$ that

$$\begin{aligned}\mathbb{E}_{\phi,i}\left[I_{\{\upsilon_H<1\}}V(X_1,\alpha(1))\right] \le&(H+C_2)\mathbb{P}_{\phi,i}\{\upsilon_H<1\}\\ \le&2(H+C_2)-(H+C_2)\mathbb{P}_{\phi,i}\{\upsilon_H<1\}.\end{aligned} \tag{11.91}$$

Let $\mathscr{C}_V := \{(\psi', j') : V(\psi, j) \le 2(H + C_2)\}$. In view of (11.89) and (11.91),

$$\begin{aligned}\mathbb{E}_{\phi,i}V(X_1,\alpha(1)) \le&\mathbb{E}_{\phi,i}\left[I_{\{\upsilon_H<1\}}V(X_1,\alpha(1))\right]+\mathbb{E}_{\phi,i}V(X_{1\wedge\upsilon_H},\alpha(1\wedge\upsilon_H))\\ \le&V(\phi,i)-\min\{C_1,H+C_2\}+2(H+C_2)I_{\{(\phi,i)\in\mathscr{C}_V\}}.\end{aligned} \tag{11.92}$$

Let $n_0 \in \mathcal{M}$ such that

$$V(\phi,i)\ge 2(2H+2C_2+C_2r)\ \text{ for any }\ \|\phi\|\vee i\ge n_0, \tag{11.93}$$

and define $\hat{\zeta}_V = \inf\{t \ge 0 : V(X_t, \alpha(t)) \ge 2(2H+2C_2+C_2r)\}$. Similar to (11.73), we have

$$\mathbb{P}_{\phi,i}\{\hat{\zeta}_V\le r\}\le\frac{1}{2}\ \text{ for }\ (\phi,i)\in\mathscr{C}_V. \tag{11.94}$$

Thus,

$$\begin{aligned}\mathbb{P}_{\phi,i}\{\|X_r\|\vee\alpha(r)\le n_0\}\ge&\ \mathbb{P}_{\phi,i}\{V(X_r,\alpha(r))\ge H+C_2r+1\}\\ \ge&\ 1-\mathbb{P}_{\phi,i}\{\hat{\zeta}_V\le r\}=\frac{1}{2},\ (\phi,i)\in\mathscr{C}_H.\end{aligned} \tag{11.95}$$

In view of (11.82) and (11.95),

$$\mathbb{P}_{\phi,i}\{(X_{k+k_0},\alpha(k+k_0))\in\widetilde{\mathcal{E}}\}\ge\hat{d}_{H,k}\hat{\mathfrak{v}}(\widetilde{\mathcal{E}}),\ \widetilde{\mathcal{E}}\in\mathfrak{B}(\mathscr{C}\times\mathcal{M}),\ \text{ if } V(\phi,i)\le H, k>r \tag{11.96}$$

for some $\hat{d}_{H,k} > 0$. Thus, the set $\{(\phi, i) \in \mathscr{C} \times \mathcal{M} : V(\phi, i) \le H\}$ is petite for $\{(X_k, \alpha(k)) : k \in \mathcal{M}\}$. Using this and (11.96), it follows from Theorem A.1 that $\lim_{n\to\infty} \|P(n, (\phi, i), \cdot) - \mu^*\|_{TV} = 0$, where $P(t, (\phi, i), \cdot)$ is the transition probability of $(X_t, \alpha(t))$ and μ^* is an invariant probability measure of the Markov chain $\{X_k, \alpha(k), k \in \mathcal{M}\}$. It is easy to show that μ^* is also an invariant probability measure of the process $\{(X_t, \alpha(t))\}$. Thus $\|P(t, (\phi, i), \cdot) - \mu^*\|_{TV}$ is decreasing in t, which leads to $\lim_{t\to\infty} \|P(t, (\phi, i), \cdot) - \mu^*\|_{TV} = 0$.

Now we show that the process $(X_t, \alpha(t))$ is positive recurrent. Similar to (11.89), we deduce from the functional Itô formula that $\mathbb{E}_{\phi,i}\upsilon_H \le C_1^{-1}V(\phi, i)$. Owing to this and (11.96), we can use the arguments in the proof of Lemma 3.8 to show that $(X_t, \alpha(t))$ is positive recurrent relative to any set of the form $\mathcal{D} \times N$ where $\mathcal{D}$ is a nonempty open set of $\mathscr{C}$ and $N \ni i^*$. □

Example 11.35 In Example 11.32, if we assume further that

$$\lim_{|x|\to\infty}\inf_{i\in\mathcal{M}}\{|x|b(x,i)\}>0, \tag{11.97}$$

then it follows from (11.76) and (11.77) that $\mathbb{L}V(\phi, i) \le -\hat{C}$ for $\phi \in \mathscr{C}, i \ge 2$, for some positive constant $\hat{C}$. This combined with (11.79) yields that (11.88) holds for the switching diffusion $(X(t), \alpha(t))$ and the function $V(\phi, i)$ in Example 11.32. Thus the conclusion of Theorem 11.34 holds for the switching diffusion in Example 11.32 with the additional condition (11.97).

Theorem 11.36 *Suppose that either* (H11.6) *or* (H11.7) *holds. Let* $V(\cdot, \cdot) \in \mathbf{F}$ *such that*

$$\lim_{n\to\infty} \inf\{V(\phi, i) : |\phi(0)| \vee i \ge n\} = \infty. \tag{11.98}$$

Suppose further that there are C_1 *and* $C_2 > 0$ *such that*

$$\mathbb{L}V(\phi, i) \le -C_1 V(\phi, i) + C_2. \tag{11.99}$$

Then, there is a unique invariant probability measure μ^* *and* $\theta > 0$ *such that for any* $(\phi, i) \in \mathscr{C} \times \mathcal{M}$

$$\lim_{t\to\infty} \exp(\theta t)\|P(t, (\phi, i), \cdot) - \mu^*\|_{TV} = 0. \tag{11.100}$$

Proof In view of Dupire's functional Itô formula

$$\begin{aligned}
&\mathbb{E}_{\phi,i} \exp\{C_1(\eta_k \wedge t)\}V(X_{\eta_k\wedge t}, \alpha(\eta_k \wedge t)) \\
&= V(\phi, i) + \mathbb{E}_{\phi,i} \int_0^{\eta_k\wedge t} e^{C_1 s}[\mathbb{L}V(X_s, \alpha(s)) + C_1 V(X_s, \alpha(s))]ds \\
&\le V(\phi, i) + C_2\mathbb{E}_{\phi,i} \int_0^{\eta_k\wedge t} e^{C_1 s}ds \\
&\le V(\phi, i) + C_1^{-1}C_2 e^{C_1 t}.
\end{aligned} \tag{11.101}$$

Letting $k \to \infty$, we obtain

$$\mathbb{E}_{\phi,i} V(X_t, \alpha(t)) \le e^{-C_1 t} V(\phi, i) + C_1^{-1}C_2. \tag{11.102}$$

Let $\gamma_1 = e^{-C_1}$ and $\gamma_2 \in (\gamma_1, 1)$. It follows from (11.102) and (11.91) that

$$\begin{aligned}
\mathbb{E}V(X_1, \alpha(1)) &\le \gamma_1 V(\phi, i) + C_1^{-1}C_2 \\
&= \gamma_2 V(\phi, i) + \left[C_1^{-1}C_2 - (\gamma_2 - \gamma_1)V(\phi, i)\right] \\
&\le \gamma_2 V(\phi, i) + [C_1^{-1}C_2]I_{\{V(\phi,i)\le H'\}},
\end{aligned} \tag{11.103}$$

where $H' = C_1^{-1}C_2(\gamma_2 - \gamma_1)^{-1}$. Similar to (11.96), the set $\{(\phi, i) \in \mathscr{C} \times \mathcal{M} : V(\phi, i) \le H'\}$ is petite, which combined with (11.103) implies the existence of $\gamma_3 \in (0,1)$ such that $\lim_{n\to\infty} \gamma_3^n \|P(n,(\phi,i),\cdot) - \mu^*\|_{TV} = 0$ due to Theorem A.1. Then (11.100) follows from (11.103) and the decreasing property of $\|P(t,(\phi,i),\cdot) - \mu^*\|_{TV}$ in t. □

Example 11.37 Suppose that

$$\begin{aligned}
&q_{12}(\phi) = 1, q_{1j}(\phi) = 0 \text{ for } j \ge 3;\\
&q_{i,1}(\phi) = 2\int_{-r}^{0} |\phi(s)|ds, q_{i,i+1}(\phi) = i\int_{-r}^{0} |\phi(s)|ds \text{ for } i \ge 2\\
&q_{ij}(\phi) = 0 \text{ for } i \ge 2, j \notin \{1, i, i+1\}.
\end{aligned}$$

and that the equation for the diffusion part is

$$dX(t) = \sigma(X(t),\alpha(t))dW(t) - b(X(t),\alpha(t))X(t)dt$$

where $\sigma(x,i), b(x,i)$ are locally Lipchitz in x and uniformly bounded in $K \times \mathcal{M}$ for each compact set $K \in \mathbb{R}$. Let $V(\phi,i)$ be defined as in Example 11.32. Similar to Examples 11.32 and 11.35, under the assumption that $b := \inf_{(x,i)\in\mathbb{R}\times\mathcal{M}}\{b(x,i)\} > 0$, one can show that (11.99) holds in this example with this function V. Thus, the conclusion of Theorem 11.31 holds for this example. if there exists $i^* \in \mathcal{M}$ such that $\sigma(x,i^*) \neq 0$ for any $x \in \mathbb{R}$.

Example 11.38 Let

$$\begin{aligned}
&q_{12}(\phi) = 1, q_{1j}(\phi) = 0 \text{ for } j \ge 3;\\
&q_{i,i-1}(\phi) = C_i + 2|\phi(0)|, q_{i,i+1} = C_i + |\phi(-r)| \text{ for } i \ge 2,\ C_i \ge 0;\\
&q_{ij}(\phi) = 0 \text{ for } i \ge 2, j \notin \{i-1, i, i+1\}.
\end{aligned}$$

Consider the general equation for diffusion (11.5), where $\sigma(x,i), b(x,i)$ are locally Lipchitz in x in $K \times \mathcal{M}$ for each compact set $K \in \mathbb{R}$. Suppose there is a function $U(x) : \mathbb{R}^n \mapsto \mathbb{R}_+$ satisfying

- $U(x)$ is twice continuously differentiable in x.
- $\lim_{|x|\to\infty} U(x,i) = \infty$.
- There are positive constants C_1, C_2, H such that

$$\mathbb{L}_i U(x) \le -C_1 U(x) + C_2. \tag{11.104}$$

Let $V(x,i) = U(x) + i + \int_0^t \exp\{\frac{\ln 2}{r}(s+r)\}ds$. By Remark 11.23,

$$\mathbb{L}V(\phi,i) = \begin{cases} \mathbb{L}_i U(x) - i - \dfrac{\ln 2}{r}\displaystyle\int_0^t \exp\left\{\dfrac{\ln 2}{r}(s+r)\right\} ds + 2 & \text{if } i > 1\\ \mathbb{L}_i U(x) - \dfrac{\ln 2}{r}\displaystyle\int_0^t \exp\left\{\dfrac{\ln 2}{r}(s+r)\right\} ds + 1 & \text{if } i = 1 \end{cases} \tag{11.105}$$

As a consequence of (11.104) and (11.105), there are C_3 and $C_4 > 0$ such that

$$\mathbb{L}V(\phi, i) \le -C_3 V(\phi, i) + C_4 \text{ for } (\phi, i) \in \mathscr{C} \times \mathcal{M}.$$

Thus, the conclusion of Theorem 11.31 holds for this example if there exists $i^* \in \mathcal{M}$ such that $A(x, i^*)$ is elliptic.

11.5.3 Characterization of Recurrent Past-Independent Switching Diffusions

This subsection, as a generalization of Section 3.3, is devoted mainly to characterizing the recurrence of $(X(t), \alpha(t))$ using the corresponding system of partial differential equations when the switching intensities of $\alpha(t)$ depends only on the current state of $X(t)$, that is, $q_{ij}(\cdot), i, j \in \mathcal{M}$ are functions on $\mathbb{R}^n$ rather than on $\mathscr{C}$. To simplify the presentation, throughout this section, we set $q_{ii}(x) = 0$ for $(x, i) \in \mathbb{R}^n \times \mathcal{M}$. Thus, $q_i(x) = -\sum_{j\in\mathcal{M}} q_{ij}(x)$. In this section, we use the following assumption.

(H11.8) Suppose that

1. either (H11.3) or (H11.4) holds with $\phi \in \mathscr{C}$ replaced by $x \in \mathbb{R}^n$;
2. for each $i \in \mathcal{M}$, $A(x, i)$ is uniformly elliptic in each compact set;
3. for any $x \in \mathbb{R}^n$, there are $\hat{q} = \hat{q}(x) > 0$ and $n_{\hat{q}} = n_{\hat{q}}(x) > 0$ such that

$$\sum_{j\le n_{\hat{p}}} q_{ij}(x) \ge \hat{q} \text{ for any } i > n_{\hat{q}}. \tag{11.106}$$

Remark 11.39 We note the following facts:

- Part 3 of (H11.8) stems from a familiar condition for uniform ergodicity of the Markov chain having a countable state space. In other word, if (11.106) holds, for each $x \in \mathbb{R}^n$, the Markov chain $\hat{\alpha}^x(t)$ with generator $Q(x)$ has a property that $\sup_{i\in\mathcal{M}} \mathbb{E}_i \varsigma < \infty$ where ς is the first time the process $\hat{\alpha}^x(t)$ jumps to $\{1, \ldots, n_0\}$ and $\mathbb{E}_i$ is the expectation with condition $\hat{\alpha}^x(0) = i$.
- Since $q_{ij}(x)$ is continuous in $x \in \mathbb{R}^n$, with the use of the Heine-Borel covering theorem, it is easy to show that for any bounded set $D \in \mathbb{R}^n$, there is $\varepsilon_0 = \varepsilon_0(D)$ and $n_0 = n_0(D)$ such that

$$\sum_{j\le n_0} q_{ij}(x) \ge \varepsilon_0 \text{ for any } i > n_0, x \in D. \tag{11.107}$$

For an open set $D \subset \mathbb{R}^n$, define $\widetilde{\tau}_D = \inf\{t \ge 0 : X(t) \notin D\}$, and $W^{2,p}_{loc}(D)$ is the set of functions $u : \bar{D} \mapsto \mathbb{R}$ that has generalized derivatives $D^\beta u$ for any multiple-index $\beta = (\beta_1, \ldots, \beta_n)$ with $|\beta| = \sum \beta_i \le 2$ satisfying $D^\beta u \in L^p_{loc}(D)$ if $|\beta| \le 2$. Let $\mathbb{H}^p(D)$ be the set of functions $u(x, i)$ in $\bar{D} \times \mathcal{M}$ satisfying that

- For each $i \in \mathcal{M}$, $u(\cdot, i) \in W^{2,p}_{loc}(D)$ and $u(\cdot, i)$ is continuous in the closure $\bar{D}$ of D.
- For any compact set $K \subset \mathbb{R}^n$, $\sup_{(x,i)\in(K\cap\bar{D})\times\mathcal{M}}\{u(x,i)\} < \infty$.

Let $\mathcal{L}_i$ be defined as in Remark 11.7. We state the two main results of this section.

Theorem 11.40 *Suppose that* (H11.8) *holds. Let D_1 be a bounded open set of $\mathbb{R}^n$ with $\partial D_1 \in C^2$ and $D = \bar{D}_1^c$ be the complement of $\bar{D}_1$. Let $p > n$. The process $X(t)$ is recurrent relative to D_1, if and only if the Dirichlet problem*

$$\begin{cases} \mathcal{L}_i u(x,i) - q_i(x)u(x,i) + \sum_{j\in\mathcal{M}} q_{ij}(x)u(x,j) = 0 \text{ in } D\times\mathcal{M} \\ u(x,i) = f(x,i) \text{ on } \partial D\times\mathcal{M}. \end{cases} \tag{11.108}$$

has a unique solution in $\mathbb{H}^p(D)$ given that $f(x,i)$ is continuous in $x \in \partial D$ and bounded in $\partial\mathcal{D}\times\mathcal{M}$.

Theorem 11.41 *Suppose that* (H11.8) *holds. Let D_1 be a bounded open set of $\mathbb{R}^n$ with boundary $\partial D_1 \in C^2$ and $D = D_1^c$ be its complement. Let $p > n$. Suppose further that for each compact set $K \in \mathbb{R}^n$, the function $u(x,i) = \mathbb{E}_{x,i}\widetilde{\tau}_D$ is bounded in $K\times\mathcal{M}$. Then $\{u(x,i)\} \in \mathbb{H}^p(D)$, $p > n$ is a strong solution to*

$$\begin{cases} \mathcal{L}_i u(x,i) - q_i(x)u(x,i) + \sum_{j\in\mathcal{M}} q_{ij}(x)u(x,j) = -1 \text{ in } D\times\mathcal{M} \\ u(x,i) = 0 \text{ on } \partial D\times\mathcal{M}. \end{cases} \tag{11.109}$$

The solution is unique in $\mathbb{H}^p(D)$ and $p > n$.

Lemma 11.42 *Let $D \subset \mathbb{R}^n$ be an open bounded set with $\partial D \in C^2$. The Dirichlet problem*

$$\begin{cases} \mathcal{L}_i u(x,i) - q_i(x)u(x,i) + \sum_{j\in\mathcal{M}} q_{ij}(x)u(x,j) = f(x,i) \text{ in } D\times\mathcal{M} \\ u(x,i)\big|_{\partial D} = \phi(x,i) \text{ on } \partial D\times\mathcal{M} \end{cases} \tag{11.110}$$

has a unique strong solution $\{u(x,i)\} \in \mathbb{H}^p(D)$, $p > n$ if $\phi(x,i)$ and $f(x,i)$ are continuous and bounded on $\partial D\times\mathcal{M}$ and $D\times\mathcal{M}$, respectively.

Proof The proof is motivated by that of [79, Proposition A]. However, because infinitely many equations are involved, significant modification is needed. Let $p > n$ and

$$\hat{M} = \sup_{(x,i)\in\partial D\times\mathcal{M}}\{|\phi(x,i)|\} + \sup_{(x,i)\in D\times\mathcal{M}}\{|f(x,i)|\} < \infty. \tag{11.111}$$

By [279, Theorem 9.1.5], for each $i \in \mathcal{M}$, there exists a strong solution $u_0(x,i) \in W^{2,p}_{loc}(D)\cap C(\bar{D})$ to

$$\begin{cases} \mathcal{L}_i u_0(x,i) - q_i(x)u(x,i) = 0 \text{ in } D\times\mathcal{M} \\ u_0(x,i)\big|_{\partial D} = \phi(x,i) \text{ on } \partial D\times\mathcal{M}. \end{cases} \tag{11.112}$$

Let $Y^{x,i}(t)$ be the solution to

$$dY(t) = b(Y(t), i)dt + \sigma(Y(t), i)dW(t), \ t \geq 0, \tag{11.113}$$

with initial condition x and $\tau_D^{x,i} = \inf\{t \geq 0 : Y^{x,i}(t) \notin \mathcal{D}\}$. In view of the Feyman-Kac formula for diffusion processes

$$\begin{aligned} u_0(x,i) =& \mathbb{E}_{x,i}\left[\phi(Y(\tau_D)), i)\exp\left(-\int_0^{\tau_D} q_i(Y(s))ds\right)\right] \\ &- \mathbb{E}_{x,i}\int_0^{\tau_D}\exp\left(-\int_0^t q_i(Y(s))ds\right) f(Y(t),i)dt. \end{aligned} \tag{11.114}$$

Note that in (11.114) and what follows, we drop the superscripts x and i in $Y^{x,i}$ and $\tau_D^{x,i}$ whenever the expectation $\mathbb{E}_{x,i}$ or probability $\mathbb{P}_{x,i}$ is used. By part (2) of (H11.8), $\sup_{x\in\mathcal{D}} \mathbb{E}_{x,i}\tau_D < \infty$ for any $i \in \mathcal{M}$. In view of (11.114), we have

$$|u_0(x,i)| \leq \sup_{x\in\partial\mathcal{D}}\{|\phi(x,i)|\} + \sup_{x\in\mathcal{D}}\{|f(x,i)|\}\sup_{x\in\mathcal{D}}\{\mathbb{E}_{x,i}\tau_D\}. \tag{11.115}$$

Let n_0 and ε_0 satisfy (11.107). In particular, for $i > n_0$, $q_i(x) \geq \varepsilon_0 > 0$ in D, we can have the following estimate from (11.114):

$$\begin{aligned} |u_0(x,i)| \leq& \mathbb{E}_{x,i}|\phi_i(Y(\tau_D))| + \mathbb{E}_{x,i}\int_0^{\tau_D}\exp\left(-\int_0^t q_i(Y(s))ds\right)|f(Y(t),i)|dt \\ \leq& \sup_{x\in\partial\mathcal{D}}\{|\phi(x,i)|\} + \sup_{x\in\mathcal{D}}\{|f(x,i)|\}\mathbb{E}_{x,i}\int_0^{\tau_D}\exp(-\varepsilon_0 t)dt \\ \leq& \sup_{x\in\partial\mathcal{D}}\{|\phi(x,i)|\} + \sup_{x\in\mathcal{D}}\{|f(x,i)|\}\mathbb{E}_{x,i}\int_0^{\tau_D}\exp(-\varepsilon_0 t)dt \\ \leq& \sup_{x\in\partial\mathcal{D}}\{|\phi(x,i)|\} + \varepsilon_0^{-1}\sup_{x\in\mathcal{D}}\{|f(x,i)|\}. \end{aligned} \tag{11.116}$$

As a result of (11.111), (11.115), and (11.116),

$$\begin{aligned} M_0 :=& \sup_{(x,i)\in D\times\mathcal{M}}|u_0(x,i)| \leq \sup_{x\in\partial\mathcal{D}, i\in\mathcal{M}}\{|\phi(x,i)|\} + \varepsilon_0^{-1}\sup_{x\in\mathcal{D}, i>n_0}\{|f(x,i)|\} \\ &+ \sup_{x\in\mathcal{D}, i\leq n_0}\{|f(x,i)|\}\sup_{x\in\mathcal{D}, i\leq n_0}\{\mathbb{E}_{x,i}\tau_D\} < \infty. \end{aligned} \tag{11.117}$$

Since $u_0(x,i)$ is continuous in $\bar{\mathcal{D}} \times \mathcal{M}$ and $q_i(x) = \sum_{j\in\mathcal{M}} q_{ij}(x)$ is continuous and bounded in $\bar{\mathcal{D}} \times \mathcal{M}$, it is easy to show that $\sum_{j\in\mathcal{M}} q_{ij}u_0(x,j)$ is continuous in $\bar{\mathcal{D}} \times \mathcal{M}$ and

$$\sup_{(x,i)\in D\times\mathcal{M}} \left| \sum_{j\in\mathcal{M}} q_{ij} u_0(x,j) \right| := \hat{M}_0 < \infty. \tag{11.118}$$

Thus, for each $i \in \mathcal{M}$, there exists a strong solution $u_1(x,i) \in W^{2,p}_{loc}(D) \cap C(\bar{D})$ to

$$\begin{cases} \mathcal{L}_i u_1(x,i) - q_i(x) u_1(x,i) = -\sum_{j\in\mathcal{M}} q_{ij}(x) u_0(x,j) \text{ in } D \times \mathcal{M} \\ u_1(x,i)\big|_{\partial D} = \phi(x,i) \text{ on } \partial D \times \mathcal{M}. \end{cases} \tag{11.119}$$

owing to [279, Theorem 9.1.5]. Similar to (11.117), we can use (11.118) to obtain that

$$\sup_{(x,i)\in D\times\mathcal{M}} |u_1(x,i)| := M_1 < \infty. \tag{11.120}$$

Continuing this way, we can define recursively $\{u_{m+1}(x,i)\} \in \mathbb{H}^{2,p}(D)$, the strong solution to

$$\begin{cases} \mathcal{L}_i u_{m+1}(x,i) - q_i(x) u_{m+1}(x,i) = -\sum_{j\in\mathcal{M}} q_{ij}(x) u_m(x,j) \text{ in } D \times \mathcal{M} \\ u_{m+1}(x,i)\big|_{\partial D} = \phi(x,i) \text{ on } \partial D \times \mathcal{M}. \end{cases} \tag{11.121}$$

By the Feyman-Kac formula,

$$\begin{aligned} u_{m+1}(x,i) =& \mathbb{E}_{x,i}\left[\phi_i(Y(\tau_D)))\exp\left(-\int_0^{\tau_D} q_i(Y(s))ds\right)\right] \\ &+ \mathbb{E}_{x,i}\int_0^{\tau_D} \exp\left(-\int_0^t q_i(Y(s))ds\right) \sum_{j\in\mathcal{M}} q_{ij}(Y(t)) u_m(Y(t),j)dt \\ &- \mathbb{E}_{x,i}\int_0^{\tau_D} \exp\left(-\int_0^t q_i(Y(s))ds\right) f(Y(t),i)dt. \end{aligned} \tag{11.122}$$

Let $\Delta_m(x,i) = u_{m+1}(x,i) - u_m(x,i)$ and $\Delta^i_m = \sup\{|\Delta_{m+1}(x,i)| : x \in \mathcal{D}\}$. It follows from (11.122) that

$$\begin{aligned} |\Delta_{m+1}(x,i)| =& \mathbb{E}_{x,i}\int_0^{\tau_D} \exp\left(-\int_0^t q_i(Y(s))ds\right) \sum_{j\in\mathcal{M}} q_{ij}(Y(t))|\Delta_m(Y(t),j)|dt \\ \le& \sup_{\{i\in\mathcal{M}\}}\{\Delta^i_m\}\mathbb{E}_{x,i}\int_0^{\tau_D} \exp\left(-\int_0^t q_i(Y(s))ds\right) q_i(Y(t))dt \\ =& \sup_{\{i\in\mathcal{M}\}}\{\Delta^i_m\}\mathbb{E}_{x,i}\left[1 - \exp\left(-\int_0^{\tau_D} q_i(Y(s))ds\right)\right]. \end{aligned} \tag{11.123}$$

Let

$$p := \max_{\{i \le n_0\}} \mathbb{E}_{x,i}\left[1 - \exp\left(-\int_0^{\tau_D} q_i(Y(s))ds\right)\right] < 1.$$

We have from (11.123) that

$$\max_{\{i \le n_0\}} \{\Delta^i_{m+1}\} \le p \sup_{\{i \in \mathcal{M}\}} \{\Delta^i_m\}. \tag{11.124}$$

It also follows from (11.123) that

$$\sup_{\{i \in \mathcal{M}\}} \{\Delta^i_{m+1}\} \le \sup_{\{i \in \mathcal{M}\}} \{\Delta^i_m\}. \tag{11.125}$$

For $i > n_0$, using (11.123) again and then using (11.124) and (11.125), we have

$$\begin{aligned}
|\Delta_{m+2}(x,i)| \le & \max_{\{i \le n_0\}} \{\Delta^i_{m+1}\} \mathbb{E}_{x,i} \int_0^{\tau_D} \exp\left(-\int_0^t q_i(Y(s))ds\right) \sum_{j \le n_0} q_{ij}(Y(t))dt \\
& + \sup_{\{i > n_0\}} \{\Delta^i_{m+1}\} \mathbb{E}_{x,i} \int_0^{\tau_D} \exp\left(-\int_0^t q_i(Y(s))ds\right) \sum_{j > n_0} q_{ij}(Y(t))dt \\
\le & p \sup_{\{i \in \mathcal{M}\}} \{\Delta^i_{m+1}\} \mathbb{E}_{x,i} \int_0^{\tau_D} \exp\left(-\int_0^t q_i(Y(s))ds\right) \sum_{j \le n_0} q_{ij}(Y(t))dt \\
& + \sup_{\{i \in \mathcal{M}\}} \{\Delta^i_{m+1}\} \mathbb{E}_{x,i} \int_0^{\tau_D} \exp\left(-\int_0^t q_i(Y(s))ds\right) \sum_{j > n_0} q_{ij}(Y(t))dt \\
\le & p \sup_{\{i \in \mathcal{M}\}} \{\Delta^i_{m+1}\} \mathbb{E}_{x,i} \int_0^{\tau_D} \exp\left(-\int_0^t q_i(Y(s))ds\right) \sum_{j \in \mathcal{M}} q_{ij}(Y(t))dt \\
& + (1-p) \sup_{\{i \in \mathcal{M}\}} \{\Delta^i_{m+1}\} \mathbb{E}_{x,i} \int_0^{\tau_D} \exp\left(-\int_0^t q_i(Y(s))ds\right) \\
& \times \sum_{j > n_0} q_{ij}(Y(t))dt.
\end{aligned} \tag{11.126}$$

Let

$$M_D = \sup_{(x,i) \in \mathcal{D} \times N} q_i(x). \tag{11.127}$$

Note that

$$\frac{\sum_{j>n_0} q_{ij}(x)}{q_i(x)} = 1 - \frac{\sum_{j\le n_0} q_{ij}(x)}{q_i(x)} \le 1 - \frac{\varepsilon_0}{M_D} := \varepsilon_1 \text{ for } i > n_0,$$

which implies for $i > n_0$ that

$$\begin{aligned}
&\mathbb{E}_{x,i} \int_0^{\tau_D} \exp\left(-\int_0^t q_i(Y(s))ds\right) \sum_{j>n_0} q_{ij}(Y(t))dt \\
\le& \varepsilon_1 \mathbb{E}_{x,i} \int_0^{\tau_D} \exp\left(-\int_0^t q_i(Y(s))ds\right) q_i(Y(t))dt \le \varepsilon_1.
\end{aligned} \tag{11.128}$$

In view of (11.126) and (11.128), we have

$$\sup_{\{i>n_0\}} \Delta^i_{m+2} \le [p + (1-p)\varepsilon_1] \sup_{\{i\in\mathcal{M}\}} \{\Delta^i_{m+1}\}. \tag{11.129}$$

By (11.124) and (11.125),

$$\sup_{\{i\le n_0\}} \{\Delta^i_{m+2}\} \le p \max_{\{i\in\mathcal{M}\}} \{\Delta^i_{m+1}\} \le \max_{\{i\in\mathcal{M}\}} \{\Delta^i_m\}. \tag{11.130}$$

By (11.130) and (11.129),

$$\sup_{\{i\in\mathcal{M}\}} \{\Delta^i_{m+2}\} \le [p + (1-p)\varepsilon_1] \max_{\{i\in\mathcal{M}\}} \{\Delta^i_m\}. \tag{11.131}$$

In view of (11.117) and (11.118), $\sup_{i\in\mathcal{M}}\{\Delta^i_1\} \le M_0 + M_1 < \infty$. Since $p + (1-p)\varepsilon_1 < 1$, it follows from (11.125) and (11.131) that the series $\sum_{m=1}^\infty \sup_{\{i\in\mathcal{M}\}}\{\Delta^i_{m+2}\}$ is convergent. Thus $u_m(x,i)$ converges uniformly in (x,i) to a function $u(x,i)$. For each $i \in \mathcal{M}$, since $q_i(x) = \sum_j q_{ij}(x)$ is continuous, the convergence $\lim_{k\to\infty} \sum_{j<k} q_{ij}(x) = q_i(x)$ is uniform. Thus, it is not difficult to show that as $m \to \infty$, $\sum_j q_{ij}(x)u_m(x,j)$ converges uniformly to $\sum_j q_{ij}(x)u(x,j)$, (which is also continuous in x). Using this uniform convergence, passing the limit in (11.122) we have

$$\begin{aligned}
u(x,i) =& \mathbb{E}_{x,i}\left[\phi_i(Y(\tau_D))) \exp\left(-\int_0^{\tau_D} q_i(Y(s))ds\right)\right] \\
&+ \mathbb{E}_{x,i} \int_0^{\tau_D} \exp\left(-\int_0^t q_i(Y(s))ds\right) \sum_{j\in\mathcal{M}} q_{ij} u(Y(t),j)dt \\
&- \mathbb{E}_{x,i} \int_0^{\tau_D} \exp\left(-\int_0^t q_i(Y(s))ds\right) f(Y(t),i)dt.
\end{aligned} \tag{11.132}$$

Since $\sum_j q_{ij}(x)u(x,j)$ is continuous in x for each $i \in \mathcal{M}$, the representation (11.132) shows that $u(x,i)$ satisfies

$$\mathcal{L}_i u(x,i) - q_i(x)u(x,i) = f(x,i) - \sum_{j\in\mathcal{M}} q_{ij}(x)u(x,j) \text{ in } D \times \mathcal{M}$$

Since $u_m(x,i) = \phi(x,i)$ on $\partial D \times \mathcal{M}$ for all $m \in \mathcal{M}$, we have $u(x,i) = \phi(x,i)$ on $\partial D \times \mathcal{M}$. The existence of solutions is therefore proved. To prove the uniqueness, it suffices to consider the uniqueness in $\mathbb{H}^p(D)$ of the system

$$\begin{cases} \mathcal{L}_i v(x,i) - q_i(x)v(x,i) + \sum_{j=1}^{\infty} q_{ij}(x)v(x,j) = 0 \text{ in } D \times \mathcal{M} \\ v(x,i)\big|_{\partial D} = 0 \text{ on } \partial D \times \mathcal{M}. \end{cases} \tag{11.133}$$

Let $\{v(x,i)\} \in \mathbb{H}^p(D)$ be a solution of (11.133). Then we have

$$v(x,i) = -\mathbb{E}_{x,i} \int_0^{\tau_D} \exp\left(-\int_0^t q_i(Y(s))ds\right) \sum_{j\in\mathcal{M}} q_{ij} v(Y(s),j)ds. \tag{11.134}$$

Similar to (11.124), it follows from (11.134) that

$$\sup_{i\le n_0, x\in D} \{|v(x,i)|\} \le p \sup_{i\in\mathcal{M}, x\in D} \{|v(x,i)|\}.$$

Similar to (11.129), the above inequality and (11.134) imply that

$$\sup_{i\in\mathcal{M}, x\in D} \{|v(x,i)|\} \le [p + (1-p)\varepsilon_1] \sup_{i\in\mathcal{M}, x\in D} \{|v(x,i)|\}.$$

Thus $\sup_{i\in\mathcal{M}, x\in D}\{|v(x,i)|\} = 0$, that is, (11.133) has a unique solution. □

Lemma 11.43 *Let D be an open and bounded set of $\mathbb{R}^n$. Let $\xi_0 = 0$ and $\xi_k = \inf\{t \ge 0 : \alpha(t) \ne \alpha(\xi_{k-1})\}, k \in \mathcal{M}$. Let $f(x,i)$ and $g(x,i)$ are bounded and measurable functions on $D \times \mathcal{M}$ and $\partial D \times \mathcal{M}$, respectively. Then*

$$\begin{aligned} &\mathbb{E}_{x,i} 1_{\{\xi_1 \le \widetilde{\tau}_D\}} f(X(\xi_1), \alpha(\xi_1)) + \mathbb{E}_x 1_{\{\xi_1 > \widetilde{\tau}_D\}} g(X(\widetilde{\tau}_D), i) \\ =&\mathbb{E}_{x,i} \int_0^{\tau_D} q_{ij}(Y(t)) f(Y(t), j) \exp\left(-\int_0^t q_i(Y(s))ds\right) \\ &+ \mathbb{E}_{x,i} g(Y(\tau_D), i) \exp\left(-\int_0^{\tau_D} q_i(Y(t))dt\right). \end{aligned} \tag{11.135}$$

Proof Define $\beta^{x,i}(t) = i + \int_0^t \int_{\mathbb{R}} h(Y_t^{x,i}, \beta^{x,i}(t-), z)\mathfrak{p}(dt, dz)$. Let $\lambda_1^{x,i} = \inf\{t \ge 0 : \beta^{x,i}(t) \ne i\}$. We have that

$$(X^{x,i}(t), \alpha^{x,i}(t)) = (Y^{x,i}(t), \beta^{x,i}(t)) \text{ up to } \lambda_1^{x,i} = \xi_1^{x,i}, \tag{11.136}$$

where $(X^{x,i}(t), \alpha^{x,i}(t))$ is the solution to (11.7) with initial value (x, i) and $\xi_1^{x,i}$ is the first moment of jump for $\alpha^{x,i}(t)$. Thus,

$$\mathbb{P}_{x,i}\{\xi_1 \wedge \widetilde{\tau}_D < \infty\} = \mathbb{P}_{x,i}\{\lambda_1 \le \tau_D\} \ge \mathbb{P}_{x,i}\{\tau_D < \infty\} = 1. \tag{11.137}$$

In view of Lemma 11.12,

$$\begin{aligned}
&\mathbb{E}_{x,i} 1_{\{\lambda_1 \le \tau_D\}} f(Y(\lambda_1), \beta(\lambda_1)) + \mathbb{E}_{x,i} 1_{\{\lambda_1 > \tau_D\}} g(Y(\tau_D), i) \\
&= \mathbb{E}_{x,i} \int_0^{\tau_D} q_{ij}(Y(t)) f(Y(t), j) \exp\left(-\int_0^t q_i(Y(s))ds\right) \\
&\quad + \mathbb{E}_{x,i} g(Y(\tau_D), i) \exp\left(-\int_0^{\tau_D} q_i(Y(t))dt\right).
\end{aligned} \tag{11.138}$$

Combining (11.136) and (11.138), we obtain (11.135). □

Lemma 11.44 *Let D be an open bounded set in $\mathbb{R}^n$. For any $\varepsilon > 0$, there is an $n_2 = n_2(\varepsilon) > 0$ such that $\mathbb{P}_{x,i}\{\xi_{n_2} \le \widetilde{\tau}_D\} < \varepsilon$ for any $(x, i) \in \mathcal{B} \times \mathcal{M}$. It leads to $\mathbb{P}_{x,i}\{\widetilde{\tau}_D < \infty\} = 1$. Moreover, for any $k > 0$, there is a $T > 0$ such that $\mathbb{P}_{x,i}\{\xi_k \wedge \widetilde{\tau}_D > T\} < \varepsilon$.*

Proof For each $i \in \mathcal{M}$, we have that $p_{i,D} := \sup_{x \in \mathcal{D}} \mathbb{E}_{x,i}\tau_D < \infty$. By Lemma 11.43, with M_D defined as in (11.127), we have

$$\begin{aligned}
\mathbb{P}_{x,i}\{\xi_1 > \widetilde{\tau}_D\} &= \mathbb{E}_{x,i} \exp\left(-\int_0^{\tau_D} q_i(Y(t))dt\right) \ge \mathbb{E}_{x,i} \exp\left(-M_D \tau_D\right) \\
&\ge \exp\left(-M_D \mathbb{E}_{x,i}\tau_D\right) \ge \exp\left(-M_D p_{i,D}\right).
\end{aligned} \tag{11.139}$$

Let $\tilde{p} := \min_{\{i \le n_0\}}\{\exp\left(-M_D p_{i,D}\right)\}$. By (11.107), for $i > n_0$, $\dfrac{\sum_{j \le n_0} q_{ij}(x)}{q_i(x)} \ge \dfrac{\varepsilon_0}{M_D} > 0, x \in D$. Applying Lemma 11.43 again, we have

$$\begin{aligned}
\mathbb{P}_{x,i}\{\xi_1 \le \widetilde{\tau}_D, \alpha(\xi_1) \le n_0\} &= \mathbb{E}_{x,i} \int_0^{\tau_B} \sum_{j \le n_0} q_{ij}(Y(t)) \exp\left(-\int_0^t q_i(Y(s))ds\right) \\
&\ge \frac{\varepsilon_0}{M_D} \mathbb{E}_{x,i} \int_0^{\tau_B} q_i(Y(t)) \exp\left(-\int_0^t q_i(Y(s))ds\right) \\
&= \frac{\varepsilon_0}{M_D} \mathbb{P}_{x,i}\{\xi_1 \le \widetilde{\tau}_D\}.
\end{aligned} \tag{11.140}$$

By the strong Markov property, (11.139), and (11.140), we have for $i > n_0$ that

$$
\begin{aligned}
\mathbb{P}_{x,i}\{\xi_1 \le \widetilde{\tau}_D, \xi_2 > \widetilde{\tau}_D\} \ge & \mathbb{P}_{x,i}\{\xi_2 < \widetilde{\tau}_D,\ \xi_1 \ge \widetilde{\tau}_D,\ \alpha(\xi_1) \le n_0\} \\
\ge & \mathbb{P}_{x,i}\{\xi_1 \le \widetilde{\tau}_D,\ \alpha(\xi_1) \le n_0\}\left[\inf_{\{y\in\mathcal{D}, j\le n_0\}} \mathbb{P}_{y,j}\{\xi_1 < \widetilde{\tau}_D\}\right] \\
\ge & \frac{\tilde{p}\varepsilon_0}{M_D}\mathbb{P}_{x,i}\{\xi_1 \le \widetilde{\tau}_D\}.
\end{aligned} \tag{11.141}
$$

Since $\tilde{p} < 1$ and $\dfrac{\varepsilon_0}{M_D} \ge 1$, we have from (11.140) that for $x \in \mathcal{D}, i > n_0$,

$$
\begin{aligned}
\mathbb{P}_{x,i}\{\xi_2 > \widetilde{\tau}_D\} \ge & \mathbb{P}\{\xi_1 > \widetilde{\tau}_D\} + \mathbb{P}_{x,i}\{\xi_1 \le \widetilde{\tau}_D, \xi_2 > \widetilde{\tau}_D\} \\
\ge & \mathbb{P}\{\xi_1 > \widetilde{\tau}_D\} + \frac{\tilde{p}\varepsilon_0}{M_D}\mathbb{P}_{x,i}\{\xi_1 \le \widetilde{\tau}_D\} \ge \frac{\tilde{p}\varepsilon_0}{M_D}.
\end{aligned} \tag{11.142}
$$

In light of (11.139),

$$
\mathbb{P}_{x,i}\{\xi_2 > \widetilde{\tau}_D\} \ge \mathbb{P}\{\xi_1 > \widetilde{\tau}_D\} \ge \tilde{p} \text{ for } x \in D, i \le n_0. \tag{11.143}
$$

Thus, for any $x \in D$ and $i \in \mathcal{M}$, we have

$$
\mathbb{P}_{x,i}\{\xi_2 > \widetilde{\tau}_D\} \ge \frac{\tilde{p}\varepsilon_0}{M_D}. \tag{11.144}
$$

Using the strong Markov property, we have from (11.144) that

$$
\mathbb{P}_{x,i}\{\xi_{2k} \le \widetilde{\tau}_D\} \le \left(1 - \frac{\tilde{p}\varepsilon_0}{M_D}\right)^k. \tag{11.145}
$$

By letting $n_2 = 2k_2 + 1$ with k_2 being sufficiently large so that $\left(1 - \dfrac{\tilde{p}\varepsilon_0}{M_D}\right)^{k_2} < \varepsilon$, we complete the proof for the first part of this lemma.

To prove the second part, note that $\mathbb{E}_{x,i}\tau_D \le p_{i,D} < \infty$, thus for any $\varepsilon' > 0$, there is $T_1 > 0$ such that

$$
\begin{aligned}
\mathbb{P}_{x,i}\{\xi_1 \wedge \widetilde{\tau}_D \le T_1\} = & \mathbb{P}_{x,i}\{\lambda_1 \wedge \tau_D \le T_1\} \\
\ge & \mathbb{P}_{x,i}\{\tau_D \le T_1\} > 1 - \varepsilon' \text{ for all } x \in \mathcal{D}, i \le n_0.
\end{aligned} \tag{11.146}
$$

For $i > n_0$, we have

$$
\begin{aligned}
\mathbb{P}_{x,i}\{\xi_1 \wedge \widetilde{\tau}_D > T\} =&\mathbb{P}_{x,i}\{\tau_D > T, \lambda_1 > T\} \\
\geq&\mathbb{E}_{x,i}\left[I_{\{\tau_D>T\}}\int_0^T q_i(Y(t))\exp\left(-\int_0^t q_i(Y(s))ds\right)dt\right] \\
=&\mathbb{E}_{x,i}\left[I_{\{\tau_D>T\}}\exp\left(-\int_0^T q_i(Y(s))ds\right)\right] \\
\leq&\mathbb{E}_{x,i}\left[I_{\{\tau_D>T\}}\exp\left(-T\varepsilon_0\right)\right] \\
&\text{(since } q_i(x) > \varepsilon \text{ if } x \in \mathcal{D}, i > n_0).
\end{aligned}
\tag{11.147}
$$

Let $T_2 > T_1$ such that $\exp(-T_2\varepsilon_0) < \varepsilon'$. We have from (11.147) that

$$
\mathbb{P}_{x,i}\{\xi_1 \wedge \widetilde{\tau}_D \leq T_2\} > 1 - \varepsilon' \text{ for } x \in \mathcal{D}, i > n_0. \tag{11.148}
$$

Using (11.146) and (11.148),

$$
\mathbb{P}_{x,i}\{\xi_1 \wedge \widetilde{\tau}_D \leq T_2\} > 1 - \varepsilon' \text{ for } x \in \mathcal{D}, i \in \mathcal{M}. \tag{11.149}
$$

Using the strong Markov property, it is easy to show that

$$
\mathbb{P}_{x,i}\{\xi_k \wedge \widetilde{\tau}_D \leq kT_2\} > (1 - \varepsilon')^k \text{ for } x \in \mathcal{D}, i \in \mathcal{M}. \tag{11.150}
$$

By choosing ε' such that $(1-\varepsilon')^k > 1-\varepsilon$, we obtain the second part of this lemma. □

Lemma 11.45 *Let $D \in \mathbb{R}^n$ be a bounded set. For $i_0 \in \mathcal{M}$, $T > 0$, $\varepsilon > 0$, there is a $k_0 = k_0(i_0, T, \varepsilon) > 0$ such that $\mathbb{P}_{x,i_0}\{\zeta_{k_0} > T\} < \varepsilon$, $x \in D$, where $\zeta_k = \inf\{t > 0 : \alpha(t) \geq k\}$.*

Proof This lemma is a direct consequence of Lemma 11.15 and the Heine-Borel covering theorem. □

To proceed, we need the following lemma, which is a weak form of Harnack's principle.

Lemma 11.46 *Let D be an open bounded set in $\mathbb{R}^n$ with $\partial D \in C^2$ and fix $(x_0, i_0) \in \mathcal{D}\times\mathcal{M}$. Let $B \subset \bar{B} \subset D$ be a ball centered at x_0. Then for any $\varepsilon > 0$, there is a $c_0 = c_0(B, i_0, \varepsilon) > 0$ satisfying $u(x, i_0) \leq c_0 u(x_0, i_0) + \varepsilon \sup_{\{(y,i)\in\partial D\times\mathcal{M}\}}\{u(y, i)\}$, $x \in \bar{B}$, where $\{u(x, i)\} \in \mathbb{H}^p(D)$ satisfies*

$$
\mathcal{L}_i u(x, i) - q_i(x)u(x, i) + \sum_{j\in\mathcal{M}} q_{ij}(x)u(x, j) = 0 \text{ in } D \times \mathcal{M}.
$$

Proof Let $\phi(x, i) = u(x, i)|_{\partial D}$ and $\zeta_k = \inf\{t > 0 : \alpha(t) \geq k\}$. Let

$$u_k(x,i) = \begin{cases} u(x,i) & \text{if } i < k \\ 0 & \text{if } i \ge k. \end{cases}$$

By Itô's formula,

$$\begin{aligned} &\mathbb{E}_{x,i} u_k(X(\widetilde{\tau}_D \wedge \zeta_k \wedge t), \alpha(\widetilde{\tau}_D \wedge \zeta_k \wedge t)) \\ &= u_k(x,i) + \mathbb{E}_{x,i} \int_0^{\widetilde{\tau}_D \wedge \zeta_k \wedge t} \mathcal{L} u_k(X(s), \alpha(s)) ds \\ &= u_k(x,i) - \mathbb{E}_{x,i} \int_0^{\widetilde{\tau}_D \wedge \zeta_k \wedge t} \sum_{j \ge k} q_{\alpha(s),j}(X(s)) u_k(X(s), j) ds. \end{aligned} \tag{11.151}$$

Letting $k \to \infty$ and then $t \to \infty$, we obtain from the dominated convergence theorem that

$$u(x,i) = \mathbb{E}_{x,i} \phi(X(\widetilde{\tau}_D), \alpha(\widetilde{\tau}_D)). \tag{11.152}$$

As a result of Lemmas 11.44 and 11.45, there is a $k_1 = k_1(i_0, \varepsilon) \in \mathcal{M}$ such that

$$\mathbb{P}_{x,i}\{\widetilde{\tau}_D > \xi_{k_1}\} < \varepsilon. \tag{11.153}$$

In view of (11.152) and (11.153),

$$\begin{aligned} u(x,i_0) = &\mathbb{E}_{x,i_0} I_{\{\widetilde{\tau}_D < \xi_{k_1}\}} \phi(X(\widetilde{\tau}_D), \alpha(\widetilde{\tau}_D)) + \mathbb{E}_{x,i_0} I_{\{\widetilde{\tau}_D > \xi_{k_1}\}} \phi(X(\widetilde{\tau}_D), \alpha(\widetilde{\tau}_D)) \\ \le &\mathbb{E}_{x,i_0} I_{\{\widetilde{\tau}_D < \xi_{k_1}\}} \phi(X(\widetilde{\tau}_D), \alpha(\widetilde{\tau}_D)) + \varepsilon \sup_{(y,j) \in \partial D \times \mathcal{M}} \{\phi(y,j)\}. \end{aligned} \tag{11.154}$$

Let $\tilde{u}(x,i) = \mathbb{E}_{x,i} I_{\{\widetilde{\tau}_D < \xi_{k_1}\}} \phi(X(\widetilde{\tau}_D), \alpha(\widetilde{\tau}_D))$ for $i < k$. The process $\{(X(t), \alpha(t)), 0 \le t < \xi_{k_1}\}$ can be considered as a switching diffusion process on $\mathbb{R}^n \times \{1, \ldots, k_1 - 1\}$ with lifetime ξ_{k_1}. Its generator is

$$\widetilde{\mathcal{L}}_i f(x,i) = \mathcal{L}_i f(x,i) - q_i(x) u(x,i) + \sum_{j < k_1} q_{ij}(x) f(x,j),$$

for $i = 1, \ldots, k-1$. Then [46, Theorem 3.6] reveals that $\tilde{u}(x,i)$ satisfying

$$\begin{cases} \widetilde{\mathcal{L}}_i \tilde{u}(x,i) = 0 \text{ in } D \times \{1, \ldots, k_1 - 1\} \\ \tilde{u}(x,i)\big|_{\partial D} = \phi(x,i) \text{ on } \partial D \times \{1, \ldots, k_1 - 1\}. \end{cases} \tag{11.155}$$

By the Harnack principle for weakly coupled elliptic systems (see, e.g., [47]), there is a $c_0 = c_0(k_1)$ such that

$$\tilde{u}(x, i_0) \le c_0 \tilde{u}(x_0, i_0) \le c_0 u(x_0, i_0) \tag{11.156}$$

The desired result follows from (11.154) and (11.156). □

Remark 11.47 In (11.114) and (11.122), we apply the Feyman-Kac formula for functions in the class $W^{2,p}_{loc}(D) \cap C(\bar{D})$ rather than $C^2(D)$. Feyman-Kac formula is proved using Itô's formula, which is usually stated for C^2-functions. However, Itô's formula also holds for diffusion processes with functions in $W^{2,p}_{loc}(D) \cap C(\bar{D})$ when $p > n$. The proof for this claim can be found in [167, Theorem 2.10.2]. With a careful consideration, we can generalize the result for diffusion processes to switching diffusion processes in which the switching has a finite state space. Thus, (11.151) holds as long as $u(\cdot, i) \in W^{2,p}_{loc}(D) \cap C(\bar{D})$.

***Proof of Theorem* 11.41** Let $k_0 \in \mathcal{M}$ sufficiently large such that $D_1 \subset \{x \in \mathbb{R}^n : |x| < k_0\}$. For $k > k_0$, define $D_k = D \cap \{x \in \mathbb{R}^n : |x| < k\}$. By (11.152), $u_k(x, i) := \mathbb{E}_{x,i} \widetilde{\tau}_{D_k}$ satisfies the equation

$$\begin{cases} \mathcal{L}_i u_k(x, i) - q_i(x) u_k(x, i) + \sum_{j \in \mathcal{M}} q_{ij}(x) u_k(x, j) = -1 \text{ in } D_k \times \mathcal{M} \\ u_k(x, i)\big|_{\partial D} = 1 \text{ on } \partial D_k \times \mathcal{M}. \end{cases} \tag{11.157}$$

Let $B_1 \subset B_2$ be two balls in D and fix $(x_0, i_0) \in B_1 \times \mathcal{M}$ and let $k_1 > k_0$ be such that $B_2 \subset D_{k_1}$. Suppose that $\mathbb{E}_{x,i} \widetilde{\tau}_D < M$ for any $(x, i) \in B_2 \times \mathcal{M}$. Then $u_k(x, i) < M$ for $k > k_0$ and $(x, i) \in B_2 \times \mathcal{M}$. Let $v_{k,m} = u_k(x, i) - u_m(x, i)$ for $k > m > k_1$, we have

$$\mathcal{L}_i v_{k,m}(x, i) - q_i(x) v_{k,m}(x, i) + \sum_{j \in \mathcal{M}} q_{ij}(x) v_{k,m}(x, j) = 0 \text{ in } B_2 \times \mathcal{M}. \tag{11.158}$$

By Lemma 11.46, for any $\varepsilon > 0$, there is a $c_0 > 0$ such that

$$\begin{aligned} v_{k,m}(x, i_0) \le & c_0 v_{k,m}(x_0, i_0) + \varepsilon \sup\{v_{k,m}(y, j) : (y, j) \in B_2 \times \mathcal{M}\} \\ \le & c_0 v_{k,m}(x_0, i_0) + M\varepsilon \text{ for any } x \in B_1. \end{aligned} \tag{11.159}$$

For any $\varepsilon > 0$, since $u_k(x_0, i_0) = \mathbb{E}_{x_0,i_0} \widetilde{\tau}_{D_k} \to \mathbb{E}_{x_0,i_0} \widetilde{\tau}_D$ as $k \to \infty$, there exists $k_2 = k_2(\varepsilon)$ such that $c_0 v_{k,m}(x_0, i_0) = c_0[u_k(x_0, i_0) - u_m(x_0, i_0)] < \varepsilon$ for any $k > m > k_2$. In view of (11.159),

$$v_{k,m}(x, i_0) \le (M + 1)\varepsilon \text{ for any } (x, i_0) \in B_1 \times \mathcal{M}, k > m > k_2 \tag{11.160}$$

Thus, $u_k(x, i_0)$ converges uniformly in each compact subset of D. The limit $u(x, i_0)$ is therefore continuous for any i_0. Now, let $\phi(x, i) = u(x, i)|_{\partial B_2}$. Since $\phi(x, i)$ is continuous and uniformly bounded, by Lemma 11.42, for each $i \in \mathcal{M}$, there exists $\{\tilde{u}(x, i)\} \in \mathbb{H}^p(B_2)$ satisfying

$$\begin{cases} \mathcal{L}_i\tilde{u}(x,i) - q_i(x)\tilde{u}_k(x,i) + \sum_{j\in\mathcal{M}} q_{ij}(x)\tilde{u}(x,j) = -1 \text{ in } B_2 \times \mathcal{M} \\ \tilde{u}(x,i)\big|_{\partial D} = \phi(x,i) \text{ on } \partial B_2 \times \mathcal{M}. \end{cases} \tag{11.161}$$

Similar to (11.152), by applying Itô's formula we have that

$$\begin{aligned} \tilde{u}(x,i) =& \mathbb{E}_{x,i}\widetilde{\tau}_{B_2} + \mathbb{E}_{x,i}\phi(X(\widetilde{\tau}_{B_2}), \alpha(\widetilde{\tau}_{B_2})) \\ =& \mathbb{E}_{x,i}\widetilde{\tau}_{B_2} + \mathbb{E}_{x,i}\mathbb{E}_{X(\widetilde{\tau}_{B_2}),\alpha(\widetilde{\tau}_{B_2})}\widetilde{\tau}_D \\ =& \mathbb{E}_{x,i}\widetilde{\tau}_D \text{ (due to the strong Markov property)} \\ =& u(x,i). \end{aligned}$$

The proof is concluded. □

***Proof of Theorem* 11.40** After having Lemma 11.46, we adapt the proof of [156, Theorem 3.10] to obtain the desired result. First, suppose that (11.108) has a unique solution in $\mathbb{H}^p(D)$ for some $p > 0$ given that $f(x,i)$ is continuous and bounded on $D \times \mathcal{M}$. We define D_k as in the proof of Theorem 11.41 and $v_k(x,i) := \mathbb{P}_{x,i}\{X(\widetilde{\tau}_{D_k}) \in \partial D\}$. By (11.152), $v_k(x,i)$ is the strong solution to

$$\begin{cases} \mathcal{L}_i v_k(x,i) - q_i(x)v_k(x,i) + \sum_{j\in\mathcal{M}} q_{ij}(x)v_k(x,j) = 0 \text{ in } D_k \times \mathcal{M} \\ v_k(x,i)\big|_{\partial D} = 1 \text{ on } \partial D \times \mathcal{M}. \\ v_k(x,i)\big|_{\partial D} = 0 \text{ on } \{y \in \mathbb{R}^n : |y| = k\} \times \mathcal{M}. \end{cases} \tag{11.162}$$

By the definition of $v_k(x,i)$, we have that

$$\lim_{k\to\infty} v_k(x,i) = v(x,i) := \mathbb{P}_{x,i}\{\widetilde{\tau}_D < \infty\} \tag{11.163}$$

On the other hand, owing to Lemma 11.46, we can use arguments in the proof of Theorem 11.41 to show that $\{v(x,i)\} \in \mathbb{H}^p(D)$ is the solution to

$$\begin{cases} \mathcal{L}_i v(x,i) - q_i(x)v(x,i) + \sum_{j\in\mathcal{M}} q_{ij}(x)v(x,j) = 0 \text{ in } D \times \mathcal{M} \\ v(x,i)\big|_{\partial D} = 1 \text{ on } \partial D \times \mathcal{M}. \end{cases} \tag{11.164}$$

Clearly, $v(x,i) \equiv 1$ is the solution to (11.164). By the uniqueness of solutions among the class $\mathbb{H}^p(D)$, we have $\mathbb{P}_{x,i}\{\widetilde{\tau}_D < \infty\} = v(x,i) \equiv 1$.

Now, suppose that $\mathbb{P}_{x,i}\{\widetilde{\tau}_D < \infty\} \equiv 1$ and (11.108) has two solutions $\{v^{(1)}(x,i)\}$ and $\{v^{(2)}(x,i)\}$ for the same $f(x,i)$ being continuous and bounded in $\partial D \times \mathcal{M}$. Let $v^{(3)}(x,i) = v^{(1)}(x,i) - v^{(2)}(x,i)$. Then $\{v^{(3)}(x,i)\} \in \mathbb{H}^p(D)$ and satisfies

$$\begin{cases} \mathcal{L}_i v^{(3)}(x,i) - q_i(x)v^{(3)}(x,i) + \sum_{j\in\mathcal{M}} q_{ij}(x)v^{(3)}(x,j) = 0 \text{ in } D \times \mathcal{M} \\ v^{(3)}(x,i)\big|_{\partial D} = 0 \text{ on } \partial D \times \mathcal{M}. \end{cases}$$

Let $M^{(3)} = \sup_{(x,i)\in D\times\mathcal{M}}\{v^{(3)}(x,i)\}$. In view of (11.151), for $k > k_0 \vee |x|$, we have

$$|v^{(3)}(x,i)| = \left|\mathbb{E}_{x,i} I_{\{|X(\widetilde{\tau}_{D_k})|=k\}} v^{(3)}(X(\widetilde{\tau}_{D_k}), \alpha(\widetilde{\tau}_{D_k}))\right|$$
$$\leq M^{(3)}[1 - \mathbb{P}_{x,i}\{X(\widetilde{\tau}_{D_k}) \in \partial D\}]$$

Letting $k \to \infty$ and using $\mathbb{P}_{x,i}\{X(\widetilde{\tau}_{D_k}) \in \partial D\} \to \mathbb{P}_{x,i}\{\widetilde{\tau}_D < \infty\} = 1$ as $k \to \infty$, we obtain $v^{(3)}(x,i) \equiv 0$. □

11.6 Stability of Past-Independent Switching Diffusions

11.6.1 Definitions and Auxiliary Results

In this section, we consider the stability of the solution to (11.5) and (11.6)- To simplify the presentation, we assume that the switching intensity q_{ij} depends only on the current state of $X(t)$, which makes $(X(t), \alpha(t))$ a Markov process. The results presented below hold true if q_{ij}'s are past-dependent; see Section 11.6.5.

Throughout this section, we assume that an assumption for existence and uniqueness of Markov-Feller solution in Sects. 11.3 and 11.4 is satisfied. We suppose further that $b(0,i) = 0$ and $\sigma(0,i) = 0$ for $i \in \mathcal{M}$ and give the following definitions of stability.

Definition 11.48 The trivial solution $X(t) \equiv 0$ is said to be

- *stable in probability*, if for any $h > 0$,

$$\lim_{x\to 0}\inf_{i\in\mathcal{M}} \mathbb{P}_{x,i}\big\{X(t) \leq h\,\forall t \geq 0\big\} = 1.$$

- *asymptotic stable in probability*, if it is stable in probability and

$$\lim_{x\to 0}\inf_{i\in\mathcal{M}} \mathbb{P}_{x,i}\left\{\lim_{t\to\infty} X(t) = 0\right\} = 1.$$

Next, we state a general result that can be proved by arguments in the proofs of Lemmas 6.5, 6.6 and Remark 6.8.

Theorem 11.49 *Let D be a neighborhood of $0 \in \mathbb{R}^n$. Suppose there exist three functions $V(x,i) : D \times \mathbb{Z} \mapsto \mathbb{R}_+$, $\mu_1(x) : D \mapsto \mathbb{R}_+$, $\mu_2(x) : D \mapsto \mathbb{R}_+$ such that*

- *$\mu_1(x), \mu_2(x)$ are continuous on D, $\mu_k(x) = 0$ if and only if $x = 0$ for $k = 1, 2$;*
- *$V(x,i)$ is continuous on D and twice continuously differentiable in $\mathcal{D} \setminus \{0\}$ for each $i \in \mathcal{M}$;*
- *$\mu_1(x) \leq V(x,i)$ for any $(x,i) \in D \times \mathcal{M}$.*

Then the following conclusions hold.

- *if* $\mathcal{L}V(x,i) \le 0$ *for any* $(x,i) \in D \times \mathcal{M}$, *the trivial solution is stable in probability.*
- *if* $\mathcal{L}V(x,i) \le -\mu_2(x)$ *for any* $(x,i) \in D \times \mathcal{M}$, *the trivial solution is asymptotically stable in probability.*

Let $\hat{\alpha}(t)$ be the Markov chain with bounded generator $Q(0)$ and transition probability $\hat{p}_{ij}(t)$.

Definition 11.50 The Markov chain $\hat{\alpha}(t)$ is said to be:

- *ergodic*, if it has an invariant probability measure $\boldsymbol{\nu} = (\nu_1, \nu_2, \dots)$ and

$$\lim_{t\to\infty} \hat{p}_{ij}(t) = \nu_j \text{ for any } i, j \in \mathcal{M}$$

or equivalently,

$$\lim_{t\to\infty} \sum_{j\in\mathcal{M}} |\hat{p}_{ij}(t) - \nu_j| = 0 \text{ for any } i \in \mathcal{M},$$

- *strongly ergodic*, if

$$\lim_{t\to\infty} \sup_{i\in\mathcal{M}} \left\{ \sum_{j\in\mathcal{M}} |\hat{p}_{ij}(t) - \nu_j| \right\} = 0.$$

- *strongly exponentially ergodic*, if there exist $C > 0$ and $\lambda > 0$ such that

$$\sum_{j\in\mathcal{M}} |\hat{p}_{ij}(t) - \nu_j| \le Ce^{-\lambda t} \text{ for any } i \in \mathcal{M}, t \ge 0. \tag{11.165}$$

We refer to [3] for some properties and sufficient conditions for the aforementioned ergodicity.

11.6.2 Certain Practical Conditions for Stability and Instability

For each $h > 0$, denote by $B_h \subset \mathbb{R}^n$ the open ball centered at 0 with radius h. Throughout this section, let D be a neighborhood of 0 satisfying $D \subset B_1$. We also denote by $\hat{\alpha}(t)$ the continuous-time Markov chain with generator $Q(0)$. We first state a theorem, which generalizes [164, Theorem 4.3], a result for switching diffusions when the switching takes values in a finite set.

Theorem 11.51 *Suppose that the Markov chain $\hat{\alpha}(t)$ is strongly exponentially ergodic with invariant probability measure $\boldsymbol{\nu} = (\nu_1, \nu_2, \ldots)$ and that*

$$\sup_{i \in \mathcal{M}} \sum_{j \neq i} |q_{ij}(x) - q_{ij}(0)| \to 0 \text{ as } x \to 0. \tag{11.166}$$

Let D be a neighborhood of 0 and $V : D \mapsto \mathbb{R}_+$ satisfying that $V(x) = 0$ if and only if $x = 0$ and that $V(x)$ is continuous on D, twice continuously differentiable in $D \setminus \{0\}$. Suppose that there is a bounded sequence of real numbers $\{c_i : i \in \mathcal{M}\}$ such that

$$\mathcal{L}_i V(x) \le c_i V(x) \, \forall x \in D \setminus \{0\}. \tag{11.167}$$

Then, if $\sum_{i \in \mathcal{M}} c_i \nu_i < 0$, the trivial solution is asymptotic stable in probability.

The proof is based on the following lemma whose proof is almost identical to that of Lemma A.12.

Lemma 11.52 *If the Markov chain $\hat{\alpha}(t)$ is strongly exponentially ergodic with generator $\hat{Q}$ and invariant probability measure $\boldsymbol{\nu} = (\nu_1, \nu_2, \ldots)'$, then if $\mathbf{b} = (b_1, b_2, \ldots)'$ is bounded satisfying $\sum \nu_i b_i = 0$, then, there exists a bounded vector $\mathbf{c} = (c_1, c_2, \ldots)'$ such that $b_i = \sum \hat{q}_{ji} c_j$.*

***Proof of Theorem* 11.51** Let $\lambda = -\sum_{i \in \mathcal{M}} c_i \nu_i$. Since $\sum_{i \in \mathcal{M}} \nu_i = 1$, we have $\sum_{i \in \mathcal{M}} (c_i + \lambda) \nu_i = 0$. Since $\hat{\alpha}(t)$ is strongly exponentially ergodic, it follows from Lemma 11.52 that there exists a bounded sequence of real numbers $\{\gamma_i : i \in \mathcal{M}\}$ such that

$$\sum_{j \in \mathcal{M}} q_{ij}(0) \gamma_j = \lambda + c_i \text{ for any } i \in \mathcal{M} \tag{11.168}$$

Since $\sum_{j \in \mathcal{M}} q_{ij}(0) = 0$ for any $i \in \mathcal{M}$, it follows from (11.168) that

$$\sum_{j \in \mathcal{M}} q_{ij}(0) \gamma_j = \sum_{j \in \mathcal{M}} q_{ij}(0)(1 - p\gamma_j) = -p(\lambda + c_i) \text{ for any } i \in \mathcal{M} \tag{11.169}$$

Since $\{\gamma_i\}$ is bounded, we can choose $p \in (0, 1)$ such that

$$p|\gamma_i| \le \min\{0.25\lambda, 0.5\} \tag{11.170}$$

In view of (11.166) and (11.170), there is an $h > 0$ sufficiently small such that

$$\sum_{j \in \mathcal{M}} (1 - p\gamma_j) |q_{ij}(x) - q_{ij}(0)| < \frac{p\lambda}{4} \quad \forall x \in B_h. \tag{11.171}$$

Define the function $U(x,i) : B_h \times \mathcal{M} \mapsto \mathbb{R}_+$ by $U(x,i) = (1 - p\gamma_i)V^p(x)$. By Itô's formula, (11.166), (11.169), and (11.171), we have

$$\begin{aligned}
\mathcal{L}U(x,i) =& p(1-p\gamma_i)V^{p-1}\mathcal{L}_i V(x) - \frac{p(1-p)}{2}V^{p-2}|V_x(x)\sigma(x,i)|^2 \\
&+ V^p(x)\sum_{j\in\mathcal{M}}(1-p\gamma_j)q_{ij}(x) \\
\le& c_i p(1-p\gamma_i)V^{p-1} + V^p(x)\sum_{j\in\mathcal{M}}(1-p\gamma_j)q_{ij}(0) \\
&+ V^p(x)\sum_{j\in\mathcal{M}}(1-p\gamma_j)|q_{ij}(x)-q_{ij}(0)| \\
\le& c_i p(1-p\gamma_i)V^{p-1} - p(\lambda + c_i)V^p(x) \\
&+ V^p(x)\sum_{j\in\mathcal{M}}(1-p\gamma_j)|q_{ij}(x)-q_{ij}(0)| \\
\le& p(-\lambda - p\gamma_i)V^p(x) + V^p(x)\sum_{j\in\mathcal{M}}(1-p\gamma_j)|q_{ij}(x)-q_{ij}(0)| \\
\le& -0.75p\lambda V^p(x) + 0.25p\lambda V^p(x) = -0.5p\lambda V^p(x) \ \text{ for } (x,i)\in B_h\times\mathcal{M}.
\end{aligned} \tag{11.172}$$

By Theorem 11.49, it follows from (11.172) that the trivial solution is asymptotically stable. □

The hypothesis of this theorem seems to be restrictive. It requires the strongly exponential ergodicity of $Q(0)$ and the uniform convergence to 0 of the sum $\sum_{j\neq i}|q_{ij}(x) - q_{ij}(0)|$. To treat cases in which $Q(0)$ is strongly ergodic (not exponentially ergodic) or even only ergodic, as well as to relax the condition (11.166), we need a more complicated method. Our method, which is inspired by the idea in [18], utilizes the ergodicity of $Q(0)$ and the analysis of the Laplace transform. Similar techniques of using the Laplace transform can also be seen in the large deviations theory and related applications [14, 325]. We also take a step further by estimating the pathwise rate of convergence of solutions.

Let Γ be a family of increasing and differentiable functions $g : \mathbb{R}_+ \mapsto \mathbb{R}_+$ such that $g(y) = 0$ iff $y = 0$ and $\frac{dy}{dy}$ is bounded on [0, 1]. Since $\frac{dg}{dy}(y)$ is bounded on [0, 1] and $g(0) = 0$, it is easy to show that the function

$$G(y) := -\int_y^1 \frac{dz}{g(z)} \quad \text{on } [0,1] \tag{11.173}$$

is non-positive and strictly decreasing and $\lim_{y\to 0} G(y) = -\infty$. Its inverse G^{-1} : $(-\infty, 0] \mapsto (0, 1]$ satisfies $\lim_{t\to\infty} G^{-1}(-t) = 0$. We state some assumptions to be used in what follows; we will also provide some lemmas whose proofs are relegated to Section 11.7.3

(H11.9) There are functions $g \in \Gamma$, $V : D \mapsto \mathbb{R}_+$ such that

- $V(x) = 0$ if and only if $x = 0$
- $V(x)$ is continuous on D and twice continuously differentiable in $D \setminus \{0\}$.
- there is a bounded sequence of real numbers $\{c_i : i \in \mathcal{M}\}$ such that

$$\mathcal{L}_i V(x) \le c_i g(V(x)) \,\forall x \in D \setminus \{0\}. \tag{11.174}$$

Lemma 11.53 *Assume* (H11.9). *For any $\varepsilon, T, h > 0$, there exists an $\tilde{h} = \tilde{h}(\varepsilon, T, h)$ such that*

$$\mathbb{P}_{x,i}\{\tau_h \le T\} < \varepsilon, \quad \textit{for all } (x, i) \in B_{\tilde{h}} \times \mathcal{M}$$

where $\tau_h = \inf\{t \ge 0 : |X(t)| \ge h\}$.

Lemma 11.54 *Under the assumption* $b(0, i) = 0, \sigma(0, i) = 0, i \in \mathcal{M}$, *we have* $\mathbb{P}_{x,i}\{X(t) = 0 \textit{ for some } t \ge 0\} = 0$ *for any* $x \ne 0, i \in \mathcal{M}$.

With the auxiliary results above, we can prove our main results.

Theorem 11.55 *Suppose that the Markov chain $\hat{\alpha}(t)$ is ergodic with invariant probability measure $\boldsymbol{\nu} = (\nu_1, \nu_2, \ldots)$ and* (H11.9) *is satisfied with additional conditions:*

$$\limsup_{i\to\infty} c_i < 0, \tag{11.175}$$

and

$$M_g := \sup_{x\in D, i\in\mathcal{M}} \left\{\left|\frac{V_x(x)\sigma(x, i)}{g(V(x))}\right|\right\} < \infty. \tag{11.176}$$

Then, if $\sum_{i\in\mathcal{M}} c_i \nu_i < 0$, the trivial solution is asymptotic stable in probability, that is, for any $h > 0$ such that $B_h \subset D$, and $\varepsilon > 0$, there exists $\delta = \delta(h, \varepsilon) > 0$ such that

$$\mathbb{P}_{x,i}\left\{X(t) < h \,\forall t \ge 0, \textit{ and } \lim_{t\to\infty} X(t) = 0\right\} > 1 - \varepsilon \textit{ for any } (x, i) \in B_\delta \times \mathcal{M}.$$

Moreover, there is a $\lambda > 0$ such that

$$\mathbb{P}_{x,i}\left\{\lim_{t\to\infty} \frac{V(X(t))}{G^{-1}(-\lambda t)} \le 1\right\} > 1 - \varepsilon \textit{ for any } (x, i) \in B_\delta \times \mathcal{M}. \tag{11.177}$$

Remark 11.56 Before proceeding to the proof of the theorem, let us make a brief comment. In addition to providing sufficient conditions for asymptotic stability, a significant new element here is the rate of convergence given in (11.177). Although there are numerous treatment of stochastic stability by a host of authors for diffusions and switching diffusions. The rate result in Theorem 11.55 appears to the first one of its kind.

Proof The proof is divided into two steps. We first show the trivial solution is stable in probability and then we prove asymptotic stability and estimate the path-wise convergence rate.

Step 1: Stability. Shrinking D if necessary, we can assume without loss of generality that $V(x) \le 1$ in D. Let $h > 0$ such that $B_h \subset D$. Since $\{c_i\}$ is bounded,

$$\lim_{k\to\infty} \sum_{i\le k} c_i \nu_i = \sum_{i\in\mathcal{M}} c_i \nu_i < 0. \tag{11.178}$$

This and (11.175) show that there exists $k_0 \in \mathcal{M}$ such that

$$-\lambda_1 := \sum_{i\le k_0} c_i \nu_i < 0 \text{ and } -2\lambda_2 := \sup_{i>k_0} c_i < 0.$$

Define $G(y) = -\int_y^1 g^{-1}(z)dz$. In view of Lemma 11.54, if $X(0) \neq 0$, then $X(t) \neq 0$ a.s, which leads to $g(V(X(t)) \neq 0$ a.s. Thus, we have from Itô's formula and the increasing property of $g(\cdot)$ that

$$\begin{aligned} G\big(V(X(\tau_h \wedge t))\big) =& G(V(x)) + \int_0^{\tau_h\wedge t} \frac{\mathcal{L}_{\alpha(s)} V(X(s))}{g(V(X(s)))} ds \\ &- \int_0^{\tau_h\wedge t} \frac{\frac{dg}{dy}(V(X(s)))\Big|V_x(X(s))\sigma(X(s),\alpha(s))\Big|^2}{2g^2(V(X(s)))} ds \\ &+ \int_0^{\tau_h\wedge t} \frac{V_x(X(s))\sigma(X(s),\alpha(s))}{g(V(X(s)))} dW(s) \le G(V(x)) + H(t), \end{aligned} \tag{11.179}$$

where

$$H(t) = \int_0^{\tau_h\wedge t} c_{\alpha(s)} ds + \int_0^{\tau_h\wedge t} \frac{V_x(X(s))\sigma(X(s),\alpha(s))}{g(V(X(s)))} dW(s).$$

Because of the boundedness (11.176) and $\{c_i, i \in \mathbb{Z}\}$, $M_H := H(t) - \mathbb{E}_{x,i} \int_0^{\tau_h\wedge t} c_{\alpha(s)} ds$ is a martingale with bounded quadratic variance $[M_H]_t \le K_H t$ for some constant t independent of x, i, t. In view of [148, Lemma 19.22], $e^{\theta M_H(t) - \frac{\theta^2}{2}[M_H]_t}$ is a martingale for $\theta > 0$. As a result,

$$\begin{aligned}\mathbb{E}_{x,i}e^{\theta H(t)} =&\mathbb{E}_{x,i}\left(e^{\theta M_H(t)}e^{\theta \mathbb{E}_{x,i}\int_0^{\tau_h\wedge t}c_{\alpha(s)}ds}\right)\\ \le& e^{\frac{\theta^2 K_H t}{2}}e^{\theta \mathbb{E}_{x,i}\int_0^{\tau_h\wedge t}c_{\alpha(s)}ds}\mathbb{E}_{x,i}e^{\theta M_H(t)-\frac{\theta^2}{2}[M_H]_t}\\ \le& e^{\frac{\theta^2 K_H t}{2}}e^{\theta \mathbb{E}_{x,i}\int_0^{\tau_h\wedge t}c_{\alpha(s)}ds}\end{aligned} \tag{11.180}$$

On the other hand, we have

$$\begin{aligned}\mathbb{E}_{x,i}H(t) &\le \mathbb{E}_{x,i}\int_0^{\tau_h\wedge t}c_{\alpha(s)}ds\\ &\le \mathbb{E}_{x,i}\int_0^t c_{\alpha(s)}ds - \mathbb{E}_{x,i}\int_{\tau_h\wedge t}^t c_{\alpha(s)}ds\\ &\le \mathbb{E}_{x,i}\int_0^t c_{\alpha(s)}ds + t\bar{c}\mathbb{P}_{x,i}\{\tau_h < t\}.\end{aligned} \tag{11.181}$$

Because of the ergodicity of $\hat{\alpha}(t)$, there exists a $T > 0$ depending on k_0 such that

$$\mathbb{E}_{0,i}\int_0^t c_{\alpha(s)}ds = \mathbb{E}_i\int_0^t c_{\hat{\alpha}(s)}ds \le -\frac{3\lambda_1}{4}t \quad \forall t \ge T, i \le k_0. \tag{11.182}$$

Let $\bar{c} = \sup_{i\in\mathcal{M}}|c_i|$ and m_0 be an positive integer satisfying $m_0\lambda_2 > \bar{c} + K_H + 1$. By the Feller property of $(X(t), \alpha(t))$, there exists an $h_1 \in (0, h)$ such that

$$\mathbb{E}_{x,i}\int_0^t c_{\alpha(s)}ds \le -\frac{\lambda_1}{2}t \quad \forall t \in [T, T_2], |x| \le h_1, i \le k_0, \tag{11.183}$$

where $T_2 = (m_0 + 1)T$. In view of Lemma 11.53, there exists an $h_2 \in (0, h_1)$ such that

$$\bar{c}\mathbb{P}_{x,i}\{\tau_h < m_0T + T\} \le \frac{\lambda_1}{4} \text{ provided } |x| \le h_2, i \in \mathcal{M}. \tag{11.184}$$

Applying (11.183) and (11.184) to (11.181), we obtain

$$\mathbb{E}_{x,i}\int_0^t c_{\alpha(s)}ds \le -\frac{\lambda_1}{4}t \text{ if } 0 < |x| \le h_2, i \le k_0, t \in [T, T_2]. \tag{11.185}$$

With $\theta \in (0, 0.5 \wedge \frac{\lambda_1}{4K_H}]$ we have from (11.180) and (11.185) that

$$\mathbb{E}_{x,i}e^{\theta H(t)} \le \exp\left\{-\frac{\theta\lambda_1 t}{8}\right\} \text{ for } 0 < |x| < h_2, i \le k_0, t \in [T, T_2]. \tag{11.186}$$

In what follows, we fix a $\theta \in (0, 0.5 \wedge \frac{\lambda_1}{4K_H} \wedge \frac{\lambda_2}{M_g}]$. Exponentiating both sides of the inequality $G(V(X(\tau_h \wedge t))) \le G(V(x)) + H(t)$, we have for $0 < |x| < h_2, i \le k_0, t \in [T, T_2]$ that

$$\mathbb{E}_{x,i}U(X(\tau_h \wedge t)) \leq U(x)\mathbb{E}_{x,i}e^{\theta H(t)} \leq U(x)\exp\left\{-\frac{\theta\lambda_1 t}{8}\right\}. \tag{11.187}$$

where $U(x) = \exp(\theta G(V(x)))$. Since $\lim_{x\to 0} G(V(x)) = -\infty$,

$$\lim_{x\to 0} U(x) = 0. \tag{11.188}$$

Using the inequality $G(V(X(\tau_h \wedge t))) \leq G(V(x)) + H(t)$ and (11.180), we have

$$\mathbb{E}_{x,i}U(X(\tau_h \wedge t)) \leq U(x)\exp\{\theta[\bar{c} + K_H]t\}, \text{ for all } (x,i) \in B_h \times \mathcal{M}, t \geq 0. \tag{11.189}$$

Now, let $\Delta = \inf\{U(x) : h_2 \leq |x| \leq h\} > 0$. Define stopping times

$$\xi = \inf\{t \geq 0 : \alpha(t) \leq k_0\}, \quad \text{ and } \zeta = \inf\{t \geq 0 : U(X(t)) \geq \Delta\}.$$

Clearly, if $X(0) \in B_h$ then $\zeta \leq \tau_h$ and if $t < \zeta$ then $|X(t)| < h_2$. By computation and (11.179), we have

$$\begin{aligned}\mathcal{L}_i U(x) &\leq \theta U(x)\left[c_i + [\theta - \dot{g}(V(x))]\frac{|V_x(x)\sigma(x,i)|^2}{g(V(x))}\right] \leq \theta(-2\lambda_2 + \theta M_g)U(x)\\ &\leq -\theta\lambda_2 U(x), \quad \text{for } 0 < |x| < h, i > k_0.\end{aligned}$$

It follows from Itô's formula that

$$\begin{aligned}&\mathbb{E}_{x,i}e^{\theta\lambda_2(t\wedge\xi\wedge\zeta)}U(X(t\wedge\xi)\\ &= U(x) + \mathbb{E}_{x,i}\int_0^{t\wedge\xi\wedge\zeta} e^{\lambda_2 s}\left[\theta\lambda_2 U(X(s)) + \mathcal{L}_{\alpha(t)}U(X(s))\right]ds \leq U(x),\end{aligned} \tag{11.190}$$

for $0 < |x| < h, i \in \mathcal{M}$. We have the following estimate for $0 < |x| < h, i > k_0$:

$$\begin{aligned}&\mathbb{E}_{x,i}e^{\theta\lambda_2(T_2\wedge\xi\wedge\zeta)}U(X(T_2\wedge\xi\wedge\zeta))\\ &\quad = \mathbb{E}_{x,i}I_{\{\xi\wedge\zeta<m_0T\}}e^{\theta\lambda_2(T_2\wedge\xi\wedge\zeta)}U(X(T_2\wedge\xi\wedge\zeta))\\ &\qquad + \mathbb{E}_{x,i}I_{\{m_0T\leq\xi\wedge\zeta<T_2\}}e^{\theta\lambda_2(T_2\wedge\xi\wedge\zeta)}U(X(T_2\wedge\xi\wedge\zeta))\\ &\qquad + \mathbb{E}_{x,i}I_{\{\xi\wedge\zeta\geq T_2\}}e^{\theta\lambda_2(T_2\wedge\xi\wedge\zeta)}U(X(T_2\wedge\xi\wedge\zeta))\\ &\quad \geq \mathbb{E}_{x,i}I_{\{\xi\wedge\zeta\leq m_0T\}}U(X(\xi\wedge\zeta))\\ &\qquad + e^{\theta\lambda_2 m_0 T}\mathbb{E}_{x,i}I_{\{m_0T\leq\xi\wedge\zeta<T_2\}}U(X(\xi\wedge\zeta))\\ &\qquad + e^{\theta\lambda_2 T_2}\mathbb{E}_{x,i}I_{\{\xi\geq T_2\}}U(X(T_2)).\end{aligned} \tag{11.191}$$

Since $\mathbb{P}_{x,i}\{\zeta = 0\} = 1$ if $i \leq k_0$, (11.191) holds for $0 < |x| < h, i \in \mathcal{M}$. Noting that $U(x) \wedge \Delta \leq \Delta$ for any $x \in B_h$, we have

$$\mathbb{E}\Big[U(X(T_2 \wedge \tau_h)) \wedge \Delta \Big| \zeta < m_0 T, \zeta \le \xi\Big] \le \Delta \le U(X(\zeta)) = U(X(\xi \wedge \zeta)).$$

If $\xi < \zeta$, then $U(X(\xi)) < \Delta$. By strong Markov property of $(X(t), \alpha(t))$ and (11.187), and (11.189), we have

$$\begin{aligned}\mathbb{E}\Big[U(X(T_2 \wedge \tau_h)) \wedge \Delta \Big| \xi < m_0 T \wedge \zeta\Big] &\le U(X(\xi)) \\ &= U(X(\xi \wedge \zeta)),\end{aligned}$$

and

$$\begin{aligned}\mathbb{E}\Big[U(X(T_2 \wedge \tau_h)) \wedge \Delta \Big| m_0 T \le \xi < T_2 \wedge \zeta\Big] &\le U(X(\xi))e^{\theta(\bar{c}+M_H)T} \\ &= U(X(\xi \wedge \zeta))e^{\theta(\bar{c}+M_H)T}.\end{aligned}$$

From the three estimates above, we have

$$\begin{aligned}\mathbb{E}_{x,i} I_{\{\xi \wedge \zeta \le m_0 T\}}\big[U(X(T_2 \wedge \tau_h)) \wedge \Delta\big] &= \mathbb{E}_{x,i} I_{\{\zeta < m_0 T, \zeta \le \xi\}}\big[U(X(T_2 \wedge \tau_h)) \wedge \Delta\big] \\ &\quad + \mathbb{E}_{x,i} I_{\{\xi < m_0 T \wedge \zeta\}}\big[U(X(T_2 \wedge \tau_h)) \wedge \Delta\big] \\ &\le \mathbb{E}_{x,i} I_{\{\xi \wedge \zeta < m_0 T\}} U(X(\xi \wedge \zeta)),\end{aligned} \tag{11.192}$$

and

$$\begin{aligned}&\mathbb{E}_{x,i} I_{\{m_0 T \le \xi \wedge \zeta < T_2\}}\big[U(X(T_2 \wedge \tau_h)) \wedge \Delta\big] \\ &\quad \le e^{\theta(\bar{c}+M_H)T} \mathbb{E}_{x,i} I_{\{m_0 T \le \xi \wedge \zeta < T_2\}} U(X(\xi \wedge \zeta)) \\ &\quad \le e^{\theta \lambda_2 m_0 T} \mathbb{E}_{x,i} I_{\{m_0 T \le \xi \wedge \zeta < T_2\}} U(X(\xi \wedge \zeta)),\end{aligned} \tag{11.193}$$

where the last line follows from $m_0 \lambda_2 > \bar{c} + M_H + 1$. Applying (11.192) and (11.193) to (11.191), we obtain

$$\mathbb{E}_{x,i}\big[U(X(T_2 \wedge \tau_h)) \wedge \Delta\big] \le U(x) \text{ for any } (x, i) \in B_h \times \mathbb{Z}.$$

Since $\mathbb{E}_{x,i}\big[U(X(T_2 \wedge \tau_h)) \wedge \Delta\big] \le \Delta$, we have

$$\mathbb{E}_{x,i}\big[U(X(T_2 \wedge \tau_h)) \wedge \Delta\big] \le U(x) \wedge \Delta \text{ for any } (x, i) \in B_h \times \mathbb{Z}. \tag{11.194}$$

This together with the Markov property of $(X(t), \alpha(t))$ implies that $\{M(k) := \big[U(X(kT_2 \wedge \tau_h)) \wedge \Delta\big], k \in \mathcal{M}\}$ is a super-martingale. Let $\eta = \inf\{k \in \mathcal{M} : M(k) = \Delta\}$. Clearly, $\{\eta < \infty\} \supset \{\tau_h < \infty\}$. For any $\varepsilon > 0$, if $U(x) < \varepsilon\Delta$ we have that

$$\mathbb{P}_{x,i}\{\eta < k\} \le \frac{\mathbb{E}_{x,i} M(\eta \wedge k)}{\Delta} \le \frac{U(x)}{\Delta} \le \varepsilon.$$

Letting $k \to \infty$ yields

$$\mathbb{P}_{x,i}\{\tau_h < \infty\} \le \mathbb{P}_{x,i}\{\eta < \infty\} \le \varepsilon, \text{ if } U(x) < \varepsilon\Delta. \tag{11.195}$$

We complete the proof of this step by noting that $\{x : U(x) < \varepsilon\Delta\}$ is a neighborhood of x due to the fact that $\lim_{x\to 0} U(x) = 0$.

Step 2: Asymptotic stability and pathwise convergence rate. To prove the asymptotic stability in probability, we fix $h > 0$ and define $U(x), T_2, m_0, \Delta$ depending on h as in the first step. By virtue of (11.191), we have

$$\begin{aligned}
\mathbb{E}_{x,i} e^{\theta\lambda_2(T_2\wedge\xi\wedge\zeta)} U(X(T_2\wedge\xi\wedge\zeta)) \ge & \mathbb{E}_{x,i} I_{\{\xi\wedge\zeta<m_0T\}} U(X(\xi\wedge\zeta)) \\
& + e^{\theta\lambda_2 m_0 T}\mathbb{E}_{x,i} I_{\{m_0T\le\xi\wedge\zeta<T_2\}} U(X(\xi\wedge\zeta)) \\
& + e^{\theta\lambda_2 T_2}\mathbb{E}_{x,i} I_{\{\xi\wedge\zeta\ge T_2\}} U(X(T_2)) \\
\ge & \mathbb{E}_{x,i} I_{\{\xi<m_0T,\zeta>\xi\}} U(X(\xi)) \\
& + e^{\theta\lambda_2 m_0 T}\mathbb{E}_{x,i} I_{\{m_0T\le\xi<T_2,\zeta>\xi\}} U(X(\xi)) \\
& + e^{\theta\lambda_2 T_2}\mathbb{E}_{x,i} I_{\{\xi\wedge\zeta\ge T_2\}} U(X(T_2)).
\end{aligned} \tag{11.196}$$

Recalling that $\zeta \le \tau_h$ and $X(t) < h_2$ if $t < \zeta$, we have from (11.186) and (11.189) that

$$\begin{aligned}
\mathbb{E}_{x,i} I_{\{\zeta\ge T_2\}} I_{\{\xi<m_0T\}} U(X(T_2)) = & \mathbb{E}_{x,i} I_{\{\zeta\ge T_2\}} I_{\{\xi<m_0T\}} U(X(T_2\wedge\tau_h)) \\
\le & \mathbb{E}_{x,i} I_{\{\xi<\zeta\}} I_{\{\xi<m_0T\}} U(X(T_2\wedge\tau_h)) \\
\le & \mathbb{E}_{x,i}\left[I_{\{\xi<m_0T\wedge\zeta\}} U(X(\xi)) \exp\left\{-\theta\frac{\lambda}{8}(T_2-\xi)\right\}\right] \\
\le & \exp\left\{-\frac{\theta\lambda T}{8}\right\}\mathbb{E}_{x,i}\left[I_{\{\zeta\ge\xi\}} I_{\{\xi<m_0T\}} U(X(\xi))\right]
\end{aligned} \tag{11.197}$$

and

$$\begin{aligned}
& \mathbb{E}_{x,i} I_{\{m_0T\le\xi<T_2,\zeta\ge T_2\}} U(X(T_2)) \\
& \le \mathbb{E}_{x,i} I_{\{m_0T\le\xi<T_2\wedge\zeta\}} U(X(T_2\wedge\tau_h)) \\
& \le \mathbb{E}_{x,i}\left[I_{\{m_0T\le\xi<T_2\wedge\zeta\}} U(X(\xi)) \exp\{\theta(\bar{c}+M_H)(T_2-\xi)\}\right] \\
& \le \exp\{\theta(\bar{c}+M_H)T\}\mathbb{E}_{x,i}\left[I_{\{m_0T\le\xi<T_2\wedge\zeta\}} U(X(\xi))\right] \\
& \le \exp\{-\theta T\}\exp\{\theta\lambda_2 m_0 T\}\mathbb{E}_{x,i}\left[I_{\{m_0T\le\xi<T_2\wedge\zeta\}} U(X(\xi))\right].
\end{aligned} \tag{11.198}$$

On the other hand, we can write

$$\mathbb{E}_{x,i} I_{\{\xi\wedge\zeta\ge T_2\}} U(X(T_2)) = e^{-\theta\lambda_2 T_2} e^{\theta\lambda_2 T_2}\mathbb{E}_{x,i} I_{\{\xi\wedge\zeta\ge T_2\}} U(X(T_2)). \tag{11.199}$$

Letting $p = \max\left\{\exp\left\{-\frac{\theta\lambda T}{8}\right\}, \exp\{-\theta T\}, \exp\{-\theta\lambda_2 T_2\}\right\} < 1$ and adding (11.197), (11.198), and (11.198) side by side and then using (11.196), we have

$$\mathbb{E}_{x,i} I_{\{\zeta\ge T_2\}} U(X(T_2)) \le pU(x), \text{ for } (x,i) \in B_h \times \mathcal{M}.$$

By the strong Markov property of the process $(X(t), \alpha(t))$,

$$\begin{aligned}\mathbb{E}_{x,i} I_{\{\zeta\ge 2T_2\}} U(X(2T_2)) =&\mathbb{E}_{x,i}\left[I_{\{\zeta\ge T_2\}}\mathbb{E}_{X(T_2),\alpha(T_2)} I_{\{\zeta\ge T_2\}} U(X(T_2))\right]\\ \le& p\mathbb{E}_{x,i} I_{\{\zeta\ge T_2\}} U(X(T_2))\\ \le& p^2 U(x), \text{ for } (x,i) \in B_h \times \mathcal{M}.\end{aligned}$$

Continuing in this fashion, we have

$$\mathbb{E}_{x,i} I_{\{\zeta\ge kT_2\}} U(X(kT_2)) \le p^k U(x), \text{ for } (x,i) \in B_h \times \mathcal{M}.$$

Since $2\theta < 1$, we have from (11.189) that $\mathbb{E}_{x,i} e^{2\theta H(s)} \le e^{2\theta[\bar c + M_H]s}$. This and the Burkholder-Davis-Gundy inequality imply

$$\begin{aligned}&\mathbb{E}_{x,i}\sup_{t\le T_2}\left|\int_0^{t\wedge\tau_h} e^{\theta H(s)}\frac{V_x(X(s))\sigma(X(s),\alpha(s))}{g(V(X(s)))}dW(s)\right|\\ &\le\left[\mathbb{E}_{x,i}\int_0^{T_2\wedge\tau_h} e^{2\theta H(s)}\frac{|V_x(X(s))\sigma(X(s),\alpha(s))|^2}{g^2(V(X(s)))}ds\right]^{\frac12}\\ &\le\left[M_g^2\mathbb{E}_{x,i}\int_0^{T_2} e^{2\theta H(s)}ds\right]^{\frac12}\le\left[M_g^2\int_0^{T_2} e^{2\theta[\bar c+M_H]s}ds\right]^{\frac12} := \widetilde K_1.\end{aligned} \tag{11.200}$$

On the other hand,

$$\begin{aligned}&\mathbb{E}_{x,i}\sup_{t\le T_2}\left|\int_0^{t\wedge\tau_h} e^{\theta H(s)}\left[\theta c_{\alpha(s)} + \frac{\theta^2}{2}\frac{|V_x(X(s))\sigma(X(s),\alpha(s))|^2}{2g^2(V(X(s)))}\right]ds\right|\\ &\le(\bar c + M_H)\mathbb{E}_{x,i}\int_0^{T_2\wedge\tau_h} e^{\theta H(s)}ds \le (\bar c + M_H)\int_0^{T_2} e^{\theta[\bar c+M_H]}ds := \widetilde K_2.\end{aligned} \tag{11.201}$$

It follows from (11.200) and (11.201) that

$$\mathbb{E}_{x,i}\sup_{t\le T_2} U(X(t\wedge\tau_h)) = U(x)\mathbb{E}_{x,i}\sup_{t\le T_2} e^{\theta H_t} \le U(x)\widetilde K_3, \tag{11.202}$$

where $\widetilde K_3 = 1 + \widetilde K_1 + \widetilde K_2$. By the strong Markov property of $(X(t), \alpha(t))$, we derive from (11.202) that

$$
\begin{aligned}
&\mathbb{E}_{x,i} I_{\{\zeta=\infty\}} \sup_{t\in[kT_2,(k+1)T_2]} U(X(t\wedge\tau_h)) \\
&\leq \mathbb{E}_{x,i} I_{\{\zeta\geq kT_2\}} \sup_{t\in[kT_2,(k+1)T_2]} U(X(t\wedge\tau_h)) \\
&\leq \widetilde{K}_3 \mathbb{E}_{x,i} I_{\{\zeta\geq kT_2\}} U(X(kT_2) \leq \widetilde{K}_3 U(x)\rho^k,
\end{aligned}
\tag{11.203}
$$

which combined with the Markov inequality leads to

$$
\begin{aligned}
&\mathbb{P}_{x,i}\left\{ I_{\{\zeta=\infty\}} \sup_{t\in[kT_2,(k+1)T_2]} U(X(t\wedge\tau_h)) > (\rho+\widetilde{\varepsilon})^k \right\} \\
&\leq \frac{1}{(\rho+\widetilde{\varepsilon})^k} \mathbb{E}_{x,i}\left[I_{\{\zeta=\infty\}} \sup_{t\in[kT_2,(k+1)T_2]} U(X(t\wedge\tau_h)) \right] \\
&\leq \widetilde{K}_3 U(x) \frac{\rho^k}{(\rho+\widetilde{\varepsilon})^k} \quad k\in\mathcal{M},
\end{aligned}
\tag{11.204}
$$

where $\widetilde{\varepsilon}$ is any number in $(0, 1-\rho)$. In view of the Borel-Cantelli lemma, for almost all $\omega\in\Omega$, there exists random integer $k_1 = k_1(\omega)$ such that

$$
I_{\{\zeta=\infty\}} \sup_{t\in[kT_2,(k+1)T_2]} U(X(t)) < (\rho+\widetilde{\varepsilon})^k \text{ for any } k \geq k_1.
$$

Thus, for almost all $\omega \in \{\zeta=\infty\}$, we have

$$
G(V(X(t))) \leq [t/T_2]\ln(\rho+\widetilde{\varepsilon}) \leq -\lambda t \ \text{ for } t \geq k_1 T_2. \tag{11.205}
$$

where $[t/T_2]$ is the integer part of t/T_2 and $\lambda = -\dfrac{\ln(\rho+\widetilde{\varepsilon})}{2T_2} > 0$. Since $G(y)$ is decreasing and maps $(0,h]$ onto $(-\infty,0]$, (11.177) follows from (11.195) and (11.205). The proof is complete. □

In Theorem 11.55, under the condition that $\alpha(t)$ is merely ergodic, we need and additional condition (11.175) to obtain the stability in probability of the system. If $\alpha(t)$ is strongly ergodic, the condition (11.175) can be removed.

Theorem 11.57 *Suppose that*

- *for any $T>0$ and a bounded function $f:\mathcal{M}\mapsto\mathbb{R}$, we have*

$$
\lim_{x\to 0}\sup_{i\in\mathcal{M}}\left\{\left|\mathbb{E}_{x,i}\int_0^T f(\alpha(s))ds - \mathbb{E}_i\int_0^T f(\hat{\alpha}(s)ds\right|\right\} = 0. \tag{11.206}
$$

- (H11.9) *is satisfied*
- *the Markov chain $\hat{\alpha}(t)$ is strongly ergodic with invariant probability measure $\boldsymbol{\nu}=(\nu_1,\nu_2,\ldots)$.*

Suppose further that (11.176) *is satisfied and* $\sum_{i\in\mathcal{M}} c_i\nu_i < 0$. *Then the conclusion of Theorem 11.55 holds.*

Remark 11.58 Lemma 11.68 in Section 11.7.3 implies that (11.206) holds if (H11.9) and (11.166) hold.

***Proof of Theorem* 11.57** Let $\lambda = -\sum_{i\in\mathcal{M}} c_i\nu_i$. Because of the uniform ergodicity of $\hat{\alpha}(t)$, there exists a $T > 0$ such that

$$\mathbb{E}_{0,i}\int_0^t c_{\alpha(s)}ds = \mathbb{E}_i\int_0^t c(\hat{\alpha}(s))ds \le -\frac{3\lambda}{4}t \quad \forall t \ge T, i \in \mathcal{M}. \tag{11.207}$$

By (11.206), there exists an $h_1 \in (0, h)$ such that

$$\mathbb{E}_{x,i}\int_0^T c_{\alpha(s)}ds \le -\frac{\lambda}{2}T \quad \forall\, |x| \le h_1, i \in \mathcal{M}. \tag{11.208}$$

In view of Lemma 11.53, there exists an $h_2 \in (0, h_1)$ such that

$$\bar{c}\mathbb{P}_{x,i}\{\tau_h < T\} \le \frac{\lambda}{4} \text{ provided } |x| \le h_2, i \in \mathcal{M}. \tag{11.209}$$

Applying (11.208) and (11.209) to (11.181), we have

$$\mathbb{E}_{x,i}H(T) \le -\frac{\lambda}{4}T \text{ if } 0 < |x| \le h_2, i \in \mathcal{M}. \tag{11.210}$$

Using (11.210), we can use arguments in the proof Theorem 11.55 to show that

$$\mathbb{E}_{x,i}e^{\theta H(T)} \le \exp\left\{-\frac{\theta\lambda T}{8}\right\} \text{ for } 0 < |x| < h_2 \tag{11.211}$$

for a sufficiently small $\theta > 0$. This implies that

$$\mathbb{E}_{x,i}U\big(X(T\wedge\tau_h)\big) \le \exp\left\{-\frac{\theta\lambda T}{8}\right\}U(x), \tag{11.212}$$

where $U(x) = \exp(\theta G(V(x)))$. Thus, $\{M_k := U\big(X\big((kT)\wedge\tau_h\big)\big), k = 0, 1, \ldots\}$ is a bounded supermartingale. Then we can easily obtain the stability in probability of the trivial solution. Moreover, proceeding as in Step 2 of the proof of Theorem 11.55, we can obtain the asymptotic stability as well as the rate of convergence. The arguments are actually simpler because (11.212) holds uniformly in $i \in \mathcal{M}$, rather than $i \in \{1, \ldots, k_0\}$ in the proof of Theorem 11.55. □

Remark 11.59 Consider the special case $g(y) \equiv y$. With this function, we have $U(X(t)) = V(X(t))$. Thus, if (H11.9) holds with $g(y) \equiv y$, then the conclusion on stability in Theorems 11.55 and 11.57 are still true without the condition (11.176) because we still have $\mathbb{E}V(X(t \wedge \tau_h)) \le V(x)e^{\bar{c}t}$, which can be used in place of (11.189). However, in order to obtain asymptotic stability and rate of convergence, (11.176) is needed. In that case, if the initial value is sufficiently closed to 0, $V(X(t))$ will converges exponentially fast to 0 with a large probability.

Theorem 11.60 *Consider the case that the state space of $\alpha(t)$ is finite, say $\mathcal{M} = \{1, \ldots, m_0\}$ for some positive integer m_0, rather than $m_0 = \infty$. Suppose that $Q(0)$ is irreducible and let $\boldsymbol{\nu}$ be the invariant probability measure of the Markov chain with generator $Q(0)$. If $\sum_{i\in\mathcal{M}} c_i\nu_i < 0$ then the trivial solution is asymptotically stable in probability, and for any $\varepsilon > 0$, there are $\lambda > 0$, $\delta > 0$ such that*

$$\mathbb{P}_{x,i}\left\{\lim_{t\to\infty}\frac{V(X(t))}{G^{-1}(-\lambda_3 t)} \le 1\right\} > 1-\varepsilon \textit{ for any } (x,i) \in B_\delta \times \mathcal{M}.$$

We now provide some conditions for instability in probability.

Theorem 11.61 *Suppose that the Markov chain $\hat{\alpha}(t)$ is ergodic with invariant probability measure $\boldsymbol{\nu} = (\nu_1, \nu_2, \ldots)$ and that there are functions $g \in \Gamma$, $V : D \mapsto \mathbb{R}_+$ such that*

- *$V(x) = 0$ if and only if $x = 0$*
- *$V(x)$ is continuous on D and twice continuously differentiable in $D \setminus \{0\}$.*
- *there is a bounded sequence of real numbers $\{c_i : i \in \mathcal{M}\}$ such that*

$$\mathcal{L}_i V(x) \ge c_i g(V(x))\,\forall x \in D \setminus \{0\}. \tag{11.213}$$

If (11.176) is satisfied and if $\sum_{i\in\mathcal{M}} c_i\nu_i < 0$ and $\limsup_{i\to\infty} c_i < 0$, then the trivial solution is unstable in probability.

Proof Define $G(y) = -\int_y^1 g^{-1}(z)dz$ as in Theorem 11.55. We have from Itô's formula,

$$\begin{aligned}
-G\big(V(X(\tau_h \wedge t))\big) &= -G(V(x)) - \int_0^{\tau_h\wedge t}\frac{\mathcal{L}_{\alpha(s)}V(X(s))}{g(V(X(s)))}ds \\
&\quad + \int_0^{\tau_h\wedge t}\frac{\dot{g}(V(X(s)))|V_x(X(s))\sigma(X(s),\alpha(s))|^2}{2g^2(V(X(s)))}ds \\
&\quad - \int_0^{\tau_h\wedge t}\frac{V_x(X(s))\sigma(X(s),\alpha(s))}{g(V(X(s)))}dW(s) \\
&\le -G(V(x)) + \widetilde{H}(t)
\end{aligned} \tag{11.214}$$

where

$$\widetilde{H}(t) = -\int_0^{\tau_h\wedge t} c_{\alpha(s)}ds - \int_0^{\tau_h\wedge t}\frac{V_x(X(s))\sigma(X(s),\alpha(s))}{g(V(X(s)))}dW(s).$$

Then using (11.214) and proceeding in the same manner as in the proof of Theorem 11.55 with $H(t)$ replaced with $\widetilde{H}(t)$, we have can find a sufficiently small $\widetilde{\theta}, \widetilde{\Delta} > 0$ and a sufficiently large $T_3 > 0$ such that

$$\mathbb{E}_{x,i} I_{\{\widetilde{\zeta} \geq kT_3\}} \widetilde{U}(X(kT_2)) \leq p^k \widetilde{U}(x), \text{ for } (x, i) \in B_h \times \mathcal{M}.$$

where $\widetilde{U}(x) = \exp\{-\widetilde{\theta} G(V(x))\}$, and $\widetilde{\zeta} = \inf\{k \geq 0 : U(X(kT_3)) \leq \widetilde{\Delta}^{-1}\}$. Note that, unlike $U(x)$, we have $\lim_{x \to 0} \widetilde{U}(x) = \infty$. Since $U(X(kT_3)) \geq \widetilde{\Delta}^{-1}$ if $\widetilde{\zeta} \geq k$, we have that

$$\mathbb{P}_{x,i}\{\widetilde{\zeta} = \infty\} = \lim_{k \to \infty} \mathbb{P}_{x,i}\{\widetilde{\zeta} \geq k\} = 0.$$

□

Similarly, we can obtain a counterpart of Theorem 11.57 for instability.

Theorem 11.62 *Suppose that the Markov chain $\hat{\alpha}(t)$ is strongly ergodic with invariant probability measure $\boldsymbol{\nu} = (\nu_1, \nu_2, \dots)$ and that there are functions $g \in \Gamma$, $V : D \mapsto \mathbb{R}_+$ such that*

- *$V(x) = 0$ if and only if $x = 0$*
- *$V(x)$ is continuous on D and twice continuously differentiable in $D \setminus \{0\}$.*
- *there is a bounded sequence of real numbers $\{c_i : i \in \mathcal{M}\}$ such that*

$$\mathcal{L}_i V(x) \geq c_i g(V(x)) \, \forall x \in D \setminus \{0\}. \tag{11.215}$$

If (11.176) *and* (11.166) *are satisfied and if $\sum_{i \in \mathcal{M}} c_i \nu_i > 0$ then the trivial solution is unstable in probability.*

11.6.3 Stability and Instability of Linearized Systems

Suppose that (11.206) is satisfied and that $\hat{\alpha}(t)$ is a strongly ergodic Markov chain.

(H11.10) Suppose that for $i \in \mathcal{M}$, there exist $b(i), \sigma_k(i) \in \mathbb{R}^{n \times n}$ bounded uniformly for $i \in \mathcal{M}$ such that

$$\xi_i(x) := b(x, i) - b(i)x, \quad \zeta_i(x) := \sigma(x, i) - (\sigma_1(i)x, \dots, \sigma_d(i)x)$$

satisfying

$$\lim_{x \to 0} \sup_{i \in \mathcal{M}} \left\{ \frac{|\xi_i(x)| \vee |\zeta_i(x)|}{|x|} \right\} = 0. \tag{11.216}$$

For $i \in \mathcal{M}$, $k \in \{1, \ldots, n\}$, let $\Lambda_{1,i}$ and $\Lambda_{2,i,k}$ be the maximum eigenvalues of $\dfrac{b(i) + b'(i)}{2}$ and $\sigma_k(i)\sigma_k'(i)$ respectively. Similarly, denote by $\lambda_{1,i}$ and $\lambda_{2,i,k}$ be the minimum eigenvalues of $\dfrac{b(i) + b'(i)}{2}$ and $\sigma_k(i)\sigma_k'(i)$ respectively.

Suppose that $\Lambda_{1,i}$ and $\Lambda_{2,i,k}$ are bounded in $i \in \mathcal{M}$ then we claim that if

$$\sum_{i \in \mathcal{M}} \nu_i \left(\Lambda_{1,i} + \frac{1}{2} \sum_{k=1}^{n} \Lambda_{2,i,k} \right) < 0,$$

then the trivial solution is asymptotic stable.

To show that, let $\varepsilon > 0$ be sufficiently small such that

$$\sum_{i \in \mathcal{M}} \nu_i \left(\varepsilon + \Lambda_{1,i} + \frac{1}{2} \sum_{k=1}^{n} \Lambda_{2,i,k} \right) < 0. \tag{11.217}$$

Define $V(x) = |x|^p$, carry out the calculation and obtain the estimates as in that of [164, Theorem 4.3], we can find a sufficiently small $p > 0$ and $\hbar > 0$ such that

$$\mathcal{L}_i V(x) \le p \left(\varepsilon + \Lambda_{1,i} + \frac{1}{2} \sum_{k=1}^{n} \Lambda_{2,i,k} \right) V(x) \text{ for } 0 < |x| < \hbar. \tag{11.218}$$

(Note that the existence of such p and $\hbar$ satisfying (11.218) uniformly for $i \in \mathcal{M}$ is due to (11.218) and the boundedness of $\Lambda_{1,i}$ and $\Lambda_{2,i,k}$.)

By (11.217) and (11.218), it follows from Theorem 11.57 that the trivial solution is asymptotic stable and for any $\varepsilon > 0$, there exists $\delta > 0$, $\lambda > 0$ such that

$$\mathbb{P}_{x,i} \left\{ \lim_{t \to \infty} e^{\lambda t} |X(t)| \le 1 \right\} \ge 1 - \varepsilon \text{ for } (x, i) \in B_\delta \times \mathcal{M}.$$

Similarly, if $\sum_{i \in \mathcal{M}} \nu_i \left(\lambda_{1,i} + \frac{1}{2} \sum_{k=1}^{n} \lambda_{2,i,k} \right) > 0$, and if $\lambda_{1,i}$ and $\lambda_{2,i,k}$ are bounded in $i \in \mathcal{M}, k = 1, \ldots, n$, the trivial solution is unstable. To sum up, we have the following result.

Proposition 11.63 *Let* (H11.10) *be satisfied. We claim that*

- *if* $\Lambda_{1,i}$ *and* $\Lambda_{2,i,k}$ *are bounded in* $i \in \mathcal{M}$ *and* $\sum_{i \in \mathcal{M}} \nu_i \left(\Lambda_{1,i} + \frac{1}{2} \sum_{k=1}^{n} \Lambda_{2,i,k} \right) < 0$, *then the trivial solution is asymptotically stable in probability;*
- *if* $\lambda_{1,i}$ *and* $\lambda_{2,i,k}$ *are bounded in* $i \in \mathcal{M}, k = 1, \ldots, n$, *and*

$$\sum_{i\in\mathcal{M}} \nu_i \left(\lambda_{1,i} + \frac{1}{2}\sum_{k=1}^{n} \lambda_{2,i,k}\right) > 0,$$

then the trivial solution is unstable in probability.

11.6.4 Examples

This subsection provides several examples.

Example 11.64 Consider a real-valued switching diffusion

$$dX(t) = b(\alpha(t))X(t)[|X(t)|^{\gamma} \vee 1]dt + \sigma(\alpha(t))\sin^2 X(t)dW(t),\ \ 0 < \gamma < 1, \tag{11.219}$$

where $a \vee b = \max(a, b)$ for two real numbers a and b, and $Q(x) = \big(q_{ij}(x)\big)_{\mathcal{M}\times\mathcal{M}}$ with

$$q_{ij}(x) = \begin{cases} -\check{p}_1(x) & \text{if } i = j = 1 \\ \check{p}_1(x) & \text{if } i = 1,\ j = 2 \\ -\hat{p}_i(x) - \check{p}_i(x) & \text{if } i = j \geq 2 \\ \hat{p}_i(x) & \text{if } i \geq 2,\ j = i - 1 \\ \check{p}_i(x) & \text{if } i \geq 2,\ j = i + 1. \end{cases}$$

Note that the drift grow faster than linear and the diffusion coefficient is locally like x^2 near the origin for the continuous state. Suppose that $b(i)$, $\sigma(i)$, $\check{p}_i(x)$, $\hat{p}_i(x)$ are bounded for $(x, i) \in \mathbb{R} \times \mathcal{M}$ and $\check{p}_i(x)$, $\hat{p}_i(x)$ are continuous in $\mathbb{R}^n$ for each $i \in \mathcal{M}$. It is well-known (see [3, Chapter 8]) that if

$$\nu^* := \sum_{k=2}^{\infty}\prod_{\ell=2}^{k} \frac{\check{p}_{\ell-1}(0)}{\hat{p}_{\ell}(0)} < \infty,$$

then $\hat{\alpha}(t)$ is ergodic with the invariant measure $\boldsymbol{\nu}$ given by

$$\nu_1 = \frac{1}{\nu^*},\ \nu_k = \frac{1}{\nu^*}\prod_{\ell=2}^{k} \frac{\check{p}_{\ell-1}(0)}{\hat{p}_{\ell}(0)}, k \geq 2.$$

We suppose that

$$\sum b(i)\nu_i < 0, \quad \text{and} \quad \limsup_{i\to\infty} b(i) < 0.$$

we will show that the trivial solution is stable. Let $0 < \varepsilon < -\sum b(i)\nu_i$, then $\sum[b(i) + \varepsilon]\nu_i < 0$. Let $V(x) = x^2$. We have $\mathcal{L}_i V(x) = 2b(i)|x|^{2+2\gamma} + \sigma^2(i)\sin^4(x)$. Since $\gamma < 1$ and $\sigma(i)$ is bounded, there exists an $\hbar > 0$ such that $\sigma^2(i)\sin^4(x) \leq \varepsilon|x|^{2+2\gamma}$ given that $|x| \leq \hbar$. then

$$\mathcal{L}_i V(x) \leq [2b(i) + \varepsilon]|x|^{2+2\gamma} = [2b(i) + \varepsilon]V^{1+\gamma}(x) \text{ in } [-\hbar, \hbar] \times \mathcal{M}.$$

By Theorem 11.55, the trivial solution is asymptotically stable in probability. Moreover, for the function $g(y) = y^{1+\gamma}$,

$$G(y) := -\int_y^1 \frac{1}{g(z)} ds = 1 - y^{-\gamma},\ y \in (0, 1] \text{ and } G^{-1}(-t) = \frac{1}{[t+1]^{1/\gamma}},\ \text{for } t \geq 0$$

Thus, for any $\varepsilon > 0$, there exists a $\delta > 0$ such that if $(x, i) \in [0, \delta] \times \mathcal{M}$, then, there exists a $\lambda > 0$ satisfying

$$\mathbb{P}_{x,i}\left\{\limsup_{t\to\infty} t^{1/\gamma} X^2(t) \leq \lambda\right\} > 1 - \varepsilon.$$

Example 11.65 This example considers a random-switching linear systems of differential equations:

$$dX(t) = A(\alpha(t))X(t)dt, \tag{11.220}$$

where $A(i) \in \mathbb{R}^{n\times n}$ satisfying $\sup_{i\in\mathcal{M}}\{|\lambda_i| \vee |\Lambda_i|\} < \infty$ with λ_i, Λ_i being the minimum and maximum eigenvalues of $A(i)$, respectively. Let

$$Q(x) = \begin{pmatrix} -1-\sin|x| & 1+\sin|x| & 0 & 0 & 0 & \cdots \\ 1+\sin|x| & -2-2\sin|x| & 1+\sin|x| & 0 & 0 & \cdots \\ 1+\sin|x| & 0 & -2-2\sin|x| & 1+\sin|x| & 0 & \cdots \\ 1+\sin|x| & 0 & 0 & -2-2\sin|x| & 1+\sin|x| & \cdots \\ \vdots & \vdots & \vdots & \vdots & \vdots & \ddots \end{pmatrix}.$$

By [3, Proposition 3.3], it is easy to verify that the Markov chain $\hat{\alpha}(t)$ with generators $Q(0)$ is strongly ergodic. Solving the system of equations $\boldsymbol{\nu}Q(0) = 0$ and $\sum \boldsymbol{\nu}_i = 1$, we obtain that the invariant measure of $\hat{\alpha}(t)$ is $(\nu_i)_{i=1}^{\infty} = (2^{-i})_{i=1}^{\infty}$. Thus, if $\sum \lambda_i 2^{-i} > 0$ the trivial solution to (11.220) is unstable. In case $\sum \Lambda_i 2^{-i} < 0$ the trivial solution to (11.220) is asymptotically stable in probability. In particular, suppose that $n = 2$ and $A(i)$ are upper triangle matrices, that is, $A(i) = \begin{pmatrix} a_i & b_i \\ 0 & c_i \end{pmatrix}$. If a_i and c_i are positive for $i \geq 2$, then the system $dX(t) = A(i)X(t)dt$ is unstable. However, if $a_1, c_1 < -\sup_{i\geq 2}\{a_i, c_i\}$, then $\sum(a_i \vee c_i)2^{-i} < 0$. Thus, the switching differential system is asymptotically stable. The stability of the system at state 1 and the switching process become a stabilizing factor.

On the other hand, if $a_i \wedge c_i$ is negative for $i \geq 2$, then the system $dX(t) = A(i)X(t)dt$ is asymptotically stable. Suppose further that $a_1, c_1 > \sup_{i\geq 2}\{-(a_i \wedge c_i)\}$, then $\sum(a_i \vee c_i)2^{-i} > 0$. Under this condition, the switching differential system is unstable.

11.6.5 Further Remarks

Using a new method, we provide sufficient conditions for stability and instability in probability of a class of regime-switching diffusion systems with switching states belonging to a countable set. The conditions are based on the relation of a "switching-independent" Lyapunov function and the generator of the switching part.

Although the systems under consideration are memoryless, the main results of this section hold if we assume that the switching intensities of $\alpha(t)$ is past-dependent, that is, (11.6) is satisfied. The main reason is that whether q_{ij}'s depend on $X(t)$ only or they depend on X_t, when considering the stability of the trivial solution, we will approximate $q_{ij}(\cdot)$ by $q_{ij}(0)$, the switching intensities of $\alpha(t)$ when $X(t) \equiv 0$.

Our method can also be applied to regime-switching jump diffusion processes. The results obtain by using our method will generalize existing results (e.g., [282, 299]) to the case of regime-switching jump diffusions with countable regimes.

On the other hand, there is a gap between sufficient conditions for stability and instability in Proposition 4.1. To overcome the difficulty, we need to use a polar coordinate transformation to decompose of $X(t)$ into the radial part $r(t) = |X(t)|$ and the angular part $Y(t) = X(t)/r(t)$. Then, the Lyapunov exponents with respect to invariant measures of the linearized process of $(Y(t), \alpha(t))$ will determine whether or not the system is stable. This approach has been used to treat many linear and linearized stochastic systems (e.g., [8, 14, 156]). In our setting, the switching $\alpha(t)$ take values in a noncompact space; thus, it is more difficult to examine invariant measures. We will address this problem together with necessary conditions of stability in a subsequent paper.

11.7 Proofs of Several Results

This section is devoted to the proofs of some technical results. To simplify the notation, we denote by $\mathbb{P}_{\phi,i}$ the probability measure conditional on the initial value (ϕ, i), that is, for any $t > 0$,

$$\begin{aligned}\mathbb{P}_{\phi,i}\{(X_t,\alpha(t))\in\cdot\} &= \mathbb{P}\{(X_t^{\phi,i},\alpha^{\phi,i}(t))\in\cdot\},\\ \mathbb{P}_{\phi,i}\{(Y_t,\beta(t))\in\cdot\} &= \mathbb{P}\{(Y_t^{\phi,i},\beta^{\phi,i}(t))\in\cdot\}, \quad \text{and}\\ \mathbb{P}_{\phi,i}\{(Z_t,\gamma(t))\in\cdot\} &= \mathbb{P}\{(Z_t^{\phi,i},\gamma^{\phi,i}(t))\in\cdot\}.\end{aligned}$$

Let $\mathbb{E}_{\phi,i}$ be the expectation associated with $\mathbb{P}_{\phi,i}$.

11.7.1 Proofs of Some Results in Sects. 11.3 and 11.4

First, we prove the following result.

Lemma 11.66 *Let either* (H11.1) *or* (H11.3) *combined with* (ii) *of* (H11.1) *be satisfied. Assume further that* $q_i(\cdot), q_{ij}(\cdot)$ *are continuous functions in* $\mathscr{C}$ *for each* $i, j \in \mathcal{M}$. *Then the solution* $(X_t, \alpha(t))$ *to* (11.7) *satisfies* (11.5) *and* (11.6).

Proof It is clear that the solution $(X_t, \alpha(t))$ to (11.7) satisfies (11.5). Fix $\phi \in \mathscr{C}$, $i, j \in \mathcal{M}, i \neq j$. Applying the generalized Itô formula (11.9) to the function $V(x,k) := I_{\{k=j\}}$, $(x,k) \in \mathbb{R}^n \times \mathcal{M}$, we have

$$\mathbb{P}_{\phi,i}\{\alpha(\Delta)=j\} = \mathbb{E}_{\phi,i}V(X(\Delta),\alpha(\Delta)) = \mathbb{E}_{\phi,i}\int_0^\Delta q_{\alpha(t-),j}(X_t)dt, \quad \text{for } \Delta > 0.$$

where $q_{ii}(\phi) := -q_i(\phi) = -\sum_{j\neq i} q_{ij}(\phi)$. Since $\alpha(t)$ is càdlàg and $X(t)$ is continuous, $\lim_{t\to 0^+}\alpha(t) = i$ and $\lim_{t\to 0^+} X_t = \phi$ $\mathbb{P}_{\phi,i}$-a.s. In light of the continuity of $q_{ij}(\cdot)$, we obtain $\lim_{t\to 0^+} q_{\alpha(t-),j}(X_t) = q_{ij}(\phi)$ $\mathbb{P}_{\phi,i}$-a.s. which implies that

$$\lim_{\Delta\to 0^+}\frac{1}{\Delta}\int_0^\Delta q_{\alpha(t-),j}(X_t)dt = q_{ij}(\phi) \;\; \mathbb{P}_{\phi,i}\text{-a.s.}$$

Note that $q_{ij}(\cdot)$ is uniformly bounded, so is $\frac{1}{\Delta}\int_0^\Delta q_{\alpha(t-),j}(X_t)dt$. By virtue of the Lebesgue dominated convergence theorem, we have

$$\lim_{\Delta\to 0^+}\frac{\mathbb{P}_{\phi,i}\{\alpha(\Delta)=j\}}{\Delta} = \lim_{\Delta\to 0^+}\mathbb{E}_{\phi,i}\left(\frac{1}{\Delta}\int_0^\Delta q_{\alpha(t-),j}(X_t)dt\right) = q_{ij}(\phi). \tag{11.221}$$

Likewise, using the function $V(x,k) = I_{\{k\neq i\}}$ and the generalized Itô formula, we obtain

$$\lim_{\Delta\to 0^+}\frac{1-\mathbb{P}_{\phi,i}\{\alpha(\Delta)=i\}}{\Delta} = q_i(\phi). \tag{11.222}$$

The proof is complete by noting that (11.6) follows from (11.221), (11.222), and the Markov property of $(X(t), \alpha(t))$. □

***Proof of Lemma* 11.12** To prove claim (i), we apply the generalized Itô formula to $V(j) = I_{\{j=i\}}$:

$$V(\beta(\lambda_1 \wedge t)) = 1 - \int_0^{\lambda_1 \wedge t} q_i(Y_s)ds + \int_0^{\lambda_1 \wedge t} \int_{\mathbb{R}} \Big(V\big(i + h(Y_t, i, z)\big) - 1\Big)\widetilde{\mathfrak{p}}(ds, dz).$$

Since $W(\cdot)$ is independent of the Poisson random measure, taking the conditional expectation with respect to $\mathcal{F}_T^W$ yields

$$\begin{aligned}\mathbb{E}_{\phi,i}\big[I_{\{\lambda_1 > t\}}\big|\mathcal{F}_T^W\big] =& \mathbb{E}_{\phi,i}\big[V(\beta(\lambda_1 \wedge t))\big|\mathcal{F}_T^W\big] = -\mathbb{E}_{\phi,i}\big[\int_0^{\lambda_1 \wedge t} q_i(Y_s)ds\big|\mathcal{F}_T^W\big] + 1 \\ =& - \mathbb{E}_{\phi,i}\big[\int_0^t q_i(Y_s)ds I_{\{\lambda_1 > s\}}\big|\mathcal{F}_T^W\big] + 1 \\ =& - \int_0^t q_i(Y_s)\mathbb{E}_{\phi,i}\big[I_{\{\lambda_1 > s\}}\big|\mathcal{F}_T^W\big]ds + 1.\end{aligned}$$

Hence,

$$\frac{d}{dt}\mathbb{E}_{\phi,i}\big[I_{\{\lambda_1 > t\}}\big|\mathcal{F}_T^W\big] = -q_i(Y_t)\mathbb{E}_{\phi,i}\big[I_{\{\lambda_1 > t\}}\big|\mathcal{F}_T^W\big].$$

Since $\mathbb{E}_{\phi,i}\big[I_{\{\lambda_1 > 0\}}\big|\mathcal{F}_T^W\big] = 1$, we obtain

$$\mathbb{P}_{\phi,i}\big(\{\lambda_1 > t\}\big|\mathcal{F}_T^W\big) = \mathbb{E}_{\phi,i}\big[I_{\{\lambda_1 > t\}}\big|\mathcal{F}_T^W\big] = \exp\Big(- \int_0^t q_i(Y_s)ds\Big). \tag{11.223}$$

Next we proceed to prove claim (ii). First, we try to find the distribution of (λ_1, β_1) conditioned on $\mathcal{F}_T^W$ when $\lambda_1 \in [0, T]$. Fix $j \neq i$ and let $f(t, k) : [0, T] \times \mathcal{M} \to \mathcal{M}$ be any bounded measurable function satisfying $f(t, k) = 0$ if $k \neq j$. Applying the generalized Itô formula, we obtain

$$\begin{aligned}f(\lambda_1 \wedge T, \beta(\lambda_1 \wedge T)) =& \int_0^{\lambda_1 \wedge T} q_{ij}(Y_t) f(t, j)dt \\ &+ \int_0^{\lambda_1 \wedge T} \int_{\mathbb{R}} \Big(f\big(s, i + h(Y_t, i, z)\big) - f(Y_t, i)\Big)\widetilde{\mathfrak{p}}(ds, dz).\end{aligned}$$

Since $W(\cdot)$ is independent of the Poisson random measure, taking the conditional expectation with respect to $\mathcal{F}_T^W$, we have

$$\mathbb{E}_{\phi,i}\big[f(\lambda_1 \wedge T, \beta(\lambda_1 \wedge T))\big|\mathcal{F}_T^W\big] = \mathbb{E}_{\phi,i}\Big[\int_0^{\lambda_1 \wedge T} q_{ij}(Y_t) f(t, j)dt\Big|\mathcal{F}_T^W\Big]$$

$$
\begin{aligned}
&= \mathbb{E}_{\phi,i}\Big[\int_0^T q_{ij}(Y_t) f(t,j) dt I_{\{\lambda_1 > t\}} dt \Big| \mathcal{F}_T^W\Big] \\
&= \int_0^T q_{ij}(Y_t) f(t,j) \mathbb{E}\big[I_{\{\lambda_1 > t\}} | \mathcal{F}_T^W\big] dt \\
&= \int_0^T q_{ij}(Y_t) f(t,j) \exp(-\int_0^t q_i(Y_s) ds) dt.
\end{aligned}
$$

As a result, for $t \in [0, T]$,

$$
\mathbb{P}_{\phi,i}\{\lambda_1 \in dt, \beta(\lambda_1) = j | \mathcal{F}_T^W\} = q_{ij}(Y_t) \exp(-\int_0^t q_i(Y_s) ds) dt.
$$

Thus,

$$
\begin{aligned}
&\mathbb{E}_{\phi,i}\Big[g(Y_{(1)}, \lambda_1, \beta_1) I_{\{\lambda_1 \le T\}} \Big| \mathcal{F}_T^W\Big] \\
&\quad = \sum_{j=1, j\neq i}^{\infty} \int_0^T g(Y_t, t, j) \mathbb{P}_{\phi,i}\{\lambda_1 \in dt, \beta(\lambda_1) = j | \mathcal{F}_T^W\} \\
&\quad = \sum_{j=1, j\neq i}^{\infty} \int_0^T g(Y_t, t, j) q_{ij}(Y_t) \exp(-\int_0^t q_i(Y_s) ds) dt
\end{aligned}
$$

as desired. □

***Proof of Proposition* 11.13** First, we prove (11.25) for the case $l = 0$. Since $(X_t, \alpha(t)) = (Y_t, \lambda(t))$ up to the moment $\alpha_1 = \lambda_1$, we have

$$
\begin{aligned}
&\mathbb{E}_{\phi,i}\Big[f(X_T, \alpha(T)) I_{\{\tau_1 > T\}}\Big] = \mathbb{E}_{\phi,i}\Big[f(Y_T, i) I_{\{\lambda_1 > T\}}\Big] \\
&= \mathbb{E}_{\phi,i}\Big[\mathbb{E}_{\phi,i}\big(f(Y_T, i) I_{\{\lambda_1 > T\}} | \mathcal{F}_T^W\big)\Big] = \mathbb{E}_{\phi,i}\Big[f(Y_T, i) \mathbb{E}_{\phi,i}\big(I_{\{\lambda_1 > T\}} | \mathcal{F}_T^W\big)\Big] \\
&= \mathbb{E}_{\phi,i}\Big[f(Y_T, i) \exp\Big(-\int_0^T q_i(Y_s) ds\Big)\Big],
\end{aligned}
\tag{11.224}
$$

where the last equality is consequence of (i) of Lemma 11.12. Since $\gamma(\cdot)$ and $Y(\cdot)$ are independent, $Z_t = Y_t$ up to the moment θ_1 and $\mathbb{P}_{\phi,i}\{\theta_1 > T\} = \exp(-T)$, we obtain

$$
\begin{aligned}
&\mathbb{E}_{\phi,i}\Big(f(Z_T, \gamma(T)) I_{\{\theta_1 > T\}} \exp\{-\int_0^T q_i(Z_s) ds\}\Big) \\
&\quad = \mathbb{E}_{\phi,i}\Big(f(Y_T, i) I_{\{\theta_1 > T\}} \exp\{-\int_0^T q_i(Y_s) ds\}\Big)
\end{aligned}
$$

$$
\begin{aligned}
&= \mathbb{P}_{\phi,i}\{\theta_1 > T\}\mathbb{E}_{\phi,i}\Big(f(Y_T, i)\exp\{-\int_0^T q_i(Y_s)ds\}\Big) \\
&= \exp(-T)\mathbb{E}_{\phi,i}\Big(f(Y_T, i)\exp\{-\int_0^T q_i(Y_s)ds\}\Big). \qquad (11.225)
\end{aligned}
$$

From (11.224) and (11.225), we have for $t \in [0, T]$ that

$$
\mathbb{E}f(X_T, \alpha(T))I_{\{\tau_1 > T\}} = \exp(T)\mathbb{E}\Big(f(Z_T, \gamma(T))I_{\{\theta_1 > T\}}\exp\{-\int_0^T q_i(Z_s)ds\}\Big). \qquad (11.226)
$$

We now prove (11.25) for $l = 1$. Let $g(\phi, t, i), \tilde{g}(\phi, t, i) : \mathscr{C} \times [0, \infty) \times \mathcal{M} \to \mathbb{R}$ be bounded measurable functions and $g(\phi, t, i) = \tilde{g}(\phi, t, i) = 0$ if $t > T$. It follows from (ii) of Lemma 11.12 that

$$
\begin{aligned}
&\mathbb{E}_{\phi,i}g(X_{(1)}, \tau_1, \alpha_1) = \mathbb{E}_{\phi,i}g(Y_{(1)}, \lambda_1, \beta_1) \\
&= \sum_{i_1 \neq i}\int_0^T \mathbb{E}_{\phi,i}\Big(g(Y_t, t, i_1)q_{ii_1}(Y_t)\exp(-\int_0^t q_i(Y_s)ds)\Big)dt. \qquad (11.227)
\end{aligned}
$$

On the other hand,

$$
\begin{aligned}
&\mathbb{E}_{\phi,i}\Big[\tilde{g}(Z_{(1)}, \theta_1, \gamma_1)\exp(-\int_0^{\theta_1} q_i(Z_s)ds)\Big] \\
&= \mathbb{E}_{\phi,i}\Big[\tilde{g}(Y_{(1)}, \theta_1, \gamma_t)\exp(-\int_0^{\theta_1} q_i(Y_s)ds)\Big] \\
&= \sum_{i_1 \neq i}\int_0^T \mathbb{E}_{\phi,i}\Big(\tilde{g}(Y_t, t, i_1)\exp(-\int_0^t q_i(Y_s)ds)\Big)\mathbb{P}_{\phi,i}\{\theta_1 \in dt, \gamma_1 = i_1\} \\
&= \sum_{i_1 \neq i}\int_0^T \mathbb{E}_{\phi,i}\Big(\tilde{g}(Y_t, t, i_1)\exp(-\int_0^t q_i(Y_s)ds)\Big)\rho_{ii_1}\exp(-t)dt. \qquad (11.228)
\end{aligned}
$$

Substituting $\tilde{g}(\phi, t, i) = g(\phi, t, i)\exp(t) \times \dfrac{q_{ii}(\phi)}{\rho_{ii}}$ into (11.228), we have

$$
\begin{aligned}
&\mathbb{E}_{\phi,i}\Big[g(Z_{(1)}, \theta_1, \gamma_1)\exp(\theta_1) \times \frac{q_{i\gamma_1}(Z_s)}{\rho_{i\gamma_1}}\exp(-\int_0^{\theta_1} q_i(Z_s)ds)\Big] \\
&= \sum_{i_1 \neq i}\int_0^T \mathbb{E}_{\phi,i}\Big[g(Y_t, t, i_1)\exp(t)\frac{q_{ii_1}(Y_t)}{\rho_{ii_1}}\exp(-\int_0^t q_i(Y_s)ds)\Big]\rho_{ii_1}\exp(-t)dt \\
&= \sum_{i_1 \neq i}\int_0^T \mathbb{E}_{\phi,i}\Big[g(Y_t, t, i_1)q_{ii_1}(Y_t)\exp(-\int_0^t q_i(Y_s)ds)\Big]dt. \qquad (11.229)
\end{aligned}
$$

It follows from (11.227) and (11.229) that

$$\begin{aligned}&\mathbb{P}_{\phi,i}\{\tau_1 \in dt, \alpha_1 = i_1, X_{(1)} \in d\phi_1\}\\ &\quad = \mathbb{E}_{\phi,i}\Big[I_{\{\theta_1 \in dt, \gamma_1 = i_1, Z_{(1)} \in d\phi_1\}} \exp(t) \times \frac{q_{ii_i}(Z_t)}{\rho_{ii_1}} \exp(-\int_0^t q_i(Z_s)ds)\Big].\end{aligned} \tag{11.230}$$

We now use the strong Markov property of $(X_t, \alpha(t))$ and $(Z_t, \gamma(t))$, (11.230) as well as (11.226) with ϕ, i, T replaced by $\phi_1, i_1, T-t$, respectively;

$$\begin{aligned}&\mathbb{E}_{\phi,i} f(X_T, \alpha(T)) I_{\{\tau_1 \le T < \tau_2\}} I_{\{\alpha_1 = i_1\}}\\ &= \int_0^T \int_{\mathscr{C}} \Big[\mathbb{P}\{\tau_1 \in dt, \alpha_1 = i_1, X_{(1)} \in d\phi_1\} \times \mathbb{E}_{\phi_1,i_1} f(X_{T-t}, \alpha(T-t)) I_{\{\tau_1 > T-t\}}\Big]\\ &= \int_0^T \int_{\mathscr{C}} \Big[\mathbb{E}_{\phi,i}\Big(I_{\{\theta_1 \in dt, \gamma_1 = i_1, Z_{(1)} \in d\phi_1\}} \exp(t) \times \frac{q_{ii_i}(Z_s)}{\rho_{ii_1}} \exp(-\int_0^t q_i(Z_s)ds)\Big)\\ &\qquad \times \exp(T-t)\mathbb{E}_{\phi_1,i_1} f(Z_{T-t}, \gamma(T-t)) I_{\{\theta_1 > T-t\}} \exp\{-\int_0^{T-t} q_{i_1}(Z_s)ds\}\Big]\\ &= \exp(T)\mathbb{E}_{\phi,i}\Big[f(Z_T, \gamma(T)) I_{\{\theta_1 \le T < \theta_2\}} I_{\{\gamma_1 = i_1\}} \exp\Big(-\int_{\theta_1}^T q_{i_1}(Z_s)ds\Big)\\ &\qquad \frac{q_{ii_1}(Z_{(1)})}{\rho_{ii_1}} \exp\Big(-\int_0^{\theta_1} q_i(Z_s)ds\Big)\Big].\end{aligned} \tag{11.231}$$

We have already proved (11.25) for $l = 0, 1$. Using the same argument, the induction, and the strong Markov property of $(X_t, \alpha(t))$ and $(Z_t, \gamma(t))$, we can obtain (11.25) for any $l \in \mathcal{M}$. □

Proof of Lemma **11.15** By (11.19), we can find $m \in \mathcal{M}$ such that

$$\mathbb{P}_{\phi,i_0}\big\{\tau_{m+1} < T\big\} < \frac{\Delta}{2}, \ \forall (\phi, i) \in \mathscr{C} \times \mathcal{M}. \tag{11.232}$$

Now, let $\varepsilon = \varepsilon(\Delta) > 0$ (to be specified later). In view of [201, Theorem 4.3, p. 61], for each $i \in \mathcal{M}$, there is a constant C_i such that

$$\mathbb{E}_{\phi,i_0}|Y(t) - Y(s)|^6 \le C_i|t-s|^3 \,\forall t, s \in [0, T], \ \forall \|\phi\| \le R+1. \tag{11.233}$$

By the Kolmogorov-Chentsov theorem (see [149, Theorem 2.8]), there is a positive random variable $h_i^{\phi}(\omega)$ such that

$$\mathbb{P}_{\phi,i_0}\Big\{\sup_{t,s\in[0,T], 0<t-s<h_i^{\phi}(\omega)} \frac{|Y^{\phi,i}(t) - Y(s)|}{(t-s)^{0.25}} \le 4\Big\} = 1.$$

Since C_i in (11.233) does not depend on $\phi \in \{\psi : \|\psi\| \le R+1\}$, it can be seen from the proof of the Kolmogorov-Chentsov theorem that for any $\varepsilon > 0$, there is a constant $h_i > 0$ satisfying

$$\mathbb{P}_{\phi,i_0}\left\{ \sup_{t,s\in[0,T],0<t-s<h_i} \frac{|Y(t)-Y(s)|}{(s-t)^{0.25}} \le 4\right\} > 1-\varepsilon,\ \forall\, \|\phi\| \le R+1. \quad (11.234)$$

Without loss of generality, we can choose $h_{i+1} < h_i, \forall i \in \mathcal{M}$. Let

$$h_{i,T} = \Big\{\psi(\cdot) \in \mathscr{C}([0,T],\mathbb{R}) : \|\psi\| \le R+1 \text{ and } \sup_{t,s\in[0,T],0<t-s<h_i} \frac{|\psi(s)-\psi(t)|}{(s-t)^{0.25}} \le 4\Big\},$$

and

$$h_i = \Big\{\psi(\cdot) \in \mathscr{C} : \|\psi\| \le R+1 \text{ and } \sup_{t,s\in[-r,0],0<t-s<h_i} \frac{|\psi(s)-\psi(t)|}{(s-t)^{0.25}} \le 4\Big\}.$$

Hence $h_{i+1,T} \supset h_{i,T}$ and $h_{i+1} \supset h_i$. For $d > 0$ and a compact set $\mathcal{K} \subset \mathscr{C}$, we define

$$\mathcal{K}_d := \{\psi \in \mathscr{C} : \exists \phi \in \mathcal{K} \text{ such that } \|\psi - \phi\| < d\}.$$

Define $\mathcal{K}^0 = \{\psi(\cdot) = \phi_0(\cdot) + c : c \in \mathbb{R}^n, |c| \le 1\}$, which is compact, and $\mathcal{K}^1 = \mathcal{K}^0 \uplus h_{i_0}$. For each $\phi \in \mathcal{K}^1$, there is $n_{\phi,i_0} > i_0$ such that

$$\sum_{k=n_{\phi,i_0}+1}^{\infty} q_{i_0,k}(\phi) = q_{i_0}(\phi) - \sum_{k=1,k\ne i_0}^{n_{\phi,i_0}} q_{i_0k}(\phi) < \frac{\varepsilon}{2}.$$

By the continuity of q_{i_0} and $q_{i_0k}(\phi)$, there is a $d_{\phi,i_0} > 0$ such that

$$\sum_{k=n(\phi)+1}^{\infty} q_{i_0,k}(\phi') = q_{i_0}(\phi') - \sum_{k=1,k\ne i_0}^{n_{\phi,i_0}} q_{i_0k}(\phi') < \varepsilon\ \forall\, \|\phi' - \phi\| < d_{\phi,i_0}.$$

Since $\mathcal{K}^1$ is compact, there exist $n_1 > 0$ and $d_1 > 0$ such that

$$\sum_{k=n_1+1}^{\infty} q_{i_0,k}(\phi) < \varepsilon\ \forall \phi \in \mathcal{K}^1_{d_1}.$$

Define $\mathcal{K}^2 = \mathcal{K}^1 \uplus h_{n_1}$. Using the compactness of $\mathcal{K}^2$, there exist $n_2 > n_1$ and $d_2 \in (0, d_1]$ such that

$$\sum_{k=n_2+1}^{\infty} q_{i,k}(\phi) < \varepsilon \,\forall i \in N_{n_1}, \phi \in \mathcal{K}^2_{d_2}.$$

Continuing this way, for $\mathcal{K}^m = \mathcal{K}^{m-1} \uplus h_{n_{m-1}}$, there exists $n_m > n_{m-1}$ and $d_m \in (0, d_{m-1}]$ such that

$$\sum_{k=n_m+1}^{\infty} q_{i,k}(\phi) < \varepsilon \,\forall i \in N_{n_{m-1}}, \phi \in \mathcal{K}^m_{d_m}.$$

Set $\mathcal{K}^{\phi,1} = \{\phi\} \uplus h_{i_0}$ and $\mathcal{K}^{\phi,k} = \mathcal{K}^{\phi,k-1} \uplus h_{n_{k-1}}$ for $\phi \in \mathscr{C}$ and $k = 2, \ldots, m$. It is not difficult to verify that

$$\mathcal{K}^{\phi,k} \subset \mathcal{K}^k_{d_k} \,\forall k = 1, \ldots, m, \text{ for } \|\phi - \phi_0\| < \frac{d_m}{2}. \tag{11.235}$$

Denote by $\{Y(\cdot) \in h_{n_0,T}\}$ the event $\{t \in [0, T] \mapsto Y(t)$ being a function belonging to $h_{n_0,T}\}$. Clearly, if $Y(\cdot) \in h_{i_0,T}$, then $Y_t \in K^{\phi,1} \,\forall t \in [0, T]$. Thus, we can proceed as follows:

$$\begin{aligned}
&\mathbb{P}_{\phi,i_0}\big\{\tau_1 \le T, \big(X_{(1)}, \alpha_1\big) \notin \mathcal{K}^{\phi,1} \times N_{n_1}\big\} \\
&= \mathbb{P}_{\phi,i_0}\big(\{\tau_1 \le T, \alpha_1 > n_1\} \cup \{\tau_1 \le T, X(\tau_1) \notin \mathcal{K}^{\phi,1}\}\big) \\
&= \mathbb{P}_{\phi,i_0}\big(\{\lambda_1 \le T, \beta_1 > n_1\} \cup \{\lambda_1 \le T, Y(\lambda_1) \notin \mathcal{K}^{\phi,1}\}\big) \\
&\le \mathbb{P}_{\phi,i_0}\{\lambda_1 \le T, Y(\cdot) \in h_{n_0,T}, \beta_1 > n_1\} + \mathbb{P}\{Y(\cdot) \notin h^T_{n_0}\} \\
&\le \mathbb{E}_{\phi,i_0}\big[\mathbb{E}\big(I_{\{Y(\cdot)\in h_{n_0,T}\}} I_{\{\lambda_1 \le T, \beta_1 > n_1\}} | \mathcal{F}^W_T\big)\big] + \varepsilon \\
&= \mathbb{E}_{\phi,i_0}\big[I_{\{Y(\cdot)\in h_{n_0,T}\}} \mathbb{E}\big(I_{\{\lambda_1 \le T, \beta_1 > n_1\}} | \mathcal{F}^W_T\big)\big] + \varepsilon \\
&= \mathbb{E}_{\phi,i_0}\Big[I_{\{Y(\cdot)\in h_{n_0,T}\}} \int_0^T \sum_{i>n_1} q_{i_0,i}(Y_t) \exp\Big(-\int_0^t q_{i_0}(Y_s)ds\Big)dt\Big] + \varepsilon \\
&\le \mathbb{E}_{\phi,i_0}\Big[I_{\{Y(\cdot)\in h_{n_0,T}\}} \int_0^T \varepsilon dt\Big] + \varepsilon \le (T+1)\varepsilon.
\end{aligned} \tag{11.236}$$

Similarly, if $(\phi_1, i_1) \in \mathcal{K}^{\phi,1} \times N_1$, then $\mathbb{P}_{\phi_1,i_1}\big\{\tau_1 \le T, \big(X_{(1)}, \alpha_1\big) \notin \mathcal{K}^{\phi,2} \times N_{n_2}\big\} \le (T + 1)\varepsilon$. Using the strong Markov property of $(X_t, \alpha(t))$, we obtain

$$
\begin{aligned}
&\mathbb{P}_{\phi,i_0}\Big\{\tau_1 < T,\, \big(X_{(1)},\alpha_1\big) \in \mathcal{K}^{\phi,1} \times N_{n_1},\, \tau_2 \le T,\, \big(X_{(2)},\alpha_2\big) \notin \mathcal{K}^{\phi,2} \times N_{n_2}\Big\}\\
&\le \mathbb{P}_{\phi,i_0}\big\{\tau_1 < T,\, \big(X_{(1)},\alpha_1\big) \in \mathcal{K}^{\phi,1} \times N_{n_1}\big\}\\
&\times \mathbb{P}_{\phi,i_0}\Big[\tau_2 \le T+\tau_1,\, \big(X_{(2)},\alpha_2\big) \notin \mathcal{K}^{\phi,2} \times N_{n_2} \Big|\, \tau_1 < T,\, \big(X_{(1)},\alpha_1)\big) \in \mathcal{K}^{\phi,1} \times N_{n_1}\Big]\\
&\le \sup_{(\phi_1,i_1)\in\mathcal{K}_1^{\phi}\times N_{n_1}} \mathbb{P}_{\phi_1,i_1}\big\{\tau_1 \le T,\, \big(X_{(1)},\alpha_1\big) \notin \mathcal{K}^{\phi,2}\times N_{n_2}\big\} \le (T+1)\varepsilon.
\end{aligned}
$$

Continuing this way, we can show that for any $k = 1,\ldots,m$ that

$$
\begin{aligned}
&\mathbb{P}_{\phi,i_0}\Big\{\tau_k \le T,\, \big(X_{\tau_k},\alpha_k\big) \notin \mathcal{K}_k^{\phi} \times N_{n_k},\, \big(X_{(j)},\alpha_j\big) \in \mathcal{K}_j^{\phi} \times N_{n_j},\, j = 1,\ldots,k-1\Big\}\\
&\qquad \le (T+1)\varepsilon.
\end{aligned}
\tag{11.237}
$$

Consequently,

$$
\mathbb{P}_{\phi,i_0}\Big\{\exists k = 1,\ldots,m : \tau_k \le T \text{ and } \big(X_{(k)},\alpha_k\big) \notin \mathcal{K}_k^{\phi} \times N_{n_k}\Big\} \le (T+1)m\varepsilon.
$$

Hence, if we choose $\varepsilon = \dfrac{1}{2m(T+1)}\Delta$, then

$$
\begin{aligned}
&\mathbb{P}_{\phi,i_0}\Big\{\forall k = 1,\ldots,m : \tau_k > T \text{ or } \alpha_k \in N_{n_k}\Big\}\\
&\ge \mathbb{P}\Big\{\forall k = 1,\ldots,m : \tau_k > T \text{ or } \big(X_{(k)},\alpha_k\big) \in \mathcal{K}_k^{\phi} \times N_{n_k}\Big\} \ge 1 - \frac{\Delta}{2}.
\end{aligned}
\tag{11.238}
$$

It follows from (11.232) and (11.238) that

$$
\mathbb{P}_{\phi,i_0}\Big(\{\tau_{m+1} > T\} \cap \Big\{\forall k = 1,\ldots,m : \tau_k > T \text{ or } \alpha_k \in N_{n_k}\Big\}\Big) \ge 1 - \Delta.
$$

It is easily verified that if $\omega \in \{\tau_{m+1} > T\} \cap \{\forall k = 1,\ldots,m : \tau_k > T \text{ or } \alpha_k \in N_{n_k}\}$, then $\alpha(t) \in N_{n_m}$, $\forall t \in [0,T]$. The assertion of the lemma is proved. □

11.7.2 Proofs of Some Results in Section 11.5

Let $Y^{x,i}(t)$ be the solution to

$$
dY(t) = b(Y(t),i)dt + \sigma(Y(t),i)dW(t),\ t \ge 0 \tag{11.239}
$$

with initial condition $(x, i) \in \mathbb{R}^n \times \mathcal{M}$. For $(\phi, i) \in \mathscr{C} \times \mathcal{M}$, we denote by $Y^{\phi,i}(t), t \geq -r$ the process satisfying $Y^{\phi,i}(t) = \phi$ if $t \in [-r, 0]$ and $Y^{\phi,i}(t)$ solves (11.239) for $t > 0$. Clearly $Y^{\phi,i}(t) = Y^{\phi(0),i}(t)$ for $t \geq 0$. Let $\beta^{\phi,i}$ be the solution to

$$\beta^{\phi,i}(t) = i + \int_0^t \int_{\mathbb{R}} h(Y_t^{\phi,i}, \beta^{\phi,i}(t-), z)\mathfrak{p}(dt, dz), t \geq 0 \tag{11.240}$$

satisfying $Y^{\phi,i}(t) = \phi(t)$ in $[-r, 0]$ and $\beta^{\phi,i}(0) = i$. Let $\xi_1^{\phi,i}(t)$ and $\lambda_1^{\phi,i}(t)$ be the first jump times of $\alpha^{\phi,i}(t)$ and $\beta^{\phi,i}(t)$, respectively. Clearly, we have that

$$X^{\phi,i}(t) = Y^{\phi,i}(t),\ \alpha^{\phi,i}(t) = \beta^{\phi,i}(t) \text{ up to } \xi_1^{\phi,i}(t) = \lambda_1^{\phi,i}(t). \tag{11.241}$$

***Proof of Lemma* 11.24** Since $q_{ij}(\cdot)$ is continuous, there is an $\varepsilon \in (0, 1)$ such that $q_{ij}(\psi) > 0$ given that $\|\psi - \phi\| < \varepsilon$. Let $M_\phi = \sup_{\psi \in \mathscr{C}, \|\psi-\phi\|<1}\{q_i(\psi)\} < \infty$. Let $\delta_1 > 0$ such that

$$|\phi(s) - \phi(s')| < \frac{\varepsilon}{5} \text{ provided } |s - s'| < \delta_1, s, s' \in [-r, 0]. \tag{11.242}$$

Under either (H11.3) or (H11.4), standard arguments show that there exists a sufficiently small $\delta_2 \in (0, \delta_1]$ satisfying

$$\mathbb{P}_{\psi,i}\left\{|Y(t) - \psi(0)| \leq \frac{\varepsilon}{5}\ \forall t \in [0, \delta_2]\right\} \geq \frac{1}{2},\ \forall \psi \in \mathscr{C}, \|\psi - \phi\| < \varepsilon, \tag{11.243}$$

and

$$\mathbb{P}_{\psi',j}\left\{|Y(t) - \psi'(0)| \leq \frac{\varepsilon}{5}\ \forall t \in [0, \delta_2]\right\} \geq \frac{1}{2},\ \forall \psi' \in \mathscr{C}, \|\psi' - \phi\| < \varepsilon. \tag{11.244}$$

In view of (11.242), it can be checked that

$$\|Y_t^{\psi,i} - \phi\| \leq \frac{3\varepsilon}{5}\ \forall t \in [0, \delta_2] \text{ if } |Y^{\psi,i}(t) - \psi(0)| \leq \frac{\varepsilon}{5}\ \forall t \in [0, \delta_2] \text{ and } \|\psi - \phi\| < \frac{\varepsilon}{5} \tag{11.245}$$

and

$$\|Y_t^{\psi',j} - \phi\| \leq \varepsilon\ \forall t \in [0, \delta_2] \text{ if } |Y^{\psi',j}(t) - \psi'(0)| \leq \frac{\varepsilon}{5}\ \forall t \in [0, \delta_2] \text{ and } \|\psi' - \phi\| < \frac{3\varepsilon}{5}. \tag{11.246}$$

By virtue of (11.243), (11.245), and Lemma 11.12, for $\psi \in \mathscr{C}$, $\|\psi - \phi\| < \dfrac{\varepsilon}{5}$, we have

$$
\begin{aligned}
&\mathbb{P}_{\psi,i}\left\{\|Y_{\lambda_1}-\phi\| \le \frac{3\varepsilon}{5} \text{ and } \lambda_1 < \delta_2, \beta(\lambda_1)=j\right\}\\
&= \mathbb{E}_{\psi,i}\int_0^{\delta_2} I_{\{\|Y_t-\phi\|\le\frac{3\varepsilon}{5}\}} q_{i,j}(Y_t)\exp\left(-\int_0^t q_i(Y_s)ds\right)dt\\
&\ge \mathbb{E}_{\psi,i}\left[I_{\{|Y(u)-\psi(0)|\le\frac{\varepsilon}{5}\ \forall u\in[0,\delta_2]\}}\int_0^{\delta_2} q_{i,j}(Y_t)\exp\left(-\int_0^t q_i(Y_s)ds\right)dt\right]\\
&\ge \frac{\delta_2}{2}\inf_{\phi'\in\mathscr{C}:\|\phi'-\phi\|\le\frac{3\varepsilon}{5}}\{q_{i,j}(\phi')\}\times\inf_{\phi'\in\mathscr{C}:\|\phi'-\phi\|\le\frac{3\varepsilon}{5}}\left\{\exp\left(-\int_0^{\delta_2} q_i(\phi'(s))ds\right)dt\right\}\\
&:= p_1 > 0.
\end{aligned}
\tag{11.247}
$$

Now, we have from the Markov property that

$$
\begin{aligned}
&\mathbb{P}_{\psi,i}\{\|X_{\delta_2}-\phi\|<\varepsilon, \alpha(\delta_2)=j\}\\
&\quad\ge \mathbb{P}_{\psi,i}\left\{\xi_1<\delta_2, \alpha_1=j, \|X_{\xi_1}-\phi\|<\frac{3\varepsilon}{5}\right\}\\
&\qquad\times \mathbb{P}_{\psi,i}\left\{\|X_{\delta_2}-\phi\|<\varepsilon, \xi_2>\delta_2 \,\Big|\, \xi_1<\delta_2, \alpha_1=j, \|X_{\xi_1}-\psi\|<\varepsilon\right\}.
\end{aligned}
\tag{11.248}
$$

By (11.243) and (11.246), if $\|\psi'-\phi\|\le\frac{3\varepsilon}{5}$, then

$$
\begin{aligned}
&\mathbb{P}_{\psi',j}\{\|X_t-\phi\|<\varepsilon\,\forall t\in[0,\delta_2], \xi_1>\delta_2\}\\
&\quad= \mathbb{P}_{\psi',j}\{\|Y_t-\phi\|<\varepsilon\,\forall t\in[0,\delta_2], \lambda_1>\delta_2\}\\
&\quad\ge \mathbb{E}_{\psi',j}\left[1_{\{\|Y_t-\phi\|<\varepsilon\,\forall t\in[0,\delta_2]\}}\exp\left(-\int_0^{\delta_2} q_j(Y_s)ds\right)\right]\\
&\quad\ge \mathbb{P}_{\psi',j}\left\{|Y(t)-\psi'(0)|\le\frac{\varepsilon}{5}\,\forall t\in[0,\delta_2]\right\}\\
&\qquad\times\inf_{\phi'\in\mathscr{C}:\|\phi'-\phi\|\le\varepsilon}\left\{\exp\left(-\int_0^{\delta_2} q_i(\phi'(s))ds\right)\right\}\\
&\quad:= p_2 > 0.
\end{aligned}
\tag{11.249}
$$

By the strong Markov property of $(X_t, \alpha(t))$, applying estimates (11.247) and (11.249) to (11.248), we obtain

$$
\sup_{\psi\in\mathscr{C}:\|\psi-\phi\|\le\frac{\varepsilon}{5}}\mathbb{P}_{\psi,i}\{\|X_{\delta_2}-\phi\|<\varepsilon, \alpha(\delta_2)=j\} > p_1p_2 > 0.
$$

□

***Proof of Lemma* 11.25** Using the Kolmogorov-Chentsov theorem [149, Theorem 2.8], for each $i \in \mathcal{M}$ and $R > 0$, there exists an $h_{i,R} > 0$ such that

$$\mathbb{P}_{\phi,i}\left\{\sup_{t,s\in[0,r],0<t-s<h_{i,R}} \frac{|Y(t)-Y(s)|}{(s-t)^{0.25}} \le 4\right\} > \frac{1}{2}, \ \forall \|\phi\| \le R. \tag{11.250}$$

Let

$$\mathcal{A} = \left\{\phi \in \mathcal{C} : |\phi(-r)| \le R, \sup_{t,s\in[-r,0],0<t-s<h_{i,R}} \frac{|\phi(t)-\phi(s)|}{(s-t)^{0.25}} \le 4\right\}.$$

Let $R' > R$ such that $\|\phi\| < R'$ for any $\phi \in \mathcal{A}$ and $M_{R'} = \sup_{\|\phi\|<R'}\{q_i(\phi)\} < \infty$. For $\|\phi\| \le R$, we have that

$$\begin{aligned}\mathbb{P}_{\psi,i}\{X_r \in \mathcal{A}, \xi_1 > r\} =&\mathbb{P}_{\psi,i}\{Y_r \in \mathcal{A}, \lambda_1 > r\}\\ =&\mathbb{E}_{\psi,i}\left[I_{\{Y_r\in\mathcal{A}\}}\int_0^r \exp(-q_i(Y_t))dt\right]\\ \ge&\mathbb{E}_{\psi,i} I_{\{Y_r\in\mathcal{A}\}} \exp(-rM_{R'})\\ \ge&0.5\exp(-rM_{R'}),\end{aligned}$$

which implies (11.51).

To prove (11.52), note that $\sup_{(x,i)\in\mathbb{R}^n\times\mathcal{M}} \mathbb{E}_{x,i}\tau_k < \infty$, where $\tau_k = \inf\{t \ge 0 : |Y(t)| \ge k\}$, (e.g., [156, Corrolary 3.3] or [321, Theorem 3.1]). Thus, there is a $T > 0$ such that

$$\mathbb{P}_{x,i}\left\{\tau_k < T\right\} > \frac{1}{2}, \ \forall x \in \mathbb{R}^n.$$

Denote $\widetilde{\tau}_k = \inf\{t \ge 0 : \|X_t\| \ge k\}$. For $\phi \in \mathcal{C}$ with $\|\phi\| \le R < k$ we have from (11.241) and Lemma 11.12 that

$$\begin{aligned}\mathbb{P}_{\phi,i}\left\{\widetilde{\tau}_k < T\right\} \ge&\mathbb{P}_{\phi,i}\left\{\widetilde{\tau}_k < T, \alpha(t) = i \ \text{ for } t \in [0,\widetilde{\tau}_k)\right\}\\ =&\mathbb{P}_{\phi,i}\left\{\tau_k < T, \beta(t) = i \ \text{ for } t \in [0,\tau_k)\right\}\\ =&\mathbb{E}_{\phi,i}\left[I_{\{\tau_k<T\}}\exp\left(-\int_0^{\tau_k} q_i(Y_s)ds\right)\right]\\ \ge&\exp(-M_kT)\,\mathbb{E}_{\phi,i} I_{\{\tau_k<T\}} \ge 0.5\exp(-M_kT),\end{aligned}$$

where $M_k = \sup_{\|\phi\|<k}\{q_i(\phi)\} < \infty$. The proof is therefore complete. □

We need an auxiliary lemma to obtain Lemma 11.26.

Lemma 11.67 *Fix $i \in \mathcal{M}$ and suppose $A(x, i)$ is elliptic uniformly in each compact subset of $\mathbb{R}^n$. For D be a bounded open set in $\mathbb{R}^n$ and K_1, K_2 be open sets whose closures are contained in D. Then*

$$\inf_{\{\phi\in\mathscr{C}:\phi(0)\in K_1\}} \mathbb{P}_{\phi,i}\big(\{Y(T) \in K_2\} \cap \{Y(t) \in D \forall t \in [0, T]\}\big) > 0, \tag{11.251}$$

and there is a measure $\boldsymbol{\nu}$ on $\mathfrak{B}(\mathscr{C})$ such that $\mathbb{P}_{\phi,i}\{Y_{T+r} \in \mathcal{B}, Y(t) \in D, t \in [0, T + r]\} \geq \boldsymbol{\nu}(\mathcal{B})$. Moreover, if $\mathcal{B} \subset \{\phi \in \mathscr{C} : \phi(t) \in D, t \in [-r, 0]\}$ is an open set of $\mathscr{C}$, then $\boldsymbol{\nu}(\mathcal{B}) > 0$.

Proof For a bounded continuous function $f(x) : \mathbb{R}^n \mapsto \mathbb{R}$ vanishing outside K_2, let $u_f(t, x)$ be the solution to

$$\begin{cases} \dfrac{\partial u}{\partial t} + \mathcal{L}_i u = 0 \text{ in } D \times [0, T) \\ u\,(T, x) = f\,(x) \text{ on } D, \\ u(t, x) = 0 \text{ on } \partial D \times [0, T]\,. \end{cases} \tag{11.252}$$

It is well-known (see, e.g., [201, Theorem 2.8.2]) that

$$u_f(t, x) = \mathbb{E}_{x,i}\left[f\big(Y(T - t)\big) I_{\{Y(s)\in D\,\forall s\in[0,T-t]\}}\right].$$

Let g be a continuous function in D such that $0 \leq g(x) \leq 1 \,\forall x \in D$, $g(x) = 0$ outside K_2 and $g(x) > 0$ for some $x \in K_2$. By the strong maximum principle for parabolic equations (see [82, Theorem 7.12]), $u_g(0, x) > 0$ for all $x \in D$, which implies that

$$u_C := \inf\{u_g(0, x) : x \in K_1\} > 0. \tag{11.253}$$

By the definition of $g(\cdot)$, we can obtain that

$$\mathbb{P}_{x,i}\,\{Y(T) \in K_2, Y(s) \in D\,\forall s \in [0, T]\} \geq u_g(0, x)\,\forall x \in D. \tag{11.254}$$

The first desired result follows from (11.253) and (11.254). Moreover, in view of Harnack's inequality (see [82, Theorem 7.10]), there is $\tilde{\rho}_i > 0$ such that $u_f(y, T) \geq \tilde{\rho}_i u_f(x_0, \frac{T}{2})$ for all $y \in K_2$ and f being bounded continuous. Thus

$$\mathbb{E}_{x,i}\left[f\big(Y(T)\big) I_{\{Y(s)\in D\,\forall s\in[0,T-t]\}}\right] \geq \rho_i \mathbb{E}_{x_0,i}\left[f\,(Y\,(0.5T))\, I_{\{Y(s)\in D\,\forall s\in[0,T-t]\}}\right]$$

for any bounded and continuous function f. Thus, we obtain that

$$\begin{aligned}&\mathbb{P}_{x,i}\left\{Y(T)\in B \text{ and } Y(s)\in D\,\forall s\in[0,T]\right\}\\&\geq\tilde{\rho}_i\mathbb{P}_{x_0,i}\left\{Y\left(0.5T\right)\in B,\ \text{ and } Y(s)\in D\,\forall s\in[0,0.5]\right\}\\&\geq\rho_i\tilde{\nu}(B)\end{aligned}$$

for any Borel set B, where

$$\nu(\cdot)=\mathbb{P}_{x_0,i}\left\{Y(0.5T)\in\cdot,\ \text{ and } Y(s)\in D\,\forall s\in[0,0.5T]\right\}$$

and $\rho_i = \tilde{\rho}_i\mathbb{P}_{x_0,i}\left\{Y(s)\in D\,\forall s\in[0,0.5T]\right\}$, which is positive due to (11.251). Denote $\hat{\mathcal{D}} = \{\phi \in \mathscr{C} : \phi(t) \in D, t \in [-r, 0]\}$. For any Borel set $\mathcal{B} \subset \mathscr{C}$, we have from the Markov property of $Y^i(t)$ that

$$\begin{aligned}&\mathbb{P}_{x_0,i}\left\{Y_{T+r}\in\mathcal{B} \text{ and } Y(s)\in D\,\forall s\in[0,T+r]\right\}\\&=\mathbb{P}\left\{Y_{T+r}\in\mathcal{B}\cap\hat{\mathcal{D}}\middle|Y(T)=y\right\}\mathbb{P}_{\phi_0,i}\{Y(T)\in dy \text{ and } Y(s)\in D\,\forall s\in[0,T+r]\}\\&\geq\rho_i\int_{y\in D}\mathbb{P}\left\{Y_{T+r}\in\mathcal{B}\cap\hat{\mathcal{D}}\middle|Y(T)=y\right\}\tilde{\nu}(dy)\\&=\boldsymbol{\nu}(\mathcal{B}\cap\hat{\mathcal{D}}).\end{aligned}$$

Now, let $\mathcal{B}$ be an open subset of $\hat{\mathcal{D}}$. Denote $B = \{\phi(-r) : \phi \in \mathcal{B}\}$. Then B is an open subset of D. By the support theorem (see [266, Theom 3.1])),

$$\mathbb{P}_{y,i}\left\{Y_{T+r}\in\mathcal{B}\right\}>0 \text{ for any } y\in B. \tag{11.255}$$

In light of (11.251),

$$\tilde{\nu}(B)=\mathbb{P}_{\phi_0,i}\left\{Y(0.5T)\in B,\ \text{ and } Y(s)\in D\,\forall s\in[0,0.5T]\right\}>0. \tag{11.256}$$

In view of (11.255) and (11.256),

$$\boldsymbol{\nu}(\mathcal{B})\geq\int_{y\in B}\mathbb{P}_{y,i}\left\{Y_{T+r}\in\mathcal{B}\right\}\tilde{\nu}(dy)>0$$

if $\mathcal{B}$ is an open subset of $\hat{\mathcal{D}}$. □

***Proof of Lemma* 11.26** Let $D = \{x \in \mathbb{R}^n : |x| < R + 1\}$, $K_1 = K_2 = \{x \in \mathbb{R}^n : |x| < R\}$ and $M_{R,i} = \sup\{q(\phi, i) : \|\phi\| \leq R + 1\} < \infty$. In view of Lemma 11.67,

$$\mathbb{P}_{\phi,i^*}\left\{Y_{T+r}\in\mathcal{B}, Y(t)<R+1, t\in[0,T+r]\right\}\geq\boldsymbol{\nu}(\mathcal{B})$$

where $\boldsymbol{\nu}(\cdot)$ is defined as in Lemma 11.67 with i replaced by i^*. Thus,

$$\begin{aligned}
&\mathbb{P}_{\phi,i^*}\{X_{T+r}\in\mathcal{B}\text{ and }\alpha(T+r)=i^*\}\\
&\geq\mathbb{P}_{\phi,i^*}\{X_{T+r}\in\mathcal{B}\text{ and }\alpha(t)=i^*,\ \|X_t\|<R+1\,\forall t\in[0,T+r]\}\\
&\geq\mathbb{E}_{\phi,i^*}\left[I_{\{Y_{T+r}\in\mathcal{B},\|Y_t\|<R\,\forall t\in[0,T+r]\}}\exp\left(-\int_0^{T+r}q_{i^*}(Y_t)dt\right)\right]\\
&\geq\exp(-M_{R,i}(T+r))\boldsymbol{\nu}(\mathcal{B}).
\end{aligned}$$

The proof is complete. □

11.7.3 Proofs of Some Results in Section 11.6

***Proof of Lemma* 11.53** Since $V(0)=0$ and $V(x)$ is continuous on D, we can find $h_*>0$ such that $B_{h_*}\subset D$ and $V(x)\leq 1$ for any $x\in B_{h_*}$. Because $\tau_{h_1}\leq\tau_{h_2}$ if $h_1\leq h_2$, it suffices to prove the Lemma for any $h\leq h_*$.

Since g is continuously differentiable and $g(0)=0$, there is $K_g>0$ such that $g(z)\leq K_g|z|$ for $|z|\leq 1$. Thus, we have

$$\mathcal{L}_i V(x)\leq K_g\sup_{i\in\mathcal{M}}\{|c_i|\}V(x),\ \ (x,i)\in B_{h_*}\times\mathcal{M}$$

Letting $\tilde{K}=K_g\sup_{i\in\mathcal{M}}\{|c_i|\}$, by Itô's formula,

$$\begin{aligned}
\mathbb{E}_{x,i}V(X(t\wedge\tau_h))\leq& V(x)+\tilde{K}\mathbb{E}_{x,i}\int_0^{t\wedge\tau_h}V(X(s))ds\\
\leq& V(x)+\tilde{K}\mathbb{E}_{x,i}\int_0^{t}\mathbb{E}_{x,i}V(X(s\wedge\tau_h))ds.
\end{aligned}$$

By the Gronwall inequality, we can easily obtain

$$\mathbb{E}_{x,i}V(X(T\wedge\tau_h))\leq V(x)e^{KT}.$$

Let $v_h=\inf\{V(x):|x|=h\}>0$. An application of Markov's inequality yields that

$$\mathbb{P}_{x,i}\{\tau_h\leq T\}\leq\frac{V(x)e^{KT}}{v_h}.$$

Since $V(0)=0$ and V is continuous on D, there is a $\tilde{h}>0$ such that

$$\mathbb{P}_{x,i}\{\tau_h\leq T\}\leq\frac{V(x)e^{KT}}{v_h}\leq\varepsilon$$

for any $x \in B_{\tilde{h}}$ as desired. □

***Proof of Lemma* 11.54** The proof, which is analogous to that of Lemma 6.1, is omitted. □

Lemma 11.68 *Suppose that* (H11.9) *and* (11.166) *hold. Then for any* $T > 0$ *and a bounded function* $f : \mathcal{M} \mapsto \mathbb{R}$*, we have*

$$\lim_{x \to 0} \sup_{i \in \mathcal{M}, t \in [0,T]} \left\{ \left| \mathbb{E}_{x,i} f(\alpha(t)) - \mathbb{E}_i f(\hat{\alpha}(t)) \right| \right\} = 0. \tag{11.257}$$

Proof Let $\Xi(x,k) := \sum_{j \in \mathcal{M}, j \neq k} |q_{kj}(x) - q_{kj}(0)|$. Similar to (2.118), we have $\mathbb{P}_{x,i,i}\{\vartheta \le T \wedge \tau_h\} = \mathbb{E}_{x,i,i} \int_0^{\vartheta \wedge T \wedge \tau_h} \Xi(X(t), \alpha(t))dt$. For any $\varepsilon > 0$, let $h > 0$ such that $B_h \in D$ and $\sup_{(x,k) \in B_h \times \mathcal{M}} \Xi(x,k) < \frac{\varepsilon}{2T}$ we obtain that

$$\begin{aligned} \mathbb{P}_{x,i,i}\{\vartheta \le T \wedge \tau_h\} &= \mathbb{E}_{x,i,i} \int_0^{\vartheta \wedge T \wedge \tau_h} \Xi(X(t), \alpha(t))dt \\ &\le T \sup_{(x,i) \in B_h \times \mathcal{M}} \Xi(x,k) \le \frac{\varepsilon}{2}. \end{aligned} \tag{11.258}$$

In view of Lemma 11.53, there is $\delta > 0$ such that $\mathbb{P}_{x,i,i}\{\tau_h \le T\} \le \frac{\varepsilon}{2}$. This and (11.258) derive

$$\mathbb{P}_{x,i,i}\{\vartheta \wedge \tau_h \le T\} \le \mathbb{P}_{x,i,i}\{\vartheta \le T \wedge \tau_h\} + \mathbb{P}_{x,i,i}\{\tau_h \le T\} \le \varepsilon.$$

We have that

$$\begin{aligned} \left| \mathbb{E}_{x,i} f(\alpha(t)) - \mathbb{E}_{0,i} f(\alpha(t)) \right| &= \left| \mathbb{E}_{x,i,i} \left[f(\alpha(t)) - f(\hat{\alpha}(t)) \right] \right| \\ &= \left| \mathbb{E}_{x,i,i} I_{\{\vartheta \wedge \tau_h \le t\}} \left[f(\alpha(t)) - f(\hat{\alpha}(t)) \right] \right| \\ &\le 2M_f \mathbb{P}_{x,i,i}\{\vartheta \wedge \tau_h \le t\} \le 2M_f \varepsilon, \ \text{ for } t \in [0,T]. \end{aligned}$$

where $M_f = \sup_{i \in \mathcal{M}} |f(i)|$. The lemma is proved. □

11.8 Notes

This chapter focuses on switching diffusions where the random switching depends not only on the current state of the continuous component but also on its past history. This extension broadens the applicability of switching diffusions. The basic properties such as existence and uniqueness of solutions, recurrence, ergodicity, stability, and Feller properties are presented. Also included is Dupire's functional Itô formula. Appropriate references are provided for further reading.

Chapter 12
Population Dynamics Modeled by Switching Diffusion

This chapter introduced population and infectious disease systems under random environments modeled by switching diffusions. It is organized as follows. Section 12.1 presents a general theory for coexistence and persistence of population dynamics driven by switching diffusion together with some applications to Lotka-Volterra models. Then in the subsequent sections, we consider some specific models for which some conditions proposed in Section 12.1 are not satisfied. In particular, Section 12.2 deals with a chemostat model and Section 12.3 considers an infectious disease model with vaccination.

12.1 General Stochastic Kolmogorov Systems Under Regime-Switching

12.1.1 Introduction

Populations in nature do not evolve in isolation. Consequently, much of ecology focuses on understanding the traits that enable two species to coexist or lead one species to dominate the habitat of another. It is crucial to determine the fate of an invading species: will it successfully establish itself or perish in the attempt? If it does invade, will it coexist with the native species? Mathematical models of invasibility have greatly enhanced our understanding of infectious disease outbreaks [57] and ecological processes [183] and [36]. Empirical evidence indicates that heterogeneity, due to abiotic factors (such as precipitation, temperature, sunlight) or biotic factors (such as competition, predation), plays a significant role in determining invasibility; see [59] and [241]. Environmental fluctuations make population dynamics inherently stochastic.

The combined effects of biotic interactions and environmental fluctuations are crucial for understanding species richness. Biotic effects can sometimes lead to

H.-D. Nguyen et al., *Hybrid Switching Diffusions*, Probability Theory and Stochastic Modelling 63, https://doi.org/10.1007/978-3-031-93303-5_12

species extinction, but the addition of environmental randomness might reverse extinction into coexistence. Conversely, deterministic systems that predict coexistence can lead to extinction when environmental fluctuations are considered. A successful method to study this interplay involves modeling populations as discrete or continuous time Markov processes and examining their long-term behavior; see [17, 19, 30, 37, 38, 48, 83, 84, 180, 252, 253] and references therein.

A natural approach to analyzing species coexistence is to examine the average per-capita growth rate of a population when it is rare. Intuitively, if this growth rate is positive, the population increases when its density is rare and can invade, whereas if it is negative, the population decreases and goes extinct. In the case of only two populations, coexistence is ensured if each population can invade when it is rare while the other population remains stationary [49, 83, 270].

There exists a general theory for coexistence in deterministic models [125, 126, 133]. According to this theory, a sufficient condition for persistence is the presence of a fixed set of weights associated with the interacting populations such that this weighted combination of the populations' invasion rates is positive for any invariant measure supported by the boundary (i.e., associated with a sub-collection of populations) [125].

Recent studies have explored the impact of environmental stochasticity on continuous-time models. For instance, [17] found that if a deterministic continuous-time model meets the persistence criterion mentioned above, then under certain weak assumptions, the corresponding stochastic differential equation with a small diffusion term has a positive stationary distribution concentrated on the positive global attractor of the deterministic system. Additionally, for general stochastic difference and differential equations with arbitrary levels of noise in a *compact* state space, sufficient conditions for persistence are provided in [252].

There has been much progress in theory of stochastic Kolmogorov system. Reference [16] provided a general theory for persistence of stochastic population models. In [117], a Kolmogorov system driven by a Brownian motion was studied systematically; sharp conditions for persistence and existence were obtained. Subsequently, reference [118] fully classified models in [117] for three species. In what follows, we present the results corresponding to [117] in a setting of Kolmogorov systems modeled by a switching diffusion

$$dX_i(t) = X_i(t) f_i(\mathbf{Z}(t))dt + X_i(t) g_i(\mathbf{Z}(t)) dE_i(t), \ i = 1, \ldots, n, \tag{12.1}$$

where $\mathbf{Z}(\cdot) = (\mathbf{X}(\cdot), \alpha(\cdot))$, $\mathbf{X}(t) = (X_1(t), \ldots, X_n(t))$, and $\alpha(t)$ is an irreducible Markov chain on a finite state space $\mathcal{M} = \{1, \cdots, m_0\}$ with generator $Q = (q_{ij})_{m_0 \times m_0}$, that is,

$$\begin{aligned} &\mathbb{P}\{\alpha(t+\Delta) = j | \alpha(t) = i, \alpha(s), s \le t\} = q_{ij}\Delta + o(\Delta) \text{ if } i \ne j \text{ and} \\ &\mathbb{P}\{\alpha(t+\Delta) = i | \alpha(t) = i, \alpha(s), s \le t\} = 1 + q_{ii}\Delta + o(\Delta). \end{aligned} \tag{12.2}$$

We use the following notation for simplicity $\mathbb{R}^n_+ = [0,\infty)^n, \mathbb{R}^{n,\circ}_+ = (0,\infty)^n, \partial\mathbb{R}^n_+ = \mathbb{R}^n_+ \setminus \mathbb{R}^{n,\circ}_+, \mathbb{S} = \mathbb{R}^n_+ \times \mathcal{M}, \mathbb{S}^\circ = \mathbb{R}^{n,\circ}_+ \times \mathcal{M}, \partial\mathbb{S} = \partial\mathbb{R}^n_+ \times \mathcal{M}$, $\mathbf{x} = (x_1,\ldots,x_n)$, $\mathbf{z} = (\mathbf{x},\alpha)$. Denote $\mathbf{E}(t) = (E_1(t),\ldots,E_n(t))' = \Gamma'\mathbf{B}(t)\,\Gamma$ is an $n\times n$ matrix such that $\Gamma'\Gamma = \Sigma = (\sigma_{ij})_{n\times n}$ and $\mathbf{B}(t) = (B_1(t),\ldots,B_n(t))$ is a vector of independent standard Brownian motions adapted to the filtration $\{\mathcal{F}_t\}_{t\geq 0}$. System (12.1) describes the dynamics of n interacting populations $(X_1(t),\ldots,X_n(t))$.

We write $\mathcal{L}$, the generator of the process $(\mathbf{Z}(t)) := ((X_1(t),\ldots,X_n(t),\alpha(t))$ as

$$
\begin{aligned}
[\mathcal{L}F](\mathbf{z}) = \sum_i x_i f_i(\mathbf{z})\frac{\partial F}{\partial x_i}(\mathbf{z}) + \frac{1}{2}\sum_{i,j}\sigma_{ij}x_ix_jg_i(\mathbf{z})g_j(\mathbf{z})\frac{\partial^2 F}{\partial x_i\partial x_j}(\mathbf{z})\\
+\sum_{j\in\mathcal{M}} q_{mj}F(\mathbf{x},j).
\end{aligned}
$$

and use the norm $\|\mathbf{x}\| = \sum_{i=1}^n |x_i|$ in $\mathbb{R}^n$.

(H12.1) Assume the following hold.

1. $\operatorname{diag}(g_1(\mathbf{z}),\ldots,g_n(\mathbf{z}))\Gamma^\top\Gamma\operatorname{diag}(g_1(\mathbf{z}),\ldots,g_n(\mathbf{z})) = (g_i(\mathbf{z})g_j(\mathbf{z})\gamma\sigma_{ij})_{n\times n}$ is a positive definite matrix for any $\mathbf{z}\in\mathbb{S}$.
2. $f_i(\cdot,\alpha), g_i(\cdot,\alpha) : \mathbb{R}^n_+ \to \mathbb{R}$ are locally Lipschitz functions for any $\alpha\in\mathcal{M}$.
3. There are $\delta_0\in(0,1)$, $M>0$, $H>0$ and a function $\mathcal{W} : \mathbb{R}^n_+ \mapsto [1,\infty)$ such that

$$
\mathcal{W}(\mathbf{x}) \text{ is twice differentiable in } \mathbf{x} \text{ and } \liminf_{\mathbf{x}\to\infty}\frac{\mathcal{W}(\mathbf{x})}{\ln\|\mathbf{x}\|} > n. \tag{12.3}
$$

$$
[\mathcal{L}\mathcal{W}](\mathbf{z}) + 2\delta_0\hat{\Lambda}(\mathbf{z}) + \sum_{i=1}^n(|f_i(\mathbf{z})| + \sigma_{ii}^2 g_i^2(\mathbf{z}) \leq HI_{\{\|\mathbf{x}\|\leq M\}} - 2 \tag{12.4}
$$

where

$$
\hat{\Lambda}(\mathbf{z}) := \sum_{i,j}\big((|\mathcal{W}_{x_i}(\mathbf{x})x_i| + 1)(|\mathcal{W}_{x_j}(\mathbf{x})x_j| + 1)|g_i(\mathbf{z})g_j(\mathbf{z})||\sigma_{ij}|\big) \tag{12.5}
$$

Remark 12.1 Parts (2) and (3) of (H12.1) guarantee the existence and uniqueness of strong solutions to (12.1). We need part (1) of (H12.2) to ensure that the solution to (12.1) is a nondegenerate diffusion. Moreover, we show later that (12.4) implies the tightness of the family of transition probabilities of the solution to (12.1). Finally, note that for a function $U : \mathbb{R}^n_+ \to \mathbb{R}$ of $\mathbf{x}$, $\mathcal{L}U$ is a function of $\mathbf{z} = (\mathbf{x},\alpha)$. To emphasize the variable-dependence, we write $U(\mathbf{x})$ and $[\mathcal{L}U](\mathbf{z})$.

12.1.2 Invariant Measures, Lyapunov Exponents

For $\mathbf{p} = (p_1, \cdots, p_n) \in \mathbb{R}^n_+$, define the function $V_\mathbf{p} : \mathbb{S}^\circ \to \mathbb{R}$ by

$$V_\mathbf{p}(\mathbf{x}) = \mathcal{W}(\mathbf{x}) - \sum_{i=1}^{n} p_i \ln X_i, \text{ and} \tag{12.6}$$

$$\Lambda_\mathbf{p}(\mathbf{z}) := \sum_{i,j=1}^{n} \big(\mathcal{W}_{x_i}(\mathbf{x})x_i g_i(\mathbf{z}) + p_i g_i(\mathbf{z})\big)\big(\mathcal{W}_{x_j}(\mathbf{x})x_j g_j(\mathbf{z}) + p_j g_j(\mathbf{z})\big)\sigma_{ij} \tag{12.7}$$

Lemma 12.2 *For any $\mathbf{z} \in \mathbb{S} = \mathbb{R}^n_+ \times \mathcal{M}$, there exists a pathwise unique strong solution $\mathbf{Z}(t)$ to* (12.1) *with initial value $\mathbf{z}$. Moreover, for any $I \subset \{1, \cdots, n\}$ and $\mathbf{x} \in \mathbb{R}^{I,\circ}_+$ where $\mathbb{R}^{I,\circ}_+ = \{\mathbf{x} \in \mathbb{R}^n_+ : x_i = 0 \text{ if } i \notin I \text{ and } x_i > 0 \text{ if } i \in I\}$, the solution $\mathbf{Z}(t)$ with initial value $\mathbf{z}$ will stay in $\mathbb{R}^{I,\circ}_+$ with probability 1. For $\mathbf{z} \in \mathbb{R}^{n,\circ}_+$ and $V(\cdot)$ defined by* (12.6), *we have*

$$\mathbb{E}_\mathbf{z} e^{\delta_0 V(\mathbf{Z}(t))} \le \exp(\delta_0 H t) e^{\delta_0 V(\mathbf{x})}. \tag{12.8}$$

Proof We restrict our proof for the existence and uniqueness of the solution with initial value $\mathbf{z} \in \mathbb{S}^\circ$. If $\mathbf{x} \in \mathbb{R}^{I,\circ}_+$ for any $I \subset \{1, \cdots, n\}$, the proof carries over. Let $V_\mathbf{p}(\cdot)$ be defined by (12.6).

$$\lim_{m\to\infty} \inf\{V_\mathbf{p}(\mathbf{x}) : x_i \vee x_i^{-1} > m \text{ for some } i = 1, \ldots, n\} = \infty. \tag{12.9}$$

Note that

$$\mathcal{L}e^{\delta_0 V_\mathbf{p}} t(\mathbf{z}) = \delta_0 e^{\delta_0 V_\mathbf{p}(\mathbf{x})}\left[[\mathcal{L}\mathcal{W}](\mathbf{z}) - \sum\left(p_i f_i(\mathbf{z}) - \frac{p_i g_i^2(\mathbf{z})\sigma_{ii}}{2}\right) - \frac{\delta_0}{2}\Lambda_\mathbf{p}(\mathbf{z})\right]. \tag{12.10}$$

In view of (12.4) and the fact that $\Lambda_\mathbf{p}(\mathbf{z}) \le \hat{\Lambda}(\mathbf{z})$, we have

$$\begin{aligned}\mathcal{L}e^{\delta_0 V_\mathbf{p}} t(\mathbf{z}) \le& \delta_0\left(H I_{\|\mathbf{x}\|\le M} - 2\right) e^{\delta_0 V_\mathbf{p}(\mathbf{x})} - \delta_0 e^{\delta_0 V(\mathbf{x})} \sum\left(|f_i(\mathbf{z})| + \frac{\sigma_{ii}^2}{2} g_i^2(\mathbf{z})\right)\\ \le& \delta_0 H e^{\delta_0 V_\mathbf{p}(\mathbf{x})}.\end{aligned} \tag{12.11}$$

Since the coefficients of (12.1) are locally Lipschitz, using (12.9) and (12.10), it follows from Theorem 2.12 that (12.1) has a unique solution $\mathbf{Z}(t)$ remaining in $\mathbb{S}^\circ$ almost surely if the initial value $\mathbf{z} \in \mathbb{S}^\circ$. The estimate (12.8) can also be derived from the proof of Theorem 2.12. □

Lemma 12.3 *Let $H_1 = H_0 \sup_{\|\mathbf{x}\|\le M} \mathcal{W}(\mathbf{x})$. For any $\mathbf{z} \in \mathbb{S}$.*

$$\mathbb{E}_{\mathbf{z}} e^{\delta_0 \mathcal{W}(\mathbf{X}(t))} \le \delta_0 H_1 + e^{\delta_0 \mathcal{W}(\mathbf{x})} e^{-\delta_0 t} \tag{12.12}$$

and

$$\mathbb{E}_{\mathbf{z}} \int_0^t e^{\delta_0 \mathcal{W}(\mathbf{X}(s))} \left[1 + \sum (|f_i(\mathbf{Z}(s))| + |g_i(\mathbf{Z}(s))|^2)\right] ds \le H_1 \left(\frac{e^{\delta_0 \mathcal{W}(\mathbf{z})}}{\delta_0} + t \right). \tag{12.13}$$

Moreover, the solution process $\mathbf{Z}(t)$ *is a Markov-Feller process in* $\mathbb{R}^n_+$*.*

Proof Similar to (12.11), it follows from (12.4) that

$$[\mathcal{L}e^{\delta_0 \mathcal{W}}](\mathbf{z}) \le \delta_0 \left(H I_{\{\|\mathbf{x}\| \le M\}} - 2\right) e^{\delta_0 \mathcal{W}(\mathbf{x})} - \delta_0 e^{\delta_0 V(\mathbf{x})} \sum \left(|f_i(\mathbf{z})| + \frac{1}{2} g_i^2(\mathbf{z}) \right). \tag{12.14}$$

Letting $\eta_k = \inf\{t > 0 : \|\mathbf{X}(t)\| \ge k\}$, by applying Dynkin's formula to the function $\varphi(x, t) = e^{\delta_0(t + \mathcal{W}(\mathbf{x}))}$ together with the stopping time $\eta_k \wedge t$, and making use of (12.14), we have that

$$\begin{aligned} &\mathbb{E}_{\mathbf{z}} e^{\delta_0(\eta_k \wedge t)} e^{\delta_0 \mathcal{W}(\mathbf{X}(\eta_k \wedge t))} \\ &\quad = e^{\delta_0 \mathcal{W}(\mathbf{x})} + \mathbb{E}_{\mathbf{z}} \int_0^{\eta_k \wedge t} e^{\delta_0 s} \left(2\delta_0 e^{\delta_0 \mathcal{W}(\mathbf{X}(s))} + [\mathcal{L}e^{\delta_0 \mathcal{W}}](\mathbf{Z}(s)) \right) ds \\ &\quad \le e^{\delta_0 \mathcal{W}(\mathbf{x})} + \delta_0 H_1 \mathbb{E}_{\mathbf{z}} \int_0^{\eta_k \wedge t} e^{\delta_0 s} ds \le e^{\delta_0 \mathcal{W}(\mathbf{x})} + H_1 e^{\delta_0 t}. \end{aligned} \tag{12.15}$$

Letting $k \to \infty$ in (12.15) together with Fatou's lemma yields $e^{\delta_0 t} \mathbb{E}_{\mathbf{z}} e^{\delta_0 \mathcal{W}(\mathbf{X}(t))} \le e^{\delta_0 \mathcal{W}(\mathbf{x})} + \delta_0 H_1 e^{\delta_0 t}$, which in turn implies

$$\mathbb{E}_{\mathbf{z}} e^{\delta_0 \mathcal{W}(\mathbf{X}(t))} \le \delta_0 H_1 + e^{\delta_0 \mathcal{W}(\mathbf{x})} e^{-\delta_0 t}$$

as required. Another application of Dynkin's formula combined with (12.14) yields

$$\begin{aligned} &\mathbb{E}_{\mathbf{z}} e^{\delta_0 \mathcal{W}(\mathbf{X}(\eta_k \wedge t))} \\ &\quad = e^{\delta_0 \mathcal{W}(\mathbf{x})} + \mathbb{E}_{\mathbf{z}} \int_0^{\eta_k \wedge t} [\mathcal{L}e^{\delta_0 \mathcal{W}}](\mathbf{Z}(s)) ds \\ &\quad \le e^{\delta_0 \mathcal{W}(\mathbf{x})} + \delta_0 H_1 \mathbb{E} \int_0^{\eta_k \wedge t} ds - \delta_0 \mathbb{E}_{\mathbf{z}} \int_0^{\eta_k \wedge t} e^{\delta_0 \mathcal{W}(\mathbf{x})} \\ &\qquad \times \left[1 + \sum (|f_i(\mathbf{Z}(s)| + |g_i(\mathbf{Z}(s)|^2)\right] ds, \end{aligned}$$

which implies

$$\delta_0 \mathbb{E}_{\mathbf{z}} \int_0^{\eta_k \wedge t} e^{\delta_0 \mathcal{W}(\mathbf{X}(s))} \left[1 + \sum (|f_i(\mathbf{Z}(s)| + |g_i(\mathbf{Z}(s)|^2)\right] ds \le e^{\delta_0 \mathcal{W}(\mathbf{x})} + \delta_0 H_1 t.$$

Letting $k \to \infty$, we obtain (12.13).

Finally, since $\lim_{\|\mathbf{x}\|\to\infty} e^{\delta_0 \mathcal{W}(\mathbf{x})} = \infty$, it follows easily from (12.15) that $\lim_{k\to\infty} \mathbb{P}_{\mathbf{z}}\{\eta_k > t\} = 0$ uniformly in each compact set of $\mathbb{R}^n_+$. The above facts coupled with the assumption that $f_i(\cdot), g_i(\cdot), i = 1, \ldots, n$ are locally Lipschitz allow us to follow the proof of Theorem 2.23 by a truncation argument in order to get the Markov-Feller property of $\mathbf{Z}(t)$. □

For $\mathbf{z} \in \mathbb{R}^n_+ \times \mathcal{M}$, define the occupation measure

$$\Pi^{\mathbf{z}}_t(\cdot) := \frac{1}{t}\int_0^t \mathbb{P}_{\mathbf{z}}\{(\mathbf{Z}(s) \in \cdot\}ds,\ t > 0.$$

Lemma 12.4 *Let μ be an invariant probability measure of $(\mathbf{Z}(t))$. Then*

$$\int_{\mathbb{S}} e^{\delta_0 \mathcal{W}(\mathbf{x})}\left(1 + \sum_i (|f_i(\mathbf{z})| + |g_i(\mathbf{z})|^2)\right) d\mu \le H_1$$

and

$$\int_{\mathbb{S}} [\mathcal{L}\mathcal{W}](\mathbf{z}) d\mu = 0. \tag{12.16}$$

Furthermore,

$$\mathbb{E}_{\mathbf{z}} e^{\delta_0 \int_0^t \hat{\Lambda}(\mathbf{Z}(s))ds} \le \exp\left(\delta_0 \mathcal{W}(\mathbf{x}) + H_1 \delta_0 t\right). \tag{12.17}$$

Proof Let $F(\mathbf{z}) := e^{\delta_0 \mathcal{W}(\mathbf{x})}\left(1 + \sum(|f_i(\mathbf{z})| + |g_i(\mathbf{z})|^2)\right)$. Because μ is an invariant measure, we have

$$\int_{\mathbb{S}} k \wedge F(\mathbf{z})\mu(d\mathbf{z}) = \int_{\mathbb{S}} \left(\frac{1}{t}\mathbb{E}_{\mathbf{z}_0} \int_0^t (k \wedge F(\mathbf{Z}(s)))\, ds\right) \mu(d\mathbf{z}_0).$$

For any $\mathbf{z}_0 \in \mathbb{R}^n_+ \times \mathcal{M}$, we have

$$\limsup_{t\to\infty} \frac{1}{t}\mathbb{E}_{\mathbf{z}_0} \int_0^t (k \wedge F(\mathbf{Z}(s)))\, ds \le H.$$

By Lebesgue's dominated convergence theorem for bounded functions, we have

$$\int_{\mathbb{S}} k \wedge F(\mathbf{z})\mu(d\mathbf{z}) = \limsup_{t\to\infty} \int_{\mathbb{S}} \left(\frac{1}{t}\mathbb{E}_{\mathbf{z}_0} \int_0^t (k \wedge F(\mathbf{Z}(s)))\, ds\right) \mu(d\mathbf{z}_0) \le H.$$

Letting $k \to \infty$ and making use of Fatou's Lemma, we get

$$\int_{\mathbb{S}} F(\mathbf{z})\mu(d\mathbf{z}) \le H_1.$$

Next, we will prove (12.16). Letting $\eta_k = \inf\{t > 0 : \mathbf{X}(t) \ge k\}$, we have from Dynkin's formula that

$$0 \le \mathbb{E}_{\mathbf{z}} \mathcal{W}(\mathbf{X}(\eta_k \wedge t)) = \mathcal{W}(\mathbf{x}) + \mathbb{E}_{\mathbf{z}} \int_0^{\eta_k \wedge t} [\mathcal{L}\mathcal{W}](\mathbf{Z}(s))ds.$$

Again, we deduce from the upper boundedness of $-[\mathcal{L}\mathcal{W}](\mathbf{z})$ and $\int_{\mathbb{S}} \mathcal{W}(\mathbf{x})$ that

$$\mathbb{E}_{\mathbf{z}} \mathcal{W}(\mathbf{X}(t)) - \mathbb{E}_{\mathbf{x}} \int_0^t [\mathcal{L}\mathcal{W}](\mathbf{Z}(s))ds = \mathcal{W}(\mathbf{x}).$$

Since μ is an invariant probability measure,

$$\mathbb{E}_{\mu} \mathcal{W}(\mathbf{X}(t)) = \int_{\mathbb{S}} \mathcal{W}(\mathbf{x})\mu(d\mathbf{z}), \mathbb{E}_{\mu}[\mathcal{L}\mathcal{W}](\mathbf{Z}(s)) = \int_{\mathbb{S}} [\mathcal{L}\mathcal{W}](\mathbf{z})\mu(d\mathbf{z}).$$

As a result,

$$\int_{\mathbb{S}} [\mathcal{L}\mathcal{W}](\mathbf{z})\mu(d\mathbf{z}) = 0,$$

$$\begin{aligned} 0 \le \mathcal{W}(\mathbf{X}(t)) =& \mathcal{W}(\mathbf{x}) + \int_0^t [\mathcal{L}\mathcal{W}](\mathbf{Z}(s))ds + M_{\mathcal{W}}(t) \\ \le& \mathcal{W}(\mathbf{x}) + \int_0^t (H - 2\delta_0 \hat{\Lambda}(\mathbf{Z}(s)))ds + M_{\mathcal{W}}(t) \end{aligned}$$

with

$$M_{\mathcal{W}}(t) = \sum_{\ell=1}^n \int_0^t \mathcal{W}_{x_i}(\mathbf{X}(s)) X_i(s) g_i(\mathbf{Z}(s)) dE_\ell(s). \tag{12.18}$$

Multiplying the inequality by δ_0 and exponentiating both sides, we have

$$\begin{aligned} &\exp\left(\delta_0 \int_0^t \hat{\Lambda}(\mathbf{Z}(s))ds\right) \\ &\le \exp\left(\delta_0 \mathcal{W}(\mathbf{X}(t)) + \delta_0 \int_0^t \hat{\Lambda}(\mathbf{Z}(s))ds\right) \\ &\le \exp\left(\delta_0 \mathcal{W}(\mathbf{x}) + H\delta_0 t\right) \times \exp\left(\delta_0 M_{\mathcal{W}}(t) - \delta_0 \int_0^t \hat{\Lambda}(\mathbf{Z}(s))ds\right). \end{aligned} \tag{12.19}$$

Note that $M_W(t)$ is a local martingale with quadratic variation

$$[M_W(t)] = \int_0^t \Lambda_0(\mathbf{Z}(s))ds \le \int_0^t \hat{\Lambda}(\mathbf{Z}(s))ds.$$

Since $e^{\delta_0 M_{\mathcal{W}}(t) - \frac{\delta_0^2}{2}[M_{\mathcal{W}}(t)]}$ is a martingale due to [148, Lemma 19.22], and $\delta_0 \le \frac{1}{2}$ then, $\exp\left(\delta_0 M_{\mathcal{W}}(t) - \delta_0 \int_0^t \hat{\Lambda}(\mathbf{Z}(s))ds\right)$ is a supermartingale, which means

$$\mathbb{E}_{\mathbf{z}} \exp\left(\delta_0 M_{\mathcal{W}}(t) - \delta_0 \int_0^t \hat{\Lambda}(\mathbf{Z}(s))ds\right) \le 1.$$

Taking expectation both sides and use the inequality above, we obtain (12.17). □

Lemma 12.5 *Let $\{\mathbf{z}_k := (\mathbf{x}_k, \alpha_k)\}$ be a sequence satisfying the following conditions:*

- $\|\mathbf{x}_k\| \le M$, $T_k > 1$ *for all* $k \in \mathbb{N}$ *and* $\lim_{k\to\infty} T_k = \infty$.
- $\Pi_{T_k}^{\mathbf{z}_k}$ *converges weakly to a probability measure* π.
- $h : \mathbb{S} \to \mathbb{R}$ *is any continuous function satisfying* $|h(\mathbf{z})| < K_h e^{\delta \mathcal{W}(\mathbf{x})}(1 + \sum(|f_i(\mathbf{z})| + |g_i(\mathbf{z})|^2))$, $\mathbf{z} \in \mathbb{S}$, *for some* $K_h \ge 0$, $\delta < \delta_0$.

Then π is an invariant measure of $\{\mathbf{Z}(\cdot)\}$ and

$$\int_{\mathbb{S}} h(\mathbf{z})\Pi_{T_k}^{\mathbf{z}_k}(d\mathbf{z}) \to \int_{\mathbb{S}} h(\mathbf{z})\pi(d\mathbf{z}).$$

Proof The fact that any weak-limit of a tight family of occupation measures is an invariant measure is well-known, see [77, 252]. Let $V_M = \sup\{e^{\delta_0 \mathcal{W}(\mathbf{x})} : \|\mathbf{x}\| \le M, m \in \mathcal{M}\}$ and fix $\varepsilon > 0$. Pick $l_\varepsilon \in \mathbb{N}$ such that

$$\frac{\varepsilon e^{(\delta_0 - \delta)\mathcal{W}(\mathbf{x})}}{K_h H_1(1 + V_M)} > 1 \text{ for any } \|\mathbf{x}\| \ge l_\varepsilon.$$

Let $\phi_l(\cdot) : \mathbb{R}_+^n \to [0, 1]$ be a continuous function with compact support satisfying $\phi_l(\mathbf{x}) = 1$ if $\|\mathbf{x}\| \le l_\varepsilon$. One gets the following sequence of inequalities:

$$\begin{aligned}
&\int_{\mathbb{S}} (1 - \phi_l(\mathbf{x}))\, |h(\mathbf{z})| \Pi_{T_k}^{\mathbf{z}_k}(d\mathbf{z}) \\
\le& K_h \int_{\mathbb{S}} (1 - \phi_k(\mathbf{z}))\, e^{\delta \mathcal{W}(\mathbf{x})} \left(1 + \sum(|f_i(\mathbf{z})| + |g_i(\mathbf{z})|^2)\right) \Pi_{T_k}^{\mathbf{z}_k}(d\mathbf{z}) \\
\le& \frac{K_h \varepsilon}{K_h H_2(1 + V_M)} \int_{\mathbb{S}} (1 - \phi_l(\mathbf{x}))\, e^{\delta_0 \mathcal{W}(\mathbf{x})} \left(1 + \sum(|f_i(\mathbf{z})| + |g_i(\mathbf{z})|^2)\right) \Pi_{T_k}^{\mathbf{z}_k}(d\mathbf{z}) \\
\le& \varepsilon.
\end{aligned} \tag{12.20}$$

where the last inequality follows by (12.12). Similarly, we have

$$\int_{\mathbb{S}} (1 - \phi_l(\mathbf{x}))\, |h(\mathbf{z})| \pi(d\mathbf{z}) \le \varepsilon \tag{12.21}$$

Since $\Pi_{T_k}^{\mathbf{z}_k}$ converges weakly to π, we have

$$\lim_{k\to\infty} \int_{\mathbb{S}} \phi_l(\mathbf{x}) h(\mathbf{z}) \Pi_{T_k}^{\mathbf{z}_k}(d\mathbf{z}) = \int_{\mathbb{R}_+^n} \phi_l(\mathbf{x}) h(\mathbf{z}) \pi(d\mathbf{z}). \tag{12.22}$$

As a consequence of (12.20), (12.21), and (12.22)

$$\limsup_{k\to\infty} \left| \int_{\mathbb{S}} h(\mathbf{z}) \Pi_{T_k}^{\mathbf{z}_k}(d\mathbf{z}) - \int_{\mathbb{S}} h(\mathbf{z}) \pi(d\mathbf{z}) \right| \le 2\varepsilon. \tag{12.23}$$

The desired result follows by letting $\varepsilon \to 0$. To proceed, we need the following lemma, whose proof is analogous to that of Lemma 11.33. □

Lemma 12.6 *For any $T > 0$, the Markov chain $\{(\mathbf{Z}(kT), k \in \mathbb{N}\}$ on $\mathcal{S}^\circ$ is irreducible and aperiodic. Moreover, every compact set $\mathcal{S}^\circ$ is petite.*

12.1.3 Permanence

This section is devoted to finding conditions under which $\mathbf{Z}$ converges to a unique invariant probability measure supported on $\mathbb{S}^\circ$. Let $\mathbb{M}$ be the set of ergodic invariant probability measures of $(\mathbf{Z}(t))$ supported on the boundary $\partial\mathbb{S}$. The condition for coexistence is:

(H12.2) For any $\mu \in \mathbb{M}$,

$$\max_{\{i=1,\ldots,n\}} \left\{ \int_{\partial\mathbb{S}} \left(f_i(\mathbf{z}) - \frac{\sigma_{ii} g_i^2(\mathbf{z})}{2} \right) \mu(d\mathbf{z}) \right\} > 0.$$

(In view of Lemma 12.4, $\int_{\partial\mathbb{S}} \left(f_i(\mathbf{z}) - \frac{\sigma_{ii} g_i^2(\mathbf{z})}{2} \right) \mu(d\mathbf{z})$ is well defined.)

It is proved in [252, Lemma 4] by the minmax principle that (H12.2) is equivalent to the existence of $\mathbf{p} > 0$ such that

$$\min_{\mu\in\mathbb{M}} \left\{ \sum_{i=1}^{n} p_i \left(\int_{\partial\mathbb{S}} \left(f_i(\mathbf{z}) - \frac{\sigma_{ii} g_i^2(\mathbf{z})}{2} \right) \mu(d\mathbf{z}) \right) \right\} := 2\rho^* > 0. \tag{12.24}$$

which together with (12.16) implies that

$$\min_{\mu\in\mathbb{M}}\left\{\left(\int_{\partial\mathbb{S}}\left(-\mathcal{L}\mathcal{W}(\mathbf{x})+\sum_{i=1}^{n}p_i\left(f_i(\mathbf{z})-\frac{\sigma_{ii}g_i^2(\mathbf{z})}{2}\right)\mu(d\mathbf{z})\right)\right)\right\}=2\rho^*. \tag{12.25}$$

Because $\mathbb{M}$ has finite cardinality and $\Phi(\mathbf{z}) := -\mathcal{L}\mathcal{W}(\mathbf{x}) + \sum_{i=1}^{n} p_i\left(f_i(\mathbf{z})-\frac{\sigma_{ii}g_i^2(\mathbf{z})}{2}\right)$ is lower bounded due to (12.4), there exists $R > 0$ such that

$$\min_{\mu\in\mathbb{M}}\left\{\left(\sum_{i\in\mathcal{M}}\int_{\partial\mathbb{S}}(\Phi(\mathbf{z})\wedge R)\,d\mu\right)\right\}\geq\frac{3}{2}\rho^*. \tag{12.26}$$

Note that by rescaling if necessary, we can assume that $\|\mathbf{p}\| = \delta_0$.

Lemma 12.7 *Assume that* (H12.2) *holds. There is a* $T^* > 0$ *such that for any* $T > T^*$, $\mathbf{z} \in \partial\mathbb{S}$, $\|\mathbf{x}\| \leq M$,

$$\frac{1}{T}\int_0^T \mathbb{E}_{\mathbf{z}}\left(R\wedge\Phi(\mathbf{Z}(t))\right)dt \geq \rho^* \tag{12.27}$$

Proof Suppose that the conclusion of this lemma were not true. Then, we could find $\mathbf{z}_k \in \partial\mathbb{S}$, $\|\mathbf{x}_k\| \leq M$ and $T_k > 0$, $\lim_{k\to\infty} T_k = \infty$ such that

$$\frac{1}{T_k}\int_0^{T_k}\mathbb{E}_{\mathbf{z}_k}\left(R\wedge\Phi(\mathbf{Z}(t))\right)dt < \rho^*,\ k\in\mathbb{N}. \tag{12.28}$$

It follows from Lemma 12.3 that $\left\{\Pi_{T_k}^{\mathbf{z}_k} : k \in \mathbb{N}\right\}$ is tight. As a result it has a convergent subsequence in the weak*-topology. Without loss of generality, we can suppose that $\left\{\Pi_{T_k}^{\mathbf{z}_k} : k \in \mathbb{N}\right\}$ is a convergent sequence in the weak*-topology whose limit is an invariant probability measure μ of $(\mathbf{Z}(t))$. As a consequence of the weak convergence,

$$\lim_{k\to\infty}\frac{1}{T_k}\int_0^{T_k}\mathbb{E}_{\mathbf{z}_k}\left(R\wedge\Phi(\mathbf{Z}(t))\right)dt=\int_{\mathbb{S}}(R\wedge\Phi(\mathbf{z}))\,d\mu.$$

In view of Lemma 12.4 and (12.26), we get that

$$\lim_{k\to\infty}\frac{1}{T_k}\int_0^{T_k}\mathbb{E}_{\mathbf{z}_k}\left(R\wedge\Phi(\mathbf{Z}(t))\right)dt=\int_{\mathbb{S}}(R\wedge\Phi(\mathbf{z}))\,d\mu\geq\frac{3}{2}\rho^*,$$

which contradicts (12.28). □

From now on let $n^* \in \mathbb{N}$ such that

$$\rho^*(n^*-1) > H_1 + 1. \tag{12.29}$$

Proposition 12.8 *Let* $V(\cdot)$ *be defined by* (12.6) *with* $\mathbf{p}$ *satisfying* (12.26) *and*

$$T^* > \frac{2}{H} \sup_{\|\mathbf{x}\| \le M} \{\mathcal{W}(\mathbf{x})\}$$

satisfying the assumptions from Lemma (12.7). *There is* $\theta \in \left(0, \frac{\delta_0}{2}\right)$, $K_\theta > 0$, *such that for any* $T \in [T^*, n^*T^*]$ *and* $\mathbf{z} \in \mathbb{R}_+^{n,\circ} \times \mathcal{M}$, $\|\mathbf{x}\| \le M$,

$$\mathbb{E}_\mathbf{z} U_+(\mathbf{X}(T)) \le \exp(-0.5\theta\rho^* T)U_+(\mathbf{x}) + K_\theta$$

where $U_+(\mathbf{x}) := e^{\theta V_\mathbf{p}(\mathbf{x})}$.

Proof We have from Itô's formula that

$$\begin{aligned}
V_\mathbf{p}(\mathbf{X}(T)) =& V_\mathbf{p}(\mathbf{x}) - \int_0^T \Phi(\mathbf{Z}(t))dt + M_\mathcal{W}(t) + M_\mathbf{p}(t)\\
\le& V_\mathbf{p}(\mathbf{x}) - \int_0^T (R \wedge \Phi(\mathbf{Z}(t)))\, dt + M_{V_\mathbf{p}}(T)\\
\le& V_\mathbf{p}(\mathbf{x}) - \mathbb{E}_\mathbf{z} \int_0^T (R \wedge \Phi(\mathbf{Z}(t)))\, dt + M_\mathcal{W}(t) + M_\mathbf{p}(t)\\
&+ \left(\mathbb{E}_\mathbf{z} \int_0^T R \wedge \Phi(\mathbf{Z}(t))dt - \int_0^T (R \wedge \Phi(\mathbf{Z}(t)))\, dt\right)
\end{aligned} \tag{12.30}$$

where

$$M_\mathbf{p} = \sum_{i=1}^n \int_0^T \left[\mathcal{W}_{x_i} X_i(t) g_i(\mathbf{Z}(t)) + p_i\right] g_i(\mathbf{Z}(t)) dE_i(t). \tag{12.31}$$

Note that $\tilde{M}(T) := \left(\int_0^T R \wedge \Phi(\mathbf{Z}(t))\right) dt$ is a bounded martingale (in particular, $|\tilde{M}(T)| \le c_R T$ for some constant c_R depending on R. Since $\exp\left(\theta \tilde{M}(T) - \frac{\theta^2}{2}[\tilde{M}(T)]\right)$ is a martingale, we have

$$\mathbb{E}_\mathbf{z} \exp\left(\theta \tilde{M}(T) - c_R^2 T^2 \frac{\theta^2}{2}\right) \le \mathbb{E}_\mathbf{z} \quad \exp\left(\theta \tilde{M}(T) - \frac{\theta^2}{2}[\tilde{M}(T)]\right) \le 1$$

which implies

$$\mathbb{E}_\mathbf{z} \exp\left(\theta \tilde{M}(T)\right) \le \exp\left(c_R^2 T^2 \frac{\theta^2}{2}\right). \tag{12.32}$$

Analogously, since $M_{V_{\mathbf{p}}}(T)$ is a local martingale with quadratic variation $\int_0^T \Lambda_{\mathbf{p}}(\mathbf{Z}(t))dt$ and $\Lambda_{\mathbf{p}}(\mathbf{z}) \leq \hat{\Lambda}(\mathbf{z})$,

$$\begin{aligned}&\mathbb{E}\exp\left(\theta M_{V_{\mathbf{p}}}(T) - \frac{\theta^2}{2}\int_0^T \hat{\Lambda}(\mathbf{Z}(t))dt\right)\\ \leq&\mathbb{E}\exp\left(\theta M_{V_{\mathbf{p}}}(T) - \frac{\theta^2}{2}[M_{V_p}(T)]\right) \leq 1 \text{ for } 0 < \theta \leq 1.\end{aligned} \tag{12.33}$$

Applying Holder's inequality, we have for $\theta \leq \frac{1}{2} \wedge \delta_0$ that

$$\begin{aligned}&\mathbb{E}_{\mathbf{z}}\exp\left(2\theta M_{V_{\mathbf{p}}}(T)\right)\\ =&\mathbb{E}_{\mathbf{z}}\left[\exp\left(\theta M_{V_{\mathbf{p}}}(T) - \theta^2\int_0^T \hat{\Lambda}(\mathbf{Z}(t))dt\right)\exp\left(\theta^2\int_0^T \hat{\Lambda}(\mathbf{Z}(t))dt\right)\right]^2\\ \leq&\mathbb{E}_{\mathbf{z}}\exp\left(2\theta M_{V_{\mathbf{p}}}(T) - 2\theta^2\int_0^T \hat{\Lambda}(\mathbf{Z}(t))dt\right)\mathbb{E}_{\mathbf{z}}\exp\left(2\theta^2\int_0^T \hat{\Lambda}(\mathbf{Z}(t))dt\right)\\ \leq&\exp\left(2\theta^2(\mathcal{W}(\mathbf{x}) + HT)\right),\end{aligned} \tag{12.34}$$

where the last inequality follows from (12.17) and (12.33). Now, using Holder's inequality again yields

$$\begin{aligned}\mathbb{E}_{\mathbf{z}}\exp\left(\theta(\tilde{M}(T) + M_{V_{\mathbf{p}}}(T))\right) \leq& \left(\mathbb{E}_{\mathbf{z}}\exp\left(2\theta\tilde{M}(T)\right)\mathbb{E}_{\mathbf{z}}\exp\left(\theta M_{V_{\mathbf{p}}}(T)\right)\right)^{\frac{1}{2}}\\ \leq&\exp\left(\theta^2(\mathcal{W}(\mathbf{x}) + HT) + c_R^2T^2\theta^2\right),\end{aligned} \tag{12.35}$$

where the last inequality comes from (12.32) and (12.34). Next, in view of Lemma 12.7 and the Feller property of $(\mathbf{Z}(t))$, there exists $\tilde{\delta} > 0$ such that if $\|\mathbf{x}\| \leq M$, $\text{dist}(\mathbf{x}, \partial\mathbb{R}^n_+) < \tilde{\delta}$ and $T \in [T^*, n^*T^*]$ then

$$\mathbb{E}_{\mathbf{z}}\int_0^T (R \wedge \Phi(\mathbf{Z}(t)))\,dt \geq \frac{7}{8}\rho^*T. \tag{12.36}$$

We have from (12.30), (12.35), (12.36), and the inequality $\sup_{\|\mathbf{x}\|\leq M}\mathcal{W}(\mathbf{x}) \leq \frac{T}{8}$ that

$$\begin{aligned}
&\mathbb{E}_{\mathbf{z}} \exp\left(\theta V_{\mathbf{p}}(\mathbf{X}(T))\right) \\
&\leq \exp\left(\theta V_{\mathbf{p}}(\mathbf{x}) - \theta \mathbb{E}_{\mathbf{z}} \int_0^T (R \wedge \Phi(\mathbf{Z}(t)))\, dt\right) \times \mathbb{E}_{\mathbf{z}} \exp\left(\theta(\tilde{M}(T) + M_{V_{\mathbf{p}}}(T))\right) \\
&\quad \leq \exp\left(\theta V_{\mathbf{p}}(\mathbf{x}) - \theta \mathbb{E}_{\mathbf{z}} \int_0^T (R \wedge \Phi(\mathbf{Z}(t)))\, dt\right) \\
&\quad \exp\left(\theta^2(\mathcal{W}(\mathbf{x}) + HT) + c_R^2 T^2 \theta^2\right) \\
&\quad \leq \exp\left(\theta V_{\mathbf{p}}(\mathbf{x})\right) \exp\left(-\theta \frac{5\rho^*}{8} T + c_R^2 T^2 \theta^2\right) \\
&\quad \leq \exp\left(\theta V_{\mathbf{p}}(\mathbf{x})\right) \exp\left(-\frac{\theta \rho^*}{2} T\right) \text{ for } T \in [T^*, n^* T^*]
\end{aligned} \tag{12.37}$$

if $\theta \leq \min\left\{\delta_0, \frac{1}{2}, \frac{\rho^*}{8H}, \frac{1}{8c_R^2 n^* T^*}\right\}$

In view of (12.8), we have for $\mathbf{x}$ satisfying $\|\mathbf{x}\| \leq M, \operatorname{dist}(\mathbf{x}, \partial \mathbb{R}_+^n) \geq \tilde{\delta}$ and $T \in [T^*, n^* T^*]$ that

$$\mathbb{E}_{\mathbf{z}} e^{\theta V_{\mathbf{p}}(\mathbf{X}(T))} \leq e^{\theta n^* T^* H} \sup_{\|\mathbf{x}\| \leq M, \operatorname{dist}(\mathbf{x}, \partial \mathbb{R}_+^n) \geq \tilde{\delta}, m \in \mathcal{M}} \{e^{\theta V_{\mathbf{p}}(\mathbf{x})}\} =: K_\theta < \infty. \tag{12.38}$$

The desired result follows from (12.37) and (12.38). □

Theorem 12.9 *Suppose that* (H12.2) *holds. Let θ be as in Proposition* 12.8. *There are $\kappa = \kappa(\theta, T^*) \in (0, 1)$, $\tilde{K} = \tilde{K}(\theta, T^*) > 0$ such that*

$$\mathbb{E}_{\mathbf{z}} U_+(\mathbf{X}(n^* T^*)) \leq \kappa U_+(\mathbf{x}) + \tilde{K} \text{ for all } \mathbf{z} \in \mathcal{S}^\circ := \mathbb{R}_+^{n,\circ} \times \mathcal{M}. \tag{12.39}$$

As a result, $\mathbf{Z}(t)$ has a unique invariant probability measure π^* in $\mathcal{S}^\circ$ and

$$\lim_{t \to \infty} \|P_{\mathbf{Z}}(t, \mathbf{z}, \cdot) - \pi^*(\cdot)\|_{TV} = 0, \ \mathbf{z} \in \mathcal{S}^\circ, \tag{12.40}$$

exponentially fast where $\|\cdot, \cdot\|_{TV}$ is the total variation norm and $P_{\mathbf{Z}}(t, \mathbf{z}, \cdot)$ is the transition probability of $(\mathbf{Z}(t))_{t \geq 0} = (\mathbf{X}(t), \alpha(t))_{t \geq 0}$. Moreover, for any initial value $\mathbf{z} \in \mathcal{S}^\circ$ and any π^*-integrable function f, we have

$$\mathbb{P}_{\mathbf{z}}\left\{\lim_{T \to \infty} \frac{1}{T} \int_0^T f(\mathbf{Z}(t))\, dt = \int_{\mathcal{S}^\circ} f(\mathbf{u}) \pi(d\mathbf{u})\right\} = 1. \tag{12.41}$$

Proof From (12.11), we have

$$\mathcal{L} U_+(\mathbf{x}) \leq -\theta U_+(\mathbf{x}) \text{ if } \|\mathbf{x}\| \geq M. \tag{12.42}$$

Define

$$\tau = \inf\{t \geq 0 : \|\mathbf{X}(t)\| \leq M\}. \tag{12.43}$$

In view of (12.42), we can obtain from Dynkin's formula that

$$\begin{aligned}
&\mathbb{E}_{\mathbf{z}}\left[\exp\left(\theta(\tau \wedge n^*T^*)\right) U_+(\mathbf{X}(\tau \wedge n^*T^*))\right] \\
&\leq U_+(\mathbf{x}) + \mathbb{E}_{\mathbf{z}} \int_0^{\tau \wedge n^*T^*} \exp(\theta s)[\mathcal{L}U_+(\mathbf{X}(s)) + \theta U_+(\mathbf{X}(s))]ds \\
&\leq U_+(\mathbf{x}).
\end{aligned}$$

Thus,

$$\begin{aligned}
U_+(\mathbf{x}) \geq &\mathbb{E}_{\mathbf{z}}\left[\exp\left(\theta(\tau \wedge n^*T^*)\right) U_+(\mathbf{X}(\tau \wedge n^*T^*))\right] \\
= &\mathbb{E}_{\mathbf{z}}\left[I_{\{\tau \leq (n^*-1)T^*\}} \exp\left(\theta(\tau \wedge n^*T^*)\right) U_+(\mathbf{X}(\tau \wedge n^*T^*))\right] \\
&+ \mathbb{E}_{\mathbf{z}}\left[I_{\{(n^*-1)T^* < \tau < n^*T^*\}} \exp\left(\theta(\tau \wedge n^*T^*)\right) U_+(\mathbf{X}(\tau \wedge n^*T^*))\right] \\
&+ \mathbb{E}_{\mathbf{z}}\left[I_{\{\tau \geq n^*T^*\}} \exp\left(\theta(\tau \wedge n^*T^*)\right) U_+(\mathbf{X}(\tau \wedge n^*T^*))\right] \\
\geq &\mathbb{E}_{\mathbf{z}}\left[I_{\{\tau \leq (n^*-1)T^*\}} U_+(\mathbf{X}(\tau))\right] \\
&+ \exp\left(\theta(n^*-1)T^*\right) \mathbb{E}_{\mathbf{z}}\left[I_{\{(n^*-1)T^* < \tau < n^*T^*\}} U_+(\mathbf{X}(\tau))\right] \\
&+ \exp\left(\theta n^*T^*\right) \mathbb{E}_{\mathbf{z}}\left[I_{\{\tau \geq n^*T^*\}} U_+(\mathbf{X}(n^*T^*))\right].
\end{aligned} \tag{12.44}$$

By the strong Markov property of $(\mathbf{X}(t), \alpha(t))$ and Proposition 12.8, we obtain

$$\begin{aligned}
&\mathbb{E}_{\mathbf{z}}\left[I_{\{\tau \leq (n^*-1)T^*\}} U_+(\mathbf{X}(n^*T^*))\right] \\
&\quad \leq \mathbb{E}_{\mathbf{z}}\left[I_{\{\tau \leq (n^*-1)T^*\}} \exp\left(-0.5\theta p^*(n^*T^* - \tau)\right) \left(U_+(\mathbf{X}(\tau)) + K_\theta\right)\right] \\
&\quad \leq \exp(-0.5\theta\rho^*T^*) \mathbb{E}_{\mathbf{z}}\left[I_{\{\tau \leq (n^*-1)T^*\}} \left(U_+(\mathbf{X}(\tau)) + K_\theta\right)\right]
\end{aligned} \tag{12.45}$$

The strong Markov property of $(\mathbf{X}(t), \alpha(t))$ and (12.8) lead to

$$\begin{aligned}
&\mathbb{E}_{\mathbf{z}}\left[I_{\{(n^*-1)T^* < \tau < n^*T^*\}} U_+(\mathbf{X}(n^*T^*))\right] \\
&\quad \leq \mathbb{E}_{\mathbf{z}}\left[I_{\{(n^*-1)T^* < \tau < n^*T^*\}} \exp\left(\theta H(n^*T^* - \tau)\right) \left(U_+(\mathbf{X}(\tau))\right)\right] \\
&\quad \leq \exp(\theta H T^*) \mathbb{E}_{\mathbf{z}}\left[I_{\{(n^*-1)T^* < \tau < n^*T^*\}} \left(U_+(\mathbf{X}(\tau))\right)\right]
\end{aligned} \tag{12.46}$$

Applying (12.45) and (12.46) to (12.44) yields

$$
\begin{aligned}
U_+(\mathbf{x}) \geq\, & \mathbb{E}_{\mathbf{z}}\left[I_{\{\tau\leq(n^*-1)T^*\}}U_+(\mathbf{X}(\tau))\right] \\
&+\exp\left(\theta(n^*-1)T^*\right)\mathbb{E}_{\mathbf{z}}\left[I_{\{(n^*-1)T^*<\tau<n^*T^*\}}U_+(\mathbf{X}(\tau))\right] \\
&+\exp\left(\theta n^*T^*\right)\mathbb{E}_{\mathbf{z}}\left[I_{\{\tau\geq n^*T^*\}}U_+(\mathbf{X}(n^*T^*))\right] \\
\geq\, & \exp(0.5\theta\rho^*T^*)\mathbb{E}_{\mathbf{z}}\left[I_{\{\tau\leq(n^*-1)T^*\}}U_+(\mathbf{X}(n^*T^*))\right]-K_\theta\mathbb{P}_{\mathbf{z}}\{\tau\leq(n^*-1)T^*\} \\
&+\exp(-\theta HT^*)\exp\left(\theta(n^*-1)T^*\right)\mathbb{E}_{\mathbf{z}}\left[I_{\{(n^*-1)T^*<\tau<n^*T^*\}}U_+(\mathbf{X}(n^*T^*))\right] \\
&+\exp\left(\theta n^*T^*\right)\mathbb{E}_{\mathbf{z}}\left[I_{\{\tau\geq n^*T^*\}}U_+(\mathbf{X}(n^*T^*))\right] \\
\geq\, & \exp(m\theta T^*)\mathbb{E}_{\mathbf{z}}U_+(\mathbf{X}(n^*T^*))-K_\theta
\end{aligned}
\tag{12.47}
$$

where $m^* = \min\{0.5\rho^*, (n^*-1)-H\} > 0$ by (12.29). The proof of (12.39) is complete by taking $\kappa = \exp(-m\theta T^*)$ and $\tilde{K} = \exp(-m^*\theta T^*)K_\theta$.

By Lemma 12.6, the Markov chain $\{\mathbf{Z}(kn^*T^*) : k \in \mathbb{N}\}$ is irreducible and aperiodic. Moreover, each compact set of $\mathcal{S}^\circ$ is petite. As an application of Theorem A.1, we deduce from (12.47) that

$$
\|P_{\mathbf{Z}}(kn^*T^*, \mathbf{z}, \cdot) - \pi^*(\cdot)\|_{TV} < C_{\mathbf{z}}r^k \tag{12.48}
$$

where π^* is an invariant probability measure of $\{\mathbf{Z}(kn^*T^*) : k \in \mathbb{N}\}$ on $\mathcal{S}^\circ$, for some $r \in (0,1)$ and $C_{\mathbf{z}} > 0$ a constant depending on $\mathbf{z} \in \mathcal{S}^\circ$.

We also have that for any compact set $K \subset \mathbb{R}^{n,\circ}_+$, $\mathbb{E}_{\mathbf{z}}\tau^*_K < \infty$ where τ^*_K is the first time $\{\mathbf{X}(kn^*T^*)\}$ enters K.

Thus, the process $\mathbf{Z}(\cdot)$ is a positive recurrent diffusion, or equivalently, $\mathbf{Z}(\cdot)$ has a unique invariant probability measure on $\mathcal{S}^\circ$; see Theorem 4.3. Because of (12.48), the unique invariant probability measure of the process $\mathbf{X}$ must be π^*. Moreover, it is well-known that $\|P_{\mathbf{Z}}(t, \mathbf{z}, \cdot) - \pi^*(\cdot)\|_{TV}$ is decreasing in t (it can be shown easily using the Kolmogorov-Chapman equation). We therefore obtain an exponential upper bound for $\|P_{\mathbf{Z}}(t, \mathbf{z}, \cdot) - \pi^*(\cdot)\|_{TV}$. □

12.1.4 Extinction

This section is devoted to the study of conditions under which some of the species will go extinct with probability 1. Consider $\mu \in \mathcal{M}$. Since the diffusion is nondegenerate in each subspace, there exist $0 < n_1 < \cdots < n_k \leq n$ (if $k = 0$, there are no $n_1, \ldots, n_k$) such that $\text{supp}(\mu) = \mathcal{S}^\mu := \mathbb{R}^\mu_+ \times \mathcal{M}$ where $\mathbb{R}^\mu_+ := \{(x_1, \ldots, x_n) \in \mathbb{R}^n_+ : x_i = 0 \text{ if } i \in I^c_\mu\}$ for $I_\mu := \{n_1, \ldots, n_k\}$ and $I^c_\mu := \{1, \ldots, n\} \setminus \{n_1, \ldots, n_k\}$. Let

$$
\mathbb{R}^{\mu,\circ}_+ := \{(x_1, \ldots, x_n) \in \mathbb{R}^n_+ : x_i = 0 \text{ if } i \in I^c_\mu \text{ and } x_i > 0 \text{ if } x_i \in I_\mu\}
$$

and $\partial\mathbb{R}^\mu_+ := \mathbb{R}^\mu_+ \setminus \mathbb{R}^{\mu,\circ}_+$, $\partial\mathcal{S}^\mu := \partial\mathbb{R}^\mu_+ \times \mathcal{M}$, $\mathcal{S}^{\mu,\circ} := \mathbb{R}^{\mu,\circ}_+ \times \mathcal{M}$.

Lemma 12.10 *For any* $\mu \in \mathbb{M}$,

$$\int_{\partial \mathbb{R}^n_+} \left(f_i(\mathbf{z}) - \frac{\sigma_{ii} g_i^2(\mathbf{z})}{2} \right) \mu(d\mathbf{z}) = 0\,,\ i \in I_\mu$$

Proof Let $\mathbf{Z}(t)$ be a stationary solution with distribution μ. In view of Itô's formula,

$$\frac{\ln X_i(t)}{t} = \frac{\ln X_i(0)}{t} + \frac{1}{t}\int_0^t \left[f_i(\mathbf{Z}(s)) - \frac{g_i^2(\mathbf{Z}(s))\sigma_{ii}}{2} \right] ds + \frac{1}{t}\int_0^t g_i(\mathbf{Z}(s)) dE_i(s)$$

By the strong law of large numbers (Theorem 4.4), if $\mathbf{Z}(0) = \mathbf{z}_0 \in \mathcal{S}^{\mu,\circ}$ and $i \in I_\mu$, then

$$\begin{aligned} &\lim_{t\to\infty} \frac{1}{t}\int_0^t \left[f_i(\mathbf{Z}(s)) - \frac{g_i^2(\mathbf{Z}(s))\sigma_{ii}}{2} \right] ds \\ &= \int_{\partial \mathcal{S}} \left(f_i \mathbf{z}) - \frac{\sigma_{ii} g^2(\mathbf{z})}{2} \right) \mu(d\mathbf{z}) \ \ \mathbb{P}_{\mathbf{x}_0}\text{- a.s.} \end{aligned}$$

and

$$\lim_{t\to\infty} \frac{1}{t}\int_0^t g_i^2(\mathbf{Z}(s)) ds = \int_{\partial \mathcal{S}} \frac{\sigma_{ii} g_i^2(\mathbf{z})}{2} \mu(d\mathbf{z}) \ \ \mu\text{- a.s.}$$

In view of the strong law of large numbers for local martingales, the latter display implies that $\lim_{t\to\infty} \frac{1}{t}\int_0^t g_i(\mathbf{Z}(s)) dE_i(s) = 0 \quad \mu$- a.s. On the other hand, $X_i(t), i \in I_\mu$ goes to neither 0 nor ∞ as $t \to \infty$. Thus $\lim_{t\to\infty} \frac{\ln X_i(t)}{t} = 0, i \in I_\mu$, which implies the desired result. □

The condition for extinction is given next.

(H12.3) There exists $\mu \in \mathcal{M}$ such that

$$\max_{i \in I_\mu^c} \int_{\partial \mathcal{S}} \left(f_i \mathbf{z}) - \frac{\sigma_{ii} g_i \mathbf{z})^2}{2} \right) \mu(d\mathbf{z}) < 0. \tag{12.49}$$

If $\mathbb{R}^\mu_+ \neq \{\mathbf{0}\}$, suppose further that for any $\nu \in \mathbb{M}_\mu := \{\widetilde{\nu}' \in \mathcal{M} : \mathrm{supp}(\widetilde{\nu}') \subset \partial \mathbb{R}^\mu_+\}$, we have

$$\max_{i \in I_\mu} \int_{\partial \mathcal{S}} \left(f_i \mathbf{z}) - \frac{\sigma_{ii} g_i \mathbf{z})^2}{2} \right) \nu(d\mathbf{z}) > 0 \tag{12.50}$$

The condition (12.50) is equivalent to the existence of $0 < \hat{p}_i < 1, i \in I_\mu$ such that for any $\nu \in \mathbb{M}_\mu$, we have

$$\sum_{i\in I_\mu} \hat{p}_i \int_{\partial\mathcal{S}} \left(f_i\mathbf{z}) - \frac{\sigma_{ii}g_i\mathbf{z})^2}{2} \right) \nu(d\mathbf{z}) > 0$$

Thus, there is $\check{p} \in (0, 1)$ sufficiently small such that

$$\begin{aligned}&\sum_{i\in I_\mu} \hat{p}_i \int_{\partial\mathcal{S}} \left(f_i\mathbf{z}) - \frac{\sigma_{ii}g_i\mathbf{z})^2}{2} \right) \nu(d\mathbf{z}) \\ &\quad - \check{p} \max_{i\in I_\mu^c} \int_{\partial\mathcal{S}} \left(f_i\mathbf{z}) - \frac{\sigma_{ii}g_i\mathbf{z})^2}{2} \right) \nu(d\mathbf{z}) > 0 \text{ for any } \nu \in \mathbb{M}_\mu\end{aligned} \tag{12.51}$$

Let $\tilde{\mathbf{p}} = (\tilde{p}_1, \ldots, \tilde{p}_n) \in \mathbb{R}^n$ where $\tilde{p}_i = \hat{p}_i$ if $i \in I_\mu$, $\tilde{p}_i = -\check{p}$ if $i \in I_\mu^c$. In view of (12.51), (12.49) and Lemma 12.10, there is $\rho_e > 0$ such that for any $\nu \in \mathbb{M}_\mu \cup \{\mu\}$,

$$\sum_{i\in I_\mu} \hat{p}_i \int_{\partial\mathcal{S}} \left(f_i\mathbf{z}) - \frac{\sigma_{ii}g_i\mathbf{z})^2}{2} \right) d\nu - \check{p} \max_{i\in I_\mu^c} \left\{ \int_{\partial\mathcal{S}} \left(f_i\mathbf{z}) - \frac{\sigma_{ii}g_i\mathbf{z})^2}{2} \right) d\nu \right\} > 3\rho_e. \tag{12.52}$$

Lemma 12.11 *Suppose that* (H12.3) *holds. Let M, H be as in* (12.4). *Let $n_e \in \mathbb{N}$ such that $\rho_e(n_e - 1) > H + 1$. There are $T_e \geq 0$, $\delta_e > 0$ such that, for any $T \in [T_e, n_eT_e]$, $\|\mathbf{x}\| \leq M$, $x_i < \delta_e, i \in I_\mu^c$, we have*

$$\begin{aligned}&\frac{1}{T}\int_0^T \mathbb{E}_\mathbf{z}\left([\mathcal{L}W](\mathbf{Z}(t))\right) dt - \sum_{i\in I_\mu} \hat{p}_i \frac{1}{T}\int_0^T \mathbb{E}_\mathbf{z}\left(f_i(\mathbf{X}(t), i) - \frac{\sigma_{ii}g_i(\mathbf{X}(t), i)^2}{2} \right) dt \\ &\quad + \check{p} \max_{i\in I_\mu^c} \left\{ \frac{1}{T}\int_0^T \mathbb{E}_\mathbf{z}\left(f_i\mathbf{Z}(t)) - \frac{\sigma_{ii}g_i\mathbf{Z}(t))^2}{2} \right) dt \right\} \leq -\rho_e.\end{aligned} \tag{12.53}$$

Proof Analogous to Lemma 12.7, using (12.52), one can show there exists $T_e > 0$ such that for any $T > T_e$, $\mathbf{x} \in \mathbb{R}_+^\mu$, $\|\mathbf{x}\| \leq M$, we have

$$\begin{aligned}&\frac{1}{T}\int_0^T \mathbb{E}_\mathbf{z}\left([\mathcal{L}W](\mathbf{Z}(t))\right) dt - \sum_{i\in I_\mu} \hat{p}_i \frac{1}{T}\int_0^T \mathbb{E}_\mathbf{z}\left(f_i\mathbf{Z}(t)) - \frac{\sigma_{ii}g_i\mathbf{X}(t), i)^2}{2} \right) dt \\ &\quad + \check{p} \max_{i\in I_\mu^c} \left\{ \frac{1}{T}\int_0^T \mathbb{E}_\mathbf{z}\left(f_i\mathbf{Z}(t)) - \frac{\sigma_{ii}g_i\mathbf{Z}(t))^2}{2} \right) dt \right\} \leq -2\rho_e\end{aligned} \tag{12.54}$$

By the Feller property of $\mathbf{Z}(t)$ and compactness of the set $\{\mathbf{x} \in \mathbb{R}^{\mu}_{+}, \|\mathbf{x}\| \le M\}$, there is $\delta_e > 0$ such that for any $T \in [T_e, n_e T_e]$, $\|\mathbf{x}\| \le M, x_i < \delta_e, i \in I^c_{\mu}$, the estimate (12.53) holds. □

Proposition 12.12 *Suppose that* (H12.3) *holds. Let* $\delta_0 > 0$ *be as in* (12.4). *There is* $\theta \in (0, \delta_0)$ *such that for any* $T \in [T_e, n_e T_e]$ *and* $\mathbf{x} \in \mathbb{R}^{n,\circ}_{+}$ *satisfying* $\|\mathbf{x}\| \le M$, $x_i < \delta_e$, $i \in I^c_{\mu}$

$$\mathbb{E}_{\mathbf{z}} U_{-}(\mathbf{X}(T)) \le \exp(-0.5\theta\rho_e T) U_{-}(\mathbf{x}).$$

where

$$U_{-}(\mathbf{x}) = \sum_{i \in I^c_{\mu}} \left[e^{\mathcal{W}(\mathbf{x})} \frac{x_i^{\check{p}}}{\prod_{j \in I_{\mu}} x_j^{\hat{p}_j}} \right]^{\theta}, \mathbf{x} \in \mathcal{S}^{\circ}.$$

Proof For $i \in I^c_{\mu}$, let $U_{i,-}(\mathbf{x}) = e^{\mathcal{W}(\mathbf{x})} \dfrac{x_i^{\check{p}}}{\prod_{j \in I_{\mu}} x_j^{\hat{p}_j}} = e^{V_{\mathbf{p}_i}(\mathbf{x})}$ where $\mathbf{p}_i = (p_1, \cdots, p_n)$ with $p_i = \check{p}, p_j = \hat{p}_j, j \in I_{\mu}$ and $p_j = 0$ elsewhere. Similar to Proposition 12.8, by making use of Lemma 12.11, one can find $\theta > 0$ such that

$$\mathbb{E} U^{\theta}_{i,-}(\mathbf{X}(T)) \le \exp(-0.5\theta\rho_e T) U^{\theta}_{i}(\mathbf{z})$$

if $T \in [T_e, n_e T_e]$ and $\mathbf{x} \in \mathbb{R}^{n,\circ}_{+}$ satisfies $\|\mathbf{x}\| \le M$, $x_i < \delta_e$. The proof is complete by noting that $U_{-}(\mathbf{x}) = \sum_{i \in I^c_{\mu}} U^{\theta}_{i,-}(\mathbf{x})$. □

Lemma 12.13 *For* $\theta \in [0, \delta_0]$, *we have*

$$\mathbb{E}_{\mathbf{z}} U_{-}(\mathbf{X}(t)) \le \exp(\theta H t) U_{-}(\mathbf{x}) \ \mathbf{z} \in \mathcal{S}^{\circ}$$

Proof By the arguments from the proof of (12.8), for $\theta \le \delta_0$, $i \in I^c_{\mu}$ we have

$$\mathbb{E}_{\mathbf{z}} U^{\theta}_{i,-}(\mathbf{X}(t)) \le \exp(\theta H t) U^{\theta}_{i,-}(\mathbf{z}) \, , \mathbf{z} \in \mathcal{S}^{\circ}.$$

From this estimate, we can take the sum over I^c_{μ} to obtain the desired result. □

Theorem 12.14 *Under* (H12.3) *for any* $\delta < \delta_0$ *and any* $\mathbf{z} \in \mathbb{S}^{\circ}$ *we have*

$$\lim_{t \to \infty} \mathbb{E}_{\mathbf{z}} \left(\bigwedge_{i=1}^{n} X_i(t) \right)^{\delta} = 0, \tag{12.55}$$

where $\bigwedge_{i=1}^{n} X_i(t) := \min\{X_1(t), \ldots, X_n(t)\}$.

Proof Just as in (12.42), we have

$$\mathcal{L}U_-(\mathbf{x}) \leq -\theta U_-(\mathbf{x}) \text{ if } \|\mathbf{x}\| \geq M. \tag{12.56}$$

Let

$$\varsigma := \frac{\delta_e^{\check{p}\theta}}{C_U^\theta}, \quad C_U := \sup\left\{\frac{\prod_{i\in I_\mu} x_i^{\hat{p}_i}}{e^{W(\mathbf{x})}} : \mathbf{z} \in \mathcal{S}^\circ\right\} < \infty,$$

and

$$\xi := \inf\{t \geq 0 : U_-(\mathbf{X}(t)) \geq \varsigma\}.$$

Clearly, if $U_-(\mathbf{x}) < \varsigma$, then $\xi > 0$ and for any $i \in I_\mu^c$, we get

$$X_i(t) \leq \delta_e\,, t \in [0, \xi). \tag{12.57}$$

Let

$$\tilde{U}_-(\mathbf{x}) := \varsigma \wedge U_-(\mathbf{x}).$$

We have from the concavity of $x \mapsto x \wedge \varsigma$ that

$$\mathbb{E}_\mathbf{z}\tilde{U}_-(\mathbf{X}(T)) \leq \varsigma \wedge \mathbb{E}U_-(\mathbf{X}(T)).$$

Let τ be defined by (12.43). By (12.56) and Dynkin's formula, we have that

$$\begin{aligned}
&\mathbb{E}_\mathbf{z}\left[\exp\left(\theta(\tau \wedge \xi \wedge n_e T_e)\right) U_-(\mathbf{X}(\theta(\tau \wedge \xi \wedge n_e T_e)))\right] \\
&\quad \leq U_-(\mathbf{x}) + \mathbb{E}_\mathbf{z}\int_0^{\theta(\tau\wedge\xi\wedge n_e T_e)} \exp(\theta s)\left([\mathcal{L}U_-](\mathbf{Z}(s)) + \theta U_-(\mathbf{X}(s))\right) ds \\
&\quad \leq U_-(\mathbf{x}).
\end{aligned}$$

As a result,

$$\begin{aligned}
U_-(x, m) \geq &\mathbb{E}_\mathbf{z}\left[\exp\left(\theta(\tau \wedge \xi \wedge n_e T_e)\right) U_-(\mathbf{X}(\tau \wedge \xi \wedge n_e T_e))\right] \\
\geq &\mathbb{E}_\mathbf{z}\left[I_{\{\tau\wedge\xi\wedge(n_e-1)T_e=\tau\}} U_-(\mathbf{X}(\tau))\right] \\
&+ \mathbb{E}_\mathbf{z}\left[I_{\{\tau\wedge\xi\wedge(n_e-1)T_e=\xi\}} U_-(\mathbf{X}(\xi))\right] \\
&+ \exp\left(\theta(n_e - 1)T_e\right)\mathbb{E}_\mathbf{z}\left[I_{\{(n_e-1)T_e<\tau\wedge\xi<n_e T_e\}} U_-(\mathbf{X}(\tau \wedge \xi))\right] \\
&+ \exp\left(\theta n_e T_e\right)\mathbb{E}_\mathbf{z}\left[I_{\{\tau\wedge\xi\geq n_e T_e\}} U_-(\mathbf{X}(n_e T_e))\right].
\end{aligned} \tag{12.58}$$

By the strong Markov property of $\mathbf{Z}(t)$ and Proposition 12.12 (which we can use because of (12.57)),

$$\begin{aligned}
&\mathbb{E}_{\mathbf{z}}\left[I_{\{\tau\wedge\xi\wedge(n_e-1)T_e=\tau\}}U_-(\mathbf{X}(n_eT_e))\right]\\
&\quad\leq \mathbb{E}_{\mathbf{z}}\left[I_{\{\tau\wedge\xi\wedge(n_e-1)T_e=\tau\}}\exp\left(-\frac{1}{2}\theta\rho_e(n_eT_e-\tau)\right)U_-(\mathbf{X}(\tau))\right]\\
&\quad\leq \mathbb{E}_{\mathbf{z}}\left[I_{\{\tau\wedge\xi\wedge(n_e-1)T_e=\tau\}}U_-(\mathbf{X}(\tau))\right].
\end{aligned}\tag{12.59}$$

Similarly, by the strong Markov property of $\mathbf{Z}(t)$ and Lemma 12.13, we obtain

$$\begin{aligned}
&\mathbb{E}_{\mathbf{z}}\left[I_{\{(n_e-1)T_e<\tau\wedge\xi<n_eT_e\}}U_-(\mathbf{X}(n_eT_e))\right]\\
&\quad\leq \mathbb{E}_{\mathbf{z}}\left[I_{\{(n_e-1)T_e<\tau\wedge\xi<n_eT_e\}}\exp\left(\theta H(n_eT_e-\tau\wedge\xi)\right)U_-(\mathbf{X}(\tau\wedge\xi))\right]\\
&\quad\leq \exp(\theta HT_e)\mathbb{E}_{\mathbf{z}}\left[I_{\{(n_e-1)T_e<\tau\wedge\xi<n_eT_e\}}U_-(\mathbf{X}(\tau\wedge\xi))\right].
\end{aligned}\tag{12.60}$$

If $U_-(\mathbf{x})<\varsigma$ then applying (12.59), (12.60), and the inequality

$$\tilde{U}_-(\mathbf{X}(n_eT_e),\alpha(n_eT_e))\leq U_-(\mathbf{X}(n_eT_e),\alpha(n_eT_e))$$

to (12.58) yields

$$\begin{aligned}
\tilde{U}_-(\mathbf{x})&=U_-(\mathbf{x})\\
&\geq \mathbb{E}_{\mathbf{z}}\left[I_{\{\tau\wedge\xi\wedge(n_e-1)T_e=\tau\}}U_-(\mathbf{X}(\tau))\right]\\
&\quad+\mathbb{E}_{\mathbf{z}}\left[I_{\{\tau\wedge\xi\wedge(n_e-1)T_e=\xi\}}U_-(\mathbf{X}(\xi))\right]\\
&\quad+\exp\left(\theta(n_e-1)T_e\right)\mathbb{E}_{\mathbf{z}}\left[I_{\{(n_e-1)T_e<\tau\wedge\xi<n_eT\}}U_-(\mathbf{X}(\tau\wedge\xi))\right]\\
&\quad+\exp\left(\theta n_eT_e\right)\mathbb{E}_{\mathbf{z}}\left[I_{\{\tau\wedge\xi\geq n_eT_e\}}U_-(\mathbf{X}(n_eT_e))\right]\\
&\geq \mathbb{E}_{\mathbf{z}}\left[I_{\{\tau\wedge\xi\wedge(n_e-1)T_e=\tau\}}U_-(\mathbf{X}(n_eT_e))\right]\\
&\quad+\mathbb{E}_{\mathbf{z}}\left[I_{\{\tau\wedge\xi\wedge(n_e-1)T_e=\xi\}}\tilde{U}_-(\mathbf{X}(n_eT_e))\right]\\
&\quad+\exp\left(\theta(n_e-1)T_e-\theta HT_e\right)\mathbb{E}_{\mathbf{z}}\left[I_{\{(n_e-1)T_e<\tau\wedge\xi<n_eT_e\}}U_-(\mathbf{X}(n_eT_e))\right]\\
&\quad+\exp\left(\theta n_eT_e\right)\mathbb{E}_{\mathbf{z}}\left[I_{\{\tau\wedge\xi\geq n_eT_e\}}U_-(\mathbf{X}(n_eT_e))\right]\\
&\geq \mathbb{E}_{\mathbf{z}}\tilde{U}_-(\mathbf{X}(n_eT_e))\quad(\text{since }\tilde{U}_-(\cdot)\leq U_-(\cdot)).
\end{aligned}\tag{12.61}$$

Clearly, if $U_-(\mathbf{x})\geq\varsigma$ then

$$\mathbb{E}_{\mathbf{z}}\tilde{U}_-(\mathbf{X}(n_eT_e))\leq\varsigma=\tilde{U}_-(\mathbf{x}).\tag{12.62}$$

As a result of (12.61) and (12.62), and the Markov property of $(\mathbf{Z}(t))$, the sequence $\{Y(k):k\in\mathbb{N}\}$ where $Y(k):=\tilde{U}_-(\mathbf{X}(kn_eT_e))$ is a supermartingale. For $\lambda\in(0,\varsigma)$ and $\varepsilon\in(0,1)$, let $\zeta_\lambda:=\inf\{k\in\mathbb{N}:Y(k)\geq\lambda\}$. If $U_-(\mathbf{x})\leq\lambda\varepsilon$ we have

$$\mathbb{E}_{\mathbf{z}}Y(k \wedge \zeta_\lambda) \le \mathbb{E}_{\mathbf{z}}Y(0) = U_-(\mathbf{x}) \le \lambda\varepsilon \quad \text{for all } k \in \mathbb{N}. \tag{12.63}$$

Subsequently, (12.63) combined with the Markov inequality yields

$$\mathbb{P}_{\mathbf{z}}\{\zeta_\lambda < k\} \le \lambda^{-1}\mathbb{E}_{\mathbf{z}}Y(k \wedge \zeta_\lambda) \le \varepsilon \quad \text{for all } k \in \mathbb{N}, \text{if } U_-(\mathbf{x}) \le \lambda\varepsilon.$$

Next, let $k \to \infty$ to get

$$\mathbb{P}_{\mathbf{z}}\{\zeta_\lambda < \infty\} \le \varepsilon \quad \text{if } U_-(\mathbf{x}) \le \lambda\varepsilon. \tag{12.64}$$

Next, we show by contradiction that $(\mathbf{Z}(t))$ is transient. If the process $(\mathbf{Z}(t))$ is recurrent in $\mathcal{S}^\circ$, then $\mathbf{Z}(t)$ will enter $\mathcal{K} \times M$ in a finite time almost surely given that $\mathbf{X}(0) \in \mathbb{R}^{n,\circ}_+$ and $\mathcal{K}$ is a compact set of $\mathbb{R}^{n,\circ}_+$. We have from (12.11) and the proof of Theorem 2.12 that

$$\inf_{\mathbf{z} \in \mathcal{K} \times \mathcal{M}} \mathbb{P}_{\mathbf{z}}\{V_{\mathbf{p}}(\mathbf{X}(t)) \le n \text{ for all } t \in [0, n_e T_e]\} \to 1 \text{ as } n \to \infty \text{ if } \|\mathbf{p}\| \le 1, \mathbf{p} > 0. \tag{12.65}$$

By the strong Markov property, for sufficiently small $\lambda > 0$,

$$\mathbb{P}_{\mathbf{z}}\{X_i(kn_eT_e) \ge \lambda,\ i = 1, \ldots, n \text{ for some } k \in \mathbb{N}\} > 1 - \varepsilon,\ \mathbf{z} \in \mathcal{S}^\circ. \tag{12.66}$$

If $\mathbf{z} \in \mathcal{S}^\circ$ is such that $U_-(\mathbf{x})$ is sufficiently small then both (12.64) and (12.66) hold, a contradiction. Thus, $\mathbf{Z}$ is transient.

As a result, any weak*-limit of $P_{\mathbf{Z}}(t, \mathbf{z}, \cdot)$ is a probability measure concentrated on $\partial\mathcal{S}$. Suppose that (12.55) does not hold. Then there exist $\mathbf{z}_0 \in \mathcal{S}^\circ$, $\hat{\varepsilon} > 0$ and a sequence $t_k \uparrow \infty$ as $k \to \infty$ such that

$$\int_{\mathcal{S}} \left(\bigwedge_{i=1}^n x_i^\delta\right) P_{\mathbf{Z}}(t_k, \mathbf{z}_0, d\mathbf{z}) > \hat{\varepsilon} \text{ for any } k \in \mathbb{N}. \tag{12.67}$$

Since the family $\{P_{\mathbf{Z}}(t, \mathbf{z}_0, \cdot), t \ge 0\}$ is tight in $\mathcal{S}$ due to Lemma 12.3, there exists a subsequence $\{t_k\}$ (still denoted by $\{t_k\}$ for convenience) such that $P_{\mathbf{Z}}(t_k, \mathbf{z}_0, \cdot)$ converges weakly to π as $k \to \infty$. Similar computations to the ones from Lemma 12.5 show that and $h(\cdot)$ is a continuous function on $\mathbb{R}^n_+$ such that for all $\mathbf{z} \in \mathcal{S}$ we have $|h(\mathbf{z})| < K(1 + \|\mathbf{x}\|)^\delta$, $\delta < \delta_0$ then $h(\cdot)$ is π-integrable and $\int_{\mathcal{S}} h(\mathbf{z}) P_{\mathbf{Z}}(t_k, \mathbf{z}_0, d\mathbf{z}) \to \int_{\mathbb{R}^n_+} h(\mathbf{z})\pi(d\mathbf{z})$. Because π with $\text{supp}(\pi) \subset \partial\mathbb{R}^n_+$, we have

$$\int_{\mathcal{S}} \left(\bigwedge_{i=1}^n x_i^\delta\right) \pi(d\mathbf{z}) = 0, \text{ and } \left(\bigwedge_{i=1}^n x_i^\delta\right) \le K(1 + \|\mathbf{x}\|)^\delta.$$

These facts imply

$$\lim_{k\to\infty}\int_{\mathcal{S}}\left(\bigwedge_{i=1}^{n}x_i^{\delta}\right)P_{\mathbf{Z}}(t_k,\mathbf{z}_0,d\mathbf{z})=0,$$

which contradicts (12.67). As a result, (12.55) is proved. □

Let $\mathbb{M}^1$ be the set of $\mu\in\mathcal{M}$ satisfying (H12.3) and $\mathbb{M}^2=\mathbb{M}\setminus\mathbb{M}^1$. For a subset $\widetilde{\mathbb{M}}\subset\mathbb{M}$, denote by $\text{Conv}(\widetilde{\mathbb{M}})$ the convex hull of $\widetilde{\mathbb{M}}$, that is, the set of probability measures π of the form $\pi(\cdot)=\sum_{\mu\in\widetilde{\mathbb{M}}}p_\mu\mu(\cdot)$ with $p_\mu>0,\sum_{\mu\in\widetilde{\mathbb{M}}}p_\mu=1$. To characterize the extinction of subspecies, we need some additional conditions.

(H12.4) Suppose that (12.4) is strengthened to

$$\begin{aligned}&[\mathcal{L}\mathcal{W}](\mathbf{z})+\delta_0e^{\delta_1\mathcal{W}(\mathbf{x})}\sum_{i,j}\left((|\mathcal{W}_{x_i}(\mathbf{x})x_i|+1)(|\mathcal{W}_{x_j}(\mathbf{x})x_j|+1)|g_ig_j||\sigma_{ij}|\right)\\&\quad+\sum_{i=1}^{n}(|f_i(\mathbf{z})|+g_i^2(\mathbf{z})\le-2\text{ if }\|\mathbf{x}\|\ge M.\end{aligned}\tag{12.68}$$

Assumption (H12.4) forces the growth rates of $g_i^2(\cdot)$ to be slightly lower than those of $|f_i(\cdot)|$. This is needed in order to suppress the diffusion part so that we can obtain the tightness of the random normalized occupation measures

$$\widetilde{\Pi}_t(\cdot):=\frac{1}{t}\int_0^t I_{\{\mathbf{X}(s)\in\cdot\}}ds,\ t>0$$

as well as the convergence of $\int_{\mathcal{S}}\left(f_i(\mathbf{z})-\frac{g_i^2(\mathbf{z})}{2}\right)\widetilde{\Pi}_{t_k}(d\mathbf{z})$ to $\int_{\mathcal{S}}\left(f_i(\mathbf{z})-\frac{g_i^2(\mathbf{z})}{2}\right)$ $\pi(d\mathbf{z})$ given that $\widetilde{\Pi}_{t_k}$ converges weakly to π. Having these properties, we can analyze the asymptotic behavior of sample paths of the solution.

(H12.5) Suppose that either $\mathbb{M}^2=\emptyset$ or for any $\nu\in\mathbb{M}^2$,

$$\max_{\{i=1,\dots,n\}}\left\{\int_{\partial\mathcal{S}}\left(f_i(\mathbf{z})-\frac{\sigma_{ii}g_i^2(\mathbf{z})}{2}\right)\nu(d\mathbf{z})\right\}>0.$$

Theorem 12.15 *Suppose that* (H12.4) *and* (H12.5) *are satisfied and* $\mathbb{M}^1\neq\emptyset$. *Then for any* $\mathbf{x}\in\mathbb{R}_+^{n,\circ}$

$$\sum_{\mu\in\mathbb{M}^1}P_{\mathbf{z}}^{\mu}=1\tag{12.69}$$

where

$$P_{\mathbf{z}}^{\mu} := \mathbb{P}_{\mathbf{z}}\left\{\lim_{t\to\infty}\frac{\ln X_i(t)}{t} = \int_{\partial\mathcal{S}}\left(f_i(\mathbf{z}) - \frac{\sigma_{ii}g_i^2(\mathbf{z})}{2}\right)\mu(d\mathbf{z}) < 0, i \in I_{\mu}^c\right\}.$$

Moreover $P_{\mathbf{z}}^{\mu} > 0$ *for each* $\mu \in \mathbb{M}^1$.

The proof is divided into several auxiliary lemmas.

Lemma 12.16 *Suppose that* (H12.4) *is satisfied. Let* $\mathbf{z}$ *be any element in* $\mathcal{S}$. *Then there is a* $\hat{K} > 0$ *such that*

$$\mathbb{P}_{\mathbf{z}}\left\{\limsup_{t\to\infty}\frac{1}{t}\int_0^t e^{\delta_1 \mathcal{W}(\mathbf{x})}\left(1 + \sum(|f_i(\mathbf{Z}(s))| + |g_i(\mathbf{Z}(s))|^2)\right)ds \le \hat{K}\right\} = 1. \tag{12.70}$$

Moreover,

$$\mathbb{P}_{\mathbf{x}}\left\{\lim_{t\to\infty}\frac{1}{t}\int_0^t\sum_{i=1}^n \mathcal{W}_{x_i}(\mathbf{X}(s))X_i(s)g_i((\mathbf{Z}(s))dE_i(s) = 0\right\} = 1, \tag{12.71}$$

$$\mathbb{P}_{\mathbf{x}}\left\{\lim_{t\to\infty}\frac{1}{t}\int_0^t g_i((\mathbf{Z}(s))dE_i(s) = 0, i = 1, \cdots, n\right\} = 1. \tag{12.72}$$

and there is $\hat{K}_1 > 1$ *such that*

$$\mathbb{P}_{\mathbf{z}}\left\{\liminf\frac{1}{t}\int_0^t I_{\{\|\mathbf{X}(s)\|\le\hat{K}_1\}}ds \ge \frac{1}{2}\right\} = 1, \ \mathbf{z} \in \mathcal{S}. \tag{12.73}$$

Proof Computation similar to those used to prove (12.14) gives us

$$\begin{aligned}\mathcal{L}e^{\delta_1 \mathcal{W}}t(\mathbf{z}) \le& \delta_1\left(HI_{\|\mathbf{x}\|\le M} - 2\right)e^{\delta_1\mathcal{W}(\mathbf{x})} - \delta_1 e^{2\delta_1\mathcal{W}(\mathbf{x})}\hat{\Lambda}(\mathbf{z}) \\ &- \delta_1 e^{\delta_1\mathcal{W}(\mathbf{x})}\sum_{i=1}^n(|f_i(\mathbf{z})| + g_i^2(\mathbf{z})) \\ \le& \tilde{H}_2 - \delta_1 e^{2\delta_1\mathcal{W}(\mathbf{x})}\hat{\Lambda}(\mathbf{z}) - \delta_1 e^{\delta_1\mathcal{W}(\mathbf{x})}\sum_{i=1}^n(|f_i(\mathbf{z})| + g_i^2(\mathbf{z})\end{aligned} \tag{12.74}$$

where $\tilde{H}_2 = \delta_1 \sup_{\|\mathbf{x}\|\le M, m\in\mathcal{M}}\left\{e^{\delta_1\mathcal{W}(\mathbf{x})}\right\}$.

Equation (12.74) together with Itô's formula implies $e^{\delta_1\mathcal{W}(\mathbf{X}(t))}$

$$\begin{aligned}
\frac{e^{\delta_1 \mathcal{W}(\mathbf{X}(t))}}{t} =&\frac{e^{\delta_1 \mathcal{W}(\mathbf{X}(0))}}{t} + \frac{1}{t}\int_0^t [\mathcal{L}e^{\delta_1 \mathcal{W}}](\mathbf{Z}(s))ds \\
&+ \frac{\delta_1}{t}\int_0^t e^{\delta_1 \mathcal{W}(\mathbf{X}(s))}\sum_{i=1}^n \mathcal{W}_{x_i}(\mathbf{X}(s))X_i(s)g_1((\mathbf{Z}(s))dE_i(s) \\
\leq& \frac{e^{\delta_1 \mathcal{W}(\mathbf{X}(0))}}{t} + \tilde{H}_2 - \frac{\delta_1}{t}\int_0^t e^{\delta_1 \mathcal{W}(\mathbf{X}(s))}\sum_{i=1}^n (\delta_1 + |f_i(\mathbf{Z}(s))| \\
&+ g_i^2(\mathbf{Z}(s)))ds - \delta_1 \frac{1}{t}\int_0^t e^{2\delta_1 \mathcal{W}(\mathbf{X}(s))}\hat{\Lambda}(\mathbf{Z}(s))ds \\
&+ \frac{\delta_1}{t}\int_0^t e^{\delta_1 \mathcal{W}(\mathbf{X}(s))}\sum_{i=1}^n \mathcal{W}_{x_i}(\mathbf{X}(s))X_i(s)g_1((\mathbf{Z}(s))dE_i(s)
\end{aligned}$$

Since $\frac{e^{\delta_1 \mathcal{W}(\mathbf{X}(t))}}{t} \geq 0$, we have

$$\begin{aligned}
&\frac{\delta_1}{t}\int_0^t e^{\delta_1 \mathcal{W}(\mathbf{X}(s))}\sum_{i=1}^n (\delta_1 + |f_i(\mathbf{Z}(s))| + g_i^2(\mathbf{Z}(s)))ds \\
&\leq H_2 - \frac{\delta_1}{t}\int_0^t e^{2\delta_1 \mathcal{W}(\mathbf{X}(s))}\hat{\Lambda}(\mathbf{Z}(s))ds \qquad (12.75) \\
&+ \frac{\delta_1}{t}\int_0^t e^{\delta_1 \mathcal{W}(\mathbf{X}(s))}\sum_{i=1}^n \mathcal{W}_{x_i}(\mathbf{X}(s))X_i(s)g_1((\mathbf{Z}(s))dE_i(s)
\end{aligned}$$

Let $M_3(t) := \int_0^t e^{\delta_1 \mathcal{W}(\mathbf{X}(s))}\sum_{i=1}^n \mathcal{W}_{x_i}(\mathbf{X}(s))X_i(s)g_1((\mathbf{Z}(s))dE_i(s)$ has quadratic variation

$$[M_3(t)] = \int_0^t e^{2\delta_1 \mathcal{W}(\mathbf{X}(s))}\Lambda_{\mathbf{0}}(\mathbf{Z}(s))ds \leq K_3\int_0^t e^{2\delta_1 \mathcal{W}(\mathbf{X}(s))}\hat{\Lambda}(\mathbf{Z}(s))ds$$

By the strong law of large numbers for martingales (see [201, Theorem 1.3.4], we obtain that

$$\lim_{t\to\infty}\frac{M_3(t)}{t + \int_0^t e^{2\delta_1 \mathcal{W}(\mathbf{X}(s))}\hat{\Lambda}(\mathbf{Z}(s))ds} = 0 \text{ a.s.},$$

which implies

$$\limsup_{t\to\infty}\left[-1 - \frac{\delta_1}{2}\frac{1}{t}\int_0^t e^{2\delta_1 \mathcal{W}(\mathbf{X}(s))}\hat{\Lambda}(\mathbf{Z}(s))ds + \frac{M_3(t)}{t}\right] \leq 0 \qquad (12.76)$$

Applying (12.76) to (12.75), we get

$$\limsup_{t\to\infty}\frac{\delta_1}{t}\int_0^t e^{\delta_1 \mathcal{W}(\mathbf{X}(s))}\sum_{i=1}^n(\delta_1+|f_i(\mathbf{Z}(s))|+g_i^2(\mathbf{Z}(s)))ds\le H_2+1 \ \mathbb{P}_{\mathbf{z}}\text{ - a.s.}$$

which proves (12.70). From (12.75), we also have

$$\begin{aligned}\frac{\delta_1}{2t}&\int_0^t e^{2\delta_1 \mathcal{W}(\mathbf{X}(s))}\hat{\Lambda}(\mathbf{Z}(s))ds\\ \le& H_2-\frac{\delta_1}{2t}\int_0^t e^{2\delta_1 \mathcal{W}(\mathbf{X}(s))}\hat{\Lambda}(\mathbf{Z}(s))ds\\ &+\frac{\delta_1}{t}\int_0^t e^{\delta_1 \mathcal{W}(\mathbf{X}(s))}\sum_{i=1}^n \mathcal{W}_{x_i}(\mathbf{X}(s))X_i(s)g_1((\mathbf{Z}(s))dE_i(s)\end{aligned}\tag{12.77}$$

and using (12.76) again we have

$$\limsup_{t\to\infty}\frac{\delta_1}{2t}\int_0^t e^{2\delta_1 \mathcal{W}(\mathbf{X}(s))}\hat{\Lambda}(\mathbf{Z}(s))ds\le H_2+1 \ \mathbb{P}_{\mathbf{z}}\text{ - a.s.}$$

Note that $M_W(t)$ is a local martingale with quadratic variation

$$[M_{\mathcal{W}}(t)]=\int_0^t \Lambda_{\mathbf{0}}(\mathbf{Z}(s))ds\le\frac{1}{2}\int_0^t\hat{\Lambda}(\mathbf{Z}(s))ds.$$

By the strong law of large numbers for martingales (see [201, Theorem 1.3.4], we obtain that

$$\lim_{t\to\infty}\frac{M_{\mathcal{W}}(t)}{t+\int_0^t\hat{\Lambda}(\mathbf{Z}(s))ds}=0\text{ a.s.}$$

Since $\lim_{t\to\infty}\frac{1}{t}\int_0^t\hat{\Lambda}(\mathbf{Z}(s))ds\le\hat{K}$ we can easily obtain that $\lim_{t\to\infty}\frac{M_{\mathcal{W}}(t)}{t}=0$ a.s., that is, (12.71). Likewise, (12.72) is proved similarly.

On the other hand, since $\inf_{\mathbf{z}\in\mathcal{S}:\|\mathbf{x}\|\ge n}e^{\delta_1\mathcal{W}(\mathbf{x})}\to\infty$ as $n\to\infty$, a simple application of Markov's inequality to (12.70) will result in (12.73). □

Lemma 12.17 *Suppose we have a sample path of* $\{\mathbf{X}(t)\}$ *satisfying*

$$\limsup_{t\to\infty}\frac{1}{t}\int_0^t e^{\delta_1 \mathcal{W}(\mathbf{X}(s))}\sum_{i=1}^n(1+|f_i(\mathbf{Z}(s))|+g_i^2(\mathbf{Z}(s)))ds\le\hat{K}$$

and that $\widetilde{\Pi}_{T_k}(\cdot)$ *converges weakly to an invariant probability measure* π *of* $\{\mathbf{X}(t)\}$ *with* $\lim_{k\to\infty}T_k=\infty$. *Then for this sample path, we have* $\int_{\mathbb{R}^n_+}h(\mathbf{z})\widetilde{\Pi}^{\mathbf{x}_k}_{T_k}(d\mathbf{z})\to$

$\int_{\mathbb{R}^n_+} h(\mathbf{z})\pi(d\mathbf{z})$ *for any continuous function* $h : \mathbb{R}^n_+ \to \mathbb{R}$ *satisfying* $|h(\mathbf{z})| < K_h e^{\delta \mathcal{W}(\mathbf{x})}(1 + \sum(|f_i(\mathbf{z})| + |g_i(\mathbf{z})|^2))$, $\mathbf{x} \in \mathbb{R}^n_+$, *with* K_h *a positive constant and* $\delta \in [0, \delta_1)$.

Proof The proof is almost the same as that of Lemma 12.5 and is left for the reader. □

Lemma 12.18 *For any initial condition* $\mathbf{z} \in \mathcal{S}$*, the family* $\{\widetilde{\Pi}_t(\cdot), t \geq 1\}$ *is tight in* $\mathcal{S}$*, and its weak*-limit set (denoted by* $\mathcal{U} = \mathcal{U}(\omega)$*) is a set of invariant probability measures of* $\{\mathbf{X}(t)\}$ *with probability 1.*

Proof The tightness follows from Lemma 12.16. The property of the weak*-limit set of normalized occupation measures was first proved in [252, Theorems 4, 5] for compact state spaces and then generalized to a locally compact state space in [83, Theorem 4.2]. Similar results for general Markov processes can be found in [16]. □

Lemma 12.19 *Let* $\mu \in \mathbb{M}^1$*. For any* $\mathbf{z} \in \mathcal{S}^\circ$,

$$\mathbb{P}_{\mathbf{z}}\Big\{\mathcal{U}(\omega) \subset \text{Conv}(\mathbb{M}_\mu \cup \{\mu\})\Big\} = \mathbb{P}_{\mathbf{z}}\Big\{\mathcal{U}(\omega) = \{\mu\}\Big\}$$

Proof Since μ satisfies (H12.3), it follows from (12.50) that, there are $p_i^\mu \in (0, 1)$, $i \in I_\mu$ such that

$$\sum_{i \in I_\mu} p_i^\mu \int_{\mathbb{S}} \left(f_i(\mathbf{z}) - \frac{\sigma_{ii} g_i^2(\mathbf{z})}{2} \right) d\nu > 0, \ \nu \in \mathbb{M}_\mu. \tag{12.78}$$

As a result of Lemmas 12.4 and 12.18,

$$\mathbb{P}_{\mathbf{x}} \left\{ \lim_{t \to \infty} \frac{1}{t} \int_0^t [\mathcal{L}\mathcal{W}](\mathbf{Z}(s))ds = 0 \right\} = 1$$

In light of Itô's formula and (12.71), for any initial value $\mathbf{z} \in \mathcal{S}^\circ$,

$$\limsup_{t \to \infty} \frac{\mathcal{W}(\mathbf{X}(t))}{t} = \lim_{t \to \infty} \frac{1}{t} \int_0^t [\mathcal{L}\mathcal{W}](\mathbf{Z}(s))ds + \lim_{t \to \infty} \frac{M_{\mathcal{W}}(t)}{t} = 0 \text{ a.s.} \tag{12.79}$$

Since $\liminf_{\mathbf{x} \to \infty} \frac{\mathcal{W}(\mathbf{x})}{\ln \|\mathbf{x}\|} > n$, (12.79) is followed by

$$\mathbb{P}_{\mathbf{x}} \left\{ \limsup_{t \to \infty} \frac{\ln X_i(t)}{t} \leq 0, \ i = 1, \ldots, n \right\} = 1, \ \mathbf{z} \in \mathcal{S}^\circ. \tag{12.80}$$

In view of (12.71) and (12.80), to prove the lemma, it suffices to show that if $\mathcal{U}(\omega) \subset \text{Conv}(\mathbb{M}_\mu \cup \{\mu\})$ and

$$\lim_{t\to\infty}\frac{1}{t}\int_0^t g_i(\mathbf{Z}(s))dE_i(s)=0,\ i=1,\ldots,n \tag{12.81}$$

and

$$\limsup_{t\to\infty}\frac{\ln X_i(t)}{t}\le 0,\ i=1,\ldots,n \tag{12.82}$$

then $\mathcal{U}(\omega)=\{\mu\}$.

We argue by contradiction. Assume there is a sequence $\{t_k\}$ with $\lim_{k\to\infty}t_k=\infty$ such that $\widetilde{\Pi}_{t_k}(\cdot)$ converges weakly to an invariant probability of the form $\pi=(1-\rho)\pi_1+\rho\mu$ where $\rho\in(0,1]$ and $\pi_1\in\text{Conv}(\mathbb{M}_\mu)$. It follows from Lemma 12.16 that

$$\begin{aligned}
&\lim_{k\to\infty}\frac{1}{t_k}\sum_{i\in I_\mu}p_i^\mu\int_0^{t_k}\left(f_i(\mathbf{Z}(s))-\frac{\sigma_{ii}g_i^2(\mathbf{Z}(s))}{2}\right)ds\\
&=\sum_{i\in I_\mu}p_i^\mu\int_{\mathbb{R}^n_+}\left(f_i(\mathbf{z})-\frac{\sigma_{ii}g_i^2(\mathbf{z})}{2}\right)\pi(d\mathbf{z})\\
&=(1-\rho)\sum_{i\in I_\mu}p_i^\mu\int_{\mathbb{R}^n_+}\left(f_i(\mathbf{z})-\frac{\sigma_{ii}g_i^2(\mathbf{z})}{2}\right)\pi(d\mathbf{z})\\
&\quad+\rho\sum_{i\in I_\mu}p_i^\mu\int_{\mathbb{R}^n_+}\left(f_i(\mathbf{z})-\frac{\sigma_{ii}g_i^2(\mathbf{z})}{2}\right)\mu(d\mathbf{z})\\
&=(1-\rho)\sum_{i\in I_\mu}p_i^\mu\int_{\mathbb{R}^n_+}\left(f_i(\mathbf{z})-\frac{\sigma_{ii}g_i^2(\mathbf{z})}{2}\right)\pi(d\mathbf{z})\ \text{(by Lemma 12.10)}\\
&>0\ \text{(by virtue of (12.78))}.
\end{aligned} \tag{12.83}$$

As a result of (12.81), (12.83), and Itô's formula

$$\begin{aligned}
&\lim_{k\to\infty}\sum_{i\in I_\mu}p_i^\mu\frac{\ln X_i(t_k)}{t_k}\\
&=\lim_{k\to\infty}\frac{1}{t_k}\sum_{i\in I_\mu}p_i^\mu\int_0^{t_k}\left[\left(f_i(\mathbf{Z}(s))-\frac{\sigma_{ii}g_i^2(\mathbf{Z}(s))}{2}\right)ds+\sum_{i\in I_\mu}g_i(\mathbf{Z}(s))dE_i(s)\right]\\
&>0,
\end{aligned}$$

which contradicts (12.82). This completes the proof. □

Lemma 12.20 *Let $\mu \in \mathbb{M}^1$. For any $k \in \mathbb{N}$, $\varepsilon > 0$, there is a $\Delta > 0$ such that*

$$\mathbb{P}_{\mathbf{z}}\left\{\lim_{t\to\infty}\frac{\ln X_i(t)}{t} = \int_{\partial S}\left(f_i(\mathbf{z}) - \frac{\sigma_{ii}g_i^2(\mathbf{z})}{2}\right)\mu(d\mathbf{z}) < 0, i \in I_\mu^c\right\} > 1-\varepsilon$$

if $\mathbf{x} \in \mathcal{K}_\mu^{k,\Delta}$ where

$$\mathcal{K}_\mu^{k,\Delta} = \{\mathbf{x} \in \mathbb{R}_+^{n,\circ}, k^{-1} \le x_i \le k \text{ for } i \in I_\mu, x_i < \Delta \text{ for } i \in I_\mu^c\}.$$

Proof Let $\tilde{U}_-(\mathbf{x})$ be the function defined as in the proof of Theorem 12.14. In view of (12.65), there is a $\Delta_1 > 0$ such that

$$\mathbb{P}_{\mathbf{z}}\left\{\max_{i\in I_\mu^c}\{X_i(t)\} > \Delta_1 \,\forall t \in [0, n_eT_e]\right\} > 0.5,\ \mathbf{z} \in \mathcal{H}_\mu \times \mathcal{M}, \tag{12.84}$$

where

$$\mathcal{H}_\mu = \left\{\mathbf{x} \in \mathbb{R}_+^n : \|\mathbf{x}\| \le \hat{K}_1,\ \max_{i\in I_\mu^c}\{x_i\} \ge 1\right\}.$$

It can be seen that $\nu(\mathcal{H}_\mu) > 0$ for any $\nu \in \mathcal{M} \setminus (\mathbb{M}_\mu \cup \{\mu\})$. By the definition of $\tilde{U}_-(\cdot)$, there is a $\Delta > 0$ sufficiently small such that

$$\sup_{\mathbf{x}\in\mathcal{K}_\mu^{k,\Delta}, m\in\mathcal{M}}\{\tilde{U}_-(\mathbf{x})\} \le \frac{2}{\varepsilon}\inf_{\mathbf{y}\in\mathbb{R}_+^{n,\circ}, x_i\ge\Delta_1, i\in I_\mu^c, j\in\mathcal{M}}\{\tilde{U}(\mathbf{y}, j)\}. \tag{12.85}$$

In view of (12.85), since $\{Y(j) := \tilde{U}_\theta(\mathbf{X}(jn_eT_e)) : j \in \mathbb{N}\}$ is a supermartingale, similar to (12.64), we can obtain

$$\mathbb{P}_{\mathbf{z}}\left\{\max_{i\in I_\mu^c}\{X_i(jn_eT_e)\} < \Delta_1 \text{ for all } j \in \mathbb{N}\right\} > 1 - \frac{\varepsilon}{2} \text{ if } \mathbf{x} \in \mathcal{K}_\mu^{k,\Delta}. \tag{12.86}$$

Let $\tau_{\mathcal{H}_\mu} = \inf\{t > 0 : \mathbf{X}(t) \in \mathcal{H}_\mu\}$. Suppose that there is an $\mathbf{x} \in \mathcal{K}_\mu^{k,\Delta}$ such that

$$\mathbb{P}_{\mathbf{z}}\left\{\limsup_{t\to\infty}\frac{1}{t}\int_0^t I_{\{\mathbf{X}(s)\in\mathcal{H}_\mu\}}ds > 0\right\} > \varepsilon. \tag{12.87}$$

Then

$$\mathbb{P}_{\mathbf{z}}\{\tau_{\mathcal{H}_\mu} < \infty\} > \varepsilon. \tag{12.88}$$

By the strong Markov property of $\{\mathbf{X}(t) : t \in \mathbb{R}_+\}$, it follows from (12.84) and (12.88) that

$$\mathbb{P}_{\mathbf{z}}\left(\{\tau_{\mathcal{H}_\mu} < \infty\} \cap \left\{\max_{i \in I_\mu^c}\{X_i(t)\} \geq \Delta_1 \text{ for } t \in [\tau_{\mathcal{H}_\mu}, \tau_{\mathcal{H}_\mu} + n_e T_e]\right\}\right) > 0.5\varepsilon,$$

which contradicts (12.86). Thus, (12.87) does not hold, that is, we have

$$\mathbb{P}_{\mathbf{z}}\left\{\lim_{t\to\infty} \frac{1}{t}\int_0^t I_{\{\mathbf{X}(s)\in\mathcal{H}_\mu\}} ds = 0\right\} > 1 - \varepsilon,\ \mathbf{z} \in \mathcal{K}_\mu^{k,\Delta} \times \mathcal{M}. \tag{12.89}$$

If for an $\omega \in \Omega$, and a sequence $\{t_j\}$ with $\lim_{j\to\infty} t_j = \infty$, $\widetilde{\Pi}_{t_j}(\cdot)$ converges weakly to an invariant probability of the form $\pi = (1-\rho)\pi_1 + \rho\pi_2$ where $\rho \in (0, 1]$ and $\pi_1 \in \text{Conv}(\mathbb{M}_\mu \cup \{\mu\})$, $\pi_2 \in \text{Conv}(\mathcal{M} \setminus (\mathbb{M}_\mu \cup \{\mu\}))$ then

$$\limsup_{j\to\infty} \frac{1}{t_j}\int_0^{t_j} I_{\{\mathbf{X}(s)\in\mathcal{H}_\mu\}} ds \geq \pi(\mathcal{H}_\mu) \geq \rho\pi_2(\mathcal{H}_\mu) > 0.$$

This inequality, combined with Lemma 12.18 and (12.89), implies that

$$\mathbb{P}_{\mathbf{z}}\left\{\mathcal{U}(\omega) \subset \text{Conv}(\mathbb{M}_\mu \cup \{\mu\})\right\} > 1 - \varepsilon,\ \mathbf{z} \in \mathcal{K}_\mu^{k,\Delta} \times \mathcal{M}.$$

In view of Lemma 12.19,

$$\mathbb{P}_{\mathbf{z}}\Big\{\mathcal{U}(\omega) = \{\mu\}\Big\} > 1 - \varepsilon, \mathbf{z} \in \mathcal{K}_\mu^{k,\Delta} \times \mathcal{M}. \tag{12.90}$$

In view of Lemma 12.16 and (12.90), we have for $\mathbf{z} \in \mathcal{K}_\mu^{k,\Delta} \times \mathcal{M}$ and for each $i = 1, \ldots, n$ that

$$\begin{aligned}\mathbb{P}_{\mathbf{z}}&\left\{\frac{1}{t}\int_0^t \left(f_i(\mathbf{Z}(s)) - \frac{\sigma_{ii} g_i^2(\mathbf{Z}(s))}{2}\right) ds\right.\\ &\left.= \int_{\partial\mathcal{S}} \left(f_i(\mathbf{z}) - \frac{\sigma_{ii} g_i^2(\mathbf{z})}{2}\right) \mu(d\mathbf{z})\right\} > 1 - \varepsilon.\end{aligned} \tag{12.91}$$

The claim of this lemma follows from (12.91) and (12.72) and Itô's formula. □

***Proof of Theorem* 12.15.** First, suppose that (H12.5) is satisfied with nonempty $\mathbb{M}^2$. Then, there is $\mathbf{q} = (q_1, \ldots, q_n) \in \mathbb{R}_+^{n,\circ}$ such that $\|\mathbf{q}\| \leq 1$ and

$$\min_{\nu\in\mathbb{M}^2} \sum q_i \left\{\int_{\partial\mathcal{S}} \left(f_i(\mathbf{z}) - \frac{\sigma_{ii} g_i^2(\mathbf{z})}{2}\right) \nu(d\mathbf{z})\right\} > 0. \tag{12.92}$$

Using (12.92) and a proof by contradiction like Lemma 12.19, we can show that with probability 1, $\mathcal{U}(\omega)$ is a subset of $\text{Conv}(\mathcal{M}) \setminus \text{Conv}(\mathbb{M}^2)$. In other words, each invariant probability $\pi \in \mathcal{U}(\omega)$ has the form $\pi = (1-\rho)\pi_1 + \rho\pi_2$ where $\rho \in [0, 1)$, $\pi_1 \in \text{Conv}(\mathbb{M}^1)$, $\pi_2 \in \text{Conv}(\mathbb{M}^2)$. Let $k_0 > 1$ and for each $\mu \in \mathbb{M}^1$ define

$$\mathcal{K}_\mu^0 = \left\{\mathbf{x} \in \mathbb{R}_+^\mu : x_i \wedge x_i^{-1} \le k_0, i \in I_\mu^c\right\}.$$

By Lemma 12.20, there are $k > k_0$ and $\Delta > 0$ such that

$$\mathbb{P}_{\mathbf{z}}\left\{\lim_{t\to\infty} \frac{\ln X_i(t)}{t} = \int_{\partial S}\left(f_i(\mathbf{z}) - \frac{\sigma_{ii}g_i^2(\mathbf{z})}{2}\right)\mu(d\mathbf{z}) < 0, i \in I_\mu^c\right\} > 1 - \varepsilon \tag{12.93}$$

for all $\mu \in \mathbb{M}^1$ and $\mathbf{z} \in \mathcal{K}_\mu^{k,\Delta} \times \mathcal{M}$. Let $\psi(\cdot) : \mathbb{R}_+^n \to [0, 1]$ be a continuous function satisfying

$$\psi(\mathbf{x}) = \begin{cases} 1 \text{ if } \mathbf{x} \in \bigcup_{\mu\in\mathbb{M}^1} \mathcal{K}_\mu^0 \\ 0 \text{ if } \mathbf{x} \in \mathbb{R}_+^{n,\circ} \setminus \left(\bigcup_{\mu\in\mathbb{M}^1} \mathcal{K}_\mu^{k,\Delta}\right). \end{cases}$$

Since $\pi_1(\bigcup_{\mu\in\mathbb{M}^1} \mathcal{K}_\mu^0) > 0$ for any $\pi_1 \in \text{Conv}(\mathbb{M}^1)$ and $\mathcal{U}(\omega)$ is a subset of $\text{Conv}(\mathcal{M}) \setminus \text{Conv}(\mathbb{M}^2)$ with probability 1, then we have from Lemma 12.18 that

$$\mathbb{P}_{\mathbf{z}}\left\{\liminf_{t\to\infty} \frac{1}{t}\int_0^t \psi(\mathbf{X}(s))ds > 0\right\} = 1, \ \mathbf{z} \in \mathcal{S}^\circ. \tag{12.94}$$

Since $\phi(\mathbf{x}) = 0$ if $\mathbf{x} \in \mathbb{R}_+^{n,\circ} \setminus \left(\bigcup_{\mu\in\mathbb{M}^1} \mathcal{K}_\mu^{k,\Delta}\right)$, we deduce from (12.94) that

$$\mathbb{P}_{\mathbf{z}}\left\{\liminf_{t\to\infty} \frac{1}{t}\int_0^t I_{\left\{\mathbf{X}(s)\in\bigcup_{\mu\in\mathbb{M}^1} \mathcal{K}_\mu^{k,\Delta}\right\}}ds > 0\right\} = 1, \ \mathbf{z} \in \mathcal{S}^\circ. \tag{12.95}$$

Thus, if $\mathbf{X}(0) \in \mathbb{R}_+^{n,\circ}$, $\{\mathbf{X}(s)\}$ will enter $\bigcup_{\mu\in\mathbb{M}^1} \mathcal{K}_\mu^{k,\Delta}$ with probability 1. This fact, combined with (12.93) and the strong Markov property of $\{\mathbf{Z}(s)\}$, implies that

$$\sum_{\mu\in\mathbb{M}^1} P_{\mathbf{z}}^\mu > 1 - \varepsilon, \ \mathbf{z} \in \mathcal{S}^\circ,$$

where

$$P_{\mathbf{z}}^\mu = \mathbb{P}_{\mathbf{z}}\left\{\lim_{t\to\infty} \frac{\ln X_i(t)}{t} = \int_{\partial S}\left(f_i(\mathbf{z}) - \frac{\sigma_{ii}g_i^2(\mathbf{z})}{2}\right)\mu(d\mathbf{z}) < 0, i \in I_\mu^c\right\}, \ \mathbf{z} \in \mathcal{S}^\circ.$$

Letting $\varepsilon \to 0$ we obtain (12.69). The positivity of $P_{\mathbf{z}}^{\mu}$ follows from (12.93) and the fact that $\{\mathbf{X}(s)\}$ will visit $\mathcal{K}_{\mu}^{k,\Delta}$ with a positive probability due to the nondegeneracy of the diffusion.

Now we consider the case $\mathbb{M}^1$ being nonempty and $\mathbb{M}^2 = \emptyset$. Then, we claim that $\mathbb{M}^1 = \{\boldsymbol{\delta}^* \times \pi\}$ where $\boldsymbol{\delta}^*$ be the Dirac measure concentrated on the origin $\mathbf{0}$. Indeed, if $\mathbb{M}^1$ contains a measure μ with $\mathbb{R}_+^{\mu} \neq \{\mathbf{0}\}$, then $\boldsymbol{\delta}^* \times \pi \in \mathbb{M}_{\mu}$. Since μ satisfies (H12.3), in view of (12.50), $\boldsymbol{\delta}^* \in \mathbb{M}^2$ which results in a contradiction. Thus, $\mathbb{M} = \mathbb{M}^1 = \{\boldsymbol{\delta}^*\}$. As a result, $\mathcal{U}(\omega) = \{\boldsymbol{\delta}^* \times \pi\}$ with probability 1. Then, we can easily deduce with probability 1 that $\lim_{t\to\infty} \dfrac{\ln X_i(t)}{t} = \sum_{\alpha\in\mathcal{M}} \left(\pi_\alpha (f_i(\mathbf{0}, \alpha) - \dfrac{\sigma_{ii} g_i^2(\mathbf{0}, \alpha)}{2}\right) < 0, i = 1, \ldots, n$ since $\boldsymbol{\delta}^* \times \pi$ satisfies (12.49).

□

12.1.5 Examples

Example 12.21 Consider a stochastic Lotka-Volterra competitive model

$$
\begin{aligned}
dX_1(t) &= X_1(t)[a_1(\alpha(t)) - b_1(\alpha(t))X_1(t) - c_1(\alpha(t))X_2(t)]dt \\
&\quad + X_1(t)\sigma_1(\alpha(t))dB_1(t) \\
dX_2(t) &= X_2(t)[a_2(\alpha(t)) - b_2(\alpha(t))X_2(t) - c_2(\alpha(t))X_1(t)]dt \\
&\quad + X_2(t)\sigma_2(\alpha(t))dE_2(t)
\end{aligned}
\tag{12.96}
$$

where $b_i(\alpha), c_i(\alpha) > 0, i = 1, 2$.

For this model, we can choose $\mathcal{W}(x_1, x_2) = C_{\mathcal{W}}(1 + x_1 + x_2)$ for sufficiently large $C_{\mathcal{W}}$. One can easily verify that the conditions in (H12.1) and (H12.4) are satisfied. Moreover, it is straightforward to see that

$$
\lambda_i(\boldsymbol{\delta}^* \times \pi) = \sum_{m\in\mathcal{M}} \left(a_i(\alpha) - \frac{\sigma_i^2(\alpha)}{2}\right)\pi_\alpha, i = 1, 2.
$$

If $\lambda_1(\boldsymbol{\delta}^* \times \pi) < 0$, (resp. $\lambda_2(\boldsymbol{\delta}^* \times \pi) < 0$) there is no invariant probability measure on $\mathbb{R}_{1+}^{\circ} \times \mathcal{M} := \{(x_1, 0) : x_1 > 0\} \times \mathcal{M}$ (resp. $\mathbb{R}_{2+}^{\circ} := \{(0, x_2) : x_2 > 0\} \times \mathcal{M}$) in view of Theorem 12.14.

Using Theorems 12.9 and 12.14, we have the following classification:

- If $\lambda_1(\boldsymbol{\delta}^* \times \pi) > 0, \lambda_2(\boldsymbol{\delta}^* \times \pi) > 0$ and $\lambda_1(\mu_2) > 0, \lambda_2(\mu_1) > 0$, any invariant probability measure in $\partial\mathbb{R}_+^2$ has the form $\mu = p_0\boldsymbol{\delta}^* \times \pi + p_1\mu_1 + p_2\mu_2$ with $0 \le p_0, p_1, p_2$, and $p_0 + p_1 + p_2 = 1$. It can easily be verified that $\max_{i=1,2}\{\lambda_i(\mu)\} > 0$ for any μ having the form above. As a result there is a unique invariant probability measure π^* on $\mathbb{R}_+^{2,\circ} \times \mathcal{M}$ and $P(t, \mathbf{z}, \cdot), \mathbf{z} \in \mathbb{R}_+^{2,\circ} \times \mathcal{M}$ converges to π^* in total variation exponentially fast.

- If $\lambda_i(\boldsymbol{\delta}^* \times \pi) < 0, i = 1, 2$ then $X_i(t)$ converges to 0 almost surely with the exponential rate $\lambda_i(\boldsymbol{\delta}^* \times \pi)$ for any initial condition $(x_1, x_2, \alpha) \in \mathbb{R}_+^{2,\circ} \times \mathcal{M}$.
- If $\lambda_i(\boldsymbol{\delta}^* \times \pi) > 0, \lambda_j(\boldsymbol{\delta}^* \times \pi) < 0$ for one $i \in \{1, 2\}$ and $j \in \{1, 2\} \setminus \{i\}$, then $\lambda_j(\mu_i) < 0$ and $X_j(t)$ converge to 0 almost surely with the exponential rate $\lambda_j(\mu_i)$ for any initial condition $(x_1, x_2, \alpha) \in \mathbb{R}_+^{2,\circ} \times \mathcal{M}$, and the randomized occupation measure converges weakly to μ_i almost surely.
- If $\lambda_i(\boldsymbol{\delta}^* \times \pi) > 0, i \in \{1, 2\}$ and $\lambda_1(\mu_2) < 0, \lambda_2(\mu_1) < 0$, then $p_i^{\mathbf{x}} > 0, i = 1, 2$ and $p_1^{\mathbf{z}} + p_2^{\mathbf{z}} = 1$ where

$$p_i^{\mathbf{z}} = \mathbb{P}_{\mathbf{z}} \left\{ \mathcal{U}(\omega) = \{\mu_i\} \text{ and } \lim_{t \to \infty} \frac{\ln X_j(t)}{t} = \lambda_j(\mu_i), j \in \{1, 2\} \setminus \{i\} \right\}.$$

- If $\lambda_1(\boldsymbol{\delta}^* \times \pi) > 0, \lambda_2(\boldsymbol{\delta}^* \times \pi) > 0, \lambda_j(\mu_i) < 0, \lambda_i(\mu_j) > 0$ for $i, j \in \{1, 2\}, i \neq j$, then $X_j(t)$ converges to 0 almost surely with the exponential rate $\lambda_j(\mu_i)$, and the randomized occupation measure converges weakly to μ_i almost surely for any initial condition $(x_1, x_2, \alpha) \in \mathbb{R}_+^{2,\circ} \times \mathcal{M}$.

Example 12.22 Consider a stochastic Lotka-Volterra model with two predators competing for one prey:

$$\begin{aligned}
dX_1(t) &= X_1(t)[a_1(\alpha(t)) - b_1 X_1(t) - c_{21}(\alpha(t))X_2(t) - c_{31}(\alpha(t))X_3(t)]dt \\
&\quad + \sigma_1(\alpha(t))dB_1(t) \\
dX_2(t) &= X_2(t)[-a_2(\alpha(t)) - b_2(\alpha(t))X_2(t) + c_{12}(\alpha(t))X_1(t) \\
&\quad - c_{32}(\alpha(t))X_3(t)]dt + \sigma_2(\alpha(t))dB_2(t) \\
dX_3(t) &= X_3(t)[-a_3(\alpha(t)) - b_3(\alpha(t))X_3(t) + c_{13}(\alpha(t))X_1(t) \\
&\quad - c_{23}(\alpha(t))X_2(t)]dt + \sigma_3(\alpha(t))dB_3(t)
\end{aligned} \tag{12.97}$$

Assume that $a_i(\alpha), b_i(\alpha) > 0, i = 1, 3, c_{21}(\alpha), c_{31}(\alpha), c_{12}(\alpha), c_{31}(\alpha), c_{32}(\alpha) \geq 0$. Note that if $c_{23}(\alpha) > 0$ and $c_{32}(\alpha) > 0$, then (12.97) describes an interacting population of two predators (X_2, X_3) competing for one prey X_1. If $c_{23}(\alpha) < 0$ and $c_{32}(\alpha) > 0$, then (12.97) is a model of a tri-trophic food chain where X_3 is the top predator and X_2 is the intermediate species.

For this model, we use

$$\mathcal{W} := \ln[C_{\mathcal{W}}(1 + c_{\mathcal{W}} x_1 + x_2 + x_3)], \text{ where}$$

$$c_{\mathcal{W}} = 1 + \max_{\alpha \in \mathcal{M}} \left\{ \frac{|c_{12}(\alpha)| + |c_{13}(\alpha)|}{b_1(\alpha)} \right\}$$

For sufficiently large $C_{\mathcal{W}}$, one can easily verify that the conditions in Assumptions (H12.1) and (H12.4) are satisfied. In order to analyze this model, first, consider the equation on the boundary $\{(0, x_2, x_3) : x_2, x_3 \geq 0\}$. Since $\lambda_i(\boldsymbol{\delta}^* \times \pi) = \sum_{\alpha \in \mathcal{M}} \left(-a_i(\alpha) - \frac{\sigma_i(\alpha)}{2} \right) \pi_\alpha < 0, i = 2, 3$, an application of Theorem 12.14 to the space $\{(0, x_2, x_3) : x_2, x_3 \geq 0\}$ shows that there is only one invariant probability measure on $\{(0, x_2, x_3) : x_2, x_3 \geq 0\}$, which is $\boldsymbol{\delta}^*$. It indicates that without the prey, both predators die out.

Now, consider the equation on the boundaries $\mathbb{R}_{12+} := \{(x_1, x_2, 0) : x_1, x_2 \geq 0\}$ and $\mathbb{R}_{13+} := \{(x_1, 0, x_3) : x_1, x_3 \geq 0\}$. If $\lambda_1(\boldsymbol{\delta}^* \times \pi) = \sum_{\alpha\in\mathcal{M}}\left(a_1(\alpha) - \frac{\sigma_1(\alpha)}{2}\right)\pi_\alpha < 0$, $\boldsymbol{\delta}^*$ is the unique invariant probability measure on $\mathbb{R}^3_+$ by virtue of Theorem 12.14. If $\lambda_1(\boldsymbol{\delta}^* \times \pi) > 0$, there is an invariant probability measure μ_1 on $\mathbb{R}^\circ_{1+} := \{(x_1, 0, 0) : x_1 > 0\} \times \mathcal{M}$.

If $\lambda_1(\boldsymbol{\delta}^* \times \pi) > 0$ and $\lambda_i(\mu_1) < 0, i = 2, 3$, by Theorem 12.14, there is no invariant probability measure on $\mathbb{R}^\circ_{1i+} \times \mathcal{M}$.

If $\lambda_1(\boldsymbol{\delta}^* \times \pi) > 0$ and $\lambda_2(\mu_1) > 0$, by Theorem 12.9, there is an invariant probability measure μ_{12} on $\mathbb{R}^\circ_{12+} \times \mathcal{M}$.

By the ergodic decomposition theorem, every invariant probability measure on $\partial\mathbb{R}^3_+$ is a convex combination of $\delta^*, \mu_1, \mu_{12}, \mu_{13}$ (when these measures exist). Some computations for the Lyapunov exponents with respect to a convex combination of these ergodic measures together with an application of Theorem 12.14 show that $P(t, \mathbf{x}, \cdot), \mathbf{x} \in \mathbb{R}^{3,\circ}_+$ converges exponentially fast to an invariant probability measure π^* on $\mathbb{R}^{3,\circ}_+ \times \mathcal{M}$ if one of the following is satisfied:

- $\lambda_1(\boldsymbol{\delta}^* \times \pi) > 0, \lambda_2(\mu_1) > 0, \lambda_3(\mu_1) < 0$ and $\lambda_3(\mu_{12}) > 0$.
- $\lambda_1(\boldsymbol{\delta}^* \times \pi) > 0, \lambda_2(\mu_1) < 0, \lambda_3(\mu_1) > 0$ and $\lambda_2(\mu_{13}) > 0$.
- $\lambda_1(\boldsymbol{\delta}^* \times \pi) > 0, \lambda_2(\mu_1) > 0, \lambda_3(\mu_1) > 0, \lambda_3(\mu_{12}) > 0$, and $\lambda_2(\mu_{13}) > 0$.

As an application of Theorem 12.15, we have the following classification for extinction for an initial condition $\mathbf{z} \in \mathbb{R}^{3,\circ}_+ \times \mathcal{M}$:

- If $\lambda_1(\boldsymbol{\delta}^* \times \pi) < 0$ then $X_1(t), X_2(t), X_3(t)$, converge to 0 almost surely with the exponential rates $\lambda_i(\boldsymbol{\delta}^* \times \pi), i = 1, 2, 3$, respectively.
- If $\lambda_1(\boldsymbol{\delta}^* \times \pi) > 0, \lambda_i(\mu_1) < 0, i = 2, 3$, then $X_i(t), i = 2, 3$ converge to 0 almost surely with the exponential rate $\lambda_i(\mu_1), i = 2, 3$, respectively, and the occupation measure converges almost surely to μ_1.
- If $\lambda_1(\boldsymbol{\delta}^* \times \pi) > 0, \lambda_i(\mu_1) > 0, \lambda_j(\mu_{1i}) < 0, \lambda_j(\mu_1) < 0$ for $i, j \in \{2, 3\}, i \neq j$, then $X_j(t)$ converges to 0 almost surely with the exponential rate $\lambda_j(\mu_{1i})$ and the occupation measure converges almost surely to μ_{1i}.
- If $\lambda_1(\boldsymbol{\delta}^* \times \pi) > 0, \lambda_2(\mu_1) > 0, \lambda_3(\mu_1) > 0, \lambda_j(\mu_{1i}) < 0, \lambda_i(\mu_{1j}) > 0$ for $i, j \in \{2, 3\}, i \neq j$, then $X_j(t)$ converges to 0 almost surely with the exponential rate $\lambda_j(\mu_{1i})$, and the occupation measure converges almost surely to μ_{1i}.
- If $\lambda_1(\boldsymbol{\delta}^* \times \pi) > 0, \lambda_2(\mu_1) > 0, \lambda_3(\mu_1) > 0, \lambda_2(\mu_{13}) < 0, \lambda_3(\mu_{12}) < 0$, then $p_i^{\mathbf{z}} > 0, i = 2, 3$ and $p_2^{\mathbf{z}} + p_3^{\mathbf{z}} = 1$ where

$$p_i^{\mathbf{z}} = \mathbb{P}_{\mathbf{z}}\left\{\mathcal{U}(\omega) = \{\mu_{1i}\} \text{ and } \lim_{t\to\infty}\frac{\ln X_i(t)}{t} = \lambda_i(\mu_{1j}), i \in \{2, 3\} \setminus \{j\}\right\}.$$

12.2 Nonlinear Stochastic Systems: Chemostat Models

In this section, we deal with a specific model where part 3 of (H12.1) is not satisfied, but we can still obtain an exponential convergence rate.

12.2.1 Introduction

We consider a class of systems of nonlinear stochastic differential equations. The motivation stems from the pressing need of the treatment of chemostat models that are a laboratory apparatus used for the continuous culture of microorganism, which is a technique introduced by Novick and Szilard in [224]. This technique plays an important role in microbiology, biotechnology, and population biology and is perhaps the best laboratory idealization of nature for population studies [275]. Chemostats are also used as microcosms in ecology [15, 232] and evolutionary biology [75, 144], as well as in wastewater treatment based on chemostat models [25, 98, 119, 145, 312], which has led to numerous research inventions. Since the 1950s, much attention has been devoted to modeling and analyzing chemostat problems; see [95, 120, 128, 264] and references therein.

The dynamics of the process can be modeled in the general form by a system of ordinary differential equations

$$\begin{cases} \dfrac{dS(t)}{dt} =\dfrac{S_0 - f_0(S(t))}{\theta} - X(t)f_1\big(S(t), X(t)\big), \\ \dfrac{dX(t)}{dt} =X(t)\left(f_2\big(S(t), X(t)\big) - k_d - \dfrac{1+R}{\theta}\right), \end{cases} \tag{12.98}$$

where $S(t), X(t)$ are the substrate concentration and the bacterial concentration, respectively; S_0 and $f_0(S)$ are the input concentration and the decay rate of the substrate, respectively; $\frac{1}{\theta}$ is the dilution rate (or equivalently, θ is the mean residence time), k_d is the death rate of X, and R is the recycle ratio; $f_1(S, X)$ is the consumption rate, and $f_2(S, X)$ is the growth rate of the bacteria. The formulation in (12.98) is much more general than the existing models. The readers can find works on the specialized forms of the models with specific forms of the functions f_0, f_1, f_2 in [33, 128, 245, 315, 318] and references therein.

To better reflect the reality, effort has been devoted to stochastic systems to take into account the effect of environmental perturbations [136, 273]. A fundamental problem is the long-term behavior of the system. However, the dynamic behaviors have not been fully understood to the best our knowledge. The asymptotic features of the systems and important information such as the *wash-out time* have not been fully understood to date. In contrast to the existing work, we develop new approaches to carefully analyze the corresponding systems.

Considering the system in a fluctuating environment consisting of both small fluctuations and abrupt changes, we have a stochastic hybrid system modeled by

$$\begin{cases} dS(t) = \left[\dfrac{S_0 - f_0(S(t), \alpha(t))}{\theta} - X(t) f_1(S(t), X(t), \alpha(t))\right] dt \\ \qquad + \sigma_1(\alpha(t)) S(t) dW_1(t), \\ dX(t) = X(t) \big(f_2\big(S(t), X(t), \alpha(t)\big) - \widetilde{k}_d(\alpha(t))\big) dt + \sigma_2(\alpha(t)) X(t) dW_2(t), \end{cases} \tag{12.99}$$

where $\widetilde{k}_d(\alpha(t)) := k_d(\alpha(t)) + \frac{1+R}{\theta}$, $\alpha(t)$ is a Markov chain with state space $\mathcal{M} = \{1, \ldots, m_0\}$ and generator $Q = (q_{kl})_{m_0 \times m_0}$, independent of the Brownian motion. While this model is similar to (12.1), the third condition of (H12.1) is not satisfied for this model. The proofs for persistence and extinction therefore require some delicate treatment of the behavior of the process near infinity.

To proceed, in Section 12.2.2, we prove the existence and uniqueness of positive solutions to (12.99) and (12.2). Then a complete characterization of the asymptotic behavior of the system under consideration is provided. We show that the long-term properties of the system can be classified by using a real-valued parameter λ. If $\lambda \le 0$, the bacteria will die out; if $\lambda > 0$, the system has an invariant probability measure and the transition probability of the solution process converges to the invariant measure. One of the distinct contributions is that the critical case $\lambda = 0$ is also considered.

12.2.2 *Characterization of Long-Time Behavior*

This section is devoted to asymptotic properties of the switching diffusion models. We show that the long-term properties of the system can be completely classified by using a real parameter λ. In particular, if $\lambda \le 0$, the bacteria will die out, and we refer to such case as the system being not permanent because it does not work in long term and the efficiency goes to 0. If $\lambda > 0$, the system has an invariant probability measure, and the transition probability of the solution process converges to the invariant measure. This is what we refer to as permanence (where the terminology is inherited from the study in biological and ecological models). Moreover, we obtain the rate of convergence.

One of the highlights here is that we derive sufficient and necessary conditions for permanence. First, the process under consideration is jointly a Markov Feller process. By using the Lyapunov exponent of $X(t)$, we then establish the existence of the invariant measure of $(S(t), \alpha(t))$. Furthermore, under suitable conditions, we obtain the exponential error bounds of the difference of the transition function and that of the invariant measure in the total variation norm.

Throughout this paper, we use the lowercase letters s, x, and i to denote the initial values of $S(t)$, $X(t)$, and $\alpha(t)$, respectively. Note the distinction of

s and the input concentration of the substrate S_0. To simplify the notation, let $\hat{k}_d = \max_{i\in\mathcal{M}}\{\widetilde{k}_d(i)\}$; $\check{k}_d = \min_{i\in\mathcal{M}}\{\widetilde{k}_d(i)\}$, and $\hat{\sigma}_k = \max_{i\in\mathcal{M}}\{\sigma_k(i)\}$ and $\check{\sigma}_k = \min_{i\in\mathcal{M}}\{\sigma_k(i)\}, k = 1, 2$. We denote by $\mathbb{R}_+ = [0,\infty)$, $\mathbb{R}_+^\circ = (0,\infty)$, $\mathbb{R}_+^2 = [0,\infty)\times[0,\infty)$, and $\mathbb{R}_+^{2,\circ} = (0,\infty)\times(0,\infty)$. The operator associated with the process $(S(t), X(t), \alpha(t))$, solving (12.99) and (12.2), is given by

$$\mathcal{L}V(\phi,i) = V_\phi(\phi,i)\widetilde{f}(\phi,i) + \frac{1}{2}\mathrm{tr}[\widetilde{g}(\phi,i)\widetilde{g}'(\phi,i)V_{\phi\phi}(\phi,i)] + \sum_{j\in\mathcal{M}} q_{ij}V(\phi,j), \tag{12.100}$$

where A' denotes the transpose of A; $\phi = (s,x)$, $V_\phi(\phi,i)$, and $V_{\phi\phi}(\phi,i)$ are the gradient and Hessian of $V(\cdot,i)$ with respect to ϕ; and $\widetilde{f}$ and $\widetilde{g}$ are the drift and diffusion coefficients of (12.99), respectively. That is,

$$\begin{aligned}\widetilde{f}(\phi,i) &= \left(\frac{S_0 - f_0(s,i)}{\theta} - xf_1(s,x,i),\ \ x\Big(f_2(s,x,i) - \widetilde{k}_d(i)\Big)\right)^\top,\\ \widetilde{g}(\phi,i) &= \mathrm{diag}\big(\sigma_1(i)s, \sigma_2(i)x\big) \in \mathbb{R}^{2\times 2},\end{aligned}$$

where diag(a,b) denotes the diagonal matrix with entries a and b. Note that the particular structure of $\widetilde{g}$ implies that $\widetilde{g}\widetilde{g}^\top = \widetilde{g}^2$. In what follows, we write $V(\phi,i)$ and $V(s,x,i)$ interchangeably, whichever is more convenient. We also assume the following assumption.

(H12.6) Suppose that

- $W_1(t)$ and $W_2(t)$ are independent real-valued standard Brownian motions that are independent of the Markov chain $\alpha(t)$.
- $S_0, \theta, \widetilde{k}_d(i), \sigma_1(i), \sigma_2(i)$ are positive constants for each $i \in \mathcal{M}$.
- $f_0(s,i)$, $f_1(s,x,i)$, and $f_2(s,x,i)$ are locally Lipschitz; $f_0(0,i) = f_1(0,x,i) = 0$ satisfying $\lim_{s\to\infty} f_0(s,i) = \infty$; $0 \le f_2(s,x,i) \le \kappa_0\big(f_1(s,x,i) \wedge f_0(s,i)\big)$, $f_1(s,x,i) \le \kappa_0(1+s)$ for some $\kappa_0 > 1$. Moreover, for each $i \in \mathcal{M}$, $f_2(s,x,i)$ is uniformly continuous at $x = 0$ (i.e., $\lim_{x\to 0}\sup\{|f_2(s,x,i) - f_2(s,0,i)|\} = 0$).
- The Markov chain or its generator $Q = (q_{ij})_{m_0\times m_0}$ is irreducible, that is, for any $i, j \in \mathcal{M}$, there exist $i = i_0, i_1, \ldots, i_n = j$ such that $q_{i_{k-1},i_k} > 0$ for $k = 1,\ldots,n$.

To start, we state the following theorem, which provides the preliminary results, namely, the existence and uniqueness of the global solution, and the positivity of the solution. The proof is postponed to Section 12.2.3.

Theorem 12.23 *For any $(s,x,i) \in \mathbb{R}_+^2 \times \mathcal{M}$, there exists a unique global solution to the system* (12.99) *and* (12.2) *with initial value (s,x,i). The three-component process $\{(S(t), X(t), \alpha(t)), t \ge 0\}$ is a Markov-Feller process. Moreover, we have $\mathbb{P}_{s,x,i}\{S(t) > 0, t > 0\} = 1$ and $\mathbb{P}_{s,x,i}\{X(t) = 0, t > 0\} = 1$ if $x = 0$, $\mathbb{P}_{s,x,i}\{X(t) > 0, t > 0\} = 1$ if $x > 0$.*

To proceed, we examine the boundary by letting $X(t) = 0$. Let $\hat{S}(t)$ be the solution to (12.99) with $X(t) = 0$, i.e.,

$$d\hat{S}(t) = \frac{S_0 - f_0(\hat{S}(t), \alpha(t))}{\theta} dt + \sigma_1(\alpha(t))\hat{S}(t) dW_1(t). \tag{12.101}$$

Since the drift of (12.101) is negative when $\hat{S}(t)$ is sufficiently large and is positive when $\hat{S}(t)$ is sufficiently small, we can see

$$\left[\hat{\mathcal{L}}\left(s + \ln\frac{s}{s+1}\right)\right](s, i) \le -1 \text{ if } s < \hat{s} \text{ or } s > \frac{1}{\hat{s}},$$

for some $\hat{s} > 0$ being sufficiently small, where $\hat{\mathcal{L}}$ is the operator associated with (12.101). Thus, the hybrid diffusion (12.101) is positive recurrent due to its nondegeneracy and Theorem 3.26. Then there exists a unique invariant measure ν_0 of (12.101) thanks to Theorem 4.3. Moreover, by Theorem A.1 and the decreasing monotonicity of $\|\hat{P}(t, s, i, \cdot) - \nu_0(\cdot)\|_{TV}$, we have

$$\lim_{t\to\infty} \|\hat{P}(t, s, i, \cdot) - \nu_0(\cdot)\|_{TV} = 0, \ (s, i) \in (0, \infty) \times \mathcal{M}, \tag{12.102}$$

where $\hat{P}(t, s, i, \cdot)$ is the transition probability of $(\hat{S}(t), \alpha(t))$. Since $\mathbb{P}_{0,i}\{\hat{S}(t) > 0, t > 0\} = 1$, (12.102) holds even when $s = 0$.

When $X(t)$ is small, $S(t)$ can be approximated by $\hat{S}(t)$. Due to Itô's formula and the ergodicity of $(\hat{S}(t), \alpha(t))$, when $X(t)$ is small, the long-term growth rate $\frac{\ln X(t)}{t}$ is approximated by the critical value

$$\lambda := \sum_{i\in\mathcal{M}} \int_{\mathbb{R}_+^\circ} \left(f_2(s, 0, i) - \widetilde{k}_d(i) - \frac{\sigma_2^2(i)}{2} \right) \nu_0(ds, i). \tag{12.103}$$

As a result, the sign of λ determines whether or not $X(t)$ converges to 0. The main results of this section is provided in the next theorem to follow. First, we state an assumption that will be used in the main theorem.

(H12.7) Denoting by π the invariant measure of $\alpha(t)$, assume that one of the following conditions holds.

(1) For each $i \in \mathcal{M}$, $f_2(s, x, i)$ is non-decreasing in s and non-increasing in x for any $(s, x) \in [0, \infty) \times (0, \infty)$;

(2)

$$\sum_{i\in\mathcal{M}} \left(\limsup_{s\to\infty} f_2(s, 0, i) - \widetilde{k}_d(i) - \frac{\sigma_2^2(i)}{2} \right) \pi(i) < 0; \tag{12.104}$$

(3)

$$\sum_{i\in\mathcal{M}}\left(\liminf_{s\to\infty} f_2(s,0,i)-\widetilde{k}_d(i)-\frac{\sigma_2^2(i)}{2}\right)\pi(i)>0. \tag{12.105}$$

Theorem 12.24 *The following claims hold.*

- *If* $\lambda < 0$, *we have* $\lim_{t\to\infty}\dfrac{\ln X(t)}{t}=\lambda$ *a.s. and the distribution of* $\big(S(t),\alpha(t)\big)$ *converges weakly to the unique invariant probability measure* ν_0 *if either condition* (1) *or* (2) *of* (H12.7) *holds.*
- *If* $\lambda=0$ *and condition* (1) *of* (H12.7) *holds then*

$$\lim_{T\to\infty}\frac{1}{T}\mathbb{E}_{s,x,i}\int_0^T S(t)dt=S_0 \text{ and } \lim_{T\to\infty}\frac{1}{T}\mathbb{E}_{s,x,i}\int_0^T X(t)dt=0 \tag{12.106}$$

under additional conditions that for any $i\in\mathcal{M}$, $\frac{\partial f_0(s,i)}{\partial s}$ *and* $\frac{\partial f_2(s,0,i)}{\partial s}$ *exist and are positive, continuous, and* $\frac{\partial f_0(s,i)}{\partial s}$ *is bounded below by a positive constant.*
- *If* $\lambda>0$, *then there exists an invariant probability measure* μ^* *on* $\mathbb{R}_+^{2,\circ}\times\mathcal{M}$ *and* $\lim_{t\to\infty}\|P(t,s,i,\cdot)-\mu^*(\cdot)\|_{TV}=0$. *We assume further that condition* (3) *of* (H12.7) *holds, and* $\liminf_{s\to\infty}\frac{f_0(s,i)}{s^q}>0, i\in\mathcal{M}$ *for some* $q\in(0,1]$. *Then,*

  (i) *in case* $q<1$,

$$\lim_{t\to\infty} t^{\beta-1}\|P(t,s,x,i,\cdot)-\mu^*(\cdot)\|_{TV}=0,$$
$$\text{for any } 1\le\beta<\frac{1}{1-q},\ (s,x,i)\in\mathbb{R}_+\times\mathbb{R}_+^\circ\times\mathcal{M};$$

  (ii) *if* $q=1$, *there exists a* $\widetilde{\gamma}>0$ *such that*

$$\lim_{t\to\infty} e^{\widetilde{\gamma}t}\|P(t,s,x,i,\cdot)-\mu^*(\cdot)\|_{TV}=0,\ (s,x,i)\in\mathbb{R}_+\times\mathbb{R}_+^\circ\times\mathcal{M}, \tag{12.107}$$

  where $P(t,s,x,i,\cdot)$ *is the transition probability of* $\big(S(t),X(t),\alpha(t)\big)$.

Remark 12.25 If $\lambda>0$, it is easy to see that condition (1) of (H12.7) implies the condition (3) in (H12.7). Moreover, most models in the literature consider the case $f_0(s)=C_is$ while the coefficients $xf_1(s,x,i), xf_2(s,x)$ are linear functional response ($f_j(s,x,i)=c_{ij}x, j=1,2$), Holling type II response ($f_j(s,x,i)=\frac{c_{ij}s}{m_{ij}+a_{ij}s}, j=1,2$), Holling type III response ($f_j(s,x,i)=\frac{c_{ij}s^2}{(m_{ij1}+a_{ij1}s)(m_{ij2}+a_{ij2}s)}, j=1,2$), and Beddington-DeAngelis functional response ($f_j(s,x,i)=\frac{c_{ij}s}{m_{ij}+a_{ij}s+b_{ij}x}, j=1,2$) etc.; see, e.g., [245, 315, 318] and the references therein. It is clear that these functions satisfy (H12.6) and part (1) of (H12.7).

Proof for the Case $\lambda = 0$ We argue by contradiction. Suppose $\big(S(t), X(t), \alpha(t)\big)$ has an invariant probability measure μ^* on $\mathbb{R}_+^{2,\circ} \times \mathcal{M}$. Then, it follows from the ergodicity that for μ^*-almost every initial value (s, x, i),

$$\lim_{T\to\infty} \frac{1}{T}\int_0^T h\big(S(t), X(t), \alpha(t)\big)dt = \sum_{i'\in\mathcal{M}} \int_{\mathbb{R}_+^{2,\circ}} h(s', x', i')\mu^*(ds', dx', i'), \tag{12.108}$$

for any measurable function h that is μ^*-integrable. Since the transition probability density is continuous and positive, the invariant measure μ^* is unique and equivalent to $\widetilde{m} \times \pi$ where $\widetilde{m}$ is the Lebesgue measure on $\mathbb{R}_+^{2,\circ}$. As a result, if $\mu^*(A) = 1$, then $\mathbb{P}_{s,x,i}\big((S(t), X(t), \alpha(t)) \in A)\big) = 1$ for every $(s, x, i) \in \mathbb{R}_+^{2,\circ} \times \mathcal{M}, t > 0$, which implies that (12.108) holds for any initial value $(s, x, i) \in \mathbb{R}_+^{2,\circ} \times \mathcal{M}$. By the comparison theorem (see, e.g., [99]), we have $\hat{S}(t) \geq S(t)$ with probability 1 given that $\hat{S}(0) = S(0)$. Hence, it follows (12.99), (12.101) and positivity of $S(t)$, $\hat{S}(t)$ and additional assumption on $f_0(\cdot,\cdot)$ (non-decreasing in s) that

$$\limsup_{T\to\infty} \frac{1}{T}\mathbb{E}_{s,x,i}\int_0^T f_0(S(t), \alpha(t))dt \leq \limsup_{T\to\infty} \frac{1}{T}\mathbb{E}_{s,i}\int_0^T f_0(\hat{S}(t), \alpha(t))dt \leq S_0. \tag{12.109}$$

On the other hand, since $\dfrac{\partial f_0(s,i)}{\partial s}$ is bounded below by a positive constant and $f_0(0, i) = 0$, there exists a $\bar{g}_1 > 0$ such that

$$\frac{f_0(s,i)}{\theta} \geq \bar{g}_1 s, \ \forall s \geq 0, i \in \mathcal{M}. \tag{12.110}$$

Let $0 < \bar{p} < \frac{\bar{g}_1}{2\hat{\sigma}_1^2}$ and $0 < \bar{g}_2 < \frac{(1+\bar{p})\bar{g}_1}{4}$. By Itô's formula, we obtain that

$$\begin{aligned}
&\mathbb{E}_{s,i} e^{\bar{g}_2 T}\hat{S}^{1+\bar{p}}(T)\\
=&s^{1+\bar{p}} + \mathbb{E}_{s,i}\int_0^T e^{\bar{g}_2 t}(1+\bar{p})\hat{S}^{\bar{p}}(t)\Big(\frac{S_0 - f_0\big(\hat{S}(t), \alpha(t)\big)}{\theta} + \frac{\bar{p}\sigma_1^2(\alpha(t))}{2}\hat{S}(t)\\
&+ \frac{\bar{g}_2}{1+\bar{p}}\hat{S}(t)\Big)dt\\
\leq& s^{1+\bar{p}} + \mathbb{E}_{s,i}\int_0^T e^{\bar{g}_2 t}(1+\bar{p})\hat{S}^{\bar{p}}(t)\left(\frac{S_0}{\theta} - \frac{\bar{g}_1}{2}\hat{S}(t)\right)dt\\
\leq& s^{1+\bar{p}} + \frac{\bar{g}_3 e^{\bar{g}_2 T}}{\bar{g}_2}, \ \forall T > 0, \text{ where } \bar{g}_3 := \frac{(1+\bar{p})S_0}{\theta}\left(\frac{2S_0}{\theta\bar{g}_1}\right)^{\bar{p}}.
\end{aligned}$$

As a consequence, we have

$$\limsup_{T\to\infty} \mathbb{E}_{s,x,i} S^{1+\bar{p}}(T) \leq \limsup_{T\to\infty} \mathbb{E}_{s,i}\hat{S}^{1+\bar{p}}(T) \leq \frac{\bar{g}_3}{\bar{g}_2}. \tag{12.111}$$

Moreover, we also obtain that

$$\mathbb{E}_{s,x,i}\big(S(T)+X(T)\big)^{1+\bar{p}} \text{ is uniformly bounded in } T. \tag{12.112}$$

Using Itô's formula again, we have

$$\begin{aligned}\frac{\mathbb{E}_{s,i}\hat{S}^{1+\bar{p}}(T)}{T} &= \frac{s^{1+\bar{p}}}{T} + \frac{(1+\bar{p})S_0}{\theta T}\mathbb{E}_{s,i}\int_0^T \hat{S}^{\bar{p}}(t)dt \\ &\quad - \frac{1+\bar{p}}{\theta T}\mathbb{E}_{s,i}\int_0^T \hat{S}^{\bar{p}}(t)f_0\big(\hat{S}(t),\alpha(t)\big)dt \\ &\quad + \frac{\bar{p}(1+\bar{p})}{2T}\mathbb{E}_{s,i}\int_0^T \sigma_1^2(\alpha(t))\hat{S}^{1+\bar{p}}(t)dt,\end{aligned}$$

which together with (12.111) implies that

$$\limsup_{T\to\infty}\frac{1}{T}\mathbb{E}_{s,i}\int_0^T \hat{S}^{\bar{p}}(t)f_0\big(\hat{S}(t),\alpha(t)\big)dt \le \bar{g}_4,\ \text{ for some } \bar{g}_4<\infty. \tag{12.113}$$

We have from (12.101) and (12.111) that

$$\lim_{T\to\infty}\frac{1}{T}\mathbb{E}_{s,i}\int_0^T\left(\frac{S_0}{\theta}-\frac{f_0\big(\hat{S}(t),\alpha(t)\big)}{\theta}\right)dt = \lim_{T\to\infty}\frac{\mathbb{E}_{s,i}\hat{S}(T)-s}{T}=0.$$

Hence

$$\lim_{T\to\infty}\frac{1}{T}\mathbb{E}_{s,i}\int_0^T f_0\big(\hat{S}(t),\alpha(t)\big)dt = S_0. \tag{12.114}$$

From (12.99), with standard arguments, we have

$$\mathbb{E}_{s,x,i}\int_0^T X(t)f_1(S(t),X(t),\alpha(t))dt \le S_0T+s,\ \ T\ge 0.$$

Moreover, due to the uniform boundedness (12.112), the linear growth bound of f_0 and the almost sure convergence (12.108), it follows from the dominated convergence theorem that

$$\begin{aligned}&\lim_{T\to\infty}\frac{1}{T}\mathbb{E}_{s,x,i}\int_0^T\left(\frac{S_0-f_0\big(S(t),\alpha(t)\big)}{\theta}\right)dt \\ &= \sum_{i'\in\mathcal{M}}\int_{\mathbb{R}_+^{2,\circ}}\frac{S_0-f_0(s',x',i')}{\theta}\mu^*(ds',dx',i') =: \hat{g}.\end{aligned} \tag{12.115}$$

From (12.99) and (12.112), we deduces that

$$\lim_{T\to\infty}\frac{1}{T}\mathbb{E}_{s,x,i}\int_0^T X(t)f_1(S(t),X(t),\alpha(t))dt$$

exists (because the two others in (12.116) exist) and

$$\begin{aligned}&\lim_{T\to\infty}\frac{1}{T}\mathbb{E}_{s,x,i}\int_0^T\left(\frac{S_0-f_0\big(S(t),\alpha(t)\big)}{\theta}-X(t)f_1\big(S(t),X(t),\alpha(t)\big)\right)dt\\&=\lim_{T\to\infty}\frac{\mathbb{E}_{s,x,i}S(T)-s}{T}=0,\end{aligned}\tag{12.116}$$

where, due to Fatou's lemma,

$$\begin{aligned}&\lim_{T\to\infty}\frac{1}{T}\mathbb{E}_{s,x,i}\int_0^T X(t)f_1\big(S(t),X(t),\alpha(t)\big)dt\\&=\hat{g}\geq\sum_{i'\in\mathcal{M}}\int_{\mathbb{R}^{2,\circ}_+}x'f_1(s',x',i')\mu^*(ds',dx',i')>0.\end{aligned}$$

We have from (12.114) and (12.115) that

$$\lim_{T\to\infty}\frac{1}{T}\mathbb{E}_{s,x,i}\int_0^T\Big(f_0\big(\hat{S}(t),\alpha(t)\big)-f_0\big(S(t),\alpha(t)\big)\Big)dt=\theta\hat{g},\ s,x\geq 0,i\in\mathcal{M}.\tag{12.117}$$

Now, let $H>1$ be a sufficiently large constant satisfying that $\frac{\bar{g}_4}{H^{\bar{p}}}\leq\frac{\theta\hat{g}}{2}$. Combining with (12.113), we obtain

$$\begin{aligned}&\limsup_{T\to\infty}\frac{1}{T}\mathbb{E}_{s,x,i}\int_0^T I_{\{\hat{S}(t)>H\}}\Big(f_0\big(\hat{S}(t),\alpha(t)\big)-f_0\big(S(t),\alpha(t)\big)\Big)dt\\&\qquad\leq\limsup_{T\to\infty}\frac{1}{T}\mathbb{E}_{s,x,i}\int_0^T\frac{\hat{S}^{\bar{p}}(t)}{H^{\bar{p}}}f_0\big(\hat{S}(t),\alpha(t)\big)dt\\&\qquad\leq\frac{\bar{g}_4}{H^{\bar{p}}}\leq\frac{\theta\hat{g}}{2}.\end{aligned}\tag{12.118}$$

A consequence of (12.117) and (12.118) is that

$$\liminf_{T\to\infty}\frac{1}{T}\mathbb{E}_{s,x,i}\int_0^T I_{\{\hat{S}(t)\leq H\}}\Big(f_0\big(\hat{S}(t),\alpha(t)\big)-f_0\big(S(t),\alpha(t)\big)\Big)dt\geq\frac{\theta\hat{g}}{2}.\tag{12.119}$$

Since $\frac{\partial f_0(s,0,i)}{\partial s}$ is continuous and positive, there exists a $\bar{g}_5^H<\infty$ such that

$$f_0(s_1,i)-f_0(s_2,i)\leq\bar{g}_5^H(s_1-s_2),\ \forall 0\leq s_2\leq s_1\leq H,i\in\mathcal{M}.$$

Thus, (12.119) implies that

$$\liminf_{T\to\infty}\frac{1}{T}\mathbb{E}_{s,x,i}\int_0^T I_{\{\hat{S}(t)\le H\}}\Big(\hat{S}(t)-S(t)\Big)\,dt \ge \frac{\theta\hat{g}}{2\bar{g}_5^H}. \tag{12.120}$$

Similarly, there exists a $\bar{g}_6^H > 0$ such that

$$f_2(s_1,0,i)-f_2(s_2,0,i)\ge \bar{g}_6^H(s_1-s_2),\ \text{ for any } 0\le s_2\le s_1\le H, i\in\mathcal{M}.$$

Hence, combining with $f_2(s,x,i)$ being non-increasing in x, we obtain

$$f_2(s_1,0,i)-f_2(s_2,x,i)\ge \bar{g}_6^H(s_1-s_2)\ \forall x\ge 0, 0\le s_2\le s_1\le H, i\in\mathcal{M}. \tag{12.121}$$

As a consequence of (12.120) and (12.121),

$$\begin{aligned}
&\lim_{T\to\infty}\frac{1}{T}\mathbb{E}_{s,x,i}\int_0^T \Big(f_2\big(\hat{S}(t),0,\alpha(t)\big)-f_2\big(S(t),X(t),\alpha(t)\big)\Big)\,dt\\
&\quad\ge \lim_{T\to\infty}\frac{1}{T}\mathbb{E}_{s,x,i}\int_0^T I_{\{\hat{S}(t)\le H\}}\Big(f_2\big(\hat{S}(t),0,\alpha(t)\big)-f_2\big(S(t),X(t),\alpha(t)\big)\Big)\,dt\\
&\quad\ge \bar{g}_6^H\liminf_{T\to\infty}\frac{1}{T}\mathbb{E}_{s,x,i}\int_0^T I_{\{\hat{S}(t)\le H\}}\Big(\hat{S}(t)-S(t)\Big)\,dt\\
&\quad\ge \frac{\bar{g}_6^H\theta\hat{g}}{2\bar{g}_5^H}.
\end{aligned}$$

Therefore, we obtain that

$$\begin{aligned}
&\sum_{i'\in\mathcal{M}}\int_{\mathbb{R}_+^{2,\circ}}\left(f_2(s',x',i')-\widetilde{k}_d(i)-\frac{\sigma_2^2(i)}{2}\right)\mu^*(ds',dx',i')\\
&\quad=\lim_{T\to\infty}\frac{1}{T}\mathbb{E}_{s,x,i}\int_0^T\left(f_2\big(S(t),X(t),\alpha(t)\big)-\widetilde{k}_d(\alpha(t))-\frac{\sigma_2^2(\alpha(t))}{2}\right)dt\\
&\quad\le\lim_{T\to\infty}\frac{1}{T}\mathbb{E}_{s,x,i}\int_0^T\left(f_2\big(\hat{S}(t),0,\alpha(t)\big)-\widetilde{k}_d(\alpha(t))-\frac{\sigma_2^2(\alpha(t))}{2}\right)dt-\frac{\bar{g}_6^H\theta\hat{g}}{2\bar{g}_5^H}\\
&\quad\le\lambda-\frac{\bar{g}_6^H\theta\hat{g}}{2\bar{g}_5^H}=-\frac{\bar{g}_6^H\theta\hat{g}}{2\bar{g}_5^H}.
\end{aligned} \tag{12.122}$$

By (12.99), Itô's formula, the ergodicity, and (12.122), we have

$$\begin{aligned}\lim_{T\to\infty}\frac{\ln X(T)}{T} &= \lim_{T\to\infty}\frac{1}{T}\left(X(0)+\int_0^T \sigma_2(\alpha(t))dW_2(t)\right)\\ &+\lim_{T\to\infty}\frac{1}{T}\int_0^T\left(f_2\big(S(t),X(t),\alpha(t)\big)-\widetilde{k}_d(\alpha(t))-\frac{\sigma_2^2(\alpha(t))}{2}\right)dt\\ &\le -\frac{\bar{g}_6^H\theta\hat{g}}{2\bar{g}_5^H}<0 \text{ a.s.}\end{aligned}$$

As a result, $\mathbb{P}_{s,x,i}\left\{\lim_{T\to\infty}X(T)=0\right\}=1$, which contradicts the assumption that the process has an invariant probability measure on $\mathbb{R}_+^{2,\circ}\times\mathcal{M}$. Moreover, $\nu_0\times\delta^*$ is the unique invariant measure of $\big(S(t),\alpha(t),X(t)\big)$ on $\mathbb{R}_+\times\mathcal{M}\times\mathbb{R}_+$, where δ^* is the Dirac measure with mass at 0. Consider the empirical measure

$$\Pi_t^{s,x,i}(\cdot)=\frac{1}{t}\int_0^t \mathbb{P}_{s,x,i}\{(S(s),X(s),\alpha(s))\in\cdot\}ds.$$

In view of (12.112), the family $\{\Pi_t^{s,x,i}(\cdot),t\ge 0\}$ is tight for each $(s,x,i)\in\mathbb{R}_+^2\times\mathcal{M}$. By virtue of Lemma 12.18, any weak limit of $\Pi_t^{s,x,i}(\cdot)$ as $t\to\infty$ is an invariant probability measure of $\big(S(t),\alpha(t),X(t)\big)$. Since $\nu_0\times\delta^*$ is the unique invariant probability measure and we have the boundedness of $\mathbb{E}_{s,x,i}\big(S(t)+X(t)\big)^{1+\bar{p}}$ in (12.112), we can easily obtain (12.106). □

Proof for the Case $\lambda>0$ Proof of the convergence in total variation of transition probability to an invariant measure is straightforward. Applying [16, Theorem 4.4], with $W(s,x,i)=s+x$ and $V(s,x,i)=\ln\frac{x}{x+1}$, we can show that

$$\begin{aligned}&\mathbb{E}_{s,x,i}[W(S(t_0),X(t_0),\alpha(t_0))+V(S(t_0),X(t_0),\alpha(t_0))]\\ &\le -c_1+c_2 I_{\{W(s,x,i)+V(s,x,i)\le k\}}\end{aligned}$$

for some $t_0,c_1,c_2,k>0$.

Because of the nondegeneracy of the diffusion, the Markov chain $\{(S(nt_0), X(nt_0),\alpha(nt_0),n\in\mathbb{N}\}$ is irreducible and aperiodic, and $\{W(s,x,i)+V(s,x,i)\le k\}$ is petite; see Lemma 12.6.

As a result, $\lim_{n\to\infty}\|P(nt_0,(s,x,i),\cdot)-\mu^*\|_{TV}=0$ for an invariant probability measure of $\{(S(nt_0), X(nt_0), \alpha(nt_0), n\in\mathbb{N}\}$. It is then easy to show that μ^* is an invariant probability measure of the process $\{S(t),X(t),\alpha(t)\}$ and $\lim_{t\to\infty}\|P(t,s,i,\cdot)-\mu^*(\cdot)\|_{TV}=0$.

Now, we consider the rate of convergence where conditions in (H12.1) are not easy to verify for our model. We assume that condition (3) of (H12.7) holds. In view of (12.105), there exists an $\hat{H}>0$ such that

$$\sum_{i\in\mathcal{M}}\left(\inf_{s\ge\hat{H}}\{f_2(s,0,i)\}-\widetilde{k}_d(i)-\frac{\sigma_2^2(i)}{2}\right)\pi(i)$$

Therefore, by the uniform continuity at $x = 0$ of $f_2(s, x, i)$, there exists an $\varepsilon_1 > 0$ such that

$$4\widetilde{\lambda} := \sum_{i\in\mathcal{M}} \hat{h}_i\pi_i > 0, \text{ where } \hat{h}_i := \inf_{(s,x)\in[\hat{H},\infty)\times[0,\varepsilon_1]} \{f_2(s, x, i)\} - \widetilde{k}_d(i) - \frac{\sigma_2^2(i)}{2}. \tag{12.123}$$

Since $\sum_{i\in\mathcal{M}}(4\widetilde{\lambda} - \hat{h}_i)\pi_i = 0$, an application of the Fredholm alternative implies the existence of $\gamma_i > 0$ with $i \in \mathcal{M}$ such that $\sum_{j\in\mathcal{M}} q_{ij}\gamma_j = 4\widetilde{\lambda} - \hat{h}_i$ for all $i \in \mathcal{M}$. Let ρ_1 be sufficiently small satisfying

$$\rho_1\gamma_i(-4\widetilde{\lambda} + \hat{h}_i) < \widetilde{\lambda}(1 - \rho_1\gamma_i), \ \rho_1\sigma_2^2(i) < 4\widetilde{\lambda} \text{ and } \rho_1\gamma_i < 1, \forall i \in \mathcal{M},$$

and define $V_3(s, x, i) = (1 - \rho_1\gamma_i)x^{-\rho_1}$. By directed calculations, we obtain that

$$\begin{aligned}\mathcal{L}V_3(s, x, i) =& \rho_1 V_3(s, x, i)\left(-f_2(s, x, i) + \widetilde{k}_d(i) + \frac{\rho_1 + 1}{2}\sigma_2^2(i)\right)\\ &+ \rho_1(-4\widetilde{\lambda} + \hat{h}_i)V_3(s, x, i) + \rho_1^2\gamma_i(-4\widetilde{\lambda} + \hat{h}_i)x^{-\rho_1}\\ \le& - 2\rho_1\widetilde{\lambda}V_3(s, x, i) + \rho_1^2\gamma_i(-4\widetilde{\lambda} + \hat{h}_i)x^{-\rho_1}\\ \le& - \rho_1\widetilde{\lambda}V_3(s, x, i) \ \forall i \in \mathcal{M}, s \ge \hat{H}, x \le \varepsilon_1.\end{aligned} \tag{12.124}$$

Since $\left[\mathcal{L}\frac{1}{x}\right](s, x, i) \le \frac{\widetilde{k}_d(i)+\sigma_2^2(i)}{x} \le \frac{\hat{k}_d+\hat{\sigma}_2^2}{x}$, it is easily verified that

$$\mathbb{E}_{s,x,i}X^{-1}(t) \le e^{(\hat{k}_d+\hat{\sigma}_2^2)t}x^{-1}, t \ge 0, (s, x, i) \in \mathbb{R}_+^{2,\circ} \times \mathcal{M}. \tag{12.125}$$

On the other hand, by (12.103), we obtain that there is a $T_1 > 1$ satisfying

$$- \ln(1 - \rho_1\gamma_i) < \frac{\rho_1\lambda T_1}{4}, \forall i \in \mathcal{M}, \tag{12.126}$$

such that for $i \in \mathcal{M}, T \ge T_1, s \le \hat{H}$,

$$\frac{1}{T}\mathbb{E}_{s,0,i}\int_0^T \left(f_2\big(S(t), 0, \alpha(t)\big) - \widetilde{k}_d(\alpha(t)) - \frac{\sigma_2^2(\alpha(t))}{2}\right)dt > \frac{3\lambda}{4}. \tag{12.127}$$

Let $n_* \in \mathbb{N}$ and $n_* > \frac{\hat{k}_d+\hat{\sigma}_2^2}{\lambda} + 1$. Then following the argument in the proof of 12.8 to obtain the existence of some $\rho_2 \in (0, \rho_1)$ and $\varepsilon_2 \in (0, \varepsilon_1)$ satisfying

$$\mathbb{E}_{s,x,i}X^{-\rho_2}(t) \le e^{-\frac{\rho_2\lambda t}{2}}x^{-\rho_2} \text{ for } i \in \mathcal{M}, t \in [T_1, n_*T_1], s \le \hat{H}, x < \varepsilon_2. \tag{12.128}$$

With such ρ_2, using (12.128) and (12.126), we have for any $i \in \mathcal{M}, s \le \hat{H}, x < \varepsilon_2$ that

$$\begin{aligned}&\mathbb{E}_{s,x,i} V_3^{\frac{\rho_2}{\rho_1}}\big(S(t), X(t), \alpha(t)\big)\\ &\le \mathbb{E}_{s,x,i} X^{-\rho_2}(t) \le e^{-\frac{\rho_2\lambda t}{2}} x^{-\rho_2} = e^{-\frac{\rho_2\lambda t}{2}} V_3^{\frac{\rho_2}{\rho_1}}(s,x,i)(1-\rho_1\gamma_i)^{-\frac{\rho_2}{\rho_1}}\\ &\le e^{-\frac{\rho_2\lambda t}{2}} V_3^{\frac{\rho_2}{\rho_1}}(s,x,i) e^{\frac{\rho_2\lambda t}{4}} \le e^{-\frac{\rho_2\lambda t}{4}} V_3^{\frac{\rho_2}{\rho_1}}(s,x,i), \forall t \in [T_1, n_* T_1].\end{aligned} \tag{12.129}$$

Since $\rho_2 < \rho_1$, we can apply Itô's formula to obtain from (12.124) that

$$\mathcal{L} V_3^{\frac{\rho_2}{\rho_1}}(s,x,i) \le -\rho_2 \widetilde{\lambda} V_3^{\frac{\rho_2}{\rho_1}}(s,x,i) \ \forall i \in \mathcal{M}, s \ge \hat{H}, x \le \varepsilon_2. \tag{12.130}$$

Estimates (12.129) and (12.130) allow us to estimate $\mathbb{E}_{s,x,i} V_3^{\frac{\rho_2}{\rho_1}}\big(S(t), X(t), \alpha(t)\big)$ through $V_3^{\frac{\rho_2}{\rho_1}}(s,x,i)$ when x is sufficiently small and $s \ge 0$. Using (12.129), (12.130), and (12.125), we can follow the proof of Theorem 12.9 to show the existence of $q_* \in (0,1)$ and $K_* > 0$ such that

$$\mathbb{E}_{s,x,i} V_3^{\frac{\rho_2}{\rho_1}}\big(S(n_*T_1), X(n_*T_1), \alpha(n_*T_1)\big) \le q^* V_3^{\frac{\rho_2}{\rho_1}}(s,x,i) + K_*. \tag{12.131}$$

Moreover, if $\liminf_{s\to\infty} \frac{f_0(s,i)}{s^q} > 0, i \in \mathcal{M}$ for some $q \in (0,1]$, we have

$$[\mathcal{L}(\kappa_0 s + x)](s,x,i) \le \frac{S_0\kappa_0}{\theta} - \frac{\kappa_0 f_0(s,i)}{\theta} - \widetilde{k}_d(i)x \le \widetilde{c}_1 - \widetilde{c}_2(\kappa_0 s + x)^q, \tag{12.132}$$

for some $\widetilde{c}_1, \widetilde{c}_2 > 0$.

Since the drift term of $S(t)$ is positive when $S(t) = 0$, a standard argument shows that the compact set $\widetilde{\mathcal{H}}$ of $[0,\infty) \times (0,\infty)$ is petite (see, e.g., [63]). Using this fact and (12.132), we derive from Theorem A.1 that $\lim_{k\to\infty}(k+1)^{\beta-1}\|P(kn_*T_1, s, x, i, \cdot) - \mu^*(\cdot)\|_{TV} = 0$. An application of some standard arguments (e.g., the argument at the end of Theorem 12.9), we have

$$\lim_{t\to\infty} t^{\beta-1}\|P(t,s,x,i,\cdot) - \mu^*(\cdot)\|_{TV} = 0, \text{ for any } 1 \le \beta < \frac{1}{1-q}.$$

If $q = 1$, we obtain from (12.131) and (12.132) that

$$\begin{aligned}&\mathbb{E}_{s,x,i}\left[V_3^{\frac{\rho_2}{\rho_1}}\big(S(n_*T_1), X(n_*T_1), \alpha(n_*T_1)\big) + \kappa_0 S(n_*T_1) + X(n_*T_1)\right]\\ &\le \widetilde{q}^*\left(V_3^{\frac{\rho_2}{\rho_1}}(s,x,i) + \kappa_0 s + x\right) + \widetilde{K}_*,\end{aligned}$$

for some $\widetilde{q}^* \in (0, 1)$ and $\widetilde{K}_* > 0$. Theorem A.1 implies that

$$\lim_{k\to\infty} e^{\widetilde{\gamma}k}\|P(kn_*T_1, s, x, i, \cdot) - \mu^*(\cdot)\|_{TV} = 0, \text{ for some } \widetilde{\gamma} > 0.$$

Then, we can obtain the exponential rate of convergence. □

Proof for the Case $\lambda < 0$ If condition (1) of (H12.7) holds, the proof can be carried out using comparison arguments by comparing $S(t)$ and $\hat{S}(t)$. We have from a comparison argument that $S(t) \le \hat{S}(t)\,\forall t \ge 0$ with probability 1 given that $S(0) \le \hat{S}(0)$. Due to (12.111) and the ergodicity of $(\hat{S}(t), \alpha(t))$ (see Theorem 4.4), we have

$$\lim_{t\to\infty}\int_0^t \hat{S}^p(u)du = \sum_{i\in\mathcal{M}}\int_{\mathbb{R}_+} s^p\nu_0(ds, i) < \infty, \text{ for } 0 < p < 1 + \bar{p}. \tag{12.133}$$

In view of Itô's formula,

$$\begin{aligned}\frac{\ln X(t)}{t} &= = \frac{1}{t}\int_0^t \left(f_2(S(u), X(u), \alpha(u) - \widetilde{k}_d(\alpha(u))du - \frac{1}{t}\frac{\sigma_2^2(\alpha(u))}{2}\right)du\\ &\quad + \frac{1}{t}\int_0^t \sigma_2(\alpha(u))dW_2(u)\\ &\le \frac{1}{t}\int_0^t \left(f_2(\hat{S}(u), 0, \alpha(u) - \widetilde{k}_d(\alpha(u))du - \frac{\sigma_2^2(\alpha(u))}{2}\right)du\\ &\quad + \frac{1}{t}\int_0^t \sigma_2(\alpha(u))dW_2(u)\end{aligned} \tag{12.134}$$

In view of (12.133) and the strong law of large numbers for martingales, we have

$$\begin{aligned}&\frac{1}{t}\int_0^t \left(f_2(\hat{S}(u), 0, \alpha(u) - \widetilde{k}_d(\alpha(u))du - \frac{\sigma_2^2(\alpha(u))}{2}\right)du\\ &\quad + \frac{1}{t}\int_0^t \sigma_2(\alpha(u))dW_2(u) = \lambda\end{aligned} \tag{12.135}$$

Combining (12.134) and (12.135), we obtain that $\limsup_{t\to\infty}\frac{\ln X(t)}{t} \le \lambda$. Finally, using the fact that any weak limit of random occupation measures is an invariant probability measure (see Section 12.1.4), we can obtain the convergence rate (Lyapunov exponent) $\lim_{t\to\infty}\frac{\ln X(t)}{t} = \lambda$.]

Now, we assume that condition (2) of (H12.7) holds. Analogous to the proof in the case $\lambda > 0$, we have

$$\mathbb{E}_{s,x,i} V_5^{\frac{\rho_4}{\rho_3}}\left(S(n^*T_2), X(n^*T_2), \alpha(n^*T_2)\right) \le \hat{q}^* V_5^{\frac{\rho_4}{\rho_3}}(s, x, i)$$

$$+ \hat{K}_*; i \in \mathcal{M}, s \geq 0, x \leq \varepsilon_3,$$

where $V_5(s, x, i) = (1 - \rho_3\gamma_i)x^{\rho_3}$ for suitable constants $\rho_3, \rho_4, n^*, T_2, \hat{q}^* \in (0, 1), \hat{K}_*, \varepsilon_3$. Then we can mimic the proofs of 12.14 and 12.15 to obtain the desired result. □

Remark 12.26 Before proceeding further, let us make the following remarks:

- The techniques of handling the critical case can be applied to such stochastic models in epidemiology as SIR, SEIR, and SEIS models and predator-prey models, when the system exhibits certain monotone properties so that the contradiction arguments in the proof for the case $\lambda = 0$ can be applied.
- Constructing Lyapunov function for switching diffusions is practically more difficult than that for diffusions. In non-critical cases, we showed how to construct the pair of Lyapunov functions $(\widetilde{V}, \widetilde{H})$ satisfying [16, Proposition 4.13] for systems involving Markovian switching if the construction is possible but not obvious (in our model, the requirement is (2) or (3) of (H12.6) together with $\lim_{s\to\infty} \frac{f_0(s,i)}{s^q} > 0, q < 1$ being satisfied). We also showed that when it is practically impossible to find $(\widetilde{V}, \widetilde{H})$ satisfying [16, Proposition 4.13], we can obtain a sub-geometric convergence rate under certain conditions. Our techniques combined with the main approaches in Section 12.1 can work for some stochastic Kolmogorov systems under relaxed conditions.

12.2.2.1 Proof of Theorem 12.23

***Proof of Theorem* 12.23.** Since the coefficients of (12.99) are locally Lipschitz continuous in $(s, x, i) \in \mathbb{R}^2_+ \times \mathcal{M}$, the system given by (12.99) and (12.2) has a unique continuous solution up to explosion time τ_e, where $\tau_e = \inf\{t \geq 0 : S(t) \vee X(t) = \infty\}$, with the convention $\inf \emptyset = \infty$. The solution is also a strong Markov process; see Chapter 2 or [201]. If we define $\tau_k = \inf\Big\{t \geq 0 : S(t) \vee X(t) > k\Big\}$, then $\tau_e = \lim_{k\to\infty} \tau_k$. Consider $\hat{V}_1(s, x, i) = \kappa_0 s + x$ then, by the generalized Itô formula, we have

$$\begin{aligned}
\mathcal{L}\hat{V}_1(s, x, i) &= \frac{S_0\kappa_0}{\theta} - \frac{\kappa_0}{\theta} f_0(s, i) - \widetilde{k}_d(i)x + x(f_2(s, x, i) - \kappa_0 f_1(s, x, i)) \\
&\leq \frac{S_0\kappa_0}{\theta}.
\end{aligned}$$

Hence,

$$\mathbb{E}_{s,x,i}\hat{V}_1\big(S(\tau_k \wedge t), X(\tau_k \wedge t), \alpha(\tau_k \wedge t)\big) \leq \hat{V}_1(s, x, i) + \frac{S_0\kappa_0 t}{\theta},$$

which implies that

$$\mathbb{P}_{s,x,i}\{\tau_k < t\} \le \mathbb{P}_{s,x,i}\Big\{\hat{V}_1\big(S(\tau_k \wedge t), X(\tau_k \wedge t), \alpha(\tau_k \wedge t)\big) \ge k\Big\}$$
$$\le \frac{\hat{V}_1(s,x,i) + \frac{S_0\kappa_0 t}{\theta}}{k} \to 0 \text{ as } k \to \infty.$$

Therefore, we have $\mathbb{P}_{s,x,i}\{\tau_e \le t\} = 0$ or $\mathbb{P}_{s,x,i}\{\tau_e > t\} = 1$ $\forall t > 0$. As a consequence, $\mathbb{P}_{s,x,i}\{\tau_e = \infty\} = 1$. Hence, the system given by (12.99) and (12.2) has a unique global continuous solution.

Now, we move to the part of positivity of solutions. First, suppose that $s, x > 0$. For any $n \in \mathbb{N}$, we define the following truncated functions:

$$\begin{aligned} f_0^{(n)}(s,i) &= f_0(s \wedge n, i); \\ f_1^{(n)}(s,x,i) &= f_1(s \wedge n, x \wedge n, i); \\ f_2^{(n)}(s,x,i) &= f_2(s \wedge n, x \wedge n, i), \end{aligned}$$

and let $\big(S^{(n)}(t), X^{(n)}(t)\big)$ be the solution of Eqs. (12.99) and (12.2) with f_0, f_1, f_2 replaced by $f_0^{(n)}, f_1^{(n)}, f_2^{(n)}$, respectively. Denote

$$\eta^{(n)} = \inf\big\{t \ge 0 : S^{(n)}(t) \wedge X^{(n)}(t) \le 0\big\},$$

$$\eta_k^{(n)} = \inf\Big\{t \ge 0 : S^{(n)}(t) \wedge X^{(n)}(t) < \frac{1}{k}\Big\}.$$

Then $\eta^{(n)} = \lim_{k\to\infty} \eta_k^{(n)}$. Consider

$$\hat{V}_2^{(n)}(s,x,i) = s - c_1^{(n)} - c_1^{(n)} \ln \frac{s}{c_1^{(n)}} + c_2(x - 1 - \ln x),$$

where

$$c_2 = \frac{1}{\kappa_0} \text{ and } c_1^{(n)} = \frac{c_2 \check{k}_d}{\max_{i\in\mathcal{M}} \sup_{s>0} \frac{f_1^{(n)}(s,x,i)}{s}}.$$

Let $c_2^{(n)} := \max_{i\in\mathcal{M}} \sup_{s>0} \frac{f_0^{(n)}(s,i)}{s}$. Since $f_1^{(n)}(s,x,i)$ is global Lipschitz and $f_1^{(n)}(0,x,i) = 0$, it is readily seen that $\sup\limits_{s>0} \dfrac{f_1^{(n)}(s,x,i)}{s} < \infty$. Similarly, we also obtain $\sup_{s>0} \frac{f_0^{(n)}(s,i)}{s} < \infty$, so the above constants are well defined.

By the generalized Itô formula, we have

$$\mathcal{L}^{(n)}\hat{V}_2^{(n)}(s,x,i) = \left(1 - \frac{c_1^{(n)}}{s}\right)\left(\frac{S_0}{\theta} - \frac{f_0^{(n)}(s,i)}{\theta} - x f_1^{(n)}(s,x,i)\right) + \frac{c_1^{(n)}\sigma_1^2(i)}{2}$$

$$
\begin{aligned}
&+\left(c_2-\frac{c_2}{x}\right)\left(xf_2^{(n)}(s,x,i)-x\widetilde{k}_d(i)\right)+\frac{c_2\sigma_2^2(i)}{2}\\
\le&\frac{S_0+c_1^{(n)}c_2^{(n)}}{\theta}+c_2\hat{k}_d+\frac{c_1^{(n)}\hat{\sigma}_1^2+c_2\hat{\sigma}_2^2}{2}\\
&+x(c_2f_2^{(n)}(s,x,i)-f_1^{(n)}(s,x,i))\\
&+x\left(\frac{c_1^{(n)}f_1^{(n)}(s,x,i)}{s}-c_2k_d(i)\right)\\
\le&\frac{S_0+c_1^{(n)}c_2^{(n)}}{\theta}+c_2\hat{k}_d+\frac{c_1^{(n)}\hat{\sigma}_1^2+c_2\hat{\sigma}_2^2}{2}:=K^{(n)}\text{ for }s,x>0,
\end{aligned}
$$

where the operator $\mathcal{L}^{(n)}$ is defined as $\mathcal{L}$ with f_0, f_1, and f_2 replaced by $f_0^{(n)}$, $f_1^{(n)}$, and $f_2^{(n)}$, respectively. Applying Itô's formula again, we have

$$
\begin{aligned}
&\mathbb{E}_{s,x,i}\hat{V}_2^{(n)}\big(S^{(n)}(\eta_k^{(n)}\wedge t),X^{(n)}(\eta_k^{(n)}\wedge t),\alpha(\eta_k^{(n)}\wedge t)\big)\\
=&\hat{V}_2^{(n)}(s,x,i)+\mathbb{E}_{s,x,i}\int_0^{\eta_k^{(n)}\wedge t}\mathcal{L}\hat{V}_2^{(n)}\big(S^{(n)}(u),X^{(n)}(u),\alpha(u)\big)du\\
\le&\hat{V}_2^{(n)}(s,x,i)+K^{(n)}t.
\end{aligned}
$$

Since the definition of $\hat{V}_2^{(n)}$, we observe that if $\eta_k^{(n)}<t$, then

$$
\hat{V}_2^{(n)}\big(S^{(n)}(\eta_k^{(n)}\wedge t),X^{(n)}(\eta_k^{(n)}\wedge t),\alpha(\eta_k^{(n)}\wedge t)\big)\ge\big(c_1^{(n)}\ln kc_1^{(n)}-c_1^{(n)}\big)\wedge\big(c_2\ln k-c_2\big).
$$

Hence, we obtain that

$$
\mathbb{P}_{s,x,i}\{\eta_k^{(n)}<t\}\le\frac{\hat{V}_2^{(n)}(s,x,i)+K^{(n)}t}{\big(c_1^{(n)}\ln kc_1^{(n)}-c_1^{(n)}\big)\wedge\big(c_2\ln k-c_2\big)}\to 0\text{ as }k\to\infty.
$$

As a result, for any $n\in\mathbb{N}$, $\mathbb{P}_{s,x,i}\{\eta_\infty^{(n)}=\infty\}=1$. That is,

$$
\mathbb{P}_{s,x,i}\big\{S^{(n)}(t),X^{(n)}(t)>0:\forall t>0\big\}=1\text{ for any }n\in\mathbb{N},
$$

and hence

$$
\mathbb{P}_{s,x,i}\big\{S^{(n)}(t),X^{(n)}(t)>0:\forall t>0,n\in\mathbb{N}\big\}=1.
$$

Now, for any $t>0$ and $\omega\in\{\tau_e=\infty\}\cap\big\{S^{(n)}(t),X^{(n)}(t)>0:\forall t>0,n\in\mathbb{N}\big\}$, there exists $n_0=n_0(\omega,t)$ such that $S(y)(\omega)\vee X(y)(\omega)<n_0$ $\forall 0\le y\le t$. As a

consequence, $S(t)(\omega) = S^{(n_0)}(t)(\omega) > 0$ and $X(t)(\omega) = X^{(n_0)}(t)(\omega) > 0$. This, combined with $\mathbb{P}_{s,x,i}\{\tau_e = \infty\} = 1$, implies that

$$\mathbb{P}_{s,x,i}\big\{S(t) > 0 : t > 0\big\} = \mathbb{P}_{s,x,i}\big\{X(t) > 0 : t > 0\big\} = 1 \ \forall s, x > 0. \tag{12.136}$$

If $s > 0, x = 0$, the result $\mathbb{P}_{s,x,i}\big\{S(t) > 0 : t > 0\big\} = 1$ is similar proved by choosing $c_2 = 0$. Moreover, it is obvious that $\mathbb{P}_{s,x,i}\big\{X(t) = 0 : t > 0\big\} = 1$.

Consider the case when the initial value $s = 0$ and $x \geq 0$. Let $\varepsilon > 0$ be sufficiently small such that

$$\frac{S_0 - f_0(\tilde{s}, i)}{\theta} - \tilde{x} f_1(\tilde{s}, \tilde{x}, i) \geq \frac{S_0}{2\theta}, \tag{12.137}$$

for any $(\tilde{s}, \tilde{x}, \tilde{i}) \in \mathbb{R}^2 \times \mathcal{M}$ satisfying $\tilde{s} + |\tilde{x} - x| < \varepsilon$. Let

$$\tilde{\tau}_1 = \inf\{t > 0 : S(t) + |X(t) - x| \geq \varepsilon\}.$$

By the continuity of $(S(t), X(t))$, $\mathbb{P}_{0,x,i}\{\tilde{\tau}_1 > 0\} = 1$. Using the variation of constants formula (see [201, Chapter 3]), we can write $S(t)$ in the form

$$\begin{aligned} S(t) = \Phi(t) \Bigg[\int_0^t \Phi^{-1}(u) \Bigg(&\frac{S_0 - f_0\big(S(u), \alpha(u)\big)}{\theta} \\ &- X(u) f_1\big(S(u), X(u), \alpha(u)\big)\Bigg) du \Bigg], \end{aligned} \tag{12.138}$$

for $t \in [0, \tilde{\tau}_1)$, where $\Phi(t) = \exp(-\int_0^t \frac{\sigma_1^2(\alpha(u))}{2} du + \int_0^t \sigma_1(\alpha(u)) dW_2(u))$. It follows from (12.137) that

$$\frac{S_0 - f_0\big(S(u), \alpha(u)\big)}{\theta} - X(u) f_1\big(S(u), X(u), \alpha(u)\big) > 0 \text{ if } t \in (0, \tilde{\tau}_1].$$

This and (12.138) imply that $\mathbb{P}_{0,x,i}\{S(t) > 0, t \in (0, \tilde{\tau}_1]\} = 1$, which combined with (12.136) and the strong Markov property of $(S(t), X(t), \alpha(t))$ yields that $\mathbb{P}_{0,x,i}\{S(t) > 0, t \in (0, \infty)\} = 1$. The theorem is therefore proved. □

12.3 SIRS Models with Imperfect Vaccination

12.3.1 Introduction

Motivations Because of the importance, in the past decades, much attention has been devoted to modeling epidemic systems, analyzing their dynamic behaviors, and predicting the future. In the human history, infectious diseases have been making significant impacts on the health, economics, and social life of the population. At the forefront of scientific research, the compartment-type models introduced by Kermack and McKendrick [154, 155] have received much attention. These models have been developed substantially and used widely for epidemic systems in recent years. The rational is that the population will be divided into several non-overlapping classes termed compartments. For example, the so-called SIR epidemic model divides population into susceptible, infected, and recovered classes, which is suitable for diseases with permanent immunity such as rubella, whooping cough, measles, smallpox, etc. However, for some diseases such as common colds, influenza, etc., an infected individual can be reinfected again after recovery (i.e., the individual returns to the susceptible class after recovery). This leads to the so-called SIRS model, in which the term SIRS refers for the cycle "susceptible (S) $\to$ infected (I) $\to$ recovered (R) $\to$ susceptible (S) again" in the dynamics. SIRS-type model has been recognized as one of the most important models in epidemiology and mathematical biology; see [179, 192, 194, 228, 246, 272] and the references therein.

To control the spread of diseases, it is important to develop vaccines and investigate the efficiency of the vaccinations. In recent years, there is a resurgent effort in the study of epidemic models with vaccination; see [127, 184, 188, 191, 317, 319] and the references therein. From a different angle, it has also been well-recognized that the so-called hybrid systems are more versatile and more appropriate for describing dynamic systems. Such systems feature in the coexistence of continuous dynamics and discrete events and their interactions. Taking these factors into consideration, we study the following SIRS with vaccination, termed regime-switching SVIRS epidemic model:

$$
\begin{cases}
dS(t) = \Big[(1-\widetilde{q}(\alpha(t)))\, A(\alpha(t)) - I(t)g(S(t)+\beta V(t), I(t), \alpha(t)) \\
\qquad - (\alpha_0(\alpha(t)) + \widetilde{p}(\alpha(t)))S(t) + \alpha_2(\alpha(t))R(t) + \gamma(\alpha(t))V(t)) \Big] dt \\
\qquad + \sigma_1(\alpha(t))S(t)dW_1(t), \\
dV(t) = \Big[\widetilde{q}(\alpha(t))A(\alpha(t)) + \widetilde{p}(\alpha(t))S(t) - \big(\alpha_0(\alpha(t)) + \gamma(\alpha(t))\big)V(t) \Big] dt \\
\qquad + \sigma_2(\alpha(t))V(t)dW_2(t), \\
dI(t) = \Big[I(t)g(S(t)+\beta V(t), I(t), \alpha(t)) - \big(\alpha_0(\alpha(t)) + \alpha_1(\alpha(t)) \\
\qquad + \alpha_3(\alpha(t))\big)I(t) \Big] dt + \sigma_3(\alpha(t))I(t)dW_3(t), \\
dR(t) = \Big[\alpha_3(\alpha(t))I(t) - \big(\alpha_0(\alpha(t)) + \alpha_2(\alpha(t))\big)R(t) \Big] dt + \sigma_4(\alpha(t))R(t)dW_4(t), \\
S(0) = s \ge 0,\ I(0) = i \ge 0,\ V(0) = v \ge 0,\ R(t) = r \ge 0,\ \alpha(0) = k \in \mathcal{M},
\end{cases}
\tag{12.139}
$$

Here, $S(t)$ denotes the number of susceptible individuals who are infection-prone; $I(t)$ denotes the number of the infected individuals who have already contracted the disease; $V(t)$ is the number of the individuals who are vaccinated; $R(t)$ denotes the number of individuals recovering from infection; $\{W_i(t)\}$, $i = 1, \dots, 4$, are independent standard Brownian motions representing the white noise; and $\alpha(t)$ is a finite-state continuous-time Markov chain that is independent of the Brownian motions $W_i(t)$ and that has state space $\mathcal{M} = \{1, \dots, m_0\}$. In the model, there are a number of parameters: $\widetilde{q}$ is a fraction of vaccinated newborns; A is an input of new members into the population; $1 - \beta$ is the efficiency of vaccination (i.e., β is the proportion of vaccinated individuals for who the vaccination fails to protect); $\widetilde{p}$ is the proportional coefficient of vaccinated for the susceptible; γ is the rate for immunity loss of vaccines after some time; α_0 is the nature death rate of the population; α_1 is the disease-related death rate; α_2 is the rate of recovered individual becoming susceptible again; α_3 is the rate of recovery from infection; and σ_i, $i = 1, 2, 3, 4$ are intensities of the noise. All these constants are assumed to be positive and depend on $\alpha(t)$. The rate (in time) of how many individuals in susceptible group get infected is represented by a function named "incidence rate function." Numerous types of incidence rates have been modeled and considered in the literature such as the bilinear functional response $k_0 SI$ (see [63]), the nonlinear functional response $\frac{k_0 S I^l}{1+k_1 I^h}$ (see [250, 288]), and the Beddington-DeAngelis functional response $\frac{k_0 S I}{1+k_1 S+k_2 I}$ (see [67, 69]), etc. It is clear that if there is no infection in the population, there is no disease transmission and then, the incidence rate function will be 0. As a result, the incidence rate in general can also be factorized by $I(t)$. Because of this, and also for the convenience of stating results later, we will write the incidence rate function in the form Ig for some function $g(\cdot, \cdot, \cdot)$.

Our goal in this section is to provide a complete characterization of long-time behavior of (12.139) so as to answer one of the most important questions: What is the sufficient and necessary condition that guarantee the infected class of individuals to be extinct eventually. To answer this question, we introduce a real-valued threshold λ and prove that if λ is negative, the disease will die out together with the extinction rate of the disease. On the other hand, if λ is positive, the disease becomes endemic. We shall show that the system has an invariant probability measure, and the transition probability of the solution process converges to the invariant measure. Our method to determine the thresholds is from a dynamical system theory point of view, which is interesting in its own right. This methods can be generalized to deal with many other models. The algebraic representations of the thresholds are given that is easily computable. In addition, simulation studies and numerical examples are given to illustrate our results.

12.3.2 Main Results

12.3.2.1 Preliminary Results

This section is devoted to certain preliminary results related to (12.139). Let $(\Omega, \mathcal{F}, \{\mathcal{F}_t\}_{t\geq 0}, \mathbb{P})$ be a complete filtered probability space. $W_1(t)$, $W_2(t)$, $W_3(t)$, and $W_4(t)$ are independent standard Brownian motions, and $\alpha(t)$ is a finite-state continuous-time Markov chain defined by (12.2) that is independent of the Brownian motions $W_i(t), i = 1, \ldots, 4$ and that has a finite state space $\mathcal{M} = \{1, \ldots, m_0\}$. As a result, it has a unique invariant probability measure ν. Throughout the section, we denote $\mathbb{R}^2_+ = \{(s, v) \in \mathbb{R}^2 : s \geq 0, v \geq 0\}$ with interior $\mathbb{R}^{2,\circ}_+ = \{(s, v) \in \mathbb{R}^2 : s > 0, i > 0\}$, $\mathbb{R}^4_+ = \{(s, v, i, r) \in \mathbb{R}^4 : s \geq 0, i \geq 0, v \geq 0, r \geq 0\}$ with interior $\mathbb{R}^{4,\circ}_+ = \{(s, v, i, r) \in \mathbb{R}^4 : s > 0, i > 0, v > 0, r > 0\}$, $\mathbb{R}^{4,*}_+ = \{(s, v, i, r) \in \mathbb{R}^4_+ : i > 0\}$. Moreover, to simplify notation, we denote $\mathbf{z} := (s, v, i, r, k)$ and $\mathbf{Z}(\cdot) = (S(\cdot), V(\cdot), I(\cdot), R(\cdot), \alpha(\cdot))$. We will denote $\mathbb{P}_{\mathbf{z}}$ and $\mathbb{E}_{\mathbf{z}}$ the probability and expectation corresponding to the initial condition $S(0) = s$, $I(0) = i$, $V(0) = v$, $R(0) = r$, $\alpha(0) = k$, respectively.

(H12.8) We impose the following condition throughout the rest of the section.

- $g(u, i, k)$ is locally Lipschitz and bounded by $C(1 + u)$ for some constant $C > 0$ on $\mathbb{R}^2_+ \times \mathcal{M}$.
- $g(u, i, k)$ is continuous at $i = 0$ uniformly in (u, k), that is

$$\lim_{i\to 0^+} \sup_{u\geq 0, k\in\mathcal{M}} \{|g(u, i, k) - g(u, 0, k)|\} = 0.$$

- $g(u, 0, k)$ is increasing in $u \in [0, \infty)$ for each $k \in \mathcal{M}$.

Remark 12.27 We note that almost all of the incidence rate functions used in the literature as mentioned in the introduction (such as the bilinear incidence rate, the Beddington-DeAngelis incidence rate, the Holling type II functional response, etc.) satisfy these conditions. Recall that the incidence rate in our setting is Ig rather than g.

To proceed, we first establish the existence, uniqueness, and basic properties of the solution process of (12.139).

Theorem 12.28 *For any initial value* $\mathbf{z} \in \mathbb{R}^4_+ \times \mathcal{M}$, *there exists a global solution* $(\mathbf{Z}(t), \alpha(t))$ *to* (12.139) *and* (12.2) *such that* $\mathbb{P}_{\mathbf{z}}\{\mathbf{Z}(t) \in \mathbb{R}^4_+ \times \mathcal{M},\ \forall t \geq 0\} = 1$. *Moreover, for all* $\mathbf{z} = (s, v, i, r, k)$ *with* $i = 0$ *and* $s, v, r \geq 0$,

$$\mathbb{P}_{\mathbf{z},k}\{I(t) = 0,\ \forall t \geq 0\} = \mathbb{P}_{\mathbf{z},k}\{S(t) > 0,\ V(t) > 0,\ R(t) > 0,\ \forall t \geq 0\} = 1,$$

and for all $\mathbf{z}$ *with* $i > 0, s \geq 0, v \geq 0, r \geq 0$, $\mathbb{P}_{\mathbf{z}}\{\mathbf{Z}(t) \in \mathbb{R}^{4,\circ}_+ \times \mathcal{M},\ \forall t > 0\} = 1$. *We also have that the process* $\{\mathbf{Z}(t)\}$ *is a Markov-Feller process on* $\mathbb{R}^4_+ \times \mathcal{M}$.

Remark 12.29 Together with well-posedness property of the underlying equation, the results show the positivity property of the solution. That is, all components are always non-negative provided if they start from non-negative initial conditions. Moreover, $I(t) = 0$ for all t if started at $i = 0$, and always positive if started at $i > 0$; while all other components are always positive.

Proof The proof is similar to those in Sects. 12.1 and 12.2 and therefore is omitted here. □

Next, we establish the boundedness in probability and boundedness in moment of the solution process.

Lemma 12.30 *For any $q > 0$ sufficiently small, there exist $C_q > 0$ and $D_q > 0$ such that*

$$\mathbb{E}_{\mathbf{z}}\left(S(t)+V(t)+I(t)+R(t)\right)^{1+q} \le \frac{(1+s+i+v+r)^{1+q}}{e^{D_q t}} + \frac{C_q}{D_q}, \quad \forall t \ge 0. \tag{12.140}$$

As a result, for any $H > 0$, $\varepsilon > 0$, $T > 0$, there is a constant $M_{H,\varepsilon,T} > 0$ such that

$$\mathbb{P}_{\mathbf{z}}\left\{\sup_{t\in[0,T]}\{S(t)+V(t)+I(t)+R(t)\} \le M_{H,\varepsilon,T}\right\} \ge 1-\varepsilon, \ \forall \mathbf{z} \in [0,H]^4 \times \mathcal{M}. \tag{12.141}$$

Proof Denote by $\mathcal{L}$ the operator associated with the solution process of (12.139). It is well-known that for a function U of $\mathbf{z}$ that is twice differentiable with respect to s, v, i, r, we have

$$\begin{aligned}
\mathcal{L}U(\mathbf{z}) &= U_s(\mathbf{z})\big((1-\widetilde{q}(k))A - ig(s+\beta v, i, k) \\
&\quad - (\alpha_0(k)+\widetilde{p}(k))s + \alpha_2(k)r + \gamma(k)v\big) \\
&\quad + U_i(\mathbf{z})\big(ig(s+\beta v, i, k) - (\alpha_0(k)+\alpha_3(k)+\alpha_1(k))i\big) \\
&\quad + U_v(\mathbf{z})\big(\widetilde{q}(k)A(k) + \widetilde{p}(k)s - (\alpha_0(k)+\gamma(k))v\big) \\
&\quad + U_r(\mathbf{z})\big(\alpha_3(k)i - \alpha_0(k)r - \alpha_2(k)r\big) \\
&\quad + \frac{1}{2}U_{ss}(\mathbf{z})\sigma_1^2(k)s^2 + \frac{1}{2}U_{vv}(\mathbf{z})\sigma_2^2(k)v^2 + \frac{1}{2}U_{ii}(\mathbf{z})\sigma_3^2(k)i^2 \\
&\quad + \frac{1}{2}U_{rr}(\mathbf{z})\sigma_4^2(k)r^2 \\
&\quad + \sum_{l\in\mathcal{M}} q_{kl}U(\mathbf{z}, l).
\end{aligned}$$

Therefore, by direct computations for $U = (1+s+v+i+r)^{1+q}$, we have

$$
\begin{aligned}
&\mathcal{L}(1+s+v+i+r)^{1+q}\\
=&(1+q)(1+s+v+i+r)^q\big((1-\widetilde{q}(k))A - ig(s+\beta v, i, k)\\
&-(\alpha_0(k)+\widetilde{p}(k))s+\alpha_2(k)r+\gamma(k)v\big)\\
&+(1+q)(1+s+v+i+r)^q\big(ig(s+\beta v,i,k)-(\alpha_0(k)+\alpha_3(k)+\alpha_1(k))i\big)\\
&+(1+q)(1+s+v+i+r)^q\big(\widetilde{q}(k)A(k)+\widetilde{p}(k)s-(\alpha_0(k)+\gamma(k))v\big)\\
&+(1+q)(1+s+v+i+r)^q\big(\alpha_3(k)i-\alpha_0(k)r-\alpha_2(k)r\big)\\
&+\frac{q(1+q)}{2}(1+s+v+i+r)^{q-1}\Big(\sigma_1^2(k)s^2+\sigma_2^2(k)v^2+\sigma_3^2(k)i^2+\sigma_4^2(k)r^2\Big)\\
=&(1+q)(1+s+v+i+r)^q\Big(A-\alpha_0(k)s-\alpha_0(k)v-(\alpha_0(k)+\alpha_1(k))i-\alpha_0(r)\Big)\\
&+\frac{q(1+q)}{2}(1+s+v+i+r)^{q-1}\Big(\sigma_1^2(k)s^2+\sigma_2^2(k)v^2+\sigma_3^2(k)i^2+\sigma_4^2(k)r^2\Big)\\
\le&(1+q)(1+s+v+i+r)^q\\
&\Big(2A-\breve{\alpha}(1+s+v+i+r)+q\hat{\sigma}(1+s+v+i+r)\Big),
\end{aligned}
$$

where $\breve{\alpha} := \min_{k\in\mathcal{M}}\{A, \alpha_0(k), \alpha_1(k)\} > 0$, and $\hat{\sigma} := \frac{1}{2}\max_{k\in\mathcal{M}, i=1,2,3,4}\{\sigma_i^2(k)\} < \infty$. Now, for any q, $0 < q < \frac{\breve{\alpha}}{2\hat{\sigma}}$, we have

$$
\mathcal{L}(1+s+v+i+r)^{1+q} \le C_q - D_q(1+s+v+i+r)^{1+q},
$$

for $D_q = \frac{(1+q)\breve{\alpha}}{4}$, and

$$
C_q = \sup_{s,v,i,r\ge 0}\Big\{(1+q)(1+s+v+i+r)^q\Big(2A-\frac{\breve{\alpha}}{4}(1+s+v+i+r)\Big)\Big\} < \infty.
$$

Therefore, as an application of Itô's formula, one can obtain (12.140) and (12.141); see Section 12.1.2 or [202] or [203] for this well-known argument.

□

12.3.2.2 Characterization of Long-Time Behavior

In this section, a threshold λ will be introduced. This λ will completely characterize the long-time property of system (12.139). The threshold λ will be determined from the dynamics of the system on the boundary, that is, when $I(t) = R(t) = 0$, in which, $(S(t), V(t))$ follows the system:

$$
\begin{cases}
dS(t) = \Big[(1-\widetilde{q}(\alpha(t)))\,A(\alpha(t)) - (\alpha_0(\alpha(t)) + \widetilde{p}(\alpha(t)))\,S(t) \\
\qquad\qquad + \gamma(\alpha(t)V(t))\Big]dt + \sigma_1(\alpha(t))S(t)dW_1(t), \\
dV(t) = \Big[\widetilde{q}(\alpha(t))A(\alpha(t)) + \widetilde{p}(\alpha(t))S(t) - (\alpha_0(\alpha(t)) + \gamma(\alpha(t)))\,V(t)\Big]dt \\
\qquad\qquad + \sigma_2(\alpha(t))V(t)dW_2(t).
\end{cases}
\tag{12.142}
$$

As alluded to in the introduction, to examine the boundary behavior, we need to look at two-dimensional problems, which presents an added difficult. The returning to susceptible group from recovered and vaccinated individuals is also another challenge. To ease the difficulty, we consider a perturbed system of (12.142). For each $\theta \in [-1, 1]$, let $(\widetilde{S}^\theta(t), \widetilde{V}^\theta(t))$ be solution to

$$
\begin{cases}
d\widetilde{S}^\theta(t) = \Big[(1-\widetilde{q}(\alpha(t)))\,A(\alpha(t)) - (\alpha_0(\alpha(t)) + \widetilde{p}(\alpha(t)))\,\widetilde{S}^\theta(t) \\
\qquad\qquad + \gamma(\alpha(t))\widetilde{V}^\theta(t)\Big]dt + \theta\alpha_2(\alpha(t))dt + \sigma_1(\alpha(t))\widetilde{S}^\theta(t)dW_1(t), \\
d\widetilde{V}^\theta(t) = \Big[\widetilde{q}(\alpha(t))A(\alpha(t)) + \widetilde{p}(\alpha(t)))\widetilde{S}^\theta(t) - (\alpha_0(\alpha(t)) + \gamma(\alpha(t)))\,\widetilde{V}^\theta(t)\Big]dt \\
\qquad\qquad + \sigma_2(\alpha(t))\widetilde{V}^\theta(t)dW_2(t).
\end{cases}
\tag{12.143}
$$

Denote by $\mathcal{L}_\theta$ the operator associated with the solution process of (12.143), i.e., for twice differentiable function $U((s, v, k))$ (with respect to (s, v))

$$
\begin{aligned}
\mathcal{L}_\theta U((s, v, k)) = {} & U_s((s, v, k))\big((1-\widetilde{q}(k))A - (\alpha_0(k) + \widetilde{p}(k))s + \alpha_2(k)r + \gamma(k)v \\
& + \theta\alpha_2(k)\big) + U_v((s, v, k))\big(\widetilde{q}(k)A(k) + \widetilde{p}(k)s - (\alpha_0(k) + \gamma(k))v\big) \\
& + \frac{1}{2}U_{ss}((s, v, k))\sigma_1^2(k)s^2 + \frac{1}{2}U_{vv}((s, v, k))\sigma_2^2(k)v^2 \\
& + \sum_{l\in\mathcal{M}} q_{kl}U((s, v, k)).
\end{aligned}
$$

Analogous to Theorem 12.28 and Lemma 12.30, we can easily show that for each initial value $(s, v, k) \in \mathbb{R}^2_+ \times \mathcal{M}$, there exists a global solution $(\widetilde{S}^\theta(t), \widetilde{V}^\theta(t))$ to (12.143) and that $(\widetilde{S}^\theta(t), \widetilde{V}^\theta(t)) \in \mathbb{R}^{2,\circ}_+$ for all $t > 0$ almost surely and that

$$
\mathbb{E}_{(s,v,k)}\big(\widetilde{S}^\theta(t) + \widetilde{V}^\theta(t)\big)^{1+q} \le \frac{(1+s+v)^{1+q}}{e^{\widetilde{D}_q t}} + \frac{\widetilde{C}_q}{\widetilde{D}_q}, \quad \forall t \ge 0, \tag{12.144}
$$

for some small $q > 0$, and constants $\widetilde{C}_q, \widetilde{D}_q > 0$, which can be taken uniformly for any $\theta \in [-1, 1]$.

Since the diffusion in (12.143) is nondegenerate on $\mathbb{R}_+^{2,\circ} \times \mathcal{M}$, there exists uniquely an invariant probability distribution $\bar{\mu}_\theta$ of $(\widetilde{S}^\theta(t), \widetilde{V}^\theta(t), \alpha(t))$ on $\mathbb{R}_+^{2,\circ} \times \mathcal{M}$. Due to (12.144) and [117, Lemma 3.4], we have

$$\sum_{k\in\mathcal{M}} \int_{\mathbb{R}_+^2} (s+v)^{1+q} \bar{\mu}_\theta(ds, dv, k) \le \frac{\widetilde{C}_q}{\widetilde{D}_q} < \infty.$$

Thus, we can define the following λ_θ which is a "perturbed" growth rate of $I(t)$ as its density is small:

$$\begin{aligned}\lambda_\theta :=& \sum_{k\in\mathcal{M}} \int_{\mathbb{R}_+^2} \left(g(s,0,k) - (\alpha_0(k)+\alpha_3(k)+\alpha_1(k)) - \frac{\sigma_3^2(k)}{2} \right) \bar{\mu}_\theta(ds, dv, k)\\ =& \sum_{k\in\mathcal{M}} \int_{\mathbb{R}_+^2} g(s+\beta v, 0, k) \bar{\mu}_\theta(ds, dv, k)\\ &- \sum_{k\in\mathcal{M}} \left(\alpha_0(k)+\alpha_3(k)+\alpha_1(k) + \frac{\sigma_3^2(k)}{2} \right) \pi_k.\end{aligned}$$

The equality above follows from the fact that π is the marginal distribution of the third component in $\boldsymbol{\nu}_\theta$, which is because π is the unique invariant measure of $\alpha(t)$. In particular, when $\theta = 0$, we denote

$$\begin{aligned}\lambda :=\lambda_0 =& \sum_{k\in\mathcal{M}} \int_{\mathbb{R}_+^2} \left(g(s+\beta v, 0, k) - (\alpha_0(k)+\alpha_3(k)+\alpha_1(k)) - \frac{\sigma_3^2(k)}{2} \right)\\ &\bar{\mu}_0(ds, dv, k).\end{aligned} \tag{12.145}$$

This λ will be shown to be the threshold that characterizes the long-time behavior of the disease. Note that, when $\theta = 0$, (12.143) is the solution to (12.142). Thus, we can consider $\bar{\mu}_0$ as the unique invariant measure of $((S(t), V(t), I(t), R(t)), \alpha(t))$ on the boundary $\mathbb{R}_+^{4,*} \times \mathcal{M}$ (by embedding $\mathbb{R}_+^2 \times \mathcal{M}$ to $\mathbb{R}_+^{4,*} \times \mathcal{M}$).

Lemma 12.31 $\lim_{\theta\to 0} \lambda_\theta = \lambda_0 = \lambda$.

Proof Owing to (12.144), the family $\{\boldsymbol{\nu}_\theta, \theta \in [-1, 1]\}$ is tight. As a result, for any sequence θ_n that converges to 0 as n goes to ∞, we can find a subsequence, still denoted by $\{\theta_n\}_{n=1}^\infty$ for simplicity such that $\boldsymbol{\nu}_{\theta_n}$ converges weakly to a probability measure, denoted by $\bar{\boldsymbol{\nu}}$, as n goes to ∞.

On the other hand, because $\boldsymbol{\nu}_\theta$ is the unique invariant measure of (12.143), for any (s, v)-twice differentiable function $U((s, v, k))$ with compact support, we have

$$\sum_{k\in\mathcal{M}} \int_{\mathbb{R}_+^2} \mathcal{L}_\theta U((s, v, k)) \bar{\mu}_\theta(ds, dv, k) = 0.$$

Because of the weak convergence of $\{\boldsymbol{\nu}_{\theta_n}\}_{n=1}^{\infty}$ to $\bar{\boldsymbol{\nu}}$, we have

$$\lim_{n\to\infty}\sum_{k\in\mathcal{M}}\int_{\mathbb{R}^2_+}\mathcal{L}_{\theta_n}U((s,v,k))\bar{\mu}_{\theta_n}(ds,dv,k)=0, \tag{12.146}$$

for any (s, v)-twice differentiable function $U((s, v, k))$ with compact support.

Note that $\mathcal{L}_{\theta_n}U((s,v,k)) = \mathcal{L}U((s,v,k))+\theta_n\alpha_2(k)U_s((s,v,k))$ which implies

$$\begin{aligned}
&\sum_{k\in\mathcal{M}}\int_{\mathbb{R}^2_+}\mathcal{L}U((s,v,k))\bar{\boldsymbol{\nu}}(ds,dv,k)\\
&\quad=\lim_{n\to\infty}\sum_{k\in\mathcal{M}}\int_{\mathbb{R}^2_+}\mathcal{L}U((s,v,k))\bar{\mu}_{\theta_n}(ds,dv,k)\\
&\quad=\lim_{n\to\infty}\sum_{k\in\mathcal{M}}\int_{\mathbb{R}^2_+}\mathcal{L}_{\theta_n}U((s,v,k))\bar{\mu}_{\theta_n}(ds,dv,k)\\
&\qquad-\lim_{n\to\infty}\sum_{k\in\mathcal{M}}\int_{\mathbb{R}^2_+}\theta_n\alpha_2(k)U_s((s,v,k))\bar{\mu}_{\theta_n}(ds,dv,k)\\
&\quad=0.
\end{aligned} \tag{12.147}$$

The last equality is due to (12.146), the boundedness of $U_s(s, v, k)$, and the fact that $\lim_{n\to\infty}\theta_n = 0$.

Because $\sum_{k\in\mathcal{M}}\int_{\mathbb{R}^2_+}\mathcal{L}U((s,v,k))\bar{\boldsymbol{\nu}}(ds,dv,k) = 0$ for any (s, v)-twice differentiable function $U((s, v, k))$ with compact support, $\bar{\boldsymbol{\nu}}$ must be an invariant measure of (12.142). Due to the uniqueness, $\bar{\boldsymbol{\nu}}$ is identical to $\boldsymbol{\nu}$. Therefore, we have showed that any weak limit of $\boldsymbol{\nu}_\theta$ as $\theta \to 0$ must be $\boldsymbol{\nu}$. This and the tightness of $\{\boldsymbol{\nu}_\theta, \theta \in [-1, 1]\}$ implies that $\boldsymbol{\nu}_\theta$ converges weakly to $\boldsymbol{\nu}$ as $\theta \to 0$. By the same arguments as in Lemma 12.5, the limit $\lim_{\theta\to 0}\lambda_\theta = \lambda$ follows directly from this weak convergence of $\boldsymbol{\nu}_\theta$ to $\boldsymbol{\nu}$ together with (12.144) and the linear growth rate of $g(s + \beta v, 0, k)$.

□

Now, we state our main results of this section.

Theorem 12.32 *If $\lambda > 0$, there exists a unique invariant measure $\bar{\mu}^*$ of $\{\mathbf{Z}(t)\}$ on $\mathbb{R}^{4,\circ}_+ \times \mathcal{M}$. In addition, the transition probability converges to the unique invariant measure exponentially fast. That is, there is a constant $m > 0$ such that*

$$\lim_{t\to\infty} e^{mt}\|P(t,\mathbf{z},\cdot)-\bar{\mu}^*(\cdot)\|_{TV}=0,\ \forall \mathbf{z}\in\mathbb{R}^{4,*}_+\times\mathcal{M},$$

where $P(t, \mathbf{z}, \cdot)$ is the transition probability of the process $\{\mathbf{Z}(t)\}$.

Theorem 12.33 *If $\lambda < 0$, $I(t)$ converges to 0 exponentially fast and $R(t)$ converges to 0 almost surely. More precisely,*

$$\mathbb{P}_{\mathbf{z}}\left\{\lim_{t\to\infty}\frac{\ln I(t)}{t}=\lambda,\ \textit{and}\ \lim_{t\to\infty}R(t)=0\right\}=1. \tag{12.148}$$

In addition, $(S(t), V(t), \alpha(t))$ *converges (weakly) to the invariant probability measure of the solution process to the couple* (12.142) *and* (12.2).

Algebraic Representation of λ Before proving the main results, let us provide some computable representation of the threshold λ. When $g(s+\beta v, 0, k)$ is a linear function, that is, $g(s+\beta v, 0, k) = b(k)(s+\beta v)$ with $b(\cdot) : \mathcal{M} \to (0, \infty)$, we will construct an algebraic representation for λ. This representation is very useful from a computational point of view since we can compute λ by solving a system of linear equations.

Lemma 12.34 *Assume function* g *is a linear function, i.e.,* $g(s+\beta v, 0, k) = b(k)(s+\beta v)$*. Let* $C := (c_1(1), c_1(2), \ldots, c_1(m_0), c_2(1), c_2(2), \ldots, c_2(m_0))^\top$ *be the unique solution to the linear system:*

$$\begin{cases} b(k) - (\alpha_0(k) + \widetilde{p}(k))\, c_1(k) + \widetilde{p}(k)c_2(k) + \sum\limits_{l\in\mathcal{M}} \gamma_{kl}c_1(l) = 0, \\ \beta b(k) + \gamma(k)c_1(k) - (\alpha_0(k) + \gamma(k))\, c_2(k) + \sum\limits_{l\in\mathcal{M}} \gamma_{kl}c_2(l) = 0, \quad k = 1, \ldots, m_0. \end{cases} \tag{12.149}$$

We have

$$\lambda = \sum_{k\in\mathcal{M}}\left[c_1(k)(1-\widetilde{q}(k))A(k) - \left(\alpha_0(k)+\alpha_3(k)+\alpha_1(k)+\frac{\sigma_3^2(k)}{2}\right)\right]\pi_k.$$

Proof The system (12.149) can be written in the following form:

$$\mathbf{A}C = B, \tag{12.150}$$

where $B = (b(1), b(2), \ldots, b(m_0), \beta b(1), \beta b(2) \ldots, \beta b(m_0))^\top$, and

$$\mathbf{A} = \begin{bmatrix} \check{p}(1) & \ldots & -\gamma_{1m_0} & -\widetilde{p}(1) & \ldots & 0 \\ \vdots & \vdots & \vdots & \vdots & \vdots & \vdots \\ -\gamma_{m_0 1} & \ldots & \check{p}(m_0) & 0 & \ldots & -\widetilde{p}(m_0) \\ -\gamma(1) & \ldots & 0 & \check{\gamma}(1) & \ldots & -\gamma_{1m_0} \\ \vdots & \vdots & \vdots & \vdots & \vdots & \vdots \\ \mathbf{0} & \ldots & -\gamma(m_0) & -\gamma_{m_0 1} & \ldots & \check{\gamma}(m_0) \end{bmatrix}.$$

where $\check{p}(i) = \alpha_0(i)+\widetilde{p}(i)-\gamma_{ii}$, $\check{\gamma}_i = \alpha_0(i)+\gamma(i)-\gamma_{ii}$, $i = 1, \cdots, n$ Since $-\gamma_{ii} = \sum_{j\neq i}\gamma_{ij}$, it is obvious that matrix $\mathbf{A} = [a_{ij}]_{2m_0\times 2m_0}$ is diagonally dominant, i.e.,

$|a_{ii}| \ge \sum_{j \neq i, j=1}^{2m_0} |a_{ij}|$ for any $i = 1, \ldots, m_0$. It is well-known that a diagonally dominant matrix is non-singular. Thus, there exists a unique solution C to (12.150).

Denote by $\mathcal{L}_0$ the operator associated with the solution process of (12.142) and $(\widetilde{S}^0, \widetilde{V}^0)$ be the solution to (12.142). It is readily seen that if we let $U((s, v, k)) = c_1(k)s + c_2(k)v$, then

$$\begin{aligned}\mathcal{L}_0 U(s, v, k) =& c_1(k)\left[(1 - \widetilde{q}(k))A(k) + \alpha_3(k)\theta - (\alpha_0(k) + \widetilde{p}(k))\, s + \gamma(k)v\right] \\ &+ c_2(k)\left[\widetilde{q}(k)A(k) + \widetilde{p}(k)s - (\alpha_0(k) + \gamma(k))\, v\right] \\ &+ \Big(\sum_l \gamma_{kl} c_1(l)\Big)s + \Big(\sum_l \gamma_{kl} c_2(l)\Big)v \\ =& c_1(k)\left[(1 - \widetilde{q}(k))A(k)\right] - b(k)s - \beta b(k)v.\end{aligned}$$

Then, by Dynkin's formula 2.10, we have

$$\begin{aligned}&\mathbb{E}_{s,v,k} U(\widetilde{S}^0(t), \widetilde{V}^0(t), \alpha(t)) - U((\widetilde{S}^0(0), \widetilde{V}^0(0), \alpha(0)) \\ =&\mathbb{E}_{s,v,k} \int_0^t \mathcal{L}_0 U((\widetilde{S}^0(u), \widetilde{V}^0(u), \alpha(u))du \\ =&\mathbb{E}_{s,v,k} \int_0^t c_1(\alpha(u))\left[(1 - \widetilde{q}(\alpha(u)))A(\alpha(u))\right] du \\ &- \mathbb{E}_{s,v,k} \int_0^t b(\alpha(u))\left(\widetilde{S}^0(u) + \beta \widetilde{V}^0(u)\right) du.\end{aligned} \tag{12.151}$$

Moreover, one has

$$\begin{aligned}\lim_{t\to\infty} \frac{1}{t}\mathbb{E}_{s,v,k} &\int_0^t c_1(\alpha(u))\left[(1 - \widetilde{q}(\alpha(u)))A(\alpha(u))\right] du \\ &= \sum_{k\in\mathcal{M}} c_1(k)(1 - \widetilde{q}(k))A(k)\pi_k,\end{aligned} \tag{12.152}$$

$$\lim_{t\to\infty} \frac{1}{t}\mathbb{E}_{s,v,k} \int_0^t b(\alpha(u))\left(\widetilde{S}^0(u) + \beta \widetilde{V}^0(u)\right) du = \sum_{k\in\mathcal{M}} \int_{\mathbb{R}^2_+} b(k)(s + \beta v)\nu(ds, dv), \tag{12.153}$$

$$\lim_{t\to\infty} \frac{\mathbb{E}_{s,v,k} U(\widetilde{S}^0(t), \widetilde{V}^0(t), \alpha(t)) - U(\widetilde{S}^0(0), \widetilde{V}^0(0), \alpha(0))}{t} = 0. \tag{12.154}$$

Plugging (12.152), (12.153), and (12.154) in (12.151), we have

$$\sum_{k\in\mathcal{M}} c_1(k)(1 - \widetilde{q}(k))A(k)\pi_k = \sum_{k\in\mathcal{M}} \int_{\mathbb{R}^2_+} b(k)(s + \beta v)\boldsymbol{\nu}_0(ds, dv, k),$$

which and the formula (12.145) of λ imply

$$\lambda = \lambda_0 = \sum_{k\in\mathcal{M}} c_1(k)(1-\widetilde{q}(k))A(k)\pi_k - \sum_{k\in\mathcal{M}} \left(\alpha_0(k)+\alpha_3(k)+\alpha_1(k)+\frac{\sigma_3^2(k)}{2}\right)\pi_k.$$

The proof is complete.

□

12.3.2.3 Proof of Theorem 12.32

Since $g(u,i,k)$ is continuous at $i=0$ uniformly in u, there exists $\delta_1>0$ such that

$$\sup_{u\geq 0, k\in\mathcal{M}, 0\leq i\leq\delta_1} |g(u,i,k)-g(u,0,k)| < \frac{1}{4}\lambda. \tag{12.155}$$

Since the function $g(u,0,k)$ is increasing in u, and

$$\sum_{k\in\mathcal{M}}\int_{\mathbb{R}_+^2}\left(g(s+\beta v,0,k)-(\alpha_0(k)+\alpha_3(k)+\alpha_1(k))-\frac{\sigma_2^2(k)}{2}\right)\bar{\mu}(ds,dv,k)=\lambda,$$

there exists an $H_1>0$ such that

$$\begin{aligned}&\sum_{k\in\mathcal{M}}\pi_k\left(g(H_1,0,k)-(\alpha_0(k)+\alpha_3(k)+\alpha_1(k))-\frac{\sigma_3^2(k)}{2}\right)\\ &=\inf_{u\geq H_1}\sum_{k\in\mathcal{M}}\pi_k\left(g(u,0,k)-(\alpha_0(k)+\alpha_3(k)+\alpha_1(k))-\frac{\sigma_3^2(k)}{2}\right)>\frac{3}{4}\lambda.\end{aligned} \tag{12.156}$$

In view of (12.155) and (12.156), we have

$$\begin{aligned}&\sum_{k\in\mathcal{M}}\hat{h}_k\pi_k\geq\frac{1}{2}\lambda \text{ where}\\ &\quad\hat{h}_k \coloneqq \inf_{u\geq H_1, 0\leq i\leq\delta_1}\left(g(u,i,k)-(\alpha_0(k)+\alpha_3(k)+\alpha_1(k))-\frac{\sigma_3^2(k)}{2}\right).\end{aligned} \tag{12.157}$$

Let

$$\hat{\lambda}_1 := \frac{1}{3}\sum_{k\in\mathcal{M}}\hat{h}_k\pi_k. \tag{12.158}$$

Because $\sum_{k\in\mathcal{M}}(3\hat{\lambda}_1 - \hat{h}_k)\pi_k = 0$, we deduce from the Fredholm alternative that there exist $\widetilde{c}_k \geq 0$, $k \in \mathcal{M}$ such that

$$\sum_{\ell\in\mathcal{M}} q_{k\ell}\widetilde{c}_\ell = 3\hat{\lambda}_1 - \hat{h}_k. \tag{12.159}$$

We have the following lemma.

Lemma 12.35 *For any $\rho_1 \in (0, 1)$ satisfying*

$$\rho_1\left(\frac{\sigma_3^2(k)}{2}(1-\rho_1\widetilde{c}_k) + (-3\hat{\lambda}_1 + \hat{h}_k)\right) < 1 \text{ and } \frac{1}{2} \leq 1 - \rho_1\widetilde{c}_k, \text{ for all } k \in \mathcal{M}, \tag{12.160}$$

define

$$V_1(i,k) = (1-\rho_1\widetilde{c}_k)i^{-\rho_1}.$$

We have the following conclusions:

- $[\mathcal{L}V_1](\mathbf{z}) \leq -\rho_1\hat{\lambda}_1 V_1(i,k)$ *for any* $\mathbf{z} = (s, v, i, r, k)$ *with* $s \geq H_1$, $i \leq \delta_1$.
- *Let* $\hat{c}_1 := \max_{k\in\mathcal{M}}\left\{\alpha_0(k) + \alpha_3(k) + \alpha_1(k) + \frac{\sigma_3^2(k)}{2} + 2\sum_{\ell\in\mathcal{M}}|q_{k\ell}|\right\}$, *then* $[\mathcal{L}V_1](\mathbf{z}) \leq \rho_1\hat{c}_1 V_1(i,k)$ *for any* $\mathbf{z} \in \mathbb{R}_+^{4,*} \times \mathcal{M}$. *As a result,*

$$\mathbb{E}_{\mathbf{z}} V_1(I(\tau), \alpha(\tau))e^{-\rho_1\hat{c}_1\tau} \leq V_1(i,k)$$

for any initial condition $\mathbf{z} \in \mathbb{R}_+^{4,*} \times \mathcal{M}$ *and any bounded stopping time* τ.

Proof We have the following estimate:

$$\begin{aligned}
(\mathcal{L}V_1)(\mathbf{z}) =& -\rho_1 V_1(i,k)\left(g(s+\beta v, i, k) - (\alpha_0(k) + \alpha_3(k) + \alpha_1(k)) - \frac{\sigma_3^2(k)}{2}\right) \\
&+ \rho_1^2\frac{\sigma_3^2(k)}{2}V_1(i,k) + \sum_{\ell\in\mathcal{M}} q_{k\ell}V_1(i,\ell) \\
\leq& -\rho_1 V_1(i,k)\hat{h}_k + \rho_1^2\frac{\sigma_3^2(k)}{2}V_1(i,k) + \sum_{\ell\in\mathcal{M}} q_{k\ell}V_1(i,\ell),
\end{aligned} \tag{12.161}$$

where $\hat{h}_k$ is defined as in (12.157). Note that

$$
\begin{aligned}
\sum_{\ell\in\mathcal{M}} q_{k\ell}V_1(i,\ell) =& \sum_{\ell\in\mathcal{M}} q_{k\ell}(1-\rho_1\widetilde{c}_\ell)i^{-\rho_1} \\
=& -\rho_1 i^{-\rho_1}\sum_{\ell\in\mathcal{M}} q_{k\ell}\widetilde{c}_\ell \\
=&(-3\hat{\lambda}_1+\hat{h}_k)\rho_1 i^{-\rho_1} \\
=&(-3\hat{\lambda}_1+\hat{h}_k)\rho_1 V_1(i,k)+(-3\hat{\lambda}_1+\hat{h}_k)\rho_1^2 i^{-\rho_1}.
\end{aligned}
\tag{12.162}
$$

Plugging (12.162) into (12.161) and applying (12.160), we get

$$
\begin{aligned}
(\mathcal{L}V_1)(\mathbf{z}) \le& \rho_1^2\left(\frac{\sigma_3^2(k)}{2}(1-\rho_1\widetilde{c}_k)+(-3\hat{\lambda}_1+\hat{h}_k)\right)i^{-\rho_1}-3\hat{\lambda}_1\rho_1 V_1(i,k) \\
\le& \hat{\lambda}_1\rho_1 i^{-\rho_1}-3\hat{\rho}_1\lambda_1 V_1(i,k) \\
\le& \hat{\lambda}_1\rho_1\left(\frac{1}{1-\rho_1\widetilde{c}_k}-3\right)V_1(i,k) \\
\le& -\rho_1\hat{\lambda}_1 V_1(i,k) \text{ for any } \mathbf{z}=(s,i,v,r) \text{ with } s\ge H_1, i\le\delta_1.
\end{aligned}
\tag{12.163}
$$

Therefore, the first assertion is proved.

On the other hand, because

$$
\sum_{\ell\in\mathcal{M}} q_{k\ell}V_1(i,\ell)=\sum_{\ell\in\mathcal{M}} q_{k\ell}\frac{1-\rho_1\widetilde{c}_k}{1-\rho_1\widetilde{c}_\ell}V_1(i,k)\le 2\sum_{\ell\in\mathcal{M}}|q_{k\ell}|V_1(i,k),
$$

it follows from (12.161) that

$$
[\mathcal{L}V_1](\mathbf{z})\le\rho_1\hat{c}_1V_1(i,k) \text{ for any } \mathbf{z}\in\mathbb{R}_+^{4,*}\times\mathcal{M}. \tag{12.164}
$$

We deduce from (12.164) and Dynkin's formula that

$$
\mathbb{E}_{\mathbf{z}}V_1(I(\tau),\alpha(\tau))e^{-\rho_1\hat{c}_1\tau}\le V_1(i,k) \text{ for any } \mathbf{z}\in\mathbb{R}_+^{4,*}\times\mathcal{M}, \tag{12.165}
$$

and any bounded stopping time τ. The proof is complete. □

Let $n^*>0$ be such that $\hat{c}_1-(n^*-1)\hat{\lambda}_1<0$, where $\hat{c}_1$ is as in Lemma 12.35 and $\hat{\lambda}_1$ is defined in (12.158).

Lemma 12.36 *For any $H>0$, the set $B_H:=[0,H]^2\times[H^{-1},H]\times[0,H]\times\mathcal{M}$ is petite with respect to the Markov chain $\{(\mathbf{Z}(nn^*T^*)), n\in\mathbb{N}\}$.*

Sketch of Proof Note that, there exists a compact subset K_H of $\mathbb{R}_+^{4,\circ}\times\mathcal{M}$ such that

$$\mathbb{P}_{\mathbf{z}}\{\mathbf{Z}(1) \in K_H\} \geq \frac{1}{2}, z \in B_H.$$

The reason is that the drift of $S(t)$ (resp. $V(t)$) at $s = 0$ (resp. $v = 0$) is positive. Thus, in a short time, the process will enter a compact subset of $\mathbb{R}_+^{4,\circ} \times \mathcal{M}$ with a large probability. See 12.23 for a similar proof. Since the diffusion is nondegenerate, K_H is petite, see 12.6. For a more detailed proof, we refer to the proof of [220, Lemma 5.4].

□

Lemma 12.37 *There are $T^* > \max_{k \in \mathcal{M}} \frac{8\widetilde{c}_k}{\lambda}$ sufficiently large and $\rho_1 > 0$, $\delta_2 > 0$ sufficiently small such that for any $\mathbf{z} = (s, v, i, r, k) \in \mathbb{R}_+^4 \times \mathcal{M}$ with $0 \leq i \leq \delta_2$, $0 \leq s + \beta v \leq H_1$,*

$$\mathbb{E}_{\mathbf{z}} V_1(I(T), \alpha(T)) \leq e^{-\frac{1}{8}\rho_1 \lambda T} V_1(i, k), \ \forall T \in [T^*, n^* T^*];$$

where V_1 is defined as in Lemma 12.35.

Proof To simplify notation, let

$$h(s+\beta v, 0, k) := g(s+\beta v, 0, k) - (\alpha_0(k)+\alpha_3(k)+\alpha_1(k)) - \frac{\sigma_3^2(k)}{2}, \quad (12.166)$$

Because of

$$\lambda = \sum_{k \in \mathcal{M}} \int_{\mathbb{R}_+^2} h(s + \beta v, 0, k)\bar{\mu}_0(ds, dv, k),$$

there exists $H_2 > 0$ such that

$$\sum_{k \in \mathcal{M}} \int_{\mathbb{R}_+^2} \hat{h}_2(s + \beta v, 0, k)\bar{\mu}_0(ds, dv, k) > \frac{3}{4}\lambda, \quad (12.167)$$

where

$$\hat{h}_2(s + \beta v, i, k) := H_2 \wedge h(s + \beta v, i, k).$$

Define the occupation measure in $\mathbb{R}_+^4 \times \mathcal{M}$

$$\Pi_{\mathbf{z}}^t(\cdot) := \frac{1}{t}\mathbb{E}_{\mathbf{z}} \int_0^t I_{\{(\mathbf{Z}(s)) \in \cdot\}} ds, \quad (12.168)$$

then Lemma 12.5 claims that for a bounded sequence $\{\mathbf{z}_n\} \subset \mathbb{R}_+^4 \times \mathcal{M}$ and an increasing unbounded sequence $\{t_n\}$, we have that any weak limit of $\{\Pi_{\mathbf{z}_n}^{t_n}\}$ as $n \to \infty$ must be an invariant probability measure of $\{\mathbf{Z}(t)\}$. Then, from (12.167), by a simple contradiction argument (e.g., see the details in [117, Lemma 4.1]), we can find $T^* = T^*(H) > \max_{k\in\mathcal{M}} \frac{8c_k}{\lambda}$ such that for any $T \geq T^*$, we have that for any $\mathbf{z} \in \mathbb{R}^4_+$ with $0 \leq s + \beta v \leq H_1, i = 0$, and $k \in \mathcal{M}$,

$$\mathbb{E}_{\mathbf{z}}\frac{1}{T}\int_0^T \hat{h}_2(S(t)+\beta V(t), I(t), \alpha(t)))dt \geq \frac{1}{2}\lambda.$$

Because the process $\{\mathbf{Z}(t)\}$ is a Markov-Feller process on $\mathbb{R}^4_+ \times \mathcal{M}$, we can find $\delta_2 \in (0, \delta_1)$ such that for any $\mathbf{z} \in \mathbb{R}^4_+ \times \mathcal{M}$ with $0 \leq s + \beta v \leq H_1, 0 < i < \delta_2$, and $T \in [T^*, n^*T^*]$,

$$\mathbb{E}_{\mathbf{z}}\frac{1}{T}\int_0^T \hat{h}_2(S(t)+\beta V(t), I(t), \alpha(t)))dt \geq \frac{\lambda}{4}, \tag{12.169}$$

With the same arguments used in the proof of Proposition 12.8, there are $\rho_1 > 0$ sufficiently small that

$$\begin{aligned}\mathbb{E}_{\mathbf{z}} I^{-\rho_1}(T) =& i^{-\rho_1}\mathbb{E}_{\mathbf{z}} e^{\rho_1\left(-\int_0^T h(S(u)+\beta V(u), I(u), \alpha(u))du - \int_0^T \sigma_3(\alpha(u))dW_3(u)\right)}\\ \leq& i^{-\rho_1}\exp\left\{-\frac{\lambda\rho_1 T}{4}\right\},\end{aligned}$$

for any $\mathbf{z} \in \mathbb{R}^4_+ \times \mathcal{M}$ with $0 \leq s + \beta v \leq H_1, 0 < i < \delta_2$, and $T \in [T^*, n^*T^*]$. On the other hand, we have

$$(1-\rho_1\widetilde{c}_k) \geq e^{-\frac{\rho_1\lambda T}{8}} \text{ if } \frac{\lambda T}{8} \geq \widetilde{c}_k.$$

Therefore, we deduce that

$$\begin{aligned}\mathbb{E}_{\mathbf{z}} V_1(I(T), \alpha(T)) \leq& \mathbb{E}_{\mathbf{z}} I^{-\rho_1}(T) \leq i^{-\rho_1}\exp\left\{-\frac{\lambda\rho_1 T}{4}\right\}\\ \leq& (1-\rho_1\widetilde{c}_k) i^{-\rho_1}\exp\left\{-\frac{\lambda\rho_1 T}{8}\right\} = V_1(i,k)\exp\left\{-\frac{\lambda\rho_1 T}{8}\right\},\end{aligned} \tag{12.170}$$

for any $\mathbf{z} \in \mathbb{R}^4_+ \times \mathcal{M}$ with $0 \leq s + \beta v \leq H_1, 0 < i < \delta_2$, and $T \in [T^*, n^*T^*]$. The proof is complete.

□

With Lemma 12.37, Theorem 12.32 can be proved by following the arguments in the proof of Theorem 12.9.

12.3.2.4 Proof of Theorem 12.33

By Lemma 12.34, it is readily seen that λ_θ is continuous in θ. Therefore, there exists $\theta_0 > 0$ such that

$$\lambda_{\theta_0} < \frac{\lambda_0}{2} = \frac{\lambda}{2}. \tag{12.171}$$

It is noted that we are considering the case $\lambda < 0$, so $\lambda_{\theta_0} < 0$.

We will establish the following lemma, which shows that if the solution starts from initial points very close to the boundary, $I(t) \to 0$ (exponentially fast) and $R(t) \to 0$ with high probability.

Lemma 12.38 *Let $\theta_0 > 0$ be as in (12.171). For any $\varepsilon > 0$, $H > 0$, there is a constant $\theta_2 > 0$ such that*

$$\mathbb{P}_{\mathbf{z}}\left\{\lim_{t\to\infty}\frac{\ln I(t)}{t} = \lambda < 0 \text{ and } \lim_{t\to\infty} R(t) = 0\right\} \geq 1-\varepsilon,\ \forall \mathbf{z} \in [0,H]^2 \times (0,\theta_2]^2 \times \mathcal{M}. \tag{12.172}$$

Proof We obtain from the ergodicity of $(\widetilde{S}^{\theta_0}(t), \widetilde{V}^{\theta_0}(t), \alpha(t))$ that

$$\begin{aligned}
&\lim_{t\to\infty}\frac{1}{t}\int_0^t h(\widetilde{S}^{\theta_0}(u)+\beta\widetilde{V}^{\theta_0}(u), 0, \alpha(u))du \\
&\quad = \sum_{k\in\mathcal{M}}\int_{\mathbb{R}^2_+} h(s+\beta v, 0, k)\bar{\mu}_{\theta_0}(ds, dv, k) = \lambda_{\theta_0}\ a.s,
\end{aligned} \tag{12.173}$$

where h is as in (12.166), and

$$\begin{aligned}
&\lim_{t\to\infty}\frac{1}{t}\int_0^t \left(\alpha_0(\alpha(u)) + \alpha_2(\alpha(u)) + \frac{\sigma_4^2(\alpha(u))}{2}\right)du \\
&\quad = \sum_{k\in\mathcal{M}}(\alpha_0(k)+\alpha_2(k)+\frac{\sigma_4(k)}{2})\pi_k =: \hat{c}_2.
\end{aligned} \tag{12.174}$$

By the strong law of large numbers for martingales,

$$\lim_{t\to\infty}\frac{1}{t}\int_0^t \sigma_\ell(\alpha(u))dW_\ell(u) = 0 \quad \ell = 1,2,3,4. \tag{12.175}$$

Now, denote

$$\hat{\lambda}_2 := \min\left\{\frac{1}{9}|\lambda_{\theta_0}|, \frac{1}{3}\hat{c}_2\right\}. \tag{12.176}$$

It follows from (12.173) that for any $\varepsilon > 0$, there exists a $\widetilde{T}_1 = \widetilde{T}_1(H, \varepsilon) > 0$ such that $\mathbb{P}_{H,H,k}(\Omega_1) \geq 1 - \frac{\varepsilon}{5}$, where

$$\begin{aligned}\Omega_1 &= \Big\{\omega : \frac{1}{t}\int_0^t h(\widetilde{S}^{\theta_0}_{H,H,k}(u) + \beta \widetilde{V}^{\theta_0}_{H,H,k}(u), 0, \alpha(u))du \leq \lambda_{\theta_0} \\ &\quad + \hat{\lambda}_2 \leq -8\hat{\lambda}_2 \forall t \geq \widetilde{T}_1\Big\}. \end{aligned} \tag{12.177}$$

In the above, the subscript in $\mathbb{P}_{H,H,k}$, $(\widetilde{S}^{\theta_0}_{H,H,k}(t), \widetilde{V}^{\theta_0}_{H,H,k}(t))$ shows the initial value of $(\widetilde{S}^{\theta_0}(t), \widetilde{V}^{\theta_0}(t), \alpha(t))$.

Because of the comparison theorem [136], we have $S_{\mathbf{z}}(u) \leq \widetilde{S}^{\theta}_{(H,H),k}(u)$, $V_{\mathbf{z}}(u) \leq \widetilde{V}^{\theta}_{(H,H),i}(u)$, $\forall 0 \leq u \leq \widetilde{\tau}$ almost surely if $\mathbf{z} = (s, v, i, r, k)$ satisfies $s \leq H, v \leq H$. This and (12.174), (12.175) deduce that there exists $\widetilde{T}_2 = \widetilde{T}_2(\varepsilon) > 0$ such that $\mathbb{P}(\Omega_2) \geq 1 - \frac{\varepsilon}{5}$ and $\mathbb{P}(\Omega_3) \geq 1 - \frac{\varepsilon}{5}$ where

$$\Omega_2 = \Big\{\omega \in \Omega : \frac{1}{t}\Big|\int_0^t \sigma_\ell(\alpha(u))dW_\ell(u)\Big| \leq \hat{\lambda}_2,\ \ell = 1, 2, 3, 4\ \forall t \geq \widetilde{T}_2\Big\}, \tag{12.178}$$

and

$$\begin{aligned}\Omega_3 = \Big\{\omega \in \Omega : \hat{c}_2 - \hat{\lambda}_2 < \frac{1}{t}\int_0^t (\alpha_0(\alpha(u) + \alpha_2(\alpha(u))) + \frac{\sigma_4(\alpha(u))}{2})du < \hat{c}_2 + \hat{\lambda}_2 \\ \forall t \geq \widetilde{T}_2\Big\}. \end{aligned} \tag{12.179}$$

Let $\widetilde{T} = \max\{\widetilde{T}_1, \widetilde{T}_2\}$. On the finite interval $[0, \widetilde{T}]$, thanks to (12.141), we can find a sufficiently large $M > 0$ satisfying that $\mathbb{P}(\Omega_4) \geq 1 - \frac{\varepsilon}{5}$, if $\mathbf{z} \in [0, H]^2 \times [0, 1]^2 \times \mathcal{M}$ where

$$\begin{aligned}\Omega_4 = \Big\{\omega \in \Omega : &\int_0^{\widetilde{T}} h(S(u) + \beta V(u), I(u), \alpha(u))du \\ &+ \sup_{t \in [0, \widetilde{T}], \ell = 3,4} \Big|\int_0^t \sigma_\ell(\alpha(u))W_\ell(u)du\Big| < M\Big\}. \end{aligned} \tag{12.180}$$

It is noted that Ω_4 (and other Ω_i) depends on initial values of the solution, and it would be labeled as $\Omega_4^{\mathbf{z}}$ to indicate the corresponding initial values. However, to simplify notation, we will remove this superscript. Moreover, Doob's inequality allows us to choose M sufficiently large such that

$$\mathbb{P}(\Omega_5) \geq 1 - \frac{\varepsilon}{5}, \text{ where}$$

$$\Omega_5 = \left\{\omega \in \Omega : \left| \int_0^t \sigma_\ell(\alpha(u)) W_\ell(u) du \right| \le \frac{M}{2}, \forall t \in [0, \widetilde{T}] \text{ and } \ell = 1, 2, 3, 4 \right\}. \tag{12.181}$$

Denote

$$\check{\gamma} := \max_{k \in \mathcal{M}} \alpha_3(k) \text{ and } \hat{c}_3 := \max_{k \in \mathcal{M}} \left\{ \alpha_0(k) + \alpha_2(k) + \frac{\sigma_4^2(k)}{2} \right\}.$$

Since $h(u, i, k) = g(u, i, k) - (\alpha_0(k) + \alpha_3(k) + \alpha_1(k)) - \frac{\sigma_3^2(k)}{2}$ is continuous at $i = 0$ uniformly in u, we can pick $\theta_1 \in \left(0, \frac{\theta_0}{M(1+\check{\gamma} M e^{\hat{c}_3 T + M} T)}\right)$ such that

$$|h(u, i, k) - h(u, 0, k)| \le \hat{\lambda}_2 \text{ whenever } 0 \le i \le \theta_1. \tag{12.182}$$

Next, we pick θ_2 satisfying $0 < \theta_2 < \min\left\{\theta_1 e^{-M}, \theta_0 \left(1 + \frac{\check{\gamma}}{\hat{c}_2 - 4\hat{\lambda}_2}\right)^{-1}\right\}$.

Combining the second equation of (12.139), (12.181), and (12.180) implies that for all $\omega \in \Omega_3$, we have

$$\begin{aligned} I(t) =& I(0) \exp\left\{ \int_0^t h(S(u) + \beta V(u), I(u), \alpha(u)) du + \int_0^t \sigma_3(\alpha(u)) dW_3(u) \right\} \\ \le& I(0) e^M \le \theta_2 e^M < \theta_1 \text{ for any } t \in [0, \widetilde{T}], \text{ if } I(0) \le \theta_2. \end{aligned} \tag{12.183}$$

On the other hand, by a variation of constants formula for linear stochastic differential equations (e.g., [201, Chapter 3]), we have

$$R(t) = \frac{1}{\Phi(t)} \left(R(0) + \int_0^t \alpha_3(\alpha(u)) \Phi(u) I(u) du \right), \tag{12.184}$$

where

$$\Phi(t) = \exp\left\{ \int_0^t \left(\alpha_0(\alpha(u)) + \alpha_2(\alpha(u)) + \frac{\sigma_4^2(\alpha(u))}{2} \right) du - \int_0^t \sigma_4(\alpha(u)) dW_4(u) \right\}.$$

Provided $\mathbf{z} \in [0, H]^2 \times [0, 1]^2 \times \mathcal{M}$, for $t \le \widetilde{T}$ and $\omega \in \Omega_4$, we have from (12.180) that $\frac{1}{\Phi(t)} \le M$ (where M is defined as in (12.180)) and $\Phi(t) \le M e^{\hat{c}_3 t}$, which in combination with (12.183), (12.184), and the definition of θ_1 implies

$$R(t) \le M\left(r + M e^{\hat{c}_3 \widetilde{T}} \check{\gamma} \int_0^t I(u) du \right) \le M(\theta_1 + M e^{\hat{c}_3 \widetilde{T}} \check{\gamma} \widetilde{T} \theta_1) < \theta_0. \tag{12.185}$$

Define the stopping time

$$\widetilde{\zeta} := \inf\{t \geq 0 : R(t) \geq \theta_0 \text{ or } I(t) \geq \theta_1\}. \tag{12.186}$$

Because of (12.183) and (12.185),

$$\widetilde{\zeta} > \widetilde{T} \text{ if } \mathbf{z} \in [0, H]^2 \times [0, \theta_2]^2 \times \mathcal{M}, \text{ and } \omega \in \Omega_4. \tag{12.187}$$

It is noted again that $S_{\mathbf{z}}(t) \leq \widetilde{S}^{\theta_0}_{H,H,k}(t)$, $V_{\mathbf{z}}(t) \leq \widetilde{V}^{\theta_0}_{H,H,k}(t)$ for any $t \leq \widetilde{\zeta}$ given that $\mathbf{z} = (s, v, i, r, k)$ with $s \leq H$, $v \leq H$ (thanks to the comparison theorem [136]). Thus, from (12.183), (12.182) and non-decreasing in u property of $h(u, x, k)$, if $t \leq \widetilde{\zeta}$, one has

$$\begin{aligned}h(S(t) + \beta V(t), I(t), \alpha(t)) &\leq h(S(t) + \beta V(t), 0, \alpha(t)) + \hat{\lambda}_2\\ &\leq h(\widetilde{S}^{\theta_0}_{H,H,k}(t) + \beta \widetilde{V}^{\theta_0}_{H,H,k}(t), 0, \alpha(t)) + \hat{\lambda}_2,\end{aligned}$$

given the initial value of $(\mathbf{Z}(t))$ is in $[0, H]^2 \times [0, \theta_2] \times [0, \infty) \times \mathcal{M}$. As a result,

$$I(t) \leq I(0)e^{\int_0^t \left(h(\widetilde{S}^{\theta_0}_{H,H,k}(u)+\beta \widetilde{V}^{\theta_0}_{H,H,k}(u),0,\alpha(u)\right)du+\hat{\lambda}_2 t+\int_0^t \sigma_3(\alpha(u))dW_2(u)}. \tag{12.188}$$

Combining (12.177), (12.178), (12.183), (12.187), and (12.188) for $\omega \in \bigcap_{j=1}^5 \Omega_j$ and $\mathbf{z} \in [0, H]^2 \times [0, \theta_2]^2 \times \mathcal{M}$, we have $\widetilde{\zeta} > \widetilde{T}$ and that

$$\begin{aligned}I(t) &\leq I(0)e^{\int_0^t g(\widetilde{S}^{\theta_0}_{H,H,k}(u)+\beta \widetilde{V}^{\theta_0}_{H,H,k}(u),0,\alpha(u))du+\hat{\lambda}_2 t+\int_0^t \sigma_3(\alpha(u))dW_3(u)}\\ &\leq I(0)\exp\left\{-8\hat{\lambda}_2 t + 2\hat{\lambda}_2 t\right\} \leq I(0)\exp\{-6\hat{\lambda}_2 t\} < \theta_0,\ t \in [\widetilde{T}, \widetilde{\zeta}).\end{aligned} \tag{12.189}$$

We also have that for any $\widetilde{T} < t \leq \widetilde{\zeta}$ and $\omega \in \bigcap_{j=1}^5 \Omega_j$ and $\mathbf{z} \in [0, H]^2 \times [0, \theta_2]^2 \times \mathcal{M}$ that

$$e^{(\hat{c}_2 - 2\hat{\lambda}_2)t} \leq \Phi(t) \leq e^{(\hat{c}_2 + 2\hat{\lambda}_2)t},$$

which implies

$$\begin{aligned}R(t) =& \frac{1}{\Phi(t)}\left(R(0) + \int_0^t \alpha_3(\alpha(u))\Phi(u)I(u)du\right)\\ \leq& e^{(-\hat{c}_2+2\hat{\lambda})t}\left(r + \check{\gamma}\int_0^t e^{(\hat{c}_2+2\hat{\lambda}_2)u} i e^{-6\hat{\lambda}_2 u}du\right)\\ \leq& e^{(-\hat{c}_2+2\hat{\lambda}_2)t}\left((\theta_2 + \theta_2\frac{\check{\gamma}}{\hat{c}_2 - 2\hat{\lambda}_2}e^{(\hat{c}_2-4\hat{\lambda}_2)t}\right)\\ \leq& \theta_2 + \theta_2\frac{\check{\gamma}}{\hat{c}_2 - 4\hat{\lambda}_2} < \theta_0.\end{aligned} \tag{12.190}$$

Thanks to (12.189) and (12.190), we must have $\widetilde{\zeta} = \infty$ for all $\omega \in \bigcap_{j=1}^5 \Omega_j$, $\mathbf{z} \in [0, H]^2 \times [0, \theta_2]^2 \times \mathcal{M}$. Because of $\widetilde{\zeta} = \infty$, we have from (12.189) and (12.190) again that

$$I(t) \le I(0)e^{-6\hat{\lambda}_2} \text{ and } R(t) \le e^{(-\hat{c}_2+2\hat{\lambda}_2)t} r + i \frac{\check{\gamma}}{\hat{c}_2 - 4\hat{\lambda}_2} e^{-2\hat{\lambda}_2 t},$$

for any $t \ge \widetilde{T}, \omega \in \bigcap_{j=1}^5 \Omega_j, I(0) = i \le \theta_2$. This clearly implies that $\lim_{t\to\infty} I(t) = \lim_{t\to\infty} R(t) = 0 \forall \omega \in \bigcap_{j=1}^5 \Omega_j$, provided $\mathbf{z} \in [0, H]^2 \times [0, \theta_2]^2 \times \mathcal{M}$.

Next, we define a randomized occupation measure

$$\widetilde{\Pi}^t_{\mathbf{z}}(\cdot) := \frac{1}{t}\int_0^t I_{\{\mathbf{Z}(u)\in\cdot\}} du, \quad t > 0,$$

in which, the subscript in $\widetilde{\Pi}^t_{\mathbf{z}}(\cdot)$ indicates the initial condition. As an application of Lemma 12.38 and the comparison $S(t) \le \widetilde{S}^{\theta_0}_{H,H,k}(t)$, $V(t) \le \widetilde{V}^{\theta_0}_{H,H,k}(t)$, $t \ge 0$. On the other hand, with probability 1, any weak-limit of $\widetilde{\Pi}^t_{\mathbf{z}}(\cdot)$ as $t \to \infty$ is an invariant probability measure of the process $\{\mathbf{Z}(t)\}$ on $\mathbb{R}^4_+ \times \mathcal{M}$; see Lemma 12.18. Moreover, it is readily seen that $\bar{\mu}_0$, when regarded as an invariant measure of $\{\mathbf{Z}(t)\}$, is the unique invariant probability measure on $[0, \infty)^2 \times \{0\}^2 \times \mathcal{M}$. Therefore, for almost every $\omega \in \cap_{j=1}^5 \Omega_j$, $\widetilde{\Pi}^t_{\mathbf{z}}(\cdot)$ converges weakly to $\bar{\mu}_0$ as $t \to \infty$. As a result, we obtain from the weak convergence that

$$\begin{aligned} &\lim_{t\to\infty} \frac{1}{t}\int_0^t h(S(u) + \beta V(u), I(u), \alpha(u))du \\ &= \sum_{k\in\mathcal{M}} \int_{\mathbb{R}^2_+} h(s + \beta v, 0, k)\bar{\mu}_0(ds, dv, k) = \lambda, \end{aligned} \tag{12.191}$$

for almost every $\omega \in \cap_{j=1}^5 \Omega_j$. It is noted that the limit (12.191) is valid because of the weak convergence and the uniform integrability

$$\begin{aligned} &\limsup_{t\to\infty} \frac{1}{t}\int_0^t (S(u) + V(u))^{1+p} du \\ &\le \lim_{t\to\infty} \frac{1}{t}\int_0^t (\widetilde{S}^{\theta_0}(u) + \widetilde{V}^{\theta_0}(u))^{1+p} du \\ &= \sum_{k\in\mathcal{M}} \int_{\mathbb{R}^2_+} (s + v)^{1+p} \bar{\mu}_{\theta_0}(ds, dv, k) < \infty \text{ for some small } p > 0; \end{aligned}$$

see, e.g., Lemma 12.17. By (12.183), one has

$$\frac{\ln I(t)}{t} = \frac{\ln I(0)}{t} + \frac{1}{t}\int_0^t h(S(u) + \beta V(u), 0, \alpha(u))du + \frac{1}{t}\int_0^t \sigma_2(\alpha(u))dW_2(u). \tag{12.192}$$

Therefore, letting $t \to \infty$ in (12.192) and because of (12.175) and (12.191), we have that for almost every $\omega \in \cap_{j=1}^{5} \Omega_j$,

$$\lim_{t\to\infty} \frac{\ln I(t)}{t} = \lambda \text{ and } \lim_{t\to\infty} R(t) = 0.$$

As a result, by noticing that $\mathbb{P}(\cap_{j=1}^{5} \Omega_j) \geq 1 - \varepsilon$, the lemma is proved.

□

Now, we are in the position to prove Theorem 12.33

Proof of Theorem 12.33. Because of Lemma 12.38, the process $\{\mathbf{Z}(t)\}$ is transient in $\mathbb{R}_+^{4,\circ} \times \mathcal{M}$. This fact leads to that the process has no invariant probability measure in $\mathbb{R}_+^{4,\circ} \times \mathcal{M}$. As a result, $\bar{\mu}_0$ is the unique invariant probability measure of $\{\mathbf{Z}(t)\}$ in $[0, \infty)^2 \times \{0\}^2 \times \mathcal{M}$. Let $H > 0$ be sufficiently large such that $\bar{\mu}_0(\{s, v \in (0, H)\}) > 1 - \varepsilon$. On the other hand, the process $\{\mathbf{Z}(t)\}$ is tight (due to (12.140)). Therefore, for any initial condition $\mathbf{z} \in \mathbb{R}_+^4 \times \mathcal{M}$ the family of occupation measures $\Pi_{\mathbf{z}}^t(\cdot)$, which is defined in (12.168), is tight in $\mathbb{R}_+^4 \times \mathcal{M}$. Since any weak-limit of $\Pi_{\mathbf{z}}^t$ as $t \to \infty$ must be an invariant probability measure of $\mathbf{Z}(t)$, we have that $\Pi_{\mathbf{z}}^t$ converges weakly to $\bar{\mu}_0$ as $t \to \infty$. As a result, for any $\delta > 0$, there is a constant $\hat{T} > 0$ satisfying that $\Pi_{\mathbf{z}}^{\hat{T}}((0, H)^2 \times (0, \delta)^2 \times \mathcal{M}) > 1 - \varepsilon$, or equivalently,

$$\frac{1}{\hat{T}}\int_0^{\hat{T}} \mathbb{P}_{\mathbf{z}}\{\mathbf{Z}(t) \in (0, H)^2 \times (0, \delta)^2 \times \mathcal{M}\}dt > 1 - \varepsilon.$$

Therefore, $\mathbb{P}_{\mathbf{z}}\{\hat{\zeta} \leq \hat{T}\} > 1 - \varepsilon$, where

$$\hat{\zeta} = \inf\{t \geq 0 : \mathbf{Z}(t) \in (0, H)^2 \times (0, \delta)^2 \times \mathcal{M}\}.$$

We obtain from the strong Markov property and Lemma 12.38 that

$$\mathbb{P}_{\mathbf{z}}\left\{\lim_{t\to\infty} \frac{\ln I(t)}{t} = \lambda < 0 \text{ and } \lim_{t\to\infty} R(t) = 0\right\} \geq 1 - \varepsilon, \ \forall \mathbf{z} \in \mathbb{R}_+^{4,*} \times \mathcal{M}.$$

Because $\varepsilon > 0$ is arbitrary, we obtain (12.148). The proof of Theorem 12.33 is complete.

□

12.4 Notes

The results in this chapter are based on [117, 123, 218]. In recent years, the study on stochastic population dynamics and in particular Kolmogorov systems has witnessed tremendous progress. This chapter provides a stochastic process view point from the point of switching diffusions. The reader may also find additional and related materials in the aforementioned references.

Appendix A

Serving as a handy reference, this appendix collects a number of results that are used in the book. These results include Markov chains, martingales, diffusion processes, weak convergence, hybrid jump diffusions, and other miscellaneous results. In most of the cases, only results are presented. The detailed developments and discussions are omitted, but pointers are provided for further reading. We assume the knowledge of basic probability theory and stochastic processes. These can be found in standard textbooks for a course in probability, for example, Breiman [31] and Chow and Teicher [51], among others.

A.1 Discrete-Time Markov Chains

The theory of stochastic processes is concerned with structures and properties of families of random variables X_t, where t is a parameter belonging to an appropriate index set $\mathbb{T}$. The index set $\mathbb{T}$ may be discrete (i.e., $\mathbb{T} = \{0, 1, \ldots, \}$) or continuous (i.e., an interval of the real line). Stochastic processes associated with these index sets are said to be discrete-time processes and continuous-time processes, respectively; see [150], for instance. The random variables X_t can be either scalars or vectors. For a continuous-time stochastic processes, we use the notation $X(t)$, whereas for a discrete process, we use X_k.

A stochastic process is wide-sense (or covariance) stationary, if it has finite second moments, a constant mean, and a covariance that depends only on the time difference. The ergodicity of a discrete-time stationary sequence $\{X_k\}$ refers to the convergence of the sequence $(X_1 + X_2 + \cdots + X_n)/n$ to its expectation in an appropriate sense; see, for example, Karlin and Taylor [150, Theorem 5.6, p. 487] for a strong ergodic theorem of a stationary process. A stochastic process X_k is adapted to a filtration $\{\mathcal{F}_k\}$, if for each k, X_k is an $\mathcal{F}_k$-measurable random vector.

H.-D. Nguyen et al., *Hybrid Switching Diffusions*, Probability Theory and Stochastic Modelling 63, https://doi.org/10.1007/978-3-031-93303-5

Suppose that $\{\alpha_k\}$ is a stochastic process taking values in $\mathcal{M}$, which is at most countable (i.e., it is either finite $\mathcal{M} = \{1, 2, \ldots, m_0\}$ with m_0 being a finite positive integer or countable $\mathcal{M} = \{1, 2, \ldots\}$). We say that $\{\alpha_k\}$ is a Markov chain if it possesses the Markov property

$$\begin{aligned} p_{ij}^{k,k+1} &= \mathbb{P}(\alpha_{k+1} = j|\alpha_k = i) \\ &= \mathbb{P}(\alpha_{k+1} = j|\alpha_0 = i_0, \ldots, \alpha_{k-1} = i_{k-1}, \alpha_k = i), \end{aligned}$$

for any $i_0, \ldots, i_{k-1}, i, j \in \mathcal{M}$.

Given i, j, if $p_{ij}^{k,k+1}$ is independent of time k, that is, $p_{ij}^{k,k+1} = p_{ij}$, $\{\alpha_k\}$ is said to have stationary transition probabilities. The corresponding Markov chain is said to be stationary or time-homogeneous or temporally homogeneous or simply homogeneous. In this case, denote the transition matrix by $P = (p_{ij})$. Denote the n-step transition matrix by $P^{(n)} = (p_{ij}^{(n)})$, with

$$p_{ij}^{(n)} = \mathbb{P}(\alpha_n = j|\alpha_0 = i).$$

Then $P^{(n)} = P^n$. That is, the n-step transition matrix is simply the matrix P to the nth power. Note that

(a) $p_{ij} \geq 0$, $\sum_j p_{ij} = 1$, and
(b) $(P)^{k_1+k_2} = (P)^{k_1}(P)^{k_2}$, for $k_1, k_2 = 1, 2, \ldots$ This identity is known as the Chapman-Kolmogorov equation.

Suppose that $P = (p_{ij}) \in \mathbb{R}^{m_0 \times m_0}$ is a transition matrix. Then the spectral radius of P is $\rho(P) = 1$; see Section A.7 and also Karlin and Taylor [151, p. 3] for a definition of the spectral radius of a matrix. This implies that all eigenvalues of P are on or inside the unit circle.

For a Markov chain $\{\alpha_k\}$, state j is said to be accessible from state i if $p_{ij}^{(k)} = \mathbb{P}(\alpha_k = j|\alpha_0 = i) > 0$ for some $k \geq 0$. Two states i and j that are accessible from each other are said to communicate with each other and belong to the same equivalence class. A Markov chain is irreducible if all states communicate with each other. For $i \in \mathcal{M}$, let $d(i)$ denote the period of state i (i.e., the greatest common divisor of all $k \geq 1$ such that $\mathbb{P}(\alpha_{k+n} = i|\alpha_n = i) > 0$, define $d(i) = 0$ if $\mathbb{P}(\alpha_{k+n} = i|\alpha_n = i) = 0$ for all k). A Markov chain is aperiodic if each state has period one. In accordance with Kolmogorov's classification of states, a state i is recurrent if, starting from state i, the probability of returning to state i after some finite time is 1. A state is transient if it is not recurrent. Criteria on recurrence can be found in most standard textbooks of stochastic processes or Markov chains. In this book, we consider recurrence of switching diffusions, which, as for its diffusion process counterpart, can be considered as a generalization of Kolmogorov's classifications.

Note that (see Karlin and Taylor [151, p. 4]) if P is a transition matrix for a finite-state Markov chain, the multiplicity of the eigenvalue 1 is equal to the number of recurrent classes associated with P. A row vector $\pi = (\pi_1, \ldots, \pi_{m_0})$ with each

$\pi_i \geq 0$ is called a stationary distribution of $\{\alpha_k\}$ if it is the unique solution to the system of equations

$$\begin{cases} \pi P = \pi, \\ \sum_i \pi_i = 1. \end{cases}$$

As demonstrated in [151, p. 85], for i being in an aperiodic recurrent class, if $\pi_i > 0$, which is the limit of the n-step transition probability of starting in state i and returning to state i at the nth transition as $n \to \infty$, then for all j in the same equivalence class of i, $\pi_j > 0$, and the class is termed positive recurrent or strongly ergodic.

Theorem A.1 *Let $P = (p_{ij}) \in \mathbb{R}^{m_0 \times m_0}$ (with m_0 being a finite positive integer) be the transition matrix of an irreducible aperiodic finite-state Markov chain. Then there exist constants $0 < \lambda < 1$ and $c_0 > 0$ such that*

$$\left| P^k - \overline{P} \right| \leq c_0 \lambda^k \text{ for } k = 1, 2, \ldots,$$

where $\overline{P} = \mathbb{1}_{m_0}\pi$, $\mathbb{1}_{m_0} = (1, \ldots, 1)' \in \mathbb{R}^{m_0 \times 1}$, and $\pi = (\pi_1, \ldots, \pi_{m_0})$ is the stationary distribution of $\{\alpha_k\}$. This implies, in particular,

$$\lim_{k \to \infty} P^k = \mathbb{1}_{m_0}\pi.$$

Suppose $\{\alpha_k\}$ is a discrete-time Markov chain with transition probability matrix P. One of the ergodicity conditions of Markov chains is Doeblin's condition (see Doob [65, Hypothesis D, p. 192]; see also Meyn and Tweedie [211, p. 391]). Suppose that there is a probability measure μ with the property that for some positive integer n, $0 < \delta < 1$, and $\Delta > 0$, $\mu(A) \leq \delta$ implies that $P^n(x, A) \leq 1 - \Delta$ for all $x \in A$. In the above, $P^n(x, A)$ denotes the transition probability starting from x reaches the set A in n steps. Note that if α_k is a finite-state Markov chain that is irreducible and aperiodic, then the Doeblin condition is satisfied.

Given an $m_0 \times m_0$ irreducible transition matrix P and a vector G, consider

$$F(P - I) = G, \tag{A.1}$$

where F is an unknown vector. Note that zero is an eigenvalue of the matrix $P - I$ and the null space of $P - I$ is spanned by $\mathbb{1}_{m_0}$. Then by the Fredholm alternative (see Lemma A.11), (A.1) has a solution if and only if $G\mathbb{1}_{m_0} = 0$, where $\mathbb{1}_{m_0} = (1, \ldots, 1)' \in \mathbb{R}^{m_0 \times 1}$.

Define $Q_c = (P - I \vdots \mathbb{1}_{m_0}) \in \mathbb{R}^{m_0 \times (m_0+1)}$. Consider (A.1) together with the condition $F\mathbb{1}_{m_0} = \sum_{i=1}^{m_0} F_i = \widehat{F}$, which may be written as $FQ_c = G_c$ where $G_c = (G \vdots \widehat{F})$. Because for each t, (A.9) has a unique solution, it follows that $Q_c(t)Q_c'(t)$ is a matrix with full rank; therefore, the equation

$$F[Q_c Q_c'] = G_c Q_c' \tag{A.2}$$

has a unique solution, which is given by $F = G_c Q_c'[Q_c Q_c']^{-1}$.

A.2 Ergodicity and Convergence Rates of Discrete-Time Markov Chains

We present key concepts and results on ergodicity and convergence rates of discrete-time Markov chains. We refer to [211] for a comprehensive treatment as well as [109] for some recent progress on the topic. Let $\mathbf{\Phi} = (\Phi_0, \Phi_1, \ldots)$ be a discrete-time Markov chain with transition kernel $\mathcal{P}$ on a general state space $(E, \mathfrak{E})$, where $\mathfrak{E}$ is a countably generated σ-algebra. That is, $\mathbf{\Phi}$ is a discrete-time stochastic process satisfying

$$\mathbb{P}(\Phi_0 \in A_0, \cdots, \Phi_n \in A_n) = \int_{A_0} \mathbb{P}(dy_0) \int_{A_1} \mathcal{P}(y_0, dy_1) \cdots \int_{A_{n-1}} \mathcal{P}(y_{n-1}, A_n).$$

for any $\mathfrak{E}-$measurable sets $A_0, \cdots, A_n$. If there is a non-trivial σ-finite positive measure φ on $(E, \mathfrak{E})$ such that for any $A \in \mathfrak{E}$ satisfying $\varphi(A) > 0$, we have $\sum_{n=1}^{\infty} \mathcal{P}^n(x, A) > 0,\ x \in E$ where $\mathcal{P}^n$ is the n-step transition kernel of $\mathbf{\Phi}$; then the Markov chain $\mathbf{\Phi}$ is called *φ-irreducible*. It can be shown (see [226]) that if $\mathbf{\Phi}$ is φ-irreducible, then there exists a positive integer d and disjoint subsets $E_0, \ldots, E_{d-1}$ such that for all $i = 0, \ldots, d-1$ and all $x \in E_i$, we have $\mathcal{P}(x, E_j) = 1$ where $j = i+1 \pmod d$ and $\varphi\left(E \setminus \bigcup_{i=0}^{d-1} E_i\right) = 0$. The smallest positive integer d satisfying the above is called the period of $\mathbf{\Phi}$. An *aperiodic* Markov chain is a chain with period $d = 1$. A set $C \in \mathfrak{E}$ is called *petite* if there exists a non-negative sequence $(a_n)_{n\in\mathbb{N}}$ with $\sum_{n=1}^{\infty} a_n = 1$ and a nontrivial positive measure ν on $(E, \mathfrak{E})$ satisfying that $\sum_{n=1}^{\infty} a_n \mathcal{P}^n(x, A) \geq \nu(A),\ x \in C, A \in \mathfrak{E}$.

We have the following theorem on ergodicity and convergence rate which is extracted from [142, Theorem 3.6] and [210, Theorem 6.3].

Theorem A.1 *Let $\mathbf{\Phi}$ be φ-irreducible and aperiodic with a petite set C. Suppose there is a function $V : E \mapsto [1, \infty)$ and positive constant b, c, $0 \leq \alpha \leq 1$ such that*

$$\mathcal{P}V(x) := \int_{y\in E} V(y)\mathcal{P}(x, dy) \leq V(x) - cV^{\alpha}(x) + bI_{\{x\in C\}} \text{ for all } x \in E.$$

Then there exists an invariant probability measure π such that for any β satisfying $1 \leq \beta \leq \frac{1}{1-\alpha}$,

$$\lim_{n\to\infty} (n+1)^{\beta-1} \|\mathcal{P}^n(x, \cdot) - \pi(\cdot)\|_{TV} = 0, \quad \forall x \in E.$$

where $\|\cdot\|_{TV}$ *is the total variation norm. Moreover, if* $\alpha = 1$, *we have*

$$\lim_{n\to\infty} \widetilde{\beta}^n \|\mathcal{P}^n(x,\cdot) - \pi(\cdot)\|_{TV} = 0 \, \forall x \in E$$

for some $\widetilde{\beta} > 1$.

A.3 Continuous-Time Markov Chains

A right-continuous stochastic process with piecewise-constant sample paths is a jump process. Suppose that $\alpha(\cdot) = \{\alpha(t) : t \geq 0\}$ is a jump process defined on $(\Omega, \mathcal{F}, P)$ taking values in $\mathcal{M}$. Then $\{\alpha(t) : t \geq 0\}$ is a continuous-time Markov chain with state space $\mathcal{M}$, if

$$\mathbb{P}(\alpha(t) = i|\alpha(r) : r \leq s) = \mathbb{P}(\alpha(t) = i|\alpha(s)),$$

for all $0 \leq s \leq t$ and $i \in \mathcal{M}$, with $\mathcal{M}$ being either finite or countable.

For any $i, j \in \mathcal{M}$ and $t \geq s \geq 0$, let $p_{ij}(t,s)$ denote the transition probability $\mathbb{P}(\alpha(t) = j|\alpha(s) = i)$, and $P(t,s)$ the matrix $(p_{ij}(t,s))$. We name $P(t,s)$ the transition matrix of the Markov chain $\alpha(\cdot)$ and postulate that

$$\lim_{t\to s^+} p_{ij}(t,s) = \delta_{ij},$$

where $\delta_{ij} = 1$ if $i = j$ and 0 otherwise. It follows that for $0 \leq s \leq \varsigma \leq t$,

$$\begin{aligned}
&p_{ij}(t,s) \geq 0, \; i,j \in \mathcal{M}, \\
&\sum_{j\in\mathcal{M}} p_{ij}(t,s) = 1, \; i \in \mathcal{M}, \\
&p_{ij}(t,s) = \sum_{k\in\mathcal{M}} p_{ik}(\varsigma,s)p_{kj}(t,\varsigma), \; i,j \in \mathcal{M}.
\end{aligned}$$

The last identity is the Chapman-Kolmogorov equation as its discrete-time counterpart. If the transition probability $\mathbb{P}(\alpha(t) = j|\alpha(s) = i)$ depends only on $(t-s)$, then $\alpha(\cdot)$ is said to be time-homogeneous or it is said to have stationary transition probabilities. Otherwise, the process is nonhomogeneous or nonstationary. For time homogeneous Markov chains, we define $p_{ij}(h) := p_{ij}(s+h,s)$ for any $h \geq 0$.

Suppose that $\alpha(t)$ is a continuous-time Markov chain with stationary transition probability $P(t) = (p_{ij}(t))$. It then naturally induces a discrete-time Markov chain. For each $h > 0$, the transition matrix $(p_{ij}(h))$ is the transition matrix of the discrete-time Markov chain $\alpha_k = \alpha(kh)$, which is called an h-skeleton of the corresponding continuous-time Markov chain by Chung; see [52, p. 132].

Definition A.2 (q-Property) A matrix-valued function $Q(t) = (q_{ij}(t))$, for $t \ge 0$, satisfies the q-Property, if

(a) $q_{ij}(t)$ is Borel measurable for all $i, j \in \mathcal{M}$ and $t \ge 0$;
(b) $q_{ij}(t)$ is uniformly bounded. That is, there exists a constant K such that $|q_{ij}(t)| \le K$, for all $i, j \in \mathcal{M}$ and $t \ge 0$;
(c) $q_{ij}(t) \ge 0$ for $j \ne i$ and $q_{ii}(t) = -\sum_{j \ne i} q_{ij}(t)$, $t \ge 0$.

For any real-valued function f on $\mathcal{M}$ and $i \in \mathcal{M}$, write

$$Q(t)f(\cdot)(i) = \sum_{j \in \mathcal{M}} q_{ij}(t)f(j) = \sum_{j \ne i} q_{ij}(t)(f(j) - f(i)).$$

Let us recall the definition of the generator of a Markov chain.

Definition A.3 (Generator) A matrix $Q(t)$, $t \ge 0$, is an infinitesimal generator (or in short, a generator) of $\alpha(\cdot)$ if it satisfies the q-property, and for any bounded real-valued function f defined on $\mathcal{M}$

$$f(\alpha(t)) - \int_0^t Q(\varsigma)f(\cdot)(\alpha(\varsigma))d\varsigma \tag{A.3}$$

is a martingale.

Remark A.4 Motivated by the applications we are interested in, a generator is defined as a matrix satisfying the q-property above, where an additional condition on the boundedness of the entries of the matrix is posed. It naturally connects the Markov chain and martingale problems. Definitions including other classes of matrices may be devised as in Chung [52]. To proceed, we give an equivalent condition for a finite-state Markov chain generated by $Q(\cdot)$.

Lemma A.5 *Let $\mathcal{M} = \{1, \ldots, m_0\}$. Then $\alpha(t)$ taking values in $\mathcal{M}$ for $t \ge 0$, is a Markov chain generated by $Q(t)$ if and only if*

$$\left(I_{\{\alpha(t)=1\}}, \ldots, I_{\{\alpha(t)=m_0\}}\right) - \int_0^t \left(I_{\{\alpha(\varsigma)=1\}}, \ldots, I_{\{\alpha(\varsigma)=m_0\}}\right) Q(\varsigma)d\varsigma \tag{A.4}$$

is a martingale.

Proof For a proof, see Yin and Zhang [303, Lemma 2.4]. □

For any given $Q(t)$ satisfying the q-property, there exists a Markov chain $\alpha(\cdot)$ generated by $Q(t)$. If $Q(t) = Q$, a constant matrix, the idea of Ethier and Kurtz [81] can be utilized for the construction. For time-varying generator $Q(t)$, we need to use the piecewise-deterministic process approach as described in Davis [60], to define the Markov chain $\alpha(\cdot)$.

The discussion below is taken from that of Yin and Zhang [303], which was originated in the work of Davis [60]. Let $0 = \tau_0 < \tau_1 < \cdots < \tau_l < \cdots$ be a sequence of jump times of $\alpha(\cdot)$ such that the random variables τ_1, $\tau_2 - \tau_1$, ..., $\tau_{k+1} - \tau_k$, ... are independent. Let $\alpha(0) = i \in \mathcal{M}$. Then $\alpha(t) = i$ on the interval $[\tau_0, \tau_1)$. The first jump time τ_1 has the probability distribution

$$\mathbb{P}(\tau_1 \in B) = \int_B \exp\left\{\int_0^t q_{ii}(s)ds\right\}(-q_{ii}(t))\,dt,$$

where $B \subset [0, \infty)$ is a Borel set. The post-jump location of $\alpha(t) = j$, $j \neq i$, is given by

$$\mathbb{P}(\alpha(\tau_1) = j|\tau_1) = \frac{q_{ij}(\tau_1)}{-q_{ii}(\tau_1)}.$$

If $q_{ii}(\tau_1)$ is 0, define $\mathbb{P}(\alpha(\tau_1) = j|\tau_1) = 0$, $j \neq i$. Then $\mathbb{P}(q_{ii}(\tau_1) = 0) = 0$. In fact, if $B_i = \{t : q_{ii}(t) = 0\}$, then

$$\begin{aligned}\mathbb{P}(q_{ii}(\tau_1) = 0) &= \mathbb{P}(\tau_1 \in B_i)\\ &= \int_{B_i} \exp\left\{\int_0^t q_{ii}(s)ds\right\}(-q_{ii}(t))\,dt = 0.\end{aligned}$$

In general, $\alpha(t) = \alpha(\tau_l)$ on the interval $[\tau_l, \tau_{l+1})$. The jump time τ_{l+1} has the conditional probability distribution

$$\begin{aligned}&\mathbb{P}(\tau_{l+1} - \tau_l \in B_l|\tau_1, \dots, \tau_l, \alpha(\tau_1), \dots, \alpha(\tau_l))\\ &= \int_{B_l} \exp\left\{\int_{\tau_l}^{t+\tau_l} q_{\alpha(\tau_l)\alpha(\tau_l)}(s)ds\right\}\left(-q_{\alpha(\tau_l)\alpha(\tau_l)}(t + \tau_l)\right)dt.\end{aligned}$$

The post-jump location of $\alpha(t) = j$, $j \neq \alpha(\tau_l)$ is given by

$$\mathbb{P}(\alpha(\tau_{l+1}) = j|\tau_1, \dots, \tau_l, \tau_{l+1}, \alpha(\tau_1), \dots, \alpha(\tau_l)) = \frac{q_{\alpha(\tau_l)j}(\tau_{l+1})}{-q_{\alpha(\tau_l)\alpha(\tau_l)}(\tau_{l+1})}.$$

Theorem A.6 *Suppose that the matrix $Q(t)$ satisfies the q-property for $t \geq 0$. Then the following statements hold:*

(a) *The process $\alpha(\cdot)$ constructed above is a Markov chain.*
(b) *The process*

$$f(\alpha(t)) - \int_0^t Q(\varsigma)f(\cdot)(\alpha(\varsigma))d\varsigma \tag{A.5}$$

is a martingale for any uniformly bounded function $f(\cdot)$ on $\mathcal{M}$. Thus $Q(t)$ is indeed the generator of $\alpha(\cdot)$.

(c) *The transition matrix $P(t, s)$ satisfies the forward differential equation*

$$\begin{cases} \dfrac{\partial P(t,s)}{\partial t} = P(t,s)Q(t), \quad t \ge s, \\ P(s,s) = I, \end{cases} \tag{A.6}$$

where I is the identity matrix.

(d) *Assume further that $Q(t)$ is continuous in t. Then $P(t,s)$ also satisfies the backward differential equation*

$$\begin{cases} \dfrac{\partial P(t,s)}{\partial s} = -Q(s)P(t,s), \quad t \ge s, \\ P(s,s) = I. \end{cases} \tag{A.7}$$

Proof For (a)–(c), see Yin and Zhang [303, Theorem 2.5]. As for (d), see [50, p. 402]. □

Note that frequently, working with $s \in [0,T]$, the backward equations are written slightly differently by using reversed time $\tau = T - s$. In this case, the minus sign in (A.7) disappears. Suppose that $\alpha(t)$, $t \ge 0$, is a Markov chain generated by an $m_0 \times m_0$ matrix $Q(t)$. The notions of irreducibility and quasi-stationary distribution are given next.

Definition A.7 (Irreducibility)

(a) A generator $Q(t)$ is said to be weakly irreducible if, for each fixed $t \ge 0$, the system of equations

$$\begin{cases} \nu(t)Q(t) = 0, \\ \displaystyle\sum_{i=1}^{m_0} \nu_i(t) = 1 \end{cases} \tag{A.8}$$

has a unique solution $\nu(t) = (\nu_1(t), \ldots, \nu_{m_0}(t))$ and $\nu(t) \ge 0$ (i.e., $\nu_i(t) \ge 0$ for each $i = 1, \ldots, m_0$).

(b) A generator $Q(t)$ is said to be irreducible, if for each fixed $t \ge 0$ the system of Eqs. (A.8) has a unique solution $\nu(t)$ and $\nu(t) > 0$.

By $\nu(t) \ge 0$, we mean that for each $i \in \mathcal{M}$, $\nu_i(t) \ge 0$. A similar interpretation holds for $\nu(t) > 0$. It follows from the definitions above that irreducibility implies weak irreducibility. However, the converse is not true. For example, the generator

$$Q = \begin{pmatrix} -1 & 1 \\ 0 & 0 \end{pmatrix}$$

is weakly irreducible, but it is not irreducible because it contains an absorbing state corresponding to the second row in Q. A moment of reflection reveals that for a two-state Markov chain with generator

$$Q = \begin{pmatrix} -\lambda(t) & \lambda(t) \\ \mu(t) & -\mu(t) \end{pmatrix}$$

the weak irreducibility requires only $\lambda(t) + \mu(t) > 0$, whereas the irreducibility requires that both $\lambda(t)$ and $\mu(t)$ be positive. Such a definition is convenient for many applications (e.g., the manufacturing systems mentioned in Khasminskii, Yin, and Zhang [163, p. 292]).

Definition A.8 (Quasi-Stationary Distribution) For $t \geq 0$, $\nu(t)$ is termed a quasi-stationary distribution if it is the unique solution of (A.8) satisfying $\nu(t) \geq 0$.

Remark A.9 In the study of homogeneous Markov chains, stationary distributions play an important role. When we are interested in nonstationary (nonhomogeneous) Markov chains, stationary distributions are replaced by the corresponding quasi-stationary distributions, as defined above.

If $\nu(t) = \nu > 0$, it is a stationary distribution. In view of Definitions A.7 and A.8, if $Q(t)$ is weakly irreducible, then there is a quasi-stationary distribution. Note that the rank of a weakly irreducible $m_0 \times m_0$ matrix $Q(t)$ is $m_0 - 1$, for each $t \geq 0$. The definition above emphasizes the probabilistic interpretation. An equivalent definition using the algebraic properties of $Q(t)$ is provided next. One can verify their equivalence using the Fredholm alternative; see Lemma A.11.

Definition A.10 A generator $Q(t)$ is said to be weakly irreducible if, for each fixed $t \geq 0$, the system of equations

$$\begin{cases} f(t)Q(t) = 0, \\ \displaystyle\sum_{i=1}^{m_0} f_i(t) = 0 \end{cases} \tag{A.9}$$

has only the trivial (zero) solution.

A.4 Fredholm Alternative and Ramification

The Fredholm alternative, which provides a powerful method for establishing existence and uniqueness of solutions for various systems of equations, can be found, for example, in Hutson and Pym [134, p. 184].

Lemma A.11 (Fredholm Alternative) *Let* **B** *be a Banach space and* A *a linear compact operator defined on it. Let* I : **B** $\to$ **B** *be the identity operator. Assume* $\gamma \neq 0$. *Then one of the two alternatives holds:*

(a) *The homogeneous equation* $(\gamma I - A)f = 0$ *has only the zero solution, in which case* $\gamma \in \rho(A)$, *the resolvent set of* A, $(\gamma I - A)^{-1}$ *is bounded, and the inhomogeneous equation* $(\gamma I - A)f = g$ *also has one solution* $f = (\gamma I - A)^{-1}g$, *for each* $g \in \mathbf{B}$.

(b) *The homogeneous equation* $(\gamma I - A)f = 0$ *has a nonzero solution, in which case the inhomogeneous equation* $(\gamma I - A)f = g$ *has a solution if and only if* $\langle g, f^* \rangle = 0$ *for every solution* f^* *of the adjoint equation* $\gamma f^* = A^* f^*$.

Note that in (b) above, $\langle g, f^* \rangle$ is a pairing defined on $\mathbf{B} \times \mathbf{B}^*$ (with $\mathbf{B}^*$ denoting the dual of $\mathbf{B}$). This is also known as an "outer product" (see [134, p. 149]), whose purpose is similar to the inner product in a Hilbert space. If we work with a Hilbert space, this "outer product" is identical to the usual inner product. When one considers linear systems of algebraic equations, the lemma above can be rewritten in a simpler form.

Let B denote an $m_0 \times m_0$ matrix. For any $\gamma \neq 0$, define an operator $A : \mathbb{R}^{m_0 \times m_0} \to \mathbb{R}^{m_0 \times m_0}$ as

$$Ay = y(\gamma I - B).$$

Note that in this case, I is just the $m_0 \times m_0$ identity matrix I. Then the adjoint operator $A^* : \mathbb{R}^{m_0 \times m_0} \to \mathbb{R}^{m_0 \times m_0}$ is

$$A^* x = (\gamma I - B)x.$$

Suppose that b and $y \in \mathbb{R}^{1 \times m_0}$. Consider the system $yB = b$. If the adjoint system $Bx = 0$ where $x \in \mathbb{R}^{m_0 \times 1}$ has only the zero solution, then $yB = b$ has a unique solution given by $y = bB^{-1}$. If $Bx = 0$ has a nonzero solution x, then $yB = b$ has a solution if and only if $\langle b, x \rangle = 0$.

Suppose that the generator Q of a continuous-time Markov chain $\alpha_1(t)$ is a constant matrix and is irreducible. Then the rank of Q is $m_0 - 1$. Denote by $\mathcal{R}(Q)$ and $\mathcal{N}(Q)$ the range and the null space of Q, respectively. It follows that $\mathcal{N}(Q)$ is one-dimensional spanned by $\mathbb{1}$ (i.e., $\mathcal{N}(Q) = \text{span}\{\mathbb{1}\}$). As a consequence, the Markov chain $\alpha_1(t)$ with generator Q is ergodic. In what follows, denote the associated stationary distribution by $\nu = (\nu_1, \nu_2, \ldots, \nu_{m_0}) \in \mathbb{R}^{1 \times m_0}$. Consider a linear system of equations

$$Qc = \eta, \tag{A.10}$$

where c and $\eta \in \mathbb{R}^{m_0}$.

Lemma A.12 *The following assertions hold.*

(a) *Equation* (A.10) *has a solution if and only if* $\nu\eta = 0$.
(b) *Suppose that* c_1 *and* c_2 *are two solutions of* (A.10). *Then* $c_1 - c_2 = \gamma_0 \mathbb{1}$ *for some* $\gamma_0 \in \mathbb{R}$.
(c) *Any solution of* (A.10) *can be written as*

$$c = \gamma_0 \mathbb{1} + h_0,$$

where $\gamma_0 \in \mathbb{R}$ *is an arbitrary constant,* $\mathbb{1} = (1, \dots, 1)' \in \mathbb{R}^{m_0}$, *and* $h_0 \in \mathbb{R}^{m_0}$ *is the unique solution of* (A.10) *satisfying* $\nu h_0 = 0$.

Proof We begin with a stochastic representation. First, if (A.10) has a solution c, then $\nu Qc = \nu\eta$. But $\nu Q = 0$, and hence $\nu\eta = 0$. Next, suppose that c_1 and c_2 are two solutions of (A.10). Define $\widetilde{c} = c_1 - c_2$. Then we have $Q\widetilde{c} = 0$. Thus $\widetilde{c} \in \mathcal{N}(Q) = \text{span}\{\mathbb{1}\}$ and hence it follows that there exists a $\gamma_0 \in \mathbb{R}$ such that $\widetilde{c} = \gamma_0\mathbb{1}$.

Suppose that $\nu\eta = 0$. Using orthogonal decomposition, we can write η as $\eta = b + b_1$, where $b \in \mathcal{R}(Q')$ and $b_1 \in \mathcal{N}(Q)$. Thus, $b_1 = \beta\mathbb{1}$ for some $\beta \in \mathbb{R}$. Thus, η can be written as $\eta = b + \beta\mathbb{1}$. Since $\nu\eta = 0$, we obtain $\beta = -\nu b$. Then solving (A.10) is equivalent to that of

$$Qc = b - \nu b\mathbb{1}. \tag{A.11}$$

Denote by $\mathbb{E}_i$ the conditional expectation corresponding to the conditional probability $\mathbf{P}_i(\cdot) := \mathbb{P}(\cdot\,|\alpha_1(0) = i)$, and define a column vector $h = (h_1, h_2, \dots, h_{m_0})' \in \mathbb{R}^{m_0}$ by

$$h_i = \mathbb{E}_i \int_0^\infty (\nu b - b_{\alpha_1(t)})dt, \quad i \in \mathcal{M}. \tag{A.12}$$

It is readily seen that

$$\begin{aligned}
\mathbb{E}_i b_{\alpha_1(t)} &= \sum_{j=1}^{m_0} b_j \mathbb{P}(\alpha_1(t) = j|\alpha_1(0) = i) \\
&= \sum_{j=1}^{m_0} p_{ij}(t) b_j \to \sum_{j=1}^{m_0} \nu_j b_j = \nu b \text{ as } t \to \infty.
\end{aligned}$$

Moreover, because $\alpha_1(t)$ has a finite state space, the convergence above takes place exponentially fast; see [303, Appendix] and references therein. Thus, h is well defined. We proceed to show that h, in fact, is a solution of (A.11).

By direct calculation, it is seen that

$$\begin{aligned}
Qh &= \int_0^\infty (Q\nu b\mathbb{1} - QP(t)b)dt = -\int_0^\infty QP(t)bdt \\
&= -\int_0^\infty \frac{dP(t)}{dt} bdt = -P(t)b\Big|_0^\infty = -\mathbb{1}\nu b + b,
\end{aligned}$$

where $P(t) = (p_{ij}(t))$ is the transition matrix satisfying the Kolmogorov backward equation $(d/dt)P(t) = QP(t)$ with $P(0) = I$ and $\lim_{t\to\infty} P(t) = \mathbb{1}\nu$. Thus, the vector h satisfies Eq. (A.11) and hence (A.10). By using the result proved earlier, any solution of (A.10) can be represented by $c = h + \gamma\mathbb{1}$ for some $\gamma \in \mathbb{R}$.

Finally, we verify that any solution of (A.10) can be written as $c = h_0 + \alpha \mathbb{1}$ for $\alpha \in \mathbb{R}$, where h_0 is the unique solution of (A.10) satisfying $\nu h_0 = 0$. In fact, we have shown that h defined in (A.12) solves (A.10) and any solution of (A.10) can be represented by $c = h + \gamma \mathbb{1}$ for some $\gamma \in \mathbb{R}$. Now let $h_0 = h - \nu h \mathbb{1} \in \mathbb{R}^{m_0}$. Then we have

$$Qh_0 = Q(h - \nu h \mathbb{1}) = Qh - \nu h Q \mathbb{1} = Qh = \eta.$$

Also, we verify that

$$\nu h_0 = \nu(h - \nu h \mathbb{1}) = \nu h - (\nu h)(\nu \mathbb{1}) = 0.$$

Hence h_0 satisfies (A.10) and $\nu h_0 = 0$. Moreover, any solution of (A.10) can be represented by

$$c = h + \gamma \mathbb{1} = h_0 + (\gamma + \nu h)\mathbb{1} = h_0 + \gamma_0 \mathbb{1},$$

where $\gamma_0 = \gamma + \nu h \in \mathbb{R}$. It remains to show uniqueness. To this end, define an augmented matrix

$$Q_a = \begin{pmatrix} Q \\ \nu \end{pmatrix} \in \mathbb{R}^{(m_0+1) \times m_0}$$

and a new vector

$$\widehat{b} = \begin{pmatrix} b \\ 0 \end{pmatrix} \in \mathbb{R}^{m_0+1}.$$

Then (A.10) together with $\nu h = 0$ can be written as

$$Q_a h = \widehat{b}. \tag{A.13}$$

It can be shown that $Q_a' Q_a$ has full rank m_0 due to the irreducibility of Q and the solution of (A.13) can be represented by

$$h = (Q_a' Q_a)^{-1} Q_a' \widehat{b} = h_0.$$

This leads to the desired uniqueness.

□

Remark A.13 In the literature, (A.10) is sometimes referred to as Poisson equation (see [24]), and the results of (a) and (b) in the lemma are deemed to be well-known. They are more or less a consequence of the Fredholm alternative. The irreducibility of Q used in the lemma can be relaxed to weak irreducibility. The result still holds.

Perhaps being interesting in its own right, we illustrate how the solutions may be obtained through a stochastic representation in the proof above. This angle of view has not been extensively exploited to our knowledge. Although classifying solutions of an algebraic system by the sum of homogenous part and a particular solution is a time honored concept, characterizing the unique particular solution using orthogonality with respect to the stationary distribution of Q is useful in many applications.

Similar to Lemma A.12, when studying stability of switching diffusions, we need to examine an equation

$$Qc = b - \frac{1}{2}\sigma^2 + \beta\mathbb{1}, \tag{A.14}$$

where $\beta \in \mathbb{R}$, $b \in \mathbb{R}^{m_0}$ is a constant vector and $\sigma^2 = (\sigma_1^2, \ldots, \sigma_{m_0}^2)' \in \mathbb{R}^{m_0}$ with $\sigma_i^2 \geq 0$ for $i \in \mathcal{M}$. Then Eq. (A.14) has a solution if the Markov chain $\alpha_1(t)$ is ergodic. Let the associated stationary distribution be denoted by ν. Then β is given by

$$\beta = -\sum_{i=1}^{m_0} \nu_i \left(b_i - \frac{1}{2}\sigma_i^2 \right) = -\nu b + \frac{1}{2}\nu\sigma^2. \tag{A.15}$$

Moreover, let the column vector $h = (h_1, \ldots, h_{m_0})' \in \mathbb{R}^{m_0}$ be defined by

$$h_i = \mathbb{E}_i \int_0^\infty \left(\nu b - \frac{1}{2}\nu\sigma^2 - b_{\alpha_1(t)} + \frac{1}{2}\sigma^2_{\alpha_1(t)} \right) dt, \quad i \in \mathcal{M}. \tag{A.16}$$

Then h is well defined and is a solution of (A.14) with β given by (A.15).

A.5 Martingales, Gaussian Processes, and Diffusions

This section briefly reviews several random processes including martingales, Gaussian processes, and diffusions.

A.5.1 Martingales

Many applications involving stochastic processes depend on the concept of martingale. The definition and properties of discrete-time martingales can be found in Breiman [31, Chapter 5], Chung [52, Chapter 9], and Hall and Heyde [112], among others. This section provides a brief review.

Discrete-Time Martingales

Definition A.14 Suppose that $\{\mathcal{F}_n\}$ is a filtration and $\{X_n\}$ is a sequence of random variables. The pair $\{X_n, \mathcal{F}_n\}$ is a martingale if for each n,

(a) X_n is $\mathcal{F}_n$-measurable;
(b) $\mathbb{E}|X_n| < \infty$;
(c) $\mathbb{E}(X_{n+1}|\mathcal{F}_n) = X_n$ a.s.

It is a supermartingale (resp., submartingale) if (a) and (b) in the above hold, and

$$\mathbb{E}(X_{n+1}|\mathcal{F}_n) \le X_n \ \ (\text{resp., } \mathbb{E}(X_{n+1}|\mathcal{F}_n) \ge X_n) \ \ \text{a.s.}$$

In what follows if the sequence of σ-algebras is clear, we simply say that $\{X_n\}$ is a martingale.

Let $X_n = \sum_{j=1}^{n} Y_j$, where $\{Y_n\}$ is a sequence of independent and identically distributed (i.i.d.) random variables with zero mean. It is plain that

$$\begin{aligned}\mathbb{E}[X_{n+1}|Y_1, \ldots, Y_n] &= \mathbb{E}[X_n + Y_{n+1}|Y_1, \ldots, Y_n]\\ &= X_n + \mathbb{E}Y_{n+1} = X_n \ \ \text{a.s.}\end{aligned}$$

The above equation illustrates the defining relation of a martingale.

If $\{X_n\}$ is a martingale, we can define $Y_n = X_n - X_{n-1}$, which is known as a martingale difference sequence. Suppose that $\{X_n, \mathcal{F}_n\}$ is a martingale. Then the following properties hold:

(a) Suppose $\varphi(\cdot)$ is an increasing and convex function defined on $\mathbb{R}$. If for each positive integer n, $\mathbb{E}|\varphi(X_n)| < \infty$, then $\{\varphi(X_n), \mathcal{F}_n\}$ is a submartingale.
(b) Let τ be a stopping time with respect to $\mathcal{F}_n$ (i.e., an integer-valued random variable such that $\{\tau \le n\}$ is $\mathcal{F}_n$-measurable for each n). Then $\{X_{\tau\wedge n}, \mathcal{F}_{\tau\wedge n}\}$ is also a martingale.
(c) The martingale inequality (see Kushner [175, p. 3]) states that for each $\lambda > 0$,

$$\begin{aligned}&\mathbb{P}\left(\max_{1\le j\le n} |X_j| \ge \lambda\right) \le \frac{1}{\lambda}\mathbb{E}|X_n|,\\ &\mathbb{E}\max_{1\le j\le n} |X_j|^2 \le 4\mathbb{E}|X_n|^2, \ \ \text{if } \mathbb{E}|X_n|^2 < \infty \ \text{ for each } n.\end{aligned} \tag{A.17}$$

(d) The Doob inequality (see Hall and Heyde [112, p.15]) states that for each $p > 1$,

$$\mathbb{E}^{1/p}|X_n|^p \le \mathbb{E}^{1/p}\left(\max_{1\le j\le n} |X_j|\right)^p \le q\mathbb{E}^{1/p}|X_n|^p,$$

where $p^{-1} + q^{-1} = 1$.
(e) The Burkholder inequality (see Hall and Heyde [112, p.23]) is: For $1 < p < \infty$, there exist constants K_1 and K_2 such that

$$K_1 \mathbb{E}\left|\sum_{j=1}^{n} y_j^2\right|^{p/2} \le \mathbb{E}|X_n|^p \le K_2 \mathbb{E}\left|\sum_{i=j}^{n} y_j^2\right|^{p/2},$$

where $Y_n = X_n - X_{n-1}$.

Remark A.15 If $X(\cdot)$ is a right-continuous martingale in continuous time and $f(\cdot)$ is a nonnegative convex function, then

$$\mathbb{P}\Big[\sup_{s\le t\le T} f(X(t)) \ge \lambda \Big| \mathcal{F}_s\Big] \le \frac{\mathbb{E}\Big[f(X(t)\Big|\mathcal{F}_s\Big]}{\lambda}, \tag{A.18}$$

$$\mathbb{E}\sup_{0\le t\le T} |X(t)|^2 \le 4\mathbb{E}|X(T)|^2. \tag{A.19}$$

Consider a discrete-time Markov chain $\{\alpha_n\}$ with state space $\mathcal{M}$ (either finite or countable) and one-step transition probability matrix $P = (p_{ij})$. Recall that a sequence $\{f(i) : i \in \mathcal{M}\}$ is P-harmonic or right-regular (Karlin and Taylor [151, p. 48]), if (a) $f(\cdot)$ is a real-valued function such that $f(i) \ge 0$ for each $i \in \mathcal{M}$, and (b)

$$f(i) = \sum_{j\in\mathcal{M}} p_{ij} f(j) \quad \text{for each } i \in \mathcal{M}. \tag{A.20}$$

If the equality in (A.20) is replaced by $\ge$ (resp., $\le$), $\{f(i) : i \in \mathcal{M}\}$ is said to be P-superharmonic or right superregular (resp., P-subharmonic or right subregular). Considering $f = (f(i) : i \in \mathcal{M})$ as a column vector, (A.20) can be written as $f = Pf$. Similarly, we can write $f \ge Pf$ for P-superharmonic (resp., $g \le Pf$ for P-subharmonic). Likewise, $\{f(i) : i \in \mathcal{M}\}$ is said to be P left regular, if (b) above is replaced by

$$f(j) = \sum_{i\in\mathcal{M}} f(i) p_{ij} \quad \text{for each } j \in \mathcal{M}. \tag{A.21}$$

Similarly, left superregular and subregular functions can be defined.

There is a natural connection between a martingale and a discrete-time Markov chain; see Karlin and Taylor [150, p. 241]. Let $\{\alpha_n\}$ be a discrete-time Markov chain and $\{f(i) : i \in \mathcal{M}\}$ be a bounded P-harmonic sequence. Define $X_n = f(\alpha_n)$. Then $\mathbb{E}|X_n| < \infty$. Moreover, owing to the Markov property,

$$\begin{aligned}
\mathbb{E}(X_{n+1}|\mathcal{F}_n) &= \mathbb{E}(f(\alpha_{n+1})|\alpha_n)) \\
&= \sum_{j\in\mathcal{M}} p_{\alpha_n,j} f(j) \\
&= f(\alpha_n) = X_n \ \text{ a.s.}
\end{aligned}$$

Therefore, $\{X_n, \mathcal{F}_n\}$ is a martingale. Note that if $\mathcal{M}$ is finite, the boundedness of $\{f(i) : i \in \mathcal{M}\}$ is not needed.

As explained in Karlin and Taylor [150], one of the widely used ways of constructing martingales is through the utilization of eigenvalues and eigenvectors of a transition matrix. Again, let $\{\alpha_n\}$ be a discrete-time Markov chain with transition matrix P. Recall that a column vector f is a right eigenvector of P associated with an eigenvalue $\lambda \in \mathbb{C}$, if $Pf = \lambda f$. Let f be a right eigenvector of P satisfying $\mathbb{E}|f(\alpha_n)| < \infty$ for each n. For $\lambda \neq 0$, define $X_n = \lambda^{-n} f(\alpha_n)$. Then $\{X_n\}$ is a martingale.

Continuous-Time Martingales

Next, let us denote the space of $\mathbb{R}^r$-valued continuous functions on $[0, T]$ by $C([0, T]; \mathbb{R}^r)$ and the space of functions that are right-continuous with left-hand limits endowed with the Skorohod topology by $D([0, T]; \mathbb{R}^r)$; see Definition A.18. Consider $X(\cdot) = \{X(t) \in \mathbb{R}^r : t \geq 0\}$. If for each $t \geq 0$, $X(t)$ is an $\mathbb{R}^r$ random vector, we call $X(\cdot)$ a continuous-time stochastic process and write it as $X(t)$, $t \geq 0$, or simply $X(t)$ if there is no confusion.

A process $X(\cdot)$ is adapted to a filtration $\{\mathcal{F}_t\}$, if for each $t \geq 0$, $X(t)$ is an $\mathcal{F}_t$-measurable random variable; $X(\cdot)$ is progressively measurable if for each $t \geq 0$, the process restricted to $[0, t]$ is measurable with respect to the σ-algebra $\mathcal{B}[0, t] \times \mathcal{F}_t$ in $[0, t] \times \Omega$, where $\mathcal{B}[0, t]$ denotes the Borel sets of $[0, t]$. A progressively measurable process is measurable and adapted, whereas the converse is not generally true. However, any measurable and adapted process with right-continuous sample paths is progressively measurable.

Frequently, we need to work with a *stopping time* for applications. Consider $(\Omega, \mathcal{F}, P)$ with a filtration $\{\mathcal{F}_t\}$. A stopping time τ is a nonnegative random variable satisfying $\{\tau \leq t\} \in \mathcal{F}_t$ for all $t \geq 0$.

A stochastic process $\{X(t) : t \geq 0\}$ (real- or vector-valued) is a martingale on $(\Omega, \mathcal{F}, P)$ with respect to $\{\mathcal{F}_t\}$ if

(a) for each $t \geq 0$, $X(t)$ is $\mathcal{F}_t$-measurable,
(b) $\mathbb{E}|X(t)| < \infty$, and
(c) $\mathbb{E}[X(t)|\mathcal{F}_s] = X(s)$ a.s for all $t \geq s$.

If $\mathcal{F}_t$ is the natural filtration $\sigma\{X(s) : s \leq t\}$, we often say that $X(\cdot)$ is a martingale without specifying the filtration $\mathcal{F}_t$. The process $X(\cdot)$ is a local martingale if there exists a sequence of increasing stopping times $\{\tau_n\}$ such that $\tau_n \to \infty$ a.s as $n \to \infty$, and $X^{(n)}(t) := X(t \wedge \tau_n)$ is a martingale for each $n \in \mathbb{N}$.

A.5.2 Gaussian Processes and Diffusion Processes

A Gaussian random vector $X = (X_1, X_2, \ldots, X_r)$ is one whose characteristic function has the form

$$\phi(y) = \exp\left(\mathrm{i}y'\mu - \frac{1}{2}y'\Sigma y\right),$$

where $\mu \in \mathbb{R}^r$ is a constant vector, $y'\mu$ is the usual inner product, i denotes the pure imaginary number satisfying $\mathrm{i}^2 = -1$, and Σ is a symmetric nonnegative definite $r \times r$ matrix. In the above, μ and Σ are the mean vector and covariance matrix of X, respectively.

Consider a stochastic process $X(t)$, $t \geq 0$. It is a Gaussian process if for any $k = 1, 2, \ldots$ and $0 \leq t_1 < t_2 < \cdots < t_k$, $(X(t_1), X(t_2), \ldots, X(t_k))$ is a Gaussian vector. A random process $X(\cdot)$ has *independent increments* if for any $k = 1, 2, \ldots$ and $0 \leq t_1 < t_2 < \cdots < t_k$,

$$(X(t_1) - X(0)),\ (X(t_2) - X(t_1)),\ \ldots,\ (X(t_k) - X(t_{k-1}))$$

are independent. A sufficient condition for a process to be Gaussian is given next, whose proof can be found in Skorohod [262, p. 7].

Lemma A.16 *Suppose that the process $X(\cdot)$ has independent increments and continuous sample paths almost surely. Then $X(\cdot)$ is a Gaussian process.*

Next, we consider the notion of Brownian motions. An $\mathbb{R}^r$-valued random process $w(t)$ for $t \geq 0$ is a Brownian motion, if

(a) $w(0) = 0$ almost surely;
(b) $w(\cdot)$ is a process with independent increments;
(c) $w(\cdot)$ has continuous sample paths almost surely;
(d) for all $t, s \geq 0$, the increments $w(t) - w(s)$ have Gaussian distribution with $\mathbb{E}(w(t) - w(s)) = 0$ and $\mathrm{Cov}(w(t), w(s)) = \Sigma|t - s|$ for some nonnegative definite $r \times r$ matrix Σ, where $\mathrm{Cov}(w(t), w(s))$ denotes the covariance.

A Brownian motion $w(\cdot)$ with $\Sigma = I$ is termed a standard Brownian motion. In view of Lemma A.16, a Brownian motion is necessarily a Gaussian process. For an $\mathbb{R}^r$-valued Brownian motion $w(t)$, let $\mathcal{F}_t = \sigma\{w(s) : s \leq t\}$. Let $h(\cdot)$ be an $\mathcal{F}_t$-measurable process taking values in $\mathbb{R}^{r\times r}$ such that $\int_0^t \mathbb{E}|h(s)|^2 ds < \infty$ for all $t \geq 0$. Using $w(\cdot)$ and $h(\cdot)$, we may define a stochastic integral $\int_0^t h(s)dw(s)$ such that it is a martingale with mean 0 and

$$\mathbb{E}\left|\int_0^t h(s)dw(s)\right|^2 = \int_0^t \mathbb{E}\big[\mathrm{tr}(h(s)h'(s))\big]ds.$$

Suppose that $b(\cdot)$ and $\sigma(\cdot)$ are nonrandom Borel measurable functions. A process $X(\cdot)$ defined as

$$X(t) = X(0) + \int_0^t b(s, X(s))ds + \int_0^t \sigma(s, X(s))dw(s) \tag{A.22}$$

is called a diffusion. Then $X(\cdot)$ defined in (A.22) is a Markov process in the sense that the Markov property

$$\mathbb{P}(X(t) \in A|\mathcal{F}_s) = \mathbb{P}(X(t) \in A|X(s))$$

holds for all $0 \leq s \leq t$ and for any Borel set A. A slightly more general definition allows $b(\cdot)$ and $\sigma(\cdot)$ to be $\mathcal{F}_t$-measurable processes. Nevertheless, the current definition is sufficient for our purpose.

Associated with the diffusion process, there is an operator $\mathcal{L}$, known as the generator of the diffusion $X(\cdot)$. Let $C^{1,2}$ be the class of real-valued functions on (a subset of) $[0, \infty) \times \mathbb{R}^r$ whose first-order partial derivative with respect to t and the second-order mixed partial derivatives with respect to x are continuous. Define an operator $\mathcal{L}$ on $C^{1,2}$ by

$$\mathcal{L}f(t,x) = \frac{\partial f(t,x)}{\partial t} + \sum_{i=1}^{r} b_i(t,x)\frac{\partial f(t,x)}{\partial x_i} + \frac{1}{2}\sum_{i,j=1}^{r} a_{ij}(t,x)\frac{\partial^2 f(t,x)}{\partial x_i \partial x_j}, \tag{A.23}$$

where $A(t,x) = (a_{ij}(t,x)) = \sigma(t,x)\sigma'(t,x)$. The above may also be written in a more compact form as

$$\mathcal{L}f(t,x) = \frac{\partial f(t,x)}{\partial t} + b'(t,x)\nabla f(t,x) + \frac{1}{2}\text{tr}(\nabla^2 f(t,x)A(t,x)),$$

where ∇f and $\nabla^2 f$ denote the gradient and Hessian of f, respectively. Note that in this book, we use the notations $b'f$ and $\langle b, f\rangle$ interchangeably to represent an inner product.

The well-known Itô lemma (see Gihman and Skorohod [103], Ikeda and Watanabe [136], and Liptser and Shiryayev [190]) states that

$$df(t,X(t)) = \mathcal{L}f(t,X(t)) + \nabla f'(t,X(t))\sigma(t,X(t))dw(t),$$

or in its integral form

$$\begin{aligned} &f(t,X(t)) - f(0,X(0)) \\ &= \int_0^t \mathcal{L}f(s,X(s))ds + \int_0^t \nabla f'(s,X(s))\sigma(s,X(s))dw(s). \end{aligned}$$

By virtue of Itô's lemma,

$$M_f(t) = f(t,X(t)) - f(0,X(0)) - \int_0^t \mathcal{L}f(s,X(s))ds$$

is a local $\mathcal{F}_t$-martingale. Conversely, suppose that $X(\cdot)$ is right continuous. Using the notation of martingale problems given by Stroock and Varadhan [266], $X(\cdot)$ is said to be a solution of the martingale problem with operator $\mathcal{L}$ if $M_f(\cdot)$ is a martingale for each $f(\cdot,\cdot) \in C_0^{1,2}$ (the class of $C^{1,2}$ functions with compact support).

A.6 Weak Convergence

The notion of weak convergence is a generalization of convergence in distribution in elementary probability theory. In what follows, we present definitions and results, including tightness, tightness criteria, the martingale problem, Skorohod representation, Prohorov's theorem, and so on.

Definition A.17 (Weak Convergence) Let $\mathbb{P}$ and $\mathbb{P}_k$, $k = 1, 2, \ldots$, be probability measures defined on a metric space $\mathbb{S}$. The sequence $\{\mathbb{P}_k\}$ converges weakly to $\mathbb{P}$ if

$$\int f d\mathbb{P}_k \to \int f d\mathbb{P}$$

for every bounded and continuous function $f(\cdot)$ on $\mathbb{S}$. Suppose that $\{X_k\}$ and X are random variables associated with $\mathbb{P}_k$ and $\mathbb{P}$, respectively. The sequence X_k converges to X weakly if for any bounded and continuous function $f(\cdot)$ on $\mathbb{S}$, $\mathbb{E}f(X_k) \to \mathbb{E}f(X)$ as $k \to \infty$.

Let $D([0,\infty);\mathbb{R}^r)$ be the space of $\mathbb{R}^r$-valued functions defined on $[0,\infty)$ that are right continuous and have left-hand limits; let $\mathbb{L}$ be a set of strictly increasing Lipschitz continuous functions $\zeta(\cdot) : [0,\infty) \mapsto [0,\infty)$ such that the mapping is surjective with $\zeta(0) = 0$, $\lim_{t\to\infty} \zeta(t) = \infty$, and

$$\gamma(\zeta) := \sup_{0\le t<s} \left| \log\left(\frac{\zeta(s)-\zeta(t)}{s-t}\right)\right| < \infty.$$

Similar to $D([0,\infty);\mathbb{R}^r)$, we also use the notation $D([0,T];\mathbb{F})$ to denote the D-space of functions that take values in a metric space $\mathbb{F}$.

Definition A.18 (Skorohod Topology) For $\xi, \eta \in D([0,\infty);\mathbb{R}^r)$, the *Skorohod topology* $d(\cdot,\cdot)$ on $D([0,\infty);\mathbb{R}^r)$ is defined as

$$d(\xi,\eta) = \inf_{\zeta\in\mathbb{L}} \left\{ \gamma(\zeta) \vee \int_0^\infty e^{-s} \sup_{t\ge 0} \big(1 \wedge |\xi(t\wedge s) - \eta(\zeta(t)\wedge s)|\big) ds \right\}.$$

Analogous definitions and results are available for $D([0,T];\mathbb{F})$, where $\mathbb{F}$ is a metric space; see Ethier and Kurtz [81] and Billingsley [28] for related references. Although we frequently work with $D([0,T];\mathbb{R}^r)$ in this book, the following results are often stated with respect to the space $D([0,\infty);\mathbb{R}^r)$. This enables us to apply them to $t \in [0,T]$ for any $T > 0$.

Definition A.19 (Tightness) A family of probability measures $\mathcal{P}$ defined on a metric space $\mathbb{S}$ is *tight* if for each $\delta > 0$, there exists a compact set $K_\delta \subset \mathbb{S}$ such that

$$\inf_{\mathbb{P}\in\mathcal{P}} \mathbb{P}(K_\delta) \geq 1 - \delta.$$

The notion of tightness is closely related to compactness. The following theorem, known as Prohorov's theorem, gives such an implication. A complete proof can be found in Ethier and Kurtz [81].

Theorem A.20 (Prohorov's Theorem) *If $\mathcal{P}$ is tight, then $\mathcal{P}$ is relatively compact. That is, every sequence of elements in $\mathcal{P}$ contains a weakly convergent subsequence. If the underlying metric space is complete and separable, the tightness is equivalent to relative compactness.*

Although weak convergence techniques usually allow one to use weaker conditions and lead to a more general setup, it is often more convenient to work with probability one convergence for purely analytic reasons, however. The Skorohod representation provides us with such opportunities.

Theorem A.21 (The Skorohod Representation (Ethier and Kurtz [81, p. 174])) *Let X_k and X be random elements belonging to $D([0, \infty); \mathbb{R}^r)$ such that X_k converges weakly to X. Then there exists a probability space $(\widetilde{\Omega}, \widetilde{\mathcal{F}}, \widetilde{\mathbb{P}})$ on which are defined random elements $\widetilde{X}_k$, $k = 1, 2, \ldots$, and $\widetilde{X}$ in $D([0, \infty); \mathbb{R}^r)$ such that for any Borel set B and all $k < \infty$,*

$$\widetilde{\mathbb{P}}(\widetilde{X}_k \in B) = \mathbb{P}(X_k \in B), \quad \text{and} \quad \widetilde{\mathbb{P}}(\widetilde{X} \in B) = \mathbb{P}(X \in B)$$

satisfying

$$\lim_{k\to\infty} \widetilde{X}_k = \widetilde{X} \quad \text{a.s.}$$

Elsewhere in the book, when we use the Skorohod representation, with a slight abuse of notation, we often omit the tilde notation for convenience and notational simplicity.

Let $C([0, \infty); \mathbb{R}^r)$ be the space of $\mathbb{R}^r$-valued continuous functions equipped with the sup-norm topology and C_0 be the set of real-valued continuous functions on $\mathbb{R}^r$ with compact support. Let C_0^l be the subset of C_0 functions that have continuous partial derivatives up to the order l.

Definition A.22 Let $\mathbb{S}$ be a metric space and $\mathcal{A}$ be a linear operator on $B(\mathbb{S})$ (the set of all Borel measurable functions defined on $\mathbb{S}$). Let $X(\cdot) = \{X(t) : t \geq 0\}$ be a right-continuous process with values in $\mathbb{S}$ such that for each $f(\cdot)$ in the domain of $\mathcal{A}$,

$$f(X(t)) - \int_0^t \mathcal{A}f(X(s))ds$$

is a martingale with respect to the filtration $\sigma\{X(s) : s \le t\}$. Then $X(\cdot)$ is called a *solution of the martingale problem* with operator $\mathcal{A}$.

Theorem A.23 (Ethier and Kurtz [81, p. 174]) *A right-continuous process $X(t)$, $t \ge 0$, is a solution of the martingale problem for the operator $\mathcal{A}$ if and only if*

$$\mathbb{E}\left[\prod_{j=1}^{i} h_j(X(t_j))\left[f(X(t_{i+1})) - f(X(t_i)) - \int_{t_i}^{t_{i+1}} \mathcal{A}f(X(s))ds\right]\right] = 0$$

whenever $0 \le t_1 < t_2 < \cdots < t_{i+1}$, $f(\cdot)$ in the domain of $\mathcal{A}$, and $h_1, \ldots, h_i \in \mathcal{B}(\mathbb{S})$, the Borel field of $\mathbb{S}$.

Theorem A.24 (Uniqueness of Martingale Problems, Ethier and Kurtz [81, p. 184]) *Let $X(\cdot)$ and $Y(\cdot)$ be two stochastic processes whose paths are in $D([0, T]; \mathbb{R}^r)$. Denote an infinitesimal generator by $\mathcal{A}$. If for any function $f \in \mathcal{D}(\mathcal{A})$ (the domain of $\mathcal{A}$),*

$$f(X(t)) - f(X(0)) - \int_0^t \mathcal{A}f(X(s))ds, \ t \ge 0,$$

and

$$f(Y(t)) - f(Y(0)) - \int_0^t \mathcal{A}f(Y(s))ds, \ t \ge 0$$

are martingales and $X(t)$ and $Y(t)$ have the same distribution for each $t \ge 0$, $X(\cdot)$ and $Y(\cdot)$ have the same distribution on $D([0, \infty); \mathbb{R}^r)$.

We use a theorem from [171] on the equivalence between stochastic differential equations and martingale problems. While the generator presented in [171] is more general, we focus on a specialized form that better suits our objectives.

Theorem A.25 (Equivalence of Stochastic Equations and Martingale Problems) *Let ν be a σ-finite measure on $\mathbb{R}^k \setminus \{0\}$. Define the generator*

$$\begin{aligned}\mathcal{A}f(x) =& \frac{1}{2}\sum_{i,j=1}^{d} a_{ij}(x)\frac{\partial^2}{\partial x_i \partial x_j} f(x) + b(x) \cdot \nabla f(x)\\ &+ \int_{\mathbb{R}^k\setminus\{0\}} \big(f(x + h(x, y)) - f(x) - h(x, y) \cdot \nabla f(x)\big)\nu(dy),\end{aligned}$$

where $a(x) = \sigma(x)\sigma'(x)$. If $\sigma : \mathbb{R}^r \to \mathbb{R}, b : \mathbb{R}^{r\times d} \to \mathbb{R}, h : \mathbb{R}^r \times \mathbb{R}^k$ are measurable functions such that for each compact $K \subset \mathbb{R}^r$, we have

$$\sup_{x\in K}\left\{|\sigma(x)|+|b(x)|+\int_{\mathbb{R}^k\setminus\{0\}}|h(x,y)|^2\nu(dy)\right\}<\infty,$$

then any solution to the martingale problem in $D([0,T];\mathbb{R}^r)$ for generator $\mathcal{A}$ is a weak solution the stochastic differential equation

$$dX(t)=b(X(t))dt+\sigma(X(t))dB(t)+\int_{\mathbb{R}^k\setminus\{0\}}h(X(t-),y)\widetilde{N}(dt,dy)$$

where $B(t)$ is a d-dimensional Brownian motion and $N(dt,dy)$ is a Poisson random measure with intensity $\nu(dy)$ on $\mathbb{R}^k$ and $\widetilde{N}(dt,dy)=N(dt,dy)-\nu(dy)\,dt$.

Theorem A.26 *Let $X^\varepsilon(\cdot)$ be a solution of the differential equation*

$$\frac{dX^\varepsilon(t)}{dt}=F^\varepsilon(t),$$

and for each $T<\infty$, $\{F^\varepsilon(t):\ 0\le t\le T\}$ be uniformly integrable. If the set of initial values $\{X^\varepsilon(0)\}$ is tight, then $\{X^\varepsilon(\cdot)\}$ is tight in $C([0,\infty);\mathbb{R}^r)$.

Proof The proof is essentially in Billingsley [28, Theorem 7.3] (see also Kushner [175, p. 51, Lemma 7]). □

Define the notion of "p-lim" and an operator $\mathcal{A}^\varepsilon$ as in Ethier and Kurtz [81]. Suppose that $X^\varepsilon(\cdot)$ are defined on the same probability space. Let $\mathcal{F}_t^\varepsilon$ be the minimal σ-algebra over which $\{X^\varepsilon(s),\xi^\varepsilon(s):\ s\le t\}$ is measurable and let $\mathbb{E}_t^\varepsilon$ denote the conditional expectation given $\mathcal{F}_t^\varepsilon$. Denote

$$\overline{M}^\varepsilon=\Big\{f:\ f\text{ is real-valued with bounded support and is progressively measurable w.r.t. }\{\mathcal{F}_t^\varepsilon\},\ \sup_t\mathbb{E}|f(t)|<\infty\Big\}.$$

Let $g(\cdot), f(\cdot), f^\delta(\cdot)\in\overline{M}^\varepsilon$. For each $\delta>0$ and $t\le T<\infty$, $f=p-\lim_\delta f^\delta$ if

$$\sup_{t,\delta}\mathbb{E}|f^\delta(t)|<\infty,$$

then

$$\lim_{\delta\to 0}\mathbb{E}|f(t)-f^\delta(t)|=0\ \text{ for each }\ t.$$

The function $f(\cdot)$ is said to be in the domain of $\mathcal{A}^\varepsilon$; that is, $f(\cdot)\in\mathcal{D}(\mathcal{A}^\varepsilon)$, and $\mathcal{A}^\varepsilon f=g$, if

$$p-\lim_{\delta\to 0}\left(\frac{\mathbb{E}_t^\varepsilon f(t+\delta)-f(t)}{\delta}-g(t)\right)=0.$$

If $f(\cdot) \in \mathcal{D}(\mathcal{A}^\varepsilon)$, then Ethier and Kurtz [81] or Kushner [175, p. 39] implies that

$$f(t) - \int_0^t \mathcal{A}^\varepsilon f(u)du$$

is a martingale, and

$$\mathbb{E}_t^\varepsilon f(t+s) - f(t) = \int_t^{t+s} \mathbb{E}_t^\varepsilon \mathcal{A}^\varepsilon f(u)du \quad \text{a.s.}$$

In applications, ϕ-mixing processes frequently arise; see [81] and [175]. The assertion below presents a couple of inequalities for mixing processes. Further results on various mixing processes are in [81].

Lemma A.27 (Kushner [175, Lemma 4.4]) *Let $\xi(\cdot)$ be a ϕ-mixing process with mixing rate $\phi(\cdot)$ and let $h(\cdot)$ be $\mathcal{F}_t^\infty$-measurable and $|h| \le 1$. Then*

$$\Big|\mathbb{E}(h(\xi(t+s))|\mathcal{F}_0^t) - \mathbb{E}h(\xi(t+s))\Big| \le 2\phi(s).$$

If $t < u < v$, and $\mathbb{E}h(\xi(s)) = 0$ for all s, then

$$\Big|\mathbb{E}(h(\xi(u))h(\xi(v))|\mathcal{F}_0^t) - \mathbb{E}h(\xi(u))h(\xi(v))\Big| \le 4\Big(\phi(v-u)\phi(u-t)\Big)^{1/2},$$

where $\mathcal{F}_\tau^t = \sigma\{\xi(s) : \tau \le s \le t\}$.

Example A.28 A useful example of a mixing process is a function of a stationary Markov chain with finite state space. Let α_k be such a Markov chain with state space $\mathcal{M} = \{1, \dots, m_0\}$. Let $\xi_k = g(\alpha_k)$, where $g(\cdot)$ is a real-valued function defined on $\mathcal{M}$. Suppose the Markov chain or equivalently, its transition probability matrix, is irreducible and aperiodic. Then as proved in Billingsley [28, Example 19.3], ξ_k is a mixing process with the mixing measure decaying to 0 exponentially fast.

A crucial step in obtaining many limit problems depends on the verification of tightness of the sequences of interest. A sufficient condition known as Kurtz's criterion appears to be rather handy to use.

Lemma A.29 (Kushner [175, Theorem 3, p. 47]) *Suppose that $\{Y^\varepsilon(\cdot)\}$ is a process with paths in $D([0,\infty);\mathbb{R}^r)$, and suppose that*

$$\lim_{K_1\to\infty}\left\{\limsup_{\varepsilon\to 0}\mathbb{P}\left(\sup_{0\le t\le T}|Y^\varepsilon(t)| \ge K_1\right)\right\} = 0 \textit{ for each } T < \infty, \tag{A.24}$$

and for all $0 \le s \le \delta$, $t \le T$,

$$\begin{aligned}&\mathbb{E}_t^\varepsilon \min\left(1, |Y^\varepsilon(t+s) - Y^\varepsilon(t)|^2\right) \le \mathbb{E}_t^\varepsilon \gamma_\varepsilon(\delta),\\&\lim_{\delta\to 0}\limsup_{\varepsilon\to 0} \mathbb{E}\gamma_\varepsilon(\delta) = 0.\end{aligned} \tag{A.25}$$

Then $\{Y^\varepsilon(\cdot)\}$ *is tight in* $D([0,\infty); \mathbb{R}^r)$.

Remark A.30 In lieu of (A.24), one may verify the following condition (see Kurtz [170, Theorem 2.7, p. 10]). Suppose that for each $\eta > 0$ and rational $t \ge 0$ there is a compact set $\Gamma_{t,\eta} \subset \mathbb{R}^r$ such that

$$\inf_\varepsilon \mathbb{P}\left(Y^\varepsilon(t) \in \Gamma_{t,\eta}\right) > 1 - \eta. \tag{A.26}$$

A.7 Miscellany

Suppose that A is an $r \times r$ square matrix. Denote the collection of eigenvalues of A by Λ. Then the spectral radius of A, denoted by $\rho(A)$, is defined by $\rho(A) = \max_{\lambda\in\Lambda} |\lambda|$. Recall that a matrix with real entries is a positive matrix if it has at least one positive entry and no negative entries. If every entry of A is positive, we call the matrix strictly positive. Likewise, for a vector $x = (x_1, \dots, x_r)$, by $x \ge 0$, we mean $x_i \ge 0$ for $i = 1, \dots, r$; by $x > 0$, we mean all entries $x_i > 0$.

By a multi-index ζ, we mean a vector $\zeta = (\zeta_1, \dots, \zeta_r)$ with nonnegative integer components with $|\zeta|$ defined as $|\zeta| = \zeta_1 + \cdots + \zeta_r$. For a multi-index ζ, D_x^ζ is defined to be

$$D_x^\zeta = \frac{\partial^\zeta}{\partial x^\zeta} = \frac{\partial^{|\zeta|}}{\partial x_1^{\zeta_1} \dots \partial x_r^{\zeta_r}}; \tag{A.27}$$

see Friedman [94] or Gihman and Skorohod [101].

The following inequalities are widely used. The first of them is known as the Gronwall inequality and second one is the so-called generalized Gronwall inequality. Both of them can be found in [110, p. 36].

Lemma A.31 *If* $\gamma \in \mathbb{R}$, $\beta(t) \ge 0$, *and* $\varphi(t)$ *are continuous real-valued functions for* $a \le t \le b$, *which satisfy*

$$\varphi(t) \le \gamma + \int_a^t \beta(s)\varphi(s)ds, \ t \in [a, b],$$

then

$$\varphi(t) \le \gamma \exp\left(\int_a^t \beta(s)ds\right), \ t \in [a, b].$$

Lemma A.32 *Suppose that $\varphi(\cdot)$ and $\gamma(\cdot)$ are real-valued continuous functions on $[a, b]$, that $\beta(t) \ge 0$ is integrable on $[a, b]$, and that*

$$\varphi(t) \le \gamma(t) + \int_a^t \beta(s)\varphi(s)ds, \ t \in [a, b].$$

Then

$$\varphi(t) \le \gamma(t) + \int_a^t \beta(s)\gamma(s)\exp(\int_s^t \beta(u)du)ds, \ t \in [a, b].$$

When we treat stochastic integrals, the Burkholder-Davis-Gundy inequality is used quite often. We state it below, a proof of which may be found in [205, p. 70].

Lemma A.33 *Let $\psi : [0, \infty) \mapsto \mathbb{R}^{r\times d}$ for some positive integers r and d, and suppose that $\mathbb{E}\int_0^\infty |\psi(s)|^2 < \infty$. Define*

$$Z(t) = \int_0^t \psi(s)dw(s) \ \text{and} \ a(t) = \int_0^t |\psi(s)|^2 ds,$$

where $w(t)$ is a d-dimensional standard Brownian motion. Then for any $p > 0$, there exist positive constants c_p and C_p such that

$$c_p\mathbb{E}|a(t)|^{p/2} \le \mathbb{E}\sup_{0\le s\le t}|Z(s)|^p \le C_p\mathbb{E}|a(t)|^{p/2}, \ \text{for all} \ t \ge 0.$$

A.8 Notes

One may find a nonmeasure-theoretic introduction to stochastic processes in Ross [249]. The two volumes by Karlin and Taylor [150, 151] provide an introduction to discrete-time and continuous-time Markov chains. Advanced treatments of Markov chains can be found in Chung [52] and Revuz [247]. A book that deals exclusively with finite-state Markov chains is by Iosifescu [140]. The book of Meyn and Tweedie [211] examines Markov chains and their stability. Doob's book [65] gives an introduction to stochastic processes. Gihman and Skorohod's three-volume work [102] provides a comprehensive introduction to stochastic processes, whereas Liptser and Shiryayev's book [190] presents further topics such as nonlinear filtering.

References

1. M.S. Agranovich, Elliptic operators on closed manifolds, in *Current Problems in Mathematics: Fundamental Directions*, vol. 63 (Russian), Itogi Nauki i Tekhniki. (Akad. Nauk SSSR, Vsesoyuz. Inst. Nauchn. i Tekhn. Inform., Moscow, 1990), pp. 5–129
2. B. Anderson, J. Moore, *Optimal Control: Linear Quadratic Methods*. Dover Books on Engineering (Dover Publications, New York, 2007)
3. W. Anderson, *Continuous-time Markov Chains: An Applications-Oriented Approach* (Springer, Berlin, 2012)
4. S. Andradóttir, A stochastic approximation algorithm with varying bounds. Operations Res. **43**, 1037–1048 (1995)
5. D. Applebaum, Lévy processes and stochastic calculus, in *Cambridge Studies in Advanced Mathematics*, vol. 116, 2nd edn. (Cambridge University Press, Cambridge, 2009)
6. D. Applebaum, M. Siakalli, Asymptotic stability of stochastic differential equations driven by Lévy noise. J. Appl. Probab. **46**(4), 1116–1129 (2009)
7. A. Arapostathis, M.K. Ghosh, S.I. Marcus, Harnack's inequality for cooperative weakly coupled elliptic systems. Comm. Partial Differential Equations **24**(9–10), 1555–1571 (1999)
8. L. Arnold, W. Kliemann, E. Oeljeklaus, Lyapunov exponents of linear stochastic systems, in *Lyapunov Exponents* (Springer, Berlin, 1986), pp. 85–125
9. G. Badowski, G. Yin, Stability of hybrid dynamic systems containing singularly perturbed random processes. IEEE Trans. Automat. Control **47**(12), 2021–2032 (2002)
10. G. Barone-Adesi, R.E. Whaley, Efficient analytic approximation of American option values. J. Finance **42**(2), 301–320 (1987)
11. G.K. Basak, A. Bisi, M.K. Ghosh, Stability of a random diffusion with linear drift. J. Math. Anal. Appl. **202**(2), 604–622 (1996)
12. G.K. Basak, A. Bisi, M.K. Ghosh, Stability of degenerate diffusions with state-dependent switching. J. Math. Anal. Appl. **240**(1), 219–248 (1999)
13. R.F. Bass, Stochastic differential equations driven by symmetric stable processes, in *Séminaire de Probabilités, XXXVI*. Lecture Notes in Mathematics, vol. 1801 (Springer, Berlin, 2003), pp. 302–313
14. P. Baxendale, Invariant measures for nonlinear stochastic differential equations, in *Lyapunov Exponents* (Oberwolfach, 1990). Lecture Notes in Mathematics, vol. 1486 (Springer, Berlin, 1991), pp. 123–140
15. L. Becks, F. Hilker, H. Malchow, K. Jürgens, H. Arndt, Experimental demonstration of chaos in a microbial food web. Nature **435**, 1226–1229 (2005)
16. M. Benaïm, Stochastic persistence (2018). https://arxiv.org/abs/1806.08450

H.-D. Nguyen et al., *Hybrid Switching Diffusions*, Probability Theory and Stochastic Modelling 63, https://doi.org/10.1007/978-3-031-93303-5

17. M. Benaïm, J. Hofbauer, W.H. Sandholm, Robust permanence and impermanence for stochastic replicator dynamics. J. Biol. Dyn. **2**(2), 180–195 (2008)
18. M. Benaïm, C. Lobry, Lotka-Volterra with randomly fluctuating environments or "How switching between beneficial environments can make survival harder". Ann. Appl. Probab. **26**(6), 3754–3785 (2016)
19. M. Benaïm, S.J. Schreiber, Persistence of structured populations in random environments. Theoretical Population Biol. **76**(1), 19–34 (2009)
20. M.M. Benderskiĭ, L.A. Pastur, The asymptotic behavior of the solutions of a second order equation with random coefficients. Teor. Funkciĭ Funkcional. Anal. i Priložen. **22**, 3–14 160 (1975)
21. A. Bensoussan *Perturbation Methods in Optimal Control*. Wiley/Gauthier-Villars Series in Modern Applied Mathematics. (Wiley, Chichester; Gauthier-Villars, Montrouge, 1988). Translated from the French by C. Tomson
22. A. Bensoussan, P.-L. Lions, Optimal control of random evolutions. Stochastics **5**(3), 169–199 (1981)
23. A. Bensoussan, J.L. Menaldi, Hybrid control and dynamic programming. Dynam. Contin. Discrete Impuls. Systems **3**(4), 395–442 (1997)
24. A. Benveniste, M. Métivier, P. Priouret, *Adaptive Algorithms and Stochastic Approximations*. Applications of Mathematics, vol. 22 (New York)k (Springer, Berlin, 1990). Translated from the French by Stephen S. Wilson
25. B. Beran, F. Kargi, A dynamic mathematical model for wastewater stabilization ponds. Ecological Modelling **181**(1), 39–57 (2005)
26. B. Bercu, F. Dufour, G. Yin, Almost sure stabilization for feedback controls of regime-switching linear systems with a hidden Markov chain. IEEE Trans. Automat. Control **54**(9), 2114–2125 (2009)
27. R.N. Bhattacharya, Criteria for recurrence and existence of invariant measures for multidimensional diffusions. Ann. Probab. **6**(4), 541–553 (1978)
28. P. Billingsley, *Convergence of Probability Measures*. Wiley Series in Probability and Statistics: Probability and Statistics, 2nd edn. (Wiley, New York, 1999). A Wiley-Interscience Publication
29. G. Blankenship, G.C. Papanicolaou, Stability and control of stochastic systems with wide-band noise disturbances. I. SIAM J. Appl. Math. **34**(3), 437–476 (1978)
30. J. Blath, A. Etheridge, M. Meredith, Coexistence in locally regulated competing populations and survival of branching annihilating random walk. Ann. Appl. Probab. **17**(5-6), 1474–1507 (2007)
31. L. Breiman, *Probability*. Classics in Applied Mathematics, vol. 7 (Society for Industrial and Applied Mathematics (SIAM), Philadelphia, PA, 1992). Corrected reprint of the 1968 original
32. J. Buffington, R.J. Elliott, American options with regime switching. Int. J. Theor. Appl. Finance **5**(5), 497–514 (2002)
33. G. Butler, G. Wolkowicz, A mathematical model of the chemostat with a general class of functions describing nutrient uptake. SIAM J. Appl. Math. **45**, 137–151 (1985)
34. P.E. Caines, H.F. Chen, Optimal adaptive LQG control for systems with finite state process parameters. IEEE Trans. Automat. Control **30**(2), 185–189 (1985)
35. P.E. Caines, J.F. Zhang, On the adaptive control of jump parameter systems via nonlinear filtering. SIAM J. Control Optim. **33**(6), 1758–1777 (1995)
36. H. Caswell, *Matrix Population Models* (Wiley Online Library, New York, 2001)
37. P. Cattiaux, P. Collet, A. Lambert, S. Martínez, S. Méléard, J.S. Martín, Quasi-stationary distributions and diffusion models in population dynamics. Ann. Probab. **37**(5), 1926–1969 (2009)
38. P. Cattiaux, S. Méléard, Competitive or weak cooperative stochastic Lotka–Volterra systems conditioned on non-extinction. J. Math. Biol. **60**(6), 797–829 (2010)
39. S. Cerrai, *Second Order PDE's in Finite and Infinite Dimension: A Probabilistic Approach*. Lecture Notes in Mathematics, vol. 1762 (Springer, Berlin, 2001)

40. Z. Chao, K. Wang, C. Zhu, Y. Zhu, Almost sure and moment exponential stability of regime-switching jump diffusions. SIAM J. Control Optim. **55**(6), 3458–3488 (2017)
41. H.-F. Chen, Y. Zhu, Stochastic approximation procedures with randomly varying truncations. Sci. Sinica Ser. A **29**, 914–926 (1986)
42. M.-F. Chen, *From Markov Chains to Non-Equilibrium Particle Systems*, 2nd edn. (World Scientific Publishing Co. Inc., River Edge, NJ, 2004)
43. M.F. Chen, S.F. Li, Coupling methods for multidimensional diffusion processes. Ann. Probab. **17**(1), 151–177 (1989)
44. X. Chen, Z.-Q. Chen, K. Tran, G. Yin, Properties of switching jump diffusions: Maximum principles and harnack inequalities, Bernoulli **25**, 1045–1075 (2019)
45. X. Chen, Z.-Q. Chen, K. Tran, G. Yin, Recurrence and ergodicity for a class of regime-switching jump diffusions. Appl. Math. Optim. **80**, 415–445 (2019)
46. Z.-Q. Chen, Z. Zhao, Potential theory for elliptic systems. Ann. Probab. **24**(1), 293–319 (1996)
47. Z.-Q. Chen, Z. Zhao, Harnack principle for weakly coupled elliptic systems. J. Differential Equations **139**(2), 261–282 (1997)
48. P. Chesson, General theory of competitive coexistence in spatially-varying environments. Theoretical Population Biol. **58**(3), 211–237 (2000)
49. P.L. Chesson, S. Ellner, Invasibility and stochastic boundedness in monotonic competition models. J. Math. Biol. **27**(2), 117–138 (1989)
50. C.L. Chiang, *An Introduction to Stochastic Processes and Their Applications* (Robert E. Krieger Publishing Co., Huntington, N.Y., 1980)
51. Y.S. Chow, H. Teicher, *Probability Theory, Independence, Interchangeability, Martingales*. Springer Texts in Statistics, 3rd edn. (Springer, New York, 1997)
52. K.L. Chung, *Markov Chains with Stationary Transition Probabilities*. Die Grundlehren der mathematischen Wissenschaften, Band 104, 2nd edn. (Springer New York, Inc., New York, 1967)
53. K.L. Chung, *A Course in Probability Theory*, 3rd edn. (Academic Press Inc., San Diego, CA, 2001)
54. R. Cont, D.-A. Fournié, Change of variable formulas for non-anticipative functionals on path space. J. Funct. Anal. **259**(4), 1043–1072 (2010)
55. R. Cont, D.-A. Fournié, Functional itô calculus and stochastic integral representation of martingales. Ann. Probab. **41**(1), 109–133 (2013)
56. D.R. Cox, H.D. Miller, *The Theory of Stochastic Processes* (Wiley, New York, 1965)
57. P.C. Cross, J.O. Lloyd-Smith, P.L.F. Johnson, W.M. Getz, Duelling timescales of host movement and disease recovery determine invasion of disease in structured populations. Ecology Letters **8**(6), 587–595 (2005)
58. N.H. Dang, N.H. Du, G. Yin, Existence of stationary distributions for Kolmogorov systems of competitive type under telegraph noise. J. Differential Equations **257**(6), 386–409 (2014)
59. K.F. Davies, P. Chesson, S. Harrison, B.D. Inouye, B. Melbourne, K.J. Rice, Spatial heterogeneity explains the scale dependence of the native-exotic diversity relationship. Ecology **86**(6), 1602–1610 (2005)
60. M.H.A. Davis, *Markov Models and Optimization*. Monographs on Statistics and Applied Probability, vol. 49 (Chapman & Hall, London, 1993)
61. D.A. Dawson, Critical dynamics and fluctuations for a mean-field model of cooperative behavior. J. Statist. Phys. **31**(1), 29–85 (1983)
62. G.B. Di Masi, Y.M. Kabanov, W.J. Runggaldier, Mean-variance hedging of options on stocks with markov volatilities. Theory Probab. Appl. **39**(1), 172–182 (1994)
63. N.T. Dieu, D.H. Nguyen, N.H. Du, G. Yin, Classification of asymptotic behavior in a stochastic SIR model. SIAM J. Appl. Dyn. Syst. **15**(2), 1062–1084 (2016)
64. M.D. Donsker, S.R.S. Varadhan, Asymptotic evaluation of certain Markov process expectations for large time. I. II. Comm. Pure Appl. Math. **28**, 1–47; ibid. 28 (1975), 279–301 (1975)
65. J.L. Doob, *Stochastic Processes*. Wiley Classics Library (Wiley, New York, 1990). Reprint of the 1953 original, A Wiley-Interscience Publication

66. N.H. Du, N.H. Dang, Dynamics of kolmogorov systems of competitive type under the telegraph noise. J. Differential Equations **250**, 386–409 (2011)
67. N.H. Du, N.T. Dieu, N.N. Nhu, Conditions for permanence and ergodicity of certain SIR epidemic models. Acta Appl. Math. **160**, 81–99 (2019)
68. N.H. Du, R. Kon, K. Sato, Y. Takeuchi, Dynamical behavior of Lotka-Volterra competition systems: non-autonomous bistable case and the effect of telegraph noise. J. Comput. Appl. Math. **170**(2), 399–422 (2004)
69. N.H. Du, N.N. Nhu, Permanence and extinction of certain stochastic SIR models perturbed by a complex type of noises. Appl. Math. Lett. **64**, 223–230 (2017)
70. N.H. Du, V.H. Sam, Dynamics of a stochastic Lotka-Volterra model perturbed by white noise. J. Math. Anal. Appl. **324**(1), 82–97 (2006)
71. F. Dufour, P. Bertrand, Stabilizing control law for hybrid models. IEEE Trans. Automat. Control **39**(11), 2354–2357 (1994)
72. F. Dufresne, H.U. Gerber, Risk theory for the compound Poisson process that is perturbed by diffusion. Insurance Math. Econom. **10**(1), 51–59 (1991)
73. T. Duncan, *Probability Densities for Diffusion Processes with Applications to Nonlinear Filtering Theory and Detection Theory*. PhD thesis (Stanford University, Stanford, 1967)
74. B. Dupire, Functional Itô's calculus, in *Bloomberg portfolio research Paper No. 2009-04-FRONTIERS*. Available at SSRN: http://ssrn.com/abstract=1435551 or http://dx.doi.org/10.2139/ssrn.1435551
75. D. Dykhuizen, A. Dean, Evolution of specialists in an experimental microcosm. Genetics **167**, 2015–2026 (2004)
76. E.B. Dynkin, *Markov Processes. Vols. I, II.* Die Grundlehren der mathematischen Wissenschaften, Band 121, vol. 122 (Academic Press, Inc., Publishers, New York; Springer, Berlin-Göttingen-Heidelberg, 1965). Translated with the authorization and assistance of the author by J. Fabius, V. Greenberg, A. Maitra, G. Majone
77. L. Edelstein-Keshet, *Mathematical Models in Biology*. The Random House/Birkhäuser Mathematics Series (Random House Inc., New York, 1988)
78. S.D. Èĭdel'man, *Parabolic Systems* (North-Holland Publishing Co., Amsterdam, 1969). Translated from the Russian by Scripta Technica, London
79. A. Eizenberg, M. Freidlin, On the Dirichlet problem for a class of second order PDE systems with small parameter. Stochastics Stochastics Rep. **33**(3-4), 111–148 (1990)
80. R.J. Elliott, *Stochastic Calculus and Applications*. Applications of Mathematics (New York), vol. 18. (Springer, New York, 1982)
81. S.N. Ethier, T.G. Kurtz, *Markov Processes: Characterization and Convergence*. Wiley Series in Probability and Mathematical Statistics: Probability and Mathematical Statistics (Wiley, New York, 1986)
82. L. Evans, *Partial Differential Equations*. Graduate Studies in Mathematics, vol. 19, 2nd edn. (American Mathematical Society, Providence, RI, 2010)
83. S.N. Evans, A. Hening, S.J. Schreiber, Protected polymorphisms and evolutionary stability of patch-selection strategies in stochastic environments. J. Math. Biol. **71**(2), 325–359 (2015)
84. S.N. Evans, P.L. Ralph, S.J. Schreiber, A. Sen, Stochastic population growth in spatially heterogeneous environments. J. Math. Biol. **66**(3), 423–476 (2013)
85. S. Fang, T. Zhang, A study of a class of stochastic differential equations with non-Lipschitzian coefficients. Probab. Theory Related Fields **132**(3), 356–390 (2005)
86. W.H. Fleming, R.W. Rishel, *Deterministic and Stochastic Optimal Control*. Stochastic Modelling and Applied Probability, vol. 1 (Springer, Berlin, 1975)
87. W.H. Fleming, H.M. Soner, *Controlled Markov Processes and Viscosity Solutions*. Stochastic Modelling and Applied Probability, vol. 25, 2nd edn. (Springer, New York, 2006)
88. S.R. Foguel, Positive operators on $C(X)$. Proc. Amer. Math. Soc. **22**, 295–297 (1969)
89. J.-P. Fouque, G. Papanicolaou, K.R. Sircar, *Derivatives in Financial Markets with Stochastic Volatility* (Cambridge University, Cambridge, 2000)
90. J.-P. Fouque, G. Papanicolaou, R. Sircar, K. Solna, Multiscale stochastic volatility asymptotics. Multiscale Model. Simul. **2**(1), 22–42 (2003)

91. J.-P. Fouque, G. Papanicolaou, R. Sircar, K. Solna, Singular perturbations in option pricing. SIAM J. Appl. Math. **63**(5), 1648–1665 (2003)
92. M.D. Fragoso, O.L.V. Costa, A unified approach for stochastic and mean square stability of continuous-time linear systems with Markovian jumping parameters and additive disturbances. SIAM J. Control Optim. **44**(4), 1165–1191 (2005)
93. A. Friedman, *Partial Differential Equations of Parabolic Type* (Prentice-Hall, Inc., Englewood Cliffs, N.J., 1964)
94. A. Friedman, *Stochastic Differential Equations and Applications*, vol. I and II (Academic Press, New York, 1975)
95. C. Fritsch, J. Harmand, F. Campillo, A modeling approach of the chemostat. Ecological Modelling **229**, 1–13 (2015)
96. Z. Fu, Z. Li, Stochastic equations of non-negative processes with jumps. Stochastic Process. Appl. **120**(3), 306–330 (2010)
97. M.G. Garroni, J.-L. Menaldi, *Green Functions for Second Order Parabolic Integro-Differential Problems*. Pitman Research Notes in Mathematics Series, vol. 275 (Longman Scientific & Technical, Harlow; copublished in the United States with Wiley, New York, 1992)
98. G. Gehlert, J. Hapke, Mathematical modeling of a continuous aerobic membrane bioreactor for the treatment of different kinds of wastewater. Desalination **146**(1-3), 405–412 (2002)
99. C. Geiß, R. Manthey, Comparison theorems for stochastic differential equations in finite and infinite dimensions. Stochastic Process. Appl. **53**(1), 23–35 (1994)
100. M.K. Ghosh, A. Arapostathis, S.I. Marcus, Ergodic control of switching diffusions. SIAM J. Control Optim. **35**(6), 1952–1988 (1997)
101. I.I. Gīhman, A.V. Skorohod, *Stochastic Differential Equations*. Ergebnisse der Mathematik und ihrer Grenzgebiete [Results in Mathematics and Related Areas], vol. 72 (Springer, New York, 1972). Translated from the Russian by Kenneth Wickwire
102. I.I. Gīhman, A.V. Skorohod, *The Theory of Stochastic Processes. I*. Grundlehren der Mathematischen Wissenschaften, vol. 210, english edn. [Fundamental Principles of Mathematical Sciences] (Springer, Berlin, 1980). Translated from the Russian by Samuel Kotz
103. I.I. Gikhman, A.V. Skorokhod, *Introduction to the Theory of Random Processes*. (Dover Publications, Inc., Mineola, 1996). Translated from the 1965 Russian original, Reprint of the 1969 English translation, with a preface by Warren M. Hirsch
104. D. Gilbarg, N.S. Trudinger, *Elliptic Partial Differential Equations of Second Order*. Classics in Mathematics (Springer, Berlin, 2001). Reprint of the 1998 edition
105. I.V. Girsanov, Strong Feller processes. I. General properties. Teor. Verojatnost. i Primenen. **5**, 7–28 (1960)
106. K. Glover, All optimal Hankel-norm approximations of linear multivariable systems and their L^∞-error bounds. Internat. J. Control **39**(6), 1115–1193 (1984)
107. G.H. Golub, C.F. Van Loan, *Matrix Computations*. Johns Hopkins Series in the Mathematical Sciences, vol. 3, 2nd edn. (Johns Hopkins University Press, Baltimore, MD, 1989)
108. M. Hairer, Convergence of Markov processes. http://www.hairer.org/notes/Convergence.pdf (2016). Lecture notes, University Warwick
109. M. Hairer, J.C. Mattingly, Yet another look at Harris' ergodic theorem for Markov chains, in *Seminar on Stochastic Analysis, Random Fields and Applications VI: Centro Stefano Franscini, Ascona, May 2008* (Springer, Berlin, 2011), pp. 109–117
110. J.K. Hale, *Ordinary Differential Equations*, 2nd edn. (Robert E. Krieger Publishing Co., Inc., Huntington, 1980)
111. J.K. Hale, E. F. Infante, Extended dynamical systems and stability theory. Proc. Nat. Acad. Sci. U.S.A. **58**, 405–409 (1967)
112. P. Hall, C.C. Heyde, *Martingale Limit Theory and its Application*. Probability and Mathematical Statistics (Academic Press, Inc. [Harcourt Brace Jovanovich, Publishers], New York, 1980)
113. F.B. Hanson, *Applied Stochastic Processes and Control for Jump-Diffusions: Modeling, Analysis, and Computation*. Advances in Design and Control, vol. 13. (Society for Industrial and Applied Mathematics (SIAM), Philadelphia, 2007)

114. R.Z. Has'minskiĭ, On the principle of averaging the Itô's stochastic differential equations. Kybernetika (Prague) **4**, 260–279 (1968)
115. Q. He, G. Yin, Invariant density, Lyapunov exponent, and almost sure stability of Markovian-regime-switching linear systems. J. Syst. Sci. Complex. **24**(1), 79–92 (2011)
116. U. Helmke, J.B. Moore, *Optimization and Dynamical Systems*. Communications and Control Engineering Series (Springer-Verlag London, Ltd., London, 1994). With a foreword by R. Brockett
117. A. Hening, D.H. Nguyen, Coexistence and extinction for stochastic Kolmogorov systems. Ann. Appl. Probab. **28**(3), 1893–1942 (2018)
118. A. Hening, D.H. Nguyen, S.J. Schreiber, A classification of the dynamics of three-dimensional stochastic ecological systems. Ann. Appl. Probab. **32**(2), 893–931 (2022)
119. M. Henze, C. G. Jr., W. Gujer, A general model for single-sludge wastewater treatment systems. Water Res. **21**(5), 505–515 (1987)
120. D. Herbert, R. Elsworth, R. Telling, The continuous culture of bacteria; a theoretical and experimental study. J. Gen. Microbiol **14**, 601–622 (1956)
121. J.P. Hespanha, Stochastic hybrid systems: Application to communication networks, in *Hybrid Systems: Computation and Control*, ed. by R. Alur, G.J. Pappas (Springer, Berlin, 2004), pp. 387–401
122. J.P. Hespanha, A model for stochastic hybrid systems with application to communication networks. Nonlinear Anal. **62**(8), 1353–1383 (2005)
123. N.T. Hieu, D.H. Nguyen, N.N. Nguyen, G. Yin, Analyzing a class of stochastic sirs models under imperfect vaccination. J. Franklin Inst. **361**(3), 1284–1302 (2024)
124. D.J. Higham, X. Mao, A.M. Stuart, Strong convergence of Euler-type methods for nonlinear stochastic differential equations. SIAM J. Num. Anal. **40**(3), 1041–1063 (2002)
125. J. Hofbauer, A general cooperation theorem for hypercycles. Monatsh. Math. **91**(3), 233–240 (1981)
126. J. Hofbauer, J.W.-H. So, Uniform persistence and repellors for maps. Proc. Amer. Math. Soc. **107**(4), 1137–1142 (1989)
127. A.R. Hota, S. Sundaram, Game-theoretic vaccination against networked sis epidemics and impacts of human decision-making. IEEE Trans. Control Network Syst. **6**(4), 1461–1472 (2019)
128. S. Hsu, S. Hubbell, P. Waltman, A mathematical theory for single-nutrient competition in continuous culture of micro-organisms. SIAM J. Appl. Math. **32**, 366–383 (1977)
129. S.-B. Hsu, T.-W. Hwang, Y. Kuang, A ratio-dependent food chain model and its applications to biological control. Math. Biosci. **181**, 55–83 (2003)
130. J. Huang, H. Zhang, J. Zhang, A unified approach to diffusion analysis of queues with general patience-time distributions. Math. Oper. Res. **41**(3), 1135–1160 (2016)
131. J. Hull, *Options, Futures, and Other Derivatives*, 3rd edn. (Prentice-Hall, Upper Saddle River, NJ, 1997)
132. J. Hull, A. White, The pricing of options on assets with stochastic volatilities. J. Finance **42**(2), 281–300 (1987)
133. V. Hutson, A theorem on average Liapunov functions. Monatsh. Math. **98**(4), 267–275 (1984)
134. V. Hutson, J.S. Pym, *Applications of Functional Analysis and Operator Theory*. Mathematics in Science and Engineering, vol. 146 (Academic Press, Inc. [Harcourt Brace Jovanovich, Publishers], New York, 1980)
135. M.M. Benderskiĭ, L.A. Pastur, The spectrum of the one-dimensional Schrödinger equation with random potential. Mat. Sb. (N.S.) **82**(124), 273–284 (1970)
136. N. Ikeda, S. Watanabe, *Stochastic Differential Equations and Diffusion Processes*. North-Holland Mathematical Library, vol. 24, 2nd edn. (North-Holland Publishing Co., Amsterdam; Kodansha, Ltd., Tokyo, 1989)
137. A.M. Il' in, R.Z. Khasminskii, G. Yin, Singularly perturbed switching diffusions: rapid switchings and fast diffusions. J. Optim. Theory Appl. **102**(3), 555–591 (1999)
138. A.M. Il'in, R.Z. Khasminskii, G. Yin, Asymptotic expansions of solutions of integro-differential equations for transition densities of singularly perturbed switching diffusions: rapid switchings. J. Math. Anal. Appl. **238**(2), 516–539 (1999)

139. J. Imae, J.E. Perkins, J.B. Moore, Toward time-varying balanced realization via Riccati equations. Math. Control Signals Systems **5**(3), 313–326 (1992)
140. M. Iosifescu, *Finite Markov Processes and Their Applications*. Wiley Series in Probability and Mathematical Statistics (Wiley, Chichester; Editura Tehnică, Bucharest, 1980)
141. J. Jacod, A.N. Shiryaev, *Limit Theorems for Stochastic Processes*. Grundlehren der Mathematischen Wissenschaften, vol. 288, 2nd edn. [Fundamental Principles of Mathematical Sciences] (Springer, Berlin, 2003)
142. S.F. Jarner, G.O. Roberts, Polynomial convergence rates of markov chains. Ann. Appl. Probab. **12**(1), 224–247 (2002)
143. Y. Ji, H.J. Chizeck, Controllability, stabilizability, and continuous-time Markovian jump linear quadratic control. IEEE Trans. Automat. Control **35**(7), 777–788 (1990)
144. L. Jones, S. Ellner, Effects of rapid prey evolution on predator-prey cycles. J. Math. Biol. **55**, 541–573 (2007)
145. J. Kabouris, A. Georgakakos, Parameter and state estimation of the activated sludge process-i. model development. Water Res. **30**(12), 2853–2865 (1996)
146. I. Kac, N. Krasovskii, On the stability of systems with random parameters. J. Appl. Math. Mech. **24**(5), 1225–1246 (1960)
147. O. Kallenberg, *Foundations of Modern Probability*. Probability and its Applications (New York), 2nd edn. (Springer, New York, 2002)
148. O. Kallenberg, *Foundations of Modern Probability*, 3rd edn. (Springer Cham, 2021).
149. I. Karatzas, S.E. Shreve, *Brownian Motion and Stochastic Calculus*. Graduate Texts in Mathematics, vol. 113, 2nd edn. (Springer, New York, 1991)
150. S. Karlin, H.M. Taylor, *A First Course in Stochastic Processes*, 2nd edn. (Academic Press [A subsidiary of Harcourt Brace Jovanovich, Publishers], New York, 1975)
151. S. Karlin, H.M. Taylor, *A Second Course in Stochastic Processes* (Academic Press Inc. [Harcourt Brace Jovanovich Publishers], New York, 1981)
152. K. Kawaguchi, H. Morimoto, Long-run average welfare in a pollution accumulation model. J. Econom. Dyn. Control **31**, 703–720 (2007)
153. E. Keeler, M. Spence, R. Zeckhauser, The optimal control of pollution. J. Economic Theory **4**, 19–34 (1972)
154. W.O. Kermack, A.G. McKendrick, Contributions to the mathematical theory of epidemics. I, in *Proceeding of the Royal Society London. Series A*, Containing Papers of a Mathematical and Physical Character, vol. 115(772), 700–721 (1927)
155. W.O. Kermack, A.G. McKendrick, G.T. Walker, Contributions to the mathematical theory of epidemics. II. —The problem of endemicity. *Proceed. Royal Soc. London. Series A*, Containing Papers of a Mathematical and Physical Character **138**(834), 55–83 (1932)
156. R. Khasminskii, *Stochastic Stability of Differential Equations*. Stochastic Modelling and Applied Probability, vol. 66, 2nd edn. (Springer, New York, 2012)
157. R.Z. Khasminskii, F.C. Klebaner, Long term behavior of solutions of the Lotka-Volterra system under small random perturbations. Ann. Appl. Probab. **11**(3), 952–963 (2001)
158. R.Z. Khasminskii, G. Yin, Asymptotic series for singularly perturbed Kolmogorov-Fokker-Planck equations. SIAM J. Appl. Math. **56**(6), 1766–1793 (1996)
159. R.Z. Khasminskii, G. Yin, On transition densities of singularly perturbed diffusions with fast and slow components. SIAM J. Appl. Math. **56**(6), 1794–1819 (1996)
160. R.Z. Khasminskii, G. Yin, On averaging principles: an asymptotic expansion approach. SIAM J. Math. Anal. **35**(6), 1534–1560 (2004)
161. R.Z. Khasminskii, G. Yin, Limit behavior of two-time-scale diffusions revisited. J. Differential Equations **212**(1), 85–113 (2005)
162. R.Z. Khasminskii, G. Yin, Uniform asymptotic expansions for pricing European options. Appl. Math. Optim. **52**(3), 279–296 (2005)
163. R.Z. Khasminskii, G. Yin, Q. Zhang, Asymptotic expansions of singularly perturbed systems involving rapidly fluctuating Markov chains. SIAM J. Appl. Math. **56**(1), 277–293 (1996)
164. R.Z. Khasminskii, C. Zhu, G. Yin, Stability of regime-switching diffusions. Stochastic Process. Appl. **117**(8), 1037–1051 (2007)

165. F.C. Klebaner, *Introduction to Stochastic Calculus with Applications*, 2nd edn. (Imperial College Press, London, 2005)
166. P.E. Kloeden, E. Platen, *Numerical Solution of Stochastic Differential Equations*. Applications of Mathematics (New York), vol. 23 (Springer, Berlin, 1992)
167. N.V. Krylov, *Controlled Diffusion Processes*. Applications of Mathematics, vol. 14 (Springer-Verlag, New York, 1980). Translated from the Russian by A. B. Aries
168. V. Kulkarni, Fluid models for single buffer systems, in *Frontiers in Queueing: Models and Applications in Science and Engineering*, ed. by J.H. Dashalalow (CRC Press, New York, 1997), pp. 321–338
169. K. Kunwai, C. Zhu, On Feller and strong Feller properties and irreducibility of regime-switching jump diffusion processes with countable regimes. Nonlinear Anal. Hybrid Syst. **38**, 100946 (2020)
170. T.G. Kurtz, *Approximation of Population Processes*. CBMS-NSF Regional Conference Series in Applied Mathematics, vol. 36. (Society for Industrial and Applied Mathematics (SIAM), Philadelphia, PA, 1981)
171. T.G. Kurtz, Equivalence of stochastic equations and martingale problems, in *Stochastic analysis 2010* (Springer, Heidelberg, 2011), pp. 113–130
172. H. Kushner, On the differential equations satisfied by conditional probability densities of markov processes, with applications. J. SIAM Control Ser. A **2**, 106–119 (1964)
173. H.J. Kushner, *Stochastic Stability and Control*, in *Mathematics in Science and Engineering*, vol. 33 (Academic Press, New York, 1967)
174. H.J. Kushner, The concept of invariant set for stochastic dynamical systems and applications to stochastic stability, in *Stochastic Optimization and Control (Proc. Advanced Seminar, Madison, Wis., 1967)* (Wiley, New York, 1968), pp. 47–57
175. H.J. Kushner, *Approximation and Weak Convergence Methods for Random Processes, with Applications to Stochastic Systems Theory*. MIT Press Series in Signal Processing, Optimization, and Control, vol. 6 (MIT Press, Cambridge, MA, 1984), pp. 47–57
176. H.J. Kushner, *Weak Convergence Methods and Singularly Perturbed Stochastic Control and Filtering Problems*. Systems & Control: Foundations & Applications, vol. 3 (Birkhäuser Boston, Inc., Boston, MA, 1990)
177. H.J. Kushner, G. Yin, *Stochastic Approximation and Recursive Algorithms and Applications*. Applications of Mathematics, vol. 35, 2nd edn. (Springer, New York, 2003). Stochastic Modelling and Applied Probability
178. O.A. Ladyženskaja, V.A. Solonnikov, N.N. Ural' ceva, *Linear and Quasilinear Equations of Parabolic Type*. Translations of Mathematical Monographs, vol. 23 (American Mathematical Society, Providence, R.I., 1968). Translated from the Russian by S. Smith
179. G. Lan, Z. Lin, C. Wei, S. Zhang, A stochastic SIRS epidemic model with non-monotone incidence rate under regime-switching. J. Franklin Inst. **356**(16), 9844–9866 (2019)
180. R. Lande, S. Engen, B.-E. Saether, *Stochastic Population Dynamics in Ecology and Conservation* (Oxford University Press on Demand, Oxford, 2003)
181. J. LaSalle, S. Lefschetz, *Stability by Liapunov's Direct Method, with Applications*. Mathematics in Science and Engineering, vol. 4 (Academic Press, New York, 1961)
182. J.P. LaSalle, The extent of asymptotic stability. Proc. Nat. Acad. Sci. U.S.A. **46**, 363–365 (1960)
183. R. Law, R.D. Morton, Permanence and the assembly of ecological communities. Ecology **77**(3), 762–775 (1996)
184. D. Li, F. Wei, X. Mao, Stationary distribution and density function of a stochastic SVIR epidemic model. J. Franklin Inst. **359**(16), 9422–9449 (2022)
185. Z. Li, L. Mytnik, Strong solutions for stochastic differential equations with jumps. Ann. Inst. Henri Poincaré Probab. Stat. **47**(4), 1055–1067 (2011)
186. Z. Li, F. Pu, Strong solutions of jump-type stochastic equations. Electron. Commun. Probab. **17**(33), 1–13 (2012)
187. G.M. Lieberman, *Second Order Parabolic Differential Equations* (World Science Publishing Co., Inc., River Edge, NJ, 1996)

188. Y. Lin, D. Jiang, Threshold behavior in a stochastic sis epidemic model with standard incidence. J. Dyn. Differential Equations **26**(4), 1079–1094 (2014)
189. J.-J. Liou, Recurrence and transience of Gaussian diffusion processes. Kodai Math. J. **13**(2), 210–230 (1990)
190. R.S. Liptser, A.N. Shiryaev, *Statistics of Random Processes. I and II.* Applications of Mathematics, vol. 5 and 6, expanded edn. (Springer, Berlin, New York, 2001). General theory, Translated from the 1974 Russian original by A. B. Aries, Stochastic Modelling and Applied Probability
191. Q. Liu, D. Jiang, T. Hayat, A. Alsaedi, Threshold behavior in a stochastic delayed sis epidemic model with vaccination and double diseases. J. Franklin Inst. **356**(13), 7466–7485 (2019)
192. Q. Liu, D. Jiang, T. Hayat, A. Alsaedi, B. Ahmad, A stochastic SIRS epidemic model with logistic growth and general nonlinear incidence rate. Phys. A **551**, 124152 (2020)
193. Y.J. Liu, G. Yin, Q. Zhang, J.B. Moore, Balanced realizations of regime-switching linear systems. Math. Control Signals Systems **19**(3), 207–234 (2007)
194. Z. Liu, C. Tian, A weighted networked SIRS epidemic model. J. Differential Equations **269**(12), 10995–11019 (2020)
195. K.A. Loparo, G.L. Blankenship, Almost sure instability of a class of linear stochastic systems with jump process coefficients, in *Lyapunov Exponents* (Bremen, 1984). Lecture Notes in Mathematics, vol. 1186 (Springer, Berlin, 1986), pp. 160–190
196. R. Luck, Evaluation of natural enemies for biological control: a behavior approach. Trends Ecol. Evol. **5**, 196–199 (1990)
197. Q. Luo, X. Mao, Stochastic population dynamics under regime switching. J. Math. Anal. Appl. **334**(1), 69–84 (2007)
198. X. Mao, *Stability of Stochastic Differential Equations with Respect to Semimartingales.* Pitman Research Notes in Mathematics Series, vol. 251 (Longman Scientific & Technical, Harlow; copublished in the United States with Wiley, New York, 1991)
199. X. Mao, Stability of stochastic differential equations with Markovian switching. Stochastic Process. Appl. **79**(1), 45–67 (1999)
200. X. Mao, A note on the LaSalle-type theorems for stochastic differential delay equations. J. Math. Anal. Appl. **268**(1), 125–142 (2002)
201. X. Mao, *Stochastic Differential Equations and Applications* (Elsevier, Amsterdam, 2007)
202. X. Mao, G. Marion, E. Renshaw, Environmental brownian noise suppresses explosions in population dynamics. Stochastic Process. Appl. **97**(1), 95–110 (2002)
203. X. Mao, S. Sabanis, E. Renshaw, Asymptotic behaviour of the stochastic Lotka-Volterra model. J. Math. Anal. Appl. **287**(1), 141–156 (2003)
204. X. Mao, G. Yin, C. Yuan, Stabilization and destabilization of hybrid systems of stochastic differential equations. Automatica J. IFAC **43**(2), 264–273 (2007)
205. X. Mao, C. Yuan, *Stochastic Differential Equations with Markovian Switching* (Imperial College Press, London, 2006)
206. X. Mao, C. Yuan, G. Yin, Numerical method for stationary distribution of stochastic differential equations with Markovian switching. J. Comput. Appl. Math. **174**(1), 1–27 (2005)
207. X. Mao, C. Yuan, G. Yin, Approximations of Euler-Maruyama type for stochastic differential equations with Markovian switching, under non-Lipschitz conditions. J. Comput. Appl. Math. **205**(2), 936–948 (2007)
208. M. Mariton, *Jump Linear Systems in Automatic Control* (Marcel Dekker, Inc., New York, 1990)
209. R. Merton, *Continuous-Time Finance* (Wiley-Blackwell, Cambridge, MA, 1992)
210. S.P. Meyn, R.L. Tweedie, Stability of Markovian processes. I. criteria for discrete-time chains. Adv. Appl. Probab. **24**(3), 542–574 (1992)
211. S.P. Meyn, R.L. Tweedie, *Markov Chains and Stochastic Stability.* Communications and Control Engineering Series (Springer, London, 1993)
212. S.P. Meyn, R.L. Tweedie, Stability of Markovian processes. III. Foster-Lyapunov criteria for continuous-time processes. Adv. Appl. Probab. **25**(3), 518–548 (1993)

213. G.N. Milstein, *Numerical Integration of Stochastic Differential Equations*. Mathematics and its Applications, vol. 313. (Kluwer Academic Publishers Group, Dordrecht, 1995). Translated and revised from the 1988 Russian original
214. G.N. Milstein, M.V. Tretyakov, *Stochastic Numerics for Mathematical Physics*. Scientific Computation (Springer, Berlin, 2004)
215. C.M. Møller, Stochastic differential equations for ruin probabilities. J. Appl. Probab. **32**(1), 74–89 (1995)
216. B.C. Moore, Principal component analysis in linear systems: controllability, observability, and model reduction. IEEE Trans. Automat. Control **26**(1), 17–32 (1981)
217. R. Mortensen, Maximum-likelihood recursive nonlinear filtering. J. Optim. Theory Appl. **2**, 386–394 (1968)
218. D.H. Nguyen, N.N. Nguyen, G. Yin, General nonlinear stochastic systems motivated by chemostat models: complete characterization of long-time behavior, optimal controls, and applications to wastewater treatment. Stochastic Process. Appl. **130**(8), 4608–4642 (2020)
219. D.H. Nguyen, G. Yin, C. Zhu, Certain properties related to well posedness of switching diffusions. Stochastic Process. Appl. **127**, 3135–3158 (2017)
220. D.H. Nguyen, G. Yin, C. Zhu, Long-term analysis of a stochastic SIRS model with general incidence rates. SIAM J. Appl. Math. **80**(2), 814–838 (2020)
221. D.T. Nguyen, S.L. Nguyen, T.A. Hoang, G. Yin, Tamed-euler method for hybrid stochastic differential equations with Markovian switching. Nonlinear Anal. Hybrid Syst. **30**, 14–30 (2018)
222. S.L. Nguyen, T.A. Hoang, D.T. Nguyen, G. Yin, Milstein-type procedures for numerical solutions of stochastic differential equations with Markovian switching. SIAM J. Numer. Anal. **55**(2), 953–979 (2017)
223. J.R. Norris, *Markov Chains*. Cambridge Series in Statistical and Probabilistic Mathematics, vol. 2 (Cambridge University, Cambridge, 1998). Reprint of 1997 original
224. A. Novick, L. Szilard, Description of the chemostat. Science **112**, 715–716 (1950)
225. D. Nualart, *The Malliavin Calculus and Related Topics*, 2nd edn. (Springer, Berlin, 2006)
226. E. Nummelin, *General Irreducible Markov Chains and Nonnegative Operators*. Cambridge Tracts in Mathematics, vol. 83 (Cambridge University, Cambridge, 1984)
227. B. Øksendal, *Stochastic Differential Equations* (Springer, Berlin, 2003)
228. Q. Pan, J. Huang, H. Wang, An sirs model with nonmonotone incidence and saturated treatment in a changing environment. J. Math. Biol. **85**(3), 1–39 (2022)
229. T. Pang, A. Hussain, An application of functional itô's formula to stochastic portfolio optimization with bounded memory, in *Proceedings 2015 SIAM Conference Control and Its Application* (CT15), pp. 159–166 (2015)
230. G.C. Papanicolaou, D. Stroock, S.R.S. Varadhan, Martingale approach to some limit theorems, in *Papers from the Duke Turbulence Conference* (Duke University, Durham, N.C., 1976), Duke University Mathematics Series, vol. III, pp. Paper No. 6, ii+120 (Duke University, Durham, N.C., 1977)
231. E. Pardoux, A.Y. Veretennikov, On the Poisson equation and diffusion approximation. I. Ann. Probab. **29**(3), 1061–1085 (2001)
232. S. Pavlou, I. Kevrekidis, Microbial predation in a periodically operated chemostat: A global study of the interaction between natural and externally imposed frequencies. Math Biosci. **108**, 1–55 (1992)
233. J.E. Perkins, U. Helmke, J.B. Moore, Balanced realizations via gradient flow techniques. Syst. Control Lett. **14**(5), 369–379 (1990)
234. L. Perko, *Differential Equations and Dynamical Systems*. Texts in Applied Mathematics, vol. 7, 3rd edn. (Springer, New York, 2001)
235. S. Peszat, J. Zabczyk, Strong Feller property and irreducibility for diffusions on Hilbert spaces. Ann. Probab. **23**(1), 157–172 (1995)
236. K. Pichór, R. Rudnicki, Stability of Markov semigroups and applications to parabolic systems. J. Math. Anal. Appl. **215**(1), 56–74 (1997)

237. M. Prandini, J. Hu, J. Lygeros, S. Sastry, A probabilistic approach to aircraft conflict detection. IEEE Transac. Intelligent Transport. Syst. **1**(4), 199–220 (2000)
238. E. Priola, F.-Y. Wang, Gradient estimates for diffusion semigroups with singular coefficients. J. Funct. Anal. **236**(1), 244–264 (2006)
239. M.H. Protter, H.F. Weinberger, *Maximum Principles in Differential Equations* (Prentice-Hall, Inc., Englewood Cliffs, N.J., 1967)
240. P.E. Protter, *Stochastic Integration and Differential Equations*. Stochastic Modelling and Applied Probability, vol. 21, 2nd edn. (Springer, Berlin, 2005). Version 2.1, Corrected 3rd printing
241. P. Pyšek, P.E. Hulme, Spatio-temporal dynamics of plant invasions: linking pattern to process. Ecoscience **12**(3), 302–315 (2005)
242. H. Qian, F. Wu, G. Yin, A class of numerical algorithms for stochastic differential equations with randomly varying truncations. Discrete Continuous Dynam. Syst. Series S **18**(3), 578–602 (2024)
243. H. Qian, G. Yin, Q. Zhang, Deep filtering with adaptive learning rates. IEEE Transac. Automat. Control **68**, 3285–3299 (2023)
244. H. Qiao, Exponential ergodicity for SDEs with jumps and non-Lipschitz coefficients. J. Theoret. Probab. **27**(1), 137–152 (2014)
245. Z. Qui, J. Yu, Y. Zou, The asymptotic behavior of a chemostat model with the beddington-deangelis functional response. Math. Biosci. **187**(2), 175–187 (2004)
246. S. Rajasekar, M. Pitchaimani, Ergodic stationary distribution and extinction of a stochastic SIRS epidemic model with logistic growth and nonlinear incidence. Appl. Math. Comput. **377**, 125143 (2020)
247. D. Revuz, *Markov Chains*. North-Holland Mathematical Library, vol. 11, 2nd edn. (North-Holland Publishing Co., Amsterdam, 1984)
248. T. Rolski, H. Schmidli, V. Schmidt, J. Teugels, *Stochastic Processes for Insurance and Finance*. Wiley Series in Probability and Statistics (Wiley, Chichester, 1999)
249. S.M. Ross, *Stochastic Processes*. Lectures in Mathematics, vol. 14 (Wiley, New York, 1983)
250. S. Ruan, W. Wang, Dynamical behavior of an epidemic model with a nonlinear incidence rate. J. Differential Equations **188**(1), 135–163 (2003)
251. H. Sandberg, A. Rantzer, Balanced truncation of linear time-varying systems. IEEE Trans. Automat. Control **49**(2), 217–229 (2004)
252. S.J. Schreiber, M. Benaïm, K.A.S. Atchadé, Persistence in fluctuating environments. J. Math. Biol. **62**(5), 655–683 (2011)
253. S.J. Schreiber, J.O. Lloyd-Smith, Invasion dynamics in spatially heterogeneous environments. Amer. Naturalist **174**(4), 490–505 (2009)
254. S. Sethi, H. Zhang, Q. Zhang, *Average-Cost Control of Stochastic Manufacturing Systems*. Stochastic Modelling and Applied Probability, vol. 54. (Springer, New York, 2005)
255. S.P. Sethi, Q. Zhang, *Hierarchical Decision Making in Stochastic Manufacturing Systems*. Systems & Control: Foundations & Applications (Birkhäuser Boston Inc., Boston, MA, 1994)
256. J. Shao, Criteria for transience and recurrence of regime-switching diffusion processes. Electron. J. Probab. **20**(63), 15 (2015)
257. J. Shao, Strong solutions and strong Feller properties for regime-switching diffusion processes in an infinite state space. SIAM J. Control Optim. **53**(4), 2462–2479 (2015)
258. J. Shao, Invariant measures and Euler-Maruyama's approximations of state-dependent regime-switching diffusions. SIAM J. Control Optim. **56**(5), 3215–3238 (2018)
259. S. Shokoohi, L.M. Silverman, P.M. Van Dooren, Linear time-variable systems: balancing and model reduction. IEEE Trans. Automat. Control **28**(8), 810–822 (1983)
260. H.A. Simon, A. Ando, Aggregation of variables in dynamic systems. Econometrica **29**(2), 111–138 (1961)
261. R. Situ, *Theory of Stochastic Differential Equations with Jumps and Applications: Mathematical and Analytical Techniques with Applications to Engineering* (Springer, New York, 2005)

262. A.V. Skorokhod, *Asymptotic Methods in the Theory of Stochastic Differential Equations*. Translations of Mathematical Monographs, vol. 78 (American Mathematical Society, Providence, RI, 1989). Translated from the Russian by H. H. McFaden
263. M. Slatkin, The dynamics of a population in a markovian environment. Ecology **59**(2), 249–256 (1978)
264. H. Smith, P. Waltman, *The Theory of the Chemostat: Dynamics of Microbial Competition*, vol. 13 (Cambridge University, Cambridge, 1995)
265. R. Stratonovich, Conditional markov processes. Theory Probab. Appl. **5**, 156–178 (1960)
266. D.W. Stroock, S.R.S. Varadhan, *Multidimensional Diffusion Processes*. Grundlehren der Mathematischen Wissenschaften [Fundamental Principles of Mathematical Sciences], vol. 233 (Springer, Berlin, 1979)
267. D. Sworder, J. Boyd, *Estimation Problems in Hybrid Systems* (Cambridge University, Cambridge, 1999)
268. D.D. Sworder, V.G. Robinson, Feedback regulators for jump parameter systems with state and control dependent transition rates. IEEE Trans. Automat. Control **AC-18**(4), 355–360 (1973)
269. Y. Takeuchi, N.H. Du, N.T. Hieu, K. Sato, Evolution of predator-prey systems described by a Lotka-Volterra equation under random environment. J. Math. Anal. Appl. **323**(2), 938–957 (2006)
270. M. Turelli, Random environments and stochastic calculus. Theoretical Population Biol. **12**(2), 140–178 (1977)
271. E.I. Verriest, T. Kailath, On generalized balanced realizations. IEEE Trans. Automat. Control **28**(8), 833–844 (1983)
272. D. Wang, Y. Zhao, J. Luo, H. Leng, Simplicial SIRS epidemic models with nonlinear incidence rates. Chaos: J. Nonlinear Sci. **31**(5), 053112 (2021)
273. L. Wang, D. Jiang, A note on the stationary distribution of the stochastic chemostat model with general response functions. Appl. Math. Lett. **73**, 22–28 (2017)
274. W. Whitt, *Stochastic-Process Limits* (Springer, New York, 2002)
275. R. Williams, *Dynamics of Microbial Populations* (Academic Press, New York, 1971)
276. J.T. Wloka, B. Rowley, B. Lawruk, *Boundary Value Problems for Elliptic Systems* (Cambridge University, Cambridge, 1995)
277. W.M. Wonham, Some applications of stochastic differential equations to optimal nonlinear filtering. J. SIAM Control Ser. A **2**, 347–369 (1965)
278. W.M. Wonham, Liapunov criteria for weak stochastic stability. J. Differential Equations **2**, 195–207 (1966)
279. Z. Wu, J. Yin, C. Wang, *Elliptic and Parabolic Equations* (World Scientific, Hackensack, 2006)
280. F. Xi, Stability of a random diffusion with nonlinear drift. Statist. Probab. Lett. **68**(3), 273–286 (2004)
281. F. Xi, G. Yin, Asymptotic properties of a mean-field model with a continuous-state-dependent switching process. J. Appl. Probab. **46**(1), 221–243 (2009)
282. F. Xi, G. Yin, Almost sure stability and instability for switching-jump-diffusion systems with state-dependent switching. J. Math. Anal. Appl. **400**(2), 460–474 (2013)
283. F. Xi, G. Yin, C. Zhu, Regime-switching jump diffusions with non-Lipschitz coefficients and countably many switching states: existence and uniqueness, Feller, and strong Feller properties, in *Modeling, Stochastic Control, Optimization, and Applications*. IMA Volume Mathematics Application, vol. 164, pp. 571–599 (Springer, Cham, 2019)
284. F. Xi, C. Zhu, On Feller and strong Feller properties and exponential ergodicity of regime-switching jump diffusion processes with countable regimes. SIAM J. Control Optim. **55**(3), 1789–1818 (2017)
285. F. Xi, C. Zhu, Jump type stochastic differential equations with non-Lipschitz coefficients: non-confluence, Feller and strong Feller properties, and exponential ergodicity. J. Differential Equations **266**(8), 4668–4711 (2019)
286. T. Yamada, S. Watanabe, On the uniqueness of solutions of stochastic differential equations. J. Math. Kyoto Univ. **11**, 155–167 (1971)

287. H. Yang, G. Yin, Ruin probability for a model under Markovian switching regime, in *Probability, Finance and Insurance*, ed. by T. Lai, H. Yang, S. Yung (World Scientific Publication, River Edge, NJ, 2004), pp. 206–217
288. Q. Yang, D. Jiang, N. Shi, C. Ji, The ergodicity and extinction of stochastically perturbed SIR and SEIR epidemic models with saturated incidence. J. Math. Anal. Appl. **388**(1), 248–271 (2012)
289. D.D. Yao, Q. Zhang, X.Y. Zhou, A regime-switching model for European options, in *Stochastic Processes, Optimization, and Control Theory: Applications in Financial Engineering, Queueing Networks, and Manufacturing Systems*. International Series in Operations Research & Management Science, vol. 94, pp. 281–300 (Springer, New York, 2006)
290. G. Yin, On limit results for a class of singularly perturbed switching diffusions. J. Theoret. Probab. **14**(3), 673–697 (2001)
291. G. Yin, Asymptotic expansions of option price under regime-switching diffusions with a fast-varying switching process. Asymptot. Anal. **65**(3-4), 203–222 (2009)
292. G. Yin, S. Dey, Weak convergence of hybrid filtering problems involving nearly completely decomposable hidden Markov chains. SIAM J. Control Optim. **41**(6), 1820–1842 (2003)
293. G. Yin, V. Krishnamurthy, Least mean square algorithms with Markov regime-switching limit. IEEE Trans. Automat. Control **50**(5), 577–593 (2005)
294. G. Yin, V. Krishnamurthy, C. Ion, Regime switching stochastic approximation algorithms with application to adaptive discrete stochastic optimization. SIAM J. Optim. **14**(4), 1187–1215 (2004)
295. G. Yin, R.H. Liu, Q. Zhang, Recursive algorithms for stock liquidation: a stochastic optimization approach. SIAM J. Optim. **13**(1), 240–263(electronic) (2002)
296. G. Yin, X. Mao, K. Yin, Numerical approximation of invariant measures for hybrid diffusion systems. IEEE Trans. Automat. Control **50**(7), 934–946 (2005)
297. G. Yin, X. Mao, C. Yuan, D. Cao, Approximation methods for hybrid diffusion systems with state-dependent switching processes: numerical algorithms and existence and uniqueness of solutions. SIAM J. Math. Anal. **41**(6), 2335–2352 (2009/10)
298. G. Yin, Q.S. Song, Z. Zhang, Numerical solutions for jump-diffusions with regime switching. Stochastics **77**(1), 61–79 (2005)
299. G. Yin, F. Xi, Stability of regime-switching jump diffusions. SIAM J. Control Optim. **48**, 4525–4549 (2010)
300. G. Yin, H. M. Yan, S.X.-C. Lou, On a class of stochastic optimization algorithms with applications to manufacturing models, in *Model-Oriented Data Analysis (Petrodvorets, 1992)*, Contributions to Statistics (Physica, Heidelberg, 1993), pp. 213–226
301. G. Yin, H. Yang, Two-time-scale jump-diffusion models with Markovian switching regimes. Stoch. Stoch. Rep. **76**(2), 77–99 (2004)
302. G. Yin, H.Q. Zhang, Q. Zhang, *Applications of Two-time-scale Markovian Systems* (Science Press, Beijing, 2013)
303. G. Yin, Q. Zhang, *Continuous-Time Markov Chains and Applications: A Singular Perturbation Approach*. Applications of Mathematics, vol. 37 (Springer, New York, 1998)
304. G. Yin, Q. Zhang, *Discrete-Time Markov Chains: Two-Time-Scale Methods and Applications*. Applications of Mathematics, vol. 55 (Springer, New York, 2005)
305. G. Yin, Q. Zhang, G. Badowski, Asymptotic properties of a singularly perturbed Markov chain with inclusion of transient states. Ann. Appl. Probab. **10**(2), 549–572 (2000)
306. G. Yin, G. Zhao, F. Wu, Regularization and stabilization of randomly switching dynamic systems. SIAM J. Appl. Math. **72**(5), 1361–1382 (2012)
307. G. Yin, C. Zhu, On the notion of weak stability and related issues of hybrid diffusion systems. Nonlinear Anal. Hybrid Syst. **1**(2), 173–187 (2007)
308. G. Yin, C. Zhu, Regularity and recurrence of switching diffusions. J. Syst. Sci. Complex. **20**(2), 273–283 (2007)
309. J. Yong, X.Y. Zhou, *Stochastic Controls: Hamiltonian Systems and HJB Equations*. Applications of Mathematics, vol. 43. (Springer, New York, 1999)

310. C. Yuan, J. Lygeros, Stabilization of a class of stochastic differential equations with Markovian switching. Syst. Control Lett. **54**(9), 819–833 (2005)
311. C. Yuan, X. Mao, Asymptotic stability in distribution of stochastic differential equations with Markovian switching. Stochastic Process. Appl. **103**(2), 277–291 (2003)
312. M. Zaiat, F. Passig, E. Foresti, A mathematical model and criteria for designing horizontal-flow anaerobic immobilized biomass reactors for wastewater treatment. Bioresource Tech. **71**(3), 235–243 (2000)
313. M. Zakai, On the optimal filtering of diffusion processes. Z. Wahrsch. Verw. Gebiete **11**, 230–243 (1969)
314. Q. Zhang, Stock trading: an optimal selling rule. SIAM J. Control Optim. **40**(1), 64–87 (2001)
315. Q. Zhang, D. Jiang, Competitive exclusion in a stochastic chemostat model with holling Type II functional response. J. Math. Chem. **54**(3), 777–791 (2016)
316. Q. Zhang, G. Yin, On nearly optimal controls of hybrid LQG problems. IEEE Trans. Automat. Control **44**(12), 2271–2282 (1999)
317. X. Zhang, D. Jiang, A. Alsaedi, T. Hayat, Stationary distribution of stochastic sis epidemic model with vaccination under regime switching. Appl. Math. Lett. **59**, 87–93 (2016)
318. X. Zhang, R. Yuan, The existence of stationary distribution of a stochastic delayed chemostat model. Appl. Math. Lett. **93**, 15–21 (2019)
319. D. Zhao, T. Zhang, S. Yuan, The threshold of a stochastic sivs epidemic model with nonlinear saturated incidence. Phys. A **443**, 372–379 (2016)
320. X.Y. Zhou, G. Yin, Markowitz's mean-variance portfolio selection with regime switching: a continuous-time model. SIAM J. Control Optim. **42**(4), 1466–1482 (2003)
321. C. Zhu, G. Yin, Asymptotic properties of hybrid diffusion systems. SIAM J. Control Optim. **46**(4), 1155–1179 (2007)
322. C. Zhu, G. Yin, On competitive Lotka-Volterra model in random environments. J. Math. Anal. Appl. **357**(1), 154–170 (2009)
323. C. Zhu, G. Yin, On strong Feller, recurrence, and weak stabilization of regime-switching diffusions. SIAM J. Control Optim. **48**(3), 2003–2031 (2009)
324. C. Zhu, G. Yin, Q.S. Song, Stability of random-switching systems of differential equations. Quart. Appl. Math. **67**(2), 201–220 (2009)
325. X. Zong, F. Wu, G. Yin, Z. Jin, Almost sure and *p*th-moment stability and stabilization of regime-switching jump diffusion systems. SIAM J. Control Optim. **52**(4), 2595–2622 (2014)

Index

H.-D. Nguyen et al., *Hybrid Switching Diffusions*, Probability Theory and Stochastic Modelling 63, https://doi.org/10.1007/978-3-031-93303-5

Q

R

S

T

W

Y